CORRELATED INTERPLANETARY AND MAGNETOSPHERIC OBSERVATIONS

ASTROPHYSICS AND SPACE SCIENCE LIBRARY

A SERIES OF BOOKS ON THE RECENT DEVELOPMENTS
OF SPACE SCIENCE AND OF GENERAL GEOPHYSICS AND ASTROPHYSICS
PUBLISHED IN CONNECTION WITH THE JOURNAL
SPACE SCIENCE REVIEWS

VOLUME 42

CORRELATED INTERPLANETARY AND MAGNETOSPHERIC OBSERVATIONS

PROCEEDINGS OF THE SEVENTH ESLAB SYMPOSIUM HELD AT SAULGAU, W. GERMANY, 22–25 MAY, 1973

Edited by

D. EDGAR PAGE

Space Science Department (ESLAB),
European Space Research and Technology Centre,
Noordwijk, The Netherlands

D. REIDEL PUBLISHING COMPANY

DORDRECHT-HOLLAND / BOSTON-U.S.A.

Library of Congress Catalog Card Number 73–91433

ISBN 90 277 0429 5

Published by D. Reidel Publishing Company,
P.O. Box 17, Dordrecht, Holland

Sold and distributed in the U.S.A., Canada, and Mexico
by D. Reidel Publishing Company, Inc.
306 Dartmouth Street, Boston,
Mass. 02116, U.S.A.

Printed in The Netherlands by D. Reidel, Dordrecht

TABLE OF CONTENTS

3. MAGNETOSPHERIC PLASMA FLOW AND ELECTRIC FIELDS

4. SOLAR PARTICLE PROPAGATION IN THE INTERPLANETARY MEDIUM

5. SOLAR PARTICLE ENTRY TO THE EARTH

6. THE AUGUST 1972 SOLAR EVENTS

7. FUTURE STUDY TOPICS

PREFACE

The 1969 ESLAB symposium 'Intercorrelated Satellite Observations Related to Solar Events' was held at a time when the importance of bringing together measurements made simultaneously in different regions of space was beginning to be appreciated. To-day it is universally accepted that the major experimental steps forward in understanding the physics of the Sun-Earth relationships are likely to be made through pre-planned correlated satellite studies. Such considerations have led to the organisation of the International Magnetospheric Study and the joint ESRO/NASA International Magnetospheric Explorer, (Mother-Daughter-Heliocentric) mission.

The seventh ESLAB symposium was planned as a follow up to that of 1969 with the aim of deriving maximum benefit from those spacecraft which through good fortune found themselves simultaneously operating in different regions of the magnetosphere and interplanetary space. ESRO had launched in early 1972 its HEOS-2 satellite to investigate fields and particles in the unexplored region far above the North pole of the Earth and it became clear that the interesting new results arriving from that mission could profitably be linked with those from various American and the U.S.S.R. PROGNOZ satellites.

The book follows the order of the symposium (unfortunately the PROGNOZ contribution did not materialise) concentrating, through both review lectures and the presentation of new experimental results, on the nature of the boundaries between the interplanetary medium and the magnetosphere and the interaction of each region on the other.

The Editor gratefully acknowledges the support of his ESLAB colleagues, the magnificence of the Kleber Post hotel in Saulgau and above all the efficiency and hard work of Anne van den Eijkel.

Noordwijk, December 1973

LIST OF PARTICIPANTS

Anderson, K. A., University of California, Berkeley, Space Sciences Laboratory, Berkeley, Calif. 94720, U.S.A.

Axford, W. I., University of California, San Diego, Department of Applied Physics and Information Science, La Jolla, Calif. 92037, U.S.A.

Bahnsen, A., Danish Space Research Institute, Lyngby Division, Lundtoftevej 7, 2800 Lyngby, Denmark

Balogh, A., Imperial College of Science and Technology, Prince Consort Road, London S.W. 7, England

Beard, D. B., Department of Physics and Astronomy, The University of Kansas, Lawrence, Kan. 66044, U.S.A.

van Beek, H.F., Space Research Laboratory, Beneluxlaan 21, Utrecht, The Netherlands

Blake, J. B., Space Physics Laboratory, The Aerospace Corporation, P.O. Box 92957, Los Angeles, Calif. 90009, U.S.A.

Castagnoli, G., Istituto di Fisica Generale, Via P. Giuria 1, 10125 Torino, Italy

Chappell, C. R., Space Physics Laboratory, Dept. 52–12, Bldg. 202, Lockheed Palo Alto Res. Laboratory, 3251 Hanover Street, Palo Alto, Calif. 94304, U.S.A.

Chupp, E. L., University of New Hampshire, College of Technology, Dept. of Physics, DeMeritt Hall, Durham, N.H. 03824, U.S.A.

Decreau, Mrs. P., Groupe de Recherches Ionosphérique, CNRS, 45045 Orléans Cédex, France

Domingo, V., Space Science Department (ESLAB), European Space Research and Technology Centre, Noordwijk, The Netherlands

Durney, A. C., Space Science Department (ESLAB), European Space Research and Technology Centre, Noordwijk, The Netherlands

Engelmann, J., SEP, Centre d'Etudes Nucleaires de Saclay, Boîte Postale No. 2, 91, Gif-sur-Yvette, France

Fisk, L. A., NASA, Code 660, Goddard Space Flight Center, Greenbelt, Md. 20771, U.S.A.

Formisano, V., C.N.R., Laboratory Space Plasma, Via Galileo Galilei, P.O. Box 27, 00044, Frascati, Italy

Fritz, Th. A., Space Environment Lab., National Oceanic and Atmospheric Administration, Environmental Research Labs., Boulder, Colo. 80302, U.S.A.

Gall, R., Space Research Department, Instituto de Geofisica, Torre de Ciencias, Ciudad Universitario, Mexico 20, D.F. Mexico

Gonfalone, A., Space Science Department (ESLAB), European Space Research and Technology Centre, Noordwijk, The Netherlands

Grünwaldt, H., MPI für Physik und Astrophysik, Institut für Extraterrestrische Physik, 8046 Garching bei München, W. Germany

Haerendel, G., MPI für Physik und Astrophysik, Institut für Extraterrestrische Physik, 8046 Garching bei München, W. Germany

Hall, D. S., Appleton Laboratory, Ditton Park, Slough, Bucks., England

Harvey, C., Observatoire de Paris, Section d'Astrophysique, 92-Meudon, France

Haskell, G., European Space Research Organisation, 114 Ave. Charles de Gaulle, 92, Neuilly-sur-Seine, France

Hawkyard, A., European Space Research and Technology Centre, Noordwijk, The Netherlands

van Hollebeke, M. A., University of Maryland, Physics Department, College Park, Md., U.S.A.

Jones, D., Space Science Department (ESLAB), European Space Research and Technology Centre, Noordwijk, The Netherlands

Katz, L., Energetic Particles Branch, Space Physics Laboratory, Department of the Air Force AFSC, Laurence G. Hanscom Field, Bedford, Mass. 01730, U.S.A.

Kirsch, E., 3400 Göttingen, Saganer Weg 8, W. Germany

Kremser, G., MPI für Aeronomie, Institut für Stratosphärenphysik, D3411 – Lindau/Harz, Postfach 60, W. Germany

Lanzerotti, L. J., Bell Laboratories, 600 Mountain Avenue, Murray Hill, N.J. 07974, U.S.A.

McDonald, F. B., Laboratory for High Energy Astrophysics Code 660, NASA, Goddard Space Flight Center, Greenbelt, Md. 20771, U.S.A.

McKenzie, J. F., University of Khartoum, Faculty of Science, Dept. of Mathematics, P.O. Box 321, Khartoum, Sudan

Montgomery, M. D., MPI für Physik und Astrophysik, Institut für Extraterrestrische Physik, 8046 Garching bei München, W. Germany

Moreno, G., C.N.R., Laboratory Space Plasma, Via Galileo Galilei, P.O. Box 27, 00044, Frascati, Italy

Morfill, G. E., MPI für Physik und Astrophysik, Institut für Extraterrestrische Physik, 8046 Garching bei München, W. Germany

van den Nieuwenhof, R., Space Research Laboratory, Beneluxlaan 21, Utrecht, The Netherlands

Page, D. E., Space Science Department (ESLAB), European Space Research and Technology Centre, Noordwijk, The Netherlands

Parks, G., Geophysics Department, 202 ATS Building, University of Washington, Seattle, Wash. 98105, U.S.A.

Paschmann, G., MPI für Physik und Astrophysik, Institut für Extraterrestrische Physik, 8046 Garching bei München, W. Germany

Pedersen, A., Space Science Department (ESLAB), European Space Research and Technology Centre, Noordwijk, The Netherlands

Pfitzer, K. A., McDonnell Douglas Astronautics Company-West, 5301 Bolsa Avenue, Huntington Beach, Calif. 92647, U.S.A.

Pfotzer, G., MPI für Aeronomie, Institut für Stratosphärenphysik, D3411 – Lindau/Harz, Postfach 60, W. Germany

Pirre, M., Groupe de Recherches Ionosphérique, CNRS, 45045 Orléans Cédex, France

Riedler, W., Technische Hochschule Graz, Institut für Nachrichtentechnik, Krenngasse 37, A-8010 Graz, Austria

Roederer, J. G., Department of Physics, University of Denver, Denver, Colo. 80210, U.S.A.

Roelof, E. C., Department of Physics, University of New Hampshire, DeMeritt Hall, Durham, N.H. 03824, U.S.A.

van Rooijen, J., Space Research Laboratory, Beneluxlaan 21, Utrecht, The Netherlands

Rosen, A., Space Sciences Department, TRW Systems Group, One Space Park, Redondo Beach, Calif. 90278, U.S.A.

Rossberg, L., MPI für Aeronomie, Institut für Stratosphärenphysik, D3411 – Lindau/Harz, Postfach 60, W. Germany

Russell, C. T., University of California, Los Angeles, Inst. of Geophysics and Planetary Physics, Los Angeles, Calif. 90024, U.S.A.

Rycroft, M. J., Department of Physics, The University, Southampton, England

Sanderson, T. R., Space Science Department (ESLAB), European Space Research and Technology Centre, Noordwijk, The Netherlands

Schmerling, E. R., Physics and Astronomy Programs, NASA, Washington, D.C. 20546, U.S.A.

Scholer, M., MPI für Physik und Astrophysik, Inst. für Extraterrestrische Physik, 8046 Garching bei München, W. Germany

Sckopke, N., MPI für Physik und Astrophysik, Inst. für Extraterrestrische Physik, 8046 Garching bei München, W. Germany

Shea, M. A., Air Force Cambridge Res. Lab./PHE, L. G. Hanscom Field, Bedford, Mass. 01730, U.S.A.

Smart, D. F., Air Force Cambridge Res. Lab./PHE, L. G. Hanscom Field, Bedford, Mass. 01730, U.S.A.

Stauning, P. Ionosphere Laboratory, Technical University of Denmark, Lyngby, Denmark

Storey, L. R. O., Groupe de Recherches Ionosphériques, C.N.R.S., 4 Ave. de Neptune, 94 Saint-Maur-des-Fosses, France

Stuedemann, W., MPI für Aeronomie, Institut für Stratosphärenphysik, D3411 – Lindau/Harz, Postfach 60, W. Germany

Svalgaard, L., Institute for Plasma Research, Stanford University, Via Crespi, Stanford, Calif. 94305, U.S.A.

Thomas, G. R., Appleton Laboratory, Ditton Park, Slough, Bucks., England

Trendelenburg, E. A., Space Science Department (ESLAB), European Space Res. and Technology Centre, Noordwijk, The Netherlands

Treguer, L., S.E.P., Centre d'Etudes Nucléaires de Saclay, Boite Postale No. 2, 91, Gif-sur-Yvette, France

Vampola, A. L., Space Physics Laboratory, The Aerospace Corporation, P.O. Box 95085, Los Angeles, Calif. 90045, U.S.A.

Wenzel, K.-P., Space Science Department (ESLAB), European Space Research and Technology Centre, Noordwijk, The Netherlands

Wilken, B., MPI für Aeronomie, Institut für Stratosphärenphysik, D3411 – Lindau/ Harz, Postfach 60, W. Germany

PART 1

RESPONSE OF THE MAGNETOSPHERE TO CHANGES IN THE INTERPLANETARY MEDIUM

THE SOLAR WIND AND MAGNETOSPHERIC DYNAMICS

CHRISTOPHER T. RUSSELL
Institute of Geophysics and Planetary Physics.,
University of California, Los Angeles, Calif., U.S.A.

Abstract. Our understanding of the solar wind-magnetosphere interaction has evolved slowly but steadily since the advent of space exploration. The early picture of transient confinement of the geomagnetic field during geomagnetic storms by blast waves of solar plasma was soon replaced by a magnetosphere confined by a continuously flowing solar wind. This magnetosphere was found to be elongated in the antisolar direction forming a region behind the Earth known as the geomagnetic tail. In front of the magnetosphere, a shock front was observed at which the solar wind was decelerated, heated and deflected around the magnetosphere.

While the solar wind dynamic pressure determines the overall size and shape of the magnetosphere, recent measurements show that tangential stresses play an essential role in magnetospheric dynamics particularly in the geomagnetic tail. Much of the tangential stress on the magnetosphere arises due to the merging of the interplanetary field with the magnetospheric field. As the merging rate on the dayside changes and as the reconnection rate in the geomagnetic tail fluctuates, the polar cap expands and shrinks, the field strength in the tail increases and decreases, and the plasma sheet thins and thickens. Magnetospheric substorms apparently play an important role in this process as expansion phase onsets are correlated with a return to a dipolar configuration of the nightside field lines and with the appearance of southward fields in the plasma sheet, and are followed by a thickening of the plasma sheet. The sequence of events occurring in the tail during a substorm are readily explained by the formation of a neutral point near the Earth at the expansion phase onset followed by the eventual motion of this neutral point away from the Earth.

Finally, the control of the tangential stress on the magnetosphere by the interplanetary field can be used to explain most of the properties of the semiannual variation of geomagnetic activity. In particular, the Dst index, which is used to identify geomagnetic storms, can be quite accurately predicted with a knowledge of only the solar wind number density, velocity, and the north-south component of the interplanetary field.

1. Introduction

If there were no solar wind, the physics of the magnetosphere might be well understood, and the subject of little, if any, further research. However, there is a continual supersonic outflow of plasma from the Sun which confines and distorts the terrestrial magnetic field. Changes in the properties of the solar wind are accompanied by corresponding changes in the magnetosphere. Our present lack of a clear understanding of the physical processes occurring within the magnetosphere is not simply due to the fact that changes occur in the solar wind plasma, although, if such changes did not occur, experimental investigations would be greatly simplified. It is principally due to the fact that the strength of the interaction of the solar wind with the magnetosphere, i.e., the rate of energy input, is modulated by the interplanetary magnetic field, and to the fact the magnetosphere can store energy in one region for subsequent and often abrupt release into other regions. The modulation of this interaction and the abrupt release of the stored energy are the major topics of this review.

Before discussing the dynamical processes within the magnetosphere, we present a brief magnetospheric history. This will serve to introduce both the overall configuration of the magnetosphere and the nomenclature used in magnetospheric

D. E. Page (ed.), Correlated Interplanetary and Magnetospheric Observations, 3–47. *All Rights Reserved.*

research. Then, we discuss the evidence for a dynamic magnetosphere and review the nature of the changes in the magnetosphere during a period of enhanced solar wind magnetosphere coupling. Finally, we apply some of our understanding of this interaction to explaining the well-known cycles of geomagnetism, and to predicting geomagnetic activity using solar wind measurements.

2. A Brief History of the Magnetosphere

2.1. Early Magnetospheric Models

Geomagnetic observations have been made for centuries. An excellent history of these measurements prior to the 20th century can be found in Chapter XXVI of *Geomagnetism* by Chapman and Bartels (1940). Figure 1 shows the first magnetospheric model published in 1600 by Gilbert. Although the ionosphere was not discovered until 1883 by Balfour Stewart and the distortions in the magnetic field caused by the solar wind until much later, Gilbert wisely avoided extrapolating his field lines far above the surface of the Earth.

The next major step in the understanding of the magnetosphere was also the first major advance in the study of magnetospheric dynamics. Although the idea that solar corpuscular radiation was responsible for geomagnetic storms and aurora was popular in the early 20th century (see the review by Dessler (1967)), it was not until 1930 that a satisfactory theory was proposed.

In a series of papers, Chapman and Ferraro (1930; 1931; 1932) discussed a mechanism for generating geomagnetic storms. A principal feature of this model was a transient outflow of ionized gas from the Sun consisting of positive ions and electrons but having no net charge. Figure 2 illustrates the effect of this ionized gas, or plasma, in the case of an advancing infinite planar front. The plasma, being highly conducting, excludes field lines from its interior. This is mathematically equivalent to the creation of an equal and parallel dipole moment, at a distance behind the front the same as the distance that the Earth's dipole is ahead of the front. If the front of gas were not planar, as would necessarily occur as the front moved past the Earth, the strength of this image dipole field and its position would both change. We note that, when an approximate description of the magnetospheric field is sufficient, image dipole models are still used today because of their mathematical simplicity.

Figure 3 illustrates Chapman and Ferraro's method of creating the main phase of a geomagnetic storm, that period when the Earth's magnetic field is reduced. As the front of gas passes the Earth, charged particles leak into the magnetosphere and drift around the Earth, creating a current whose field opposes the Earth's main field. While this model is very close to that which we believe today, there was one important change to this picture made 30 yr later. As a result of studies of comet tails (Biermann, 1951; 1953; 1957) and theoretical investigation of the interplanetary gas by Parker (1958a) who also coined the term solar wind (Parker, 1958b), the realization grew that the solar wind was continually flowing. However, short term ($\sim$1 day) direct measurements (Gringauz *et al.*, 1960; Gringauz, 1961; Shklovsky

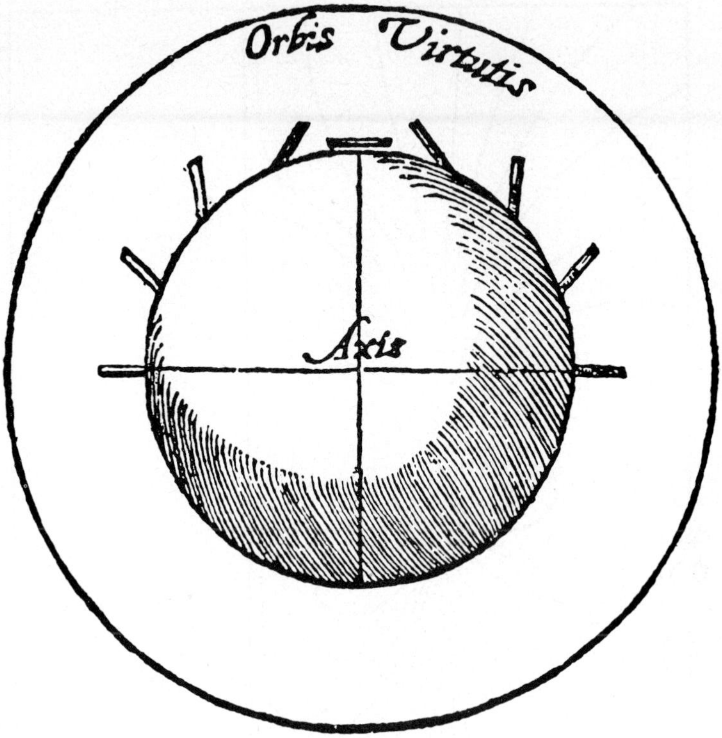

Fig. 1. Illustration of the dipolar nature of the Earth's magnetic field from Gilbert's *De Magnete* (Chapman and Bartels, 1940).

et al., 1961; Bonetti *et al.*, 1963a, b) on a variety of spacecraft were not sufficient to convince many. It was not until the continuous Mariner 2 measurements were reported (Neugebauer and Snyder, 1962; Snyder *et al.*, 1963) that the concept of a continuously flowing solar wind became commonly accepted (cf. Dessler, 1967).

While Figure 3 is suggestive of a long geomagnetic tail in the antisolar direction, the geomagnetic field would be confined to a moderately symmetric magnetic bubble by the hot, infinitely conducting solar wind plasma, if there were no viscous interaction between the solar wind and the magnetosphere. For realistic ratios of the static to dynamic pressure, Slutz and Winkelman (1964) obtain a rather short geomagnetic tail, of about 10 to 25 R_e in length. They remark that "the whole shape of the solution reminds one much more of an eyeball than of a teardrop". On the other hand, they did expect a teardrop-shaped wake region, or 'wind-tail' in the antisolar direction. Parker (1958b) pointed out that the interaction should be viscous, concluding "It is obvious that the solar wind tends to carry away lines of force of the outer geomagnetic field, just as a high wind blows smoke away from a chimney". This conclusion was not obvious to everyone, and Figure 4 illustrates the two competing models in 1960 as drawn by Piddington (1960) and Johnson (1960).

Two years later, measurements, made by Explorer 6 and Explorer 10 in the distant midnight magnetosphere, were shown to be consistent with the existence

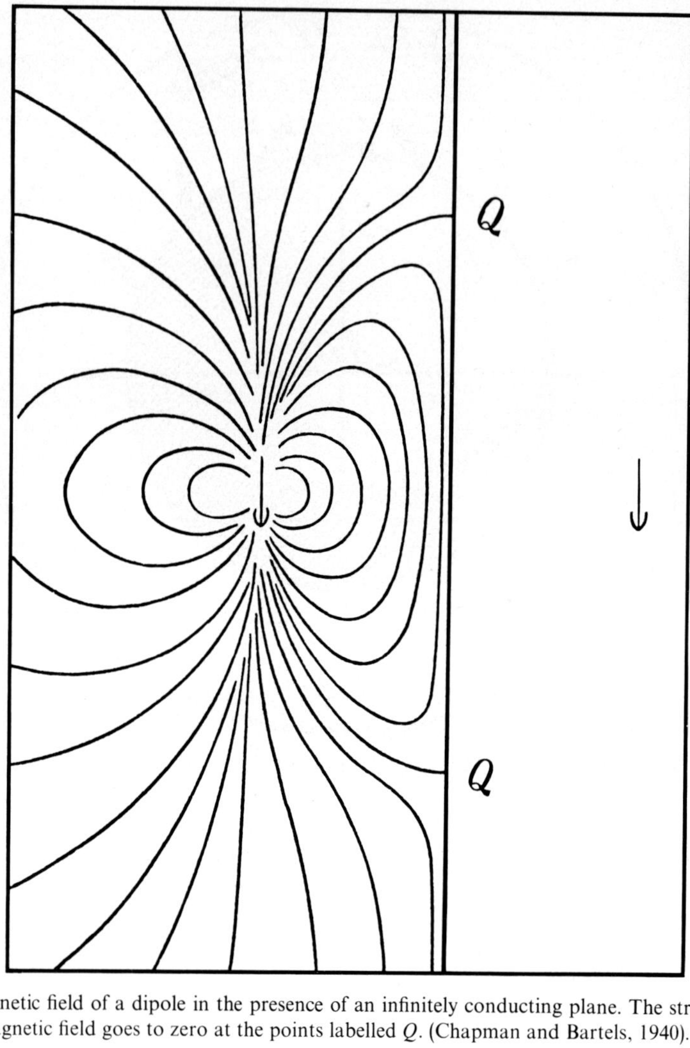

Fig. 2. Magnetic field of a dipole in the presence of an infinitely conducting plane. The strength of the magnetic field goes to zero at the points labelled Q. (Chapman and Bartels, 1940).

of an extended tail (Smith, 1962). Figure 5 shows these measurements projected into the noon-midnight meridian with the Earth's field removed (Axford *et al.*, 1965). The variation of this residual field along the satellite trajectories is qualitatively as expected for the twin solenoid current system of the magnetotail.

2.2. The Open Magnetosphere

Early models of the solar wind magnetosphere interaction ignored any possible effects of the interplanetary magnetic field. Dungey (1961), however, soon pointed out that the existence of the interplanetary magnetic field had important consequences for this interaction. Figure 6 shows Dungey's two magnetospheric models: the top panel for southward interplanetary fields and the bottom one for northward fields (Dungey, 1963). In the top model, the opposed magnetospheric and interplanetary

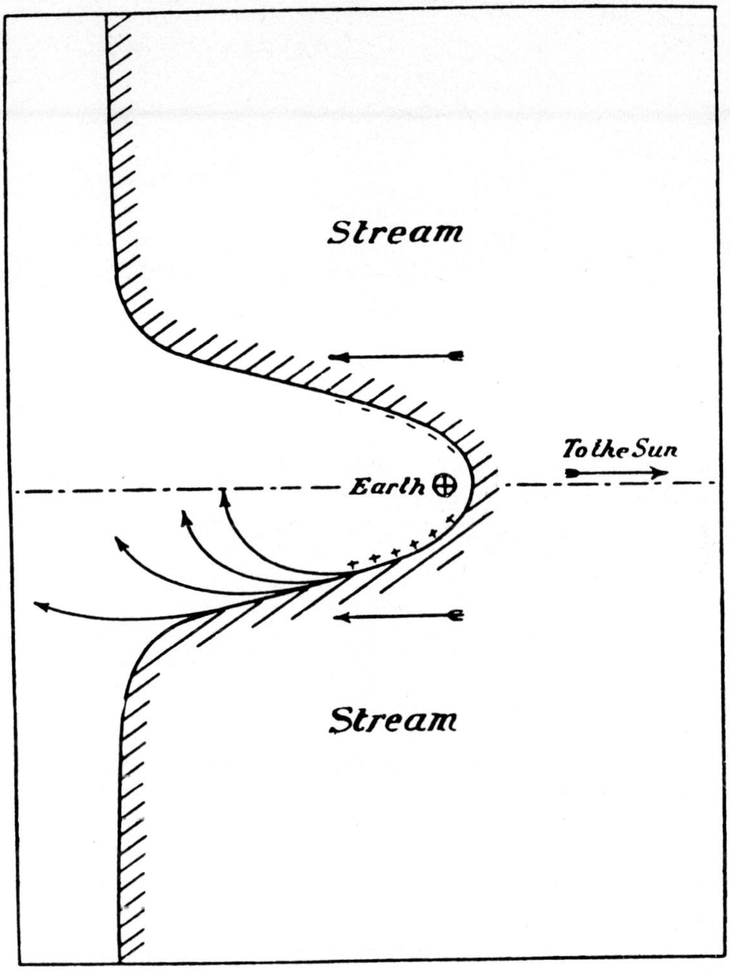

Fig. 3. Distortion of the advancing front of solar plasma (ion-electron gas) by the Earth's magnetic field. This model was proposed by Chapman and Ferraro (1930; 1931; 1932) to explain the sudden compression of the Earth's magnetic field followed by a more gradual depression of the field during a geomagnetic storm.

fields merge or reconnect at a neutral point at the front or nose of the magnetosphere. The motion of the solar wind carries the plasma on this interplanetary field line, and hence the field line itself, tailward. As the two halves of the field line move tailward in the interplanetary medium, the magnetospheric part of the lines, each of which has a foot in one polar cap, also move tailward. Eventually, the two halves of the line merge or reconnect in the tail at a neutral point, again forming a magnetospheric field line with two feet on the ground and an interplanetary field line not connected with the Earth. The newly connected field line, then moves earthward to return to the dayside merging region to repeat this process all over again.

Moving field lines, while a useful concept for some, are totally alien to others and hence the above model is viewed with some suspicion. However, it is entirely

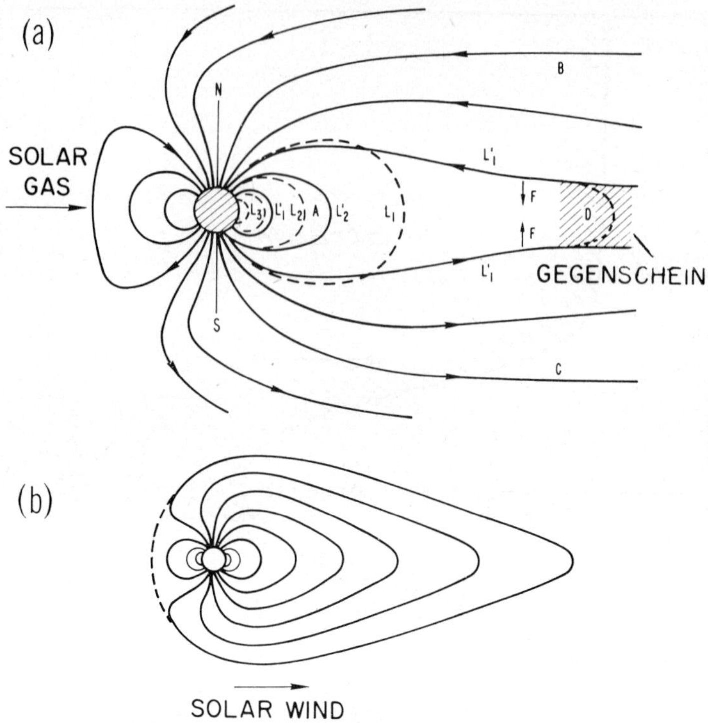

Fig. 4. Two magnetospheric models of 1960. The upper panel shows the noon-midnight meridian as drawn by Piddington (1960), and the lower panel shows the noon-midnight meridian as drawn by Johnson (1960).

equivalent to view the field lines in this model as stationary and talk entirely in terms of plasma motions. The solar wind plasma flows across the interplanetary field lines. When it encounters the magnetosphere, it mixes with an outward flow from the magnetosphere. The magnetopause is not a perfect conductor in this model and part of the interplanetary electric field (determined by the merging rate) penetrates onto the polar field lines and causes a tailward flow in this region. Flow over the north and south polar caps eventually meets in the tail. Then some of the plasma flows earthward and some tailward. We note that the electric field on closed field lines is not due to the penetration of the interplanetary electric field through the magnetopause, but represents the flow of the plasma from the reconnection region in the tail to the merging region on the dayside. Ionospheric drag on this flow will be overcome by the establishment of pressure gradients. Further, the dayside merging rate and the tail reconnection rate can differ significantly for short periods. This results in changing magnetic fields and non-conservative electric fields.

The bottom panel shows the situation for northward fields. Here field lines are 'laid' on the magnetosphere and merge at two neutral points in the tail. This process effectively adds a dayside field line and removes one from the tail. The flow on closed field lines proceeds from the dayside to the nightside. It seems improbable,

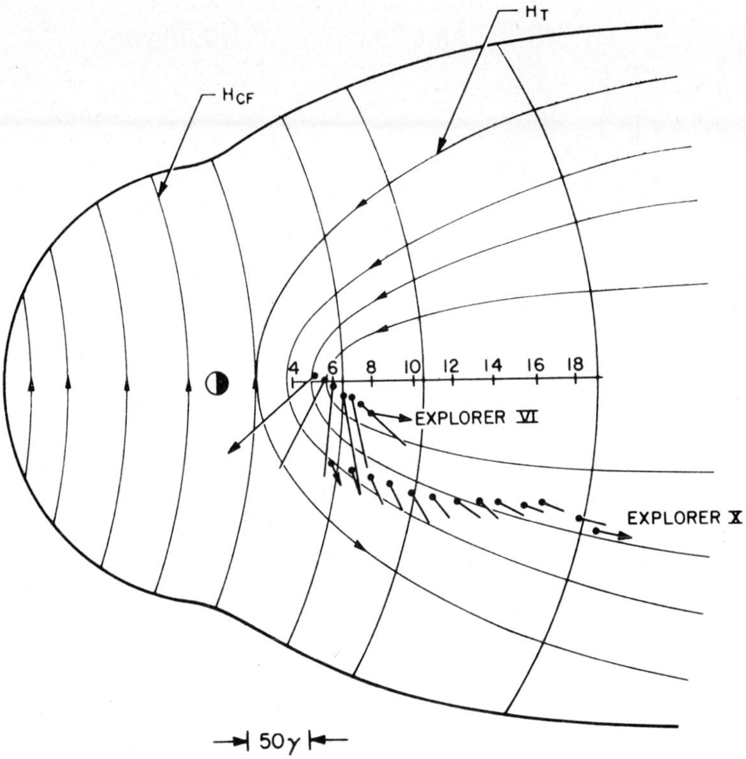

Fig. 5. Explorer VI and X magnetic field measurements in the distant midnight magnetosphere. The contribution of the Earth's main field has been removed and the residual field has been projected into the noon-midnight meridian. The field in this meridian due solely to a Chapman-Ferraro surface current, H_{CF}, and solely to a twin solenoid tail system, H_T, is also shown (Axford *et al.*, 1965).

though, that this mechanism is effective in removing flux from the tail because it requires that the same interplanetary field line become connected to the magnetosphere at two neutral points (Russell, 1972a).

Some words about the topology of Dungey's model in the presence of a southward interplanetary field are in order here. First, there are three classes of field lines: those with no feet on the ground; those with one foot on the ground; and those with two feet on the ground. These three classes are separated by two surfaces: one resembling somewhat the surface of an hour glass; and one the surface of a doughnut. These two surfaces touch at a line going all the way around from the dayside neutral point to the nightside neutral point. This line is called the separatrix. It is not necessarily neutral because field lines can be parallel to the separatrix without destroying the topology (Stern, 1972). Thus, there are only two neutral points in this model and not necessarily a neutral line, as has been emphasized by Dungey (1963).

It should be noted that Dungey drew his models to emphasize the physical mechanisms and did not attempt to draw the magnetosphere to scale. Unfortunately, this has led to some misinterpretation of Dungey's model. Dungey certainly knew

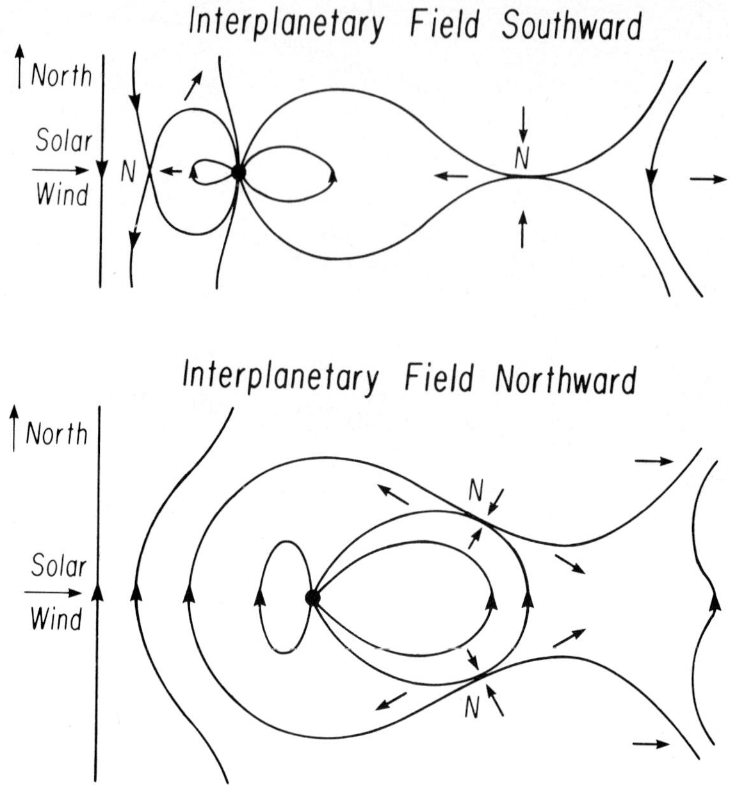

Fig. 6. Schematic illustration of the open magnetosphere, for southward, top panel, and northward, bottom panel, interplanetary fields. These sketches do not attempt to show the correct length of the tail nor the magnitude of the magnetic field normal to the magnetopause. Arrows illustrate plasma flow directions.

the correct scale. In fact, he used his model to predict the length of the tail to be 500 R_e (Dungey, 1965). This length is close to the length derived from recent studies of energetic solar particle entry into the tail (cf. Scholer, 1974; Vampola, 1974). We note that both studies of solar electrons (Lin and Anderson, 1966; Van Allen, 1970) and solar protons (Van Allen *et al.*, 1971; Morfill and Scholer, 1972; Fennell, 1973) in the magnetotail and over the polar cap leave little doubt that the magnetosphere is open.

2.3. Refining Magnetospheric Models

Figure 7 shows two models of the magnetosphere published in 1965. The model in the top panel is an open magnetosphere resembling that of Dungey but more closely approximating the correct scale (Axford *et al.*, 1965). Here the flow in the tail goes into a 'neutral' line. This line, as stated before, is the separatrix and is not necessarily neutral. The merging line would be a more appropriate name. The region labelled

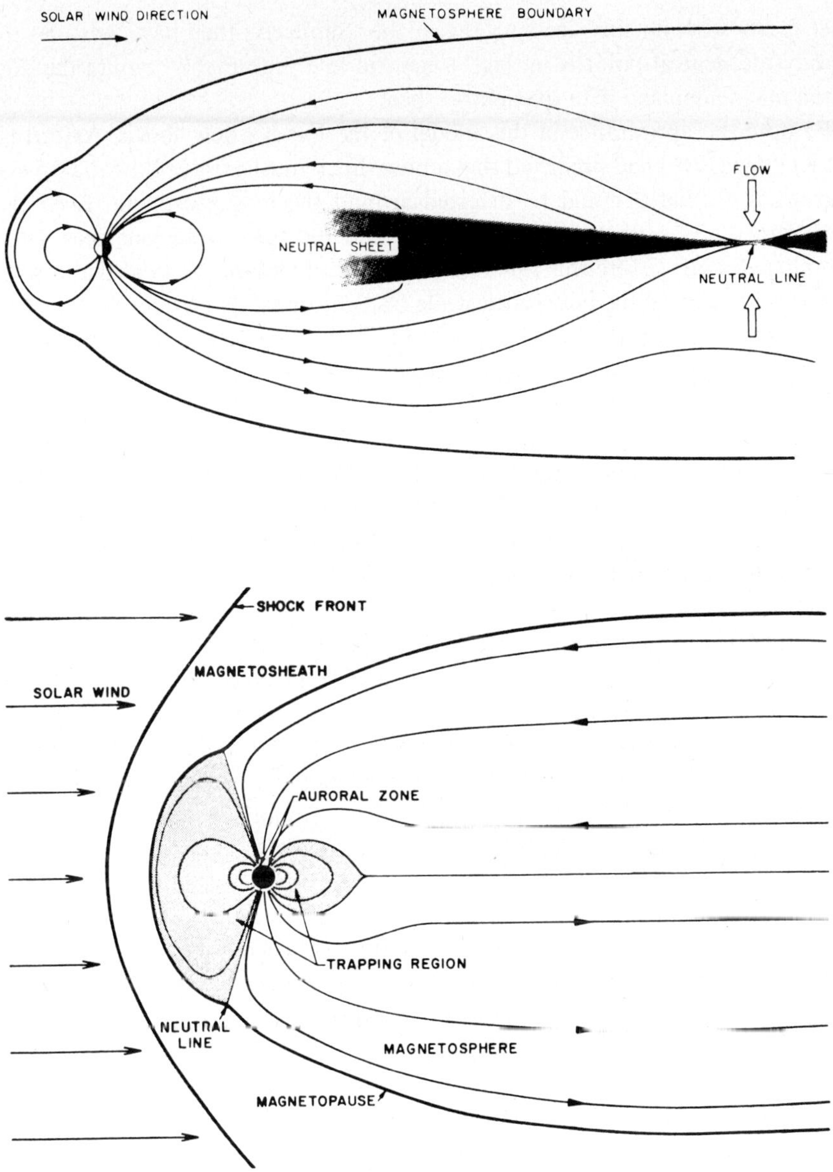

Fig. 7. Two magnetospheric models of 1965. The upper panel shows the open model of the magnetosphere with a finite tail (Axford *et al.*, 1965). The lower panel shows an open model with essentially an infinite length (Dessler and Juday, 1965).

neutral sheet is obviously not neutral, in a magnetic sense, since field lines cross this region. We would call this region the plasma sheet today, and reserve the term neutral sheet for the plane separating magnetic field in the tail pointing towards the Earth from that pointing away from the Earth.

In the lower panel, Dessler and Juday (1965) have postulated that a true neutral sheet exists with no flux crossing this plane. Similarly, they have postulated that the dayside neutral point is, in fact, a neutral line presumably joining the dimples on the magnetosphere with the neutral sheet.

We note the appearance in this model of the Earth's bow shock. Axford (1962) and Kellogg (1962) had predicted that a bow shock must exist to slow the supersonic solar wind so that it could be diverted around the magnetosphere. Even though measurements of both the magnetosheath and the solar wind had been made by several spacecraft before this time, it was not until October 7, 1962 that a satellite, Explorer 14, crossed the bow shock while transmitting data (Wolfe and Silva, 1965). Early spacecraft with their low sampling rates did little better than identify regions as either being solar wind or magnetosheath, and could not resolve the features of the bow shock itself. However, these measurements did reveal that an apparent shock transition was a permanent feature of the solar wind magnetosphere interaction (Ness *et al.*, 1964; Bridge *et al.*, 1965; Wolfe *et al.*, 1966). It was not until the OGO series of spacecraft with much higher data rates that the shock structure itself could be resolved (Holzer *et al.*, 1966; Heppner *et al.*, 1967). While the shock structure has been under intense study recently, there is yet much to be learned (Formisano, 1974).

Figure 8 shows an updating by Dessler (1968) of his magnetospheric model. This illustrates the convergence of views to that of an open tail with a neutral point (or line). However, in this model, Dessler has moved the neutral point within 30 R_e of the Earth despite the evidence presented by Mihalov *et al.* (1968) and later by Behannon (1970) that the average neutral point position is beyond the orbit of the Moon. Dessler and Hill (1970) argue the southward component tailward of the neutral point would be small and difficult to observe. Further, Hill and Dessler (1971) present a model for maintaining the plasma sheet on open field lines as this model requires with magnetic turbulence during quiet times. However, this model is not supported by observations of ULF waves in the tail (Russell, 1972b; Garrett, 1973a). Thus, there is little doubt that, at quiet times, the neutral point is beyond the orbit of the Moon. On the other hand, as we will discuss in a later section, the neutral point does move close to the Earth during geomagnetically disturbed times.

While the neutral point in the tail is very important in understanding magnetospheric processes, the so-called neutral points in the dayside magnetosphere, at the dimples in the magnetopause are also of great interest. Even in closed models of the magnetosphere, magnetosheath plasma can gain entry into the magnetosphere, and reach the ionosphere along the flux tube in the noon meridian which separates the field lines which close over the pole from those which close on the dayside. For example, in the image dipole approximation of the magnetopause, shown in Figure 2, the entry occurs at the two neutral points labelled Q.

Despite the myriad of magnetospheric missions during the 1960's, the first probing of this region was not reported until 1971. At this time encounters by four separate spacecraft with this region, called the polar cusp or cleft, were reported. At low altitudes measurements were made by ISIS 1 (Heikkila and Winningham, 1971)

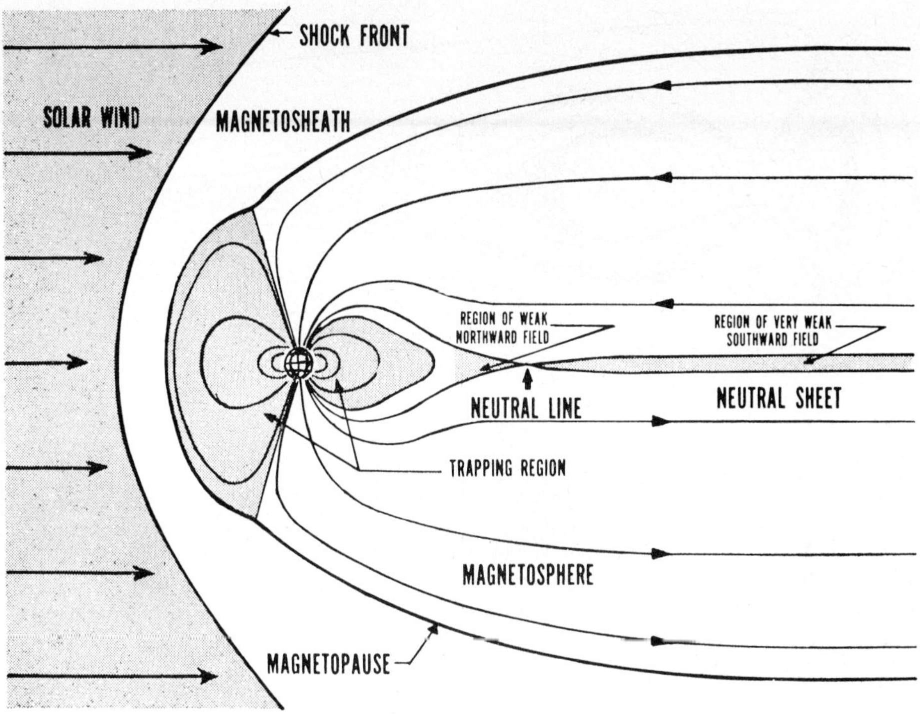

Fig. 8. The steady-state near-Earth neutral point model of Dessler (1968).

and by Injun 5 (Frank and Ackerson, 1971), and at high altitudes by Imp 5 (Frank, 1971) and by OGO 5 (Russell *et al.*, 1971). Most recently, HEOS 2 has probed the polar cusp (Bahnsen *et al.*, 1974; Paschmann *et al.*, 1974; Wenzel, 1974).

Figure 9 shows a post-polar cusp magnetospheric model by Mishin and Popov (1972). It illustrates clearly the direct penetration of the polar cusp plasma and that the cusp plasma is topologically connected to the plasma sheet boundary, since the cusp straddles the last closed field lines and first open field lines. However, the magnetopause, in the near Earth tail, is also close to the first open field lines. Thus, polar cusp plasma might be expected in this region also. Such a boundary layer has been observed by HEOS 2 (Paschmann *et al.*, 1974) and Vela satellites (Akasofu *et al.*, 1973). Haerendel (1973) has called this the polar mantle. This model also stresses that different types of precipitation are observed at different latitudes.

While this picture of the polar cusp suggests that the cusp acts mainly as a funnel to channel magnetosheath plasma into the magnetosphere and ionosphere this is far from the case. Strong currents flow in the polar cusp (Fairfield and Ness, 1972; Kivelson *et al.*, 1973; Fredricks *et al.*, 1973), and intense ULF magnetic and VLF electric oscillations are observed there (Russell *et al.*, 1971; Scarf *et al.*, 1972; Fredricks and Russell, 1973; Bahnsen *et al.*, 1974). Furthermore, both upstreaming and downstreaming particles are observed (Paschmann *et al.*, 1974).

There is one further refinement necessary in our magnetospheric model, if we are

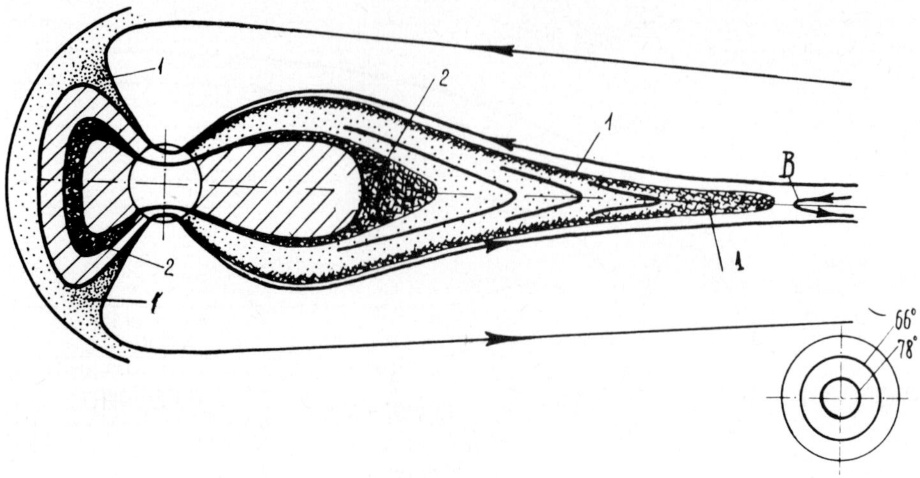

Fig. 9. The dual precipitation zone model of Mishin and Popov (1972). The polar cusp plasma is presumed to precipitate at high latitudes ($\sim 78°$) at all local times as illustrated in the sketch of the polar cap on the lower right. Similarly, there is harder low latitude precipitation from the inner edge of the plasma sheet.

to understand processes occurring in the inner magnetosphere. That refinement is the addition of the plasmasphere, a region of 'cold' ($2–10 \times 10^3$ K), dense ($\sim 10^2–10^4$ particles cm^{-3}) plasma in the inner magnetosphere, corresponding roughly to the inner cross-hatched region in Figure 9. For a recent review of the plasmasphere, see Chappell (1972). This region is simply the extension of the ionospheric plasma into the magnetosphere. The reason that the ionospheric plasma does not fill all the 'closed' flux tubes in the magnetosphere is that the filling time is long compared to the time required for cold plasma in the outer magnetosphere to drift out of the magnetosphere across the magnetopause. In a closed magnetosphere, the cold magnetospheric plasma would be circulated in cells (Axford and Hines, 1961) and not leave the system. However, in an open magnetosphere, the convection of plasma intersects the boundary of the magnetosphere except for an inner corotating core. In this inner core the density can build up and form what is called the plasmasphere. While the convection electric field and its time variations have been invoked to explain most of the observed behavior of the plasmasphere, we have only sparse data on the temporal behavior of the conservative and nonconservative electric fields in the magnetosphere (Chappell, 1974; Roederer, 1974).

3. Magnetospheric Dynamics

While the Chapman-Ferraro geomagnetic storm model proposes that there are dynamical changes in the magnetosphere during magnetic storms, it is not immediately obvious from the brief survey of the magnetosphere in the last section that the magnetosphere is almost continually undergoing significant dynamical changes, and that these changes often play a fundamental role in magnetospheric processes. In this section, we examine the evidence of magnetospheric dynamics, and the control of these dynamical processes by the interplanetary field. We, then, discuss the various dynamical changes that occur: erosion of the magnetopause, changes in the size of the polar cap, variations in the flaring angle of the tail, neutral point formation, plasma sheet motions, and the inward collapse of the midnight magnetosphere.

3.1. The Evidence for a Dynamic Magnetosphere

One of the first suggestions after the discovery of the outer radiation zone was that this zone might be the source of the electrons responsible for auroras (Van Allen *et al.*, 1959). The magnetosphere was viewed as a reservoir occasionally spilling out these electrons into the auroral regions. This was referred to as the leaky bucket model. However, O'Brien (1962) noted that the outer radiation zones would 'drain empty' of electrons in a few hours at the observed precipitation rates. He suggested instead that the magnetosphere was a 'splash catcher' trapping some of the freshly accelerated particles which caused aurora.

While it is conceivable that auroral activity occurs in a quasi-stationary pattern under which the Earth rotates, there is much evidence that auroral activity is unsteady and hence, the amount of 'splash' caught by the magnetosphere varies with time. The classic study of the aurora by Akasofu (1964) showed that there were auroral cycles which could repeat several times during the night. During these cycles, auroral arcs first brightened and then expanded poleward. On the western edge of this poleward expanding bulge, a fold in the arcs develops, called the westward travelling surge, which rapidly moves from midnight towards dusk. This period is called the expansion or break-up phase. Later, the arcs return to their pre-break-up condition during what is called the recovery phase. Chapman suggested that this cycle be called substorm and this name and the accompanying concept has become generally accepted.

In addition to the auroral substorm, there is a cycle of magnetic activity in which currents flowing in the ionosphere and the magnetosphere change in strength and in location. This cycle, called the polar magnetic substorm, is intimately related to the auroral substorm. During substorms, there are changes in the ionosphere, in particle precipitation patterns, VLF emissions, and micropulsations. These effects have been comprehensively described by Akasofu (1968), and apparently are all manifestations of one major process, the magnetospheric* substorm.

Figure 10 shows a modern splash catcher model of the magnetosphere (Davis, 1970), which summarizes much of what we know about magnetospheric processes. Starting with the pipes in the top right-hand corner, this model illustrates that there are two possible sources for magnetospheric particles: solar wind particles and atmospheric particles. The particles then flow into an 'expandable tippy thundermug', with some leakage directly into the polar cap and from the thundermug, or

* While, in retrospect, the magnetospheric nature of substorms is obvious, it was not until 1967 that the term magnetospheric substorm was suggested (McPherron *et al.*, 1967) or the magnetospheric consequences of the substorm discussed (Brice, 1967; Jelly and Brice, 1967; McPherron *et al.*, 1967).

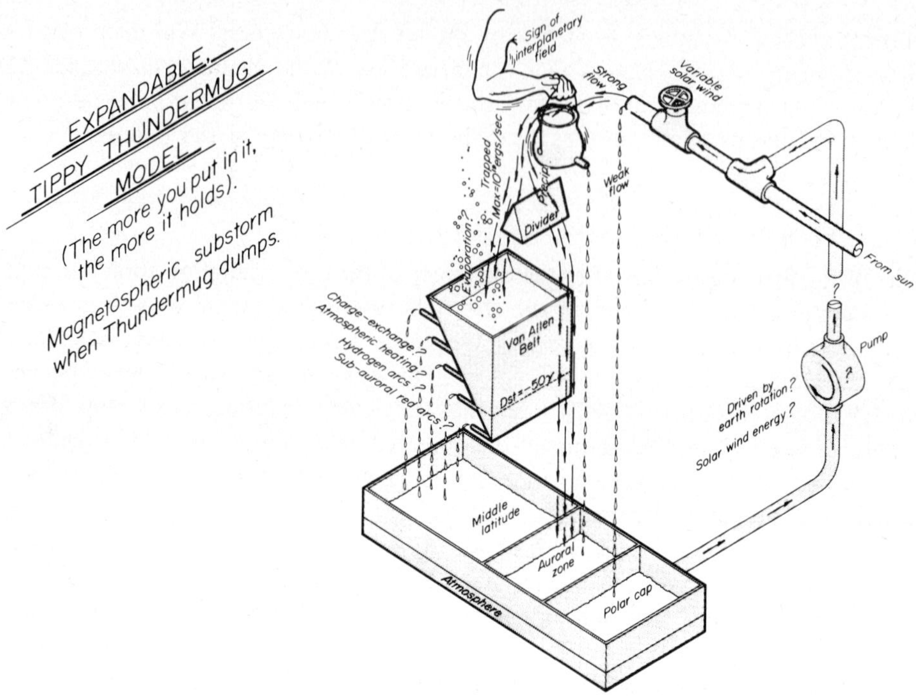

Fig. 10. The Expandable Tippy Thundermug model of substorm behavior by Davis (1970). This model continues the earlier concepts of the Van Allen belts as a leaky bucket and a splash catcher, while illustrating many of the phenomena known to occur in the magnetosphere.

pot, into the auroral zone. In this model, the pot is stable until it fills to a point such that the center of gravity is above the pivots, joining the handle to the mug. Then the pot is in a metastable state, and will tip if perturbed, say, by a southward pointing magnetic field.

When the pot tips, some of the particles are caught in the Van Allen belts and some go directly into the auroral oval. Note that it requires many tippings of the pot to fill the Van Allen belts to capacity in accord with the concept that many substorms are required to cause the main phase depression of a geomagnetic storm. There are many loss processes from the Van Allen belts; the many spigots illustrate this. The fact that the spigots are at different heights illustrates that some of these processes are thought to occur at different levels of geomagnetic activity.

At the level of analogy for which this model was intended, this model is in accord with most of what we know about the magnetosphere, with perhaps one exception. Recent studies show that the sign of the interplanetary magnetic field governs the strength of the solar wind interaction with the magnetosphere. Hence, the hand labelled 'sign of interplanetary field' would be more appropriately placed on the tap.

A recent study by Foster *et al.* (1971) emphasizes both the cyclic nature of the solar wind–magnetosphere interaction and the control of this interaction by the north-south component of the interplanetary magnetic field. Figure 11 shows their super-

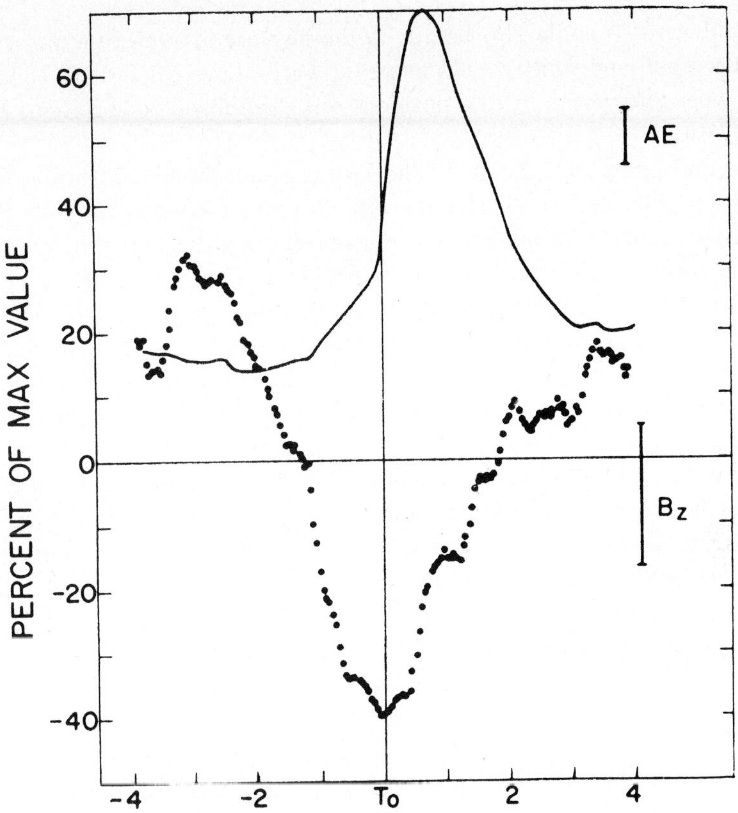

Fig. 11. Superposed epoch study of 54 isolated substorms showing the average percentage of the maximum value of the AE index and north-south component of the interplanetary magnetic field. The time scale is measured in hours from the onset of the expansion phase (Foster *et al.*, 1971).

posed epoch study of the auroral electrojet index and the north-south solar magnetospheric component of the interplanetary magnetic field for 54 isolated substorms. T_0 was determined in two ways for each event. The first determination was the time of the sharp onset of a negative bay at the auroral-zone station closest to midnight. The second determination was the time of the sharp recovery of the H-component at a low latitude premidnight station. When these two determinations agreed within 10 min, as they did in the majority of their chosen isolated events, T_0 was taken to be the average of the two determinations; otherwise the event was not used.

The AE index, which measures the strength of currents flowing in the ionosphere over selected auroral zone stations, increases with two distinct rates in this average picture. The slow increase, beginning when the north-south component *first becomes southward*, lasts about 75 min. We would identify this period with the growth phase of the substorm (McPherron, 1970). The cause of this pre-onset increase in the AE index could be due to the increase in strength of the magnetospheric convection electric field due to enhanced merging on the dayside when the interplanetary field turns southward. It could also be due to the increase in size of the polar cap due to enhanced merging which would move the auroral current systems equatorward of their quiet time positions to a position over the auroral zone stations (Akasofu *et al.*, 1973b).

The second phase, starting at T_0 and lasting about 40 min, when the AE index rises most rapidly, we would identify with the expansion phase. While the onset of this phase occurs at the minimum of B_Z, we feel this is due to the choice of isolated events for study, since events, in which the field turned northward before a substorm onset, should have a smaller probability of a subsequent onset, and since events, in which the field stayed southward long after the minimum, should be associated with multiple substorm periods and would not be included. We note that the recovery phase, when the AE index is decreasing, begins while the interplanetary field on the average is still southward. The fact that the expansion phase occurs in this average picture during a period of northward turning field suggests that the onset is not due to a strengthening of the interaction and that the increase in current measured by the AE index at this time is associated with changes internal to the magnetosphere, probably controlled by changes in the magnetic field configuration, i.e., with non-conservative electric fields.

While ground-based observations of aurora and magnetic fields have provided our earliest and most complete record of the cyclic nature of the solar wind magnetospheric interaction, controlled, as we have seen, by the interplanetary magnetic field, there is, of course, much evidence for these cycles, or substorms, in the magnetosphere itself. In the following sections we will examine the magnetospheric response in various regions of space to this time varying interaction.

3.2. The Erosion of the Magnetopause

The term 'erosion of the magnetopause' was coined by Aubry *et al.* (1970) to describe the inward motion of the dayside magnetopause in response to a southward turning of the interplanetary magnetic field even though the dynamic pressure of the solar wind remains constant. This motion can be explained in terms of current systems which arise when the interplanetary field is southward and which oppose the Earth's main field on the dayside of the Earth. Two current systems, both of which probably play a part in the field reduction, are the tail current system, which is observed to strengthen and move towards the Earth when the interplanetary field turns southward and a partial ring current associated with the enhanced convection electric field (Russell and McPherron, 1973a; Russell *et al.*, 1974). To cause the extrapolated subsolar point of the magnetopause to move from 11.7 to 9.5 R_e as the observations of Aubry *et al.* (1970) suggest occurred, requires a reduction of the dayside magnetospheric field of 17 γ.

In support of this single observation which occurred when the satellite inward motion approximately matched that of the magnetopause, Fairfield (1971) has shown that when the magnetosheath field is southward, the magnetopause is on the average one Earth radius closer to the Earth. Finally, Russell *et al.* (1971) have shown that the polar cusp moves equatorwards in the presence of a southward interplanetary magnetic field. While this motion implies that the dayside magnetosphere is shrinking, it also implies that the polar cap, at least on the dayside, is getting bigger.

3.3. The Size of the Polar Cap

If merging on the dayside proceeds at a rate faster than reconnection in the tail, the number of open field lines and hence the area of the polar cap will increase. Both the inward motion of the magnetopause and the thinning of the plasma sheet during the growth phase, which we will discuss in a later section, are indications that such is the case. Furthermore, there have been direct measurements of the equatorward motion of the polar cusp. Akasofu (1972a, b) has used all-sky camera data at local noon in winter, when such measurements can be made, to show that the midday aurora move equatorward in proportion to the accompanying geomagnetic activity. More to the point, Burch (1972) has used OGO-4 soft electron (>0.7 keV) data to measure the polar cusp position as a function of time after the southward turning of the interplanetary field, in cases of a quasi-steady southward interplanetary field with no discernible expansion phase onset. He obtains an equatorward cusp velocity of 0.1° invariant latitude per minute. While such measurements do not prove the polar cap increases in size, since it may be simultaneously moving poleward at another local time, they are, on the other hand, consistent with such an increase in size, and, hence, consistent with our interpretation of the erosion of the magnetopause.

3.4. Magnetotail Flare and Energy Density

Given that the size of the polar cap, i.e., the region of open field lines, increases in area, then the magnetic flux in the tail must increase. Far down the tail where the normal stresses on the boundary are due essentially to the thermal and magnetic pressures of the solar wind, this flux increase will result in an increase in the diameter of the tail and no change in flux density. However, closer to the Earth this is not necessarily true. The erosion of the magnetopause decreases the cross section of the magnetosphere, and hence the diameter of the near tail region, at the same time the distant tail diameter is increasing. In other words, the flaring angle of the tail boundary in the near Earth region must increase. Using the simple pressure balance model of the magnetopause (Willis and Pratt, 1972) we expect the field strength in this region to increase as the number of open field lines grow. The first observational evidence that this process was occurring came from a joint study of Pioneer 7 magnetic field and plasma data by Lazarus *et al.* (1968). While plasma sheet thinning and expansion across an observation point can lead to local increases and decreases in the field strength, this joint study separated variations in the relative energy density of particles and fields from the overall changes of energy density. Lazarus *et al.* (1968) showed that this build-up in energy density preceded a substorm, beginning about 2 h before the subsequent decrease in energy density. They also showed that one such energy density decrease was coincident with the onset of a substorm.

They did not, however, have simultaneous solar wind data and could not unambiguously determine the source of the energy density increase. It could have, for instance, been due to a change in the solar wind velocity or number density.

Camidge and Rostoker (1970) and Fairfield and Ness (1970) also confirmed this sequence with many more examples. However, neither of these studies used data to separate plasma sheet effects from strictly tail lobe changes. Aubry and McPherron (1971) showed that increases in the energy density of the tail lobes were associated with a southward turning of the interplanetary magnetic field, and not with changes in the solar wind pressure, and that the decreases were not associated with any interplanetary effect, but rather followed the onset of substorm expansion phases.

While these data are in accord with our expectations as outlined above and apprently leave no room for any interpretation except that the energy density in the near tail increases when the interplanetary field turns southward and decreases at substorm onsets, much work continues on this phenomenon. Figure 12 shows ground-based H-component magnetograms from the auroral zone near local midnight (top three traces) during an increase and subsequent decrease in the tail field strength (middle trace) at Imp 3, 24 R_e behind the Earth and 13 R_e below the expected position of the neutral sheet (Meng *et al.*, 1971). The constancy of the >40 keV electron flux (bottom trace) shows that such changes are not associated with plasma sheet effects.

Some of the disagreement in the interpretation of data such as these arises because there is auroral zone activity while the tail field is increasing. This, however, may occur during the substorm growth phase because, when the merging rate on the dayside increases, the convection electric field throughout much of the magnetosphere should increase, and thus auroral zone currents should increase. Furthermore, the changing electric and magnetic fields during the growth phase should certainly lead to increased electron precipitation, and hence to increased auroral zone conductivity, which would also lead to stronger currents.

To avoid the difficult task of interpreting auroral zone magnetograms, and since all-sky camera data are not generally available, several authors have used the sharp onset of positive bays in H-component magnetograms in the records of midlatitude observatories in the midnight sector. These studies have been quite successful. Figure 13 shows data in the lobes of the tail at a distance of from 35 to 55 R_e behind the Earth using midlatitude records in this way (Nishida and Nagayama, 1973). In every case, the midlatitude onset of magnetic activity is followed by a decrease in the lobe field strength, as measured by B_x. While this study adds six more clear examples of lobe decreases associated with onsets of magnetic activity it does not examine the simultaneous behavior of the solar wind as Aubry and McPherron (1971) have done.

To clarify further and to quantify the lobe magnetic variations, Caan *et al.* (1973a) have examined in detail the solar wind, the interplanetary magnetic field, midlatitude magnetograms, and ATS-1 magnetograms in synchronous orbit together with OGO-5 data in the lobes of the geomagnetic tail. In accord with Aubry and McPher-

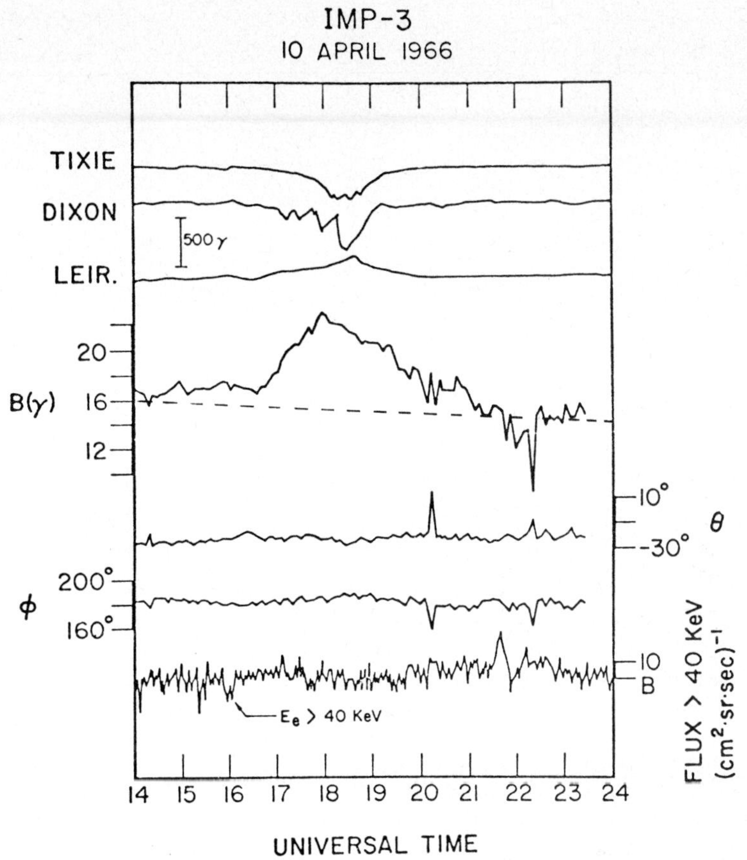

Fig. 12. Substorm signature in lobes of the tail. The top three traces show horizontal components of ground based magnetograms. The next trace shows the magnetic field strength at Imp 3 (−23.7, 12.6, −12.9) R_e in solar magnetospheric coordinates. The next two traces show the solar ecliptic latitude and longitude of the magnetic field and the lowest trace shows the flux of >40 keV electrons (Meng *et al.*, 1971).

ron they find that tail lobe energy density increases are associated with southward interplanetary fields and decreases with magnetospheric-wide substorms. However, they also find that localized, off-center or multiple-onset substorm expansion phases have little effect on the lobe magnetic field.

In an extension of this work, Caan *et al.* (1973b) have performed a superposed epoch study of the tail lobe field change using the time of the midlatitude onset of magnetic activity to order the data in the manner of the study of Foster *et al.* (1971) discussed in Section 3.1. Figure 14 shows the result of a superposition of responses in 24 cases. The rise time is about 90 min, the decay time 45 min, and the decay begins coincident with the midlatitude onset. The slight difference between pre-substorm field strength and post-substorm lobe field strength is an orbit effect. OGO 5 is typically outbound when it is high in the lobes of the tail and field strength decreases with increasing radial distance become decreases with time.

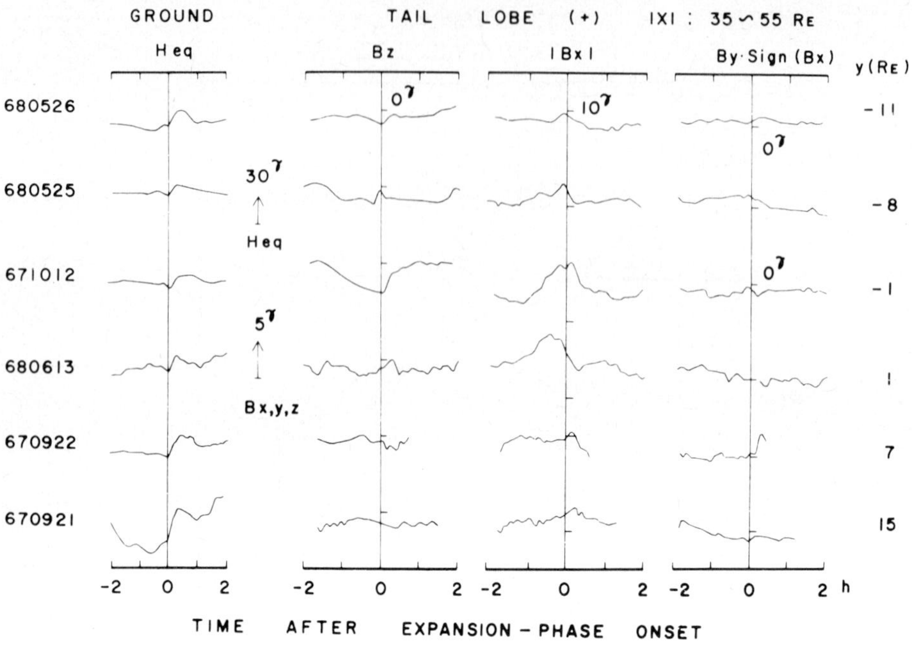

Fig. 13. Behavior of the magnetic field in the distant tail lobes ($-35>x>-55$ R_e) around the onset time of positive H-component bays at midlatitudes near midnight (Nishida and Nagayama, 1973).

In summary, the studies of the tail lobe magnetic field leave little doubt that southward turning of the interplanetary field leads to an increase in the energy density and magnetospheric-wide substorms cause the energy density in this region to decrease. Unfortunately, controversy still surrounds this latter point, not because of a lack of clarity in the reported correlations, but possibly because of a lack of understanding of the meaning of the all-sky camera and the auroral magnetogram records from which the concept of the substorm first arose. In other words, it may well be that some poleward motions of aurora do not signal expansion onsets of the magnetospheric-wide substorms and that not all auroral zone magnetic bays signal substorms. A further possibility is that auroral zone phenomena and mid-latitude bays are caused by different mechanisms, both substorm related, whose onsets are often, but not necessarily, coincident.

The increases and decreases of energy density in the tail lobes must be associated with changes in the angle of flaring of the magnetotail boundary near the Earth as the amount of magnetic flux in the tail increases and decreases. Increases in the tail flux are, of course, the result of an increase in the dayside merging rate. Similarly, as we will discuss in the next section, decreases in the magnetic flux in the tail are caused by an increase in the reconnection rate.

3.5. Neutral point formation

As mentioned in the discussion of the open magnetosphere in Section 2.2, there

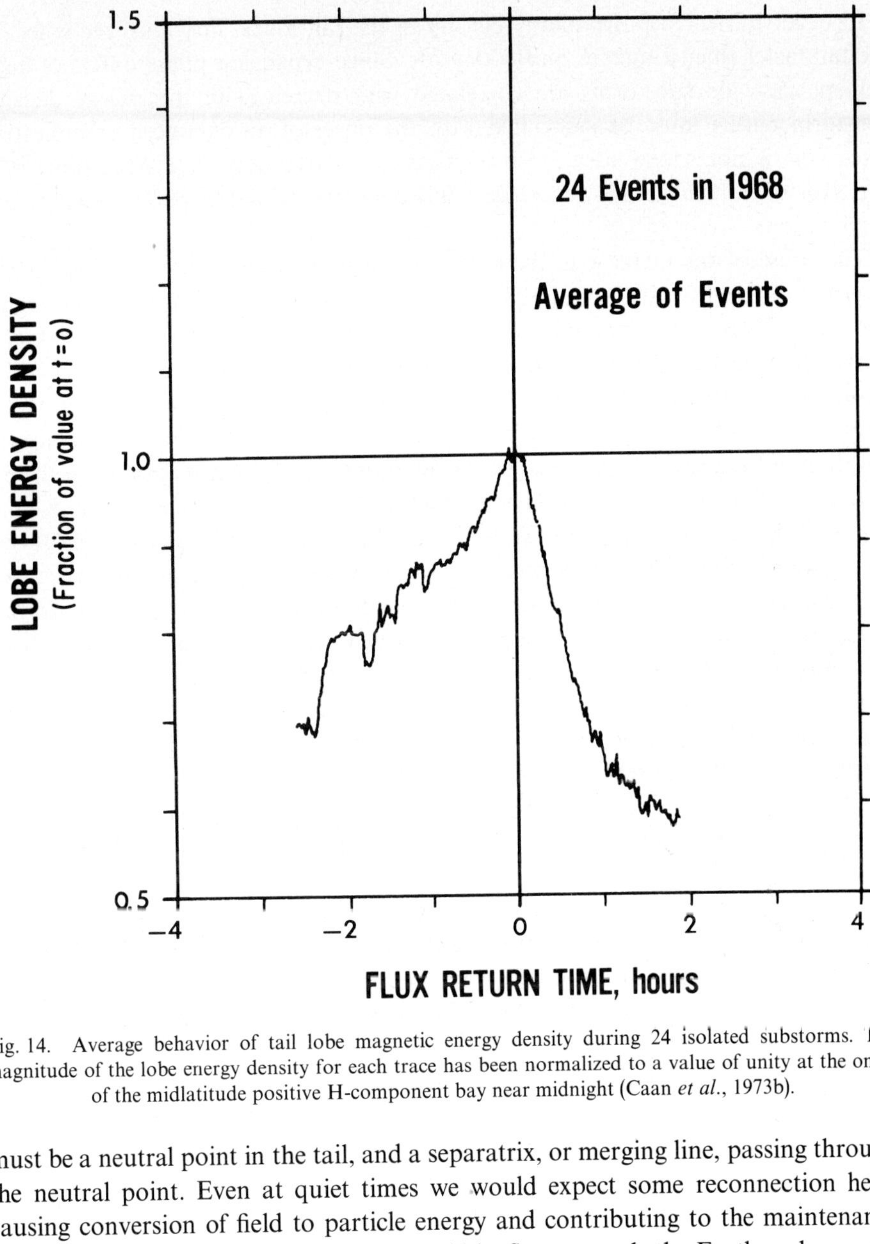

Fig. 14. Average behavior of tail lobe magnetic energy density during 24 isolated substorms. The magnitude of the lobe energy density for each trace has been normalized to a value of unity at the onset of the midlatitude positive H-component bay near midnight (Caan *et al.*, 1973b).

must be a neutral point in the tail, and a separatrix, or merging line, passing through the neutral point. Even at quiet times we would expect some reconnection here, causing conversion of field to particle energy and contributing to the maintenance of the quiet time plasma sheet. There would be flow towards the Earth and a northward magnetic field at the so called neutral sheet on the Earthward side of the neutral point, and a tailward flow and southward field on the tailward side. The existence of a southward or northward pointing magnetic field near the neutral sheet is usually an accurate indicator of whether an observation point is Earthward or tailward of the neutral point. Using this technique, quiet time observations place the neutral point beyond the orbit of the Moon (Mihalov *et al.*, 1968).

In order to decrease the energy density in the tail lobes, flux must reconnect in the tail faster than it merges on the dayside. Since expansion phase onsets of magnetospheric-wide substorms are correlated with decreases in the energy density, substorm onsets must be associated with the onset of an enhanced reconnection rate. This is not a new idea. Two of its earliest advocates were Atkinson (1966) and Rostoker (1968a). However, now tail data are available which demand this explanation.

The cause of the increase in the reconnection rate is not obvious, however. The duration of the growth phase, i.e., the delay between the enhanced dayside merging and the enhanced tail reconnection should be a clue, but it has several possible interpretations. On the one hand, Rostoker (1968b) and Coroniti and Kennel (1973) have suggested that the duration of the growth phase is the time required to propagate information to the tail neutral point to start reconnection. On the other hand, the duration of the growth phase may be the length of time for the plasma sheet to thin to near zero thickness near the Earth (McPherron, 1972; Russell, 1972a; Kennel *et al.*, 1973). Such an interpretation is also implicit in the model of Hill and Dessler (1971) who propose that the neutral point is located within the plasma sheet near the Earth at all times, but during substorms the plasma sheet disappears and merging proceeds at a more rapid rate, in a process termed vacuum merging (Dessler, 1968). However, as stated earlier this model is inconsistent with present data. More recently Hill (1973) has proposed that variations in the rate of particle entry through the magnetopause control the thickness of the plasma sheet. A more likely cause of the thinning is that during the growth phase the dayside merging rate exceeds the tail reconnection rate, causing a gradient in the flow velocity down the tail, so that near the Earth there is relatively rapid convection but far down the tail there is little or no convection (Holzer, 1971; Kennel *et al.*, 1973). To see how this causes thinning, consider a cross-sectional slab of the plasma sheet perpendicular to the Earth-Sun line. More plasma leaves this slab than enters it in the presence of such a gradient. However, the pressure in the slab is roughly constant or slightly increasing at this time, so to accommodate the net loss of plasma the slab thickness must decrease.

Plasma sheet thinning and its subsequent expansion are well established phenomena in the tail (Hones *et al.*, 1971). The thinning and expansion also affects the magnetic field observations, controls the appearance of energetic electrons and protons in tail measurements and is in fact responsible for the occurrence of islands (Russell and McPherron, 1973a). The thinning velocity is of the order of 5 to 10 km s^{-1} (Hones *et al.*, 1971; Buck *et al.*, 1973). The reported expansion velocity varies from about 20 km s^{-1} (Vasyliunas *et al.*, 1973) to about 100 km s^{-1} (Buck *et al.*, 1973). We note that one should exercise caution in interpreting what these values mean. Especially, they should not be used to calculate electric fields directly (Russell and McPherron, 1973a).

If we assume that the plasma sheet is parallel to the neutral sheet and 3 R_e in half thickness and is thinning at a rate of 5 km s^{-1}, then the plasma sheet will reach zero thickness in slightly over one hour, a time not unlike the duration of the growth phase. Furthermore, Buck *et al.* (1973) observed the plasma sheet to thin to almost zero thickness and then expand within minutes of the onset of a magnetospheric-wide substorm. Thus, the concept of a plasma sheet that thins during the growth phase until the plasma sheet reaches some critical thickness or actually disappears somewhere, thus permitting rapid merging, is in accord with observations, and forms a key part of a recent model of the growth phase (McPherron, 1972; Russell, 1972; Russell and McPherron, 1973a).

Several explanations of the thinning and thickening of the plasma sheet have been proposed in addition to convective loss and then acceleration of new plasma. The thinning and expansion could represent compression and expansion of the same volume of gas. Hones *et al.* (1971) have shown that there is little change in energy density during the thinning process ruling out the latter explanation as a significant factor. Some compression must occur as the energy density in the lobes increases but this compression is only about 25% (Caan *et al.*, 1973b) and cannot nearly account for the observed thinning.

Loss of plasma sheet particles could also occur by diffusion perpendicular to the neutral sheet onto open field lines, or by cross-tail motion into the magnetosheath. During thinning the tail is magnetically quiet, so that the former process does not appear to be important (Russell, 1972b; Garrett, 1973). On occasion plasma sheet particles have been detected in the magnetosheath (Hones *et al.*, 1972) suggesting that cross-tail flows are important in the thinning process. However, the Vela orbit lies on a spherical shell intersecting the magnetosheath near dawn and dusk. Thus, these particles could have travelled a more or less Earthward flow path before being lost to the magnetosheath.

If a new neutral point forms closer to the Earth during this thinning process, it could form when the plasma sheet thinned to zero thickness in some limited local time sector. It might also form on closed lines eventually merging more and more closed lines until it reached open field lines. How this might occur is shown schematically in Figure 15 (Vasyliunas, 1973). The top panel shows the noon-midnight meridian. The stipled region shows field lines which close on themselves. The cross-hatched region is the plasma sheet whose field lines have two feet on the ground. The middle panel shows the intersection of this bubble of field lines which close on themselves with the plane of the current sheet. As in our earlier examples, the neutral lines are not necessarily neutral. The bottom panel shows the tail cross section. This size of the bubble might increase or decrease with time depending on the external boundary conditions. We emphasize that the magnetic field lines sketched are not drawn to illustrate the Maxwell stresses expected in a near equilibrium situation.

3.6. The Location and Motion of the Neutral Point

While the data leave no doubt that enhanced reconnection does occur marking the end of the growth phase, and while it is possible to explain at least qualitatively why and how a new neutral point might form, it is not obvious *where* this neutral point

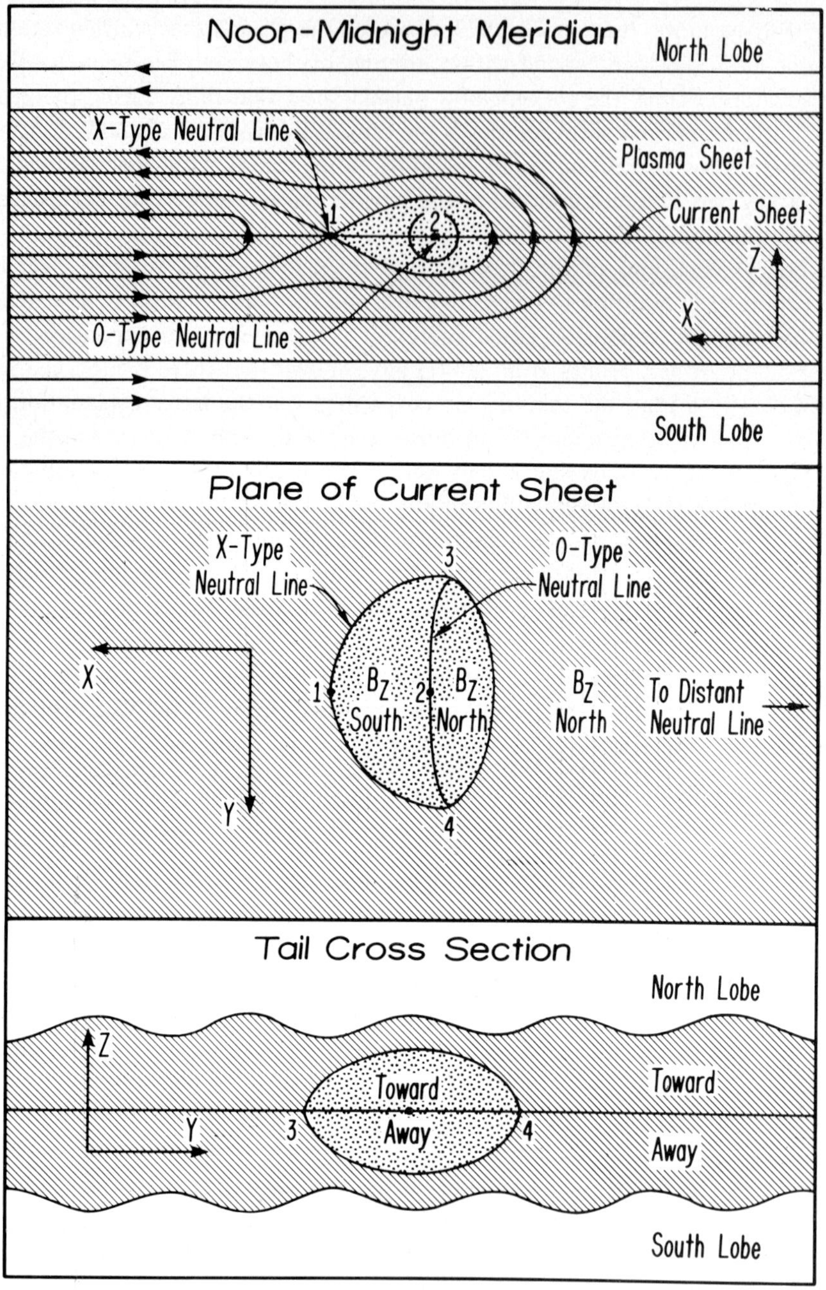

Fig. 15. Sketch of formation of X-0 neutral point pair on closed field lines as seen in three orthogonal planes.

should first form. Fortunately, the magnetic field data in the tail provide us with some limits. McPherron *et al.* (1973) infer a location within about 9 R_e for one particular substorm. Laird (1969) reports a southward component near the neutral sheet at 12 R_e. On the other hand, two years of ATS 1 2.5 min average magnetic field data near midnight reveal no southward fields. Furthermore, the sample tail data of Fairfield and Ness (1970) show many examples of transient southward fields within 34 R_e. Finally, Figure 16 shows the results of a study by Nishida and Nagayama (1973) of the tail field behavior in the plasma sheet at the onset of positive midlatitude bays near midnight. In every case for $|Y| \leqslant 10\, R_e$, the north-south

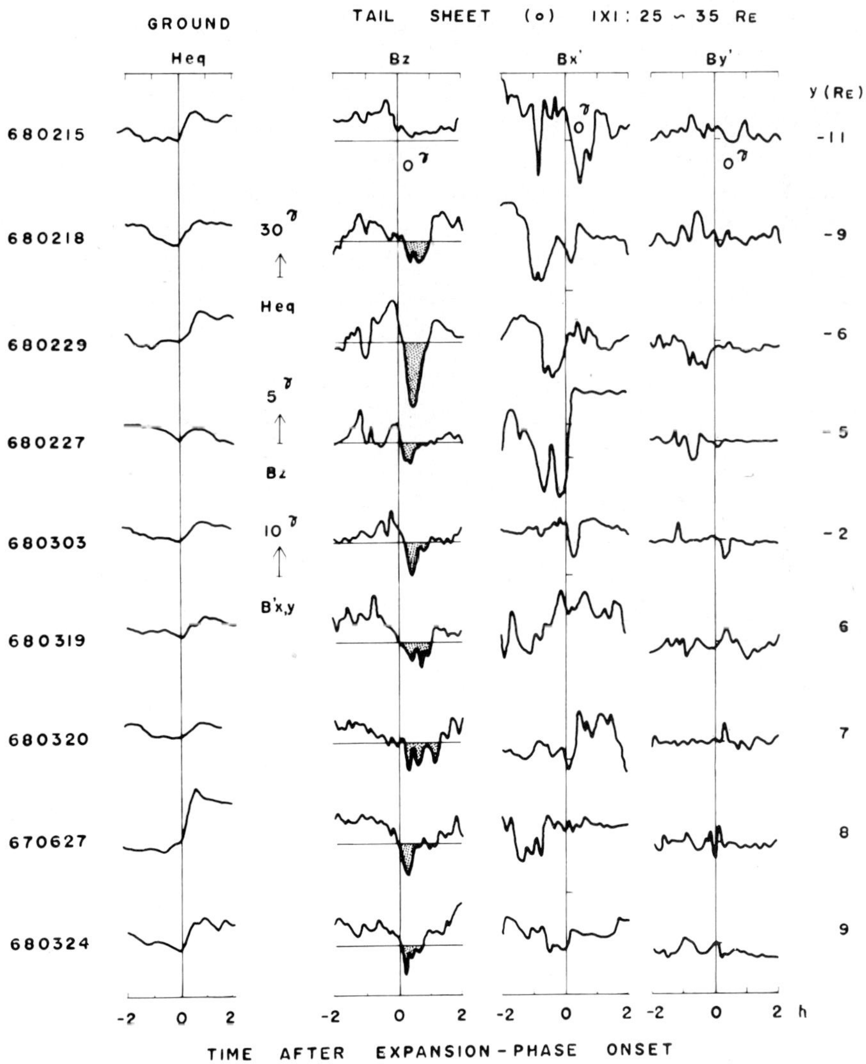

Fig. 16. Behavior of the magnetic field in the plasma sheet for $-25 < x < -35\, R_e$ around the onset time of positive H-component bays at midlatitude near midnight (Nishida and Nagayama, 1973).

component B_z becomes negative or southward immediately after the onset of the positive bay. Thus, we conclude that neutral point formation inside of 25 R_e is common for moderately large substorms, and, at least occasionally, may occur as close as 10 R_e, but it seldom if ever occurs within synchronous orbit at 6.6 R_e.

The average southward component duration in the region $-25 \geqslant x \geqslant -35$ R_e is about 1 h, according to Figure 16. The obvious reason for its disappearance after one hour is that the neutral point moved away from the Earth. This is supported by laboratory work. Merging experiments using a collisional plasma device have been performed (Bratenahl and Yeates, 1970). In an extension of this work to test the effect of boundary conditions on the merging rate, a conducting rod was placed in one of the two plasma jets coming out of the merging region. The merging region responded by moving the other way (Bratenahl, 1973). Similarly, we expect that if the tail reconnection rate significantly exceeds the dayside merging rate, pressure gradients will develop which cause the neutral point to move tailward.

In this picture, the plasma sheet is recreated on closed field lines in the merging process. Thus, as the neutral point moves away from the Earth, a satellite will observe a plasma sheet expansion. Furthermore, as convection is slowed near the Earth, the field compresses creating a stronger and more dipolar field. Alternatively, one may view the reconnection process as disrupting the tail currents. Observationally, this return to a more dipolar state is well documented and we will discuss it further in the next section.

Finally, we note that Schindler and Ness (1972) have proposed a steady state configuration of the tail field consisting of many loops similar to the loop sketched in Figure 15, which is a snapshot of a time evolving structure. While the model proposed by Schindler and Ness cannot be ruled out, there are several alternate explanations of these data. In particular, they may be observing 'quiet time' transient phenomena. For, in the one example of high resolution 'neutral point' data they show, there was a small negative bay at Narssarssuaq, Greenland near local midnight (Russell, 1973).

3.7. The Inward Moving Field Compression

Fairfield and Ness (1970) have shown that one of the major effects of a substorm is to return the nightside magnetosphere to a more dipolar state. How this return occurs is both interesting and important for several magnetospheric processes. On August 7, 1968 OGO 5 and ATS 1 were approximately lined up near the midnight meridian at the onset of a midlatitude positive bay, in excellent position to observe the near tail effects of the substorm (Russell *et al.*, 1973a). ATS 1 was at (−5.9, −2.1, 2.2) R_e and OGO 5 was at (−8.2, −2.0, 2.1) R_e in solar magnetospheric coordinates.

Figure 17 shows one second averages in solar magnetospheric coordinates of the field at both spacecraft. The onset of the midlatitude bay was about 1120 UT; a compression of the field was seen at OGO 5 at 1129:03 and then 93 s later at ATS 1. The most obvious interpretation of this event is the return of magnetic flux from the

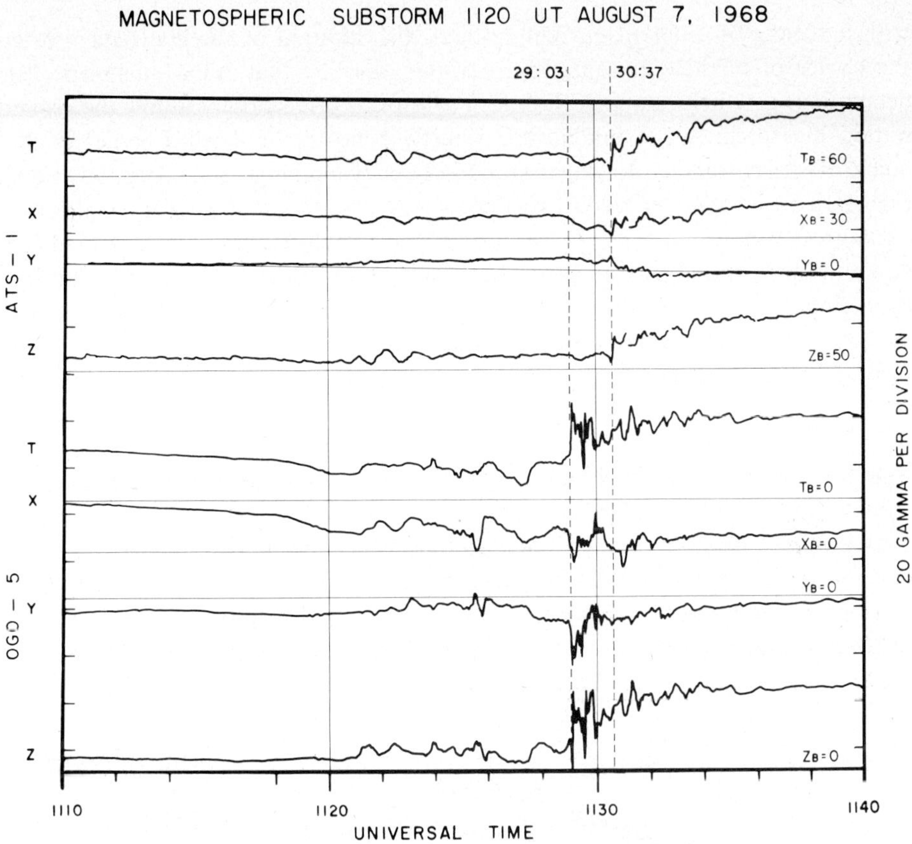

Fig. 17. One second averages of the vector magnetic field at ATS 1 and OGO 5 in solar magnetospheric coordinates and the total field strength at the two spacecraft during an isolated substorm on August 7, 1968. At this time OGO and ATS were both close to the midnight meridian. The inward moving compression reached OGO 5 about 9 min after the substorm onset as determined from midlatitude magnetograms and then reached ATS 1 94 s later (Russell *et al.*, 1973b).

tail. The average velocity of the front from OGO 5 to ATS 1 was 160 km s^{-1} which corresponds to a change in the non-conservative electric field at OGO 5 of 2.3 mV m^{-1} and at ATS 1 of 1.3 mV m^{-1}.

Such compressions of the field are not rare in this region, but such an alignment of ATS and OGO during one is. Walker and Kivelson (1972) have reported more than 10 occurrences of such events near midnight in the OGO-5 data, many of them associated with betatron acceleration of electrons and the subsequent generation of ELF chorus emissions. Further, we note that the non-conservative electric field associated with the inward moving compression is large enough to explain the sudden inward motion of whistler ducts in the outer plasmasphere at substorm onsets as reported by Carpenter and Akasofu (1972).

3.8. The Cause of Midlatitude Positive Bays

While it is possible to describe ground observations of magnetic perturbations with a variety of different current systems, the addition of satellite data restricts the choice of possible circuits. In particular, the fact that ATS-1 magnetic field fluctuations at synchronous orbit during substorms often closely follow the ground station measurements at midlatitude directly beneath the satellite indicates that much of the current is flowing beyond ATS 1 (Cummings and Coleman, 1968). Figure 18 shows a model current system consistent with both the ground and ATS-1 signatures (McPherron *et al.*, 1973a). Currents originally flowing across the tail flow down the field lines into the auroral ionosphere and then back up the field lines bypassing the region of recently merged field. These currents may be thought of as being driven by the induction of the tail and provide a mechanism for directly converting the stored energy in the tail into ionospheric heating.

3.9. Summary

The observations discussed above reveal a dynamic and constantly changing magnetosphere. Isolated substorms, those in which the end of one cycle is clearly distinguishable from the start of the next, and which are therefore easiest to study, have two clearly defined phases in space: the growth phase and the expansion phase. The neutral point motion away from the Earth may constitute a third phase, the recovery phase. In ground records, there are also two well-defined phases: the

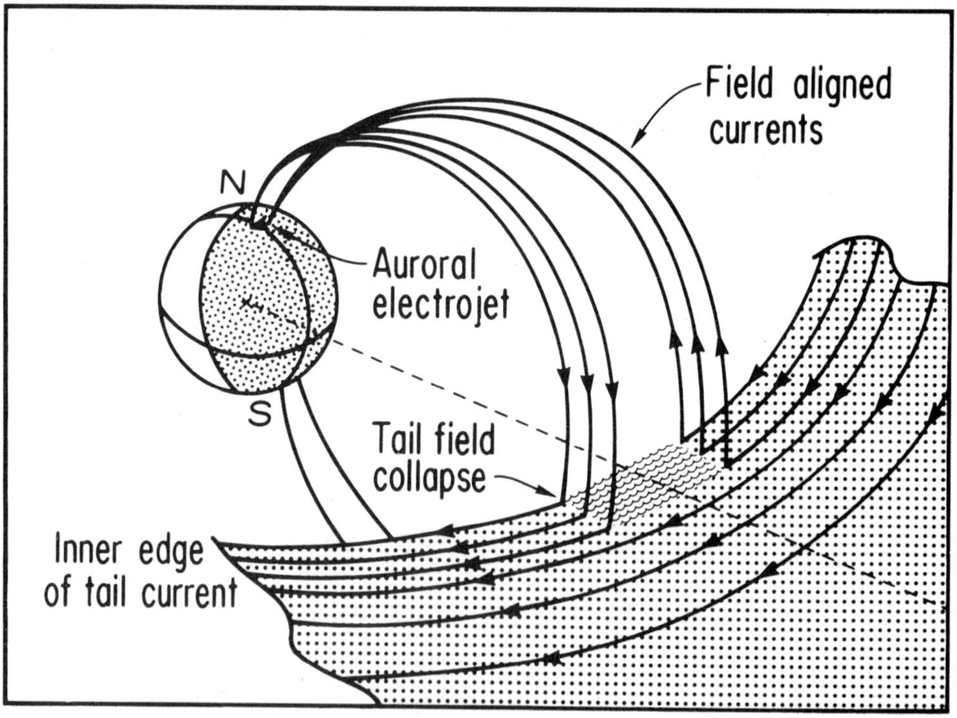

Fig. 18. Field-aligned current system, corresponding to diversion of tail current through ionosphere, which is in accord with positive H-component bays observed at midlatitudes in the midnight sector during the expansion phase of substorms (McPherron *et al.*, 1973b).

expansion phase and the recovery phase. While the growth phase has been reported in ground based data, its identification has been the subject of controversy.

Figure 19 summarizes the magnetospheric changes during the growth phase. The dayside magnetopause is eroded, the polar cusp moves equatorward, and the polar cap increases in area. The increase of magnetic flux in the polar cap, and hence in the lobes of the tail, causes an enlargement of the cross-section of the tail far away from the Earth, but erosion reduces the near-Earth cross section. The result is increased flaring of the tail boundaries and an increase in the energy density in the tail lobes near the Earth. At the same time the plasma sheet thins because merging is occurring on the dayside faster than reconnection is proceeding at night. As the convection velocity increases, we expect the plasma sheet particles to drift closer to the Earth before precipitation losses dominate (Kennel, 1969). The duration of the growth phase is probably the length of time from the onset of enhanced merging on the dayside, until the plasma sheet has become thin enough to allow the formation of a new near -Earth neutral point.

Figure 20 summarizes the sequence of tail events. Satellites such as OGO 5 whose August 15, 1968 position is shown with an *X* and Vela, denoted by an asterisk, see evidence for thinning. When the neutral point is formed (second panel), OGO 5 may leave the plasma sheet temporarily only to re-enter it rapidly. However, Vela

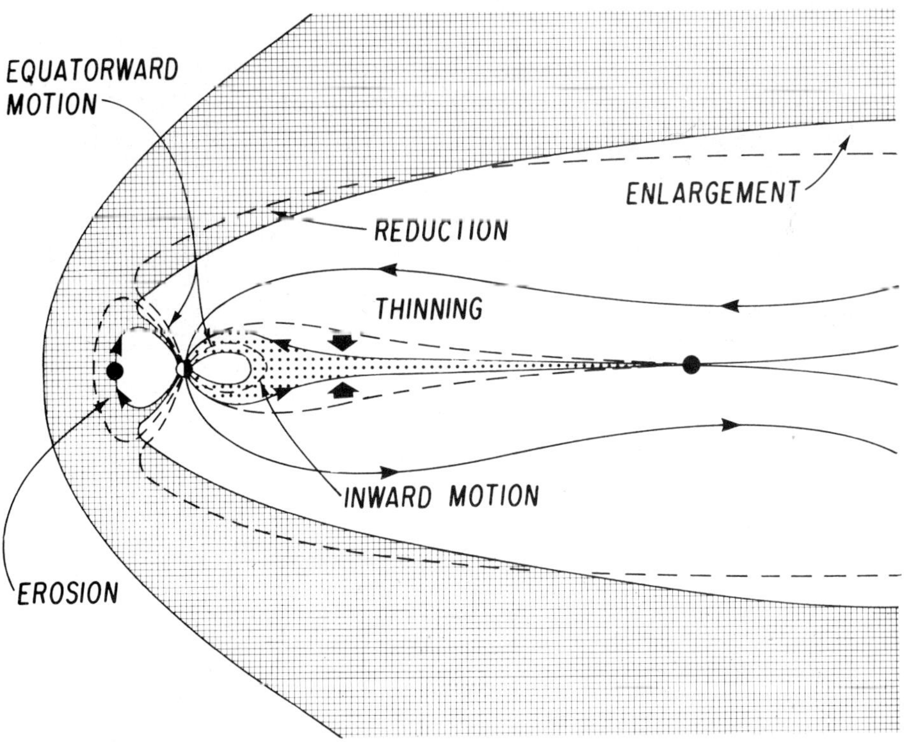

Fig. 19. Summary of the magnetospheric changes during the growth phase.

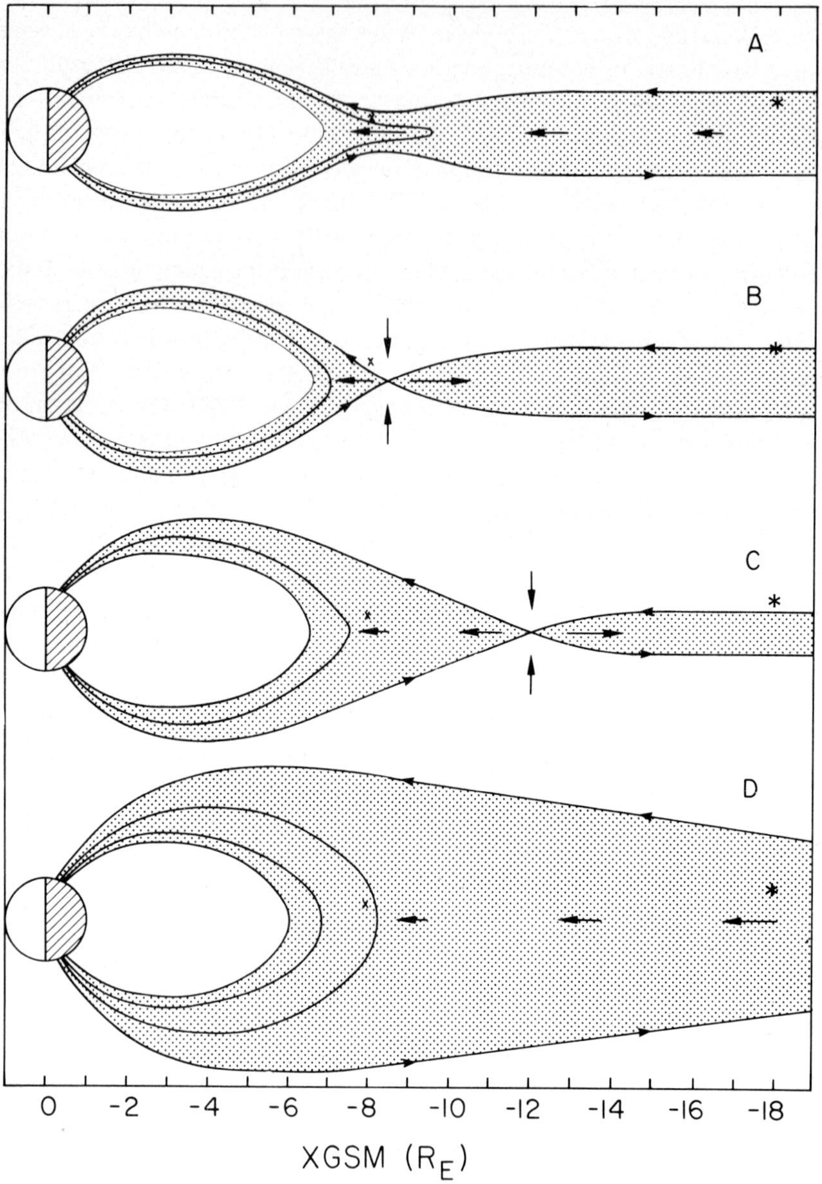

Fig. 20. Model of the expansion phase in the tail. Dotted area represents plasma sheet and length of arrows roughly proportional to convective velocity. The X marks position of OGO 5 during a substorm on August 15, 1968 studied by many authors. The asterisk marks the position of the Vela satellite during a typical event. In the upper panel flow is fastest in the near-earth plasma sheet and it thins. In the next panel a neutral point forms near the Earth, but flow towards the Earth is impeded somewhat so that the neutral point moves as shown in panel C. In the last panel the neutral point has moved past the Vela orbit, thus causing the plasma sheet to expand across Vela (after Russell, 1972a).

may continue to observe thinning. Sometime later the neutral point moves tailward past Vela, enveloping Vela in an expanding plasma sheet. When the neutral point forms, the current in the neutral sheet is disrupted, and the tail currents are diverted via a field aligned current system through the ionosphere. Finally, the return to a more dipolar field configuration in the midnight outer magnetosphere is accomplished by an inward moving compressional front.

Finally, we note that several different authors have recently proposed models which are essentially identical to this one (Nishida and Nagayama, 1973; Hones, 1974). In a field which is often characterized with almost diametrically opposed theories and many heated controversies, such convergence of opinion is indeed reassuring.

4. Geomagnetic Activity

Geomagnetic records have been kept continuously for about a century. One of the reasons for continuing to record these data, was the hope that a determination of the statistical behavior of geomagnetic activity would lead to better insight into its causes. Unfortunately, this expectation was not realized despite the discovery of well-defined diurnal, semiannual, annual, 11-yr and 22-yr cycles. On the other hand, if our understanding of the solar wind-magnetosphere interaction, and the properties of the solar wind and the Sun are correct, these cycles should be a natural consequence of our models. Thus, it is of some interest to attempt to explain these periodic variations in terms of our present knowledge. In the following section, we will briefly discuss how to explain some of these variations in terms of the control of the solar wind-magnetosphere interaction by the north-south component of the interplanetary field.

Another application we can make of our knowledge is to predict geomagnetic activity. If there is a spacecraft in the near-earth interplanetary medium, we might obtain an advanced warning of the order of an hour before the observed solar wind plasma reaches the Earth. The storage time and reaction times of the magnetosphere probably vary significantly with the strength of the interaction, but on the average these delay times might add an hour more. Thus, with real time telemetry of interplanetary data to Earth, we could possibly predict moderately accurately the conditions within the magnetosphere during the next two hours.

If we wish to extend this prediction further into the future we have to know the properties of the Earth-intersecting solar wind while it is closer to the Sun, and be able to predict how these properties will be modified in transit to the Earth. If, for example, we could infer the properties of the solar wind from solar data, from which area on the Sun the Earth-intersecting plasma came, and how these properties would change in transit, we could obtain a 2 to 5 day warning. For an even longer warning or prediction, we would have to understand how solar activity evolved, i.e., we would have to predict solar activity too.

At the present time, we are in a position to predict geomagnetic activity up to two hours in advance by using in situ interplanetary measurements. We will discuss this in Sections 4.5 and 4.6. On the other hand, it appears that any attempt to predict geomagnetic activity for longer periods in other than the most general terms is premature at the present time.

4.1. The Semiannual Variation

The semiannual variation of geomagnetic activity has been known for over half a century (Cortie, 1912). It can be seen in all indices of geomagnetic activity from storm counts to K_p. It also is a very significant modulation, clearly evident in most data even from only a few years data. Figure 21 shows the semiannual variation as determined from storm counts from data of Chapman and Bartels (1940) and by using the Dst index (Russell and McPherron, 1973b). While the phase of these variations exhibits some variation, possibly because of counting statistics, the semiannual nature and the 2 to 1 enhancement of equinoctial to solstitial storm occurrence is clearly evident. Russell and McPherron (1973b) show that such an enhancement in storm counts can be produced by a modulation of the strength of the interaction such that only 40% more energy was being extracted from the solar wind during the equinoxes than at the solstices.

Russell and McPherron pointed out that since the interplanetary field was ordered in a solar coordinate system and the interaction of the solar wind with the magnetosphere in a geomagnetic system, the varying relative orientation of these two systems could give a semiannual variation in geomagnetic activity if the magnetosphere acted as a rectifier. Since, as is well known, the magnetosphere interacts more strongly with southward fields than northward fields, the magnetosphere does act as a rectifier. Unfortunately, we do not know quantitatively the rectification laws, nor even which is the optimum coordinate system. (We do know though that the solar magnetospheric coordinate system is better than the solar ecliptic system (Hirshberg and Colburn, 1969).)

While only a limited number of rather simplistic models were tested by Russell and McPherron they did produce one which reproduced the correct phase of the semiannual variation and which could account for its amplitude. This model assumed that the solar magnetospheric system ordered the interaction, that the interaction was proportional to the strength of the southward component and that there was no interaction for northward fields.

This model, and any similar model, predicts that the semiannual variation can be separated into two annual variations, one maximizing in spring and one in fall, using the sector polarity of the interplanetary magnetic field. In particular, the activity should maximize in spring for fields in towards the Sun as measured along the Parker spiral, and in fall for fields outwards from the Sun. This prediction is confirmed by studies of the K_p index (Otaola, 1971), of the AE index (Burch, 1973) and of the C9 index (Russell and McPherron, 1973b). Thus, there can be little doubt that the semiannual variation is but another manifestation of the control of mag-

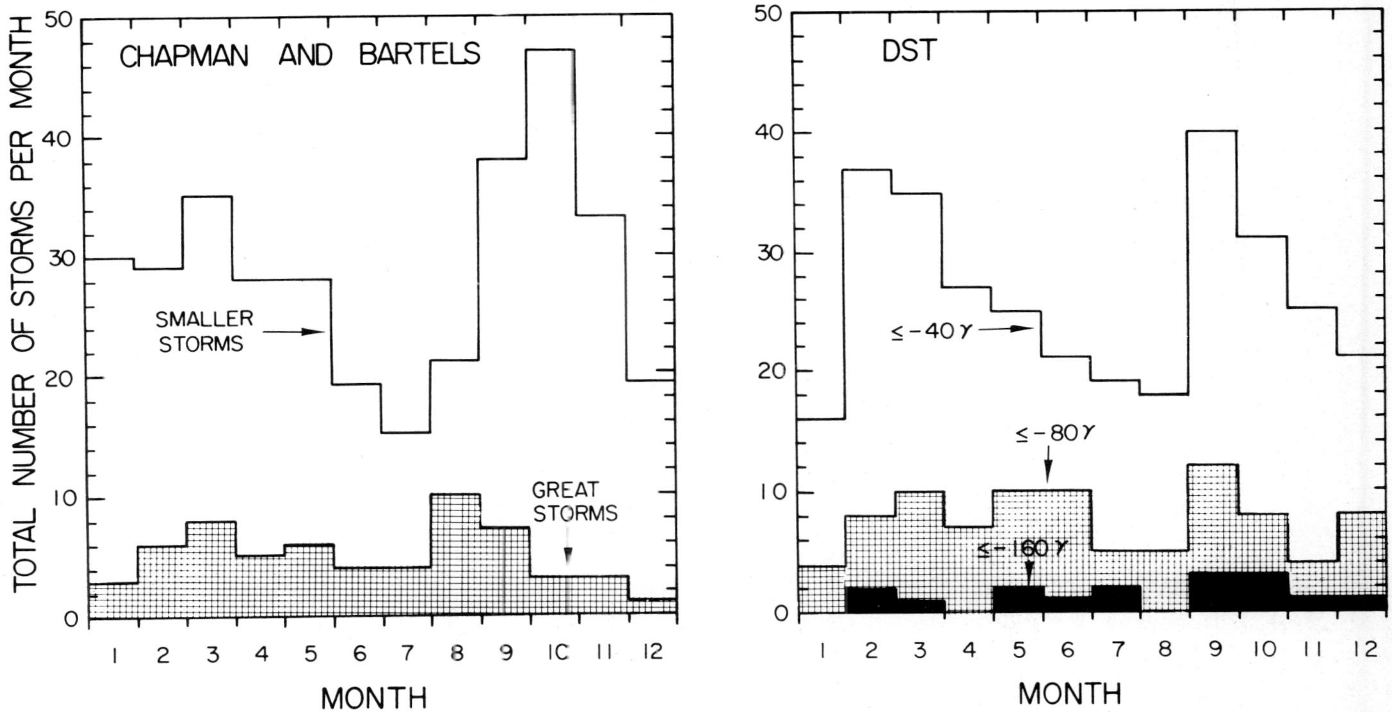

Fig. 21. The semiannual variation of the occurrence of geomagnetic storms. Left: occurrence of great storms and smaller storms from 1875 to 1927 given by Chapman and Bartels (1940). Right: occurrence of storms with Dst minimums less than -40, -80, and -160 γ during the years 1958 and 1961 through 1969 (Russell and McPherron, 1973b).

netospheric dynamics by the southward component of the interplanetary field.

4.2. The Annual Variation

While the semiannual variation is well defined in all geomagnetic indices, the annual variation is less well defined. In fact, it is so poorly defined that some authors ignore its existence by referring to the 6 month variation as the annual variation. However, by annual variation we mean here a 12 month cycle. Another source of confusion is the fact that, at any one station, there will be an annual variation due to the annual variation in the properties of the ionosphere. Furthermore, the planetary annual variation might be either of two types: an annual variation independent of the semiannual variation or an amplitude modulation of the semiannual variation. Fourier analysis of geomagnetic indices would yield an annual peak, in the former case and peaks at 0.5 and 1.5 yr in the latter case. When Fourier analysis of long continuous series of geomagnetic indices is performed there is no annual line nor is there a line at 1.5 yr (Fraser-Smith, 1972). There is, of course, a strong line at 0.5 yr corresponding to the semiannual variation.

Despite this fact, there has been a recent resurgence of interest in the annual variation in geomagnetic activity. Some of these papers seem to propose that there is an annual variation independent of the semiannual variation (Wulf, 1971; Mayaud, 1972), while others propose a modulation mechanism (Meyer, 1972; Münch, 1972). In addition, both Meyer and Münch provide a plausible explanation of why the annual variation is not usually found in harmonic dial analyses or in Fourier analyses. This is due to the variable phase of the annual modulation. It can be strongest in spring or strongest in fall, and one phase only persists a few years. This explains why the annual variation has been occasionally detected, for example, as sidebands on the 27 day peak (Coleman and Smith, 1966), or as the occasional appearance of the 1.5 yr peak (Fraser-Smith, 1972).

While the explanation of the semiannual variation proposed by Russell and McPherron does not in itself contain an annual variation, it is easy to see how such a modulation might arise. Since the spring maximum is caused by fields predominantly towards the Sun and the fall maximum by fields away from the Sun, a situation in which one of the equinoxes is more active than the other will occur when fields of one polarity are more 'active' than those of the other polarity. Since there is such a strong variation in the magnetic sector structure with latitude (Rosenberg and Coleman, 1969) it is reasonable to presume that the sector boundaries are more nearly parallel to the ecliptic than perpendicular to it. A similar picture has been proposed by Schulz (1972) on slightly different grounds. Thus, these variations in the activity of sectors with different polarities simply represent north-south asymmetries in activity on the Sun. We note that Münch (1972) reaches the same conclusion about north-south asymmetries causing the annual modulation using a non-magnetic, axial model for the interaction. Finally, we note that the reason for the reversal in the phase of this effect, i.e., which equinox has the strongest activity, is due both to the reversal of the solar magnetic field every 22 yr and to reversals in the north-south asymmetry.

4.3. The 11- and 22-yr cycles

Since solar activity follows an 11-yr cycle in which the number of active centers, i.e., sunspots, changes, as well as the positions of these active centers on the solar disk, it is not surprising that geomagnetic activity also follows an 11-yr cycle. Almost any model of the interaction predicts this. There is also a 22-yr solar cycle in overall solar magnetic polarity. This cycle is not obvious in sunspot number or position, but it does show up in the relative magnetic polarity of the leading and following spots in sunspot groups.

A corresponding 22-yr cycle in geomagnetic activity was first found by Chernosky (1966) in which a 22-yr cycle consists of 11 active years followed by 11 quiet years, with the switch between active and quiet years occurring about 2 yr after solar maximum.

The existence of this effect is in accord with our present knowledge of the solar wind-magnetosphere interaction. Furthermore, the phase of the effect and the phase of observed variations in the sector structure over the course of the 22-yr cycle are in accord with the model of Russell and McPherron (1973b), since the switching point ($\sim$2 yr after solar maximum) is when the heliographic latitude dependence of the dominant polarity of the interplanetary magnetic field switches sign (Rosenberg and Coleman, 1969; Wilcox and Scherrer, 1972). This heliographic latitude dependence either increases the amount of southward field at the equinoxes, by modulating the amount of away and towards polarity, or decreases it, according to the polarity of the solar field.

4.4. The diurnal variation

Many authors have found diurnal variations of geomagnetic activity (Nicholson and Wolf, 1955; McIntosh, 1959; Mayaud, 1967). The question at hand is whether these variations are due to ionospheric effects or whether they are controlled by the solar wind-magnetosphere interaction. In order to check whether such diurnal variations exist in so called 'planetary' indices, the diurnal variation of K_p and Dst (Russell and McPherron, 1973c) and AE (Allen, 1973) has been studied as a function of time of year. In each case, there was a diurnal variation, *but this variation was constant throughout the year* while all reasonable theories of the interaction predict a changing diurnal variation. While it is possible that in designing these planetary indices the originators were able to remove the annual cycle in the diurnal variation yet permit a constant diurnal variation to exist, it indicates that these indices and their cousins have severe limitations for short time scale variations.

Finally, we note that when treating ground station records, extreme caution must be exercised in interpreting the results of diurnal variation studies. Ground stations are very sensitive to nearby current systems in the ionosphere. Further, the interplanetary magnetic field is not the only agent producing semiannual variations on the Earth. Thus, if a semiannual variation of the diurnal pattern is found, it is not necessarily associated with the solar wind-magnetosphere interaction. The semiannual variation in the density of the upper atmosphere is well established (Cook,

1968; 1970) and can be explained in terms of EUV heating, taking care to include properly the effects of oblique incidence of the solar flux, sphericity of the atmosphere and ellipticity of the Earth's orbit (Ching and Chiu, 1972a, b). This semiannual change in the neutral density is accompanied by corresponding changes in the ionosphere (Bhatnagar, 1971). Thus, there is good reason to suspect that the semi-annual variations of the diurnal pattern of geomagnetic activity may not be a magnetospheric effect. To settle this question ground station records should be analyzed to determine how these records respond to specific interplanetary events as a function of time of day, time of year, and geomagnetic latitude.

4.5. The Cause of Geomagnetic Storms

A geomagnetic storm is defined by the existence of a 'main phase' during which the magnetic field at the Earth's surface is depressed. This depression is caused by a ring current within the magnetosphere. While we have examined magnetospheric substorms in detail, and have made the analogy of the Van Allen belts acting as a bucket catching the splash from substorms, we have not actually discussed what conditions are required to build up the ring current, and thus 'cause' a geomagnetic storm.

In one of the first studies of the cause of geomagnetic storms, using interplanetary data, Hirshberg and Colburn (1969) examined the response of the magnetosphere to four 'interplanetary' storms using three hour averages of the interplanetary magnetic field. The one interplanetary storm with only northward field produced the smallest amount of geomagnetic activity. One of these storms was examined with high resolution data, and it was found that the main phase began coincident with the arrival of the southward field. However, later they showed that the arrival of the southward field was coincident, within the resolution of their data, with the arrival of the helium-enriched driver gas (Hirshberg *et al.*, 1969; Hundhausen, 1970). We note that the solar wind dynamic pressure data were only available sporadically during the course of the storm.

Kokubun (1972) has studied several SSC storms in 1967 using hourly averages of the solar wind velocity, density and interplanetary magnetic field. His study showed that only shock waves followed by southward fields caused geomagnetic storms. He was unable to determine the functional relationship between the size of the storm and the parameters of the solar wind, possibly because of his use of low temporal resolution data.

More recently Russell *et al.* (1973a) have pursued this topic using high time resolution data. Figure 22 shows the solar wind number density and velocity, the interplanetary magnetic field strength and solar magnetospheric north-south component measured on Explorer 33 in the solar wind on March 3, 1968. In the bottom two panels are the Dst and AE indices for the same time period. The apparent cause of the strengthening of the ring current, by almost 30 γ after 1600 UT, was the strong southward component of the interplanetary magnetic field at 1500 UT. No change is seen in the solar wind number density, velocity or the magnetic field strength. On the other hand, the magnetic field is predominantly southward on this day.

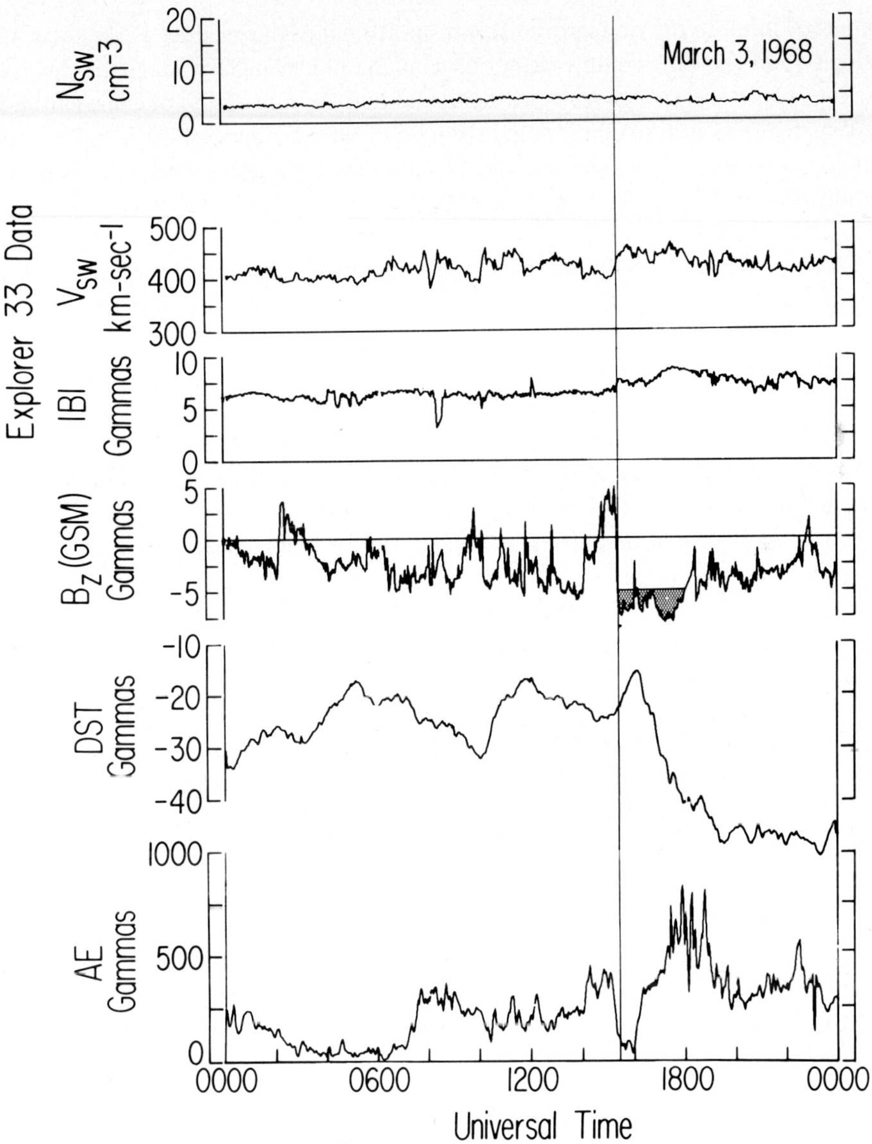

Fig. 22. Solar wind and geomagnetic activity data for March 3, 1968. From top to bottom: the solar wind number density, N_{SW}, and the solar wind velocity, V_{SW}, measured by the MIT plasma probe on Explorer 33, the interplanetary field strength, $|B|$, and its solar magnetospheric Z-component, B_Z, as measured by the Ames Research Center fluxgate magnetometer on Explorer 33; 2.5 min evaluations of the Dst and AE indices (Russell *et al.*, 1973a).

This and several similar examples led Russell *et al.* (1973a) to postulate that there was a threshold, i.e., that the southward component had to exceed a certain value, depending on solar wind velocity and pre-existing ring current strength, before further injection would occur. They postulated that this occurred because the rate of energy input to the ring current was proportional to the product of the solar wind velocity and the north-south component of the interplanetary magnetic field (i.e., the interplanetary east-west electric field) and that the rate of loss of energy from the ring current was proportional to the strength of the ring current. Thus, net injection would occur only when the southward component exceeded a threshold value, and would proceed until a new ring current strength was reached at which dissipation balanced injection.

These results indicate that the occurrence of a sufficiently strong southward interplanetary magnetic field, (say, $B_z < -5\ \gamma$), for a sufficiently long time, (several hours), will produce a moderate size geomagnetic storm, (Dst ~ -50 to $-75\ \gamma$). The stronger is the southward component and the longer the injection persists, at least until saturation is reached, the more intense the main phase minimum will be. In this model, the frequent appearance of shocks or sudden pressure increases, several hours before the main phase begins, occurs because these discontinuities are effective producers of strong magnetic fields normal to the solar equatorial plane. When these strong normal fields are southward a main phase begins to develop.

4.6. Predicting Geomagnetic Activity

As stated in the introduction to Section 4, the most we can hope to do at the present time, is to issue detailed predictions of geomagnetic activity two hours in advance using in situ measurements and real time telemetry. Accurate and moderately detailed predictions 2 to 5 days in advance may be possible in the future, but much more research has to be performed into how structures are modified in their passage from the sun to the Earth and how the characteristics of the solar corona and the near-Sun solar wind can be predicted from solar observations.

Arnoldy (1971) has studied the relationship of the interplanetary parameters to the AE index. The principal contribution to the AE index was the integrated southward magnetic field over the previous hour. In this integral northward fields are set equal to zero. Garrett (1973b) has performed a similar study and found two populations of substorms: one set which responds as Arnoldy predicted and one set which responded roughly twice as strongly to an equivalent southward magnetic field, but which was associated with a large variance in the interplanetary magnetic field. One solution to this apparent dichotomy of magnetospheric behavior is that the product of solar wind velocity times the interplanetary southward component, i.e., the east-west electric field, integrated over the previous hour controls, the AE index, and the apparent correlation with the variance is due to an association of large variances with high speed streams.

Burton *et al.* (1973), on the other hand, have concerned themselves with predicting the Dst index using in situ measurements of the solar wind velocity and number density and the interplanetary field strength. Figure 23 shows the result of this prediction for March 3, 4 and 5, 1968, a moderately disturbed period. The philosophy of this calculation is to take an initial Dst value, calculate the energy lost and the energy injected during an interval, calculate a change in Dst, add this to the initial

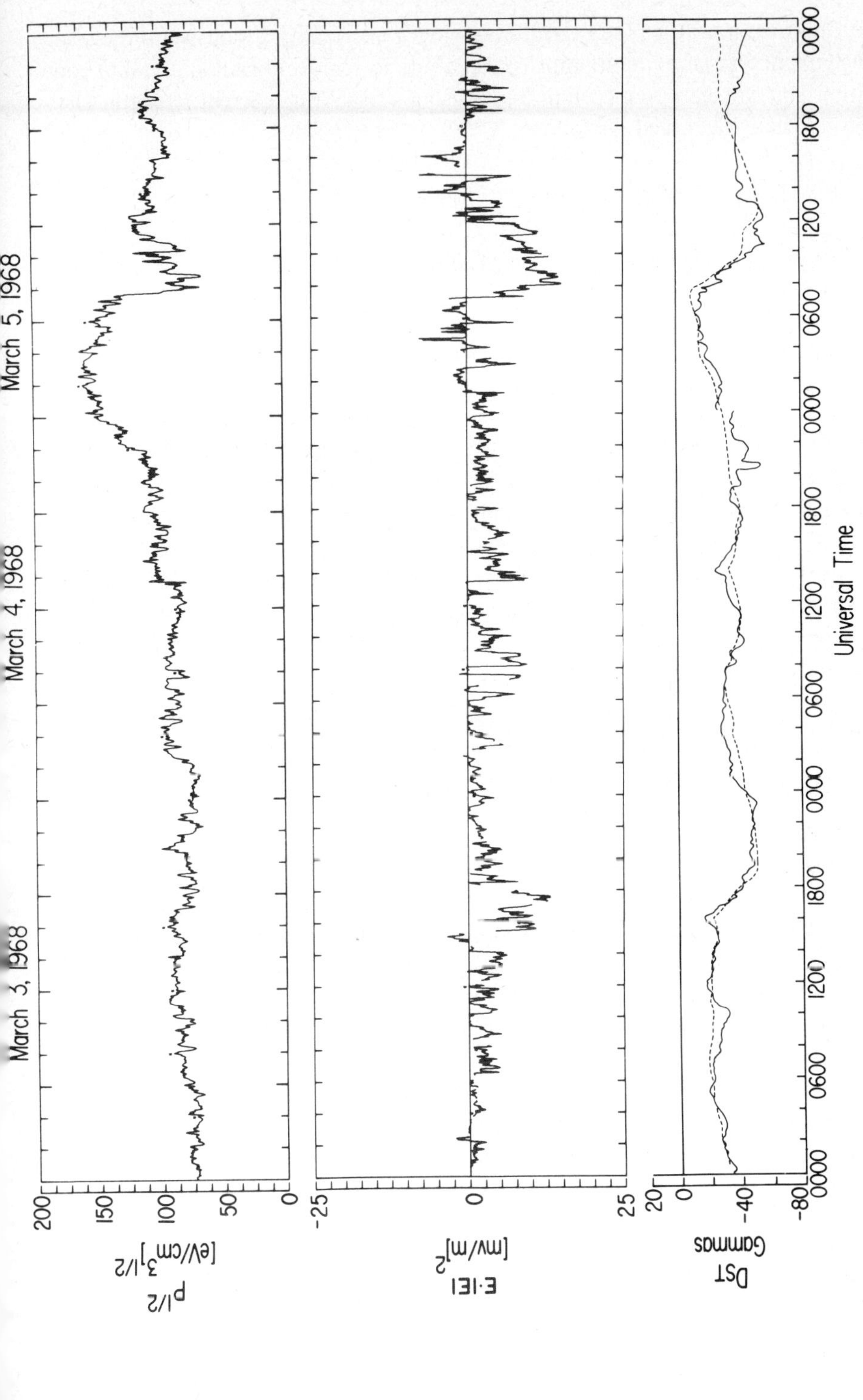

Fig. 23. Prediction of Dst using solar wind dynamic pressure and interplanetary electric field. Top panel: square root of the dynamic pressure. Middle panel: square of the westward solar magnetospheric component of the interplanetary electric field times its sign. Bottom panel: actual (solid line) and predicted (dashed line) Dst (Burton *et al.*, 1973).

value, and then repeat for the next interval. Since Dst is also responsive to the solar wind dynamic pressure, an attempt is also made to remove dynamic pressure contribution and calculate ring current losses using a 'true' ring current strength, Dst′.

The calculation proceeds as follows:

$$\text{(1)} \qquad \mathrm{Dst}'(t) = \mathrm{Dst}(t) - a\sqrt{mn(t)\,V^2(t)} + b,$$

$$\text{(2)} \qquad \frac{\partial}{\partial t}(\mathrm{Dst}'(t)) = -cE_W^2(t-\tau) - d\,\mathrm{Dst}'(t),$$

$$\text{(3)} \qquad \mathrm{Dst}(t+\Delta) = \mathrm{Dst}(t) + \Delta\frac{\partial}{\partial t}(\mathrm{Dst}'(t)),$$

where m is the proton mass, n the number density and V the velocity of the solar wind. E_W is the east-west component of the interplanetary ($\mathbf{V}\times\mathbf{B}$) electric field, set equal to zero if E_W is negative or eastward, and τ is a time lag for magnetospheric response. In these expressions, Dst is in γ, P in eV cm^{-3}, and E in mV m^{-1}. The constants used in constructing Figure 23 from Equations (1) to (3) are: a equals $0.18\ \gamma-(\text{eV cm}^{-3})^{-1/2}$; b equals $16\ \gamma$; d equals $3.5\times10^{-5}\ \text{s}^{-1}$; and c equals $4.4\times10^{-4}\ \gamma-(\text{mV m}^{-1})^{-2}\ \text{s}^{-1}$. The response time τ, of the magnetosphere to the interplanetary electric field has been held constant at 30 min in this study. While these constants work equally well on quiet days as well as disturbed, optimization of these constants is continuing using data obtained during a variety of geomagnetic conditions during several different years.

4.7. Summary

It is indeed reassuring that, with our present knowledge of the solar wind interaction with the magnetosphere and some of the properties of the solar wind and the Sun, we can provide plausible mechanisms to explain the various cycles of geomagnetic activity. It is surprising though that an explanation for the diurnal variation is not as simply derived, in view of the success of the model in explaining the semiannual, annual and 22-yr cycles. This is probably due to the fact that the diurnal variation is associated with ionospheric effects rather than due to some deficiency in our understanding of the magnetosphere.

It is also reassuring that we are now in a position to predict geomagnetic activity on the basis of interplanetary measurements. While this does not demonstrate the accuracy of our understanding of the details of magnetospheric processes, the quantitative, empirical relationships that have been derived, place well defined boundary conditions on these processes.

5. Conclusions

The history of the magnetosphere demonstrates the difficulty of achieving progress in magnetospheric research without in situ observations. This is not simply because remote sensing equipment, such as a ground based magnetometer, or an all-sky camera is insensitive to major magnetospheric processes, nor is it always because the correct theory cannot be deduced from the available data. For example, the concept of a magnetospheric cavity, a continually flowing solar wind, a geomagnetic tail, and a bow shock preceded their direct observation. However, these concepts did not gain popular acceptance until in situ measurements were available, in part because of the existence of alternate models, which were also apparently consistent with the data.

On the other hand, there are certain processes occurring in and around the magnetosphere for which remote sensing is no help. For example, the structure and nature of the bow shock and the magnetopause leave no signature in ground based data. A knowledge of the structure and the nature of the processes occurring therein is necessary, if we are to be able to extend our magnetospheric experience to astrophysical and plasma physical environments. Otherwise, the magnetospheric physicist will be just a weatherman, with a detailed knowledge of what has happened in the past and therefore what should happen in the future, but little other than this empirical knowledge.

Fortunately, the decade of the 1960's provided magnetospheric physicists with a wealth of in situ data, and our models of the magnetosphere evolved very rapidly, to a point today where there is very little controversy about the observational magnetosphere. Controversy now focuses on mechanisms and temporal behavior. For example, there are no longer any serious questions about whether the magnetosphere is open or closed. However, many arguments as to the relative timing of auroral zone phenomena, midlatitude events and magnetospheric changes still persist. Many of these arguments appear capable of resolution from presently acquired but incompletely analyzed data. However, some future data remains to be gathered to settle some timing questions, such as on the relative timing of the equatorward motion of aurora near local noon and the southward turning of the interplanetary magnetic field.

Further, there is still a need for properly conceived correlative observations, which do not involve timing so much as relative magnitudes and locations, such as comparing the magnitude of changes in the tail energy density with the strength of the plasma injection at synchronous orbit, and comparing ionospheric current systems deduced from ground station chains with currents measured in space. Finally, there are measurements that have never been made such as of the velocity, thickness and current density of magnetospheric boundaries, which require coordinated measurements from a closely spaced pair of satellites.

The experimental task is made easier though, by the fact that we know roughly what we are looking for. We know the range of parameters to be measured and we have phenomenological models for the behavior of the magnetosphere, which while they may not provide the reasons for magnetospheric behavior, they outline what this behavior is.

As the term phenomenological model implies, we have not achieved much progess in magnetospheric theory in the past few years, especially in the area of magnetospheric dynamics. For example, a successful theory might involve modelling all the stresses in the magnetosphere and showing how a perturbation of this stress balance would cause growth of some macroscopic instability of the night time magnetosphere. However, such modelling is still in its infancy. On the other hand, for many purposes, we do not need to know the explicit details of these mechanisms. For example, we can now explain many of the patterns of geomagnetic activity, and can predict the state of the magnetosphere as measured by geomagnetic indices from interplanetary measurements. Thus, while we have not solved all problems of magnetospheric dynamics, we can draw some satisfaction from the fact that we now have knowledge which can be put to use.

Acknowledgments

The work on substorms at UCLA much of which is incorporated in this manuscript is the culmination of the ideas and work of many individuals. These include M. P. Aubry, R. K. Burton, M. N. Caan, P. J. Coleman, Jr., F. V. Coroniti, T. A. Farley, R. E. Holzer, C. F. Kennel, M. G. Kivelson, R. L. McPherron and R. J. Walker. Equally important to this effort were the contributions of advice and data from our collaborators, both those associated with OGO 5: R. M. Buck, C. R. Chappell, R. W. Fredricks, M. Neugebauer, F. L. Scarf, E. J. Smith and H. I. West, Jr.; and those associated with other satellites and ground based observations: J. H. Binsack, D. L. Carpenter, D. S. Colburn, D. H. Fairfield, G. K. Parks and C. P. Sonett. Both the resources of the National Space Science Data Center and the World Data Center A were invaluable in our studies. The National Aeronautics and Space Administration provided support for the analysis of OGO-5 magnetic field data reported herein under NASA contract NAS 5-9098 and under NGR 05-007-004. The National Science Foundation provided support for the study of geomagnetic storms under NSF grant GA 34148-X.

References

Akasofu, S.-I.: 1964, *Planetary Space Sci.* **12**, 273.
Akasofu, S.-I.: 1968, *Polar and Magnetospheric Substorms*, D. Reidel, Dordrecht, Holland.
Akasofu, S.-I.: 1972a, *J. Geophys. Res.* **77**, 244.
Akasofu, S.-I.: 1972b, *J. Geophys. Res.* **77**, 2303.
Akasofu, S.-I., Hones, Jr., E. W., Bame, S. J., Asbridge, J. R., and Liu, A. T. Y.: 1973, *J. Geophys. Res.* **78**, 7257.
Akasofu, S.-I., Perreault, P. D., Yasuhara, F., and Meng, C.-I.: 1973b, *EOS* **54**, 414 (abstract).
Allen, J. H.: 1973, *EOS Trans. AGU* **54**, 242 (abstract).
Arnoldy, R. L.: 1971, *J. Geophys. Res.* **76**, 5189.
Atkinson, G.: 1966, *J. Geophys. Res.* **71**, 5157.
Aubry, M. P. and McPherron, R. L.: 1971, *J. Geophys. Res.* **76**, 4381.
Aubry, M. P., Russell, C. T., and Kivelson, M. G.: 1970, *J. Geophys. Res.* **75**, 7018.
Axford, W. I.: 1962, *J. Geophys. Res.* **67**, 3791.
Axford, W. I. and Hines, C. O.: 1961, *Can. J. Phys.* **39**, 1433.
Axford, W. I., Petschek, H. E., and Siscoe, G. L.: 1965, *J. Geophys. Res.* **70**, 1231.

Bahnsen, A., D'Angelo, N., and Ungstrup, E.: 1974, this volume, p. 237.
Behannon, K. W.: 1970, *J. Geophys. Res.* **75**, 743.
Bhatnagar, V. P.: 1971, *Planetary Space Sci.* **19**, 723.
Biermann, L.: 1951, *Z. Astrophys.* **29**, 274.
Biermann, L.: 1953, *Mem. Soc. Roy. Sci. Liege Quati. Ser.* **13**, 291 (abstract).
Biermann, L.: 1957, *Observatory* **77**, 109.
Bonetti, A., Bridge, H. S., Lazarus, A. J., Lyon, E. F., Rossi, B., and Scherb, F.: 1963a, *Space Res.* **3**, 540.
Bonetti, A., Bridge, H. S., Lazarus, A. J., Rossi, B., and Scherb, F.: 1963b, *J. Geophys. Res.* **68**, 4017.
Bratenahl, A.: 1973, personal communication.
Bratenahl, A. and Yeates, C. M.: 1970, *Phys. Fluids* **13**, 2696.
Brice, N.: 1967, in *Conjugate Point Symposium, June 13–16, 1967, Boulder, Colorado*, ESSA Technical Memorandum, IERTM-ITSA72, paper IV-2.
Bridge, H., Egidi, A., Lazarus, A., Lyon, E., and Jacobson, L.: 1965, *Space Res.* **5**, 969.
Buck, R. M., West, Jr., H. I., and D'Arcy, Jr., R. G.: 1973, *J. Geophys. Res.* **78**, 3103.
Burch, J. L.: 1972, *J. Geophys. Res.* **77**, 6696.
Burch, J. L.: 1973, *J. Geophys. Res.* **78**, 1047.
Burton, R. K., McPherron, R. L., and Russell, C. T.: 1973, *EOS Trans. AGU* **54**, 415 (abstract).
Caan, M. N., McPherron, R. L., and Russell, C. T.: 1973a, *J. Geophys. Res.*, in press.
Caan, M. N., McPherron, R. L., and Russell, C. T.: 1973b, in Proceedings of the Chapman Memorial Symposium on Magnetospheric Motions, Boulder, Colorado, June.
Camidge, F. P. and Rostoker, G.: 1970, *Can. J. Phys.* **48**, 2002.
Carpenter, D. L. and Akasofu, S.-I.: 1972, *J. Geophys. Res.* **77**, 6854.
Chapman, S. and Bartels, J.: 1940, *Geomagnetism*, vol. II, Oxford University Press, United Kingdom.
Chapman, S. and Ferraro, V. C. A.: 1930, *Nature* **126**, 129.
Chapman, S. and Ferraro, V. C. A.: 1931, *Terr. Mag.* **77**, 171.
Chapman, S. and Ferraro. V. C. A.: 1932, *Terr. Mag.* **37**, 147.
Chappell, C. R.: 1972, *Rev. Geophys. Space Phys.* **10**, 951.
Chappell, C. R.: 1974, this volume, p. 277.
Chernosky, E. J.: 1966, *J. Geophys. Res.* **71**, 965.
Ching, B. K. and Chiu, Y. T.: 1972a, *Planetary Space Sci.* **20**, 1745.
Ching, B. K. and Chiu, Y. T.: 1972b, *Planetary Space Sci.* **20**, 1761.
Coleman, P. J., Jr. and Smith, E. J.: 1966, *J. Geophys. Res.* **71**, 4685.
Cook, G. E.: 1968, *Planetary Space Sci.* **16**, 1161.
Cook, G. E.: 1970, *Planetary Space Sci.* **18**, 1573.
Coroniti, F. V. and Kennel, C. F.: 1973, *J. Geophys. Res.* **78**, 2837.
Cortie, A. L.: 1972, *Monthly Notices Roy. Astron. Soc.* **73**, 52.
Cummings, W. D. and Coleman, Jr., P. J.: 1968, *Radio Sci.* **3**, 758.
Davis, T. N.: 1970, Personal communication.
Dessler, A. J.: 1968, *J. Geophys. Res.* **73**, 209.
Dessler, A. J.: 1970, *Rev. Geophys.* **5**, 1.
Dessler, A. J. and Hill, T. W.: 1970, *J. Geophys. Res.* **75**, 7323.
Dessler, A. J. and Juday, R. D.: 1965, *Planetary Space Sci.* **13**, 63.
Dungey, J. W.: 1961, *Phys. Rev. Letters* **6**, 47.
Dungey, J. W.: 1963, in C. DeWitt, J. Hieblot, and A. Lebeau (eds.), *The Earth's Environment*, Gordon and Breach, New York.
Dungey, J. W.: 1965, *J. Geophys. Res.* **70**, 1753.
Fairfield, D. H.: 1971, *J. Geophys. Res.* **76**, 6700.
Fairfield, D. H. and Ness, N. F.: 1970, *J. Geophys. Res.* **75**, 7032.
Fairfield, D. H. and Ness, N. F.: 1972, *J. Geophys. Res.* **77**, 611.
Fennell, J. F.: 1973, *J. Geophys. Res.* **78**, 1036.
Formisano, V.: 1974, this volume, p. 187.
Foster, J. C., Fairfield, D. H., Ogilvie, K. W., and Rosenberg, T. J.: 1971, *J. Geophys. Res.* **76**, 6971.
Frank, L. A.: 1971, *J. Geophys. Res.* **76**, 5202.
Frank, L. A. and Ackerson, K. L.: 1971, *J. Geophys. Res.* **76**, 3612.
Fraser-Smith, A. C.: 1972, *J. Geophys. Res.* **77**, 4209.
Fredricks, R. W. and Russell, C. T.: 1973, *J. Geophys. Res.* **78**, 2917.
Fredricks, R. W., Scarf, F. L., and Russell, C. T.: 1973, *J. Geophys. Res.* **78**, 2133.

Garrett, H. B.: 1973a, *J. Geophys. Res.* **78**, 3799.
Garrett, H. B.: 1973b, *Planetary Space Sci.* (in press).
Gringauz, K. I.: 1961, *Space Res.* **2**, 539.
Gringauz, K. I., Bezrukikh, V. V., Ozerov, V. D., and Rybchinskiy, R. Ye.: 1960, *Soviet Phys. Doklady* **5**, 361.
Haerendel, G.: 1973, personal communication.
Heikkila, W. J. and Winningham, J. D.: 1971, *J. Geophys. Res.* **76**, 883.
Heppner, J. P., Sugiura, M., Skillman, T. L., Ledley, B. G., and Campbell, M.: 1967, *J. Geophys. Res.* **72**, 5417.
Hill, T. W.: 1973, *Planetary Space Sci.* **21**, 1307.
Hill, T. W. and Dessler, A. J.: 1971, *Planetary Space Sci.* **19**, 1275.
Hirshberg, J. and Colburn, D. S.: 1969, *Planetary Space Sci.* **17**, 1183.
Hirshberg, J., Alksne, A., Colburn, D. S., Bame, S. J., and Hundhausen, A. J.: 1970, *J. Geophys. Res.* **75**, 1.
Holzer, R. E., McLeod, M. G., and Smith, E. J.: 1966, *J. Geophys. Res.* **71**, 1481.
Holzer, T. E.: 1971, personal communication.
Hones, E. W., Jr.: 1974, this volume, p. 263.
Hones, E. W., Jr., Asbridge, J. R., and Bame, S. J.: 1971, *J. Geophys. Res.* 76, 4402.
Hones, E. W., Jr., Akasofu, S.-I., Bame, S. J., and Singer, S.: 1972, *J. Geophys. Res.* **77**, 6688.
Hundhausen, A. J.: 1972, *Ann. Geophys.* **26**, 427.
Jelly, D. and Brice, N.: 1967, *J. Geophys. Res.* **72**, 5919.
Johnson, F. S.: 1960, *J. Geophys. Res.* **65**, 3049.
Kellogg, P. J.: 1962, *J. Geophys. Res.* **67**, 3805.
Kennel, C. F.: 1969, *Rev. Geophys.* **7**, 379.
Kennel, C. F., Coroniti, F. V., McPherron, R. L., and Russell, C. T.: 1973, unpublished manuscript.
Kivelson, M. G., Russell, C. T., Neugebauer, M., Scarf, F. L., and Fredricks, R. W.: 1973, *J. Geophys. Res.* **78**, 3761.
Kokubun, S.: 1972, *Planetary Space Sci.* **20**, 1033.
Laird, M. J.: 1969, *J. Geophys. Res.* **74**, 133.
Lazarus, A. J., Siscoe, G. L., and Ness, N. F.: 1968, *J. Geophys. Res.* **73**, 2399.
Lin, R. P., and Anderson, K. A.: 1966, *J. Geophys. Res.* **71**, 4213.
Mayaud, P. N.: 1967, *Ann. Geophys.* **23**, 585.
Mayaud, P. N.: 1972, *J. Geophys. Res.* **77**, 6870.
McIntosh, D. H.: 1959, *Phil. Trans. Roy. Soc. London, A* **251**, 525.
McPherron, R. L.: 1970, *J. Geophys. Res.* **75**, 5592.
McPherron, R. L.: 1972, *Planetary Space Sci.* **20**, 1521.
McPherron, R. L., Parks, G. K., and Coroniti, F. V.: 1967, *Conjugate Point Symposium, June 13–16. 1967, Boulder, Colorado, ESSA Technical Memorandum*, IERTM-ITSA72, paper III-4.
McPherron, R. L., Aubry, M. P., Russell, C. T., and Coleman, Jr., P. J.: 1973a, *J. Geophys. Res.* **78**, 3068.
McPherron, R. L., Russell, C. T., and Aubry, M. P.: 1973b, *J. Geophys. Res.* **78**, 3131.
Meng, C.-I., Akasofu, S.-I., Hones, Jr., E. W., and Kawasaki, K.: 1971, *J. Geophys. Res.* **76**, 7584.
Meyer, J.: 1972, *J. Geophys. Res.* **77**, 3566.
Mihalov, J. D., Colburn, D. S., Currie, R. G., and Sonett, C. P.: 1968, *J. Geophys. Res.* **73**, 943.
Mishin, V. M. and Popov, G. V.: 1972, *Issled. Geomagnetizm., Aeronomii, Fiz. Soln.* **26**, 150.
Morfill, G. and Scholer, M.: 1972, *Planetary Space Sci.* **20**, 2113.
Münch, J. W.: 1972, *Planetary Space Sci.* **20**, 225.
Ness, N. F., Scearce, C. S., and Seek, J. B.: 1964, *J. Geophys. Res.* **69**, 3531.
Neugebauer, M. and Snyder, C. W.: 1962, *Science* **138**, 1095.
Nicholson, S. B. and Wolf, O. R.: 1955, *J. Geophys. Res.* **60**, 389.
Nishida, A. and Nagayama, N.: 1973, *J. Geophys. Res.* (in press).
O'Brien, B. J.: 1962, *J. Geophys. Res.* **67**, 3687.
O'Brien, B. J.: 1964, *J. Geophys. Res.* **69**, 13.
Otaola, J. A.: 1971, personal communication.
Parker, E. N.: 1958a, *Astrophys. J.* **128**, 664.
Parker, E. N.: 1958b, *Phys. Fluids* **1**, 171.
Paschmann, G., Grünwaldt, H., Montgomery, M. D., Rosenbauer, H., and Sckopke, N.: 1974, this volume, p. 249.

Piddington, J. H.: 1960, *J. Geophys. Res.* **65**, 93.
Roederer, J. G.: 1974, this volume, p. 257.
Rosenberg, R. L. and Coleman, Jr., P. J.: 1969, *J. Geophys. Res.* **74**, 5611.
Rostoker, G.: 1968a, *J. Geophys. Res.* **73**, 4217.
Rostoker, G.: 1968b, *J. Geophys. Res.* **73**, 4382.
Russell, C. T.: 1972a, The configuration of the magnetosphere, in E. R. Dyer (ed.), *Critical Problems of Magnetospheric Physics*, IUCSTP Secretariat, National Academy of Sciences, Washington, D.C., p. 1.
Russell, C. T.: 1972b, *Planetary Space Sci.* **20**, 1541.
Russell, C. T.: 1973, *J. Geophys. Res.* **78** (in press).
Russell, C. T. and McPherron, R. L.: 1973a, *Space Sci. Rev.* (in press).
Russell, C. T. and McPherron, R. L.: 1973b, *J. Geophys. Res.* **78**, 92.
Russell, C. T. and McPherron, R. L.: 1973c, *J. Geophys. Res.* (in press).
Russell, C. T., Chappell, C. R., Montgomery, M. D., Neugebauer, M., and Scarf, F. L.: 1971, *J. Geophys. Res.* **76**, 6743.
Russell, C. T., Neugebauer, M., and Kivelson, M. G.: 1974, this volume, p. 139.
Russell, C. T., McPherron, R. L., and Burton, R. K.: 1973a, *J. Geophys. Res.* (submitted).
Russell, C. T., Kivelson, M. G., McPherron, R. L., and Walker, R. J.: 1973b, *EOS Trans. AGU* **54**, 413 (abstract).
Scarf, F. L., Fredricks, R. W., Green, I. M., and Russell, C. T.: 1972, *J. Geophys. Res.* **77**, 2274.
Schindler, K. and Ness, N. F.: 1972, *J. Geophys. Res.* **77**, 91.
Scholer, M.: 1974, this volume, p. 419.
Schulz, M.: 1972, Aerospace Report No. TR-0073 (3260-20)-11.
Shklovsky, I. S., Moroz, V. I., and Kurt, V. G.: 1961, *Soviet Astron. – AJ* **4**, 871.
Slutz, R. J. and Winkelman, J. R.: 1964, *J. Geophys. Res.* **69**, 4933.
Smith, E. J.: 1962, *J. Geophys. Res.* **67**, 2045.
Snyder, C. W., Neugebauer, M., and Rao, U. R.: 1963, *J. Geophys. Res.* **68**, 6361.
Stern, D. P.: 1972, Goddard Space Flight Center preprint X-641-72-643.
Vampola, A. L.: 1974, this volume, p. 463.
Van Allen, J. A.: 1970, *J. Geophys. Res.* **75**, 29.
Van Allen, J. A., McIlwain, C. E., and Ludwig, G. H.: 1959, *J. Geophys. Res.* **64**, 271.
Van Allen, J. A., Fennell, J. F., and Ness, N. F.: 1971, *J. Geophys. Res.* **76**, 4262.
Vasyliunas, V. M.: 1973, personal communication.
Vasyliunas, V. M., Hones, Jr., E. W., Bame, S. J., and Singer, S.: 1973, *EOS* **54**, 427 (abstract).
Walker, R. J. and Kivelson, M. G.: 1972, *EOS Trans. AGU* **53**, 1097 (abstract).
Wenzel, K. P.: 1974, this volume, p. 229.
Wilcox, J. M. and Scherrer, P. H.: 1972, *J. Geophys. Res.* **77**, 5385.
Willis, D. M.: 1971, *Rev. Geophys. Space Phys.* **9**, 953.
Willis, D. M. and Pratt, R. J.: 1972, *J. Atmospheric Terrestr. Phys.* **34**, 1955.
Wolfe, J. H. and Silva, R. W.: 1965, *J. Geophys. Res.* **70**, 3575.
Wolfe, J. H., Silva, R. W., and Meyers, M. A.: 1966, *J. Geophys. Res.* **71**, 1319.
Wulf, O. R.: 1971, *J. Geophys. Res.* **76**, 1837.

INTERPLANETARY CONTROL OF MAGNETOSPHERIC DYNAMICS

G. K. PARKS*, E. H. GARDNER**, C. S. LIN†, R. E. NEWELL, and P. J. WEHRENBERG†

Space Sciences Division of Geophysics Program, University of Washington Seattle, Wash. 98195, U.S.A.

Data from the interplanetary space, the magnetosphere at the geostationary orbit, and the auroral zone have been correlated for comprehensive understanding of the solar wind-magnetospheric interaction and occurrence of magnetospheric substorms. There is evidence to show that plasma convection toward the Earth accompanies substorm activities. Properties of the convection speed and thickness of the inner edge and the phase relation between convection and substorms are derived. On several occasions, simultaneous data in the interplanetary space show that such convection is directly controlled by the interplanetary space electric field. The response of the magnetospheric convection to changes of the interplanetary space electric field is prompt, suggesting that the interplanetary space electric field has direct access to the magnetosphere. This observation also suggests that the conductivity of the magnetopause in the cross-tail direction is finite. Preliminary data analysis indicates a phase lag of the convection with respect to electric field changes. Such effects could arise from polarization drift of the magnetospheric plasma particles associated with $d\mathbf{E}/dt$.

1. Introduction

Although it has been recognized for a long time that the interaction of magnetized solar wind and the geomagnetic field might be responsible for many geophysical disturbances including the auroras and the outer radiation belts, it is not yet understood what interaction mechanisms are involved. From the experimental point of view, there is sufficient warrant to indicate that the solar wind magnetic field plays an active role. Spacecraft observations since 1966 have shown that variations in polar magnetograms are strongly correlated with the direction of the interplanetary magnetic field. Notably, Fairfield and Cahill (1966), Rostoker and Falthammer (1967) and Schatten and Wilcox (1967) showed that increasing geomagnetic activity in the auroral and polar regions is likely to be associated with periods when the solar wind magnetic field pointed in the southward direction. These observations were amplified recently by Arnoldy (1971) who showed that whether a substorm might occur or not depends on the length of time the magnetic field points in the southward direction. There are now additional correlation results that show that the polar cap magnetic

* Also affiliated with the Departments of Physics and Atmospheric Sciences.
** On leave from the U.S. Air Force Institute of Technology.
† Also affiliated with the Physics Department.

D. E. Page (ed.), Correlated Interplanetary and Magnetospheric Observations, 49–60.

field variations are intimately related to the azimuthal component of the interplanetary magnetic field (Svalgaard, 1968; Mansurov, 1969; Fris-Christensen *et al.*, 1972).

There are two models of the solar wind-magnetosphere interaction that emphasize the importance of the magnetic field direction. One is based on the 'merging' concept (it is recommended that the reader consult the articles by Dungey (1961), Parker (1958), Sweet (1958), and Petschek (1964) for the meaning of merging). The other model originally proposed by Alfvén (1968) and Alfvén and Falthammer (1971) suggests that the solar wind penetrates into the magnetosphere by the action of the $\mathbf{V} \times \mathbf{B}$ electric field. Both models favor the southward interplanetary field direction to accelerate particles and produce the auroral-type disturbances.

It is our opinion that observations have not unambiguously resolved which of the two (or whether both) are active at the magnetopause. Although most observations have been interpreted as favoring the merging model, a quantitative merging theory is still lacking. Moreover, there is a large class of observations not compatible with the merging concept. For example, the substorms that occur during periods of prolonged northward magnetic field pointing (Rostoker and Falthammer, 1967). Also the recent correlation observations that show that the outer magnetospheric particle distribution responds promptly to changes in the interplanetary magnetic field direction (Parks and Pellat, 1972).

A comprehensive model must take into account all the plasma parameters in the solar wind and the magnetosphere. In this regard, the existing models have not considered the importance of solar wind parameters like the bulk velocity, plasma kinetic temperature, and particle density. One of the important tasks is to establish from observations how such parameters are correlated with the geomagnetic activity. Then one can evaluate if the interaction mechanisms include viscous effects or Kelvin-Helmholtz instabilities at the magnetopause boundary. How these might further couple into the magnetosphere to produce substorm-like disturbances can then be studied.

Inclusion of plasma parameters is especially important when considering substorm dynamics since substorms are instabilities of the magnetosphere. In substorm studies, it is convenient to study separately the physics of substorm triggering and the mechanisms by which the magnetosphere is brought to this unstable state. More precisely, the following questions might be considered meaningful: Under what conditions does the magnetospheric plasma distribution become unstable to sustain particle acceleration and precipitation responsible for substorms? How is this unstable state brought about? What roles do the solar wind and the direction of the interplanetary magnetic field play? What physical processes are occurring at the magnetopause to couple the interplanetary space to the magnetosphere?

Precise answers to these questions are not yet available. However, it is evident that in the magnetosphere the plasma sheet is quite active. As Vasyliunas (1968) has shown, the inner edge of the plasma sheet is convected closer to earth during polar magnetic bays. In one event, the inner edge was observed near $L = 6$. These observations have since been verified by others (Schield and Frank, 1970). In addition, Chase (1969) and Hones *et al.* (1971) showed that particle energy spectra in the auroral zone resembled those of the plasma sheet.

That the plasma sheet is convected closer to earth during geomagnetic active times has also been observed by geostationary experiments. Shelley *et al.* (1972) observed enhanced plasma fluxes at the geostationary orbit from which the convection speeds were calculated. McIlwain (1972) has derived plasma flow contours in the equatorial plane on the basis of the plasma data obtained during substorms. Then there are the indirect observations from the ATS 1 by Cummings *et al.* (1968) that the magnetosphere is considerably inflated when substorms occur. The work of Lezniak and Winckler (1970) suggests that the inflation is due to plasma sheet particles. Recently, Lin and Parks (1973) used more detailed data and found new particle and magnetic field variations that support the idea that the plasma sheet is frequently present at the ATS-1 orbit and responsible for the inflation.

To understand substorm dynamics, it is important to study what causal relationship exists between the plasma sheet convection and occurrence of substorms. The ATS-1 orbit through the local evening sector is close to the inner edge of the plasma sheet and the high energy electron distribution is not contaminated by the substorm-accelerated electrons in this local time sector. Hence one can employ the energetic electrons as tracers to study the adiabatic L-shell motion with reference to the substorm activity. (This technique was used in Lezniak and Winckler, 1970.) The L-shell motion is caused by the slow changing of the local magnetic field arising from such effects as the plasma sheet motion and diamagnetism.

This approach was adopted by Lin and Parks (1973) and a considerable amount of new information was found: the thickness of the inner edge, the convection speed, and the relationship of the convection to triggering of substorms. We will summarize the principal results of this study in the next section in view of their importance to further understanding of magnetospheric substorm phenomena. Then in Section 3 the origin of the electric field that drives the plasma sheet convection is examined. Pertinent discussion follows concerning the correlation observations.

2. Plasma Sheet Convection and Magnetospheric Substorms

Figure 1 shows the 6-min averaged particle and field data from the Minnesota and UCLA experiments on the ATS-1. According to the H-component magnetic field, the magnetosphere at synchronous altitudes was very much inflated. There are also large correlated variations in the particle and field data. These are similar events previously studied by Lezniak and Winckler (1970) and the particle variations in UT interval 00-10 h are mostly adiabatic variations. Let us now examine in more detail the variations that occurred around 0400 UT.

The expanded view around 04 UT is shown in Figure 2 where data represent 15-s averages. We note that between 03 and 04 UT, the Z-component decreased at the rate of about 0.6 γ min^{-1}. Then at about 04 UT we see a sharp decrease, about 20 γ

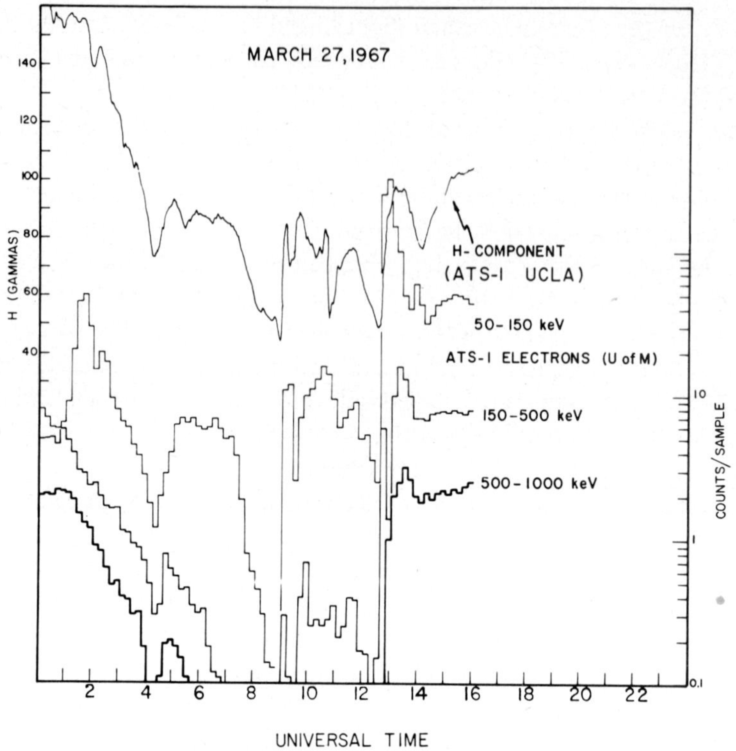

Fig. 1. Six minute averaged energetic electron and *H*-component magnetic field data obtained at the geostationary orbit. Local Time of the satellite is Universal Time – 10 h.

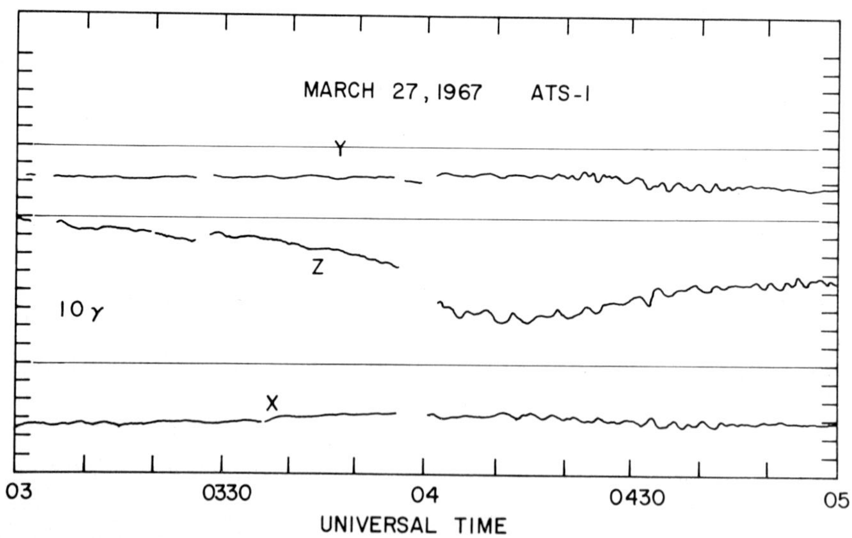

Fig. 2. The expanded view of the magnetic field data around 04 UT. The data are 15-s averages. *Z* is along the spin axis, *X* is positive toward the Sun, and *Y* completes the right-handed system.

in 5 min. Small waves were then observed persisting throughout the recovery, past 05 UT.

The two-step process, slow and abrupt changes, is a common feature at the ATS-1 orbit. Figure 3 shows more examples. The features are generally observed only in

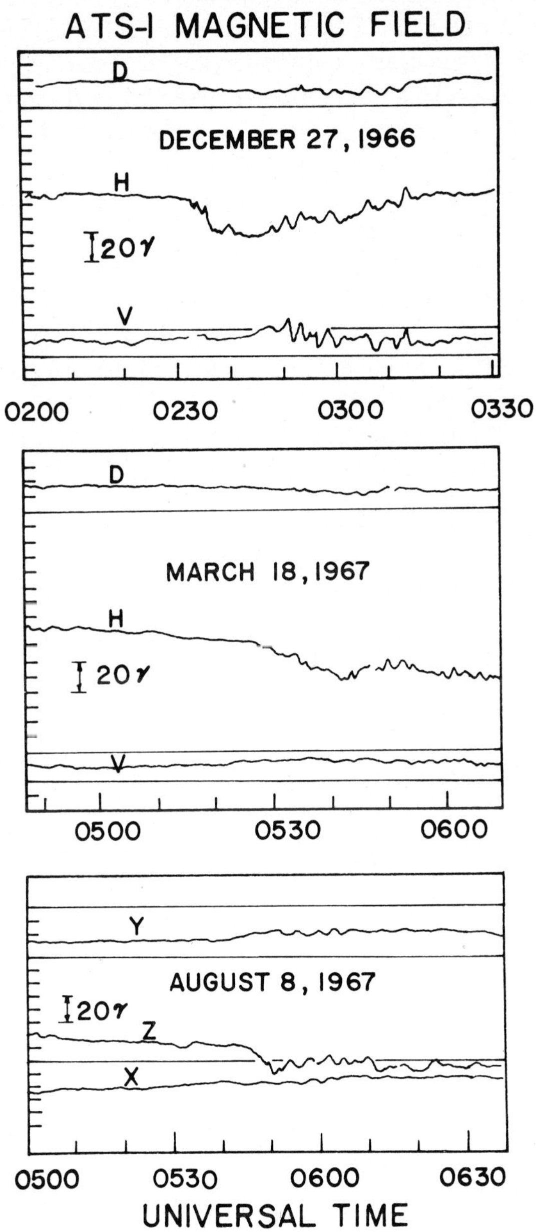

Fig. 3. More examples of the abrupt discontinuous events. Note that the discontinuities are only observed in the *H* or the *Z* component.

the magnetic field component parallel to the dipole direction (that is, in either *H* or *Z*). Lin and Parks (1973) attribute these variations to effects arising from the plasma sheet motion and the subsequent encounter of the inner edge by the satellite. The slow decrease is believed to result as the current associated with the main body of the plasma sheet is convected closer to the position of the satellite. This perturbation field is in the opposite direction from the earth's magnetic field direction and thus as the distance between the current and the satellite decrease the local magnetic field intensity decrease also. The abrupt decrease results when the inner edge encounters the satellite and when the satellite becomes immersed in the plasma sheet. The magnetic field decreases owing to plasma diamagnetic effects.

A rough estimate of the convection speed is obtained assuming the geomagnetic tail field model of Williams and Mead (1965). Using an average dB/dt during the slow decrease as about 0.5 γ min^{-1} and assuming that the inner edge is initially located at about 10 R_e, the convection speed obtained is about 0.1 R_e min^{-1}. This value is in the range of values obtained by Shelley *et al.* (1972) of convection speeds at synchronous altitudes. The model used and the calculations are shown in Figure 4.

On the basis of the interpretation given above, one can estimate the width of the inner edge. Figure 5 shows the durations of the abrupt edges observed in many regions of the local time. The average duration constructed from these observations is about

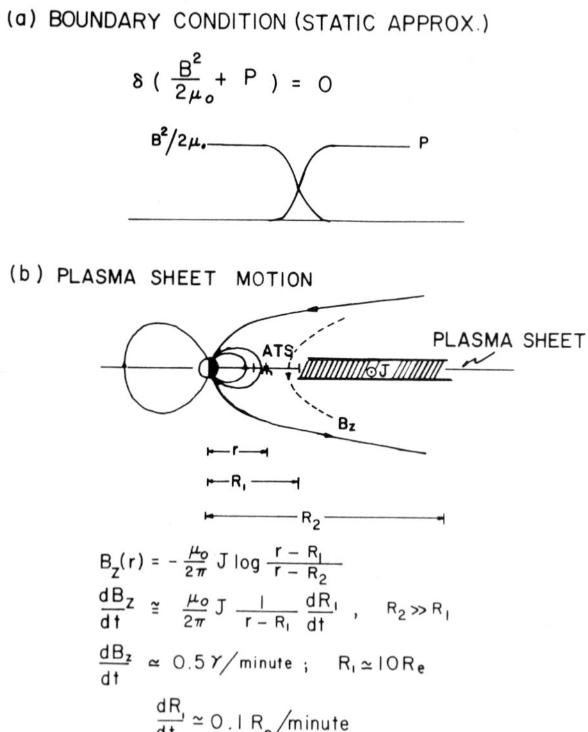

Fig. 4. Schematic diagram and examples of calculations.

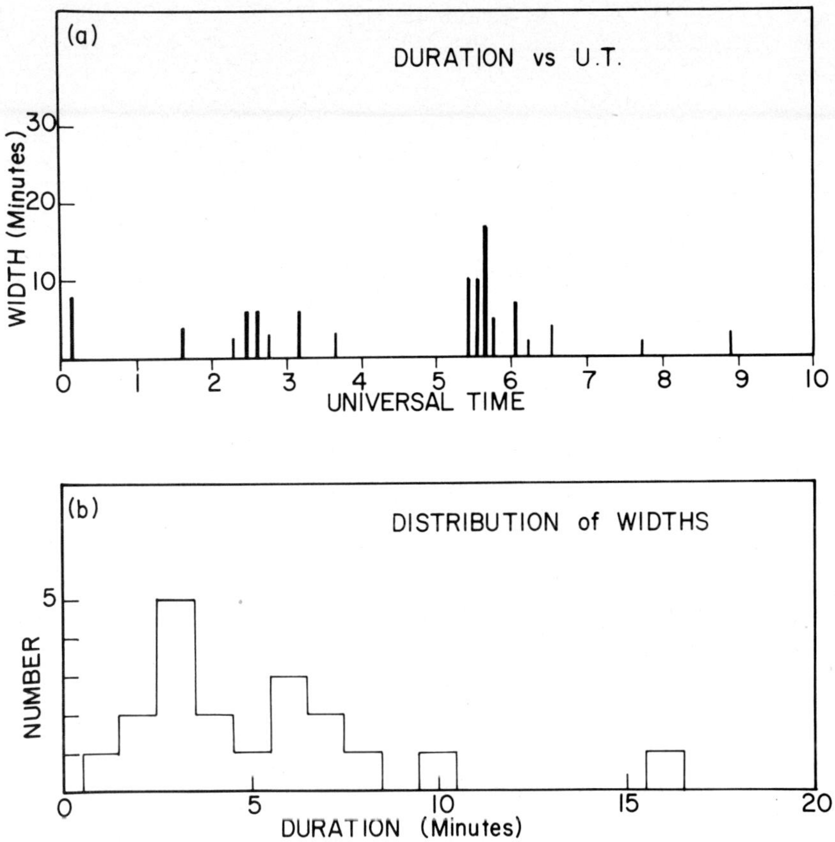

Fig. 5. The duration of the abrupt events in the UT where observed (top). Below, the distribution of the durations of the abrupt events.

5 ± 2 min. If we used 0.1 R_e min^{-1} convection speed, the inner edge width is on the average about 0.5 R_e.

An attempt was made to relate the plasma sheet motion to occurrence of substorms. Substorm onset was determined from auroral zone magnetograms as well as particle precipitation data when available. Define Δt as the time difference between the occurrence of the abrupt edge at the ATS-1 and the onset of a substorm. The results shown in Figure 6 indicate that Δt is scattered above and below the zero line which is the onset time of substorm. Here negative Δt means the abrupt edge was observed before the substorm onset and the positive Δt means the edge was observed after the auroral substorm onset.

An important conclusion can be drawn from the events with positive Δt values. According to our interpretation of the plasma sheet motion, the positive values indicate that the motion continues on inward without interruption even after a substorm onset has been recorded. This feature contrasts with models which require plasma sheet plasma to 'stack' up just prior to the substorm triggering (Vasyliunas

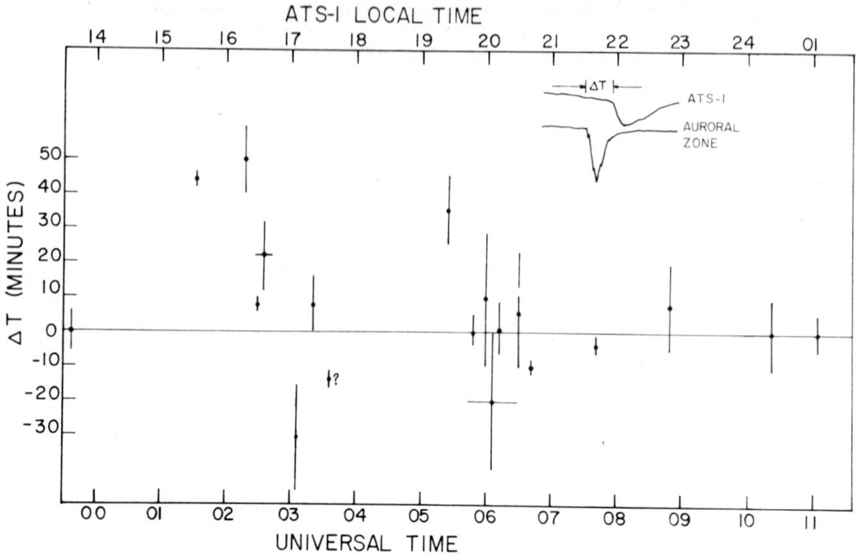

Fig. 6. Phase relationship between the onset of the abrupt edge at the ATS 1 and the onset of substorms determined from auroral magnetograms and particle precipitation data.

and Wolf, 1973). According to our data, the motion is smooth and continues on inward without interruption.

3. Plasma Sheet Convection and Interplanetary Electric Field

This section examines the origin of the electric field which drives the plasma sheet convection. For this, we have examined the behavior of the interplanetary space $\mathbf{V} \times \mathbf{B}$ electric field. Figure 7 shows the relationship between the interplanetary magnetic field and the $\mathbf{V} \times \mathbf{B}$ electric field. We have assumed for simplicity that the solar wind streams in the x-direction only. Thus, the electric field is confined in the plane perpendicular to the Earth-Sun line. The angle ξ defines the orientation of the electric field vector. When ξ is in the range $+90°$ and $-90°$, there is an electric field component directed from dawn to dusk. The definition of $\xi = \tan^{-1} B_y / -B_z$ includes both the azimuthal and polar component of the interplanetary magnetic field.

Data from the Explorer 35 have been used to construct the electric field vectors every 12 min for March 27, 1967. The original magnetic field data are shown in Figure 8. Figure 9 shows the angle ξ as a function of time plotted to the same time scale as the ATS-1 quantities (the magnitude of E was fairly constant and is left out in the figure). First to note is that the interplanetary magnetic field total intensity remained relatively constant but the direction varied considerably in both the polar and azimuthal direction. This behavior of the interplanetary B field is quite typical. The angular variations are reflected in the time variations of ξ. Examination of Figure 8 shows a striking correlation between the variation of ξ and particle and magnetic field H-component at the ATS-1. Notice that whenever ξ approaches zero, the ATS-1

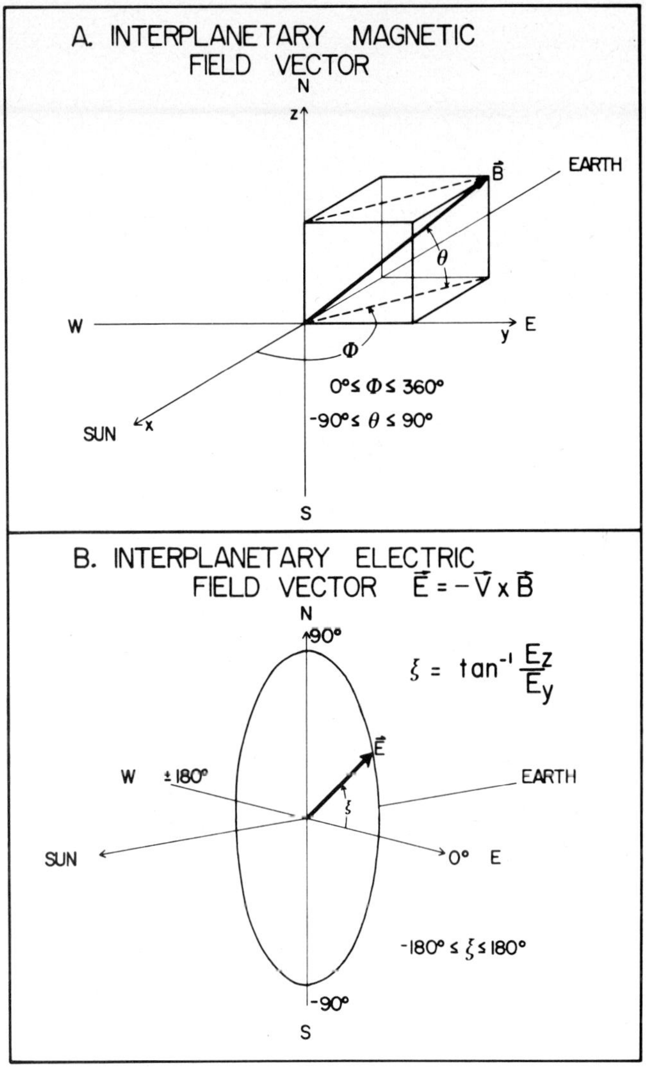

Fig. 7. Relationship between the coordinates used to define interplanetary magnetic field and the electric field.

magnetic field decreases. When ξ departs from zero toward larger angles the ATS-1 H component increases. This correlated behavior was observed repeatedly during this day. Even some of the finer details show correlation.

This type of correlation has already been reported by Parks and Pellat (1972) and the March 27, 1967 events further demonstrate that there is close coupling between the interplanetary space and the magnetosphere. As we have already indicated that changes of the ATS-1 magnetic field intensity arise from the high β plasma sheet plasma, the correlation shown in Figure 9 suggests that the interplanetary electric field penetrates into the magnetosphere and controls the plasma sheet convection. This neces-

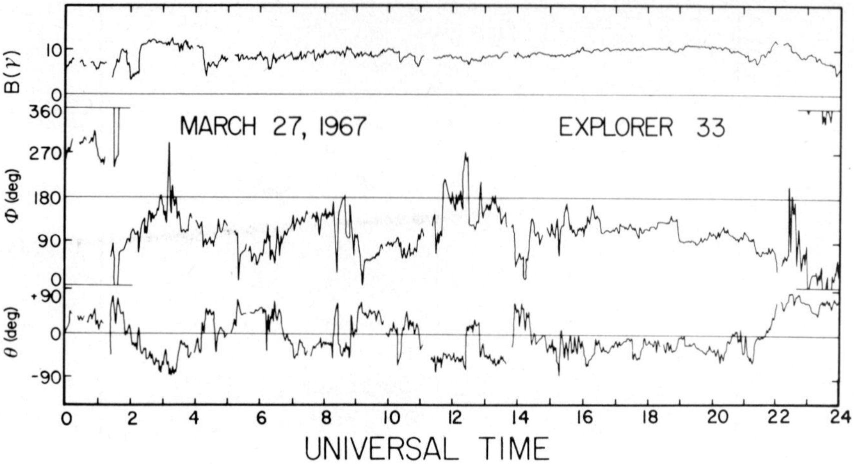

Fig. 8. Original magnetic field data in the interplanetary space measured on the Explorer 33.

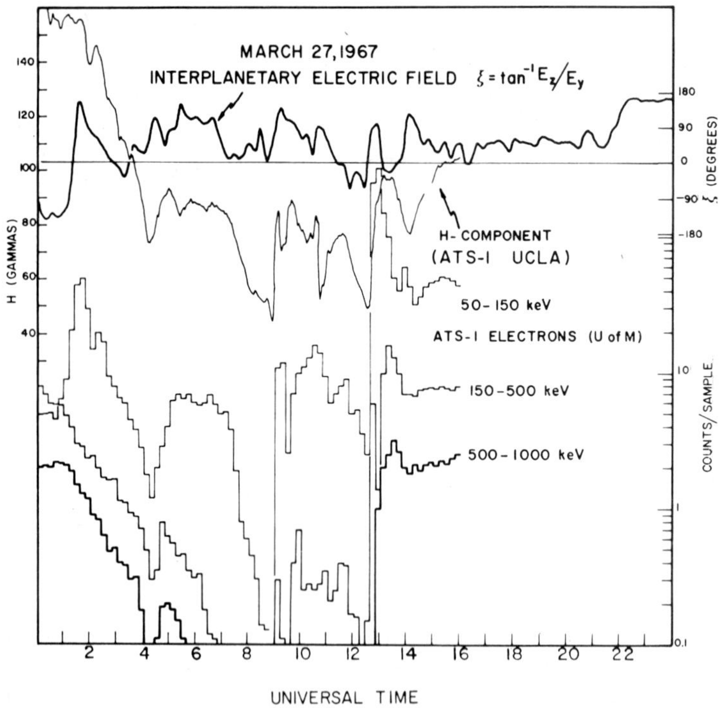

Fig. 9. Correlation of the interplanetary electric field direction and the magnetic field and energetic electrons at the ATS-1 orbit.

sarily implies that the magnetopause has a finite conductivity since otherwise the electric field would be shielded across the boundary.

The fact that the plasma sheet responds fairly promptly to the changing interplanetary electric field indicates that the polarization electric field in the plasma sheet is not fully developed in spite of the fact that the dielectric permeability is extremely high, of the order of 10^5. It is apparent that the polarization charges are quickly destroyed or allowed to flow, most probably along the magnetic field into the polar ionosphere. The pulse of current flowing is of sufficient intensity to modify the ionospheric magnetic field, and we are studying polar magnetograms for such effects. If the polarization currents do flow as described, then the variations of the polar magnetic field should be correlated to the changing interplanetary electric field and the correlation of DP2 and the interplanetary magnetic field (Nishida, 1968) might be indicating this effect.

The conceptually simple picture just described is incomplete. The final analysis must include the plasma conductivity properties of the ionosphere and the feed-back effects. The ionospheric parameters influence the rate at which the polarization currents could flow and the ionospheric electric field modifies the plasma sheet convection. (The total electric field in the magnetosphere is the sum of all fields, the primary solar wind driving field, the polarization field, and the ionospheric electric fields.) For electric field measurements in the ionosphere see Mozer (1972) and Heppner (1972). The paper of Pellat and Laval (1972) should be consulted as to how one might evaluate the self-consistent problem.

4. Conclusion

Our study of the solar wind-magnetospheric substorm problem examines the effects of the interplanetary electric field in the magnetosphere. The interplanetary electric field, defined through the angle ξ and its magnitude $|\mathbf{E}|$, takes into account both the solar wind velocity and the polar and azimuthal components of the interplanetary magnetic field. This approach has proven quite successful for the event discussed in this text. Other examples are being studied and thus far support the results shown here. The effects of the electric field intensity and phasing between the variations in the interplanetary and magnetospheric parameters are studied to better evaluate the magnetopause boundary and ionospheric effects.

Acknowledgment

We thank Professors Coleman and McPherron of UCLA for use of the ATS-1 magnetic data and Professor Winckler of Minnesota for the ATS-1 energetic electron data. The Explorer data was provided by the World Data Center and we thank Dr Vette. The research in Washington is supported in part through grants awarded by the Atmospheric Sciences Section of the National Science Foundation, Grants GA 33021X, GA 33719, and the National Science Foundation Grant GU 2655.

References

Alfvén, H.: 1968, *J. Geophys. Res.* **73**, 4379.
Alfvén, H. and Falthammer, C.-G.: 1971, *Cosmic Electrodyn.* **2**, 78.
Arnoldy, R. L.: 1971, *J. Geophys. Res.* **77**, 5189.
Chase, L. M.: 1969, *J. Geophys. Res.* **74**, 348.
Cummings, W. D., Barfield, J. N. and Coleman, Jr., P. J.: 1968, *J. Geophys. Res.* **73**, 6687.
Dungey, J. W.: 1961, *Phys. Rev. Letters* **6**, 47.
Fairfield, D. H. and Cahill, L. J.: 1966, *J. Geophys. Res.* **71**, 155.
Fris-Christensen, E., Lassen, K., Wilcox, J. M., Gonzalez, W., and Colburn, D. S.: 1972, *J. Geophys. Res.* **77**, 3371.
Heppner, J.: 1972, *J. Geophys. Res.* **77**, 4877.
Hones, E., Asbridge, J., Bame, S., and Singer, S.: 1971, *J. Geophys. Res.* **76**, 63.
Lezniak, T. W. and Winckler, J. R.: 1970, *J. Geophys. Res.* **75**, 7075.
Lin, C. S. and Parks, G. K.: 1973, in press.
Mansurov, S.: 1969, *Geomag. Aeron.* **9**, 622.
McIlwain, C.: 1972, Preprint, Physics Dept. Univ. of Calif., La Jolla, Calif.
Nishida, A.: 1968, *J. Geophys. Res.* **73**, 5549.
Mozer, F.: 1972, *J. Geophys. Res.* **77**, 6129.
Parks, G. K. and Pellat, R.: 1972, *J. Geophys. Res.* **77**, 266.
Parker, E. N.: 1958, *Astrophys. J.* **128**, 664.
Pellat, R. and Laval, G.: 1972, Proceedings of COSPAR, Madrid, Spain.
Petschek, H.: 1964, NASA publication SP-50.
Rostoker, G. and Falthammer, C.-G.: 1967, *J. Geophys. Res.* **72**, 5853.
Schatten, K. and Wilcox, J.: 1967, *J. Geophys. Res.* **72**, 5185.
Schield, M. A. and Frank, L.: 1970, *J. Geophys. Res.* **75**, 5401.
Shelley, E. G., Johnson, R. G., and Sharp, R. D.: 1972, *J. Geophys. Res.* **77**, 6104.
Svalgaard, L.: 1968, Geophys. Papers, R-6, Danish Metro. Inst. Copenhagen, Denmark.
Sweet, P. A.: 1958, Proceedings of the IAU Symp. on Electromagnetic Phenomena in Cosmic Phys., Cambridge Univ. Press, New York.
Vasyliunas, V.: 1968, *J. Geophys. Res.* **73**, 2839.
Vasyliunas, V. and Wolf, R.: 1973, *Rev. Geophys.* **11**, 181.
Williams, D. and Mead, G.: 1965, *J. Geophys. Res.* **70**, 3017.

THE RELATION BETWEEN THE AZIMUTHAL COMPONENT OF THE INTERPLANETARY MAGNETIC FIELD AND THE GEOMAGNETIC FIELD IN THE POLAR CAPS

LEIF SVALGAARD

Institute for Plasma Research, Stanford University, Stanford, Calif., U.S.A.

Abstract. The recently discovered relation between the azimuthal component of the interplanetary magnetic field and magnetic variations in the Earth's polar caps is reviewed. When the IMF azimuthal component is positive (typical of an interplanetary sector with magnetic field directed away from the Sun) geomagnetic perturbations directed away from the Earth are observed within 8° from the corrected geomagnetic pole. When the IMF azimuthal component is negative (typically within toward sectors) the geomagnetic perturbations are directed towards the Earth at both poles. These perturbations can also be described by an equivalent current flowing at a constant magnetic latitude of 80°–82° clockwise around the magnetic poles during toward sectors and counterclockwise during away sectors. This current fluctuates in magnitude and direction with the azimuthal component of the IMF, with a delay time of the order of 20 min. The importance of this effect for our understanding of both solar magnetism and magnetospheric physics is stressed in view of the possibility for investigating the solar sector structure during the last five sunspot cycles.

1. Introduction

Of all solar wind parameters, it appears to be the interplanetary magnetic field (IMF) which is most directly related to terristrial effects. Dungey (1961) suggested that a southward directed IMF may merge with the geomagnetic field and enhance the convection of magnetospheric plasma. In particular, Dungey suggested that the classical SD current system arose from such a mechanism. Many recent investigations have confirmed the existence of such a relationship (e.g. Meng *et al.*, 1973). The influence of a southward IMF component has played an important role in many theoretical investigations of the interaction between the solar wind and the terrestrial magnetosphere, especially in attempts to understand the mechanism of polar substorms. For some other aspects of magnetospheric physics, it is, however, apparent that the southward component of the IMF does not play a major role. Access of solar particles to the polar caps seems to be unaffected by the north-south component (Fennell, 1973); and the spatial configuration of the polar cap electric field as observed by Heppner (1972) is controlled by the azimuthal (east–west) component of the IMF. Finally a distinct variation of the geomagnetic field in the polar caps has been found to depend critically upon the azimuthal component of the IMF (Svalgaard, 1968, 1972, 1973a; Mansurov, 1969; Friis-Christensen *et al.*, 1971, 1972; Wilcox, 1972; Kawasaki *et al.*, 1973).

The discovery of these particular variations has sparked considerable controversy and re-thinking about the nature of polar cap magnetic variations, and will be the subject of this review. Part of the controversy is due to the fact that the influence of the north-south component was predicted, while the influence of the azimuthal component was not. But, as has been pointed out by Vasyliunas, no quantitative models

D. E. Page (ed.), Correlated Interplanetary and Magnetospheric Observations, 61–84. *All Rights Reserved.*

of the *open* magnetosphere exist at present, and therefore it is not clear what the detailed predictions of the open magnetosphere concept are. There is even some doubt as to exactly how necessary the southward component is in controlling geomagnetic activity now that it has done its job in resolving the open versus closed magnetosphere controversy.

2. Polar Cap Magnetic Field Variations

The diurnal variation of the geomagnetic field in the polar caps was traditionally considered to be simple, consisting of a rather regular sinusoidal wave in all three components. The variation of the horizontal components resembles the magnetic effects of a uniform horizontal current sheet covering the entire polar cap. This current sheet stays fixed in relation to the direction to the Sun, while the Earth rotates under it. Figure 1a shows the kind of observations leading to this picture. The average daily variations during the summer season of the North component, X, and of the East component, Y, at an observing station near the magnetic pole, are sinusoidal to within a few gammas, and their phases differ by six hours. The uniformity of the variation across the polar cap is also illustrated in the figure. Figure 1b shows that the time of maximum X is controlled by local time, i.e., the rotation of the Earth. The amplitude of the regular diurnal variation is very nearly constant over the polar cap as shown in Figure 1c, being larger during summer than in winter.

Since the westward auroral electrojet is nearest to a polar cap station in the early morning hours, an increase of the vertical component, Z, is then observed. In the late afternoon, the eastward auroral electrojet is nearest to the station and a depression of Z is observed. The variation of Z during a day is thus also of a regular character. The amplitude of this variation increases with increasing distance from the magnetic pole, since the observing station then is closer to the polar electrojets. For very high latitude stations this amplitude is in general much smaller than the amplitude of the variation in the horizontal components. This makes the Z component particularly well-suited for studies of variations unrelated to the polar electrojets.

We do not believe today that a horizontal sheet current actually flows across the polar cap. Electric field directions derived from Barium releases in the polar ionosphere show that the magnetic effects must be caused almost entirely by a source other than overhead ionospheric currents (Heppner *et al.*, 1971). Whatever the detailed mechanism may be, it is generally believed that some of the currents which close the auroral electrojets are the cause of the uniform perturbation field across the polar cap, and therefore the cause of the regular daily variations. Possibly field-aligned current sheets flowing nearly vertically into and out of the ionosphere may be of importance in producing the variations as has been suggested by Heppner *et al.* (1971) and by Kawasaki and Akasofu (1973).

By means of the polar electrojet, geomagnetic disturbances within the polar caps are tied to the substorm activity in the auroral zones. If the substorm activity is high, strong disturbances are always observed inside the polar cap. On the other hand, con-

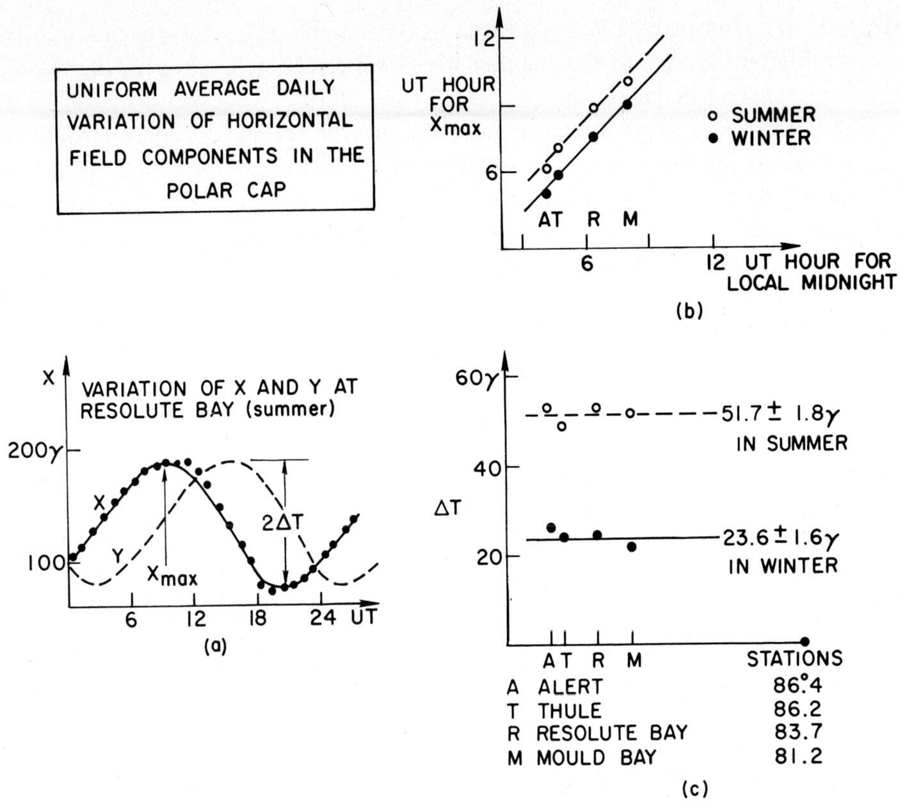

Fig. 1. Average daily variation of geomagnetic field components in the polar cap.

(a) Variation of X and Y at Resolute Bay during the summer season of 1965. The solid circles show the observed X variation. The dashed curve shows the corresponding daily variation of the Y component.

(b) Local time control of the time of observed maximum value of the north component for four northern polar cap stations.

(c) Uniformity of the amplitude of the regular daily variation of the horizontal components. The approximate invariant latitude for the four stations used is given below:

A	Alert	86°.4 N
T	Thule	86.2
R	Resolute Bay	83.7
M	Mould Bay	81.2

(After Svalgaard, 1973a).

siderable magnetic disturbance may be observed in the polar cap even if the activity at lower latitudes is very low, showing that processes particular to very high latitudes are in effect. The special polar cap disturbances differ from auroral latitude activity in several ways, the most obvious being that auroral latitude activity peaks at midnight while polar cap disturbances are most prominent around noon. The polar cap disturbances were usually described as being irregular without distinct patterns; sometimes the perturbations were positive and sometimes negative, so that they tend to cancel out when several days are averaged; in this way the average daily variation takes on its simple form as shown in Figure 1. However Svalgaard (1968) and independently Mansurov (1969) were able to classify the polar disturbances into two clearly different types and showed that they were closely related to the direction of the interplanetary magnetic field.

The two different types are most clearly seen in the vertical, Z, component at stations close to the invariant poles, such as Thule in the northern hemisphere and Vostok in the south. Figure 2 shows typical variations at these two stations. One observes either a substantial increase of the vertical component for several hours around local noon, or on other days a decrease. A change from one type of variation (e.g., increases) to the other type occurs at the same time at both stations. Typically one type is observed to persist for several days, then the variation may change, often quite abruptly, to the other type, which then is observed for the following several days, etc.

3. Interplanetary Sector Structure

The interplanetary magnetic field is an extension of the weak solar photospheric magnetic field carried out by a steady expansion of the solar corona: the solar wind. The magnetic field lines are 'frozen' into the plasma and move with it. Due to interactions between inhomogeneities in the plasma and the high speed (≈ 400 km s^{-1}) of the solar wind, this interplanetary magnetic field is highly variable on a short time scale (i.e., minutes) as observed near the Earth. If the field components are averaged over a few hours, much of this variability disappears and a larger scale field-structure emerges (Wilcox, 1968). In general, the average field is either directed away from the Sun along an Archimedean spiral line resulting from combining the radially outflowing solar wind with the rotation of the Sun, or directed towards the Sun along the spiral. Near the Earth the spiral angle is about 45°. The average field is usually in the same direction (i.e., either away from the Sun or toward the Sun) for several consecutive days. Then the average field direction changes abruptly to the opposite direction, which is then observed for the next several days. The interplanetary magnetic field can thus be divided into regions called sectors where the field direction is dominantly in the same direction either towards the Sun or away from the Sun along the spiral line. Usually there are four such sectors so that the Earth is swept by IMF with the same predominant direction for about $27/4 \approx 7$ days. It is important to note that this sector-structure only emerges if the field components are averaged over sufficiently long intervals (e.g., three hours), and that on a shorter time scale the IMF is highly variable in both magnitude and direction.

4. Polar Cap Response to IMF Polarity

The IMF polarity and the type of polar cap variation have the common property of staying the same for several days and then abruptly changing to the opposite. By closer examination it is indeed found that these changes take place simultaneously, so that when the Earth is within an away IMF sector the variation shown in Figure 2a

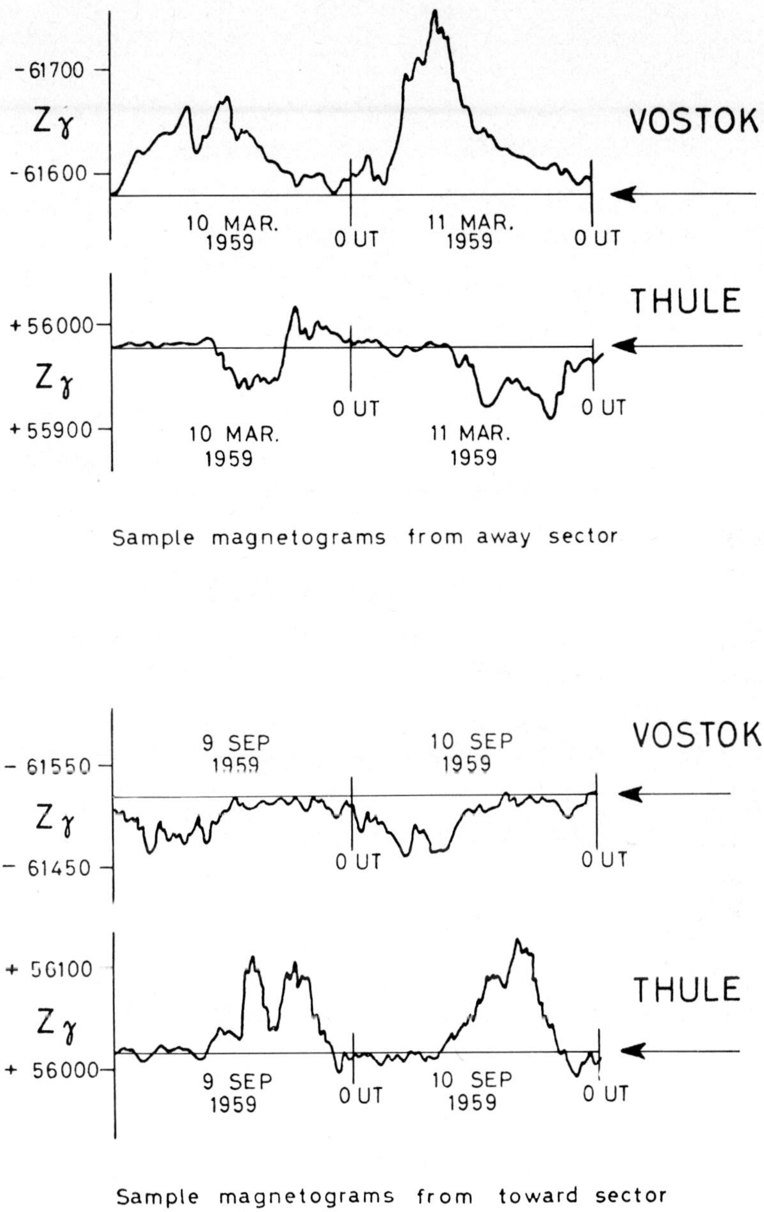

Fig. 2. Sample Z magnetograms from (a) away sector and (b) toward sector showing typical daily variations of the polar geomagnetic field observed at Vostok (84°.9 S) and Thule (86°.2 N). The arrows point to the quiet undisturbed level. (After Svalgaard, 1972).

is observed, while the other type shown in Figure 2b is seen when the Earth is immersed in a toward-the-Sun directed interplanetary field. We can summarize the description of this surprising effect and at the same time provide a useful mnemonic by saying that during an *away* sector the polar cap disturbance field is directed *away*

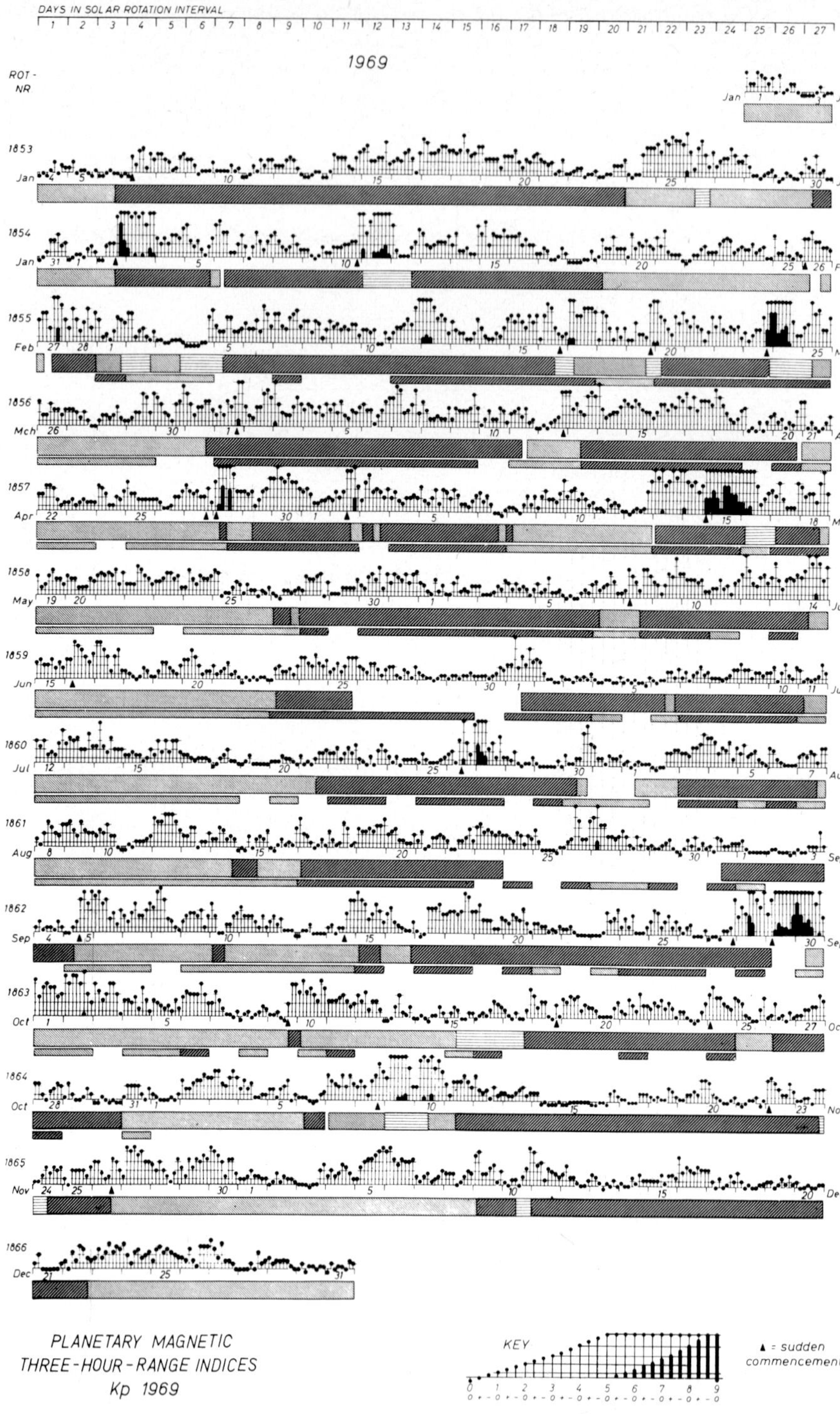

Fig. 3. Interplanetary magnetic field polarity (away from or towards the Sun) during 1969 plotted on a chart of magnetic three-hour index K_p (after J. Bartels). Light shading is field polarity away from the Sun, and dark shading is field polarity toward the Sun. The top bar on each line represents the interplanetary magnetic field polarity observed with the Ames Research Center magnetometer on the spacecraft Explorers 33 and 35, and the bottom (narrower) bar on each line represents the field polarity inferred from observations of the polar cap geomagnetic field. The shading with horizontal lines represents intervals of mixed field polarity. (After Friis-Christensen *et al.*, 1971).

from the Earth at both poles. When the Earth is within a *toward* sector, the disturbance field is directed *toward* the Earth at both poles. The magnitude of this effect varies with season, since it is largest during local summer when a disturbance field of several hundred gammas is not uncommon. It is thus important to note that the influence of the IMF polarity on the polar cap magnetic variations is not a subtle, minor or hard-to-detect effect, but is usually dominating over all other processes disturbing the polar cap field.

Therefore it is normally a straightforward procedure to infer what the IMF polarity is just by inspection of a polar cap Z magnetogram. As an unambiguous test of the effect, the IMF sector polarity was inferred from the polar cap station Thule for the years 1969 and 1970. This was done by Friis-Christensen *et al.* (1971) for 1969 and by Svalgaard (1972) for 1970. In both cases it was done before the sector polarity observed by spacecraft was available. The sector polarity was deduced by visual inspection of the magnetograms, and no scaling or computing was performed. Figure 3 shows a comparison with the observed polarity for 1969. The correspondence between the inferred and the observed sector polarities is very close. The positions of the sector boundaries usually agree to within less of a day. The same close agreement was found for the 1970 comparison. Table I summarizes the number of days with agreement and the number of days with disagreement for the two years separately.

TABLE I

Time (in days) for each combination of polarity of the interplanetary magnetic field observed with spacecraft (IMF) and inferred from observations of polar geomagnetic field (polar cap) for 1969 and 1970. Days with missing data or mixed polarity are excluded. After Friis-Christensen *et al.* (1971) and Wilcox and Colburn (1973).

Polar cap	IMF Away	IMF Toward
Away	71	11
Toward	7	85

1969: 90% correct

Polar cap	IMF Away	IMF Toward
Away	95	16
Toward	25	105

1970: 83% correct

The detailed agreement between the IMF sector polarity as observed by spacecraft near the Earth and as inferred from polar cap magnetograms leaves no doubt about the reality of the close relationship between the two fields and is also a very good argument for an open magnetosphere versus a closed one.

5. Critical Component of IMF

Even if the agreement between the sector polarity measured by spacecraft and inferred from the ground is very close ($\approx 85\%$) there are times where the polar cap magnetograms show the signature for a sector polarity opposite to the one observed

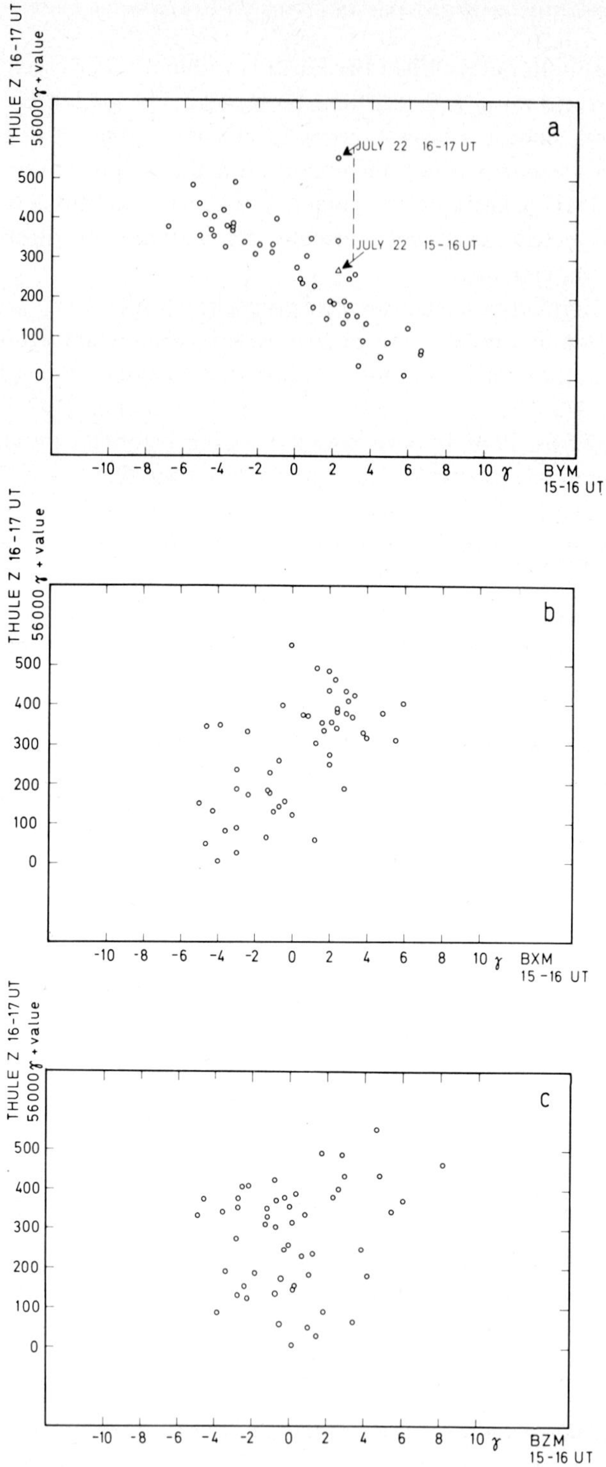

Fig. 4. Hourly mean values of the Z component ($> 56000\ \gamma$) at Thule at 1600–1700 UT versus the hourly mean values at 1500–1600 UT of the components of the interplanetary magnetic field during June–August 1969: (a) eastward component *BYM*, (b) sunward component *BXM*, and (c) N–S component *BZM*. (After Friis-Christensen *et al.*, 1972).

by spacecraft. In a study of these disagreements, Friis-Christensen *et al.* (1972) noted that on some of these days the dominant direction of the interplanetary field departed considerably from the average Archimedean spiral direction, so that the azimuthal component of the field was in the opposite direction to that expected for a spiral field. This led to the suggestion that the cause of the effects observed in the polar cap was not the polarity of the interplanetary field as such, but rather the direction of the azimuthal (east–west) component of this field. Further investigations have confirmed that this is indeed the case. However, the interplanetary field direction is almost always near enough to the spiral direction that an azimuthal component in one direction corresponds to a given polarity of the IMF.

Friis-Christensen *et al.* (1972) plotted hourly mean values of the vertical component, Z, at Thule versus hourly mean values of the components of IMF. The coordinate system used by them was the geocentric solar magnetospheric system (GSM) in which the X axis points toward the Sun from the center of the Earth, the Y axis is perpendicular to the X axis and to the Earth's magnetic dipole, so that the X-Z plane contains the dipole axis and the Z axis is positive in the northward direction. In this coordinate system the components of the IMF are BXM, BYM, and BZM. For the ideal spiral direction of IMF in an away sector BXM is negative and BYM is positive, whereas in a toward sector BXM is positive and BYM negative.

Friis-Christensen *et al.* (1972) found that if the IMF components were compared with the Thule Z component one hour later the best correlations were obtained. Figure 4 shows their result for the summer 1969. It is evident that the IMF component best correlated with Thule Z is BYM, while there seems to be little or no correlation with the two other IMF components.

That the relation between Thule Z and the IMF azimuthal component holds not only on a daily or hourly basis, but extends even to short-period fluctuations was demonstrated by Kawasaki *et al.* (1973). This is clearly seen in Figure 5 where the Thule Z-component is superposed on the IMF Y-component observed by the IMP-3 satellite. The records have been slightly shifted with respect to each other to obtain the best visually observable correlation for the short-period variations. The two records are remarkably well-correlated during the day hours (10–22 UT). Even variations of the IMF Y-component on a timescale of 20 min appear to be correlated with similar structures in the Thule Z records. On the average the variations of Thule Z are delayed about 25 min with respect to variations seen at IMP 3. This delay is considerably longer than the mean transit time (<10 min) for the solar wind from the spacecraft to the Earth, so we may conclude that the response time for the polar magnetic field to variations of the Y component of the IMF is of the order 20 min.

6. Analysis of Sector Effects in the Polar Cap

When the sector polarity dependent polar cap variations were first discovered in the vertical component, it was also noted that at somewhat lower corrected geomagnetic latitudes ($\approx 80°$) the horizontal components showed opposite perturbations depend-

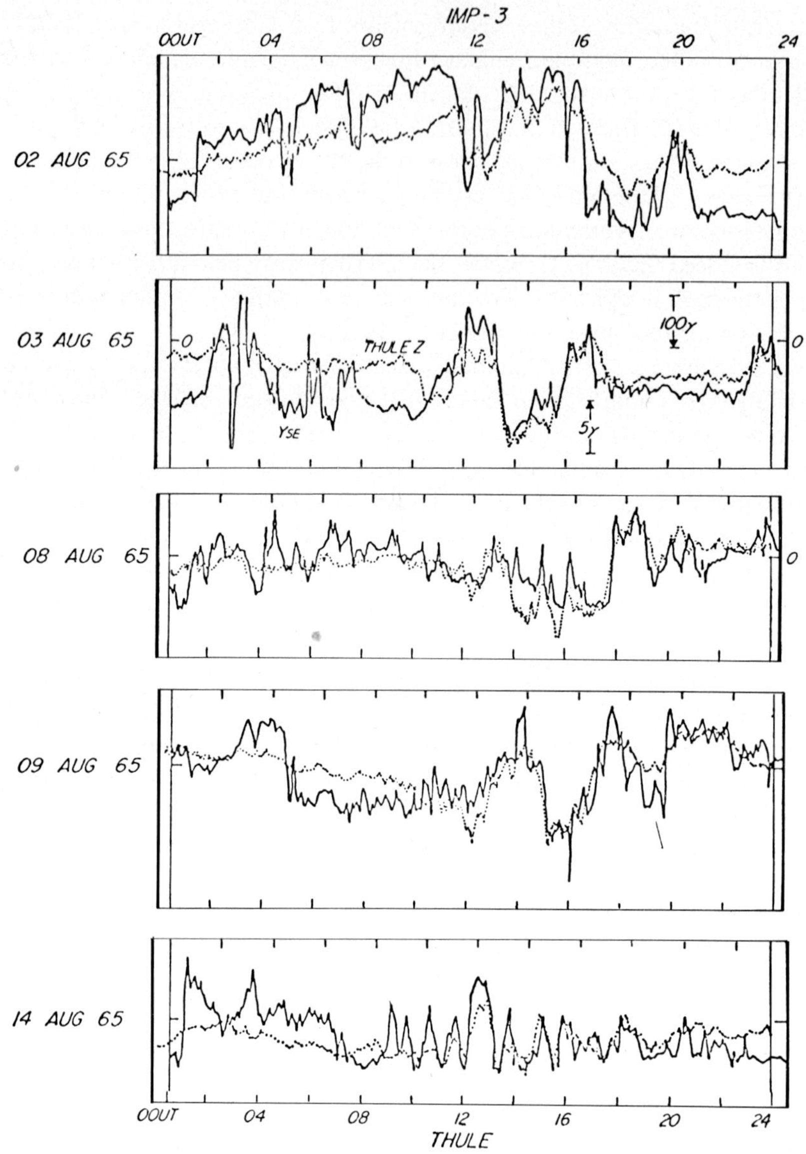

Fig. 5. Comparison of Thule Z magnetograms with the azimuthal (Y) component of the interplanetary magnetic field in the solar ecliptic (SE) coordinate system observed by the IMP-3 satellite for several days during northern local summer. The Thule Z component (dotted line) is plotted positive downward. (Adapted after Kawasaki *et al.*, 1973).

ing on the sector polarity. At Godhavn (77.5° N) the horizontal component is generally increased around noon during away sectors and decreased during toward sectors. To get a clearer picture of the nature of these polar cap variations, Svalgaard (1973a) analyzed data from nine stations in the northern hemisphere during the quiet year 1965. The geomagnetic field data was first converted to X, Y and Z components and then corrected for secular variation. All days with data gaps were excluded from the analysis. Diurnal variation curves of all three geomagnetic components were then computed for each IMF polarity (as observed by spacecraft) separately for the whole year. The average diurnal variations throughout the year were finally computed using all days.

Figure 6 shows the diurnal variation found for Z at Resolute Bay. The open circles display the variation found on days within away sectors (IMF Y component positive), while the solid dots show the average variation observed during towards sectors (IMF Y component negative). For about half of the day the sector polarity does not seem to have any marked influence on the value of the Z component. However, during the other half of the day, roughly between 12^{h} and 24^{h} UT, the variation of the Z component is strongly dependent on the IMF polarity. On days with positive IMF polarity (away from the Sun) the Z component is decreased and reaches a minimum around 18^{h} UT, whereas on days with negative IMF polarity the Z component is increased by about the same amount during the same interval of the day.

The average daily variation found by using all days of the year irrespective of sector polarity is shown by the dashed curve in Figure 6, demonstrating the almost complete cancellation of the sector polarity influence when about the same number of days with opposite IMF polarity are averaged together. In view of the detailed short-period dependence as shown in Figure 5, we would also expect this cancellation effect.

The influence of IMF polarity on the vertical component at Resolute Bay is rather typical for other polar cap stations and for other magnetic elements as well. For any given polarity of the IMF (or more strictly for a given sign of the azimuthal component), we find a certain deviation (taking place roughly between 12^{h} and 24^{h} UT in the northern hemisphere) from the simple sinusoidal daily variation for each magnetic element. For the opposite polarity about the same deviation is observed, but with the opposite sign. Figure 7 gives another example of the general behavior for the North (X) component at Resolute Bay. It seems as if the influence of the sector polarity is just superposed on the simple regular diurnal variation, and thus may be extracted by subtracting this sinusoidal variation from the observed diurnal variations. Examples of the result of the subtraction are given in the lower part of Figures 6 and 7.

7. Polar Cap Current System

Using the data and procedures discussed in the previous section, the nature of the influence of IMF on the high latitude geomagnetic field was studied by Svalgaard (1973a). The IMF effects over the northern polar cap were found to be strongest at 17^{h}–19^{h} UT. At that time it is local noon over the northern magnetic pole, and conditions seem the most favorable for direct interactions between the solar wind and the polar cap field. From very high latitudes and down to about 75° the influence of the IMF sector structure was dominating over all other effects, whereas there were only very slight, if any, indications of any IMF polarity influence at auroral latitudes (below 70°). The effects are thus well confined to the interior of the polar cap and can

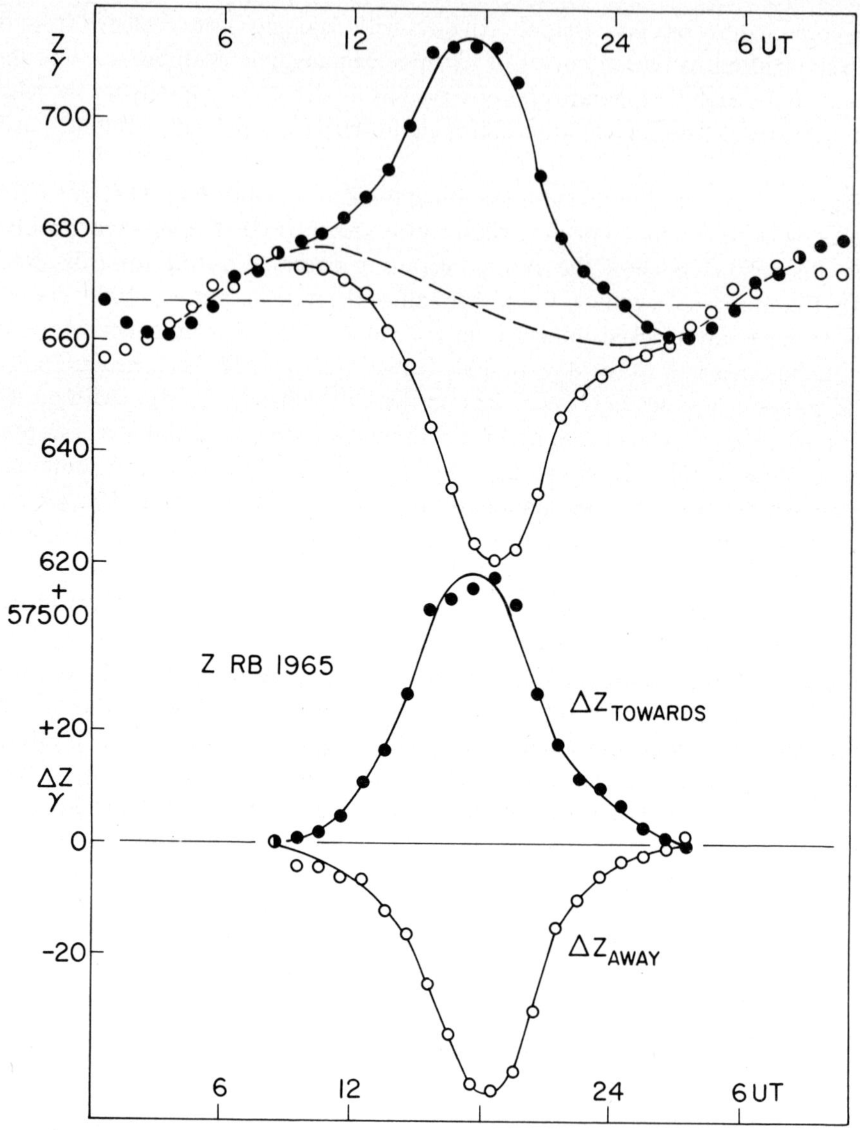

Fig. 6. Daily variations of the vertical component at Resolute Bay for 1965. Open circles show the variation on days when the Earth was immersed in interplanetary sectors of magnetic field pointing away from the Sun. Filled circles show the variation during toward sectors. The dashed curve is the average variation on all days irrespective of the sector polarity, and the straight dashed line indicates the location of the quiet undisturbed level. The average variation has been subtracted from the data in the lower panel of the figure. (After Svalgaard, 1973a).

hardly be interpreted as extensions of certain types of substorm activity. Kawasaki *et al.* (1973), analysing correlated short-period fluctuations of the two fields, also reached the conclusion that the effects depending on the sector polarity are not related to substorms.

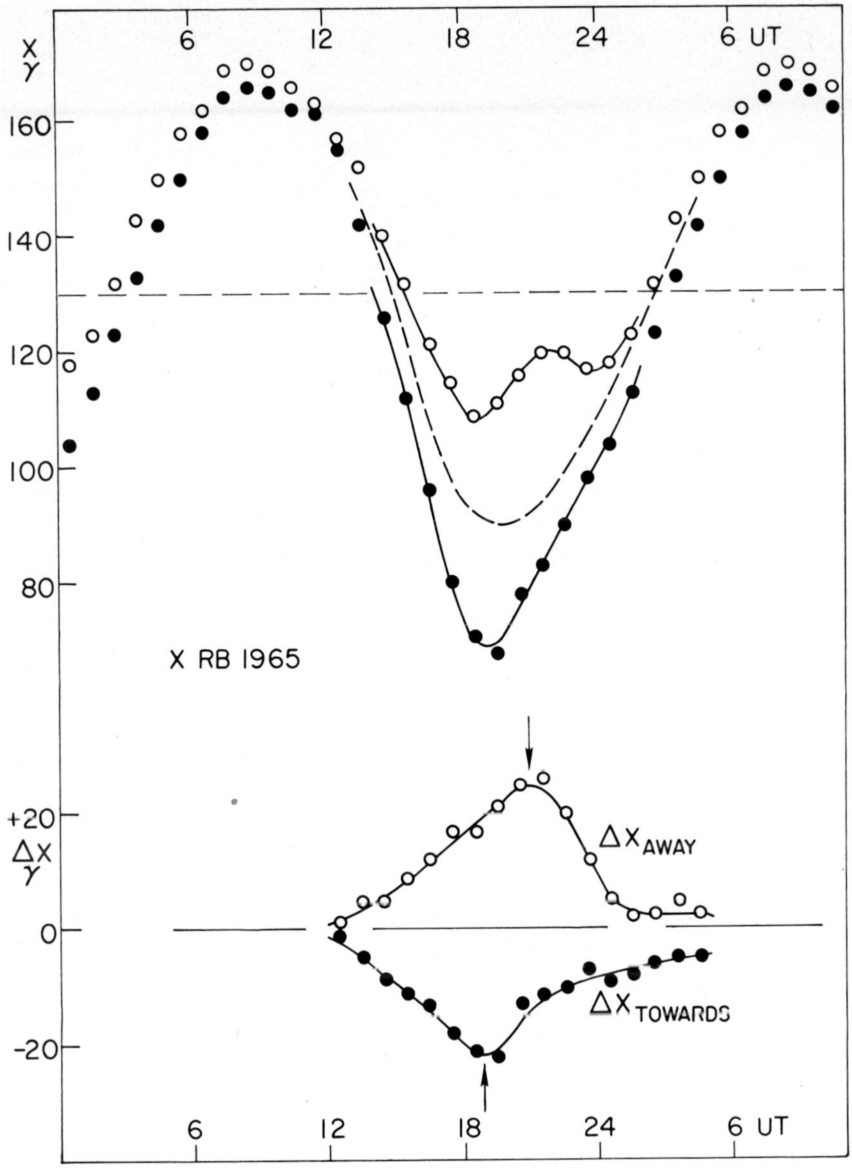

Fig. 7. Daily variation of the North component at Resolute Bay for 1965. The format is the same as used in Figure 6. (After Svalgaard, 1973a).

A synoptic presentation of the IMF sector effects as observed at 18^h UT during away and toward sectors is given in Figure 8 and 9 respectively. Horizontal perturbations are shown as vectors attached to station circles for six stations in the northern polar cap. The simultaneous vertical perturbation is given as a signed number next to the station circle. The positions of the geographical pole (GP) as well as of the magnetic pole (MP) are indicated on the figures. Traditionally the Z component is

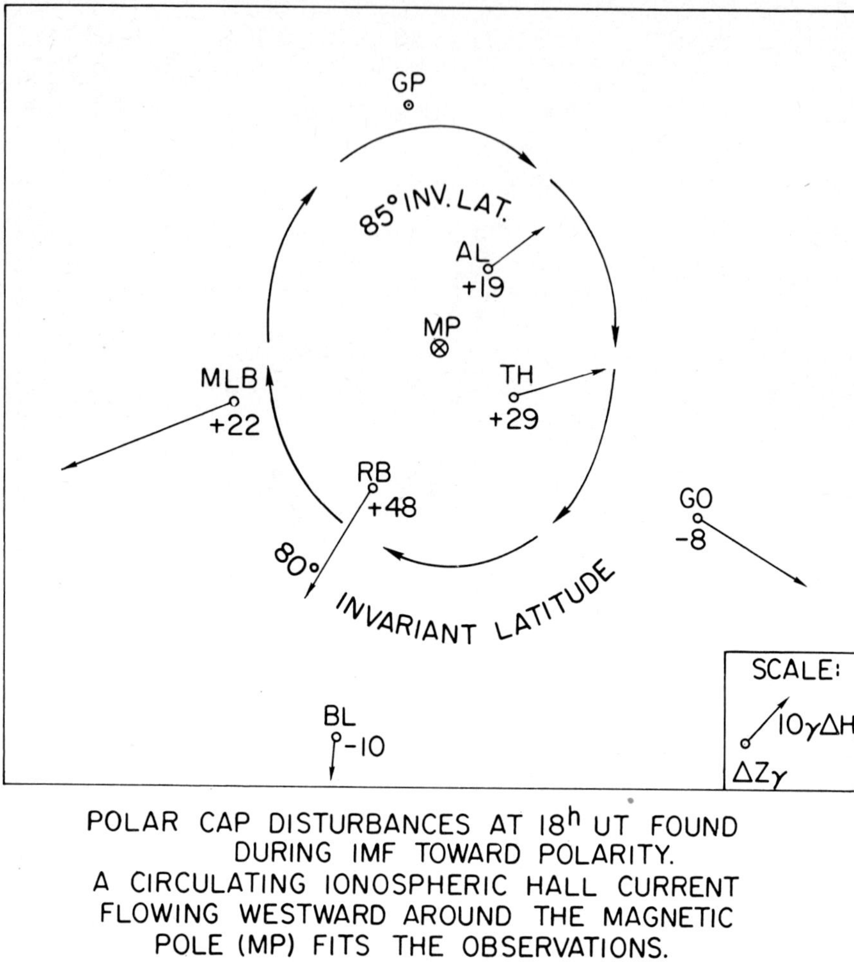

Fig. 8. Synoptic map of polar cap magnetic perturbations at 1800 UT during IMF away polarity. Horizontal perturbations are shown as vectors attached to station circles for Alert (AL), Thule (TH), Resolute Bay (RB), Mould Bay (MLB), Godhavn (GO), and Baker Lake (BL). Vertical perturbations are given as signed numbers next to the station circles. A circulating current which may produce the magnetic perturbations is indicated on the figure. (After Svalgaard, 1973a).

considered positive when directed downwards (i.e., towards the Earth); however a Z perturbation directed away from the Earth decreases the magnitude of the vertical force in the northern polar cap, but increases it in the southern polar cap.

It is evident from Figures 8 and 9 that there is a clear systematic difference in the way the geomagnetic field is disturbed during conditions with opposite IMF polarities. For away IMF polarity (Figure 8) the horizontal perturbation vectors all converge towards the magnetic pole, and vertical perturbations directed away from the Earth (negative in the northern hemisphere) are observed near the pole while vertical perturbations towards the Earth (positive in northern hemisphere) are seen below 80° invariant latitude. For towards polarity (Figure 9) the direction of all perturba-

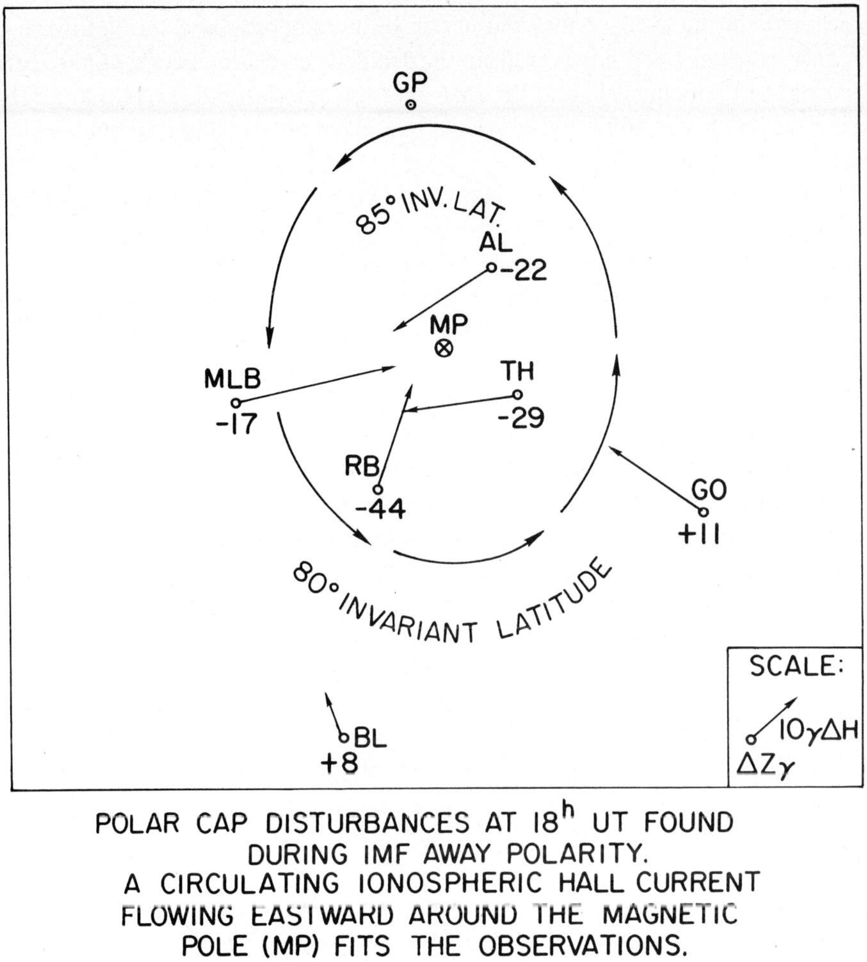

Fig. 9. Same as Figure 8, but for IMF toward polarity. The direction of all perturbations is reversed compared to the situation found during away sectors, and the polar cap current is also reversed. The position of the geographical pole (GP) and of the invariant magnetic pole (MP) are shown in both figures. (After Svalgaard, 1973a).

tions is reversed: horizontal perturbations diverge from the magnetic pole and vertical perturbations toward the Earth occur near the pole.

These magnetic effects are precisely what might be produced by a circulating ionospheric current flowing counterclockwise for away IMF polarity and clockwise for toward polarity. The location and direction of this current (presumably a Hall current) is indicated in Figures 8 and 9. From an analysis of geomagnetic data from seven stations in the northern and eleven stations in the southern hemisphere, Mansurov and Mansurova (1971) deduced essentially the same current system to account for the polarity effects observed within the polar caps. However in the interpretation of the effect they tentatively concluded that the current only appeared in the northern hemisphere during away sectors and in the southern hemisphere during toward sectors only. It seems clear, however, from the detailed correlation between short-period variations in the polar field and the IMF that the correlation is present at all times during the day hours and is *not* restricted to a given polarity (away in the north, toward in the south).

The perturbations and currents shown in Figures 8 and 9 were derived by subtracting the all-day average field values at 18^h UT from the values observed for the two IMF polarities separately. If we instead compute the mean value of the field components over a day the regular sinusoidal diurnal variation will tend to cancel out, but the IMF polarity effects will remain and influence the mean value systematically depending on sector polarity.

The mean values of the three components (X, Y, Z), corrected for secular variation, were computed for nine northern stations by Svalgaard (1973a) for three samples of data through 1965: namely (i) for all days within away sectors, (ii) for all days within toward sectors, and (iii) for all days irrespective of the sector polarity. Subtracting the average field values for the last sample from the first two samples should then show any effects related to IMF polarity. Figure 10 shows the result plotted against invariant latitude. The IMF polarity effect in the Z component is again characteristic of a current encircling the magnetic pole; the current direction again reverses when the IMF polarity reverses. The Z perturbation changes sign at about 80° invariant latitude suggesting that the current is nearly overhead at this latitude. The magnitude of perturbations of the horizontal component directed towards the invariant pole is shown in the lower part of Figure 10. The horizontal effects are strongest at about 82° invariant latitude which is nearly where the Z perturbations change sign.

Mansurov and Mansurova plotted the differences between the hourly mean centered at noon and the daily mean for the horizontal and vertical components. Their result for the southern hemisphere is shown in Figure 11. The mean for all days irrespective of sector polarity is shown by the dashed lines. Their figure is very much like Figure 10, which is for the northern hemisphere. Thus several methods of analysis all lead to the conclusion that the effect of the azimuthal component of the IMF on the geomagnetic field can be described as the magnetic effects of a current system encircling the magnetic poles at about 80° invariant latitude during the day hours. The current direction changes when the sign of the IMF azimuthal component changes.

8. Seasonal Variations of the Polar Cap Current

The magnitude of the influence of the IMF polarity on the polar geomagnetic field has a very pronounced seasonal variation, being largest during local summer. A simple measure of this change is the difference between seasonal averages of the Z component close to the pole during toward sectors and the averages during away sectors for the same season. This seasonal difference of ΔZ for three polar caps

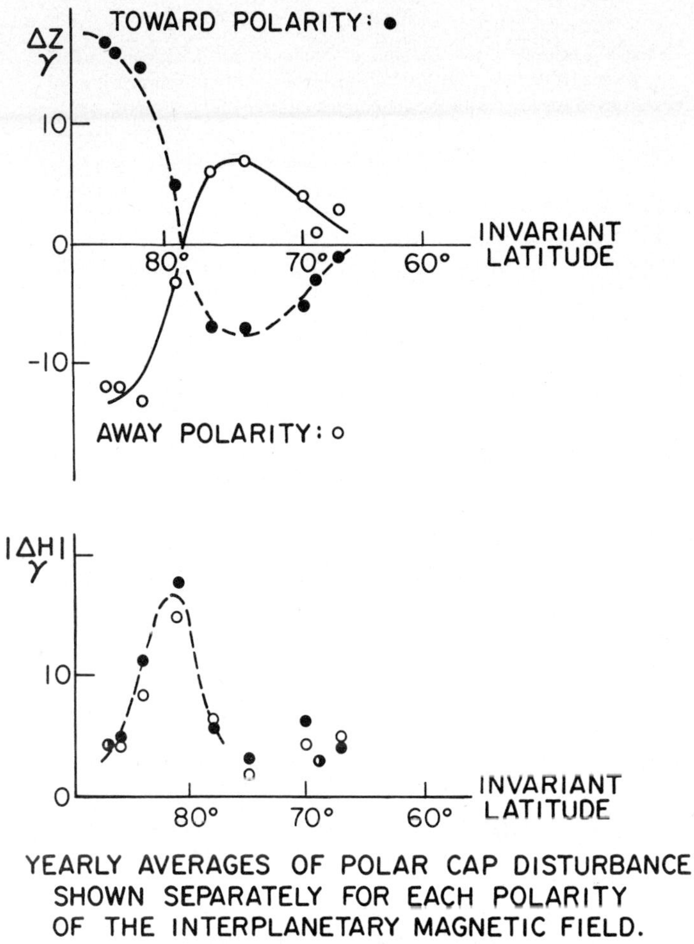

Fig. 10. Difference between the yearly mean of Z and H for both IMF polarities and the all-day average values for 1965. The differences are plotted against invariant latitude for nine stations at high northern latitudes. Open circles are used for away polarity values and solid circles for toward polarity. (After Svalgaard, 1973a).

stations is shown in Figure 12. The magnitude of the effect is very small (only a few gammas) during winter but increases sharply towards the summer. In addition, we note that the change throughout the year is nearly identical for all three stations; this seems to be a consequence of the fact that all three are well within the rather uniform perturbation field of the encircling polar cap current. There exists another seasonal effect in the time of maximum perturbation, which in the northern polar cap occurs earlier during the summer than during winter. Table II lists the time of maximum perturbation for five stations during different seasons. There is a progressive change towards earlier hours of maximum perturbation as we go from winter to summer. This effect might depend on geographical latitude; it is largest at the northernmost station Alert, and becomes smaller with decreasing geographic lati-

TABLE II

Seasonal variation of average time of maximum perturbation of polar cap geomagnetic field. The total change from winter to summer is also given together with the geographic latitude of the stations.

Station	Winter	Equinox	Summer	Change	Geogr. latitude
Alert	$18^{\mathrm{h}}.0$ UT	$14^{\mathrm{h}}.0$	$13^{\mathrm{h}}.0$ UT	$5^{\mathrm{h}}.0$	$82^{\circ}.5$
Thule	18.0	15.5	15.0	3.0	77.5
Mould Bay	21.5	19.5	19.0	2.5	76.2
Resolute Bay	18.5	18.0	17.5	1.0	74.7
Godhavn	17.0	16.5	16.5	0.5	69.2

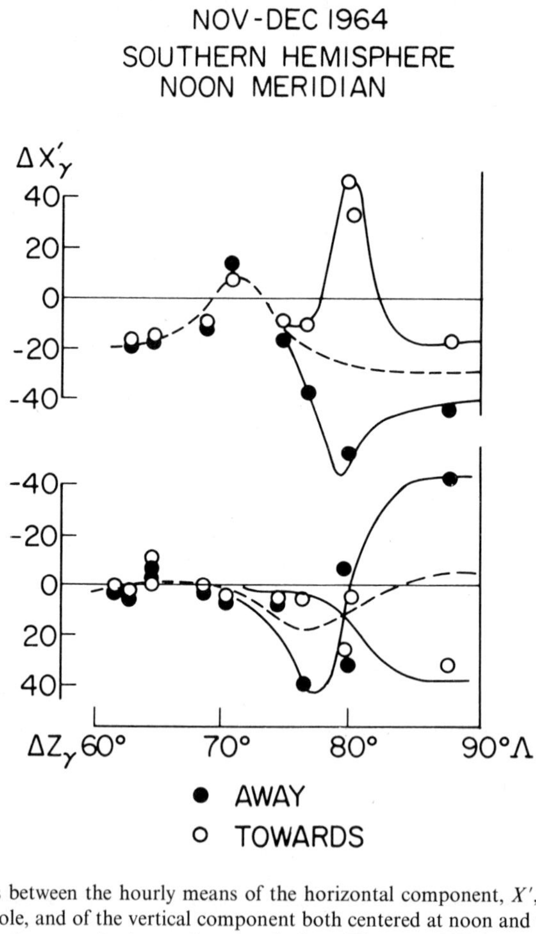

Fig. 11. Differences between the hourly means of the horizontal component, X', pointing towards the corrected magnetic pole, and of the vertical component both centered at noon and the daily mean values. The data are from local summer in the southern hemisphere. The mean difference for all days irrespective of sector polarity is shown by the dashed lines. (Adapted after Mansurov and Mansurova, 1972).

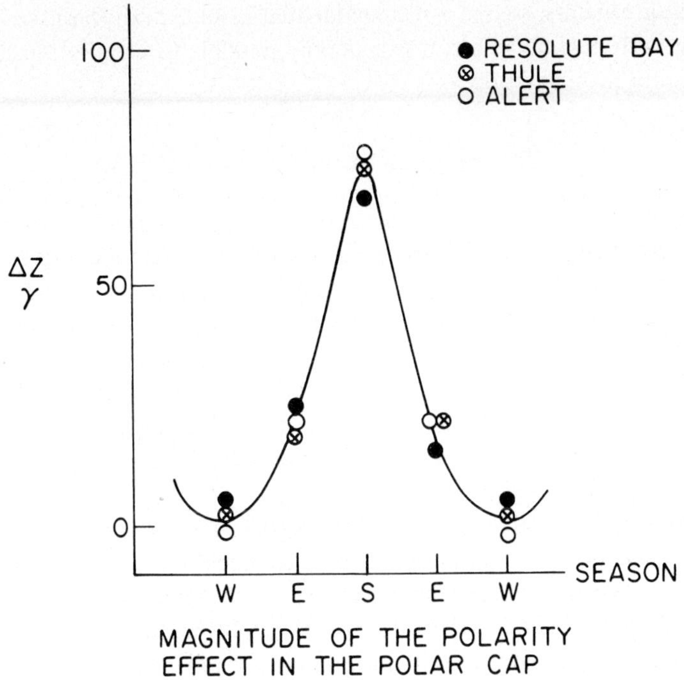

Fig. 12. Seasonal variation of the magnitude of the polar cap perturbations related to the IMF polarity. The effect is largest during local summer (S), smaller during the equinoxes (E), and smallest during winter (W). (After Svalgaard, 1973a).

tude. There is evidence from Vostok in the south that this seasonal effect is reversed in the southern polar cap, so that time of maximum perturbation changes towards later hours from winter towards local summer. However, more study of this particular effect is needed before a definitive conclusion can be reached on this point.

We have seen that significant daily and seasonal variations of the intensity of the polar cap current system exist; their nature suggests that ionospheric conductivity as well as magnetospheric geometry probably both play a role for the mechanism responsible for originating and maintaining the polar cap current system.

Even if the magnitude of the polar cap disturbances is small during winter, it is still possible to identify the signature of the IMF polarity, and therefore to infer the sector polarity. Part of the reason for this is that other disturbances tend to have small magnitudes too. It is however necessary to employ high sensitivity, temperature compensated magnetographs. Comparisons between the list of inferred polarities published by Svalgaard (1972) and the polarities measured by spacecraft (Wilcox and Colburn, 1972) made separately for the summer half-year and for the winter half-year, show very little difference in the accuracy of the inferred polarities. The accuracy of the inferred polarity is only 3% lower during winter than during summer. Also, Heckman *et al.* (1973) compared the polarity inferred from Thule during local winter with that inferred from Vostok which then had local summer and found very few cases of disagreements, so one can conclude that in spite of the large seasonal variation of the IMF polarity effect, it is generally possible to identify the effect almost independent of its magnitude.

9. Physical Cause of the Polar Cap Current

The existence of the polar cap current system is a challenge to our present understanding of magnetospheric physics, according to which the polar cap current ought not to be there at all. The very direct polar cap response to changes in the IMF polarity and also the direct access of solar flare particles to the polar caps strongly suggests that we are dealing with an open magnetosphere, where reconnection of geomagnetic and interplanetary field lines is important. The details of this reconnection process are not well understood, and in addition our understanding of the situation is hampered by the difficulty of simply mentally visualizing the resulting three-dimensional field line configuration.

In a qualitative analysis of the general properties of an open magnetosphere Stern (1972) examined the topological effects of reconnection between the geomagnetic field and an interplanetary azimuthal magnetic field. After all, the large-scale interplanetary magnetic field is oriented approximately perpendicular to the Earth's dipole axis rather than parallel to it. Stern's model is able to explain: (i) a rather uniform dawn to dusk polar cap electric field including (ii) the IMF polarity dependent asymmetry observed by Heppner, and (iii) an ionospheric polar cap current changing direction when the IMF polarity changes, and finally (iv) the familiar two-celled magnetospheric convection pattern with return flow at sub-polar cap latitudes. It may well be that Stern's model of the open magnetosphere is a significant step towards the kind of models we will be working with in the future.

In Stern's model there exist four types of field lines – closed field lines starting and ending on the Earth, open lines connected to the northern polar cap, open lines connected to the southern polar cap and unlinked lines with no connection to the dipole. Similarly, space may be divided into four regions, each containing only field lines of one type. Already Dungey showed that in models of this type the four regions generally meet along a line, called by Stern the *separatrix*. The separatrix is not a neutral line but has a field direction always pointing from south to north. Figure 13 shows the geometry of the model as seen from the Sun for the two cases of interplanetary field polarity. The critical point in Stern's model is now that interplanetary field lines when brought out to Earth by the solar wind will be swept some distance along the separatrix before breaking loose and connecting with polar field lines.

Far in space there exists a constant electric field along the vertical axis of Figure 13. Thus the electric field across the polar cap, assuming that the magnetic field lines are equipotentials, should be quite uniform. However, if the separatrix *does* have a field direction along it, it exerts a sweeping effect, most pronounced for field lines that impinge near the boundary of the polar cap: these are swept towards the northernmost neutral point and their ends on Earth are swept towards the points to which the

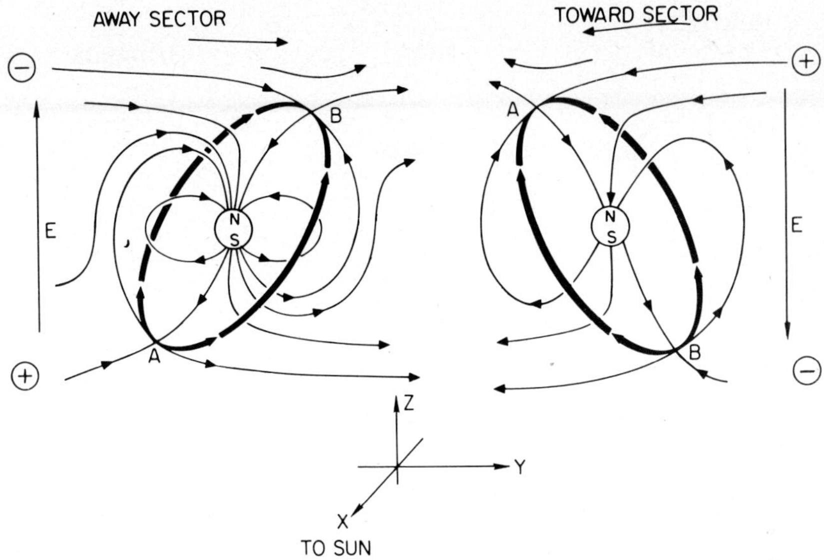

Fig. 13. Schematic view of the field of a dipole immersed in a constant field orthogonal to its axis, with the solar wind blowing into the plane of the figure. At the left is the situation for an external field parallel to the *Y* axis, and at the right for an external field antiparallel to the *Y* axis. These two cases correspond to an away and toward polarity of the IMF respectively. The separatrix is also shown schematically (see text). (After Stern, 1972).

neutral points are connected. Therefore the equipotentials become distorted and bent along the polar cap boundary.

This situation is shown in Figure 14. The left hand pattern corresponds to polar equipotentials with no sweeping effect along the separatrix. The pattern on the right corresponds to southern polar cap for an away sector. If the external field is reversed – corresponding to a toward sector – the patterns also reverse as indicated on the figure. So while the electric field in the center of the polar cap is *always* from dawn to dusk, the sweeping effect of the separatrix results in a structural property which is opposite in the two polar caps and which also reverses when the azimuthal component of the interplanetary field reverses. This is precisely what is required to explain the magnetic observations in the polar caps. The nested, crescent-shaped equipotentials are roughly aligned with the polar cap boundary, so that with enough ionospheric conductivity a Pedersen current will flow orthogonal to the boundary, but the Hall current, which is larger and therefore most important in producing magnetic variations, will run *along* the boundary. It is easily seen from Figure 14 that the Hall current has opposite directions for the two IMF polarities in the correct sense to enable one to identify it with the circulating polar cap current deduced from magnetic observations.

10. Solar Physics and Polar Geomagnetic Observations

The relation between the polarity of the interplanetary magnetic field and the polar

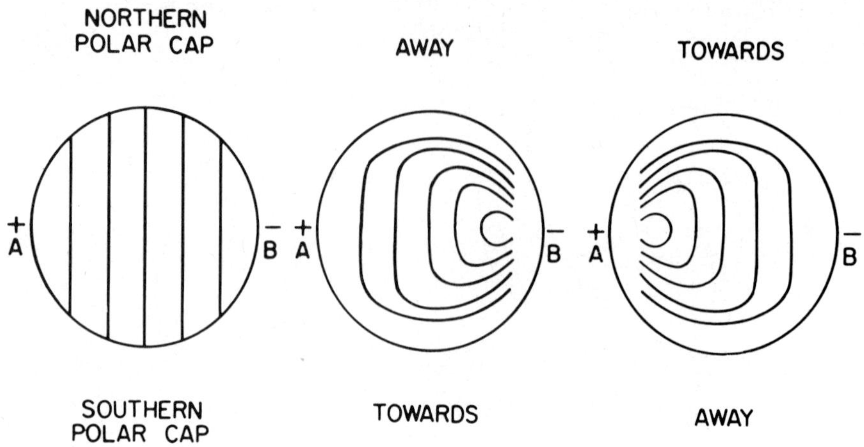

Fig. 14. Schematic view of polar cap equipotentials for (left) no sweeping along the separatrix, (middle) with field lines shifted towards the dusk side (B), and (right) with field lines swept along the separatrix towards the dawn side (A). The patterns corresponding to opposing polar caps and external field directions are indicated on the figure. (After Stern, 1972).

field of the Earth is one of the most direct solar-terrestrial relationships known. The effect is always there; it is usually very clear and sharp, and it is literally connected with the fundamental agent of solar activity: the solar magnetic field. Even in the space-age it shows the importance of using the Earth itself as an Interplanetary Monitoring Platform. Polar geomagnetic observations have been carried out for most of the present century, and therefore it has been possible to infer the sector polarity for each day back to 1926 using magnetograms from Godhavn (Svalgaard, 1972). The interplanetary magnetic field structure is very similar to the large-scale solar photospheric magnetic field (Severny *et al.*, 1970). The availability of the inferred field polarities over the past five sunspot cycles could add significantly to our knowledge of the Sun and the solar cycle.

One such application of the inferred field was the confirmation by Wilcox and Scherrer (1972) of a suggestion by Rosenberg and Coleman that when the Earth is north of the solar equatorial plane the IMF should be biased by the polarity of the northern magnetic pole of the Sun by the southern pole when the Earth is south of the solar equator. It was even possible to show that the polar regions of the Sun change polarity two to three years after sunspot maximum on the average, something which has only been directly observed one or maybe two times.

When the interplanetary sector structure was first discovered during the declining phase of sunspot cycle #19 there were four stable sectors. Solar activity in the next cycle tended to obscure the regular sector pattern and the IMF showed an evolving character. Now, in the declining portion of the present cycle a stable four-sector pattern is again observed (Figure 15). Analysis of the inferred field shows (Svalgaard, 1973b) that this four-sector pattern is a persistent feature of at least the last five sunspot cycles and that it is most clearly observed for about three years prior to sunspot

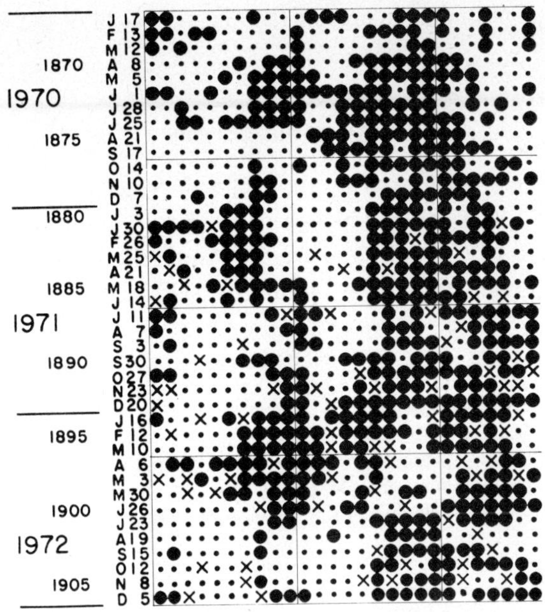

Fig. 15. Inferred interplanetary magnetic sector structure 1970–1972. For each day in the 27-day Bartels rotations the dominant IMF polarity is marked. Towards days are indicated by large filled circles, away days with a small dot; days with ambiguous or mixed field are marked by a cross. A large toward sector is prominent in the right hand side of the figure, while a less clear toward sector may be seen from about day 6 to day 12 in most rotations.

minimum. This together with other evidences hint towards a higher degree of organization of the solar magnetic field than would follow from the classical picture of solar magnetism.

The solar sector structure has proved to be a useful frame for organizing both solar activity and terrestrial responses to it. Since the structure is quasistable there are good prospects for short-term forecasting of both solar and terrestrial events if the evolution of the solar magnetic sector structure could be monitored on a real-time basis. The effect reviewed in this paper – the influence of the sector polarity on polar cap currents – is already being used for practical forecasting purposes at the Space Environment Services Center in Boulder, Colorado, in the United States and at the IZMIRAN near Moscow in the Soviet Union (Heckman *et al.*, 1973). Geomagnetic data from Thule and from Vostok are transmitted in near real-time to the two forecast centers allowing them to infer the sector polarity and to keep track of changes in the sector structure with very short notice.

With the completion of the Stanford Solar Observatory, which is dedicated to observing the large-scale solar magnetic field, it should even be possible to observe a sector boundary on the sun up to a week before it reaches the Earth. Going back in time, the many decades of polar geomagnetic observations can now be used to probe the secrets of the solar cycle and to show us new aspects of the magnetosphere and the polar cap ionosphere by demonstrating a new and very direct solar-terrestrial relationship: the continuous linkage of the magnetic field of the Earth with that of the Sun.

Acknowledgements

This work was supported in part by the Office of Naval Research under Contract N00014-67-A-0112-0068, by the National Aeronautics and Space Administration under Grant NGR 05-020-559, and by the Atmospheric Sciences Section of the National Science Foundation under Grant GA-31138.

References

Dungey, J. W.: 1961, *Phys. Rev. Letters* **6**, 47.
Fennel, J. F.: 1973, *J. Geophys. Res.* **78**, 1036.
Friis-Christensen, E., Lassen, K., Wilcox, J. M., Gonzalez, W., and Colburn, D. S.: 1971, *Nature Phys. Sci.* **233**, 48; see also *Nature Phys. Sci.* **234** (1971), 140.
Friis-Christensen, E., Lassen, K., Wilhjelm, J., Wilcox, J. M., Gonzalez, W., and Colburn, D. S.: 1972, *J. Geophys. Res.* **77**, 3371.
Heckman, G. R., Mansurov, S. M., Wilcox, J. M., and Svalgaard, L.: 1973, *Trans. AGU* **54**, 447.
Heppner, J. P.: 1972, *J. Geophys. Res.* **77**, 4877.
Heppner, J. P., Stolarik, J. D., and Wescott, E. M.: 1971, *J. Geophys. Res.* **76**, 6028.
Kawasaki, K. and Akasofu, S.-I.: 1973, *Planetery Space Sci.* **21**, 329.
Kawasaki, K., Yasuhara, F., and Akasofu, S.-I.: 1973, *Planetary Space Sci.* **21**, 1743.
Mansurov, S. M.: 1969, *Geomagnetizm i Aeronomiya* **9** (English translation) 622.
Mansurov, S. M. and Mansurova, L. G.: 1971, *Geomagnetizm i Aeronomiya* **11** (English translation) 92.
Mansurov, S. M. and Mansurova, L. G.: 1972, Preprint #30, Akademiya nauk SSSR, Institut Zemnogo magnetizma, ionosfery i rasprostraneniya radiovoln, Moscow.
Meng, C.-I., Tsurutani, B., Kawasaki, K., and Akasofu, S.-I.: 1973, *J. Geophys. Res.* **78**, 617.
Severny, A., Wilcox, J. M., Scherrer, P. H., and Colburn, D. S.: 1970, *Solar Phys.* **15**, 3.
Stern, D.: 1972, *Goddard Space Flight Center* Report X-641-72-463, 28 pp.
Svalgaard, L.: 1968, *Geophys. Papers R-6*, 11 pp. Danish Meteorol. Inst., Copenhagen.
Svalgaard, L.: 1972, *J. Geophys. Res.* **77**, 4027.
Svalgaard, L.: 1973a, *J. Geophys. Res.* **78**, 2064.
Svalgaard, L.: 1973b, *Trans. AGU* **54**, 447.
Wilcox, J. M.: 1968, *Space Sci. Rev.* **8**, 258.
Wilcox, J. M.: 1972, *Rev. Geophys. Space Phys.* **10**, 1003.
Wilcox, J. M. and Colburn, D. S.: 1972, *J. Geophys. Res.* **77**, 751.
Wilcox, J. M. and Colburn, D. S.: 1973, *Interplanetary Sector Structure 1970* (to be published).
Wilcox, J. M. and Scherrer, P. H.: 1972, *J. Geophys. Res.* **77**, 5385.

INTERPLANETARY MAGNETIC FIELD DIRECTION AND AURORAL ZONE DISTURBANCES

L. ROSSBERG

Max-Planck-Institut für Aeronomie, Institut für Stratosphärenphysik, Lindau, Harz, Germany

W. RIEDLER

Institut für Nachrichtentechnik und Wellenausbreitung, Technische Hochschule Graz, Austria and Kiruna Geophysical Observatory, Sweden

G. SKOVLI

Norwegian Defence Research Establishment, Division for Electronics, Kjeller, Norway

P. STAUNING

Technical University, Ionosphere Laboratory, Lyngby, Denmark

and

G. R. THOMAS

Appleton Laboratory, Slough, England

1. Introduction

During the last few years the demand for multi-satellite studies of outer magnetospheric phenomena has grown, since we have mounting evidence that the variability of the disturbance phenomena in this region must be better known in order to understand the physics behind them. We report such a study, using data from altogether 12 satellites and discuss a number of interesting and partly unsuspected results. Details are given in Rossberg *et al.* (1973).

2. The Study

One of our objectives is a study of the dynamic behaviour of the outer magnetosphere in response to the interplanetary magnetic field direction during a magnetic moderately disturbed period ($K_p \leqslant 3$) on 11 November 1969.

A schematic view of the positions of the different satellites with respect to the Earth is shown in Figure 1. The best data coverage is obtained near the auroral zones by 5 polar-orbiting satellites. They are supplemented by observations in the pre-midnight sector of the synchronous orbit (ATS 1 and 5), in and near the dawn flank of the tail (Vela 5B, IMP 5), near the bow shock (Vela 5A) and in the interplanetary medium (HEOS 1 and OGO 5).

Magnetic field measurements from an extensive network of ground-based observatories are used to derive an AE-index. This is shown in Figure 2 together with the latitudinal direction of the interplanetary magnetic field, the plasma-injection times observed by ATS 5, K_p and Dst. The auroral-zone magnetic activity from 00 UT to 10 UT is rather low, localized and concentrated at high latitudes as can be inferred

D. E. Page (ed.), Correlated Interplanetary and Magnetospheric Observations, 85–93.

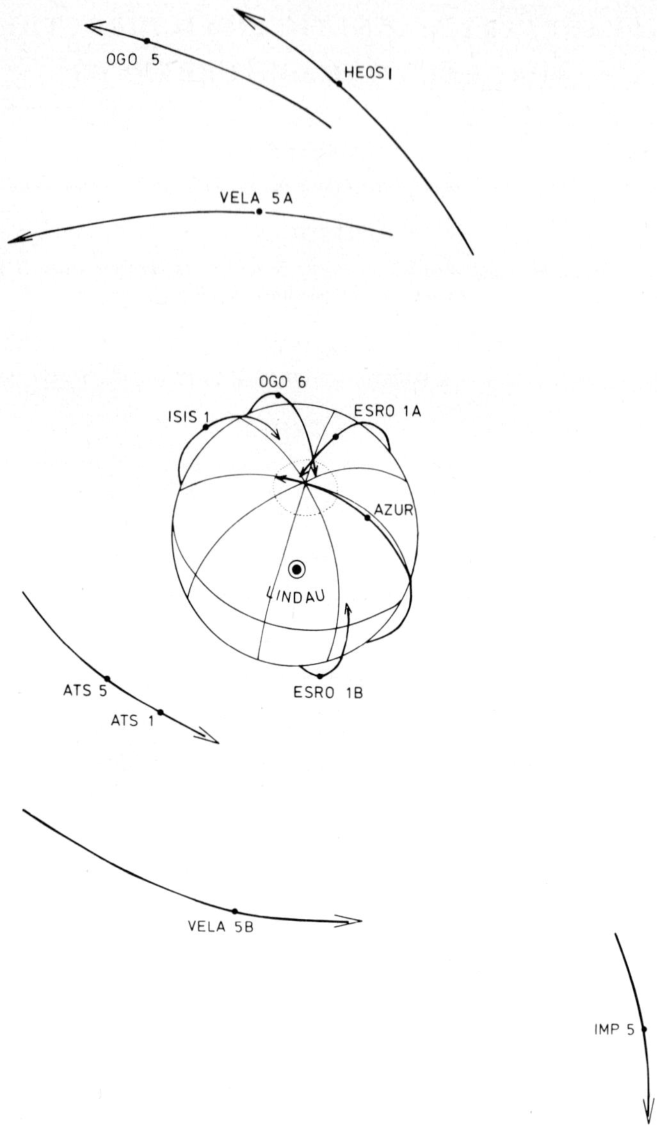

Fig. 1. Sketch of the satellites' positions involved in the 11 November 1969 multi-satellite study.

from the dominant contribution of the Fort Smith data (broken line from 06 UT to 1030 UT) and the persistent low value of K_p. The interplanetary magnetic field direction is generally northward until 09 UT; the short southward excursions have no significant effect on the low magnetic activity.

In contrast to this the polar satellite observations of energetic electrons and protons show a variety of pronounced disturbances at different local times. This can be best demonstrated by a synoptic display of the 40-keV electron observations during

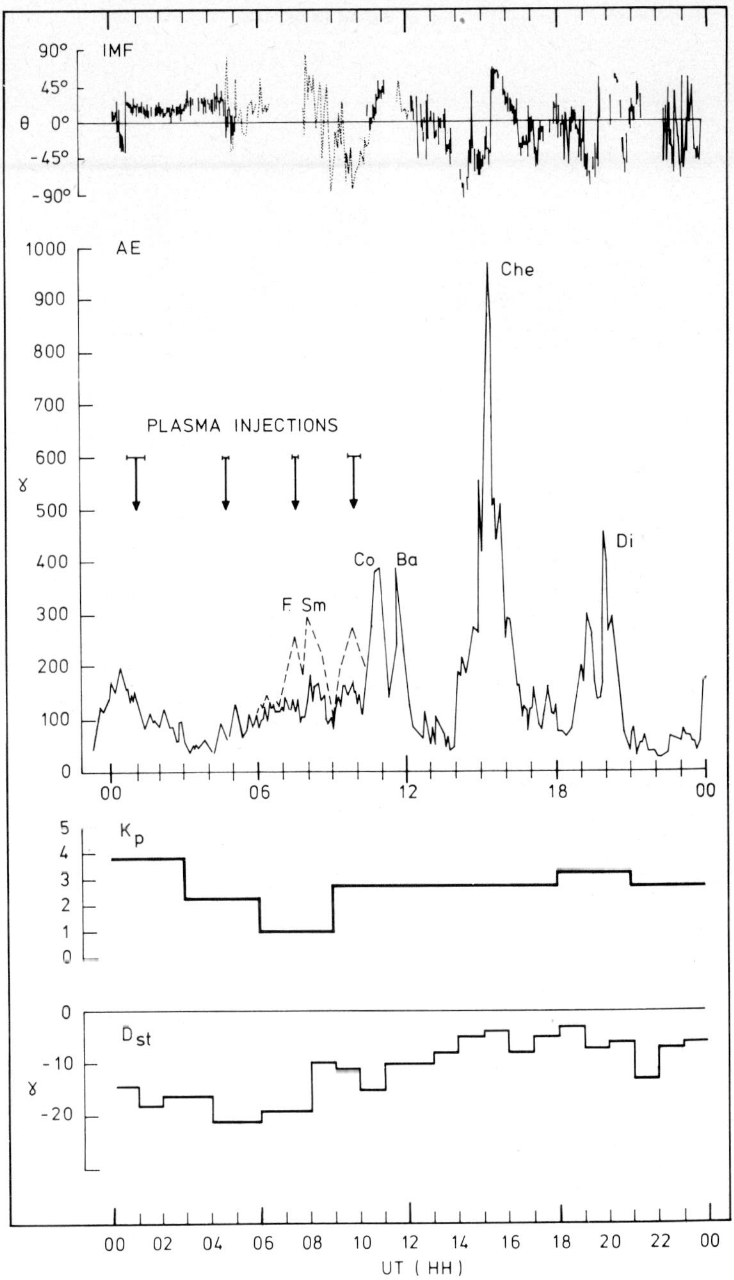

Fig. 2. The figure shows from top to bottom: – The latitudinal direction of the interplanetary field (IMF) given by the angle θ (geocentric solar ecliptic coordinates). Solid line: data from HEOS A1 (by courtesy of P. Hedgecock, Imperial College of London); dotted line: data from Explorer 35 (by courtesy of D. Colburn, NASA Ames Research Center, Moffet Field, Calif.). – The times of plasma-injection which occurred at the synchronous orbit near 23 MLT (by courtesy of C. McIlwain, University of San Diego) indicated by arrows; the horizontal bars give the uncertainty of the injection time. – The AE-index derived from magnetograms of observatories given in the Appendix using a sampling time of 5 min. A broken line is used for the contribution of the Fort Smith magnetogram because of uncertainties in the quiet day level. – The 3-hourly values of K_p. – The hourly valves of Dst.

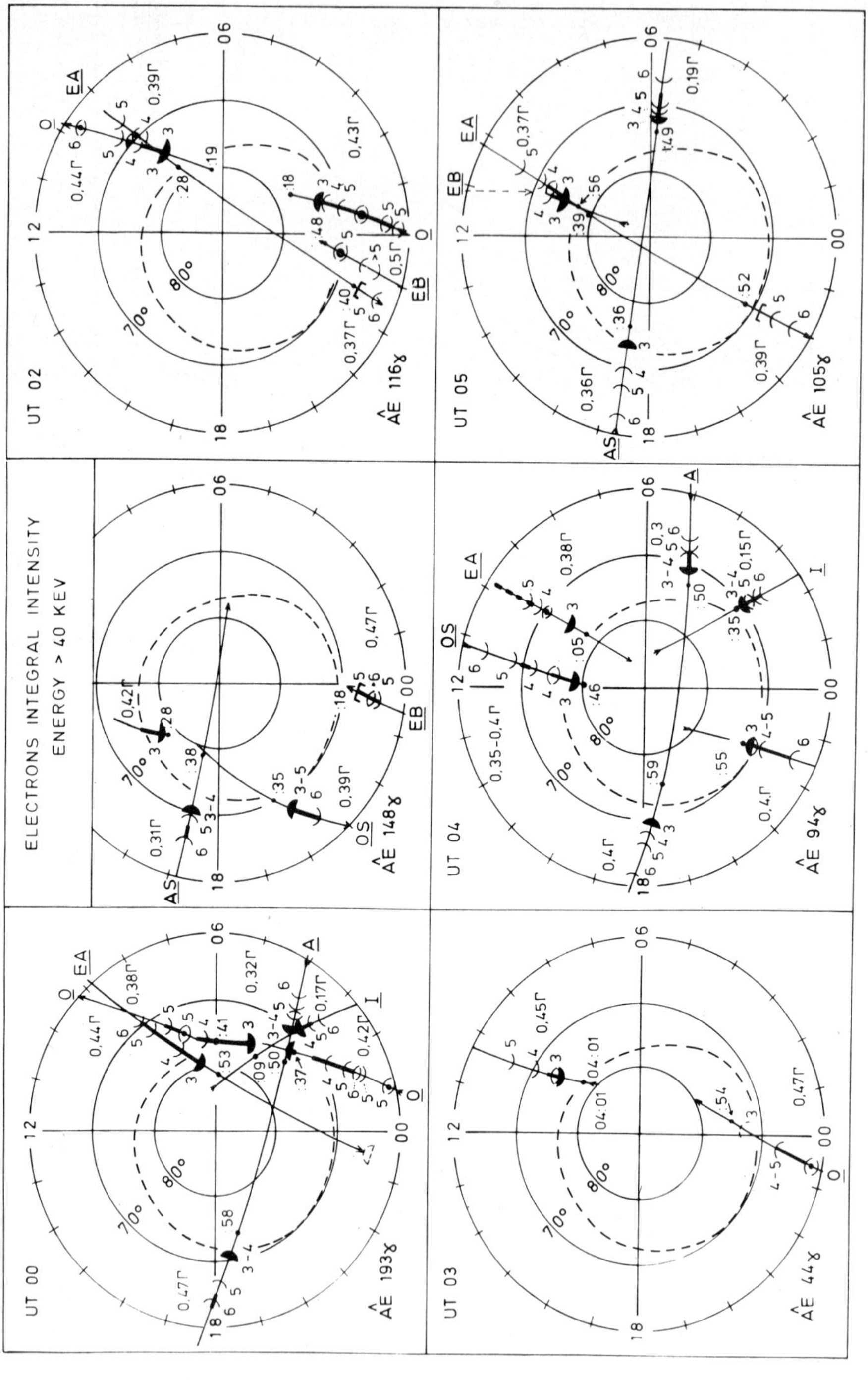

Fig. 3. Synoptic presentation of satellite observations of >40-keV electrons during the first 6 h of 11 November 1969. For details see text.

the rather quiet period from 00 UT to 06 UT (Figure 3). Shown for each hour are the satellite trajectories over the northern and southern (S) hemisphere in an INLAT-MLT coordinate system. The names of the satellites are indicated by their initial letters: *A*ZUR, *E*SRO 1*A* and *B*, *I*SIS 1 and *O*GO 6. Shown also is 'reduced' information on the trapped and precipitating electron intensities (electrons cm^2 s $sr)^{-1}$. The symbols used for this information are explained in Figure 4. Additional information is given in the dials on the magnitude of the magnetic field *B* at the satellite's position near 65° latitude and the highest value of the AE-index during the respective hour. The average position of the 35-keV background boundary (broken oval line after McDiarmid and Burrows, 1968) may be used for reference purposes.

One sees at once that considerable deviations of the boundary positions relative to the average oval form occur during the whole time period at all local times, even in the afternoon sector. Enhanced precipitation is always observed *except* in the afternoon sector, where only two precipitation events occur, spatially limited, near 65°. Note, however, that no enhanced precipitation occurs during the poleward displace-

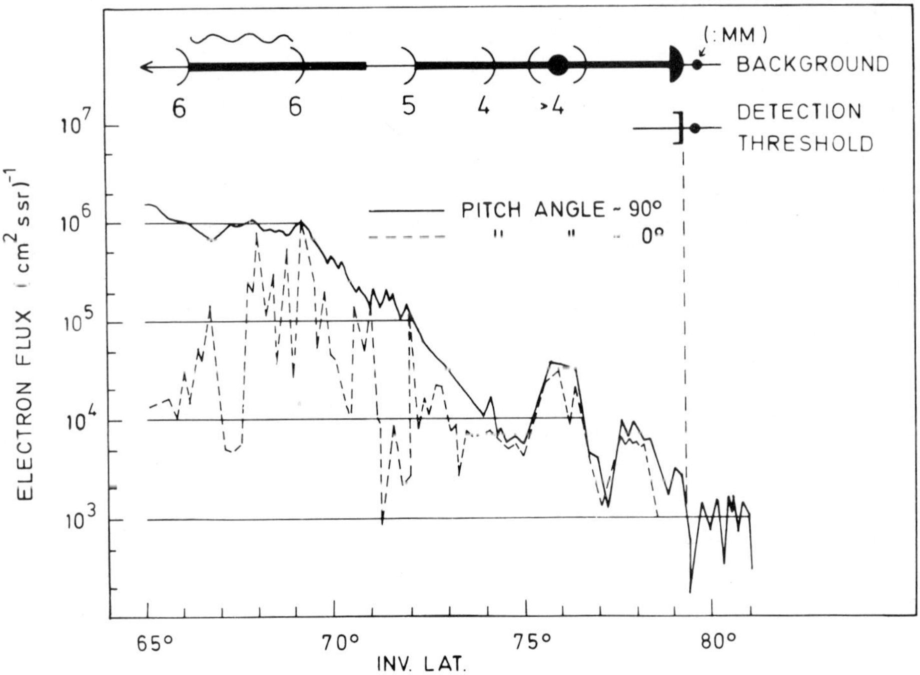

Fig. 4. The figure shows how the information in a detailed 40-keV intensity profile is reduced to the information contained in the dials. The symbols used along the trajectory indicate the positions of the intensity levels in powers of 10. The bars are bent towards increasing intensities, a dot indicates the position of a spike. The umbrella-like symbol is used at the point where the intensities reach the background level; the square-bracket symbol is used where the intensities measured by the experiment S71A of ESRO 1A/B drop below the detection threshold ($\sim 10^5$ e cm^{-2} s^{-1} sr^{-1}). The small dot poleward of these two symbols gives the time in minutes of the hour (UT). The solid line along the trajectory indicates enhanced precipitation, and the wavy line between the two flux symbols labelled '6' indicates small fluctuations about this level.

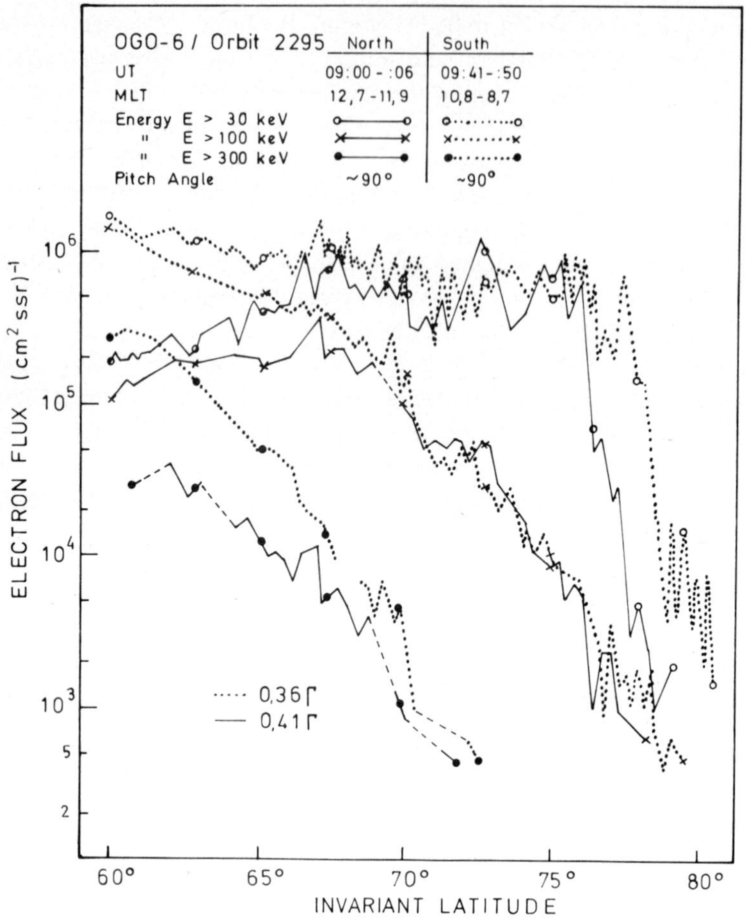

Fig. 5. Dayside observations of the electron flux distribution for $E > 30$, > 100 and > 300-keV observed during successive traversals of the northern and southern auroral zone by OGO 6 in the time intervals 0900–0906 and 0941–0950 UT. The figure shows that the steep enhancement at the boundary occurs only at the > 30-keV energy level.

ment of the background boundary at 0536 UT. This displacement is associated with an enhancement of > 250-keV protons and with the second plasma-injection shown in Figure 2 and may be the high-latitude signature of an inflated portion of the magnetosphere.

Note in particular the dramatic boundary variations from 0218 UT to 0248 UT which occurred during a period of quiet northward directed interplanetary magnetic field between the first and second plasma injection. The poleward displacement occurs also at $E > 100$ keV on the night side but not at $E > 300$ keV. The dayside boundaries show no poleward displacement.

With regard to the average position of the dayside boundaries and its dependence on the interplanetary magnetic field direction (equatorial displacements during southward interplanetary magnetic field and vice versa (Hruska *et al.*, 1972), we find that the equatorward displacement from 0053 UT to 0228 UT during quiet northward directed interplanetary magnetic field is not consistent with the average behaviour. Moreover we also observe a pronounced and persistent poleward displacement from 08 UT to 10 UT when the interplanetary magnetic field points predominantly southward. Two OGO 6 observations of >30, >100 and >300-keV trapped elec-

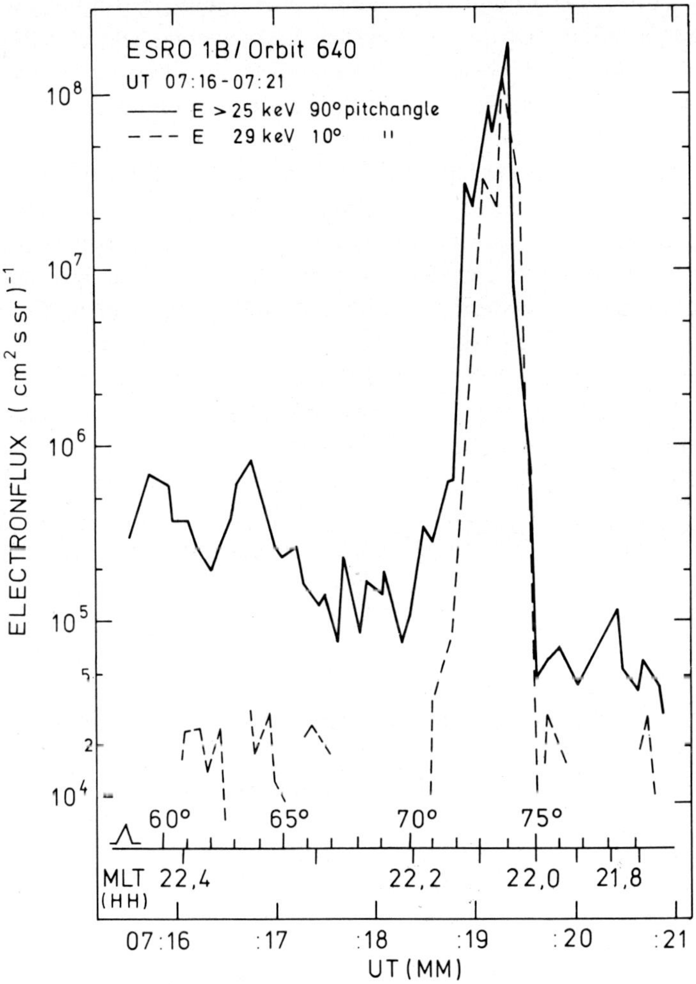

Fig. 6. Observation by ESRO 1B of a high latitude '40-keV electron' spike or burst at 0719 UT.

trons made over the southern and northern hemisphere during this period are shown in Figure 5. Note the unusual steep edge of the >30 keV intensity profiles which results in similarly unusual soft energy spectra. The poleward displacement of the boundaries, opposite to the equatorward displacement expected during southward directed field, may be explained by a compression of the dayside magnetosphere.

Such a compression is indicated by the simultaneous occurrence of sudden impulses on ground and the sudden disappearance of magnetosheath particles at the position of Vela 5A near the bow shock at 08 UT. At this time the interplanetary magnetic field is still northward but starts to turn southward. The steep edge of the 30-keV electron intensity profiles shown in Figure 5 might then be the only indication of an eroding effect of the southward directed field.

The result shows that the boundary position of energetic electrons on the dayside is not related in a unique way to the interplanetary magnetic field direction alone. It should be noted that the nightside disturbances, characterized by repeated observations of high-latitude injections of 40-keV electrons in the premidnight area, starts well before the sudden impulse and does not change its character during the occurrence of the dayside disturbance (Rossberg *et al.*, 1973) and Figure 6. Furthermore the associated magnetic nightside activity is not of the classical substorm type but can best be described as a quasi-periodic intensification of a weak electrojet centred at high latitudes in the premidnight area [Rostoker, private communication].

The onset of the first classical substorm, indicated by the sudden deepening of a bay disturbance at College (1030 UT), occurs at the end of the prolonged period of southward directed field, when the field turns northward again. Our observations provide evidence that no close relationship exists between the instantaneous IMF-direction and high-latitude disturbance phenomena and that a certain positive value of $\sum B_s t$ (summation of the IMF southward component over time) is the dominating necessary condition for the occurrence of a substorm. This is in agreement with the recent conclusion of Garrett (1973) who also demonstrated the importance of the variance of the IMF for triggering a substorm.

Acknowledgement

The authors would like to acknowledge the sponsorship of the following national institutions:

- Bundesministerium für Forschung und Technologie, Bundesrepublik Deutschland
- Danish Space Research Institute
- Royal Norwegian Council for Science and Technology
- Science Research Council, United Kingdom
- Space Research Institute of the Austrian Academy of Science
- Swedish Natural Science Research Council

The ESRO 1B data (Figure 6) are published with the permission of the Director of the Appleton Laboratory.

We would also like to thank all the people, who assisted in the design, qualification and integration of the individual experiments and in the data reduction. Finally, our special thanks go to the Directors of the Max-Planck-Institut für Aeronomie, for their support and continuous encouragement during the coordination of this study.

Appendix

Station names and their abbreviations. The magnetometer data of these stations are used for defining the AE-index

Station names	Abbreviations
Abisko	Ab
Barrow	Ba
College	Co
Cape Wellen	CW
Chelyuskin	Che
Dixon	Di
Fort Smith	F.Sm
Great Whale River	GW
Kiruna	Ki
Leirvogur	Lei
Meanook	Me
Narssarssuaq	Na
Sodankylä	So

References

Garrett, H. B.: 1973, *EOS, Trans. Am. Geophys. Union* **54**, 413.

Hruska, A., Burrows, J. R., and McDiarmid, I. B.: 1972, *J. Geophys. Res.* **77**, 2770.

McDiarmid, I. B. and Burrows, J. R.: 1968, *Can. J. Phys.* **46**, 49.

Rossberg, L., Lammers, E., Riedler, W., Skovli, G., Søraas, F., Stauning, P., Theile, B. and Thomas, G. R.: 1973, ESRO Scientific Report SR 23 (in press).

PART 2

THE NATURE OF MAGNETOSPHERIC BOUNDARIES

THE MAGNETOSPHERIC BOUNDARY

D. B. BEARD and J. Y. CHOE
University of Kansas, Lawrence, Kan., U.S.A.

1. The Magnetopause

Magnetic fields deflect charged particles and the fields are somewhat impervious to charged particles not already trapped within them. Thus, as we should expect, the Earth's magnetic field makes a bubble in the supersonic plasma flowing outward from the Sun and called the solar wind. This bubble in which the Earth's magnetic field is largely confined and compressed is called the *magnetosphere* and the boundary between the Earth's magnetic field on the inside and the turbulent interplanetary plasma and magnetic field on the outside is called *magnetopause*. The magnetopause is very thin when compared to its radius of curvature (about 10^{-3} to 10^{-5}) and can be regarded as infinitesimally thin. Within this thin boundary layer electric currents flow which are created by the magnetically deflected plasma. The effect of these currents is to cancel or terminate the geomagnetic field outside the magnetosphere and to augment the geomagnetic dipole field inside since the effect of the outside pressure is to compress the geomagnetic field.

Figure 1 illustrates the coordinate system used and proton and electron trajectories incident on a magnetic field boundary with the magnetic field perpendicular to the plane of the figure. Since the protons have more momentum they penetrate more deeply than the electrons and the resultant charge separation creates an electric field normal to the magnetopause. The electric field decelerates the protons and accelerates the electrons which travel a long way in the boundary layer as illustrated in Figure 2 calculated by Alan McMahon many years ago. The reflection is specular and the magnetic field decreases very rapidly outward as $B = B_{\max}\ e^{-x/x_0}$ where $x_0 = (m_e c^2/8\pi n_0 e^2)^{1/2}$ and m_e is the mass of an electron. For conditions at the magnetopause x_0 is of the order of a kilometer. This solution for the magnetic field at a plasma-magnetic field boundary was first derived independently by Dungey (1958) and Rosenbluth (1957) and has been developed further by several others since them. If the electric field due to charge separation is neutralized by the flow of ambient electrons the theory must be modified slightly because in this instance the protons are responsible for the current (Beard, 1968) $B = B_{\max}\ e^{-x/x_0}$ where $x_0 = (m_p c^2/8\pi n_0 e^2)^{1/2}$ and m_p is the mass of a proton. In this case the boundary layer is some 40 times thicker and is of the order of 100 km.

2. The Shape of the Magnetosphere

The problem of finding the shape of the magnetosphere has one great merit in that the problem is over determined; there exist more conditions to be satisfied than there

D. E. Page (ed.), Correlated Interplanetary and Magnetospheric Observations, 97–114.

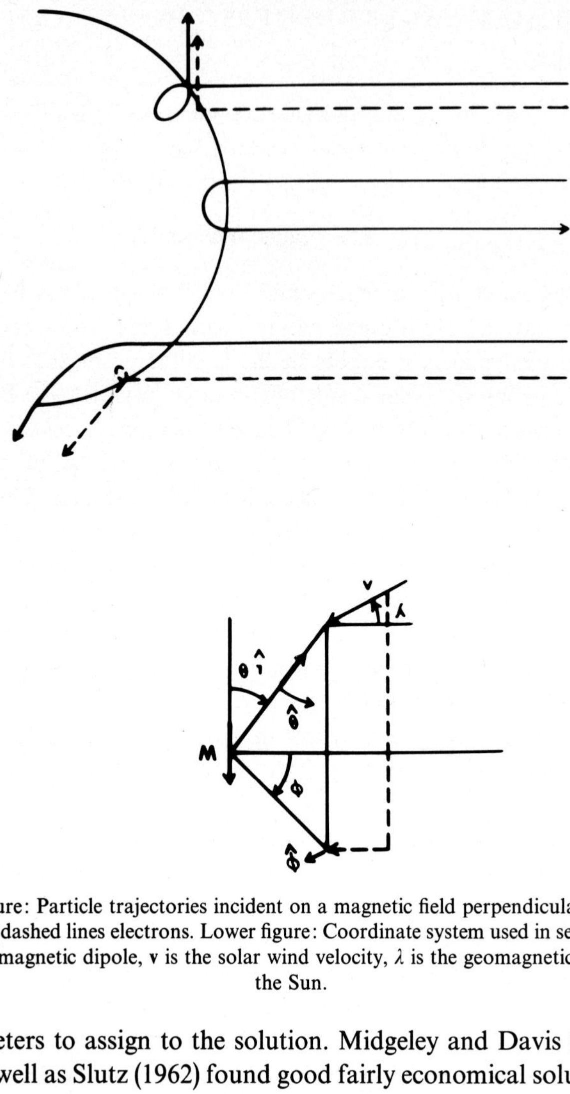

Fig. 1. Upper figure: Particle trajectories incident on a magnetic field perpendicular to plane of figure; solid lines protons, dashed lines electrons. Lower figure: Coordinate system used in self consistent calculations. **M** is the geomagnetic dipole, **v** is the solar wind velocity, λ is the geomagnetic latitude position of the Sun.

are free parameters to assign to the solution. Midgeley and Davis (1962) and Midgeley (1963) as well as Slutz (1962) found good fairly economical solutions by searching for a surface which resulted in zero magnetic field in the plasma outside. Somewhat apologetically I will describe here an older method of solution partly because it has been more generally used and because I am more familiar with it. Since the field outside is zero, the normal component of the field at the surface must also be zero for the current in the surface can produce a discontinuity in only the tangential component of the field. An additional condition which can be used is to equate the plasma pressure outside to the magnetic pressure inside, $p = B^2/8\pi$. We have then

$$B_{\text{out}} = 0$$
$$\hat{\mathbf{n}}_s \cdot \mathbf{B}_{\text{in}} = 0$$
$$B_{\text{in}} = \sqrt{8\pi p},$$

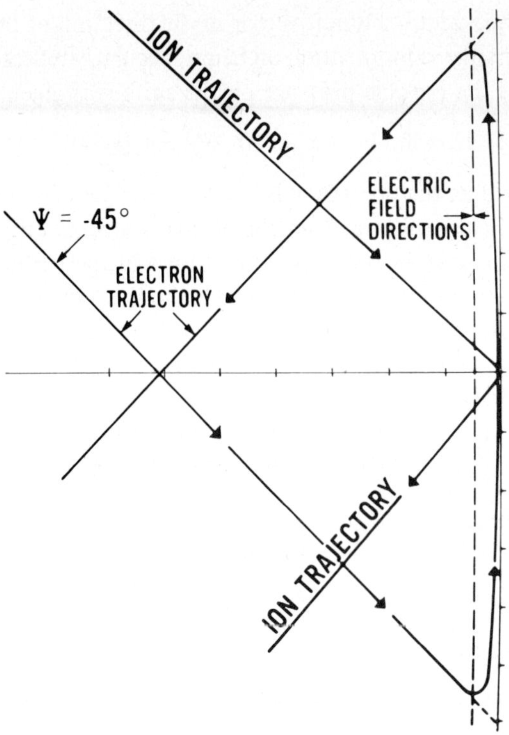

Fig. 2. Charged particle trajectories in a boundary layer between a plasma and a magnetic field as computed by McMahon. Magnetic field at right and directed perpendicular to the plane of the figure.

where $\hat{\mathbf{n}}_s$ is the unit normal to the surface. If we combine the last two equations we obtain

$$|\mathbf{n}_s \times \mathbf{B}| = \pm(8\,\pi p)^{1/2}. \tag{1}$$

Any one of these conditions may be used to obtain the correct surface but the last condition, Equation 1, has been shown to be most efficient in obtaining a rapidly convergent solution for a variety of problems (Baker *et al.*, 1964).

The magnetospheric bubble is an obstacle in the supersonic solar wind. Since the charged particles outside the surface interact with one another through the weak interplanetary magnetic field a shock structure is formed on the surface of the magnetosphere as first recognized by Kellogg (1962), Axford (1962), Lees (1964), and Spreiter and Jones (1963). This heats up the solar plasma and in the antisolar direction enhances the thermal random pressure tending to close the tail. Nevertheless other currents act to keep the tail open and on the solar side the turbulent bow shock pressure is only a small perturbation on the directed solar stream pressure on the front of the magnetosphere (Shen, 1972). Therefore as a first approach to the problem we compute the particle pressure on a surface exerted by a cold directed incident stream. The number of particles striking the surface per cm^2 per sec is $n_0 v(\hat{\mathbf{n}}_s \cdot \hat{\mathbf{v}})$ where n_0 is the number density of protons, v the speed of the stream, $\hat{\mathbf{n}}_s$ the unit normal to the surface, and $\hat{\mathbf{v}}$ the unit vector in the direction of the stream. The protons experience a change in momentum on being specularly reflected from the surface of $2\,mv\hat{\mathbf{n}}_s\cdot\hat{\mathbf{v}}$. Hence the particle pressure on the surface is given by

$$p=n_0 v\hat{\mathbf{n}}_s\cdot\hat{\mathbf{v}}2\,mv\hat{\mathbf{n}}_s\cdot\hat{\mathbf{v}}=2\,n_0 mv^2(\hat{\mathbf{n}}_s\cdot\hat{\mathbf{v}})^2=p_0(\hat{\mathbf{n}}_s\cdot\hat{\mathbf{v}})^2.$$

The magnetic field at the boundary is given by the geomagnetic dipole field, $\mathbf{B}_d$, plus the magnetic field resulting from the electric currents all over the boundary. The surface has a very large radius of curvature and the surface current field can conveniently be split up into two parts: that due to the local surface which is essentially a plane, $\mathbf{B}_{Pl}$, and contributes the major magnetic effect and that due to distant surface currents, $\mathbf{B}_{sc}$, from far away surface elements where the surface departs from the local tangential plane. The geodipole field and the field from distant surface currents do not change across the infinitesimally thin surface; the field due to the local tangential plane is equal and opposite on opposite sides of the surface. Hence

$$\mathbf{B}_{\text{inside}}=\mathbf{B}_d+\mathbf{B}_{Pl}+\mathbf{B}_{sc}$$
$$\mathbf{B}_{\text{outside}}=\mathbf{B}_d-\mathbf{B}_{Pl}+\mathbf{B}_{sc}=0$$

and therefore

$$\mathbf{B}_{\text{inside}}=2(\mathbf{B}_d+\mathbf{B}_{sc}).$$

Inserting our expressions for particle pressure and field into Equation (1), we obtain

$$|\hat{\mathbf{n}}_s\times(\mathbf{B}_d+\mathbf{B}_{sc})|=-(2\pi p_0)^{1/2}(\hat{\mathbf{n}}_s\cdot\hat{\mathbf{v}}),\tag{2}$$

where the minus sign has been taken on the right since $\hat{\mathbf{n}}_s\cdot\hat{\mathbf{v}}$ must always be negative if the stream pressure is to be exerted on the outside of the surface.

$$\mathbf{B}_d=-(M/r^3)(\sin\theta\,\hat{\boldsymbol{\theta}}+2\cos\theta\,\hat{\mathbf{r}}),$$

where M is the geomagnetic dipole moment.

The surface function

$$f(r,\theta,\phi)=r-R(\theta,\phi)=0$$

has a unit normal derivable from rules of elementary vector calculus given by

$$\hat{\mathbf{n}}_s=\nabla f/|\nabla f|=a\left[\hat{\mathbf{r}}-\frac{1}{r}\frac{\partial R}{\partial\theta}\hat{\boldsymbol{\theta}}-\frac{1}{r\sin\theta}\frac{\partial R}{\partial\varphi}\hat{\boldsymbol{\phi}}\right]$$

$$a=\left[1+\frac{1}{r^2}\left(\frac{\partial R}{\partial\theta}\right)^2+\frac{1}{r^2\sin^2\theta}\left(\frac{\partial R}{\partial\varphi}\right)^2\right]^{-1/2}$$

In the coordinate system illustrated in Figure 1 $\hat{\mathbf{v}}$ has the form

$$\begin{aligned}\hat{\mathbf{v}}=&-(\cos\theta\sin\lambda+\cos\phi\sin\theta\cos\lambda)\,\hat{\mathbf{r}}\\&-(\cos\phi\cos\theta\cos\lambda-\sin\theta\sin\lambda)\,\hat{\boldsymbol{\theta}}\\&+\sin\phi\cos\lambda\,\hat{\boldsymbol{\phi}}\end{aligned}$$

As a first approximation we can set $B_{sc}=0$. At the subsolar point for $\lambda=0$ Equation (2) becomes

$$M/r_0^3=(2\pi p_0)^{1/2}, \quad r_0=(M^2/2\pi p_0)^{1/6}.$$

It is convenient to measure all radial distances in units of r_0 in which case Equation (2) becomes

$$|\hat{n}_s \times (\mathbf{B}_d+\mathbf{B}_{sc})| = -\hat{n}_s \cdot \hat{\mathbf{v}}. \tag{3}$$

If B_{sc} is set equal to zero, Equation (3) is a two dimensional differential equation for the surface function, $f(r, \theta, \phi)$ or $R(\theta, \phi)$, which can be solved numerically. In particular we have found the Newton-Raphson Method proposed by Mead (Mead and Beard, 1964) particularly simple and convenient to use and to estimate precision. The magnetic field deduced from the calculated surface is extremely sensitively dependent on small differences of boundary shape. We have found it necessary to use six surface points in calculating derivatives and solve the surface equation in steps of one degree intervals in some places in order to obtain a satisfactory solution. A very great effort is required to obtain a precise result particularly around the neutral point. If the neutral point is inaccurate, an unreliable surface results. An overall illustration of the shape is given in Figure 3 taken from an early figure of Midgeley and Davis.

From this first surface the electrical currents everywhere on the surface are known since

$$\mathbf{j}=\frac{c}{4\pi}\,\hat{\mathbf{n}}_s \times \mathbf{B}$$

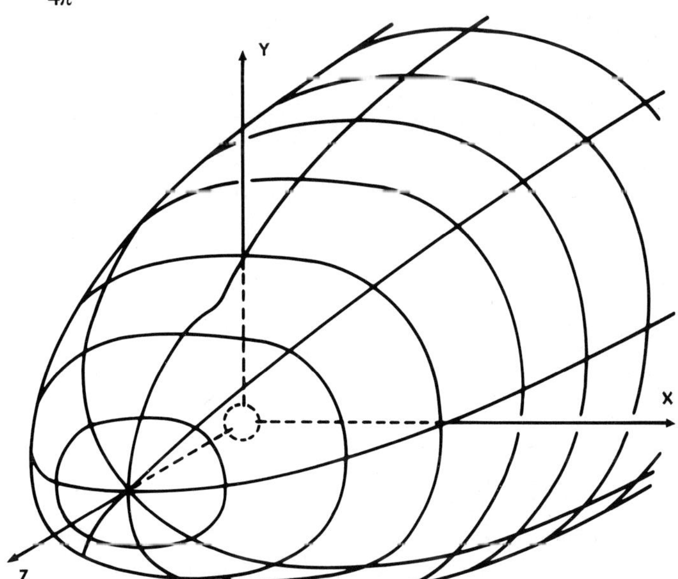

Fig. 3. Pictorial view of the magnetospheric boundary as computed by Midgeley and Davis.

and the field correction $\mathbf{B}_{sc}$ due to the departure of the surface from a local plane can be calculated by integrating the Biot Savart integral over the surface

$$\mathbf{B}^n_{sc}(r,\theta,\varphi)=\frac{1}{4\pi}\int\frac{[\hat{\mathbf{n}}_s(r',\theta',\varphi')\times\mathbf{B}^{n-1}(r',\theta',\varphi')]\times\boldsymbol{\varrho}\,\mathrm{d}s}{\varrho^3}, \tag{4}$$

where

$$\boldsymbol{\varrho}=\mathbf{r}-\mathbf{r}'$$
$$\mathbf{B}^n=\mathbf{B}_d+\mathbf{B}^n_{sc}$$

and the superscript on $\mathbf{B}$ and $\mathbf{B}_{sc}$ refers to the order of the approximation. Clearly $\mathbf{B}^0_{sc}=0$. Note that since $\mathbf{B}_{Pl}$ is already included, the point (r,θ,ϕ) at which $\mathbf{B}$ is evaluated is at the center of the boundary layer. The procedure is to solve Equation (3) with $\mathbf{B}_{sc}=0$, then find an improved value of $\mathbf{B}_{sc}$ by numerically integrating Equation (4) and using this in re-solving Equation (3). One repeats this cycle until no change in the surface is obtained on successive iterations. In each cycle the entire surface is computed successfully in 5° steps and in 1° steps near the neutral point, about 1600 points in all. The fourth iteration is usually adequate although we have iterated nine or ten times to see whether an instability will develop. A good check on the solutions is to compute the magnetic field outside the surface which should be zero. Our latest surfaces (Choe *et al.*, 1973) give an external field less than 1% of the pure dipole field at the same point *everywhere* within 20 Earth radii from the Earth.

The most recent work for $\lambda=0$, the wind blowing perpendicular to the dipole, is shown in Figures 4 and 5. A point exists in the meridian plane near the pole at which the tangential component of the magnetic field reverses. At this point called the *neutral point* the field is zero and the surface must be tangential to the solar wind velocity vector ($\hat{\mathbf{n}}_s\cdot\hat{\mathbf{v}}=0$). One critical test of a good surface solution in the meridian plane is to examine the behavior of $\hat{\mathbf{n}}_s\cdot\hat{\mathbf{v}}$ near the neutral point as shown in Figure 6. Our result behaves smoothly and does indeed reveal a tangential surface at the neutral point.

The Earth's dipole is seldom exactly perpendicular to the solar wind velocity vector. Depending on the time of day or year the dipole may be tilted as much as 37° toward or away from perpendicular to the solar wind. We have computed the surface for λ from 0 to 35° in steps of 5° and our results are shown in Figures 7–10. The neutral point region has been examined with particular care partly because the primary difference between our latest result and earlier work occurs as a consequence of our differences at the neutral point.

3. The Distortion of the Geomagnetic Field

The purpose of calculating the magnetopause shape with far greater accuracy than could ever be detected by satellite crossings is, of course, to be able to calculate the magnetic field with sufficient accuracy to fit satellite magnetic measurements and to predict particle trajectories reliably.

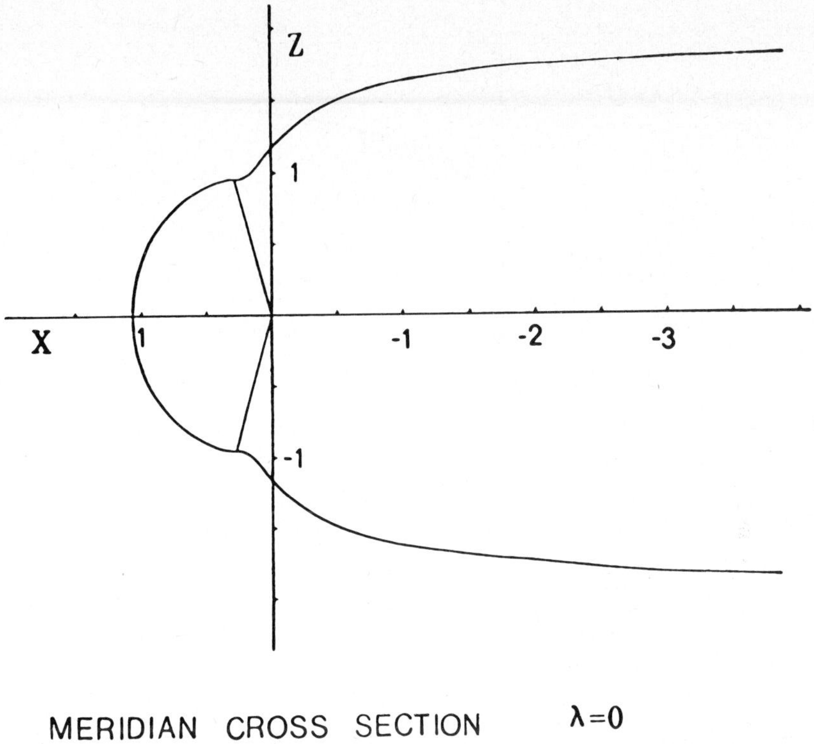

Fig. 4. Meridian cross-section of magnetosphere for dipole perpendicular to the solar wind.

Mead (1964) represented the surface current field in a particularly convenient way which has become standard. The field is represented by a scalar potential which is expanded in spherical harmonics and the coefficients are least squares fitted to the field calculated at points within the magnetosphere. We used 688 points in computing the field for each surface for the eight values of λ obtained by Choe *et al.* (1973).

$$B_s^i(r, \theta, \varphi, \lambda) = \sum_{l=1}^{l_{\max}} r^{l-1} \sum_{m=0}^{l} g_{lm}(\lambda)\, T_{lm}^i(\theta, \varphi),$$

where the superscript i refers to the r, θ, or ϕ component

$$\begin{aligned}
T_{lm}^r(\theta, \varphi) &= -lP_{lm}(\cos\theta)\cos m\varphi \\
T_{lm}^\theta(\theta, \varphi) &= -\{mP_{lm}(\cos\theta)\cos\theta + [(l+m)(l+m+1)]^{1/2} P_{l,m+1}(\cos\theta)\} \times \\
&\qquad \times \cos m\varphi \quad (m \neq 0, m \neq l) \\
&= -lP_{lm}(\cos\theta)\cos\theta\cos m\varphi \qquad (m = l) \\
&= -[l(l+1)/2]^{1/2} P_{l,1}(\cos\theta) \qquad (m = 0) \\
T_{lm}^\varphi &= m[P_{lm}(\cos\theta)(\sin\theta]\sin m\varphi
\end{aligned}$$

$P_{lm}(\cos\theta)$ is the Schmidt normalized Legendre function. Distances are expressed in earth radii and the field in Gauss. We define $G_{lm} = g_{lm} r_0^{l+2}$, $R = r/r_0$, where r_0 is the

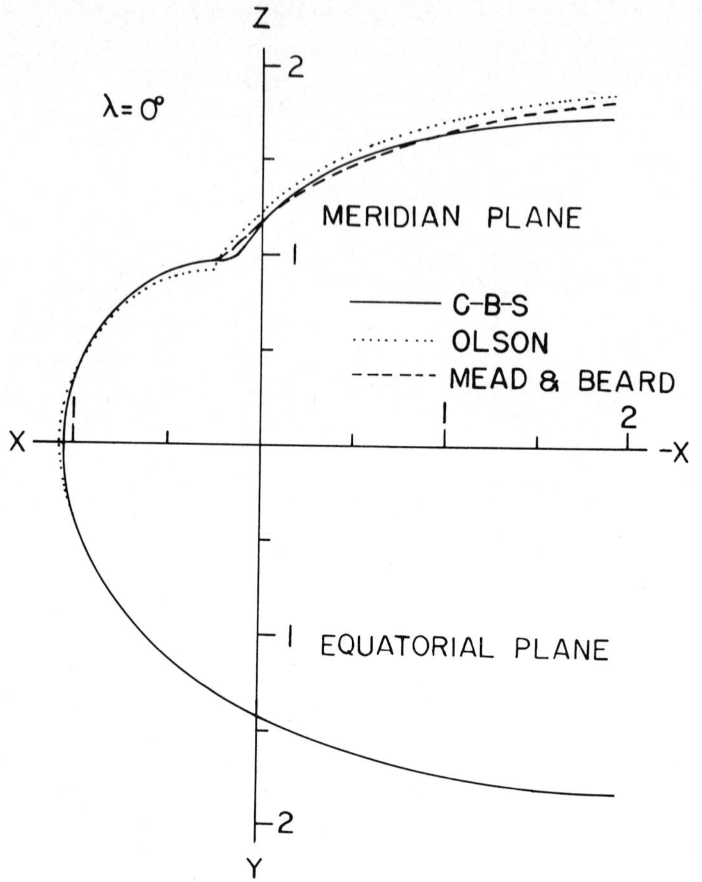

Fig. 5. Meridian plane and equatorial plane cross-sections of the magnetosphere for perpendicular incidence. (Choe *et al.*, 1973; Mead and Beard, 1964; and Olson, 1969).

first approximation perpendicular incidence subsolar point height defined earlier,

$$B_s^i(r, \theta, \varphi, \lambda) = \frac{1}{r_0^3} \sum_{l=1} R^{l-1} \sum_{m=0}^{l} G_{lm}(\lambda)\, T_{lm}^i(\theta, \varphi)$$

Table I compares our results for $\lambda = 0$ with the G_{lm} obtained by Mead (1964) and Midgeley (1964). The leading term $G_{1,0}$ giving a constant field within the magnetosphere of about 20 γ is about the same for all determinations. The advantage in repeating the calculation with the precision we sought is revealed in the second largest term G_{21} where the difference is of the order of 40%. That is, we find a field of $\pm 7\ \gamma$ instead of $\pm 10\ \gamma$ at 10 Earth radii subsolar and antisolar during quiet times.

The dependence of the G_{lm} on λ is shown for a few of the G_{lm} in Figure 11. The coefficient G_{20} which is zero for $\lambda = 0$ becomes the second most dominant term surpassing G_{21} for large λ. The projection of the field lines in the meridian plane are shown in Figures 12 and 13.

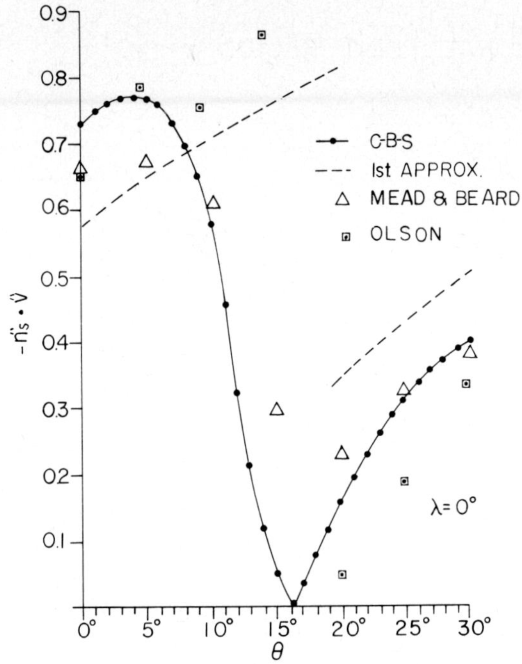

Fig. 6. Graph of $\hat{\mathbf{n}}_s \cdot \hat{\mathbf{v}}$ vs polar angle θ in the meridian plane where $\hat{\mathbf{n}}_s$ is the normal to the surface and $\hat{\mathbf{v}}$ the solar wind velocity for surfaces of Choe *et al.* (1973), Mead and Beard (1964), and Olson (1969).

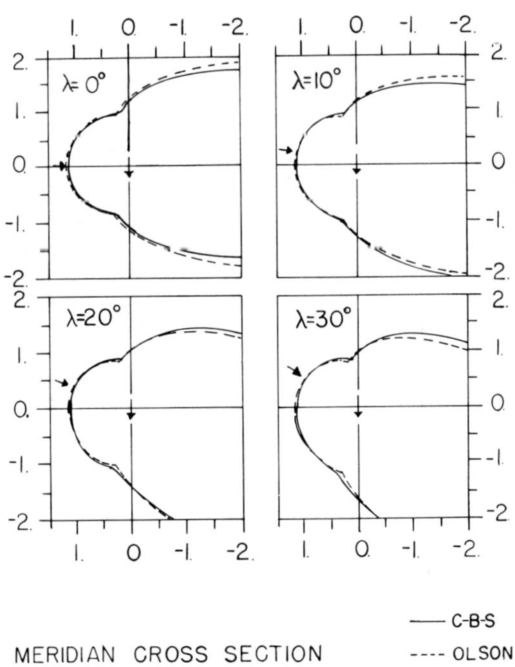

Fig. 7. Meridian cross-sections of the magnetosphere for various tilts of the Earth's dipole vector.

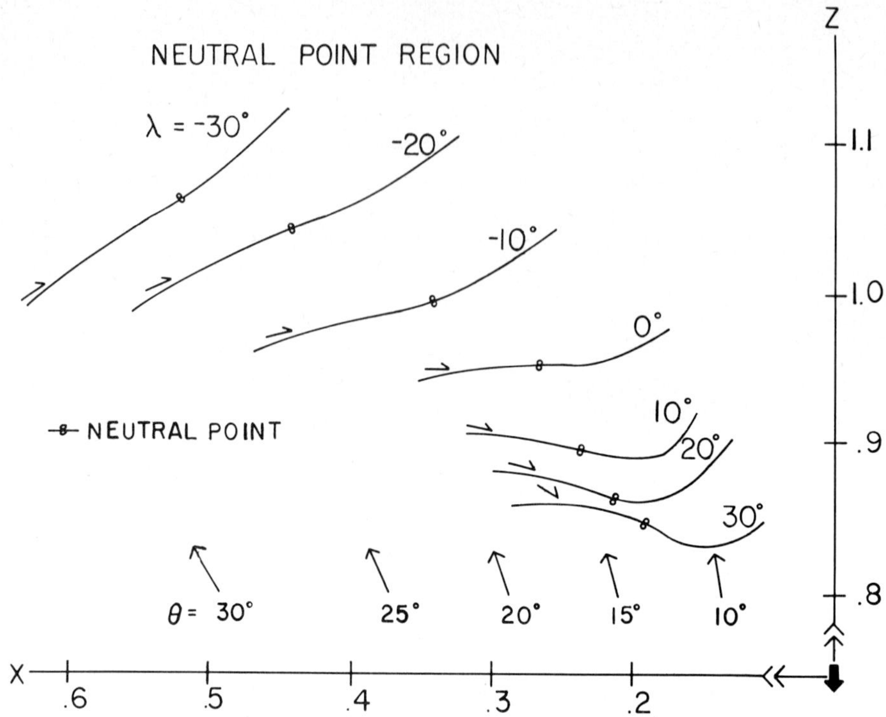

Fig. 8. Meridian plane cross-sections of the magnetosphere near the neutral point.

4. Merging of the Interplanetary Field with the Geomagnetic Field

An interplanetary magnetic field has been observed of about 6 γ at the orbit of the Earth. This arises from the pressure of the solar wind on the solar magnetic field. Looked at another way, this field is supported by a diffuse current throughout interplanetary space as a result of plasma deflection in the field. Since the interplanetary field is not parallel to the geomagnetic field, when the solar wind pushes the two fields in contact with one another they join in the reconnection process first originated by Dungey in 1953 and later applied by him to the solar wind geomagnetic field interaction (Dungey, 1961). This is illustrated in Figure 14.

Unfortunately the rate at which merging takes place on the complicated three dimensional surface of the magnetosphere has never been determined. Aside from the great complication of the geometry the essential parameters of conductivity, electric field, flow velocities etc. are uncertain. If appreciable field merging does take place there should be some component of magnetic field perpendicular to the magnetopause at the magnetopause. This has not been detected as yet. Nevertheless some correlation of geomagnetic activity with interplanetary field direction has been

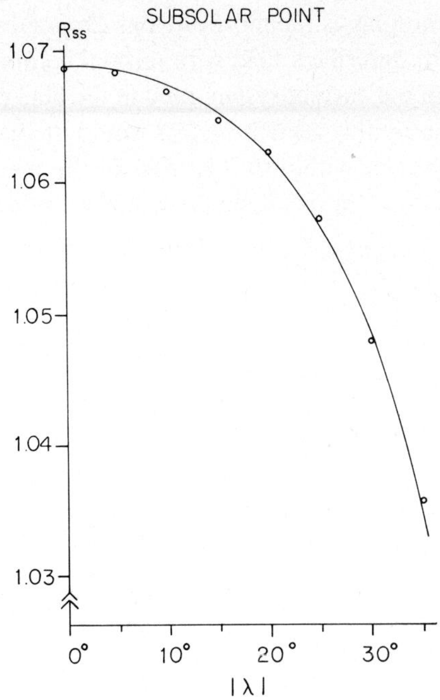

Fig. 9. The change in subsolar point with the geomagnetic latitude of the Sun.

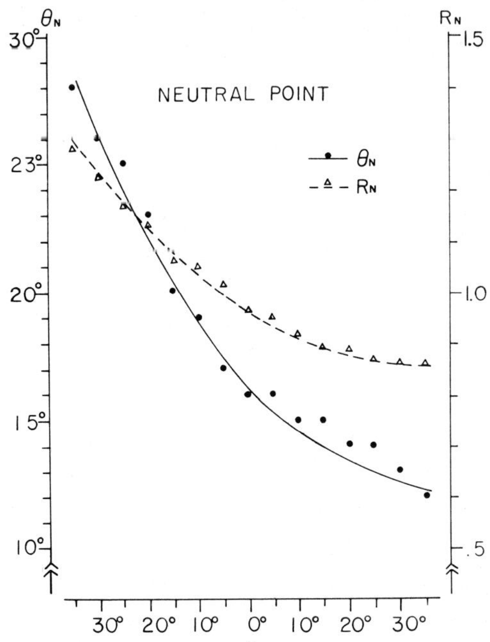

Fig. 10. The change in neutral point coordinates with the geomagnetic latitude of the Sun.

reported and field line merging is the most obvious explanation. The magnetopause current system needed to describe a field with normal components is similar to but very slightly weaker than the current system used in the field calculation in the previous section and presumably this current system would merge to a very slight extent with the large current system within the tail part of the magnetosphere. It has yet to be calculated, but it does furnish a supplementary theoretical understanding of

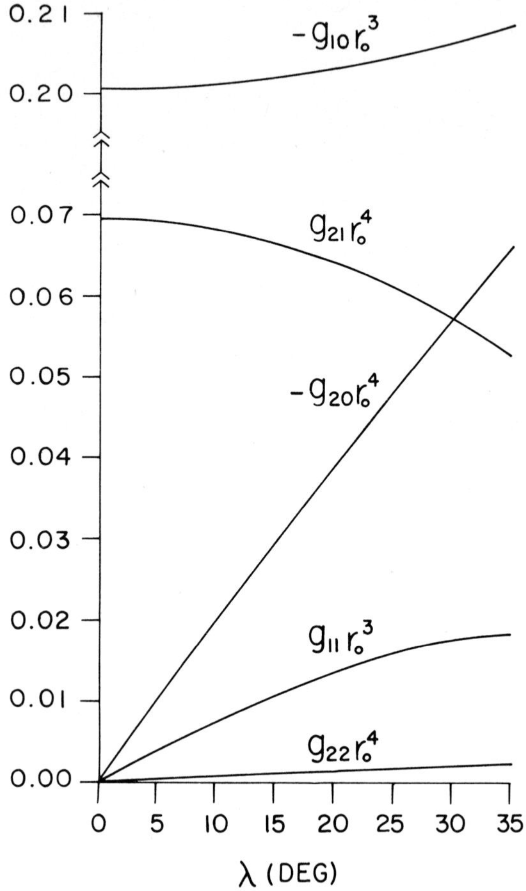

Fig. 11. The change in spherical harmonic coefficients with geomagnetic latitude of the Sun for the magnetic field expansion computed from surface currents.

the reconnection phenomenon. In effect a finite conductivity of the solar wind plasma would reduce the current in the magnetopause and at the same time create an electric field tangential to the magnetopause which would modify particle orbits within the magnetosphere. These effects would be directly related to the change in magnetic field in crossing the boundary. Thus the merging theory requiring finite plasma con-

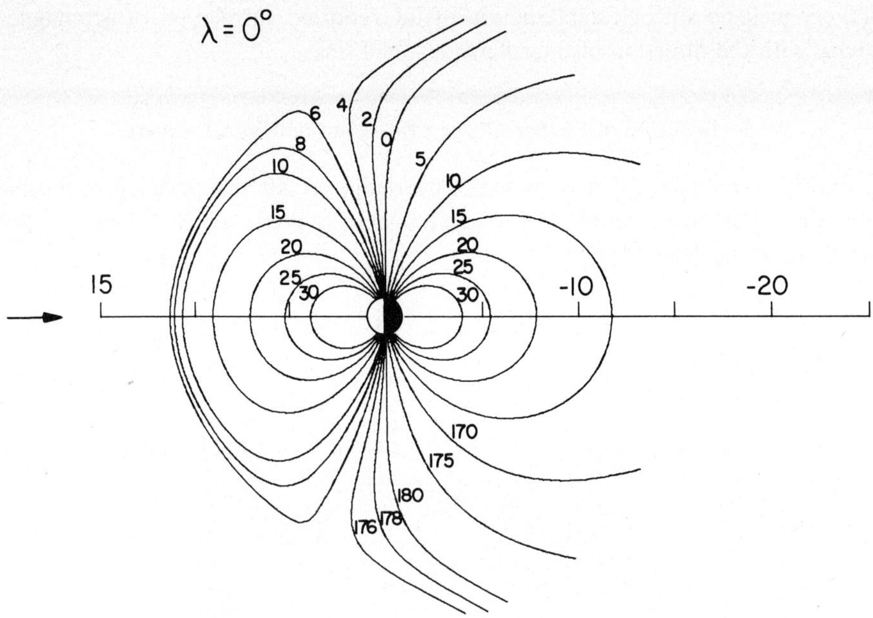

Fig. 12. Meridian plane field configuration for perpendicularly incident solar wind ($\lambda = 0$).

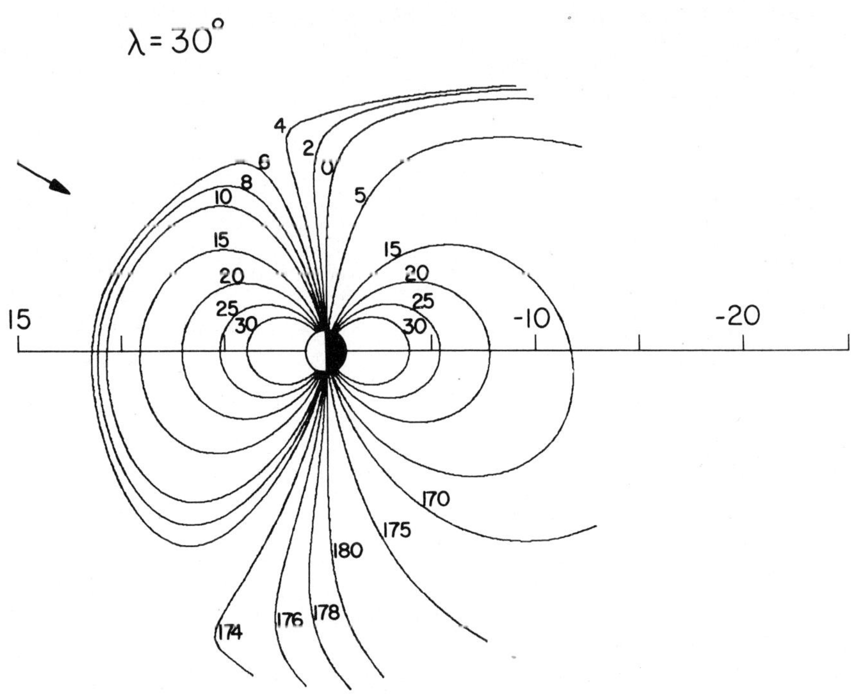

Fig. 13. Meridian plane field configuration for a solar geomagnetic latitude of 30° ($\lambda = 30°$).

ductivity yields a very clear prediction of the reported correlation of geomagnetic activity with the direction of interplanetary field lines.

5. Inclusion of Higher Order Effects in the Magnetosphere

A complete summary of our present knowledge of all the principal additional phenomena that occur within the magnetosphere is presented in Figure 15 taken from Speiser and Ness (1967).

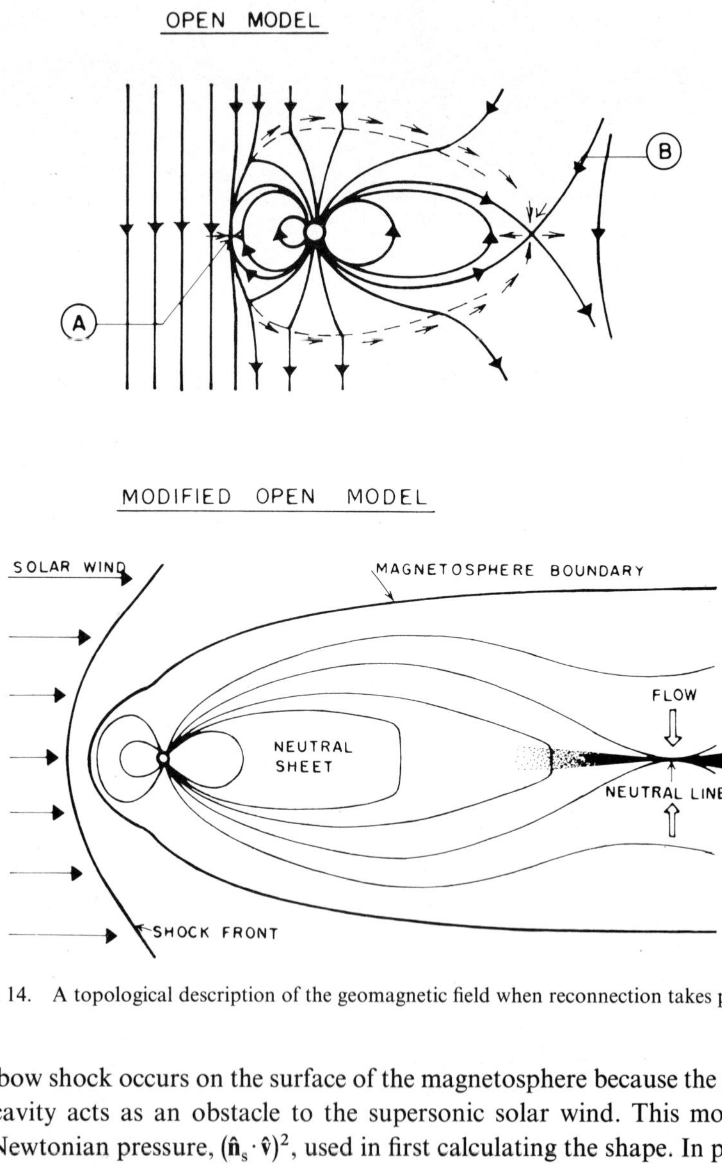

Fig. 14. A topological description of the geomagnetic field when reconnection takes place.

(i) A bow shock occurs on the surface of the magnetosphere because the magnetosphere cavity acts as an obstacle to the supersonic solar wind. This modifies the simple Newtonian pressure, $(\hat{\mathbf{n}}_s \cdot \hat{\mathbf{v}})^2$, used in first calculating the shape. In particular, the thermal pressure in the bow shock layer or *magnetosheath* drastically increases the outside pressure on the tail over that inferred from Newtonian pressure. The shape near the earth experiences little change but a more complete calculation should include the bow shock pressure. Conditions in the magnetosheath have been described by many authors (Kellogg, 1962; Axford, 1962; Lees, 1964; Spreiter and Jones, 1963 and others). Shen (1972) has included an oblique interplanetary field in obtaining the pressure at the magnetopause in the plane containing the magnetic field direction.

(ii) The plasma that leaks in and is trapped in the Van Allen radiation belts causes large approximately cylindrically symmetric currents which seriously perturb the geomagnetic field. The magnetic effect of these currents as calculated by Akasofu *et al.* (1961) is shown in Figure 16. These currents have been called ring currents and their behavior is primarily dipolar outside the current system so that the magnetosphere shape is affected very little by them. The field inside the magnetosphere is very much affected, and ring current calculations in the compressed asymmetric geomagnetic field are being prepared.

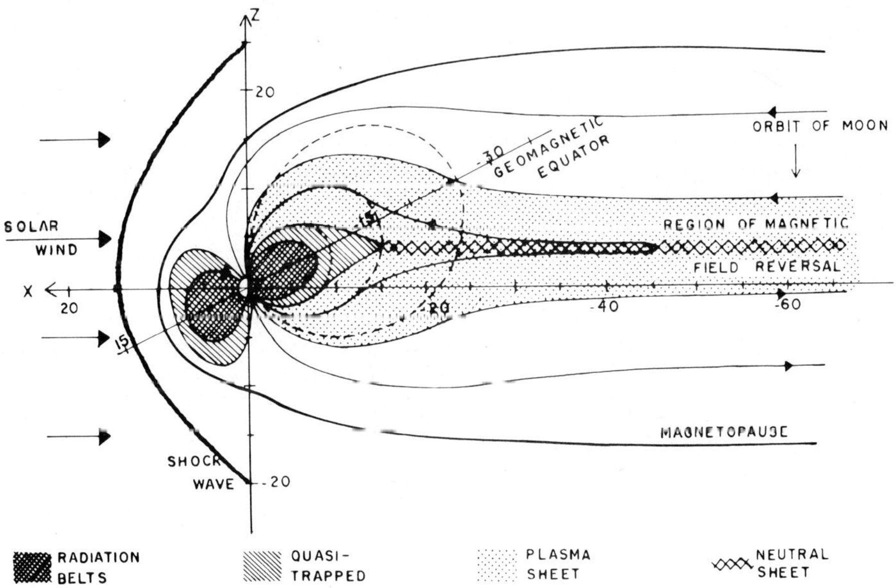

Fig. 15. An overall view of magnetospheric phenomena as presented by Speiser and Ness (1967).

(iii) Referring back to Figure 15 we can see that a fascinating and unanticipated phenomenon has been observed by satellite in the tail (Ness, 1965; Behannon, 1968; Mihalov *et al.*, 1968) which contributes to a further major distortion of the geomagnetic field. Furthermore, our theoretical study of the magnetotail indicates that everywhere in the magnetosphere but particularly in the tail the magnetopause is very porous to the entry and exit of solar wind plasma.

From about ten Earth radii in back of the Earth to far back well beyond the orbit of the Moon the magnetic field has been observed to be orders of magnitude larger than expected from the compressed dipole field. The field is aligned towards the Earth in the northern half of the tail and away from the earth in the southern half. The field reverses in a very thin equatorial plane of about one Earth radius. There is a weak

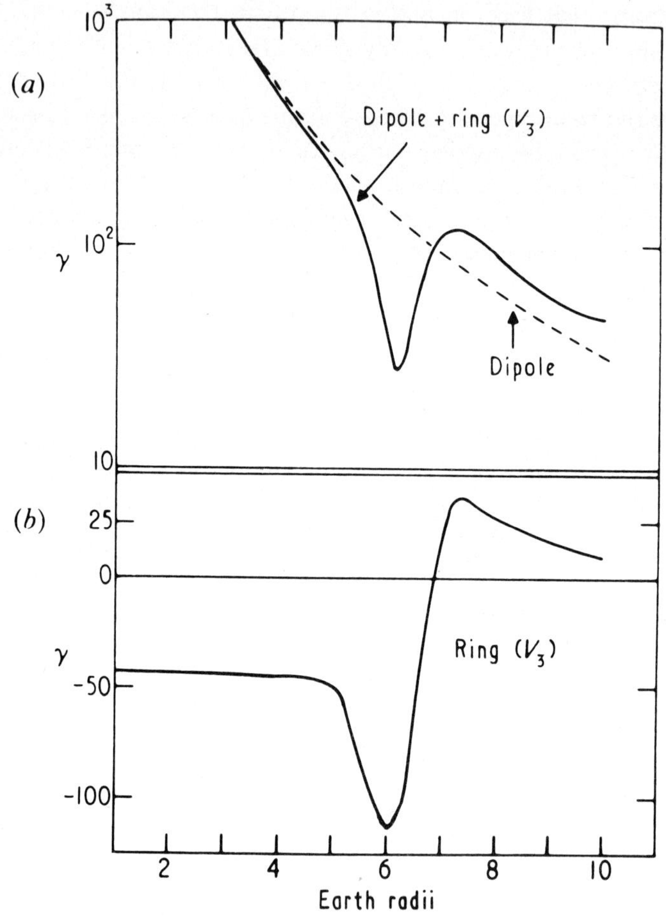

Fig. 16. The magnetic disturbance caused by the Van Allen radiation belt as computed by Akasofu *et al.* (1961).

northward field in the equatorial plane and the fields decrease slowly with distance from the Earth inversely as the 1/3 power of the distance (Behannon, 1968). The field configuration requires a large east-west current in the thin equatorial plane decreasing with distance. The enhanced magnetic pressure within the magnetosphere requires an enhanced west-east return current over the cylindrical surface of the magnetotail.

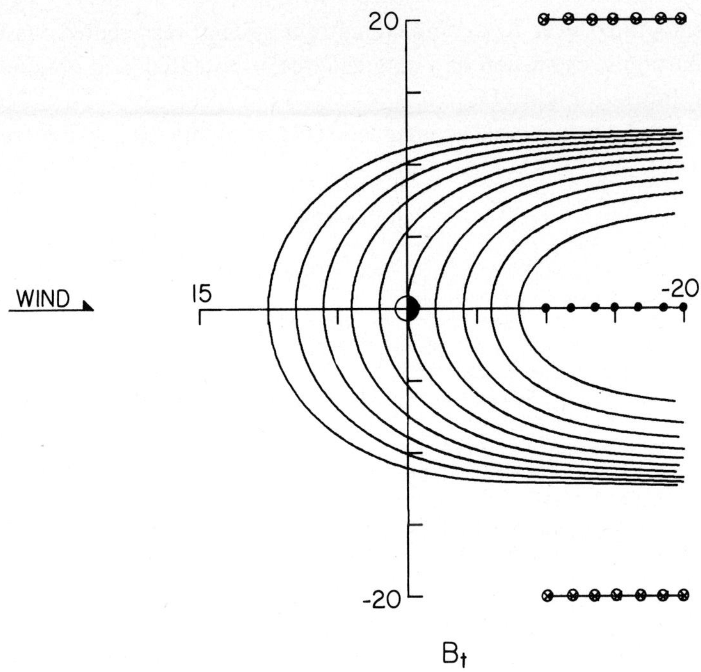

Fig. 17. The magnetic field caused by the tail current system in the meridian plane.

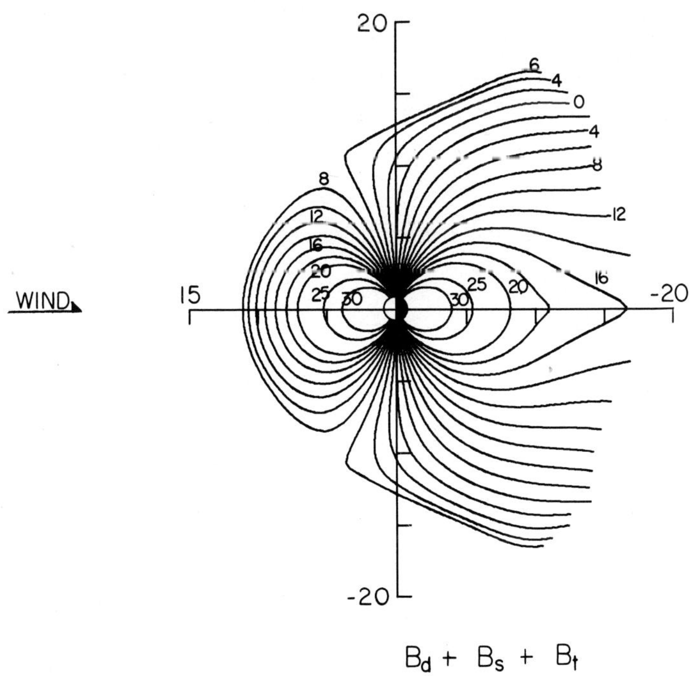

Fig. 18. The total compressed dipole field with the tail current system added in the meridian plane.

We have integrated over this complete current system, represented the result as a spherical harmonic expansion of a magnetic scalar potential, and obtained the field line traces shown in Figure 17.

If we include the compressed geomagnetic field we obtain the field line traces shown in Figure 18.

References

Akasofu, S.-I., Cain, J. C., and Chapman, S.: 1961, *J. Geophys. Res.* **66**, 4013.
Axford, W. I.: 1962, *J. Geophys. Res.* **67**, 3791.
Baker, J. C., Beard, D. B., and Young, J. C.: 1964, *Phys. Fluids* **7**, 504.
Beard, D. B.: 1968, *Planetary Space Sci.* **16**, 145.
Behannon, K. W.: 1968, *J. Geophys. Res.* **73**, 907.
Choe, J. Y., Beard, D. B., and Sullivan, E. C.: 1973, *Planetary Space Sci.* **21**, 485.
Dungey, J. W.: 1958, *Cosmic Electrodynamics*, Cambridge Univ. Press, London, p. 132.
Dungey, J. W.: 1961, *Phys. Rev. Letters* **6**, 47.
Kellogg, P. J.: 1962, *J. Geophys. Res.* **67**, 3805.
Lees, L.: 1964, *AIAAJ* **2**, 1576.
Mead, G. D.: 1964, *J. Geophys. Res.* **69**, 1181.
Mead, G. D. and Beard, D. B.: 1964, *J. Geophys. Res.* **69**, 1169.
Midgeley, J. E.: 1963, *J. Geophys. Res.* **68**, 5111.
Midgeley, J. E.: 1964, *J. Geophys. Res.* **69**, 1197.
Midgeley, J. E. and Davis, L. Jr.: 1962, *J. Geophys. Res.* **67**, 499.
Mihalov, J. D., Colburn, D. S., Currie, R. G., and Sonett, C. P.: 1968, *J. Geophys. Res.* **73**, 943.
Ness, N. F.: 1965, *J. Geophys. Res.* **70**, 2989.
Olson, W. P.: 1969, *J. Geophys. Res.* **74**, 5642.
Rosenbluth, M. V.: 1957, in R. Landshoff (ed.), *Magnetohydrodynamics*, Stanford Univ. Press, Palo Alto, p. 57.
Shen, W. W.: 1972, *Cosmic Electrodyn.* **2**, 381.
Slutz, R. J.: 1962, *J. Geophys. Res.* **67**, 505.
Speiser, T. W. and Ness, N. F.: 1967, *J. Geophys. Res.* **72**, 131.
Spreiter, J. R. and Jones, W. P.: 1963, *J. Geophys. Res.* **68**, 2121.

MAGNETOSPHERIC BOUNDARIES AND FIELDS

W. P. OLSON and K. A. PFITZER

Space Sciences Dept., McDonnell Douglas Astronautics Company-West, Huntington Beach, Calif., 92647, U.S.A.

Abstract. The quantitative determination of the magnetospheric magnetic field, **B**, is intimately related to the study of the magnetopause, plasma sheet and other boundaries in the magnetosphere. **B** is formed by the main field of the Earth and the fields from currents flowing in the magnetosphere. The boundary (or magnetoapuse) currents are formed by the shocked solar wind particles. The ions and electrons are deflected in opposite directions as they encounter the magnetospheric magnetic field. This process simultaneously forms the magnetopause and the boundary currents. Recent studies of the magnetic field in the tail of the magnetosphere also illustrate the importance of boundaries. It was possible to quantitatively model the tail field only when the cross tail currents were placed on the upper and lower boundaries of the plasma sheet. In the older models currents flow only across the center of the tail and the observed features of the tail field cannot be correctly produced. It is on the edges of the plasma sheet that the diamagnetic currents are the strongest. Drift currents also flow throughout the plasma sheet due to gradients in **B**. Success in modeling the tail field by considering diamagnetic and drift currents has led to the concept of currents distributed throughout the magnetosphere. They exist because of gradients in plasma density and **B**. The distributed currents then account for both the tail currents and the quiet time ring current. The total model magnetospheric magnetic field representation of **B** then consists of the Earth's main field and contributions from the boundary (magnetopause) and distributed currents (Olson, 1973).

The model representation of **B** has been tested in several ways using both particle and field observations. It correctly places the last closed field line at 78° magnetic latitude while older models predict 80–83°. The model yields the observed **B** values along the Earth-Sun line and the proper values of ΔB (observed total field minus observed main field) throughout the magnetosphere. Previous models do not produce the observed variations along the Earth-Sun line and ΔB contours calculated from them bear little or no resemblance to the observations. The observed decay of the tail field and other features back to lunar orbit are also correctly modeled. **B** has been approximated by expressing it analytically as a power series with exponential terms and then tested by predicting magnetospheric particle behavior. The high latitude trapping boundary for low energy particles calculated with the analytic model agrees with observations. (Previous models are in error by 4–5°.) The model was also tested by using it to compute the latitudinal cutoffs for solar cosmic rays. While older models predict cutoffs that are 6–8° in error with observations, the cutoffs computed with the present model are within 2° of the observed values.

This total magnetospheric magnetic field can also be used to redetermine the shape of the magnetopause and to study particle entry and motions in the magnetosphere. The magnetopause then exhibits the observed cusped geometry at high latitudes instead of the 'neutral points' produced by the older models. Low energy particles easily enter the magnetosphere through the dayside cusp regions and the flanks of the tail (through the dawn and dusk edges of the plasma sheet). These particles form currents across the tail and locally in the cusp region.

These studies point out the necessity of considering the low energy (<50 keV protons) plasma simultaneously with currents and fields. In order to understand and predict the formation and maintenance of magnetospheric structures (magnetopause, dayside cusp, plasma shed), it is necessary to model several current systems, the shapes and locations of boundaries, and the entry of particles. Each of these depends on all of the others. The work discussed here has been guided along these lines and the resulting models have been quite successful in predicting observed magnetospheric behavior.

1. Introduction

In order to understand and quantitatively model magnetospheric behavior it will ultimately be necessary to consider particle and field phenomena self consistently. Clearly, the low energy magnetospheric plasma is controlled by magnetic and electric

D. E. Page (ed.), Correlated Interplanetary and Magnetospheric Observations, 115–129. *All Rights Reserved.*

fields. However, the plasma forms currents and regions of charge excess which in turn modify these fields. To properly study magnetospheric dynamics it is necessary that the tilt of the Earth's axis to the solar wind direction and the various states of magnetic disturbance be included. An acceptable starting point for this task is the modeling of the magnetospheric magnetic field, with inputs from observed particle and field data. It is immediately found that the magnetospheric plasmas and fields do strongly interact and that plasma boundaries play an important role in the determination of the magnetospheric magnetic field, **B**.

The boundary of the magnetosphere, the magnetopause, is the location of strong currents formed by the deflection of the shocked solar wind plasma by the magnetospheric magnetic field. More diffuse currents flow near the plasmapause and near the edges of the plasma sheet in the tail. Boundaries are also important because their structure determines the location and entry (and exit) of plasma into the magnetosphere. Clearly, magnetosheath plasma flows throughout the dayside cusp regions. Also, in the distant tail there are indications of neutral points where the magnetosheath plasma may have easy access to the magnetosphere. Finally, particles with the proper pitch angles can penetrate the magnetopause wherever large magnetic field gradients persist parallel to the boundary just inside of the magnetopause.

2. Boundaries in the Magnetosphere

The first magnetospheric boundary to be studied in detail was the magnetopause. Its position has always been determined by assuming that at the boundary the kinetic pressure of the shocked solar wind is balanced by the energy density (pressure) of the magnetospheric magnetic field (Beard, 1960). Using this pressure balance condition, several workers have determined the shape of the magnetosphere for both perpendicular (Mead and Beard, 1964; Midgley, 1964; Slutz, 1962) and oblique (Olson, 1969; Choe *et al.*, 1973) incidence of the solar wind on the geomagnetic field. All of these models predict an infinitely long magnetospheric tail when only the kinetic pressure of solar wind is considered. These models also yield two neutral points on the high latitude dayside magnetopause. Recent observations of these regions show that they are actually characterized by a cusped geometry and the magnetosheath plasma penetrates, in these cusps, down to the ionosphere.

The magnetopause then is mutually the boundary of the shocked solar wind plasma and the geomagnetic field. The solar wind is deflected by **B** and the currents formed by this deflection act to limit the extent of **B**. The magnetopause does not completely shield the Earth from the solar wind because of its observed cusped geometry in the high latitude dayside regions and the possible X type neutral point regions in the distant tail. In addition, particles can penetrate this boundary wherever **B** just inside the magnetopause exhibits large gradients parallel to the boundary. This property is probably necessary for the formation of the plasma sheet in the tail of the magnetosphere (Pfitzer and Olson, 1973).

The tail of the magnetosphere is roughly separated into two 'lobes' where the directions of **B** are antiparallel. The lobes are separated by a region of high density plasma (the plasma pressure is approximately equal to the energy density of **B**) called the plasma sheet. The plasma sheet plays an important role in substorm dynamics. The plasma sheet is also an interesting structure in the quiet magnetosphere since its study and the study of currents flowing in the tail are intimately related. The boundaries of the plasma sheet (that separate the plasma region from the lobes of the tail) are also important because of diamagnetic currents flowing in their proximity. The final plasma structure to be considered in this paper is the plasmasphere (Carpenter, 1966). It is a region of high density plasma that extends from the ionosphere out to 3 or 4 Earth radii (R_E) geocentric distance. The plasmasphere is believed to be controlled by the corotation and cross tail electric fields. The plasma in this region however moves in the presence of a non-uniform magnetic field and is itself non-uniform in density. It therefore produces drift and diamagnetic currents that influence the structure of **B**.

3. Magnetic Fields in the Magnetosphere

Although several qualitative and two-dimensional magnetospheric magnetic field models have been published, only quantitative, three-dimensional models with some physical basis are considered here. It is necessary that the models be realistic because they will ultimately be used to self consistently study plasma structures, currents and fields.

The deflection of the solar wind by the geomagnetic field forms a current system on the magnetopause (the solar wind protons and electrons are deflected in opposite directions). In fact, it is this current system that limits the extent of the geomagnetic field. The currents flow such that their magnetic field cancels out the geomagnetic field outside of the magnetosphere and adds to it near the boundary in the magnetosphere.

These models then yielded the general shape of the magnetosphere and a rough picture of its magnetic field topology. However, observations of the magnetospheric tail (Ness, 1965) and particle data in the tail suggested that the magnetospheric magnetic field, **B** must have additional sources. Williams and Mead (1965) added a planar current sheet in the equatorial region of the tail (currents flowing from minus to plus infinity perpendicular to the Sun-Earth line) in order to separate **B** into two antiparallel 'lobe' regions. Other more physical models of currents flowing in the tail allow the currents to flow across the tail and return on the magnetopause (Schield, 1969; Olson, 1970). These models also merely try to duplicate the observed field topology and little explanation has been given for this geometry. Speiser (1965) and Beard *et al.* (1970), however have examined particle motions in the tail and suggested the types of currents that can flow.

In addition to more detailed observations of the tail field (Behanon, 1970; Mihalov *et al.*, 1968; Meng and Mihalov, 1972), recent work by Sugiura *et al.*, 1971, and Sugiura, 1972, has provided a detailed description of the near Earth magnetic field.

Sugiura's work in particular has shown that older magnetic field models do not produce the observed field geometry in the near Earth magnetosphere and that currents flow at geocentric distances as small as 3 R_E.

These observations suggest that the interaction of the magnetospheric plasma with a non-uniform magnetic field produces currents that contribute significantly to **B**. These currents are distributed throughout the magnetosphere and contribute significantly to **B**. The formation of these distributed currents and their magnetic effects are discussed in the following section.

4. A Quantitative Model of B

Both the decay of B_x (in the lobes) and B_z (in the center of the plasma sheet) with distance down the tail have been observed. It is clear that currents other than those flowing on the magnetopause (as obtained from the pressure balance condition) are necessary in order to maintain these observed fields. Initial work using the 'theta' cross section tail currents (currents flowing across the center of the tail in a sheet and returning on the boundary) was unsuccessful in modeling the observed tail field decay. It was found that in order to fit the observed tail field it is necessary to allow the currents to flow throughout the plasma sheet – not merely through its center. This behavior is expected if particle drifts and plasma diamagnetism are taken into account (Olson, 1973).

A tentative description of these tail currents is as follows. Charged particles drift across the tail due to the non-uniform magnetic field and produce a current in the dawn to dusk direction. The currents exist only in the plasma, i.e., in the plasma sheet. Although the currents flow throughout the plasma sheet, they appear to be quite strong near its boundary where the contribution from diamagnetic currents is large. These currents form charge excesses, a proton surplus near the dusk magnetopause and an electron excess near the dawn boundary. The charges in turn are most easily dissipated by causing return currents to flow through the magnetosheath. Thus, the distributed currents in the tail are similar to the older 'theta configurations' but spread throughout the plasma sheet and the magnetosheath.

The successful use of distributed currents in the tail suggested that they should also be important in the inner magnetosphere. In fact, it is believed that distributed currents account for the near earth field distortions reported by Sugiura *et al.* (1971). These currents are formed by gradients in plasma pressure in a magnetic field (diamagnetic currents) and by the drift motions of charged particles in a non-uniform magnetic field. It is difficult to determine these currents directly from particle and field observations since complete particle pitch angle and energy spectra are not available throughout the magnetosphere. However, a first order model of the distributed currents can be obtained from a general knowledge of plasma boundaries and a detailed determination of currents at a few points. The diamagnetic contribution should be largest near the plasmapause and the boundaries of the plasma sheet. The drift currents become stronger beyond the plasmapause. Plasma diamagnetism will weaken the magnetic field in the plasma regions and increase it in regions outside of the plasma. This basic property of magnetic plasmas offers a simple explanation for the gross behavior of **B** as shown by the ΔB (total field **B** minus main field) contours determined by Sugiura *et al.* (1971). General features of the near Earth distributed currents are an eastward ring-like flow around the Earth out to the vicinity of the plasmapause and beyond that a westward flow. The larger current loops become quite elliptical and extend back into the tail 10 to 15 R_E. These currents flow throughout the magnetosphere and are not restricted to the equatorial plane. In the model the distributed current system extends 3 R_E above or below the equatorial plane. The largest loop flows near the subsolar magnetopause and returns across the tail from dawn to dusk. Then there are currents that flow across the magnetosphere but return through the magnetosheath (or on the magnetopause as was done in the model, Olson, 1973).

Thus, the distributed currents form one current system that extends through the magnetosphere from the subsolar region to the distant tail. In the model the currents flow back to 212 R_E in the tail. This single system accounts for both the tail currents and the quiet time ring current. The distributed currents are completely separate from the magnetopause currents produced directly by solar wind interaction with **B**. The magnetopause currents are surface currents and quite strong. Although the current density of the distributed currents is about two orders of magnitude weaker, the total near Earth current from the distributed currents is similar in magnitude to that used in the older equatorial ring currents models of Apel *et al.* (1962), and Akasofu and Chapman (1961). These distributed currents are based on particle and field observations whenever possible and no new particle populations are necessary in order to explain the magnetic effects they produce. It will be shown that a new model of **B** that includes the contribution from these distributed currents correctly produces many observed magnetospheric particle and field features that were not obtained from the older models.

5. Tests of the Model Field

The model of the magnetospheric field, **B**, consists of a dipole representation of the main field, the field from the magnetopause currents and the field from the distributed currents. **B** was determined in the usual manner by using the Biot-Savart law to integrate over the currents. The total field model was tested by using it to calculate several observed field and particle features. **B** was computed at points between the subsolar magnetopause and the vicinity of lunar orbit in the tail. ΔB obtained from the model is shown in Figure 1. These results can be compared with the observed quiet time ΔB contours (Sugiura *et al.*, 1971) and found to be in good quantitative agreement. **B** is also determined along the Sun-Earth line (see Figure 2). It is seen that again the model agrees well with observations and that the older models (Mead, 1964; Olson, 1970) are in error by as much as 35% in the near Earth magnetosphere. The decay of the lobe (B_x) and the equatorial (B_z) fields computed from the model are in agreement with the observed decay. The model has also been used to calculate the

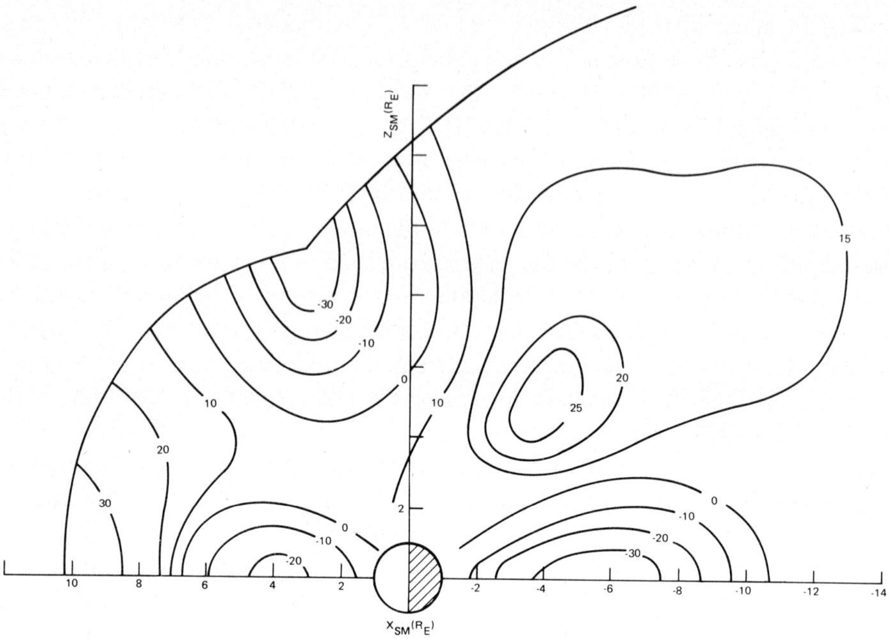

Fig. 1. ΔB contours (in gamma's) obtained from the model in the noon-midnight meridian plane. **B** was determined by direct integration over the currents.

contribution of non-ionospheric currents to the field variations at the surface of the Earth. Although these currents produce an appreciable field at the Earth's surface (the noon-midnight difference along the equator is 14 γ when the Earth's conductivity is considered), it is still believed that the dynamo currents flowing in the ionosphere are the predominant source of the S_q variations. **B** also contributes to D_{ST} which must be present even during quiet times. The contribution to D_{ST} from the boundary currents is positive (northward). The contribution from the distributed currents is divided into two parts. Those currents flowing beyond the plasmapause and across the tail contribute a large negative D_{ST} (about 40 γ) while those currents drifting eastward again produce a positive contribution to D_{ST}. The strength of the latter component is difficult to determine since there are no accurate measurements of **B** for geocentric distances less than $\sim 3\ R_E$.

The above tests were done by direct integration over the model currents. Although some accuracy is lost when **B** is expressed analytically, the great savings in computation time (and cost) dictates that an analytic representation of **B** be obtained. A generalized multivariable least squares orthonormal fitting subroutine (Pfitzer, 1973) was used to determine the coefficients of a series representation for each of the components of **B**. The analytic representation differs on the average by 9.8% from the field obtained by direct integration. The average error in the total field is less than 4%. The average difference in gammas is 0.63 γ. The averages are taken over the three components of **B**. This analytic representation of **B** is then used when **B** must be

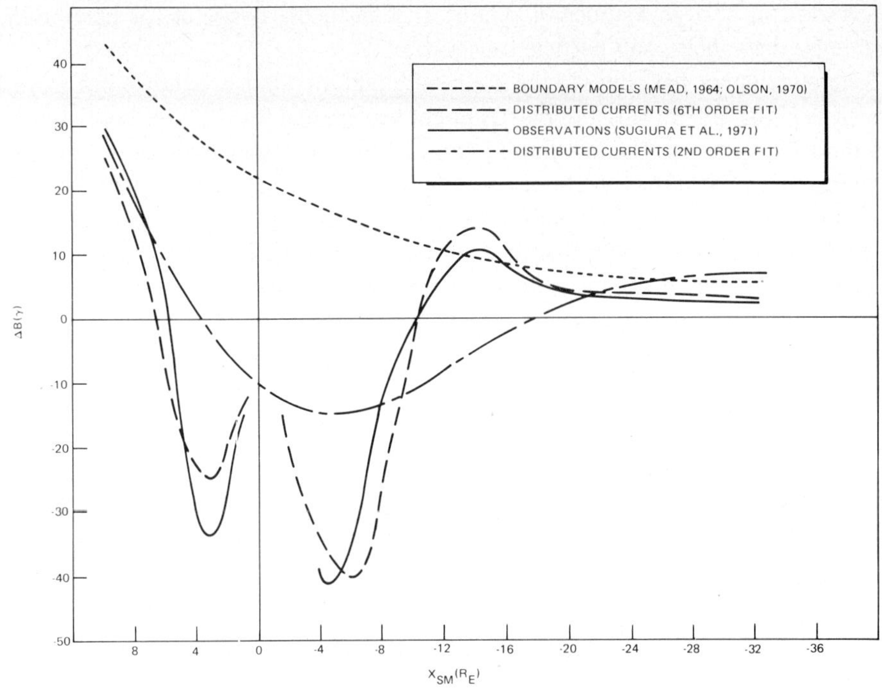

Fig. 2. ΔB along the Sun-Earth line. The observations and the values obtained by direct integration over the currents are in good agreement. The field from the magnetopause currents alone does not possess the necessary structuring in the near Earth region. The analytic representation of **B** must contain many terms in order to fit the **B** values obtained by direct integration of the currents.

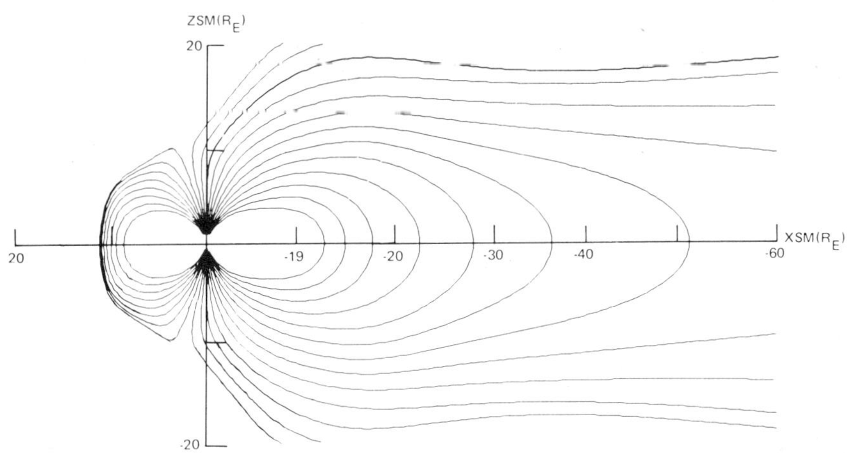

Fig. 3. Field lines in the noon-midnight meridian plane in 2° intervals from 70° to the pole. The last line to close within the dayside magnetosphere (as defined by boundary models) crosses the Earth's surface at 76°. Since the dayside cusp currents are not included in the field, the high latitude dayside magnetosphere is still not properly modeled.

determined at a large number of points. It is particularly useful for determining field lines, drift shells, and particle trajectories.

Several magnetic field lines in the noon-midnight meridian are shown in Figure 3. The lines in the tail are fairly dipolar as might be expected for the quiet configuration the model represents. The last line to close within the magnetopause on the noon meridian is at 76° although the use of the boundary field (Olson, 1970) obtained without the distributed currents forces several field lines to return beyond the magnetopause on the noon meridian instead of dropping back into the tail. The consequences of including the distributed currents in the model of **B** used to determine the shape of the magnetopause are discussed below.

Other views of field lines are shown in Figure 4. At high latitudes the longitude

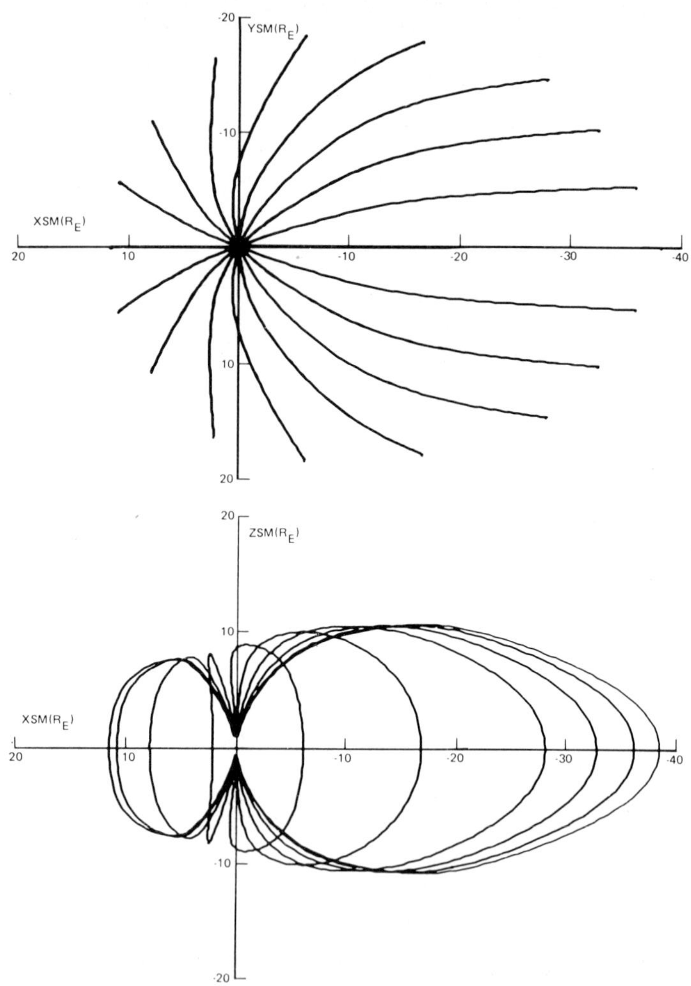

Fig. 4. Projections of field lines onto the XY and XZ planes in solar magnetospheric coordinates. The lines are shown at a constant latitude circle of 80° and a longitude spacing of 20°.

of the position of equatorial crossing of a field line can be over 100° different from the longitude of the line at its Earth intercept.

The model was also tested by using it to compute low and high energy charged particle behavior in the magnetosphere (Pfitzer, 1972). The high latitude earth intercept of the trapping boundary for 600 keV electrons calculated from the model agrees well with the observed boundary which is 4°–5° lower than that calculated from previous models. The boundary calculations were performed using the first and second adiabatic invariants and determining the last closed drift shell on which particles of a given pitch angle can exist. The boundary was experimentally determined using four years of OGO-1 and OGO-3 ion chamber and electron spectrometer data. Figure 5 shows the comparison between the experimentally determined boundary at local noon during quiet times ($K_p \leqslant 1$) and the boundary as calculated from three models. Note that the distributed current model (which includes the effect of the plasma) is in much better agreement over the entire trapping boundary.

The high latitude cutoffs for cosmic ray particles calculated from previous models have been in error with observation by 6°–8°. With the present models the calculated cutoffs are within 2° or 3° of the observed values. Cosmic ray curoffs have been

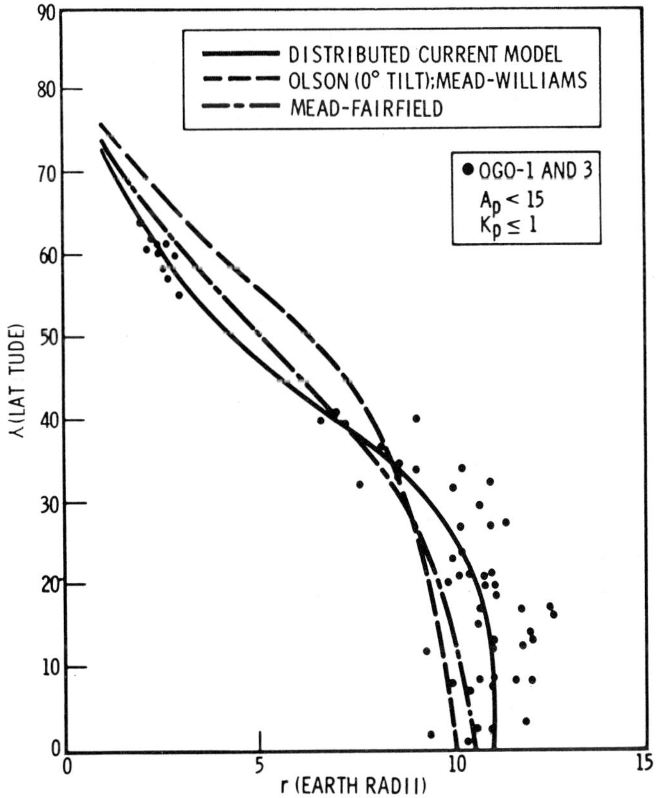

Fig. 5. Latitude dependence of the 600 keV electron trapping boundary at noon (10–14 h LT).

determined for 5, 20, and 70 MeV utilizing the distributed current model. These results, reported by Masley *et al.* (1973), are shown in Table I.

A reason for the lowering of the high latitude Earth intercept of the trapping boundary and the lowering of the cosmic ray cutoffs is the depressed field region at 4–6 R_E described above. This causes the field lines at a given latitude at the Earth's surface

TABLE I

Proton cutoffs

Comparison of observed cosmic ray cutoffs with model calculations (from Masley *et al.*, 1973). The cutoffs under distributed current are preliminary and may be further reduced when additional trajectories are calculated.

	Gall, *et al.*, 1968	Masley *et al.*, 1971				Distributed currents			OGO 6 measured
	5 MeV	5 MeV				5 MeV	20 MeV	70 MeV	5 MeV
Tilt	0	0	+30	−30	−20	0			−20
Noon (direct)	78	78	76	76	77	73	71	68	67
Noon (drift around)	70	74	–	–	–	69	68	66	
Midnight	68	69	69	70	69	~66	65	62	~63

to cross the equator at larger geocentric distances than given by previous models; and conversely, particles at a given equatorial position mirror along the field lines with lower latitude Earth intercepts.

6. Self Consistent Studies

Self consistent studies may be loosely defined as those investigations that attempt to examine and understand two or more mutually related phenomena simultaneously. Thus, studies of the magnetopause shape and magnetic field must be self consistent since the field depends on currents flowing on the magnetopause while in turn the shape of the magnetopause is determined by the total magnetospheric magnetic field (to which the magnetopause currents contribute). Self consistent problems are difficult for a number of reasons. The loop must be entered somewhere (i.e., a good guess at **B**). The determination of the magnetopause shape is then followed by the redetermination of **B**. A self consistent solution is obtained only if the set of variables converges such that, for example, **B** yields a magnetopause surface whose currents in turn produce **B**. This particular self consistent problem has been discussed in detail by many authors (Beard, 1960; Baker *et al.*, 1964; Mead and Beard, 1964; Midgley, 1964; Olson, 1969).

7. Self Consistent Magnetopause Shape and Field

There are several other self consistent problems to be solved in magnetospheric physics. Three of them are discussed here. The most simple is the extension of the magnetospheric boundary determination to include the distribution currents. This work was begun with the hope that the inclusion of the magnetic effects of the distributed currents would lower the latitude of the last closed field line and also produce a boundary geometry that is more like the observed cusped shape (instead of the dented neutral point configuration obtained in previous self consistent models) in the high latitude dayside magnetosphere. The shape of the boundary as well as the locations of the cusps are sensitive to the location and strength of the distributed currents. In order to obtain a self consistent boundary shape, several iterations of the pressure balance equations and redeterminations of current strengths on the boundary as well as the strengths of the distributed currents are required. A truly cusped geometry is not possible when only the pressure balance condition is used because the solar wind cannot directly impinge on the entire cusp surface. These changes suggest that when the cusp currents have been determined self consistently, it will be worthwhile to redetermine (self consistently) the shape of the magnetopause and the magnetopause currents.

8. Self Consistent Dayside Cusp Determination

Models of the magnetopause shape based on pressure balance yield a surface with two discontinuities along the noon meridian plane. The magnetopause currents flow in loops around these point discontinuities such that the total magnetic field at the points is zero, thus the name neutral points. It has been argued that although the zero field at those points should permit particles to enter the magnetosphere, there cannot be enough plasma flow through a point to make these neutral point regions an important source region for magnetospheric plasma. Recent observations of this region however have shown that magnetosheath plasma does indeed flow into a cusped region over an extended area. They suggest that the real magnetosphere does not exhibit neutral points and that the pressure balance condition does not adequately describe the high latitude dayside magnetopause.

It is known (Beard *et al.*, 1970; Pfitzer and Olson, 1973) that magnetosheath particles moving in the proper direction can penetrate the magnetosphere wherever there are large gradients in **B** parallel to the boundary just inside of the magnetopause. If the boundary obtained from the pressure balance condition is used as a starting point, it is clear that large gradients in **B** exist in the vicinity of the neutral points (at a latitude of about 82° in the models). All of the field lines in the magnetopause branch out from these neutral points. Thus, to self consistently study the formation of the dayside cusps it is necessary to examine the entry of particles through the region of the neutral points. The motions of the particles in the non-uniform magnetospheric field, **B**, produces drift currents. These currents in turn act to modify **B** in the region of the neutral points. This process is then repeated, plasma is allowed to enter this expanded entrance region (as determined by the new **B** and used to redetermine the magnetopause shape). This procedure is repeated until a self consistent solution is obtained. The self consistent result includes the shape and extent of the dayside cusps, the strength and direction of **B** at all points in the region of the cusps, the form of the currents flowing on the cusp boundaries and the plasma pressure everywhere within the cusp. As with every self consistent problem, a starting point must be defined. Here it was the incorrect magnetopause shape with neutral point geometry included. Work on this problem is in progress. The first iteration has transformed the cusp region from a point to an entrance area that extends several earth radii off of the noon meridian.

9. Self Consistent Plasma Sheet Determination

A similar problem on which more work has been done involves the plasma sheet in the center of the tail of the magnetosphere. It is desired to understand the formation and maintenance of the plasma sheet. Questions to be answered include: How are particles supplied to the sheet? Why is it bounded by the lobes in the tail? How does its cross section change with distance down the tail? Again, the fact that

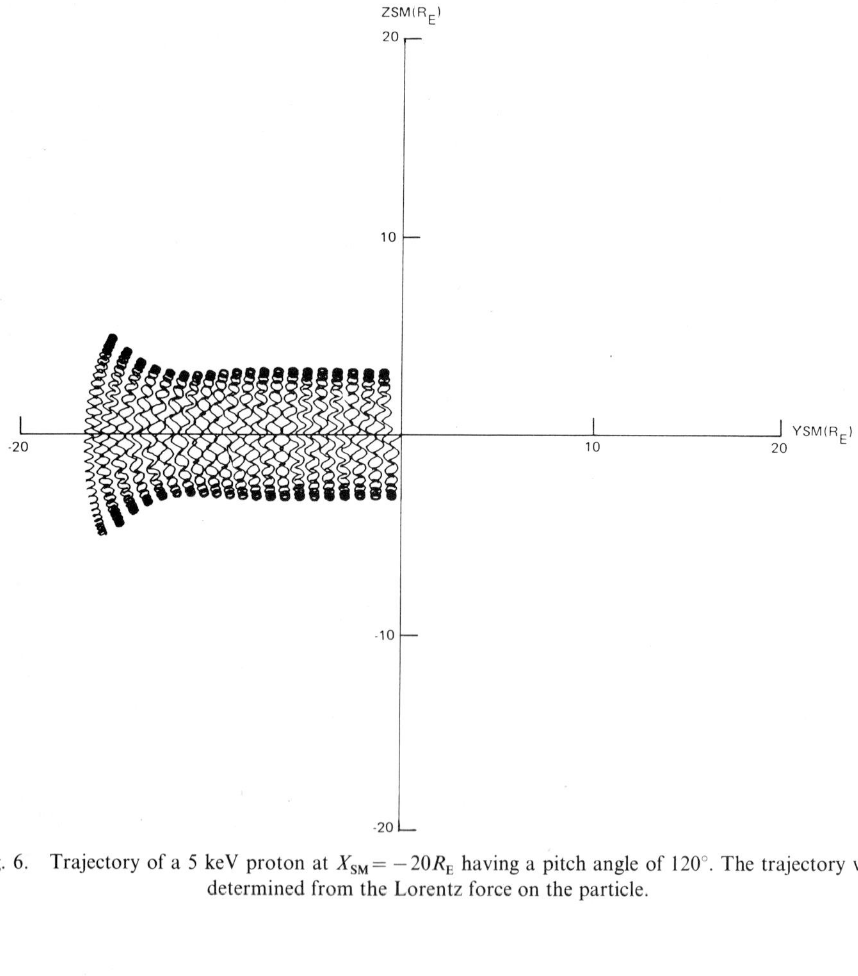

Fig. 6. Trajectory of a 5 keV proton at $X_{SM} = -20 R_E$ having a pitch angle of 120°. The trajectory was determined from the Lorentz force on the particle.

particles can enter the magnetosphere where strong field gradients persist is important. The model for **B** used here includes the distributed currents described above. With it, the field line topology can be studied along the flanks of the tail. It is seen that a weak field similar to the dayside cusp regions must persist along the flanks of the tail near the magnetopause. This geometry suggests another source region for magnetospheric particles (Olson, 1971).

Several test particles have been input to the field geometry obtained from the distributed current model at the equator on the flanks of the tail (Pfitzer and Olson, 1973). They drift in the proper direction to explain the cross tail currents (protons move from the dawn to the dusk flanks). Also, in the given field geometry they mirror farther from the equator near the flanks than they do near the center of the tail (see Figure 6). The plasma sheet is observed to have this geometry. It is noted in passing that the plasma sheet is more or less open to the magnetosheath.

To properly carry out these self consistent sudies many particle trajectories must be calculated. There must be enough particles used to adequately represent the plasma sheet plasma and simultaneously the cross tail currents. This will constitute the first step in the complete self consistent determination of the tail field and plasma properties.

10. Comments and Conclusions

The study of the distributed currents suggests modification of several concepts concerning magnetospheric structures. The distributed currents do not cross the tail in a thin sheet but rather flow throughout the plasma sheet. Thus, there may be a 'neutral sheet' where **B** is locally small, but it is only possible to fit the observed average quiet time features of the tail when distributed currents are used. The distributed currents have no sharp Earth-ward boundary and merge gradually into the near Earth currents that flow around the dayside magnetosphere. The currents in the inner magnetosphere extend out to the magnetopause and are not confined to the equatorial plane but fill the region up to 3 or 4 R_E above and below it. These currents are much weaker than those previously considered in studies of the ring current. However, because of their great volume the total current is still large and they do explain the observed magnetic effects. They are formed from low energy (<50 keV) particles and can be explained in terms of existing particle data. Because of the extended nature of these currents it is improper to speak of the 'location of the ring' (e.g., whether or not it is beyond geosynchronous orbit).

The formation of the dayside cusps and the plasma sheet also suggests that the magnetosphere should not simply be defined as a region around the Earth bounded by the magnetopause. Rather, the magnetosphere might be defined as that region of space surrounding the Earth where the Earth's main field interacts with the solar wind and its associated fields. Magnetosheath plasma flows both into the dayside cusps and across the plasma sheet in the tail. These magnetospheric features may therefore be properly thought of as extensions of the magnetosheath. It is clear that to properly understand and correctly model magnetospheric particle and field observations it is necessary to simultaneously consider both particles and fields. The work described here is an initial step in that direction. The quantitative representation of the magnetospheric magnetic field, **B**, has been extended to include the effects of currents distributed throughout the magnetosphere. These currents exist because of the mutual interaction of the magnetospheric plasma with the non-uniform magnetospheric magnetic field. The model of **B** has then been used to study cusp and plasma sheet structures. Again, the currents flowing in these regions influence **B**. The modeling done with this approach has been quite successful in predicting observed magnetospheric behavior.

Finally, it must be mentioned that all of the work described here is for quiet magnetic conditions. for steady solar wind and interplanetary parameters, that is, for a static magnetosphere. This self consistent modeling approach can also be used for the disturbed magnetosphere. Differences between quiet time and disturbed configurations can then be noted and used to determine induced electric fields in the magnetosphere. The relativistic transformation of the solar wind velocity and field also yields an electric field in a magnetospheric reference frame. Consideration of these electric fields and differences in current systems (strengths and locations) will then permit the self consistent study and modeling of magnetic storm and magnetospheric substorm processes.

Acknowledgment

This work was supported by the Office of Naval Research under Contract N00014-72-C-0233.

References

Akasofu, S. I. and Chapman, S.: 1961, *J. Geophys. Res.* **66**, 1321.

Apel, J., Singer, S. F., and Wentworth, R. C.: 1962, in H. E. Landsberg and J. V. Mieghem (eds.), *Advances in Geophysics*, vol. IX, Academic Press, New York.

Baker, J. C., Beard, D. B., and Young, J. C.: 1964, *Phys. Fluids* **7**, 504.

Beard, D. B.: 1960, *J. Geophys. Res.* **65**, 3559.

Beard, D. B., Bird, M., and Huang, Y. H.: 1970, *Planetary Space Sci.* **18**, 1349.

Behanon, K. W.: 1970, *J. Geophys. Res.* **75**, 743.

Carpenter, D. L.: 1966, *J. Geophys. Res.* **71**, 693.

Choe, J. Y., Beard, D. B., and Sullivan, E. C.: 1973, *Planetary Space Sci.* **21**, 485.

Masley, A. J., Pfitzer, K. A., and Olson, W. P.: 1973, *Trans. Amer. Geophys. Union* **54**, 408.

Mead, G. D.: 1964, *J. Geophys. Res.* **69**, 1181.

Mead, G. D. and Beard, D. B.: 1964, *J. Geophys. Res.* **69**, 1169.

Meng, C. I. and Mihalov, J. D.: 1972, *J. Geophys. Res.* **77**, 1739.

Midgley, J. E.: 1964, *J. Geophys. Res.* **69**, 1196.

Mihalov, J. D., Colburn, D. S., Currie, R. G., and Sonnet, C. P.: 1968, *J. Geophys. Res.* **73**, 943.

Ness, N. F.: 1965, *J. Geophys. Res.* **70**, 2989.

Olson, W. P.: 1969, *J. Geophys. Res.* **74**, 5642.

Olson, W. P.: 1970, A Scalar Potential Representation of the Tilted Magnetopause, McDonnell Douglas Astronautics Co. Paper WD 1332.

Olson, W. P.: 1971, Study of Particle Precipitation as an Energy Source in the Atmosphere, McDonnell Douglas Astronautics Co. Report MDC G2281.

Olson, W. P.: 1973, Distributed Magnetospheric Currents and Their Magnetic Effects, McDonnell Douglas Astronautics Co. Paper WD 2038.

Pfitzer, K. A.: 1972, *Trans. Amer. Geophys. Union* **53**, 1090.
Pfitzer, K. A.: 1973, A Multidimensional Orthonormal Analytic Function Fitting Subroutine, McDonnell Douglas Astronautics Co. Paper.
Pfitzer, K. A. and Olson, W. P.: 1973, *Trans Amer. Geophys. Union* **54**, 432.
Schield, M. A.: 1969, *J. Geophys. Res.* **74**, 1275.
Slutz, R. J.: 1962, *J. Geophys. Res.* **67**, 505.
Speiser, T. W.: 1965, *J. Geophys. Res.* **70**, 4219.
Sugiura, M.: 1972, *J. Geophys. Res.* **77**, 6093.
Sugiura, M., Ledley, B. G., Spillman, T. L., and Heppner, J. P.: 1971, *J. Geophys. Res.* **76**, 7552.
Williams, D. J. and Mead, G. D.: 1965, *J. Geophys. Res.* **70**, 3017.

A SELF CONSISTENT THEORY OF THE MAGNETOTAIL

D. B. BEARD and M. K. BIRD
University of Kansas, Lawrence, Kan., U.S.A.

The magnetic field in the tail is confined in a cylinder or slightly expanding cylinder, is aligned toward the Earth in the northern half of the cylinder and away from the Earth in the southern half, reverses across a thin equatorial plane about 1 Earth radius thick, and decreases slowly going inversely with the 1/3 power of the distance from the Earth (Behannon, 1968). The field is approximately constant above (or below) 4 or 5 Earth radii from the equatorial plane.

Now it is clear what current system is required to create this field configuration. An east west current in the thin equatorial plane is necessary to reverse the main field. This current is linearly proportional to the main field and decreases with distance as the main field does. It can be easily shown analytically that an equatorial east west current decreasing with distance gives rise to a weak northward component of field $B_Z \propto \partial B_x / \partial x$ which is about a tenth that of the main field (Beard *et al.*, 1970). The magnetic field is confined within the magnetotail. Hence there must be a current on the magnetotail surface to cause a change from magnetotail field to magnetosheath field in crossing the surface. To avoid the unessential complication of treating magnetosheath and interplanetary currents we will consider the magnetosheath field non-existent. We have then the complete magnetotail current system illustrated in Figure 1. This decreases monotonically with distance from the Earth as $1/|x|^{0.3}$ and comes in to very roughly 10 Earth radii from the Earth.

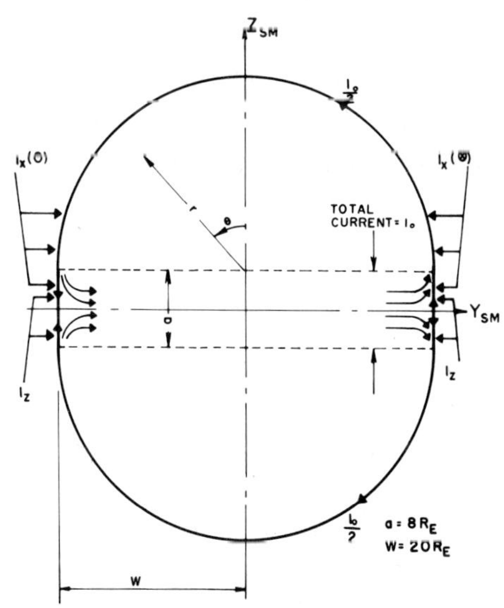

Fig. 1. The magnetotail current system in a plane perpendicular to the cylinder axis.

D. E. Page (ed.), Correlated Interplanetary and Magnetospheric Observations, 131–137. *All Rights Reserved.*

To separate the oppositely directed fields above and below the equatorial plane a hot plasma must exist within and near the equatorial plane. Indeed a hot plasma has been observed within 4 or 5 Earth radii from the equatorial plane. A hot plasma in such a field configuration will be the source of significant electrical current: (i) the gradients in the magnetic field will give rise to gradient drift currents. (ii) The field lines are sharply curved going from away from the Earth, to northward, and then towards the Earth in a distance of the order of an Earth radius. The curvature of the field lines will give rise to large curvature drift currents. (iii) The magnetization of the plasma will change with the plasma density and the strength of the field and give rise to a magnetization current. Putting all three sources of current together we arrive at the compact expression for the total plasma current

$$j = \frac{C}{B^2} \mathbf{B} X \left[\nabla P_{\mathrm{n}} + \frac{P_{\mathrm{s}} - P_{\mathrm{n}}}{B^2} (\mathbf{B} \cdot \nabla) \, \mathbf{B} \right],$$

where n and s refer respectively to the components of the tensor pressure perpendicular to the field and parallel to the field.

Inserting this current into Maxwell's equation

$$\nabla X \mathbf{B} = \frac{4\pi}{c} \mathbf{j}$$

we obtain a differential equation for $\mathbf{B}$ and P. P_{n} and P_s depend on the pitch angle distribution which is unknown. Using flux conservation $\nabla \cdot \mathbf{B} = 0$ and Liouiville's Theorem to obtain the plasma pressure along any field line we can express P in terms of $\mathbf{B}$ and eliminate P from the equation. We have integrated this equation numerically in the meridian plane assuming the spatial coordinates separable and find that only an isotropic pitch angle distribution $P_n = P_s$ will give results which fit the experimental observations. $P_s > P_n$ yields a field minimum above and below the equatorial plane while $P_s < P_n$ yields a plasma thickness that fills the entire tail and the maximum field is never achieved (Bird and Beard, 1972).

Isotropy very much simplifies our equation which can then be solved analytically in the meridian plane

$$P(x, z) = P(x_0, 0) \left| \frac{x_0}{x} \right|^{0.6} \cos^2 \omega z$$

$$B_x(x, z) = \sqrt{8\pi P(x_0, 0)} \left| \frac{x_0}{x} \right|^{0.3} \sin \omega z$$

$$B_z(x, z) = B_z(x_0, 0) \, (|x_0|/|x|)^{0.3} \cos \omega z$$

$$j_y(x, z) = \frac{c}{4\pi} \frac{P(x, 0)}{B_z^2(x, 0)/8\pi} \frac{\partial}{\partial x} B_z(x, z) \,,$$

where the x axis is towards the Sun from the Earth as origin, the z axis is northward from the equatorial plane, and y is in the equatorial plane in a right handed coordinate system.

$$\omega^2 = \frac{\alpha^2}{x^2}\left\{1 + \frac{1}{\alpha} + \frac{P}{B_z^2(x,0)/8\pi}\right\} \quad \text{with} \quad \alpha = 0.3$$

Our result is an excellent first approximation but it does not include the dipole field and the magnetic effect of distant currents in the magnetotail for which $\nabla \times \mathbf{B} = 0$. Using the current obtained from the plasma and field structure in this first approximation in the meridian plane we can complete the current system as shown in Figure 1 and integrate over its entirety to obtain a better field and plasma structure using the Biot Savart integral. This gives us an improved current system and we repeat the cycle until no further change is obtained after the third time around. Our results are illustrated in Figures 2–8.

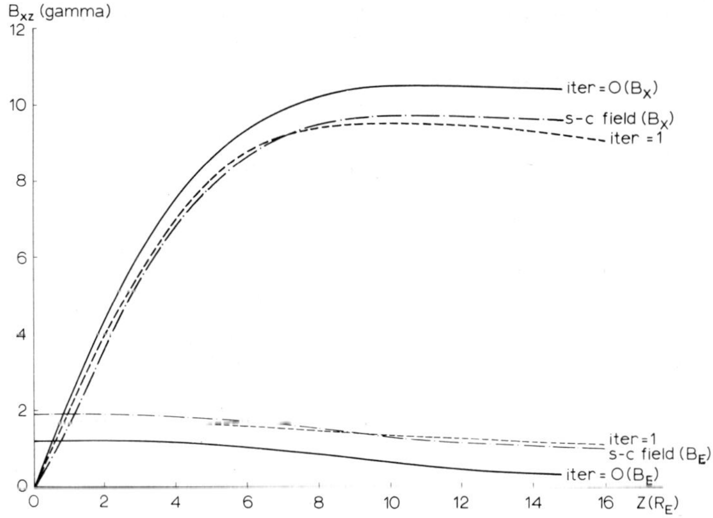

Fig. 2. Magnetic field components in the meridian plane as a function of height above the equator in Earth radii for 0, 1, and final iterations.

It is interesting to inquire what the relative importance of the various current contributions is for conditions occurring beyond the effect of the Earth's field say at the orbit of the Moon. This is shown in Figure 9. The dominant current is due to curvature drift which is cancelled in large part by the negative magnetization current in the equatorial plane. The curvature drift decreases rapidly with distance from the equatorial plane and is surpassed by the negative gradient current resulting in a weak negative net particle drift above and below the equatorial plane. The magnetization current is positive there and keeps the total electric current positive everywhere. Overall the total particle drift current is 80–90% of the total electrical current and is primarily a result of the sharp curvature near the equatorial plane.

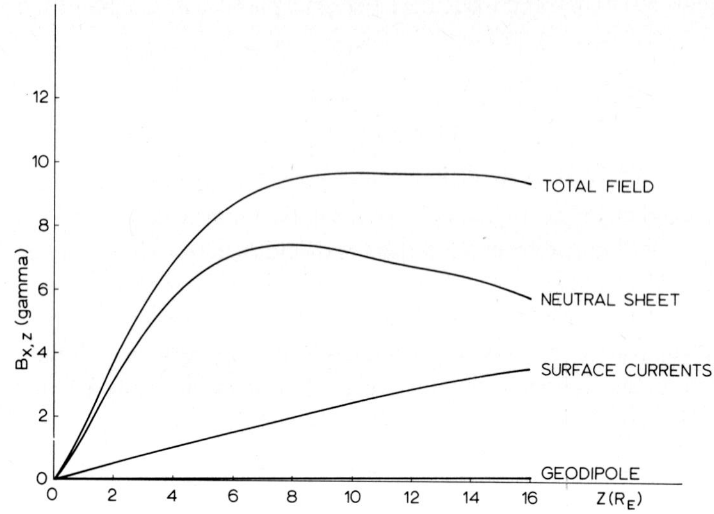

Fig. 3. Contributions to the x component of magnetic field as a function of height above the equatorial plane.

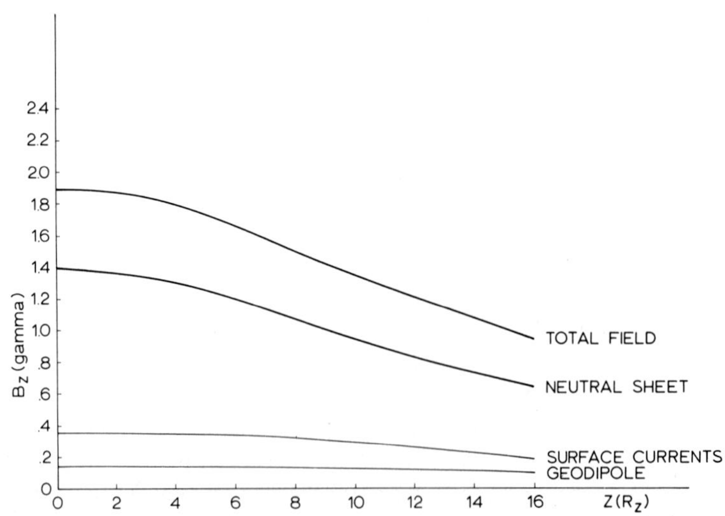

Fig. 4. Contributions to the northward or z component of magnetic field as a function of height above the equatorial plane.

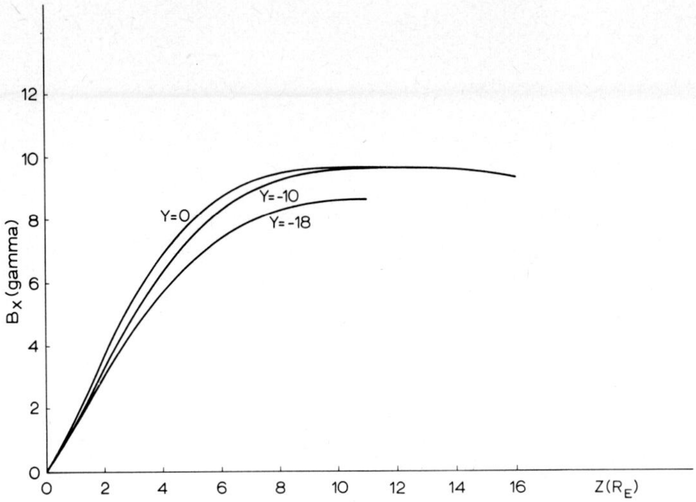

Fig. 5. The x component of magnetic field for various distances away from the meridian plane as a function of height.

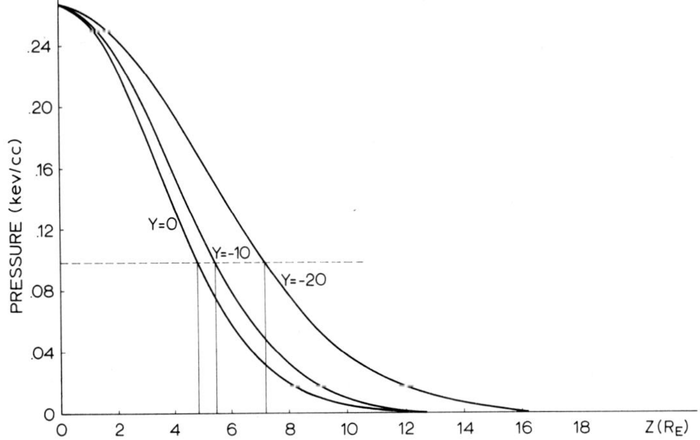

Fig. 6. Pressure for various distances away from the meridian plane as a function of height.

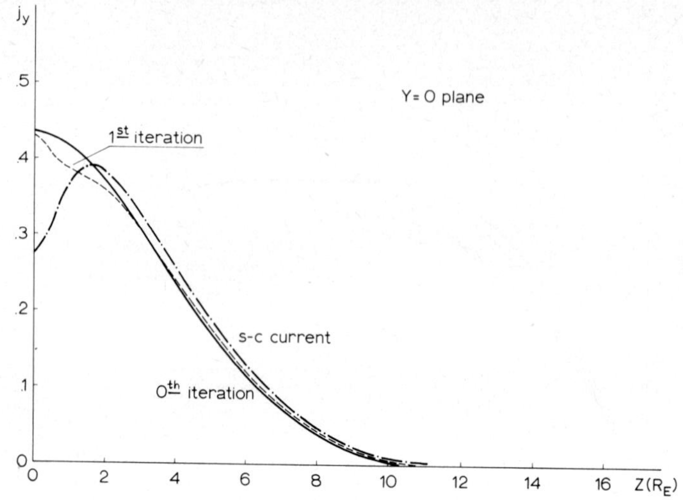

Fig. 7. Electric current as a function of height in the meridian plane for 0 and 1st order approximation and for final result.

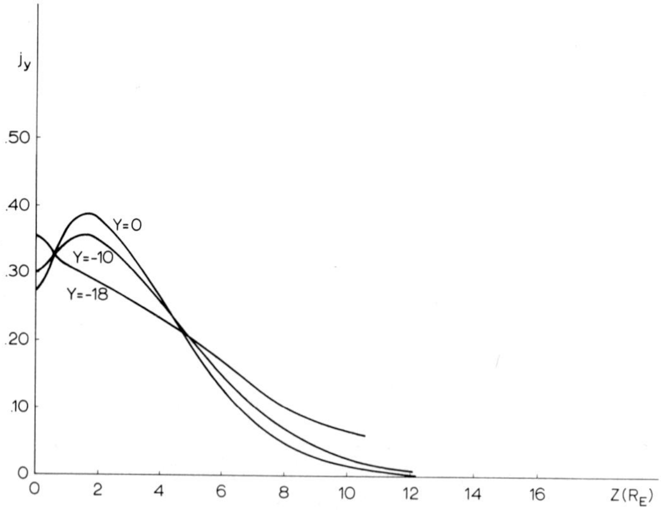

Fig. 8. Current for various distances from the meridian plane as a function of height.

This is not the entire story. The details of the magnetic and plasma structure near the magnetosheath have not been worked out exactly and we are working on that now in order to understand the details of how the plasma is caught up and injected into the current system in the magnetic field structure at the boundary. Of greater interest, we are studying the details of the near Earth termination of the tail current system

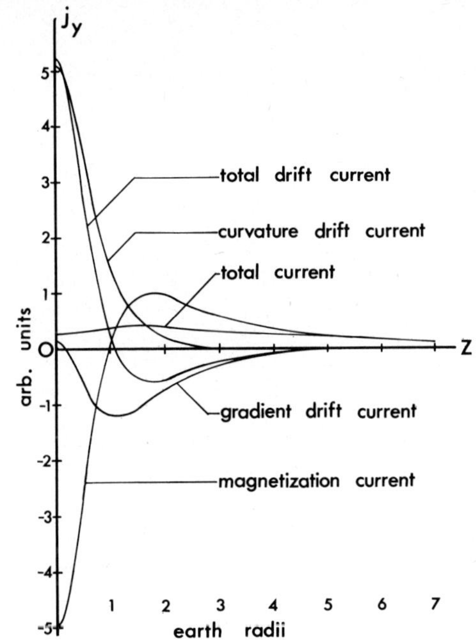

Fig. 9. Various contributions to the electric current as a function of height above the equatorial plane.

where the curvature of the dipole field is less and the field in the equatorial plane is much larger. Furthermore we have not yet included the effect of electric fields which arise from several different effects, nor studied dynamic effects.

References

Beard, D. B., Bird, M. K., and Huang, Y. H.: 1970, *Planetary Space Sci.* **18**, 1349.
Behannon, K. W.: 1968, *J. Geophys. Res.* **67**, 3791.
Bird, M. K. and Beard, D. B.: 1972, *Planetary Space Sci.* **20**, 2057.

OGO-5 OBSERVATIONS OF THE MAGNETOPAUSE

CHRISTOPHER T. RUSSELL

Institute of Geophysics and Planetary Physics, University of California, Los Angeles, Calif. 90024, U.S.A.

MARCIA NEUGEBAUER

Jet Propulsion Laboratory, Calif. Institute of Technology, Pasadena, Calif. 91103, U.S.A.

and

MARGARET G. KIVELSON

Institute of Geophysics and Planetary Physics, University of California, Los Angeles, Calif. 90024, U.S.A.

Abstract. The magnetopause is a dynamic boundary with a bewildering variety of signatures. OGO-5 observations show not only that the average boundary position moves in response to changes in dynamic pressure, but also that it moves in response to changes in the north-south component of the interplanetary field. Further, the boundary often oscillates about its average position as waves propagate along the boundary away from the nose region. The variation of the magnetic field through the boundary at times can be a simple rotation with no change in magnitude, and at other times can resemble a simple tangential discontinuity. However, it often displays complex patterns such as field enhancements on the magnetospheric side of the boundary or apparently uncorrelated field strength and direction changes. One particularly simple boundary crossing has been studied in detail with both positive ion data and magnetic field data. In this case, the electron and ion currents could be separately deduced and the structure agreed with that expected for a Chapman-Ferraro boundary with almost complete neutralization. Examination of ELF magnetic and VLF electric field data at the magnetopause revealed only occasional emissions within the field gradients.

1. Introduction

Investigations of the magnetopause roughly fall into three classifications: studies of average position, studies of motion, and studies of structure. Of the three types of investigation, the study of the average position and shape of the magnetopause has been the most popular (Cahill and Amazeen, 1963; Frank and Van Allen, 1964; Freeman, 1964; Ness *et al.*, 1964; Bridge *et al.*, 1965; Holzer *et al.*, 1966; Wolfe *et al.*, 1966; Cahill and Patel, 1967; Fairfield and Ness, 1967; Gosling *et al.*, 1967; Heppner *et al.*, 1967; Ness, 1967; Binsack, 1968; Fairfield, 1971). As a result, the average position and shape of the forward part of the magnetopause near the equatorial plane are now fairly well determined. On the other hand, little is known about the magnetopause position at high latitudes except for the recent observations of the Heos-2 satellite (Bahnsen *et al.*, 1974).

A variant on this type of study has been to determine the position of the magnetopause as a function of geomagnetic activity (Holzer *et al.*, 1966; Patel and Dessler, 1966; Cahill and Patel, 1967; Gosling *et al.*, 1967; Heppner *et al.*, 1967; Feldstein, 1970; Meng, 1970). While there is some average relationship between magnetopause position and geomagnetic activity, there are many exceptions to the general trends. Furthermore, such studies do not attack the important problem: determining the cause of geomagnetic activity and the cause of the displacement of the magnetopause from its average position.

D. E. Page (ed.), Correlated Interplanetary and Magnetospheric Observations, 139–157.

Studies of magnetopause motion have also been popular but have attracted somewhat less interest (Holzer *et al.*, 1966; Heppner *et al.*, 1967; Cummings and Coleman, 1968; Kaufmann and Konradi, 1969; Aubry *et al.*, 1970; Aubry *et al.*, 1971; Kaufmann and Konradi, 1973; Ledley, 1972). These investigations reveal oscillations with periods from seconds to tens of minutes and velocities along the boundary normal up to many tens of kilometers per second. They show further that the magnetopause can move to a new steady state position in response either to changes in the solar wind dynamic pressure or to changes in the orientation of the interplanetary magnetic field.

Finally, data from two satellites have been used to study the structure of the magnetopause: Explorer 12 (Sonnerup and Cahill, 1967; 1968; Sonnerup, 1971), and OGO 5 (Aubry *et al.*, 1971; Ogilvie *et al.*, 1971; Sonnerup *et al.*, 1973). These studies show that the magnetopause has a bewildering variety of signatures which do not always resemble magnetohydrodynamic discontinuities. At times there is a magnetic field normal to the boundary as in an open magnetosphere and at other times there is not.

In this paper we investigate the motion and structure of the boundary. In the first section we discuss the erosion of the magnetopause, i.e., the inward motion of the magnetopause when the interplanetary magnetic field turns southward. We discuss some published criticisms of the interpretation of this inward motion (Willis, 1971; Akasofu, 1972) and present counter-arguments. We next briefly discuss the oscillatory motion of the magnetopause, the high time resolution magnetic signature of the boundary. Finally, we examine the result of a joint study of the magnetopause involving the UCLA fluxgate magnetometer, the JPL solar wind probe and the UCLA/JPL search coil magnetometer.

2. Erosion of the Magnetopause

The first detailed evidence that a southward interplanetary field causes an inward displacement of the magnetopause was found in data from an inbound pass of OGO 5 at about 0900 LT on March 27, 1968 (Aubry *et al.*, 1970). Figure 1 shows one minute averages of the magnetic field in the solar magnetospheric coordinate system. OGO 5 first entered the magnetosphere at 1700 UT when the interplanetary field was northward as measured by the Ames Research Center fluxgate magnetometer on Explorer 35. Then at about 1720, the interplanetary field turned southward and the magnetopause soon ($\sim$1730 UT) moved inwards and reached the position of OGO 5. From then until 1914 the average inward velocity of OGO 5 and the magnetopause were roughly equal. During this period, the amplitude of the magnetic field was steady. Its direction fluctuated, with a southward component present through most of the interval, and dominant at the end of the interval. When OGO 5 left the region of the magnetopause at 1914, the magnetosheath field was turning northward and it remained northward for about 20 min thereafter. During this entire period there were no significant changes in the solar wind dynamic pressure.

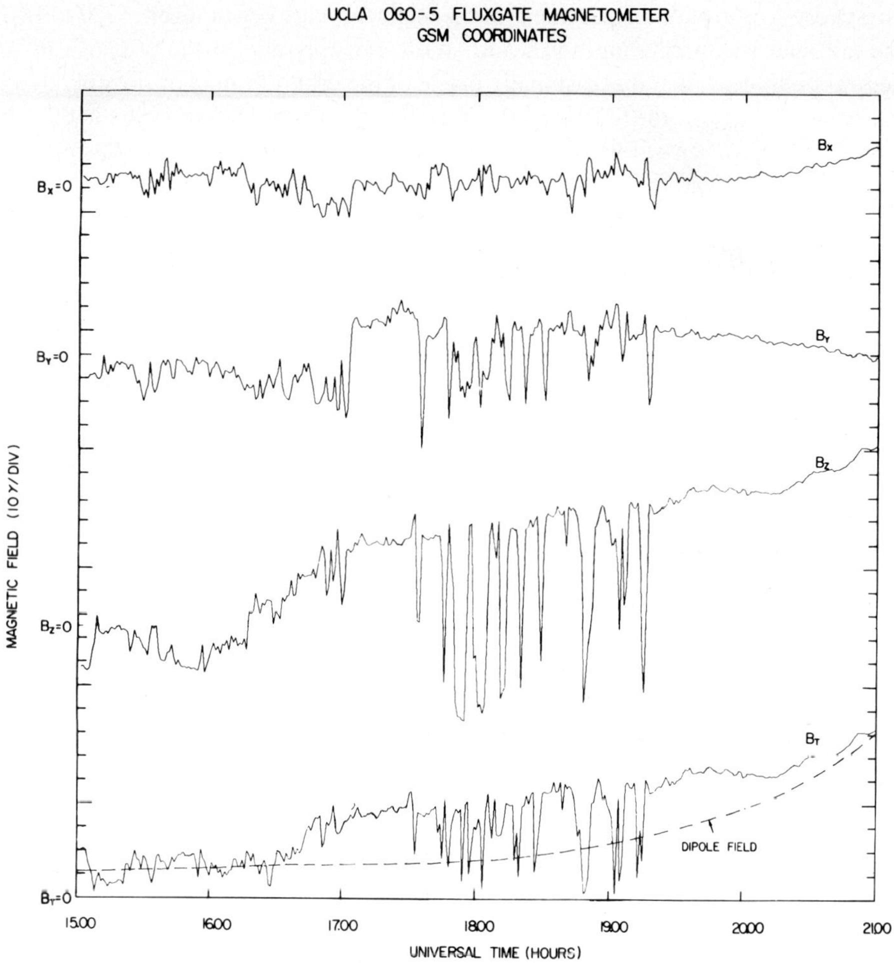

Fig. 1. One minute averages of the magnetic field in solar magnetospheric coordinates on an OGO-5 inbound pass on March 27, 1968.

Figure 2 shows the positions of the magnetopause during this pass, extrapolated from the OGO 5 crossing to the nose assuming that the magnetopause maintained a constant shape throughout the erosion process. The thick line shows the calculated position assuming an image dipole approximation and inelastic collisions. The initial position *A* is well within the range of positions which is consistent with various classical models of the solar wind-magnetosphere interaction (Schield, 1969), but the last encounter is quite inconsistent with such models. The amount of eroded magnetic flux as the boundary moved from *A* to *D* was approximately 10^{16} Maxwell which is about 15% of the magnetic flux in the tail.

Magnetic field data, from Imp 4 in the plasma sheet shown in Figure 3, reveal a decrease in the north-south component beginning at 1720 UT when the flux transport process or inward displacement of the magnetopause began. From 1720 to 1810, the tail field was oscillating north and south, possibly due to the presence of the neutral point near to and occasionally inside of the orbit of Imp 4 or possibly due to oscillatory motion of the tail about the *Y-GSM* axis. From 1810 to 1900 the tail field was northward. Then at 1900 and 1910 it went sharply southward as the plasma sheet

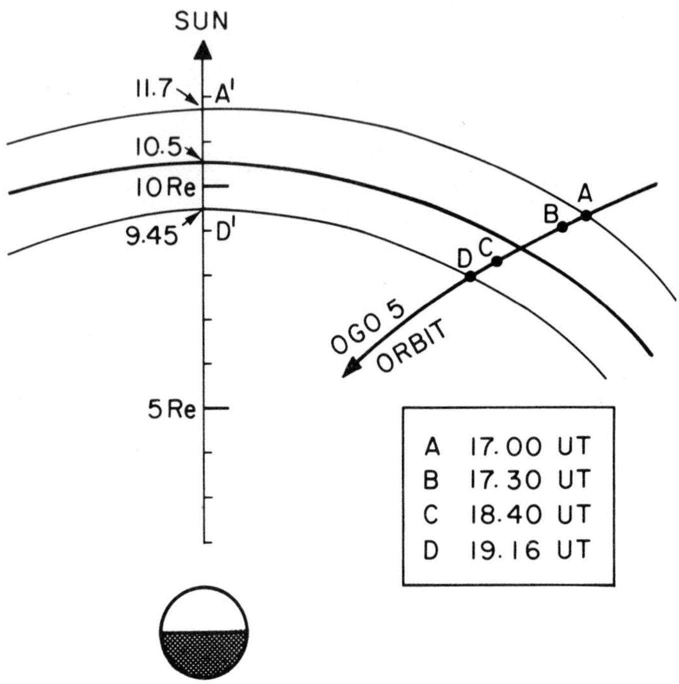

Fig. 2. Projection of the OGO-5 orbit in the solar magnetospheric equator on March 27, 1968 showing extrapolated magnetopause positions.

thinned across Imp 4. Finally at about 1940 UT, the plasma sheet expanded across Imp 4 and the tail field turned strongly northward. The reduction of the north-south component in the tail to near zero, the appearance of occasional southward components, the thinning and thickening of the plasma sheet are all tail signatures of substorm occurrence (Fairfield and Ness, 1970; Russell and McPherron, 1973). While the plasma sheet and tail field certainly changed somewhat between 1720 and 1900, it is obvious from these data that the major tail event during that period began at about 1900 UT, perhaps intensifying at 1940 UT.

Data from the auroral zone support this view. Deviations from the quiet day magnetogram traces began at Sodankyla between 1700 and 1800 UT as shown in Figure 4. However, major changes did not begin until about 1900, followed by a major intensification at about 1940 UT. Magnetic records from the 'Russian' sector, which have become available since the analysis of Aubry *et al.* (1970) are totally consistent with the Sodankyla records. The earliest auroral zone negative bay onsets occur at Cape Chelyuskin and Dixon Island at 1850 UT with an intensification and poleward expansion, seen at Heiss Island, at 1940 UT. Furthermore, a study of all 12 available midlatitude magnetograms in the midnight sector, from 20 LT to 05 LT, shows no major positive bay occurrence until 1920 UT (McPherron, 1973). While this is a clear example of the difference in onset times possible using auroral zone

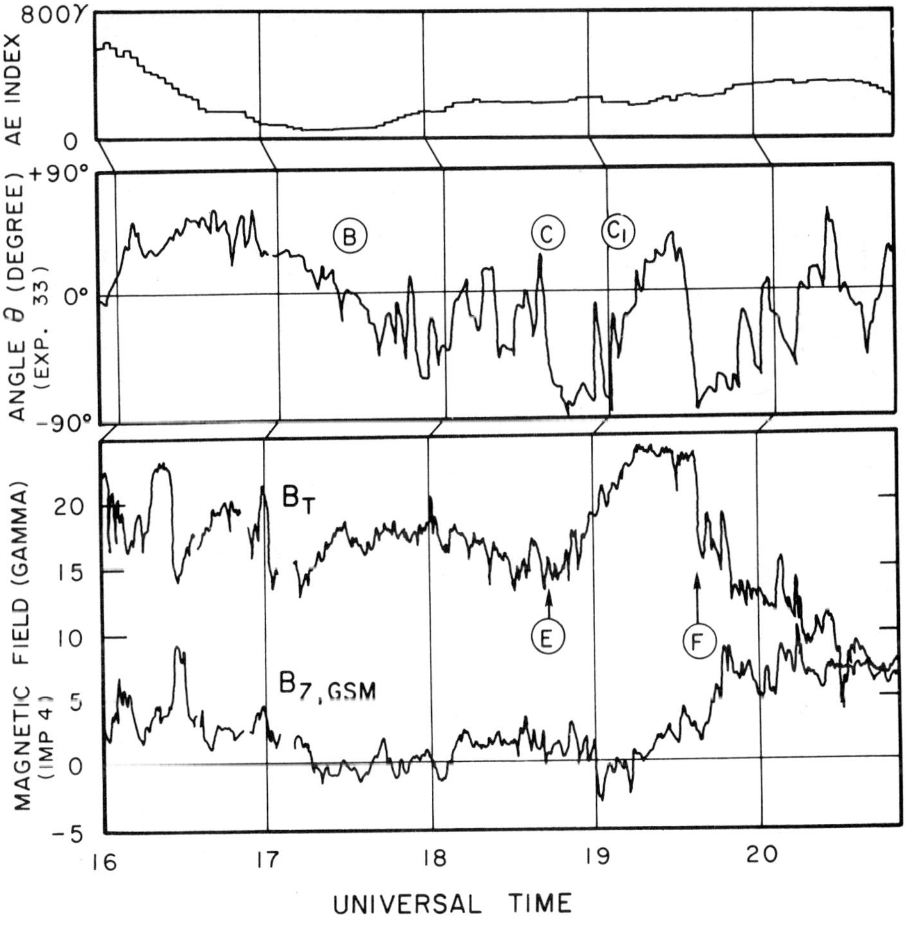

Fig. 3. Variation versus universal time of the AE index, the latitude of the magnetosheath field as measured by Explorer 33 and the total field and the Z solar magnetospheric component as measured on Imp 4 in the magnetotail at $(-24, 10, 3)$ R_E in solar magnetospheric coordinates.

and midlatitude records, it affects none of the conclusions of Aubry *et al.* (1970) who did not use the midlatitude magnetograms to identify the onset time of the substorm.

Our interpretation of this sequence of events is that when the interplanetary field turned southward, the merging rate at the dayside magnetopause was enhanced, magnetic flux was transported from the nose region to the tail and the magnetopause moved inwards. While this transport of flux certainly affected the magnetospheric convection electric field, and thus caused some ground magnetic effects, such as those identified as substorm growth phase phenomena by McPherron (1970), no major recovery occurred in the magnetotail and no significant changes occurred in the usual auroral zone magnetograms until 1850 UT. The major change, which we identify as the substorm expansion phase, we attribute to the formation of an x-type

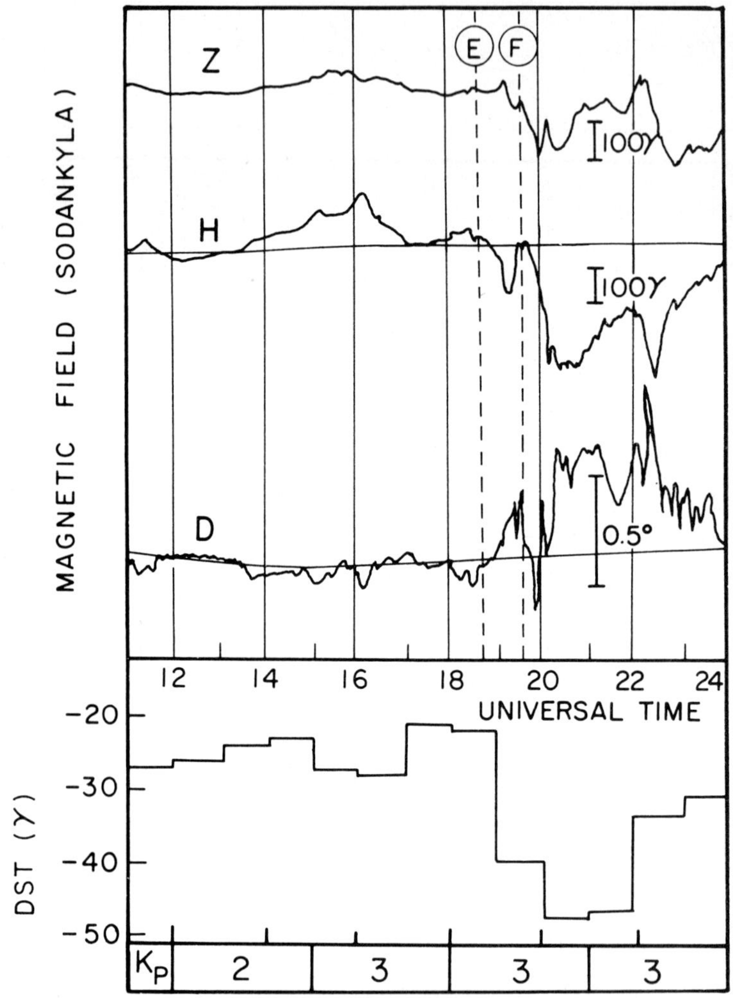

Fig. 4. Sodankyla magnetograms, 1 h averaged Dst index and three hourly average K_p. Quiet day curves for D and H are shown.

neutral point near the Earth and an enhanced reconnection rate at this point (Russell and McPherron, 1973).

These observations and interpretations have been criticized in two areas. Akasofu (1972) has stated that substorm occurrences are not related to the interplanetary field orientation and that there is not a growth phase in which energy or magnetic flux is stored in the tail. Our evidence that there exist magnetospheric responses associated with the growth phase is that the magnetopause began to move inward as soon as the interplanetary field turned southward and that the tail field began to change from a dipole-like to a tail-like orientation at the time of this inward motion. On the ground, major substorm activity, as gauged from auroral zone magnetograms and the Dst index, did not begin until almost $1\frac{1}{2}$ h later at which time the tail field began to return to a dipole-like configuration. Thus, there was a period of low geomagnetic activity during a period of strong solar wind magnetospheric interaction. This is what is implied by the concept of the growth phase.

Willis (1971) has suggested that the magnetopause cannot move inwards under constant solar wind dynamic pressure because the tangential stress does not enter the pressure balance condition which determines magnetopause position. The answer to this objection is similarly simple but in some ways more subtle. Models of the magnetopause such as that of Olson (1969) derive a magnetopause position that is a function only of solar wind dynamic pressure and dipole tilt angle. Such models ignore the viscous interaction of the solar wind on the magnetosphere, field-aligned currents, etc., and consequently do not predict the existence of or properties of the geomagnetic tail. The pressure balance maintained across the magnetopause in steady state involves the local magnetic field. The magnetic field is the sum of the contributions from *all* current systems near and distant. This is, in fact, the reason why an iterative procedure must be used to calculate magnetopause models.

Just as the presence of the ring current will cause the magnetopause to be located at greater radial distances because this current increases the field strength in the outer magnetosphere (Schield, 1969), currents which reduce the magnetic field strength in the outer magnetosphere on the dayside will cause the magnetopause to be located at lesser radial distances. The interaction of the solar wind with the magnetosphere can produce two such current systems: the tail current system and a partial ring current system associated with dayside merging. In order to produce an inward motion from 11.7 to 9.5 R_E as observed on March 27, the changes in these current systems would have to cause a net reduction of the dayside magnetic field of 17 γ.

Early models of the magnetopause such as that of Olson (1969) did not include a realistic tail current system, let alone allow the tail currents to vary in response to changes in the tangential stresses, e.g., those caused by merging. Thus, a realistic quantitative study of the response of the magnetopause position to changes in the tail currents has not been carried out. However, Unti and Atkinson (1968) have studied a two-dimensional magnetosphere with a perfect neutral sheet and find that, as expected, the magnetopause moves inwards as the magnetic flux content of the tail increases. On the other hand, we would expect that the dayside field strength reduction would only be some fraction of the tail field strength and, thus, it is improbable that it could account for the full 17 γ reduction required.

The second current system, whose existence we can only hypothesize at present, flows counter to the Chapman-Ferraro current on the magnetopause and in the outer magnetosphere, and is joined to the ionosphere along field lines. This system should be roughly centered on noon and flow eastward in the ionosphere (Coroniti and Kennel, 1973; Holzer and Reid, 1973; Reid and Holzer, 1973). Figure 5 shows schematically this current system. In the ionosphere Pedersen currents flow eastward parallel to the auroral oval wherever the convective flow is poleward, but the con-

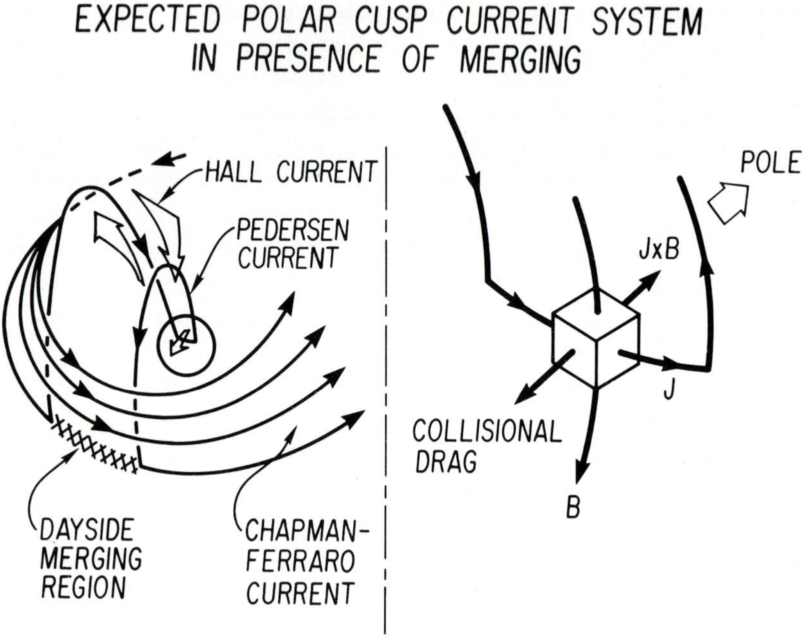

Fig. 5. Dayside polar cusp current systems associated with merging. The Pedersen current system may flow wherever there is flow poleward.

vective flow is opposed by collisions with the neutral atmosphere in the ionosphere. The **J** × **B** force associated with the Pedersen current opposes the ionospheric drag, producing the poleward flow, which is expected as merging proceeds, because merged lines are being 'dragged' tailwards and closed field lines are flowing to the dayside to replace these lines. On the magnetopause, the Chapman-Ferraro current is expected to be reduced in the merging region. In the magnetosphere, pressure gradients develop opposing the ionospheric drag on convection. Pressure gradient currents then flow westward in the magnetosphere and field-aligned currents down in the morning and up in the afternoon complete the circuit. The net effect of this system of currents is to reduce the magnetic field inside the dayside magnetosphere. We also expect Hall currents to flow in the polar cusp ionosphere, and these could also close in the magnetosphere along field lines.

Finally, we note that the magnetospheric field strength at OGO 5 was not constant as the boundary moved inwards despite the constancy of the dynamic pressure but rather increased by about 20%. This fact need not violate the pressure balance condition either. The magnetospheric magnetic field is proportional to the square root of the solar wind pressure times the cosine of the angle between the boundary normal and the solar wind flow (Willis, 1971). As the direction of the boundary normal changes, the field strength changes. If we assume a hemispherical boundary, the motion of OGO 5 closer to noon during the pass would alone have caused an increase in the field strength by 8%. An additional 10° shift in the direction of the normal towards the solar direction, i.e., from 40° to 30° from the Sun, as a result of merging could account for the rest of the observed increase in the field strength. Thus, the inward displacement of the magnetopause can be interpreted in terms of a new steady state which is consistent with the requirement of pressure balance.

In summary, the erosion of the magnetopause observed by OGO 5 on March 27, 1968 is easily understandable in terms of simple magnetospheric current systems which arise due to tangential stresses on the magnetopause, and which are not included in the usual models of magnetopause position and shape. While we have studied only this one clear example of erosion in the OGO-5 data, the effect of the southward interplanetary field on magnetopause position has been confirmed by a statistical study of Fairfield (1971) and observations of polar cusp motion in response to a southward interplanetary field (Russell *et al.*, 1971; Burch *et al.*, 1973).

3. Oscillatory Motions

Figure 1 illustrates clearly that the magnetopause often oscillates about its mean position with a period of about 10 min. This oscillation may represent the transient response of the dayside field aligned current system described above (Reid and Holzer, 1973). Occurrences of such oscillations have been used to derive amplitudes and velocities of the magnetopause motion (Holzer *et al.*, 1966; Heppner *et al.*, 1967). More rapid oscillations of the magnetopause also occur. Figure 6 shows a 9-second excursion into the magnetosheath projected into a plane perpendicular to the boundary (Aubry *et al.*, 1971). The Sun is at the top of the diagram and the nose of the magnetosphere is to the left. The projection of the magnetic field is plotted starting at point '*a*' and moving to the left one point each second. The circled numbers give the component of the field normal to the plane of the figure. The magnetopause is crossed at points *b* and *c*. The field behavior through this region is consistent with the sketch of the boundary and field line projections on either side of the boundary. This may be viewed as a snapshot of the boundary at one instant of time. Movement of this pattern with the magnetosheath flow is consistent with the features of this short section of data. We can put dimensions on this picture by assuming that the magnetosheath is flowing at 200 km s^{-1}. Then, the distance from *b* to *c* is 1800 km; and the thickness of the boundary is of the order of 200 km, or about 3 proton Larmor radii. Finally, we note that observations by Ledley (1971) and by Kaufmann and Konradi (1973) support this view of tailward moving ripples on the boundary.

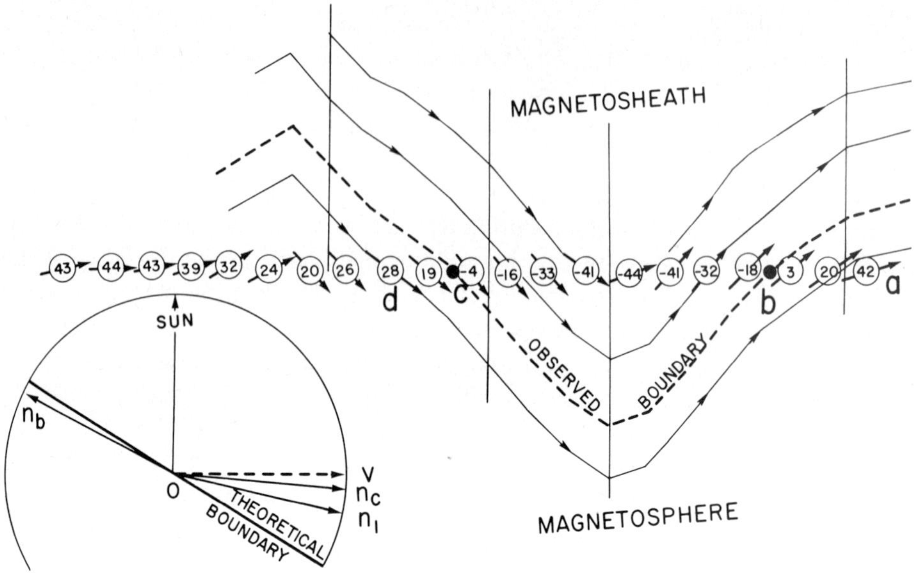

Fig. 6. Inferred boundary shape at 1860 UT on March 27, 1968. The circled numbers give the one second average field perpendicular to the plane of the figure. The short arrows give the projected field in the plane. The insert gives relevant directions in solar magnetospheric coordinates.

4. Magnetic Structure

Early Explorer 12 measurements of the magnetopause structure were interpreted as indicating basically two types of magnetopauses: a tangential discontinuity type in which there was no component of the magnetic field normal to the boundary and in which the magnetic field had a null in the center of the current sheet and a rotational type with a normal component and little change in field magnitude through the boundary. As Aubry *et al.* (1971) have pointed out, the magnetohydrodynamic solution only specifies the field change from one side of the boundary to the other and not the internal structure of the boundary.

The internal structure of the magnetopause is very varied. Figure 7 shows a hodogram of the magnetic field in the plane of the boundary for a crossing of the rotational type. Figure 8 illustrates a hybrid signature, in which the boundary has two portions in which the field magnitude changes without a change in direction (M to N and P to B) and a portion in which the change is mainly rotational, with little change in magnitude. Similar results to these have been obtained recently in a more extensive analysis of OGO-5 magnetopause crossings by Sonnerup *et al.* (1973).

One interesting feature that is occasionally observed with high time resolution data (sample rates of the order of one per second or greater) is a local maximum in the field strength on the magnetospheric side of the boundary (Neugebauer *et al.*, 1973). This phenomenon is observed for both directions of motion of the boundary and has only been observed on magnetospheric, as opposed to tail, field lines. This

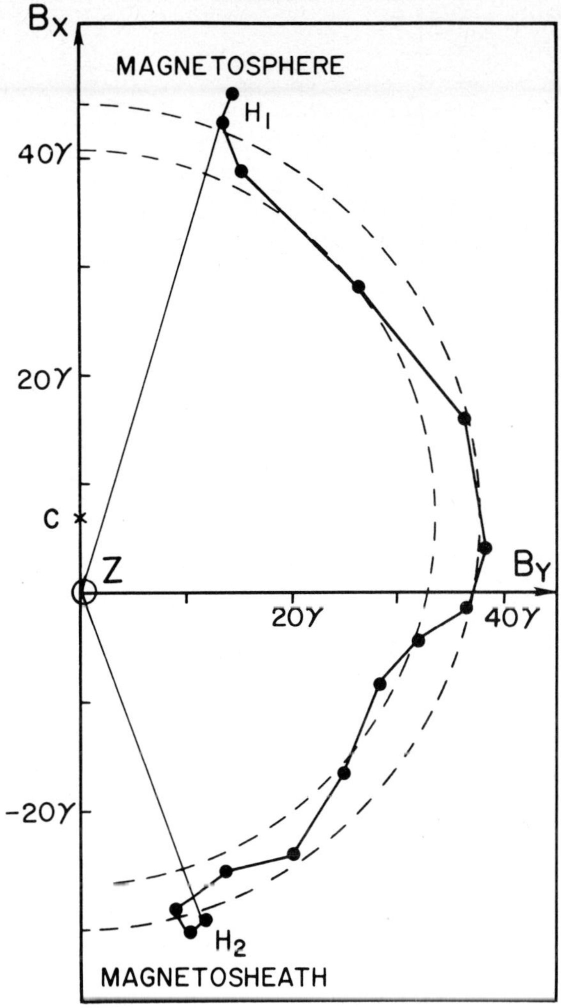

Fig. 7. Hodogram of one second averages of the magnetic field through the magnetopause at 1745 UT on March 27, 1968.

maximum is possibly a diamagnetic effect. Other complications in the boundary structure include rotations lasting much longer than gradients in the field strength, and rotations and field strength gradients occurring at different locations (Ogilvie *et al.*, 1971; Neugebauer *et al.*, 1973).

5. Ion Flux Structure

In certain regions of the magnetopause the solar oriented Faraday cup of the JPL plasma experiment on OGO 5 is directed into the magnetosheath flow. This makes possible correlated studies of the ion flux and the magnetic field through the mag-

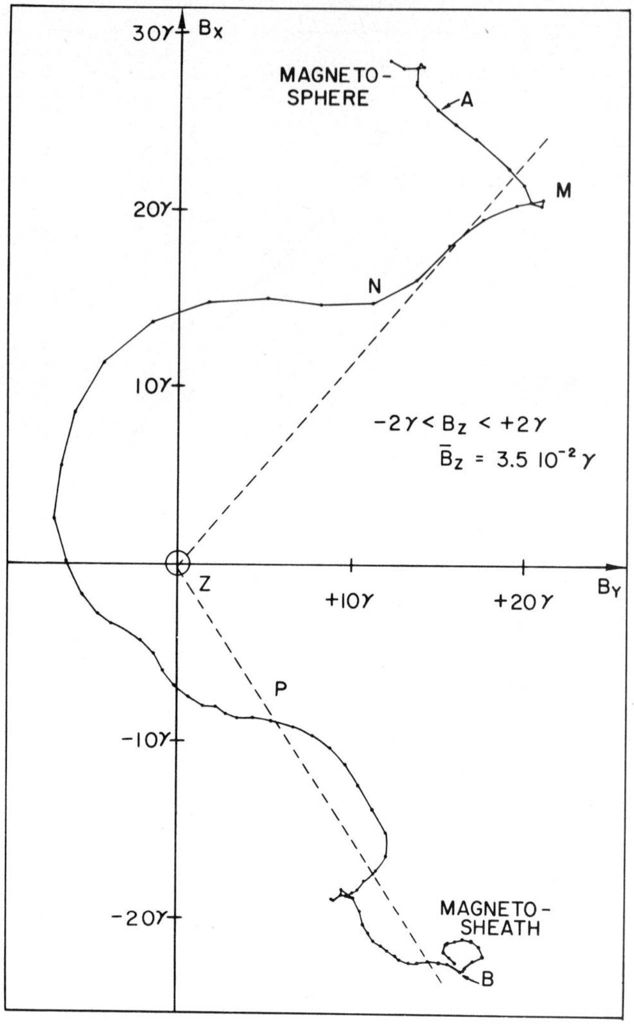

Fig. 8. Hodogram of 18 ms magnetic field data through the magnetopause at 1817 UT on March 27, 1968.

netopause. A variety of magnetopause crossings have been studied (Neugebauer *et al.*, 1973). We will briefly summarize here the properties of one crossing which had a particularly simple structure.

Figure 9 shows a plot of the total ion flux, the magnetic field strength and the longitude and latitude on the field in solar magnetospheric coordinates. The ion flux began to rise above background as soon as the magnetic field began to decrease below its magnetospheric value. Ion flux data gaps occurred when the instrument was periodically turned off. The solar magnetospheric position of OGO 5 was $(-10.0, -14.5, +12.6)\ R_E$ and the magnetic field was tail-like and pointed towards the Sun. The measured magnetosheath flow velocity was 420 ± 50 km. As sketched at the bottom

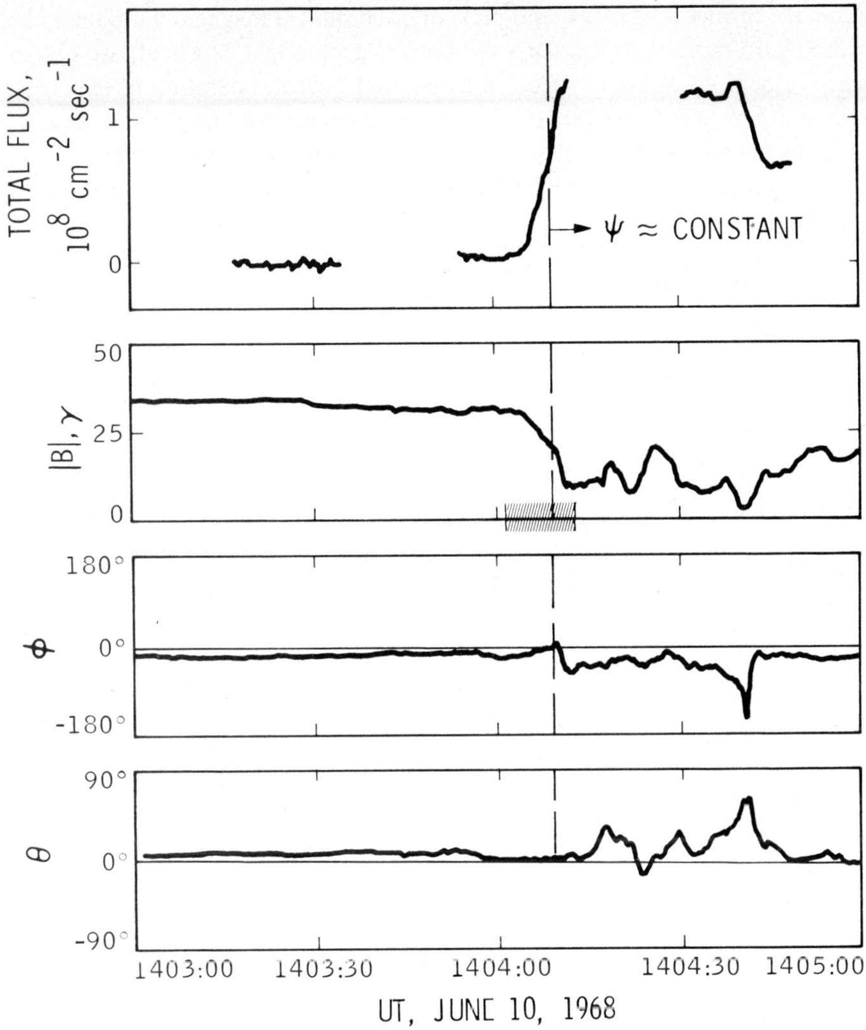

Fig. 9. Variation of charge flux, magnetic field strength and field direction with time as the magnetopause moved inward past OGO 5 near 1404 UT, June 10, 1968.

of Figure 10, the central collector of the Faraday cup is divided into three 120° sectors, the currents to each of which are plotted on the top panel. The ratio of the currents in the three detectors provides a measure of the direction of flow in the plane of the detector. After 1404:09 UT, marked with a dashed line in the figure, this direction remained roughly constant. At 1404:11, the flow direction was $13 \pm 3°$ from the Sun and its projection in the *Y-Z* plane $11 \pm 15°$ north of the solar-magnetospheric *y* axis. The normal to the magnetopause, as derived from the magnetic field data, was (0.443, −0.773, 0.453) in solar magnetospheric coordinates. The component of velocity along the shock normal was therefore 245 km s^{-1}. Either the flow in this region was not tangential to the boundary or the boundary was moving inwards at 245 km s^{-1}. If we assume the former is possible, and pick an often-quoted magnetopause thickness of 1 proton gyro radius (200 km for a 420 km s^{-1} proton in a 22 γ field), we obtain an inward velocity of 28 km s^{-1} from the observed 7-second duration of the crossing, which is not an unreasonable velocity for the crossing. Alternatively, if we use 245 km s^{-1} as the boundary velocity, the boundary is 9 proton gyroradii thick. Comparable thicknesses have been observed even closer to the nose (Kaufmann and Konradi,

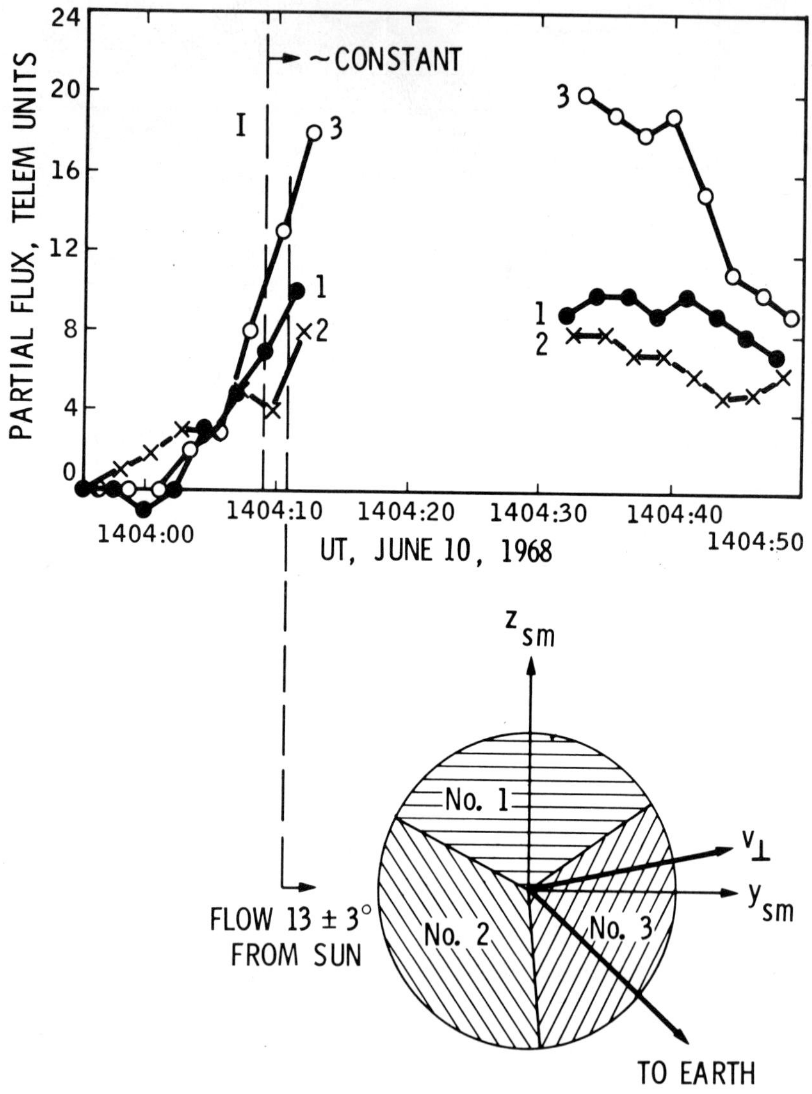

Fig. 10. Variation of the partial fluxes to each of the three sectors of the OGO Faraday cup at the magnetopause crossing shown in Figure 9. The bottom of the figure shows the orientation of each of the collector sectors as seen from the Sun, together with the projections of the plasma velocity vector and the satellite-Earth direction onto the plane of the detector.

1973). Thus, it is quite probable that the boundary was thick ($\sim$9 proton gyro radii) and that there was a rapid inward motion at this time.

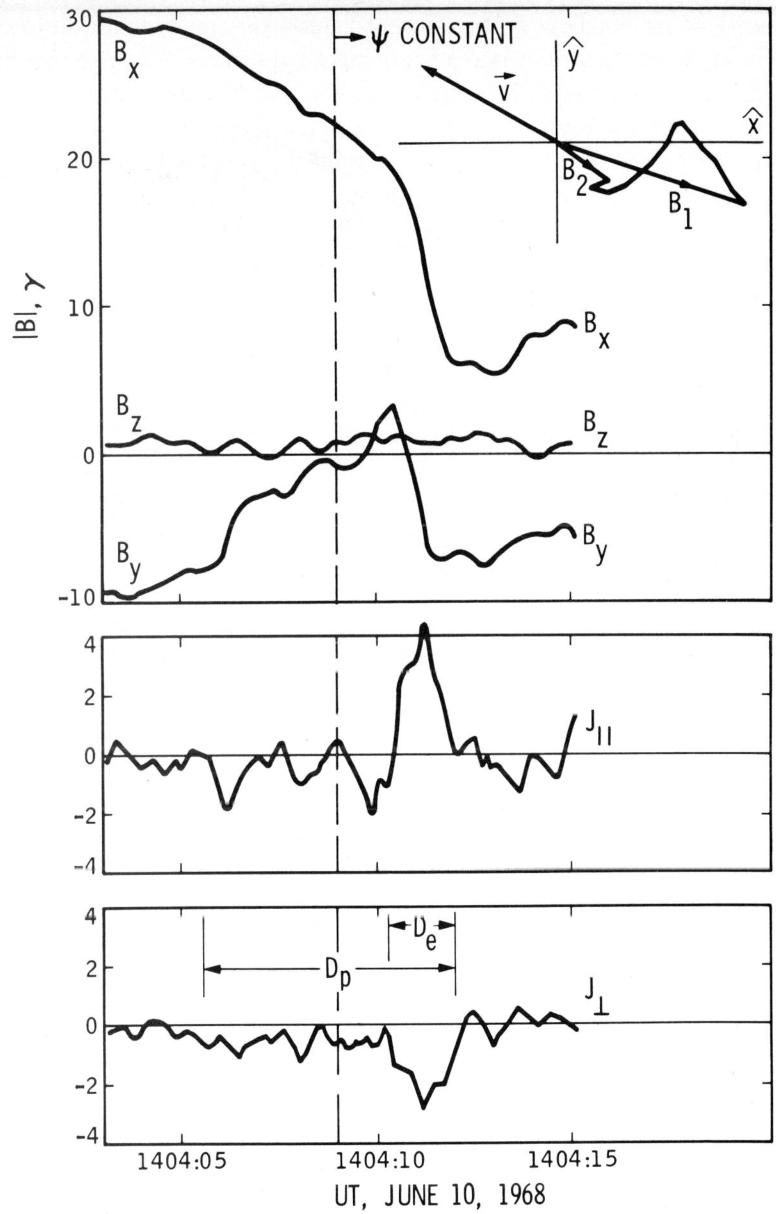

Fig. 11. Variation of the components of the magnetic field in the principal axis coordinate system for magnetopause crossing shown in Figure 9. The current parallel to the instantaneous field, $J_{\parallel}$, and that perpendicular, $J_{\perp}$, as calculated from the field variations are also shown. The inset at the upper right shows the trace of the tip of the magnetic field vector and the projection of the flow velocity onto the plane of the magnetopause.

Figure 11 shows the magnetic field in the magnetopause coordinate system; x and y are in the plane of the boundary with x along the direction of largest charge. In a planar approximation, i.e., variations only in the z direction, we can convert the observed slopes of the magnetic field to currents in arbitrary units. In the bottom panel we have rotated these currents into a component parallel and a component perpendicular to the instantaneous field direction. The parallel current consists of two components: a broad negative current on the magnetospheric side of the boundary and a narrower positive current on the magnetosheath side. Further, this current sums to zero. The perpendicular current is negative everywhere.

This boundary signature is consistent with classical theory, with the regions labeled D_p and D_e being the turning regions of the protons and electrons respectively. Figure 12 shows schematically our interpretation of this crossing. The top panel

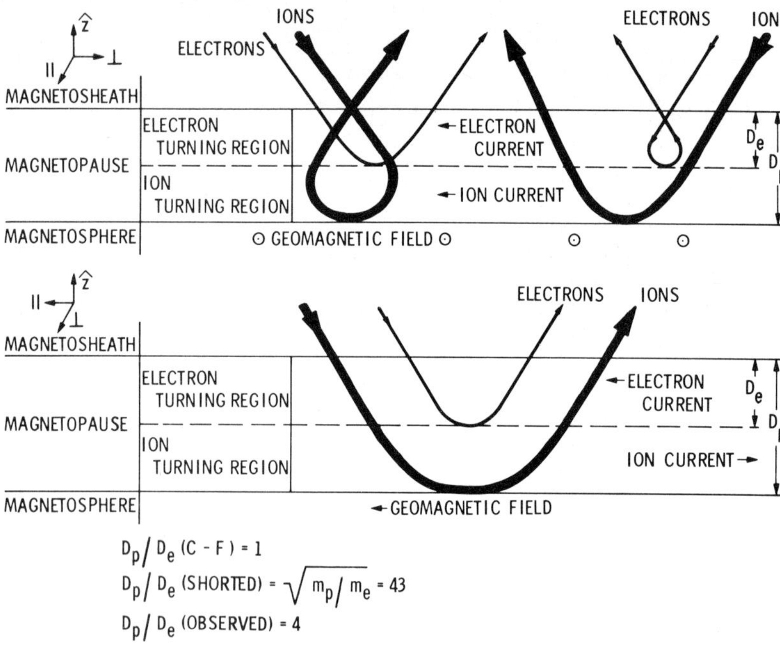

Fig. 12. An elementary magnetopause model showing ion and electron trajectories and the currents they create. The top panel shows the plane perpendicular to the magnetic field. The bottom panel shows the plane containing the boundary normal and the geomagnetic field.

shows the projection perpendicular to the geomagnetic field. Electrons and protons incident from either left or right cause a negative current throughout the boundary layer. The bottom panel shows the cross section containing the field and the boundary normal. Here electrons and protons cause currents in opposite directions along the magnetic field.

The ratio of the proton to electron turning distance is 4, as compared to unity for an unneutralized boundary with its full charge separation electric field or a ratio of 43 for a completely neutralized boundary. A formula given by Ferraro and Davies

(1968) allows us to calculate that the charge density is less than 7×10^{-5} of what it would be for an unneutralized charge-separation sheath boundary. While this simple structure permits a relatively easy comparison with theory we note that many crossings do not exhibit this simple form, and only some of the differences may be explained in terms of temporal variations.

6. Noise

A large variety of wave modes may be generated in the vicinity of the magnetopause at ULF frequencies. These wave modes commonly reveal their presence by causing multiple boundary crossings. In this section, we examine only ELF and VLF noise. Alpers (1971) noted the possible presence of the electron cyclotron instability at the magnetopause. Further, we might expect current driven instabilities similar to those seen in the bow shock (Fredricks *et al.*, 1970) and in the polar cusp (Fredricks *et al.*, 1973). However, neither electromagnetic ELF nor electrostatic VLF emissions specific to the magnetopause are common. We have examined the records of both the UCLA/JPL search coil magnetometer and the TRW electric field experiment for all crossings in which we studied the ion flux profile, and most of them are quiet. For example, no emissions were detected by either instrument during the crossing discussed in the previous section.

An example of the emissions which have been observed is shown in Figure 13.

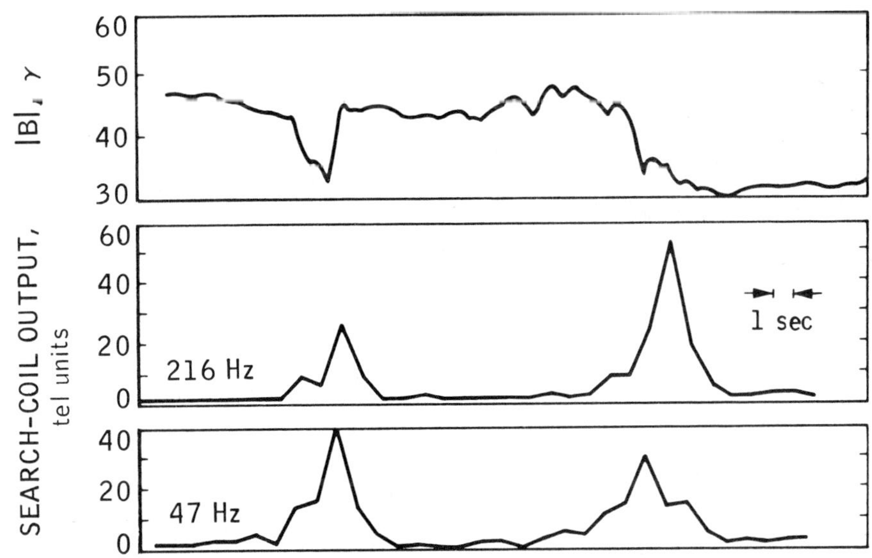

Fig. 13. Magnetic field strength and the response of 47 Hz and 216 Hz channels of the OGO-5 search coil magnetometer spectrum analyzer versus time during a magnetopause crossing at 1737 UT on December 10, 1968.

The output of two channels of the search-coil spectrum analyzer in telemetry units together with the total magnetic field at a complex magnetopause crossing are plotted. Strong emissions are present at the field strength gradients. Similar emissions have been observed on OGO 3 by Angerami (1970).

7. Summary

The magnetopause can move in response to changes in the solar wind pressure or in the north-south component of the interplanetary magnetic field. In this latter motion, pressure balance can be maintained while the magnetopause changes position and shape. As flux is transferred from the dayside to the tail, the tail currents strengthen and move towards the Earth, and dayside field aligned current systems apparently grow, reducing the magnetospheric magnetic field and permitting pressure balance to occur inward of its usual position. The boundary oscillates at a variety of frequencies and exhibits a variety of magnetic field changes through the boundary. One example of a simple boundary has been studied with both ion flux and magnetometer data. The ratio of the proton turning radius to the electron gyro radius was 4, and the boundary may have been 9 gyro radii thick. A study of ELF electromagnetic and VLF electrostatic data reveals the occasional presence of electromagnetic emissions in the field gradients of the magnetopause.

Acknowledgments

We are indebted to M. P. Aubry, R. W. Fredricks, R. L. McPherron, F. L. Scarf and E. J. Smith who have assisted us in various stages of this study. This work was supported at UCLA by the National Aeronautics and Space Administration under NASA grant NGL 05-007-004 and represents one phase of research carried out at the Jet Propulsion Laboratory, California Institute of Technology under contract NAS 7-100 sponsored by the National Aeronautics and Space Administration.

References

Akasofu, S. I.: 1972, *Scientific Design of Shuttle Auroral Observatory System*, vol. III, University of Alaska preprint, October.
Alpers, W.: 1971, *Astrophys. Space Sci.* **11**, 471.
Angerami, J. J.: 1970, *EOS* **51**, 384 (abstract).
Aubry, M. P., Russell, C. T., and Kivelson, M. G.: 1970, *J. Geophys. Res.* **75**, 7018.
Aubry, M. P , Kivelson, M. G., and Russell, C. T.: 1971, *J. Geophys. Res.* **76**, 1673.
Bahnsen, A., Iversen, I. B., and Ungstrup, E.: 1974, this volume, p. 167.
Binsack, Joseph H.: 1968, in Robert L. Carovillano, John F. McClay, and Henry R. Radoski (eds.), *Physics of the Magnetosphere*, D. Reidel Publ. Co., Dordrecht-Holland, p. 605.
Bridge, H., Egidi, A., Lazarus, A., Lyon, E., and Jacobson, L.: 1965, *Space Res.*, vol. 5, North-Holland Publ. Co., Amsterdam, p. 969.
Burch, J. L.: 1972, *J. Geophys. Res.* **77**, 6696.
Cahill, L. J. and Amazeen, P. G.: 1963, *J. Geophys. Res.* **68**, 1835.
Cahill, L. J. and Patel, V. L.: 1967, *Planetary Space Sci.* **15**, 997.
Coroniti, F. V. and Kennel, C. F.: 1973, *J. Geophys. Res.* **78**, 2837.

Cummings, W. D. and Coleman, Jr., P. J.: 1968, *J. Geophys. Res.* **73**, 5699.
Fairfield, D. H.: 1971, *J. Geophys. Res.* **76**, 6700.
Fairfield, D. H. and Ness, N. F.: 1967, *J. Geophys. Res.* **72**, 2379.
Fairfield, D. H. and Ness, N. F.: 1970, *J. Geophys. Res.* **75**, 7032.
Feldstein, Y. I.: 1970, E. R. Dyer (ed.), *Solar Terrestrial Physics*, 1970, Part III, D. Reidel Publ. Co., Dordrecht-Holland, p. 152.
Ferraro, V. C. A. and Davies, C. M.: 1968, *J. Geophys. Res.* **73**, 3606.
Frank, L. A. and Van Allen, J. A., 1964, *J. Geophys. Res.* **69**, 4923.
Fredricks, R. W., Coroniti, F. V., Kennel, C. F., and Scarf, F. L.: 1970, *Phys. Rev. Letters* **24**, 994.
Fredricks, R. W. Scarf, F. L., and Russell, C. T.: 1973, *J. Geophys. Res.* **78**, 2133.
Freeman, J. W. Jr.: 1964, *J. Geophys. Res.* **69**, 1691.
Gosling, J. T., Asbridge, J. R., Bame, S. J., and Strong, I. B.: 1967, *J. Geophys. Res.* **72**, 101.
Heppner, J. P., Sugiura, M., Skillman, T. L., Ledley, B. G., and Campbell, M.: 1967, *J. Geophys. Res.* **72**, 5417.
Holzer, T. E. and Reid, G. C.: 1973, *EOS* **54**, 414 (abstract).
Holzer, R. E., McLeod, M. G., and Smith, E. J.: 1966, *J. Geophys. Res.* **71**, 1481.
Kaufmann, R. L. and Konradi, A.: 1969, *J. Geophys. Res.* **74**, 3609.
Kaufmann, R. L. and Konradi, A.: 1973, *J. Geophys. Res.* **78**, 6549.
Ledley, B. G.: 1972, *J. Geophys. Res.* **76**, 6736.
McPherron, R. L.: 1970, *J. Geophys. Res.* **75**, 5592.
McPherron, R. L.: 1973, unpublished manuscript.
Meng, C.-I.: 1970, *J. Geophys. Res.* **75**, 3252.
Ness, N. F.: 1967, in J. W. King and W. S. Newman (eds.), *Solar Terrestrial Physics*, Academic Press, London, p. 57.
Ness, N. F., Scearce, C. S., and Seek, J. B.: 1964, *J. Geophys. Res.* **69**, 3531.
Neugebauer, M., Russell, C. T., and Smith, E. J.: 1973, *J. Geophys. Res.* (in press).
Ogilvie, K. W., Scudder, J. D., and Sugiura, M.: 1971, *J. Geophys. Res.* **76**, 3574.
Olson, W. P.: 1969, *J. Geophys. Res.* **74**, 5642.
Patel, V. L. and Dessler, A. J.: 1966, *J. Geophys. Res.* **71**, 1940.
Reid, G. C. and Holzer, T. E.: 1973, *EOS* **54**, 414 (abstract).
Russell, C. T. and McPherron, R. L.: 1973, *Space Sci. Rev.* (in press).
Russell, C. T., Chappell, C. R., Montgomery, M. D., Neugebauer, M., and Scarf, F. L.: 1971, *J. Geophys. Res.* **76**, 6743.
Schield, M. A.: 1969, *J. Geophys. Res.* **75**, 1275.
Sonnerup, B. U. O.: 1971, *J. Geophys. Res.* **76**, 6717.
Sonnerup, B. U. O. and Cahill, L. J. Jr.: 1967, *J. Geophys. Res.* **72**, 171.
Sonnerup, B. U. O. and Cahill, L. J. Jr.: 1968, *J. Geophys. Res.* **73**, 1757.
Sonnerup, B. U. O., Pierce, D. K., and Ledley, B. G.: 1973, *EOS* **54**, 428 (abstract).
Unti, T. and Atkinson, G.: 1968, *J. Geophys. Res.* **73**, 7319.
Willis, D. M.: 1971, *Rev. Geophys. Space Phys.* **9**, 953.
Wolfe, J. H., Silva, R. W., and Meyers, M. A.: 1966, *J. Geophys. Res.* **71**, 1319.

ENERGETIC ELECTRONS AT THE MAGNETOPAUSE

V. DOMINGO, D. E. PAGE, and K.-P. WENZEL

Space Science Dept. (ESLAB), European Space Research and Technology Centre, Noordwijk, The Netherlands

Abstract. We report observations of energetic and relativistic electrons made by the highly eccentric ESRO satellite HEOS 2 near the high-latitude tail magnetopause. There is generally a particle layer near the magnetopause, containing electrons up to about 1 MeV. The particle intensity shows a clear dependence on geomagnetic activity.

1. Introduction

Energetic electron spikes near the magnetopause have in the past been observed at low latitudes on the dayside and at large distances to at least about 60 R_E along the magnetotail (Meng and Anderson, 1970). First observations of energetic electrons and of the magnetic field at the high-latitude magnetopause nearly above the North pole obtained by the ESRO satellite HEOS 2 were reported by Hedgecock *et al.* (1973) and Page *et al.* (1973). In this report we present evidence of a more or less permanent layer of energetic electrons at the high-latitude magnetopause often extending to energies around 1 MeV.

2. Orbit

HEOS 2 was launched on 31 January 1972 into a highly eccentric polar orbit. The top part of Figure 1 shows a summer orbit when apogee (A) lies on the nightside of the Earth above the geomagnetic tail. Magnetopause (MP) and bow shock (BS) for the outbond and inbound passes are marked. The bottom part of the figure shows the progression of the apogee position (open circles) from launch through 1972 projected into the ecliptic plane. All outbound passes throughout the year cross the near Earth tail magnetopause almost above the North pole. The inbound passes cross the tail only during summer, whilst during winter the spacecraft enters the dayside magnetosphere at rather low magnetic latitudes directly into the outer radiation zone. In this report only magnetopause crossings through the tail are discussed.

3. Observations

Figure 2 shows the combined magnetic field and electron observations for an outbound magnetosheath pass on 26–27 February 1972. The three top plots show the magnitude $|B|$ and the polar angles Θ and φ of the magnetic field vector in GSE coordinates, kindly provided by Dr Hedgecock of Imperial College, London. The centre plots show ten-minute count rate averages of two of the electron channels (0.5–0.8 MeV, 0.8–1.1 MeV) (spin averaged) from the $dE/dX-E$ solid state detector telescope. The bottom channels show the count rates of a Geiger counter detecting electrons >40 keV, and the lowest threshold output of a solid state detector re-

D. E. Page (ed.), Correlated Interplanetary and Magnetospheric Observations, 159–165. *All Rights Reserved.*

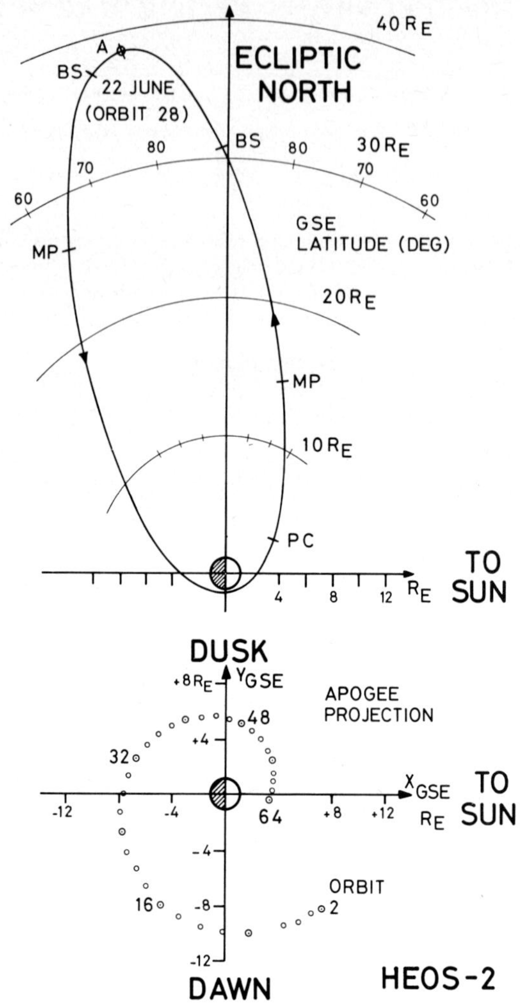

Fig. 1. HEOS-2 summer orbit (top) and progression of apogee from launch through 1972, projected into ecliptic plane (bottom).

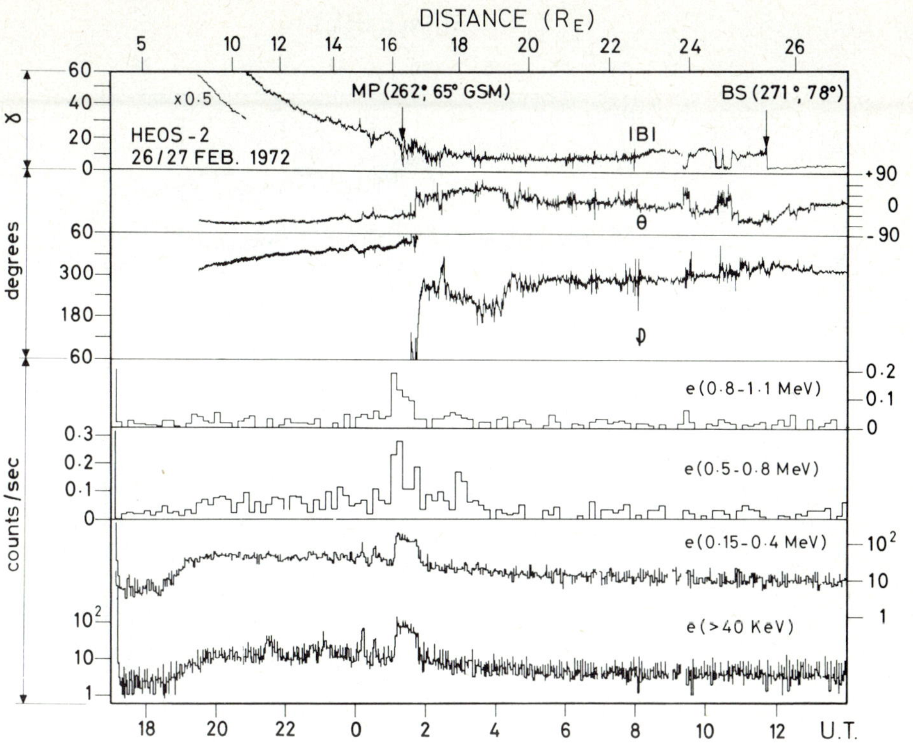

Fig. 2. Time history of magnetic field and electron observations for the outbound magnetosheath pass on 26–27 February 1972. The geocentric radial distance is given on the top scale.

sponding essentially to electrons between 0.15–0.4 MeV. Both detectors responding also to protons >0.5 MeV and 0.8–1.0 MeV respectively point perpendicular to the spin axis. The fluctuations are a measure of the anisotropy of the particles in the equatorial plane of the spacecraft, in this case in the ecliptic.

The magnetopause – taken to be located at the first major departure from the well ordered magnetic field – manifests itself on this pass first in a drop in field magnitude followed by a direction change about 20 min later. Just a few minutes prior to the field magnitude change all electron channels show an intensity increase of a factor 5–10. This increase lasts for about 1 h ($\sim 1\ R_E$) and is characterized by a rather sharp rise at the magnetosphere side and a slower drop towards the magnetosheath side. We see this intensity increase in all electron channels up to about 1 MeV. For the rest of the time in the magnetosheath the electron flux is about at its interplanetary level observed after a series of bow shock crossings around 25 R_E. This outbound pass is typical for a quiet day ($K_p = 0^-$) with low to moderate electron fluxes.

Figure 3 presents the outbound magnetosheath pass on a day of moderately large magnetic activity ($K_p = 3^-$). The magnetic field in the magnetosheath is rather disturbed. The electron flux increase starting just prior to the change in the magnetic field at 0330 UT on 3 March 1972 is more dramatic on this crossing than on the

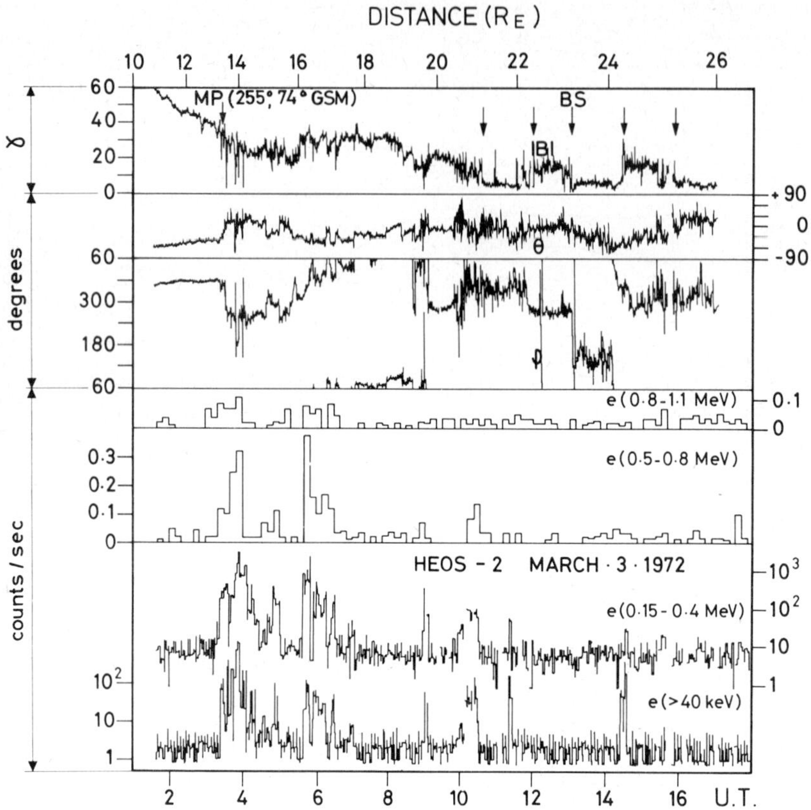

Fig. 3. Same as Figure 2 for the outbound pass on 3 March 1972.

previous one, in particular at lower energies and lasts for about 3 h. Figure 4 shows the outbound magnetopause crossing on 1–2 November 1972, on a highly disturbed day ($K_p = 5^-/7$). The same electron channels as on the last figures are plotted, plus two higher energy channels (1.1–1.8, 1.8–3.5 MeV). The magnetopause (MP) derived from magnetic field data is marked by an arrow. The feature we would point out is the intense flux increase on all electron channels, starting near the magnetopause. The energetic electron flux increases by a factor $\sim 10^2$, the relativistic electrons increase less but are clearly present at energies above 1 MeV.

Figure 5 gives an example of an inbound pass through the dawn side of the tail about 12 R_E behind the Earth. The periods in the magnetotail can clearly be identified through the regular field structure. Multiple magnetopause crossings are observed (see arrows). Again, an increased electron flux near the magnetopause is seen.

4. Discussion

We find the increased electron flux near the magnetopause on almost all of the about 70 passages through the high-latitude magnetosphere we have analyzed so far. We

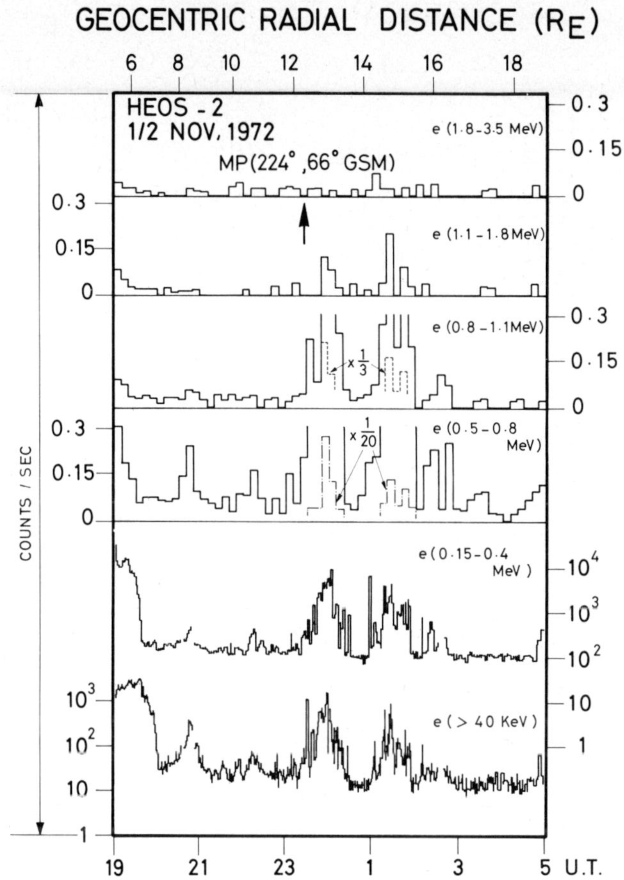

Fig. 4. Time history of electron observations for the outbound pass on 1–2 November 1972.

therefore interpret the observations as a layer of energetic electrons more or less permanently covering the high-latitude magnetopause. The layer is found on both sides of the magnetopause, if we define the latter as the first major deviation from the well ordered magnetotail. On most crossings, however, the layer extends further into the magnetosheath than into the magnetosphere. The typical thickness – assuming a stationary magnetopause – is 1–2 R_E. The electron spectrum is usually somewhat softer within the layer than outside. Preliminary analysis shows that the electrons flow essentially parallel to the magnetopause and preferably in antisolar direction down the tail.

The most interesting feature – and we think a novel feature – is that the electron fluxes near the magnetopause extend to energies of at least 1 MeV and occasionally even to about 2 MeV. This layer therefore includes electrons whose energies are considerably higher than one might have expected from the observations of Meng

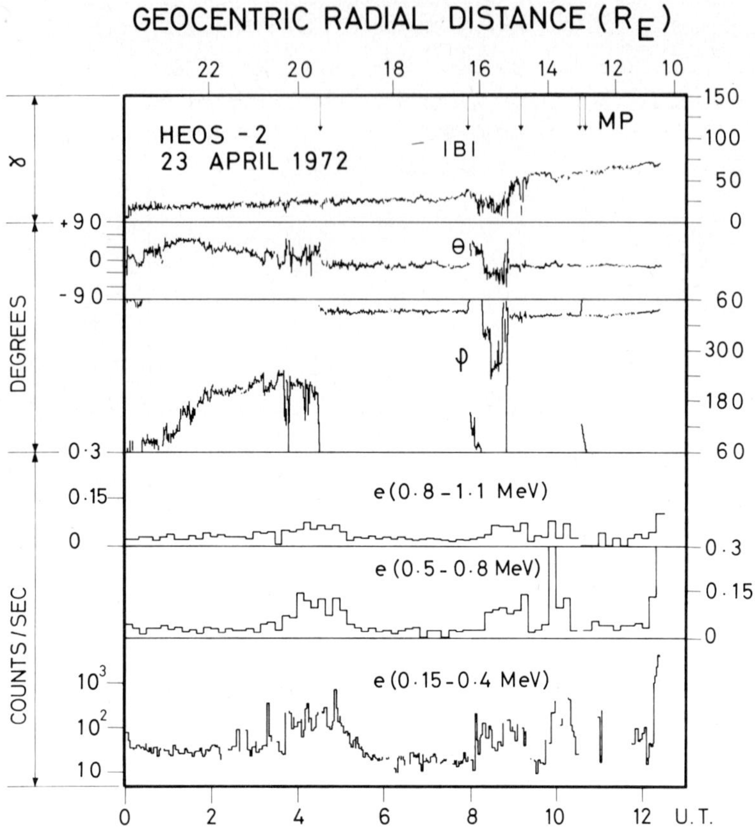

Fig. 5. Same as Figure 2 for the inbound pass on 23 April 1972.

and Anderson (1970) who reported of a layer of electrons above 40 keV near the magnetopause observed on IMP 3, IMP 1 and Explorer 35.

Figure 6 shows the correlation of the peak intensity of electrons in the magnetopause layer with geomagnetic activity. The peak count rate of 40 keV electrons is plotted against K_p. A positive correlation is clearly demonstrated.

What is the source for these electrons in the magnetopause layer? One possibility that has been mentioned by Meng and Anderson (1970) is the local energization of the low energy electrons distributed along the magnetopause by electromagnetic waves generated in the magnetosheath. Another possibility results from a model that describes the magnetopause as a tangential discontinuity. Calculations by Hill (1972), for example, show a mechanism to accelerate plasma resulting in a layer of energetic electrons at the magnetopause. Haerendel (private communication) suggested that electrons from the trapping zone on the dayside magnetosphere may be swept back onto the tail magnetosphere boundary due to reconnection of terrestrial and interplanetary magnetic fields taking place at the dayside. Further studies are clearly needed.

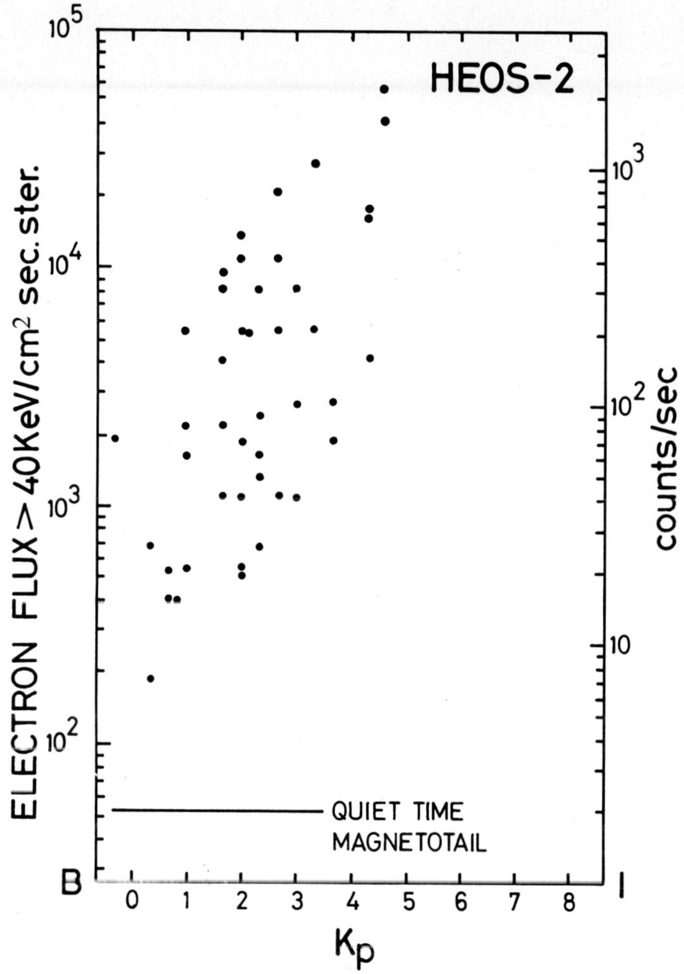

Fig. 6. Correlation of peak intensity of >40 keV electrons with K_p (B = background).

Acknowledgement

We should like to thank Dr P. C. Hedgecock, Imperial College, London, for making available to us the magnetic field data.

References

Hedgecock, P. C., Cerulli, P., Coletti, A., Egidi, A., Marconero, R., Domingo, V., Köhn, D., Page, D. E., Taylor, B. G., and Wenzel, K.-P.: 1973, *J. Geophys. Res.* **78**, 1715.

Hill, T. W.: 1972, Ph.D. thesis, Rice University, Nov.

Meng, C. I. and Anderson, K. A.: 1970, *J. Geophys. Res.* **75**, 1827.

Page, D. E., Domingo, V., Köhn, D., Taylor, B. G., Wenzel, K.-P., and Hedgecock, P. C.: 1973, *Space Res.* **13**, 631.

POLAR MAGNETOPAUSE DISTANCE

A. BAHNSEN, I. B. IVERSEN, and E. UNGSTRUP

Danish Space Research Institute, Lyngby, Denmark

Abstract. The distance from Earth to the magnetopause at very high latitudes has been measured on the HEOS-2 satellite. The wave experiment on this satellite measures magnetic and electric signals at five frequencies between 20 and 236 Hz. The transition from the magnetosphere into the sheath region is clearly indicated in these signals. Measurements over one half year show periodic variations in polar magnetopause distance around an average of 14.3 R_E. The measurements are compared to calculations of the magnetopause distance that take into account the solar wind pressure and the angle between the solar wind velocity and the Earths magnetic dipole. The periodic variations are predicted reasonably well by these calculations, but an average difference in the distance of approximately 20% exists between calculations and measurements.

The HEOS-2 satellite was launched on January 31, 1972 into a highly eccentric orbit. The orbital period is 5.2 days, and on the out-bound leg it passes over the north pole at a geocentric distance of approximately 15 R_E (Earth radii). The point of the orbit which corresponds to the average magnetopause distance, 14.3 R_E, has an ecliptic latitude of 80°, and is therefore always close to the z axis.

On this part of the orbit, the latitude varies only little with variations in the distance to the Earth, hence the satellite will always cross the magnetopause in a small region close to the North Pole. The satellite carries a complement of experiments for studying the plasma and magnetic field in the solar wind and the outer magnetosphere. One of these is an experiment to measure low frequency (20–236 Hz) magnetic and electric fields.

1. Observations

The transition from the magnetosphere into the sheath region (and through the bow shock) can be detected by the low frequency fields experiment. Data from a typical orbit arc shown on Figure 1. The magnetosheath region is characterized by large and strongly varying amplitudes of the magnetic signals, while the electric measurement shows a step-down in the noise level. Since this measurement is very sensitive to the surrounding plasma, this step is interpreted as an effect of the change in ambient plasma parameters. A detailed description of the experiment is found in Bahnsen *et al.* (1974).

We will define the distance of the magnetopause as the distance of the satellite at the time when the magnetic field amplitudes increase and/or the electric noise decreases. This determination of the magnetopause is in good agreement with the results of other experiments on the same satellite (Egidi, Hedgecock, private communication). We have plotted on Figure 2 the polar magnetopause distance r_m measured from the Earths centre in Earth radii, from February 5 to June 30, 1972. This is the top curve, marked '*a*'. In cases where the determination of r_m is ambigous, the range of possible r_m is indicated.

D. E. Page (ed.), Correlated Interplanetary and Magnetospheric Observations, 167–171.

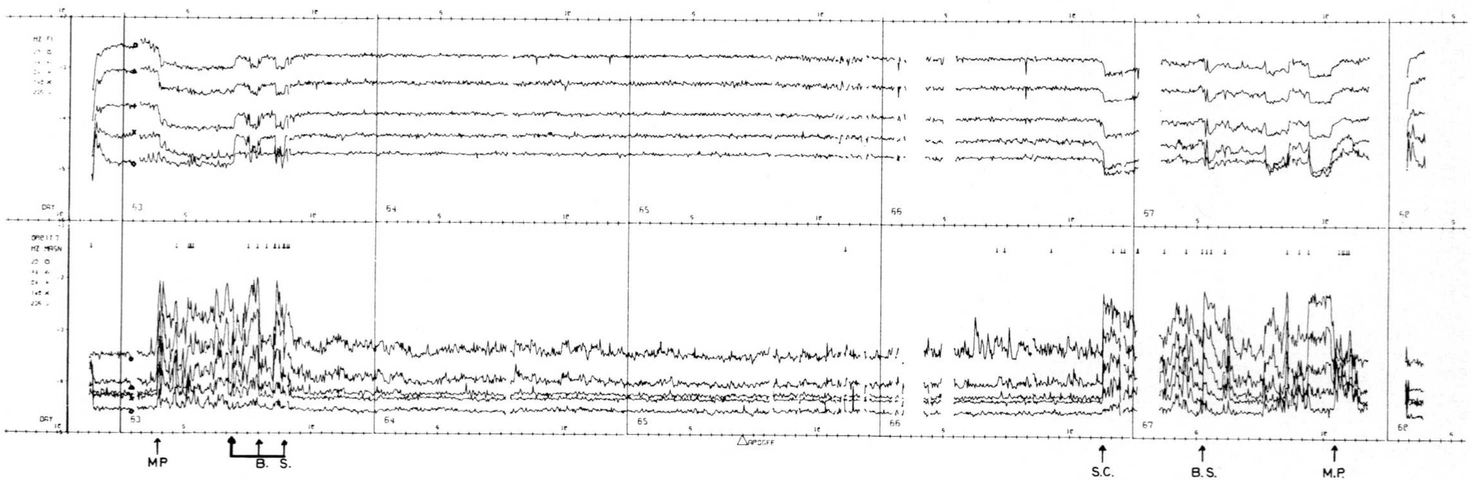

Fig. 1. Plot of amplitude of electric (top) and magnetic (bottom) signals at 20, 44, 84, 148 and 236 Hz. The abscissa is time and covers one orbit, the ordinate axes are logarithmic, for the electric signal going from 10^{-6} to 10^{-2} Vm^{-1} Hz$^{-1/2}$, for the magnetic signal from 10^{-5} to 10^{-1} γ Hz$^{-1/2}$ ($1\ \gamma = 10^{-5}$ G). Arrows indicate detection of magnetopause (M.P.), bow shock (B.S.) and a sudden commencement (S.C.).

A regular variation with a period of 26 days is evident in this curve. In the models of the solar terrestrial system used for calculations, the size and shape of the magnetopause depend mainly on two parameters: The solar wind pressure and the orientation of the Earths magnetic field, which for these calculations can be defined by a dipole of constant strength.

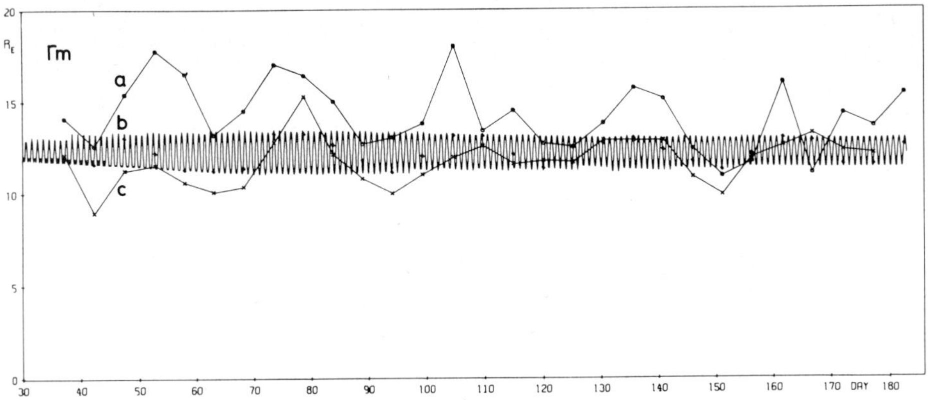

Fig. 2. Curve *a*: Experimental determination at 5.2 day intervals of magnetopause distance r_m. – Curve *b*: Daily variation of r_m calculated for a constant solar wind pressure. Points on this curve coincident with the measurements are indicated by ◇. – Curve *c*: Calculated values of r_m, using measured solar wind pressure.

The solar wind pressure p can be expected to show variations with the 27 day rotation period of the Sun. The dipole is rotating with the Earth, and will therefore have a varying angle μ to the solar wind direction, giving rise to a 24 h periodic variation of the magnetopause. Since the satellite has a 5.2 day orbital period, the Earth has rotated $0.2\ 360=72°$ between successive magnetopause crossings, and it takes 5 orbits = 26 days to cover one full rotation. There are therefore two reasons for having periodic variations in the data.

2. Magnetopause Model

In order to compare the experimental results with theory we have computed the expected distance r_m, taking into account effects of p and μ: The solar wind pressure has been included by using values of velocity and density measured on the VELA-5 satellite (*Solar Geophysical Data*, 1972). The effect of the variation in μ has been found from the work of Olson (1969). In this paper the shape of the magnetopause has been calculated for several values of the angle μ. Tables are presented of the coordinates of points on the magnetopause, in solar magnetosphere coordinate system r, θ, ϕ.

By means of these tables we have calculated the distance r_m at which the satellite would be expected to detect the magnetopause. The coordinates θ, ϕ used are of the point of the orbit where the satellite will be at 14.3 R_E, which is the average measured magnetopause distance. It should be pointed out that θ, ϕ vary quite slowly with r, on this part of the orbit.

By simple interpolation in the tables of Olson we find the daily variation in r_m; this is curve b on Figure 2. On this curve are marked points corresponding to the times that the magnetopause crossings were detected. These points then represent the values of r_m expected for a constant solar wind pressure.

Next, the effect of a varying solar wind pressure has been included according to the relation $r_m \propto p^{1/6}$. For the period of interest, 34 points of pressure were available. With the exception of a few points, the pressure varied smoothly over the whole period, making it feasible to introduce a varying solar wind pressure in the calculation of r_m for the whole period of interest. This results in curve c on Figure 2. This, then is the final result of the model calculations, to be compared with the experimental data.

3. Discussion

Comparing the two curves a and c we note the following points:

(1) The two curves have roughly the same shape, except for two cases, where irregular variations in the observed r_m are present, around day no. 110 and 160. From this we conclude that the dependence of the polar magnetopause distance on the dipole orientation is correctly represented by the model of Olson, provided that the actual solar wind pressure is taken into account.

(2) There is a displacement between curve a and c. The average difference is 2.5 R_E, or 22% of the calculated mean distance. Numerous investigations have been carried out in the past to determine the position and shape of the magnetopause. Fairfield (1971) was able to determine the shape of the magnetopause and the bow shock with good statistics, and hence make comparison with model calculations. The data presented by Fairfield cover latitudes from the ecliptic plane up to $\sim 60°$. The HEOS-2 data complement these data, since they are obtained only at high latitudes. More over Fairfield also made use of magnetopause model by Olson, which makes his discussion directly applicable to our results. Of particular interest here are three observations noted by Fairfield:

The measurements of the magnetopause distance near the subsolar point are in good agreement with theory.

The distance in the dawn-dusk meridian is generally larger than the model predicts.

The magnetopause distance is influenced by several factors such as solar wind parameters, dipole orientation and magnetospheric current systems. These combine to produce a large spread in the results.

By taking into account some of the factors that influence the magnetopause distance, we have been able to make a quantitative comparison between theory and experiment, which is in agreement with the observation made by Fairfield. We may therefore enlarge the statement made above to say that measurements made close to the dawn-dusk meridian indicate a larger magnetopause in this cross section than predicted by theory. The difference is $\sim 20\%$ at high latitudes, and is of the same order at lower latitudes.

Three factors which are neglected in most of the model calculations are given by Fairfield as reason for this discrepancy. They are: the presence of plasma in the magnetosphere, currents in the neutral sheet of the tail, and the presence of the bow shock. The region between the bow shock and the magnetopause is known to be highly turbulent, hence some of the flow energy is converted into thermal energy, and the pressure on the magnetopause is reduced. Measurements of particle flux on both sides of the polar magnetopause are necessary in order to determine the importance of these factors.

Acknowledgement

The HEOS-2 satellite was constructed, launched and operated under the responsibility of ESRO. The authors wish to express their appreciation of the effort made by ESRO to insure a successful mission. Many useful discussions with N. D'Angelo are gratefully acknowledged. The support of many staff-members of the Danish Space Research Institute in the construction and data reduction processes is very much appreciated.

References

Bahnsen, A., Iversen, I. B., Obayashi, H., Omø, K., Rasmussen, I. B., and Ungstrup, E.: 1974, Report DSRI 7301, Danish Space Research Institute.

Fairfield, D. H.: 1971, *J. Geophys. Res.* **76**, 6700.

Olson, W. P.: 1969, *J. Geophys. Res.* **74**, 5642.

Solar Geophysical Data, 1972 World Data Center A., Boulder, Colorado, U.S.A.

A STUDY OF MAGNETOSPHERIC BOUNDARIES BY THE SIMULTANEOUS OBSERVATIONS OF SOLAR AND MAGNETOSPHERIC PARTICLES

J. B. BLAKE and A. L. VAMPOLA

Space Physics Laboratory, The Aerospace Corporation Los Angeles, Calif., 90009, U.S.A.

1. Introduction

It long has been recognized that analysis of low-altitude solar-particle data can provide insight into the structure of the distant magnetosphere, the boundary with the interplanetary medium, and the dynamics of the energetic solar particles within this complex configuration of magnetic and electric fields (cf. reviews by Lanzerotti, 1972; Lanzerotti and Michel, 1972; Morfill and Scholer, 1973; Paulikas, 1973). Low energy solar protons and electrons, observed over the polar caps by low-altitude spacecraft, exhibit a very complex behavior and, unfortunately, the low-altitude data usually cannot be unequivocally interpreted. The experimental results to date have led to much heated discussion as to their interpretation.

In this paper low-altitude polar-cap observations are presented which were made during a period of rather intense precipitation of magnetospheric protons. The precipitating magnetospheric protons provided additional delineation of the magnetospheric topology. Furthermore, high-resolution proton and electron spectra and pitch angle distributions were obtained; these characteristics of the data greatly aided in their interpretation.

Extensive latitudinal structure was seen in the proton flux profiles in the region of closed field lines adjacent to the polar plateau; the protons were of both direct solar and magnetospheric origin. The structure in the proton profiles persisted well inside the stable trapping boundary of the outer-zone electrons. However solar protons did not appear to penetrate into the region of stable trapping of outer-zone protons of the *same* energy as the solar protons.

The observed local-time and latitudinal dependence of the proton energy spectra, intensities and pitch-angle distributions have some characteristics suggestive of the observations presented by Williams and Heuring (1973) and will be discussed in detail.

2. Satellite and Instrumentation

The measurements reported here were made by instruments flown aboard the satellite OV1-19 (1969-25C). The satellite was launched on 17 March 1969 into a polar orbit with an apogee of 5796 km, a perigee of 470 km and an inclination of 105°.

Electrons were measured in 24 differential energy channels with center energies from 53 keV to 5.1 MeV. A pair of 180° magnetic spectrometers employing silicon semi-conductor detectors was used. A discussion of the instrument design has been given by Vampola (1971).

D. E. Page (ed.), Correlated Interplanetary and Magnetospheric Observations, 173–186.

Two different proton spectrometers were employed. One used a single silicon semiconductor detector for energy analysis with an anticoincidence detector consisting of a plastic scintillator-photomultiplier tube combination. This instrument provided 12 differential channels from 185 keV to 2.5 MeV. A detailed description of this instrument has been published (Stevens *et al.*, 1970). The second proton spectrometer consisted of three silicon semi-conductor detectors in a telescopic array which in its low-resolution mode measured protons from 1.15 to 9.6 MeV in 3 channels.

Alpha particles were measured in 5 differential channels from 850 keV to 9.0 MeV by another three-element silicon semi-conductor detector telescope described in detail by Blake and Paulikas (1972).

All data channels were sampled one or more times per second. This data rate, coupled with the eight-second period of the satellite rotation, provided pitch-angle distributions. The pitch angles scanned in any spacecraft rotation depended upon the angle of the satellite spin vector with respect to the magnetic field vector, which changed throughout an orbit. This point is emphasized since it means that the instruments often did not scan the loss cone(s) throughout a polar pass. [If the satellite spin axis were normal to the orbital plane, then approximately all pitch angles would be scanned throughout the orbit. The OV1-19 was thus positioned at launch but perturbations quickly changed the spin axis-magnetic field orientation since the OV1-19 attitude was not actively controlled.]

However, even if the loss cones were not scanned, a normal trapped distribution ($J \propto \sin^n \alpha$) is distinguished readily from an isotropic pitch-angle distribution. Thus the stably-trapped outer-zone particles can be separated from the transient population (energy spectra provide additional evidence) without scanning the loss cone. What cannot be determined is whether particles are in the bounce loss cone since solar protons, unlike stably trapped particle distributions, exhibit rather uniform intensies outside of the loss cone with the result that no clue is given as to the loss cone intensity without direct scanning.

3. Observations

The data presented in this paper were taken between 2000 UT and 2130 UT on 14 April 1969 during a solar particle event which began ~1230 UT on 11 April 1969. At the time of the observations, a moderate magnetic disturbance was in progress with the values of D_{st} during this time period given in Table I.

Data are presented in Figure 1 for a south polar pass during the time interval from 75200 UT to 76400 UT (time in seconds) on 14 April 1969. Several electron and proton channels are shown. The figure consists of computer generated plots; no editing has been done with the exception of eliminating telemetry noise. The scanning in pitch-angle is seen most clearly in the periodic modulation of the electron countrates from 75200 UT to 75400 UT. The large scatter seen in the 230 keV and 353 keV proton channels at lower intensities is statistical in nature.

The trajectory of the spacecraft, given in the legend of Figure 1, shows that the

OV1-19 moved to larger L (higher invariant latitude) and reached the polar plateau, as delineated by the solar electrons, at dusk and at $L=6$. As the satellite moved onward towards the dayside, the L value (invariant latitude) increased very slowly. The OV1-19 moved off the polar plateau at $L \simeq 7$, at 1700 h MLT. The effect of local time upon the boundary of the polar plateau is illustrated nicely by the fact that the satellite

TABLE I
14 April 1969 magnetic index

Time (UT)	D_{st}
1900	7
2000	4
2100	−10
2200	−30
2300	−49
2400	−49

moved off the polar plateau while L was still increasing. Data acquisition terminated at 76450 UT (not shown) before the OV1-19 entered the outer zone. The polar plateau was delineated clearly by both the solar electrons and the lower energy solar protons. The boundaries of the solar plateau were so sharp that, within the one second sampling period of the two instruments, the electron fluxes declined and the proton fluxes increased.

Examination of the amplitude, spectra, and location of the peaks and other structure (Figure 1) in the proton profiles below the polar plateau revealed several features:

(1) The large peak at ~76000 UT near one boundary of the polar plateau had a substantially softer energy spectrum than the large peak near 75550 UT on the other side of the polar plateau. The softer peak disappeared at energies $\gtrsim 1$ MeV whereas the peak near 75550 UT remained visible until $\gtrsim 4$ MeV.

(2) Many small, narrow peaks were seen at the lower proton energies, especially near 75300 UT; there was a wealth of statistically significant fine structure.

(3) Since the position and amplitude of the structure varied with energy, it indicated that marked spectral variations in the proton fluxes occurred throughout the region below the polar plateau.

(4) The proton intensities in the peaks were at times more than an order-of-magnitude greater at the lower energies than the corresponding intensities over the polar plateau.

West and Vampola (1971) have shown that the electron fluxes over the polar plateau were of equal intensity to those in interplanetary space throughout this solar particle event and Fennell (1973) has shown that the proton fluxes over the central polar plateau were equal in intensity to those in interplanetary space. The fluxes in the large peaks, discussed above, were substantially larger than any uncertainty in the intercalibration of sensors onboard interplanetary and polar-orbiting spacecraft. (Such an intercalibration is based upon data comparisons over an extended time period during several solar events.) Furthermore, in the case of the lowest energy protons

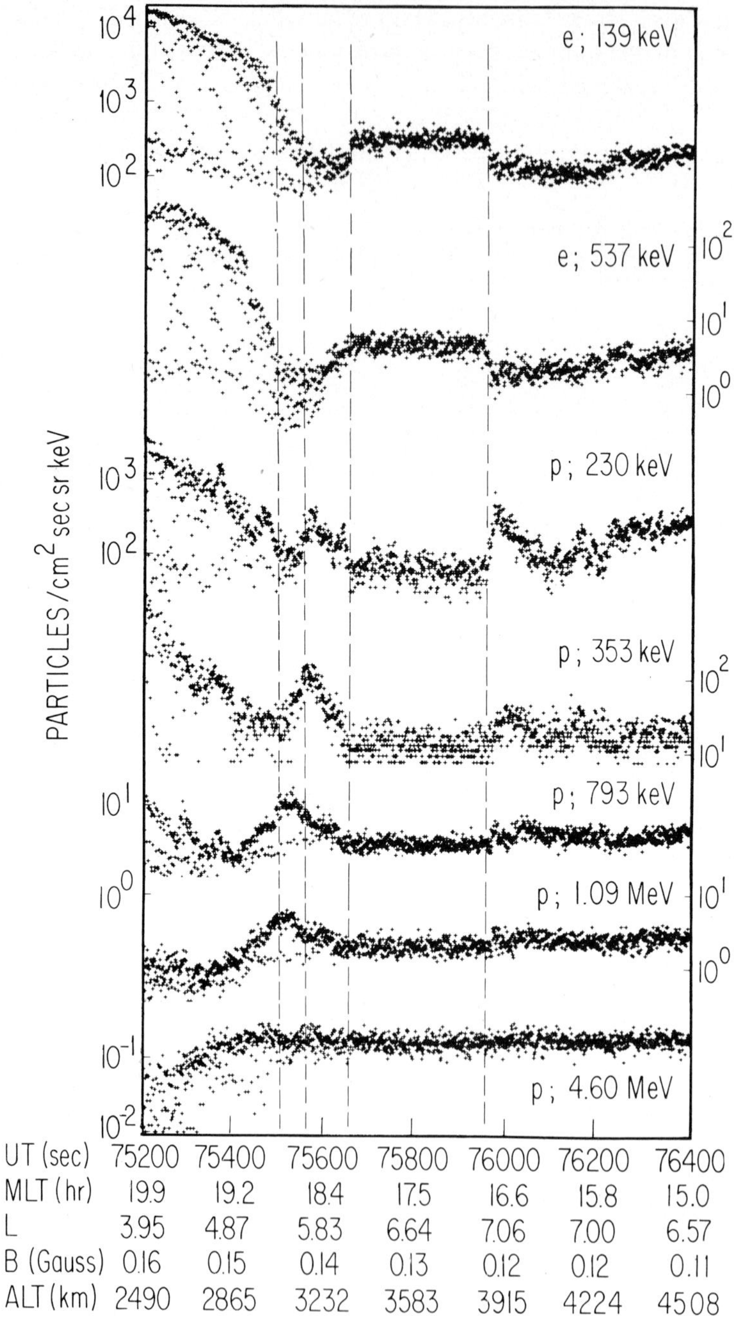

Fig. 1. Measurements of electrons and protons of several energies made between 75200 UT and 76400 UT on April 1969. At this time the OV1-19 spacecraft traversed the south polar cap. The figure consists of computer generated plots; no editing has been done with the exception of eliminating telemetry noise.

observed, the intensity in the most intense peaks was greater than the interplanetary intensity at any time during the event including the event maximum which occurred approximately 38 h before the observations discussed in this paper (Fennell, 1973). On the basis of the observations discussed above, it is clear that the intense fluxes of lower-energy protons seen at latitudes below the polar plateau are not of direct solar origin [the protons did not arrive from the sun in the vicinity of the earth at the observed energy].

Figure 2 gives pitch-angle distributions of representative proton and electron

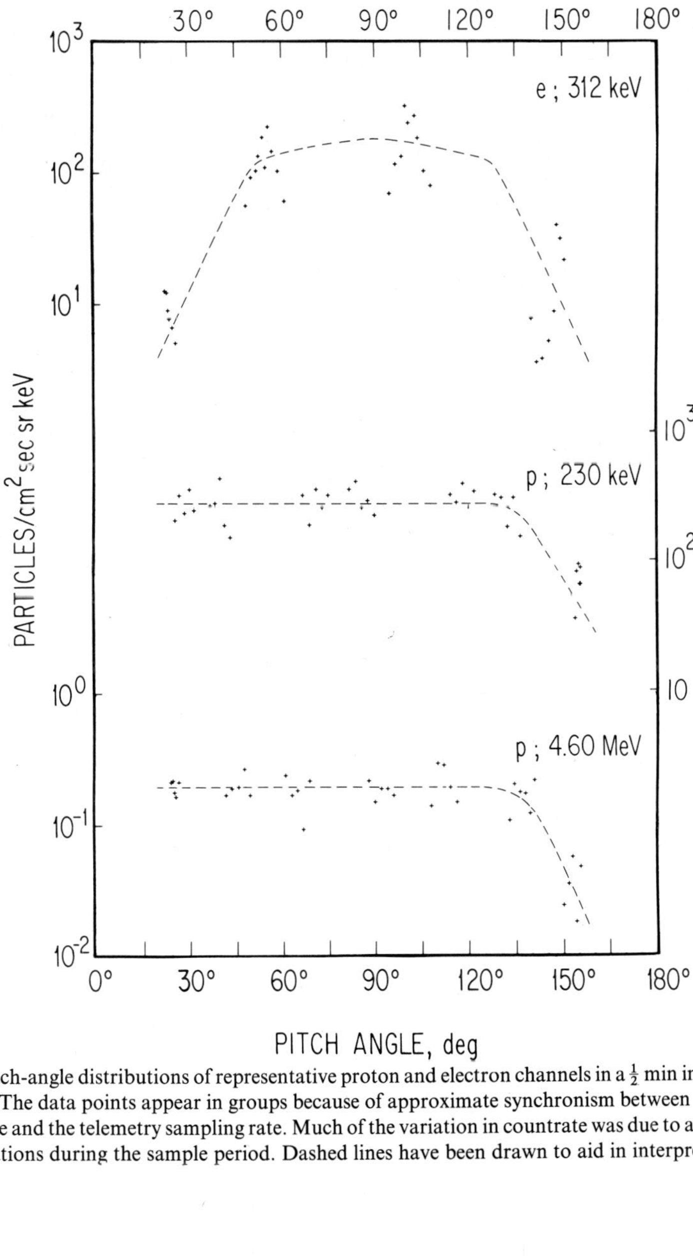

Fig. 2. Pitch-angle distributions of representative proton and electron channels in a $\frac{1}{2}$ min interval around 75470 UT. The data points appear in groups because of approximate synchronism between the spacecraft rotation rate and the telemetry sampling rate. Much of the variation in countrate was due to actual intensity variations during the sample period. Dashed lines have been drawn to aid in interpretation.

channels as observed around 75470 UT. The pitch-angle distribution in the largest proton peaks, for example around 76000 UT, are not shown because the loss cones were not well scanned at that time. The angle between the satellite spin axis and the magnetic field of the earth continually decreased during the polar traverse and, in addition, the satellite moved to higher altitude which reduced the size of the loss cones. As in the case of the polar profiles, this figure is a computer generated plot with no sorting or editing. The location of the data in clumps is due to beating between the satellite rotation rate and the data sampling rate; the sampling was not symmetric with respect to 90°. Much of the variation in intensity at a given pitch angle is due to actual intensity changes in the 0.5-min period used to generate the pitch-angle distribution. The pitch-angle distributions clearly show that the electrons have two loss cones and the protons only one; protons are precipitating into the atmosphere whereas the electrons are executing bounce motion. Examination of Figure 1 shows that these pitch-angle data were acquired well inside the outer-zone energetic-electron stable-trapping boundary. It should be noted that the electron countrates clearly showed bounce motion to ~75550 UT, although the proton pitch-angle distributions were uninformative due to the unfavorable satellite spin axis-magnetic field orientation.

Alpha-particle data were acquired during the polar traverse shown in Figure 1. The alpha-particle countrate was low due to the small geometric factor of the alpha-particle spectrometer and the low alpha-particle fluxes. However the alpha-particle fluxes were definitely higher at latitudes below the polar plateau and, in addition, the energy spectrum was much softer. The very low alpha-particle countrates preclude a quantitative comparison; however, the important point is that alpha particles, not of direct solar origin, were precipitating. Such observations are of significance in the study of the ionic composition of the radiation belts (Blake *et al.*, 1973; Fennell *et al.*, 1973) but will not be discussed further here.

Figure 3 gives proton spectra obtained over the polar plateau (open circles) and in the large low-energy proton peak observed below the polar plateau at ~75570 UT (solid circles). The straight lines are least-square fits to the data; the spectrum in the peak was divided into two components at the obvious break. The open triangles indicate the difference between the fluxes measured in the low-energy proton peak and the fluxes over the polar plateau. Above ~400 keV, the proton intensities in the region of the peak and over the polar plateau are very similar; the spectrum in the peak at low energy is exceedingly steep and bears no resemblance to that observed over the polar plateau or in interplanetary space (Fennell, 1973). Note that a spectrum at another time, say 75530 UT, would not be similar to that in the peak at 75570 UT or over the polar plateau. The observed latitudinal structure in the proton profiles results from both magnetospheric and solar particles.

Data are presented in Figure 4 of the solar profiles over the north polar cap during the time interval from 72200 UT to 73200 UT on 14 April 1969. This pass differs from the south polar cap pass shown in Figure 1 in local time (the satellite moved from morning past dawn to midnight during this traverse) and altitude. At the end

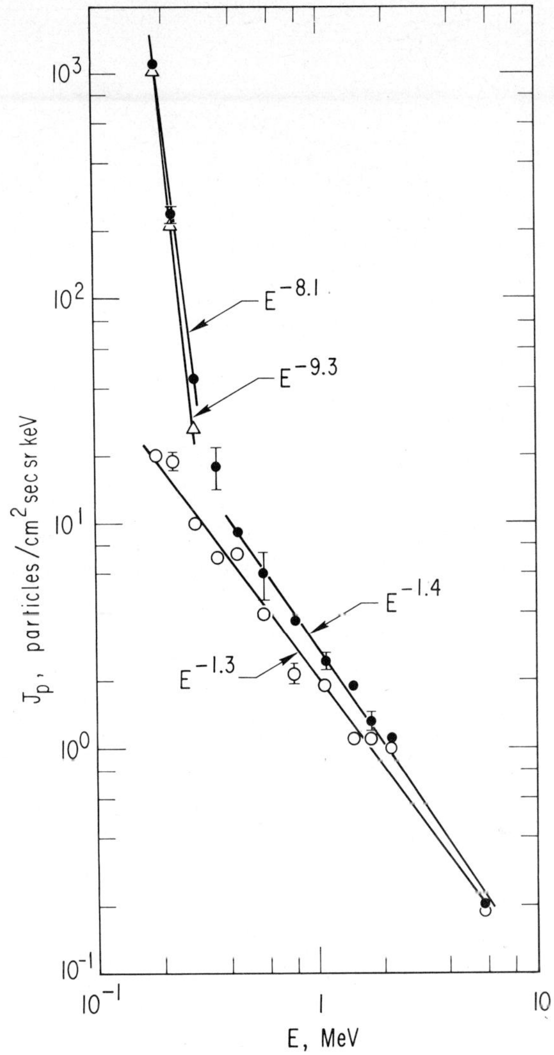

Fig. 3. Proton spectra obtained over the polar plateau (open circles) and in the large proton peak at ~75570 UT (solid circles). The straight lines are least-squares fits to the data. The spectrum in the peak was divided into two components at the obvious break. The open triangles represent the spectrum which results from subtracting the polar plateau spectrum from that observed in the peak at ~75570 UT.

of the pass (~73000 UT), the spacecraft was beneath the stable trapping region. The polar cap traverse shown in Figure 4 exhibits interesting differences and similarities with the pass given in Figure 1:

(1) No sign of precipating protons was seen on the morning side at the edge of the polar plateau at ~72650 UT. The plateau edge appeared 'normal' and quasi-trapping was seen at latitudes below the plateau edge. A small hump was seen in the lowest-

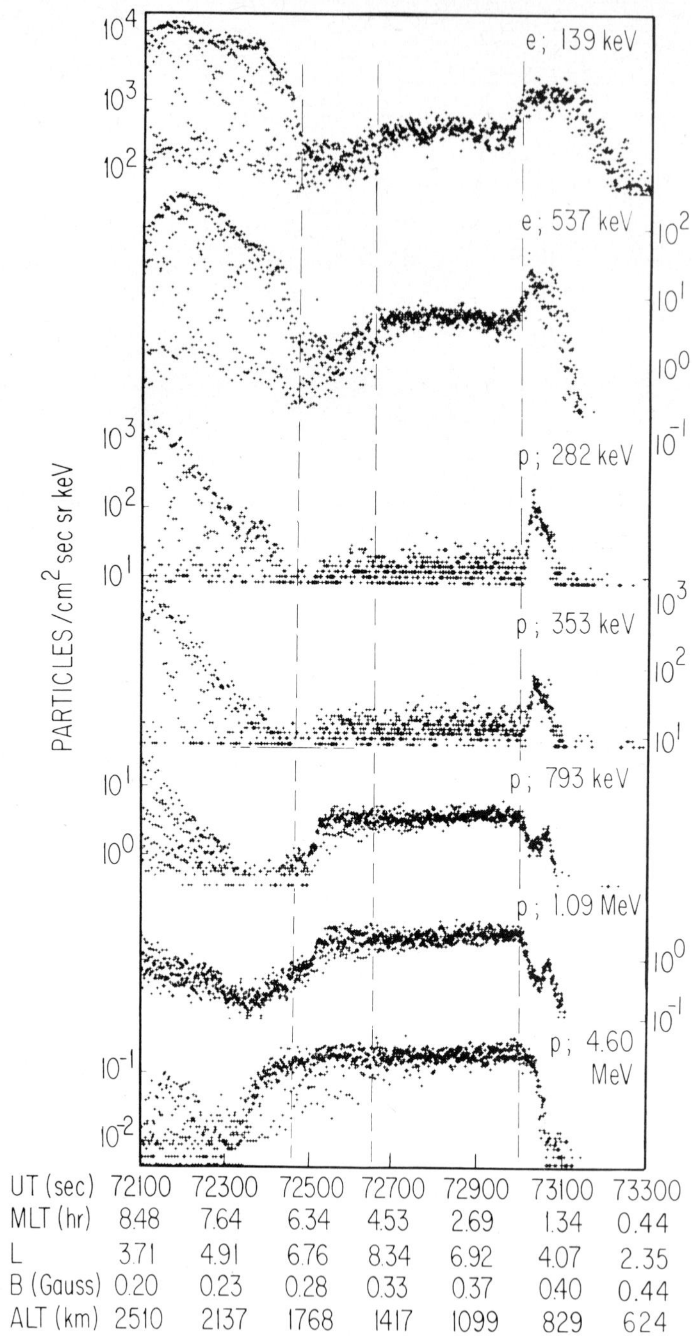

Fig. 4. The same type of plot as Figure 1 for a north polar cap traverse between 72 100 UT and 73 300 UT on 14 April 1969.

energy proton channels at $\sim$72350 UT. At that time the OV1-19 spacecraft was inside the stable trapping region as defined by the energetic electrons.

(2) On the morning side, in the quasi-trapping region as delineated by the solar electrons, the particle intensity relative to that over the polar plateau decreased with increasing particle energy in the case of electrons and increased with increasing energy in the case of protons.

(3) In contrast to the morning-side observations, enhancements of proton fluxes were seen at the midnight edge of the polar plateau. In the case of lower-energy protons the enhancements are similar in character to those seen at the evening edges of the polar plateau over the south polar cap approximately 40 min later (Figure 1). At energies above 1 MeV a peak was seen separated from the near midnight polar plateau cutoff by a region of very low fluxes.

(4) The electron profiles are complex. Figure 5 shows the region of enhancement near midnight in detail for some proton and electron channels. The electron fluxes underwent a marked change in pitch-angle distribution at higher energies at $\sim$73040 UT, as indicated by the greatly increased scatter of the electron fluxes due to the change in pitch-angle modulation. The cutoff of the 2.53 MeV proton channel can be seen to have been coincident with the change in the electron pitch-angle distribution. The presence of a spike of precipitating energetic electrons at the high-latitude boundary of the outer zone is common, particularly near midnight (Brown and Stone, 1972).

4. Discussion

The polar profiles presented in Figures 1 and 4 show a complex latitudinal structure. The high-latitude region of the polar caps was delineated by isotropic electron fluxes as is usually the case (Vampola, 1969, 1971; Paulikas *et al.*, 1970) and by electron intensities equal to those in interplanetary space at the same time (West and Vampola, 1971). During the time period under discussion here, the solar proton intensities in the plateau region were featureless and equal to the interplanetary intensities (Fennell, 1973). Thus the behavior of electrons and protons in the polar plateau region was alike. The electron intensity and pitch-angle distributions changed abruptly at the low-latitude boundaries of the plateau region. The electron polar plateau boundaries are interpreted as the high-latitude termination of closed field lines; the anisotropic pitch-angle distributions, peaked at 90°, of the electrons below the polar plateau clearly indicated locally closed field lines. In general the region immediately below the polar plateau is the quasi-trapping region; quasi-trapped because the particles cannot drift completely around the Earth, but are lost on the night side of the Earth when they drift onto open field lines. The edge of the polar plateau near midnight shows no region of quasi-trapping; the solar electron plateau ends at the stable trapping boundary.

The particle observations below the polar plateau in the region of closed field lines were characterized by intense proton precipitation without extensive electron precipitation. The proton precipitation occurred in both the quasi-trapping and the stable

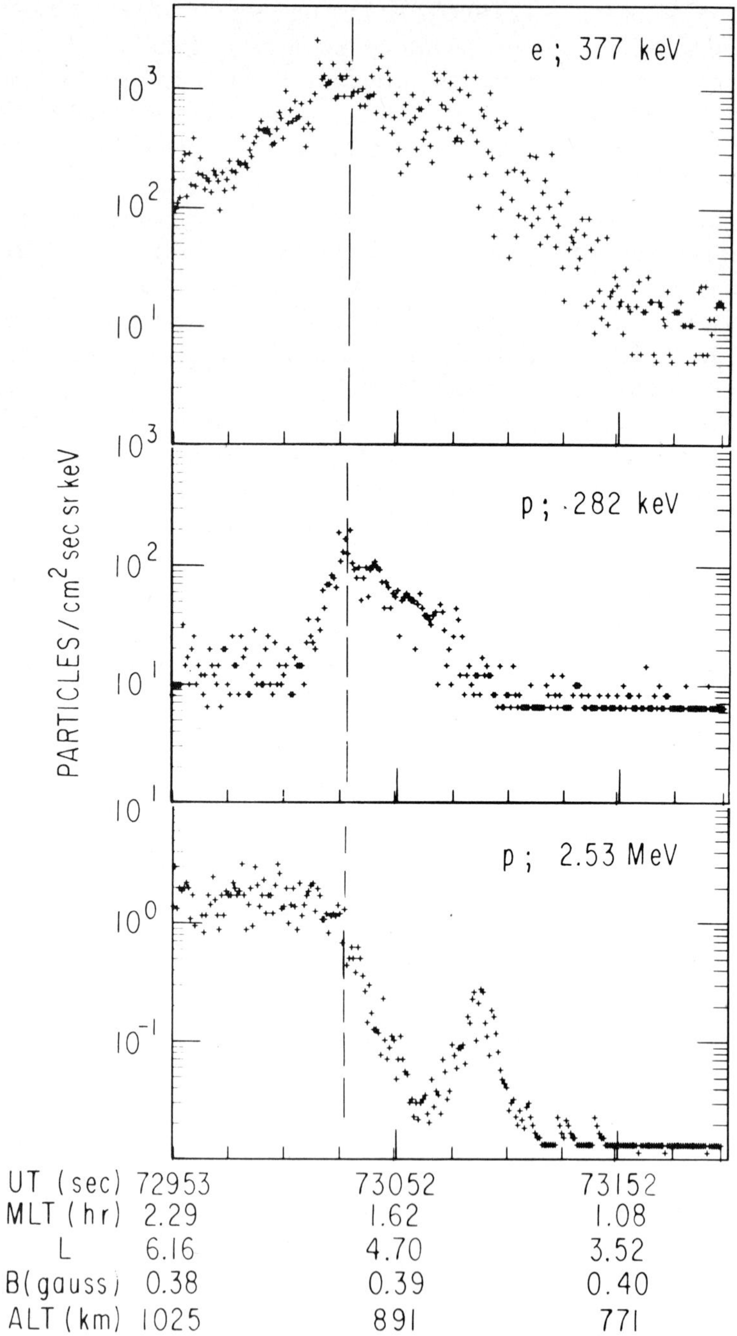

Fig. 5. An expanded portion of the polar cap traverse given in Figure 4 showing the near-midnight boundary of the polar plateau.

trapping region as defined by the energetic electrons. Some process(es) populated the bounce loss cone with both solar and magnetospheric protons, at least over the energy range 200 keV $\leqslant E \leqslant$ 4.6 MeV, and over a wide latitudinal range. At the same time, the magnetospheric electron population (above 50 keV) was not noticeably affected.

It has been suggested by Williams and Heuring (1973) that strong pitch-angle diffusion can result in isotropic solar proton pitch-angle distributions in a region of closed field-lines wherein solar protons normally exhibit anisotropic pitch-angle distributions (quasi-trapping). Morfill and Quenby (1971), Taylor and Hastie (1971) and Morfill (1973) have pointed out that breakdown of the first adiabatic invariant near the last closed field line leads to an effect indistinguishable from strong pitch-angle scattering for low-latitude particles. It is clear that the latter effect must occur and that it will be much more significant for protons with their relatively large gyro-radii. However, breakdown of the first invariant cannot be the only cause of the proton precipitation observed in the data presented here. The characteristics of the precipitation such as the intensities, spectra and latitudinal extent, discussed above, rule out breakdown of the first adiabatic invariant as the major cause of the observed proton precipitation.

The polar plateau edge near midnight, shown in Figure 4, showed a peak equatorward of the cutoff. For protons with energies above ~800 keV the peak was separated from the polar plateau by a region of relatively low flux. The energy spectrum of the protons in the peak was obviously softer than that over the polar plateau, and the latitudinal location shifted equatorward with increasing energy. A detached peak possibly could result from a change of the boundary of closed field lines. However such an origin does not readily explain the energy spectrum, intensity or energy dependence of the location of the peak. It seems more likely that these protons resulted from pitch-angle scattering of protons at high altitude. Violation of the first invariant would not be expected to have created a peak detached from the boundary of closed field lines; the most distorted field lines would be at the boundary.

Three of the four cutoffs shown in Figures 1 and 4 were in the afternoon to midnight sector. The fourth case, near dawn (at ~72600 UT) differs from the other three in that no strong proton precipitation was seen. The dawn polar plateau boundary further differs in that solar protons up to ~1 MeV are found in the quasi-trapping region (as defined by the energetic electrons). The highest energy channel, at 4.60 MeV, shows a small intensity decline at this cutoff latitude for the lower energy protons. Unfortunately the satellite spin axis-magnetic field orientation was such that the loss cone(s) were not scanned at that time. However the energy-dependent intensity decline and increased scatter in the data points suggest that the polar proton pitch-angle distribution changed.

Williams and Heuring (1973) have pointed out that solar proton pitch-angle anisotropies exhibit a time dependence during certain solar particle events. They suggest that the disappearance of anisotropic pitch-angle distributions may be the result of strong pitch-angle diffusion associated with development of the ring current. Figure

6, adapted from their paper, schematically illustrates the magnetospheric topology at the time of the OV1-19 observations presented in this paper; this topology is derived from the observed D_{st} temporal behavior around the time of the OV1-19 observations. The local and universal times of the four observations of polar plateau boundaries presented in this paper also are shown. The observation of the solar plateau edge near dawn differs from the three observations in the dusk to midnight sector in that the enhanced ring current had not reached dawn at the time of observation. Williams

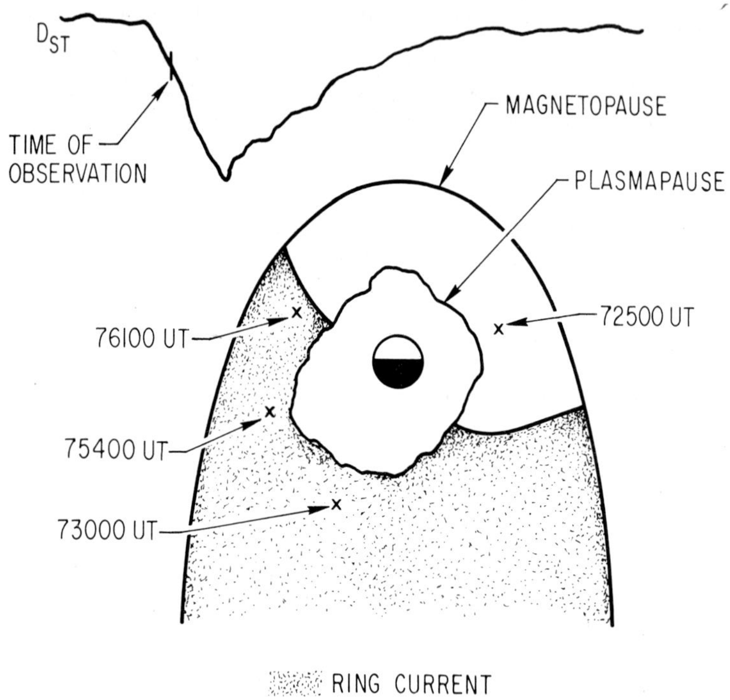

Fig. 6. A schematic illustration of the magnetospheric topography at the time of the observations presented in this paper and derived from the observed D_{st} around the time of the OV1-19 observations (adapted from Williams and Heuring, 1973). Local and universal time for the four observations of polar plateau boundaries presented in this paper are shown.

and Heuring (1973) would suggest that the protons near the dawn polar plateau edge would not be subjected to a strong diffusion process at that time. The absence of proton precipitation and the probable (but not certain) anisotropic pitch-angle distributions at ~72500 UT (dawn), in contrast to the three observations in the dusk to midnight sector are in qualitative agreement with the suggestion of Williams and Huering (1973). It should be noted that low-altitude observations such as presented here cannot verify their proposal, which is concerned with complex wave-particle interactions at high altitude, but rather are extensions of their initial observations.

The means by which solar protons penetrate deeply into the stable trapping region is still not clear (cf. discussion in Bewick *et al.*, 1973). Observations of solar protons well within the stable trapping region for electrons of a few hundred keV are shown in Figure 1 (~75 550 UT) and Figure 4 (~72 450 UT). However the stable trapping region is defined by the boundary of stably trapped electrons of a few hundred keV; these data also indicate that solar protons of a given energy did not penetrate into the region where protons of the same energy were stably trapped.

Synchronous altitude observations yield similar conclusions (cf. review by Lanzerotti, 1970). Solar protons of a few MeV reach synchronous altitude. But stably trapped protons, in the sense that protons remain when the solar source disappears, are not seen. The number of $\gtrsim 1$ MeV protons at synchronous altitude when the interplanetary fluxes are absent is essentially zero. Fillius (1968) has shown that solar protons in the 40–250 MeV range have access, at least at times, to $L \sim 4$ on the equator. His measurements also show that these solar protons did not penetrate into the region where magnetospheric protons of the same energy are found. The observations suggest that the mechanism(s) which prevents stable trapping of magnetospheric protons of a given energy on a given drift shell also can permit access of solar protons to the same drift shell. In fact, it is clear that solar protons cannot penetrate with significant efficiency into regions where they can be stably trapped; otherwise large solar proton events would have long since been observed to cause dramatic long-term enhancements in the trapped proton population. Magnetospheric loss processes cannot distinguish between a proton energized at the Sun and one energized by radial diffusion in the magnetosphere of the Earth.

Acknowledgements

The authors would like to thank Drs G. A. Paulikas, J. F. Fennell, and L. J. Lanzerotti for discussions on many aspects of this work. Dr J. R. Stevens was responsible for the design and construction of the proton spectrometer. This work was carried out under U.S. Air Force Space and Missile Systems Organization [SAMSO] Contract No. F04701-72-C-0073.

References

Bewick, A., Haskell, G. P., Hynds, R. J., and Morfill, G.: 1973, *J. Geophys. Res.* **78**, 597.
Blake, J. B. and Paulikas, G. A.: 1972, *J. Geophys. Res.* **77**, 3431.
Blake, J. B., Fennell, J. F., Schulz, M., and Paulikas, G. A.: 1973, *J. Geophys. Res.* **78**, 5498.
Brown, J. W. and Stone, E. C.: 1972, *J. Geophys. Res.* **77**, 3384.
Fennell, J. F.: 1973, *J. Geophys. Res.* **78**, 1036.
Fennell, J. F., Blake, J. B., Paulikas, G. A.: 1973, *J. Geophys. Res.* (in press).
Fillius, R. W.: 1968, *Ann. Geophys.* **24**, 821.
Lanzerotti, L. J.: 1970, in V. Manno and D. E. Page (eds.), *Intercorrelated Satellite Observations Related to Solar Events*, D. Reidel Publ. Co., Dordrecht, Holland, p. 205.
Lanzerotti, L. J.: 1972, *Rev. Geophys. Space Phys.* **10**, 379.
Lanzerotti, L. J. and Michel, F. C.: 1972, *Comments Astrophys. Space Phys.* **4**, 161.
Morfill, G.: 1973, *J. Geophys. Res.* **78**, 588.
Morfill, G. and Quenby, J. J.: 1971, *Planetary Space Sci.* **19**, 1541.
Morfill, G. and Scholer, M.: 1973, *Space Sci. Rev.* (in press).
Paulikas, G. A.: 1973, *Rev. Geophys. Space Phys.* (in press).

Paulikas, G. A., Blake, J. B., and Vampola, A. L.: 1970, in V. Manno and D. E. Page (eds.), *Intercorrelated Satellite Observations Related to Solar Events*, D. Reidel, Publ. Co., Dordrecht, Holland, p. 193.
Stevens, J. R., Martina, E. F., and White, R. S.: 1970, *J. Geophys. Res.* **75**, 5373.
Taylor, H. E. and Hastie, R. J.: 1971, *Cosmic Electrodyn.* **2**, 211.
Vampola, A. L.: 1969, *J. Geophys. Res.* **74**, 1254.
Vampola, A. L.: 1971, *J. Geophys. Res.* **76**, 36.
West, H. I., Jr. and Vampola, A. L.: 1971, *Phys. Rev. Letters* **26**, 458.
Williams, D. J. and Heuring, F. T.: 1973, *J. Geophys. Res.* **78**, 37.

THE EARTH'S BOW SHOCK FINE STRUCTURE

V. FORMISANO

Laboratorio di Ricerca e Tecnologia per lo Studio del Plasma nello Spazio,
Consiglio Nazionale delle Ricerche, CNR, Roma, Italy

1. Introduction

This paper summarizes our knowledge of the Earth's bow shock fine structure. Review papers (Sagdeev, 1967; Paul, 1968, and Friedman *et al.*, 1971) and books (Tidman and Krall, 1971) on collisionless shocks in plasmas, which have appeared so far, deal especially with laboratory produced collisionless shocks and theories to describe the corresponding observations, but only marginally with the Earth's bow shock. This is probably due to the great complexity of the bow shock structure (see Figure 1), the intermixing of the many different phenomena occurring there and the difficulties of observation. Furthermore, although the available literature on bow shock observations is now very extensive, systematic investigations were not possible until recently since in general the different parameters observed were rather uncorrelated. On many occasions for example magnetic field data on the bow shock crossings were reported without a corresponding knowledge of the plasma data and vice versa. In such cases unambiguous checking of the measurements against theories of collisionless shocks in plasmas was clearly practically impossible.

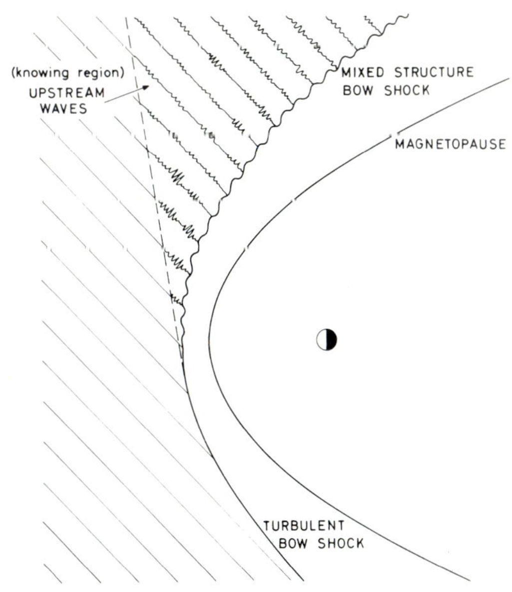

Fig. 1. Sketch in the ecliptic plane of the interaction between the solar wind and the Earth's magnetic field. The magnetopause and the bow shock are indicated together with the region where the particles reflected at the bow shock interact with the incoming plasma generating four kinds of wave phenomena (see Section 3).

D. E. Page (ed.), Correlated Interplanetary and Magnetospheric Observations, 187–223.

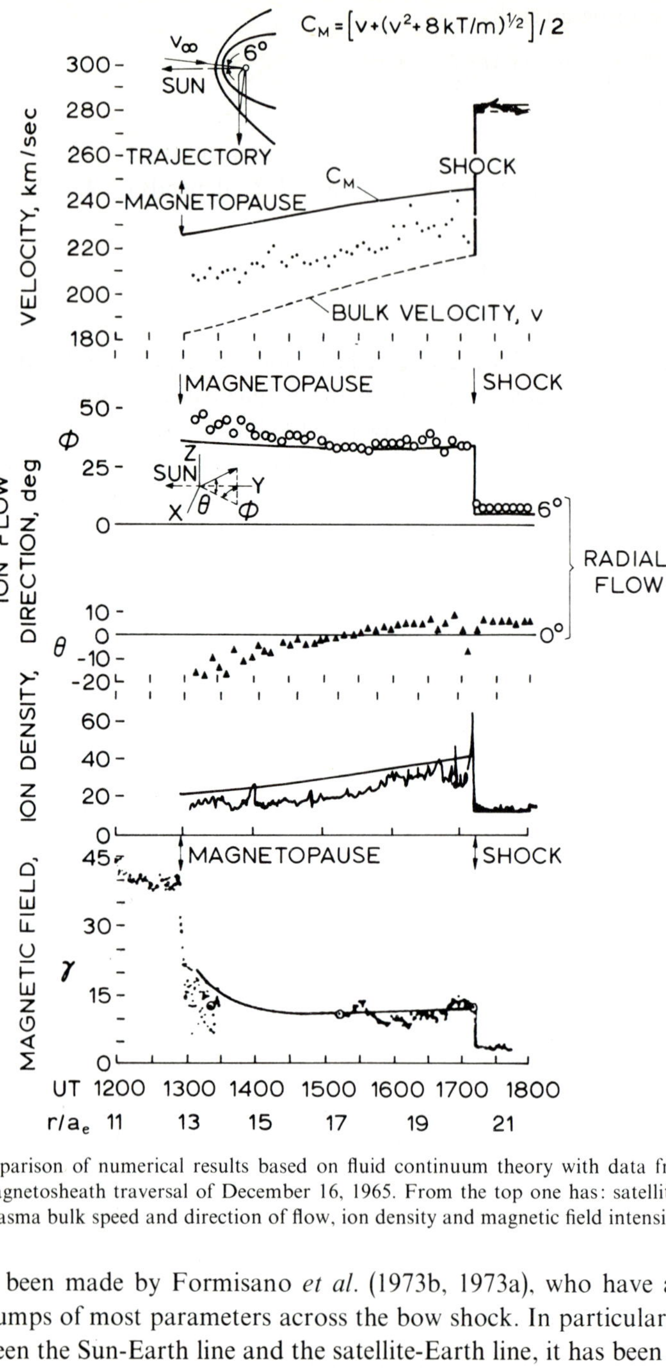

Fig. 5. Comparison of numerical results based on fluid continuum theory with data from Pioneer 6 during the magnetosheath traversal of December 16, 1965. From the top one has: satellite's trajectory, plasma bulk speed and direction of flow, ion density and magnetic field intensity.

shock, has been made by Formisano *et al.* (1973b, 1973a), who have analyzed in detail the jumps of most parameters across the bow shock. In particular, if Θ is the angle between the Sun-Earth line and the satellite-Earth line, it has been shown that the computed dependence of the quantity V_∞/V on the angle Θ agrees with the HEOS-1 data (see Figure 6). The density ratio as a function of Θ appeared not to agree with the computations of Spreiter *et al.* (1966) probably because of the approxima-

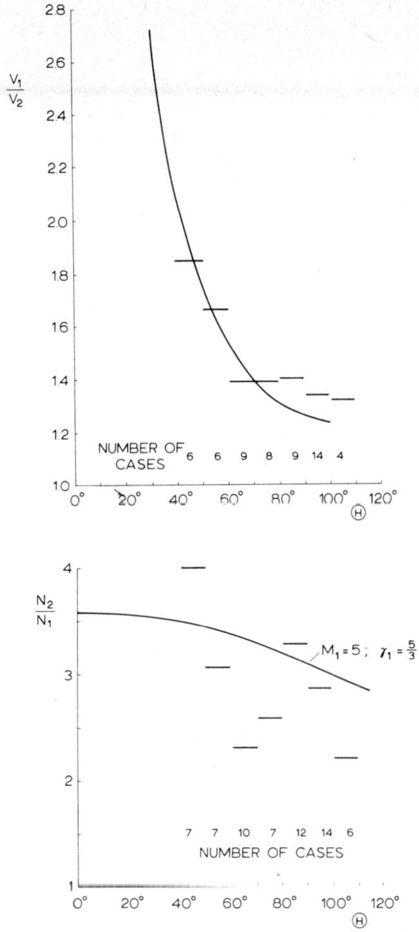

Fig. 6. Ratio of bulk speed (top) and number density (bottom), across the bow shock as function of the Sun-Earth-Satellite angle Θ observed by HEOS 1. The curves show the numerical results from Spreiter *et al.* (1966).

tions made in the theory in connection with the magnetic terms.

The dependence on the Mach number of ratios V_∞/V, T/T_∞, N/N_∞, as found in Formisano *et al.* (1973a) is shown in Figure 7. The agreement with theory is good for V_∞/V (no dependence on M_A) and for T/T_∞, but it is not satisfactory for N/N_∞. Formisano *et al.* (1973a) have tested Equation (1) using the magnetosheath proton velocity distribution given by Equation (3). It has been found that the average ratio T/T_∞ computed using Equation (1) and the measured values of M_∞, V_∞ and V, (the ratio of specific heats γ being fixed at a value of 5/3) was 0.96 times the average measured ratio T/T_∞. The magnetic field ratio B/B_∞ has been shown to be strongly dependent on M, γ, and the angle between magnetic field direction and shock normal (see Formisano *et al.*, 1973a).

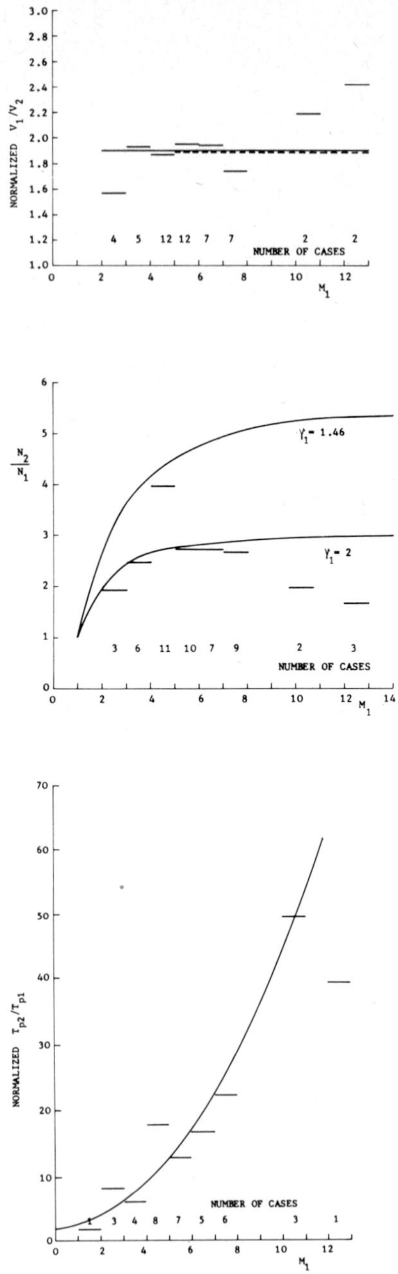

Fig. 7. Ratio of bulk speed (top), number density (middle) and temperature (bottom) across the bow shock as functions of solar wind Mach number – T_{p2}/T_{p1} and V_1/V_2 are normalized to a Sun-Earth-satellite angle $\Theta = 45°$. The curves show the numerical results from Spreiter *et al.* (1968a).

The Rankine-Hugoniot relations, which are eight in number if anisotropic fluid pressure with respect to the magnetic field direction is included, have been tested by Chao (1970, 1972) and by Formisano *et al.* (1973a) for a large number of shock crossings. In many cases it turned out to be impossible to satisfy the conservation of energy on the two sides of the discontinuity, unless it was assumed that γ upstream differed from γ downstream.

It should be noted that the computation made by Spreiter *et al.* (1966) uses a fixed value for γ ($\gamma = 5/3$) and does not consider the possibility of γ changing across the shock. In Spreiter *et al.* (1968a) the value $\gamma = 1.2$ was used, but this has not been justified and again the same value was assumed for undisturbed and disturbed solar wind. Formisano *et al.* (1973a) have shown that the undisturbed solar wind has a specific heat ratio $\gamma \geqslant 2$, γ being roughly 5/3 only when upstream waves are present. Chao (1972) has also shown that in the magnetosheath plasma the presence of turbulence results in a decrease of γ.

Shen (1971) has studied the problem of the ratio of specific heats across the Earth's bow shock, and has suggested that the ratio of specific heats can be significantly different in the upstream region of the shock front and downstream in the magnetosheath. This ratio might be larger than 2 in the undisturbed solar wind during quiet solar times and could possibly be less than 5/3 in a disturbed region such as the magnetosheath.

Another aspect of the fluid conservation equations across a shock that has been investigated with the HEOS-1 data, is the behaviour of the density ratio N_2/N_1 as compared with the ratio of the magnetic field components tangential to the surface of discontinuity. B_{T2}/B_{T1} should be rigorously equal to N_2/N_1 only if B is tangential to the shock, but for any other direction B_{T2}/B_{T1} should be roughly equal to N_2/N_1 (see Tidman and Krall, 1971). Data referring to 19 bow shock crossings are shown in Figure 8. After the first five R-H equations have been checked, the use of the coplanarity theorem allows determination of shock normal and therefore of B_{T2}/B_{T1}. It is evident from Figure 8 that B_{T2}/B_{T1} is generally different from N_2/N_1.

Concluding this section we may state that the fluid description of the interaction between the solar wind and the Earth's magnetic field appears to predict rather well the gross geometric features of the phenomenon. However, comparison between theory and experiment shows certain inadequacies in the fluid description. The magnetic field and number density jumps observed experimentally appear to be in some disagreement with the published models. The energy balance across the shock appears to be more complex than that given by the Rankine-Hugoniot relations because the fluctuation energy plays an important role on both sides of the shock. Phenomena are present that cannot be predicted in a fluid approximation.

Indeed the R-H relations have been derived under the assumption of Maxwellian velocity distribution on both sides of the discontinuity. This means, obviously, thermal equilibrium and no wave phenomena. In the actual case, however, the downstream proton velocity distribution has been found to be not Maxwellian.

The inadequacy of a pure fluid model appears even more clearly when one looks

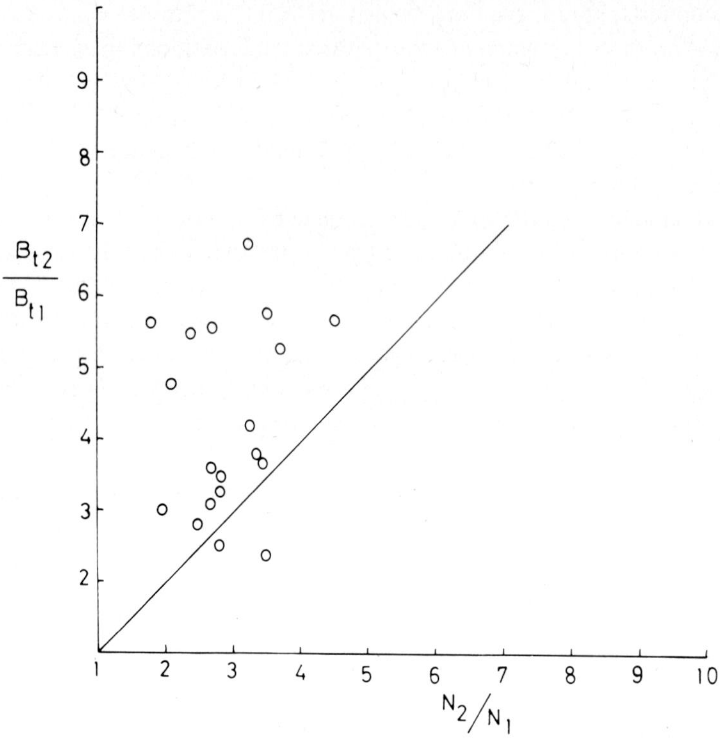

Fig. 8. Ratio of the magnetic field components tangent to the shock surface versus the ratio of the number density. The data points shown are from the HEOS-1 data. The points should be, following the fluid dynamic, all around the 45° line (also shown in figure).

in detail at the phenomena accompanying the solar wind-Earth magnetic field interaction and at the bow shock structure. The thickness of the bow shock transition appears to be of the order of a few hundred kilometers, while the collisional mean free path, in the solar wind plasma is around 1 AU. The mechanism that leads to dissipation of the directed solar wind energy thus operates on a length scale that is millions of times smaller than the distance between particle collisions and its explanation must obviously be found outside the theory of collisional fluids. Figure 1 represents a schematic picture of the bow shock and the associated phenomena. As will be discussed in detail in the remainder of the paper, very different structures of the shock transition, (indicated in the figure by turbulent bow shock and mixed structure), are seen depending on whether the angle $\widehat{Bn}$ between the interplanetary magnetic field and the shock normal is above or below 45°. For $\widehat{Bn} < 45°$, a very large region in the upstream solar wind, extending many Earth radii away from the shock transition, appears to be populated by waves ('upstream waves'), which are directly connected with the shock and which themselves influence the shock structure. This upstream wave region has been termed a 'knowing region' since the solar wind there already knows of the presence of a downstream obstacle. These structural features, and many more which result from detailed observations, are clearly not described by a fluid picture of the interaction of the solar wind with the Earth's magnetic field.

3. Shock Velocity

Single satellite observations of the bow shock necessarily give insufficient information about the shock structure unless the velocity of the satellite relative to the shock is known.

The shock speed has been estimated from single observations by a statistical study (Heppner *et al.*, 1967) of magnetosheath transitions which showed multiple bow shock crossings. The formula used, based on the consideration of an oscillatory periodic motion for the shock discontinuity is:

$$V_s = \frac{4nA}{\tau_m - \tau_1}. \tag{4}$$

Here, $\tau_m - \tau_1$ is the total time needed to observe all the crossings, m is the number of shock crossings, $n=(m-1)/2$, and $A=(r_i - r_m)/2$, r_i being the satellite radial distance at the time of the ith observation.

The range of values derived for the shock speed was 3 to 28 km s^{-1} with an average value $\sim$9 km s^{-1}. Such a value has been for a long time the bow shock and magnetopause velocity universally accepted, although magnetopause velocities higher than 75 km s^{-1} in a recession motion and higher than 156 km s^{-1} in an advance motion, were quoted by Konradi and Kaufman (1965). High velocities were also predicted by Smit (1968) for bow shock and magnetopause motion.

Formisano *et al.* (1971) suggested that in two cases seen on day 43, 1969, the shock velocity should have been roughly an order of magnitude larger than 10 km s^{-1}, in order to explain the differences between computed and observed magnetic field jumps.

Greenstadt *et al.* (1972), showed that indeed the shock velocity was on these crossings an order of magnitude larger. The shock velocity was obtained in this case from direct comparison of data taken simultaneously by the satellites HEOS 1 and OGO 5 which were at the time almost aligned with the local shock normal. High values for the bow shock speed have also been confirmed by Guha *et al.* (1972) by inspection of waves upstream of the shock and making use of the theory of Tidman and Northrop (1968) for the generation of such waves. Very recently a statistical study of the bow shock speed has been made using the mass conservation across the shock and the coplanarity theorem (Lepping and Argentiero, 1971). The shock speed is given by

$$V_s = \frac{(\bar{V}_1 \cdot \bar{n})\, N_1 - (\bar{V}_2 \cdot \bar{n})\, N_2}{N_1 - N_2}, \tag{5}$$

where $V_{1,2}$, $N_{1,2}$ are plasma velocity and density on the two sides of the discontinuity satisfying the R-H relations, and $\bar{n}$ is the shock normal obtained with a best fit technique (Chao, 1970; Lepping and Argentiero, 1971; Lepping, 1972; Linn, 1965;

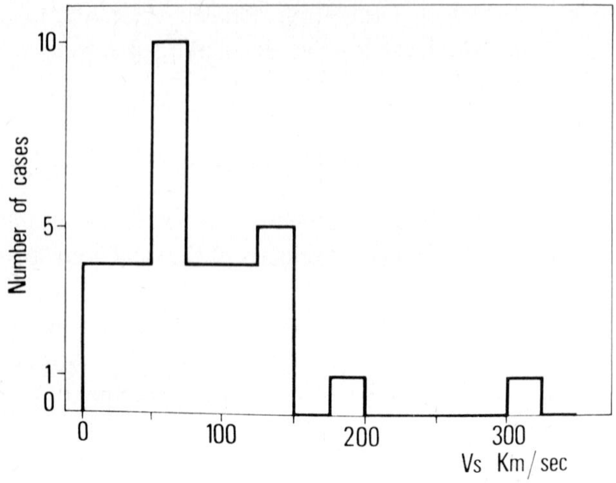

Fig. 9. Shock velocity histogram Measurements from Greenstadt *et al.* (1972), Guha *et al.* (1972) and Formisano *et al.* (1973a) are added together.

Mariani *et al.*, 1972; Abraham-Shrauner, 1972). It should be noted that the method of multiple shock crossings is found in general to give a shock speed much lower than Equation (5). For the Days 298–299 (see Figure 17) for example Equation (5) gives a shock speed around 47 km s^{-1} (Formisano *et al.*, 1973a; Formisano and Hedgecock, 1973a). Equation (4) on the other hand yields a shock speed of only 14 km s^{-1}. Figure 9 is an histogram showing the shock speed obtained in Formisano *et al.* (1973a), Greenstadt *et al.* (1972), Guha *et al.* (1972). The shock speed is, on average, 70 ~ 80 km s^{-1}, i.e. a factor 10 higher than the originally considered values. As a consequence, it appears that shock thickness values previously quoted, were also underestimated. As discussed in the following sections many theories have been compared with such small values, so that misleading conclusions have sometimes been deduced.

It is worth noting too that recently Formisano (1973) has suggested that V_s is a function of space, being larger far back in the tail because of the Mach cone shape that has to be assumed there. A change of the Mach number in the solar wind causes the tail shock to adjust itself to the new Mach cone, with a speed V_s larger than in the subsolar regions.

4. Classification of Data and the Probability of Occurrence of Certain Shock Structures

Collisionless shock structures (Sagdeev, 1967; Tidman and Krall, 1973) can be very different depending on the parameters M_A (upstream Alfvén Mach number), β (upstream ratio of plasma pressure to magnetic pressure) and shock obliqueness (as given by the angle $\widehat{Bn}$ between the magnetic field direction and the shock normal, or the angle $\widehat{Vn}$ between the plasma velocity and the shock normal). The major part of the observational work on the bow shock which has been published, does not contain all these parameters, with the consequence that an unambiguous determination of the phenomena observed may be difficult and, due to the great complexity of the situations under study, confusion may easily arise. In the following a schematic classification of the bow shock structures is introduced on the basis of the value of β and M_A. Four possibilities are considered according to the β value being much smaller than or of the order of unity and the Mach number M_A being below or above a critical value M_c (which is about $2.5 \sim 3$ for oblique shocks):

(1) $\beta \ll 1,\ M_A < M_C$ *Laminar shock* i.e. a shock for a cold plasma with low Mach number.

(2) $\beta \ll 1,\ M_A > M_C$ *Quasi-laminar shock* i.e. a shock for a cold plasma with high Mach number.

(3) $\beta \simeq 1,\ M_A < M_C$ *Quasi-turbulent shock* i.e. a shock for a warm plasma with low Mach number.

(4) $\beta \simeq 1,\ M_A > M_C$ *Turbulent shock* i.e. a shock for a hot plasma with high Mach Number.

The critical Mach number should separate shocks which are mainly resistive from those which are mainly viscous ($M > M_c$). This is deduced from laboratory plasma experiments and many theoretical works where $M < M_c$, and from space plasma physics by observation of the very different shock structures for M lower or higher than M_c.

Concerning the other parameter β a critical value $\beta_c = 0.1$ has sometimes been used to separate quasi-laminar shocks from turbulent shocks, because experimentally it has been found that the downstream amplitude of fluctuations, almost always present in the magnetosheath, goes practically to zero if upstream $\beta < \beta_c$.

Figure 10 shows the behaviour of the downstream amplitude of fluctuations as a function of upstream β for $\widehat{Bn} > 45°$. The behaviour is not affected by the Mach num ber since the same $\Delta B - \beta$ relationship has also been obtained (Formisano and Hedgecock, 1973a) by selecting only cases with upstream Alfvén Mach numbers $4 < M_A < 6$.

It appears from observations that the 'degree of obliqueness' is also important in characterizing the shock structure. It has been found that some low frequency wave features present upstream and related to the shock structure are present only when the angle between the magnetic field direction and the shock normal is small ($\widehat{Bn} < 45°$). Furthermore the downstream turbulence has an amplitude that on average is dependent on the angle $\widehat{Bn}$ between the magnetic field direction upstream and the shock normal. Figure 11 shows data from HEOS 1 (see also Formisano *et al.*. 1973b and Greenstadt and Hedgecock, 1972). ΔB downstream decreases clearly when $\widehat{Bn}$ increases, i.e. when the shock becomes locally more and more a perpendicular shock.

In the published literature one can find examples of laminar and quasi-laminar

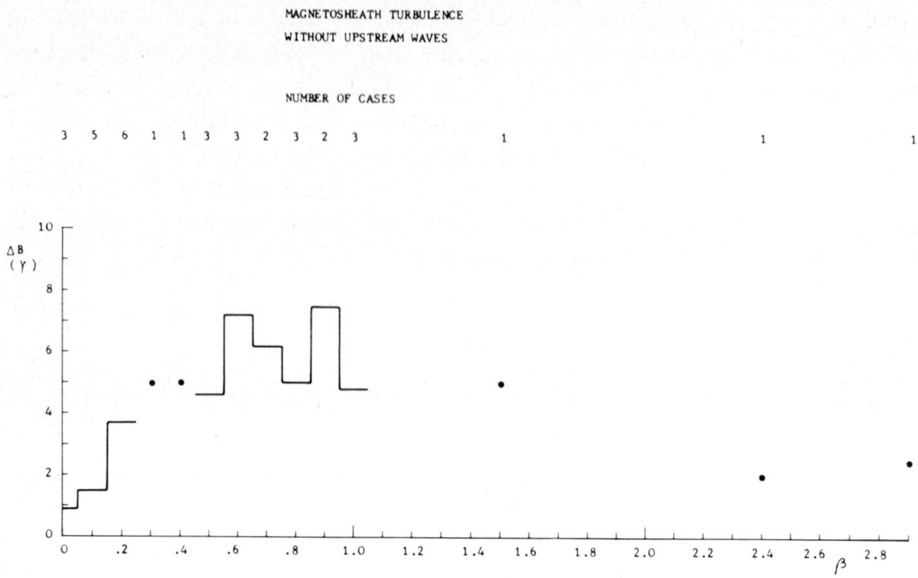

Fig. 10. Amplitude of the downstream aperiodic fluctuations in all the cases without upstream turbulence, versus β. On top of the figure are indicated the cases over which the average was made.

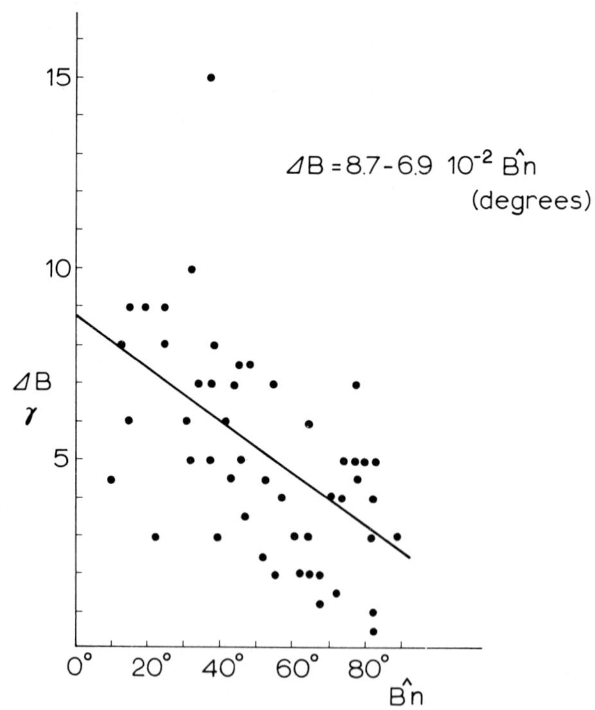

Fig. 11. Amplitude of downstream fluctuations observed in the magnetic field as function of the angle between the magnetic field direction and the shock normal. A linear best fit to the data is also given.

shocks with $\widehat{Bn}>45°$ and of turbulent shocks with both $\widehat{Bn}>45°$ and $\widehat{Bn}<45°$. One single example of a quasi-turbulent shock with $\widehat{Bn}>45°$ has been reported (Neugebauer *et al.*, 1971). The turbulent shocks with $\widehat{Bn}<45°$ shows so many new features that this class has been called '*mixed structure*' (Tidman and Krall, 1971; Formisano and Hedgecock, 1973a) or '*pulsating*' (Greenstadt *et al.*, 1967; Greenstadt *et al.*, 1970a; Greenstadt *et al.*, 1970b; Greenstadt *et al.*, 1970c) shocks.

What is the probability of finding the bow shock in any one of these conditions? We first note that, given the magnetic field direction in the solar wind, the geometry of the interaction is such that there are almost always both regions with $\widehat{Bn}>45°$ and regions with $\widehat{Bn}<45°$.

For the other two parameters β, M, we have the four possibilities listed above.

The frequency of occurrence for each of the four cases in the solar wind data of HEOS 1 has been respectively 1.9%, 1.6%, 4%, 92.5%. However, the actual laminar crossings observed by HEOS 1 were more than 2% because for an eccentric orbit satellite the probability of crossing the bow shock when it has a laminar structure is increased by the fact that a low β low M_A shock usually moves toward the region where the satellite spends most of the time of each orbit. The same is true but less significant for the other two structures (cases 2 and 3).

In concluding this section it should be pointed out that the classification proposed here is not by any means complete. Indeed for some theories the ratio of electron temperature to proton temperature is equally important in determining the presence or absence of some features. Furthermore the smaller or larger 'degree of obliqueness' discussed above ($\widehat{Bn}$ being $<45°$ or $>45°$) was related only to one, out of four, possible upstream wave phenomena. It is possible that other criteria for shock obliqueness will have to be introduced to classify the shock structure with respect to these other phenomena.

5. 'Unclassified' Data

As already mentioned, much data concerning the bow shock structure has been published without all the information needed to obtain β, M_A and $\widehat{Bn}$. It is in some cases possible nevertheless to obtain important information of general validity from such 'unclassified' data. This is true mainly for experiments, (such as the magnetic field experiment on board OGO-A), which were able to take measurements with a very high time resolution and so to investigate the fine structure of the bow shock. Heppner *et al.* (1967) found a typical magnetic shock profile in a large number of OGO-A shock crossings which is shown schematically in Figure 12. Three instants of time are indicated: τ_0 is the time when B first deviates from the interplanetary value, τ_1 the time when the field stops its rapid increase and τ_2 the time when it reaches the final downstream value. Using a shock speed of 9 km s^{-1}, the shock thickness was estimated to be 50–100 km. The time interval, $\tau_2-\tau_0$, required for the field to change

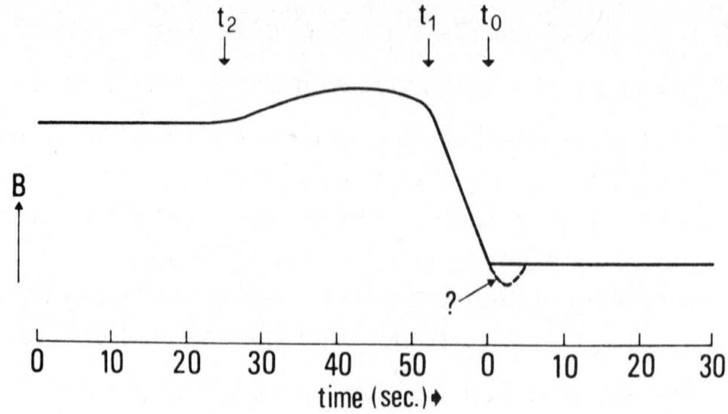

Fig. 12. Model profile of the average magnetic field during the 'most typical' shock crossings, as determined by the OGO-1 fluxgate magnetometer. The model profile is characterized by three times or time intervals.

from its initial interplanetary level to the final magnetosheath level, was estimated to be about 35 s. A very qualitative classification for the bow shock structure (see Figure 13), adopted by Heppner *et al.* (1967), was based on whether or not wave phenomena were observed at the shock and, if so, what type of wave was dominant. In Figure 13, structure (*a*) depicts a shock without the presence of waves and is similar to the mean profile discussed in Figure 12. Structure (*b*) shows the appearance of the shock transition when coherent waves with frequencies near 1 Hz were seen. In structure (*b*) these waves were present only within the B gradient; in structure (b') they were observed both upstream and downstream. Structure (*c*) gives the appearance of shock transitions when higher frequency waves ($F > 3$ Hz) were seen. Case (*c*) is with these waves present only within the gradient of the magnetic intensity B; case (c') is with the waves on both sides of the discontinuity. This classification is of course only of a qualitative and pictorial nature. In the absence of plasma data, it would be difficult to attach any theoretical interpretation to the different types of structures shown. Furthermore, the picture of the shock structure one arrives at, can be altered by the use of an erroneous shock velocity.

Besides the measurements of Heppner *et al.* (1967), (without plasma data), one must recall also detailed studies of the plasma behaviour within the shock (Greenstadt *et al.* 1967; Greenstadt *et al.*, 1968b; Greenstadt *et al.*, 1968a; Greenstadt, 1972b; Binsack, 1970). These showed features different from those seen in the magnetosheath data and in solar wind ion spectra. Particle fluxes rapidly varying in energy and direction were observed, together with large amplitude, (10–25 γ) and irregular magnetic oscillations.

6. Laminar Structures

The bow shock has been observed to be occasionally laminar (Heppner *et al.*, 1967; Formisano *et al.*, 1971; Greenstadt *et al.*, 1972; Greenstadt, 1972b). The laminar structures are characterized not only by $M_A < 3$, $\beta < 0.1$, $\widehat{Bn} > 45°$, but also by the

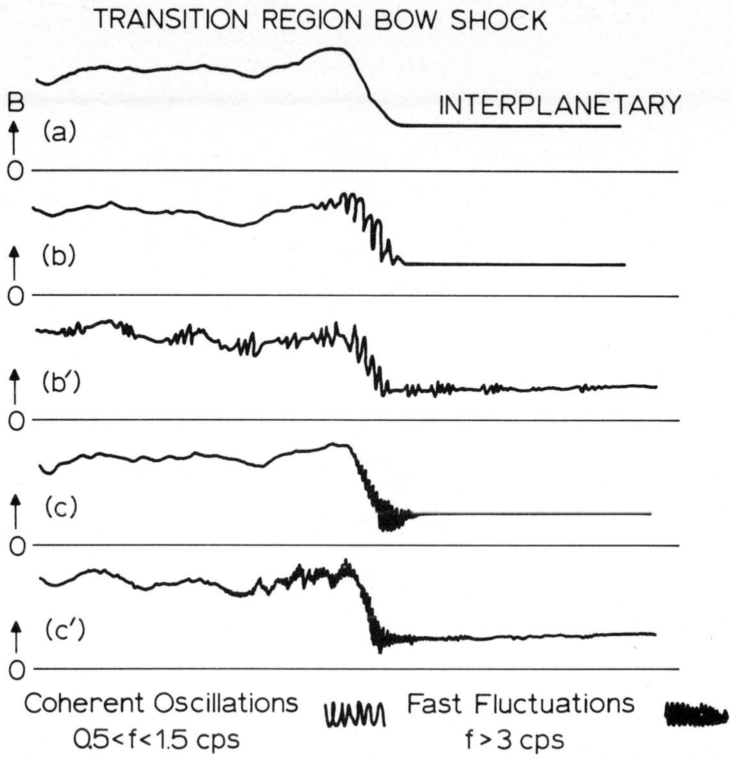

Fig. 13. Superposition of various magnetic wave observations on the shock transition profile of Figure 12. Qualitative classification of bow shock structures from OGO-1 magnetic data.

absence of downstream magnetic turbulence. This property has been studied by Formisano *et al.* (1971), Formisano *et al.* (1973b), Heppner *et al.* (1967), Fairfield (1971). Multiple laminar bow shock crossings are shown in Figure 14 as observed in plasma and magnetic data. The absence of magnetic turbulence is evident from Figure 15 which shows magnetic field intensity profiles with high time resolution for the short period indicated in Figure 14. The magnetic field jump across a laminar shock has been checked against the expression (Formisano *et al.*, 1971)

$$\frac{H_T^*}{H_T} = -\tfrac{1}{2} + \sqrt{\tfrac{1}{4} + \frac{8\pi N v^2 m_\mathrm{p}}{H_T^2}}, \tag{6}$$

where H_T^* and H_T are respectively the tangential components of the magnetic field in the magnetosheath and in the interplanetary medium; v is the velocity of the shock wave with respect to the unperturbed plasma; m_p is the proton mass. For this kind of shock structure magnetic low frequency upstream waves are not observed. Owing to the current interpretation of the low frequency upstream waves as being generated by reflected protons (as it will be discussed later), one must conclude that, in this case, the population of protons reflected backwards by the shock is negligible. Another feature, as found by Formisano *et al.* (1973b), is that the proton velocity distribution

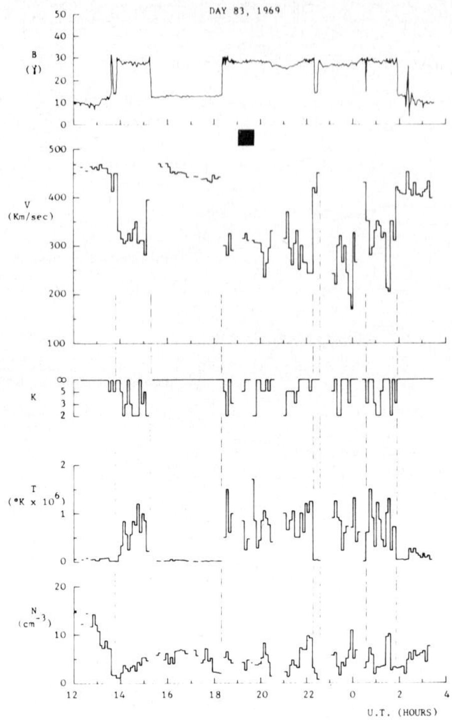

Fig. 14. Multiple laminar bow shock crossings observed by HEOS-1 on Day 83, 1969. From the top one has: Magnetic Field intensity B, b>lk speed V, the distribution function parameter K, temperature T_p and density N_p.

function downstream of a laminar shock is Maxwellian, i.e. similar to the upstream distribution.

Although predicted theoretically, no clear observation of downstream damped oscillations has so far been published. We notice also that, from the plasma data of Figure 14, it appears that the plasma parameters are not as steady as the magnetic field intensity. This indicates the presence of electrostatic fluctuations, although no electric field experiment has so far given data at the time of a laminar shock crossing.

Measurements of the shock thickness have not been published yet. A very rough indication can be obtained from the fine resolution data obtained by Greenstadt (1972b) and shown in Figure 16. These data refer to one of 13 laminar crossings observed by the two satellites OGO 5 and HEOS 1 on Day 43, 1969 and studied in Formisano *et al.* (1971), Greenstadt *et al.* (1972), Greenstadt (1972b). The average shock velocity for these crossings has been measured to be $V_s \sim 60$ km s^{-1}. If we use this velocity with the data of Figure 16, we see that a magnetic shock thickness of $\sim 6 \div 8$ s on the time scale, gives $360 \div 480$ km in length, i.e. $14 \div 19$ R_i (the upstream ion thermal gyroradius R_i being ~ 25 km). We may then conclude that the laminar shock thickness can be very roughly estimated to be $\sim 10 \div 20$ R_i. It has been shown recently

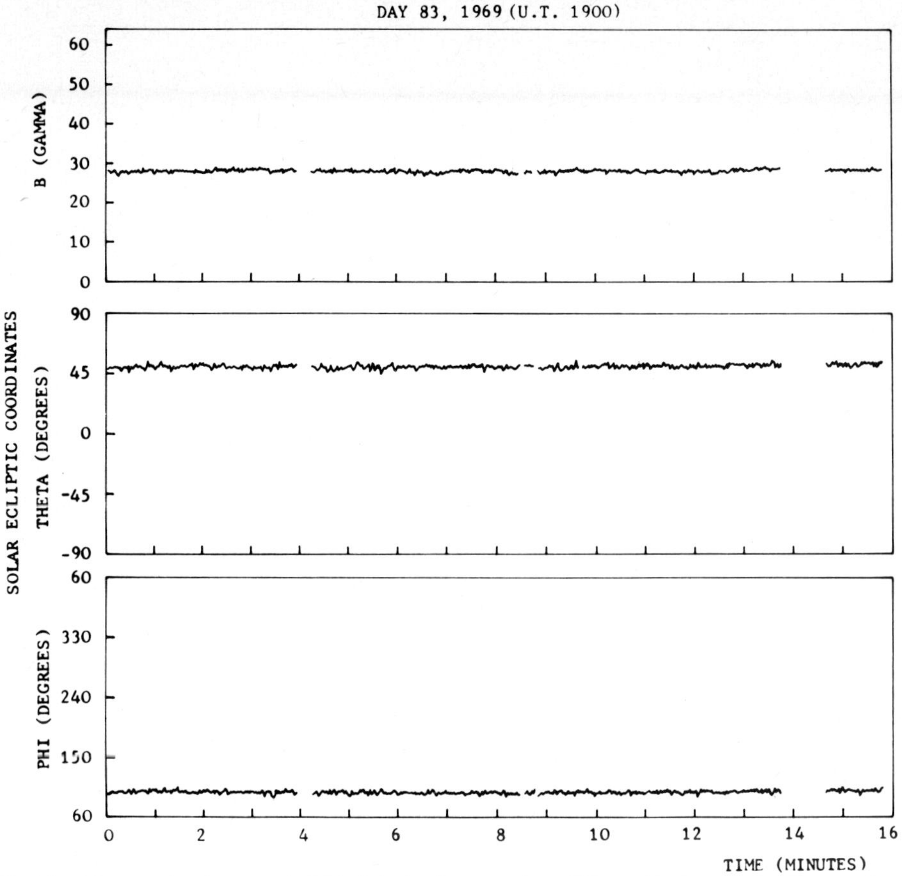

Fig. 15. Magnetosheath magnetic field of the laminar crossing of Figure 14, showing the absence of any fluctuations or turbulence.

by Greenstadt *et al.* (1973) that indeed the laminar shock thickness is of the order of hundreds km, i.e. about c/ω_{pi} and very much larger than c/ω_{pe}.

7. Quasi-Laminar Structures

Shock structures of this kind have been sporadically observed by HEOS 1. It corresponds to $M_A > 3, \beta < 0.1, \widehat{Bn} > 45°$. As in the laminar case, the magnetosheath magnetic field shows a very steady behaviour as illustrated for example in Figure 17 taken from Formisano and Hedgecock (1973a). The downstream distribution has been found to be not Maxwellian in this case. A high energy tail is in fact observed at energies higher than the energies of the incoming solar wind particles. This high energy tail is therefore made up of particles that have increased their energy through the bow shock. This feature appears to be connected with the value of the Mach number and it is present also for high β (turbulent structures). Reflected particles and, cor-

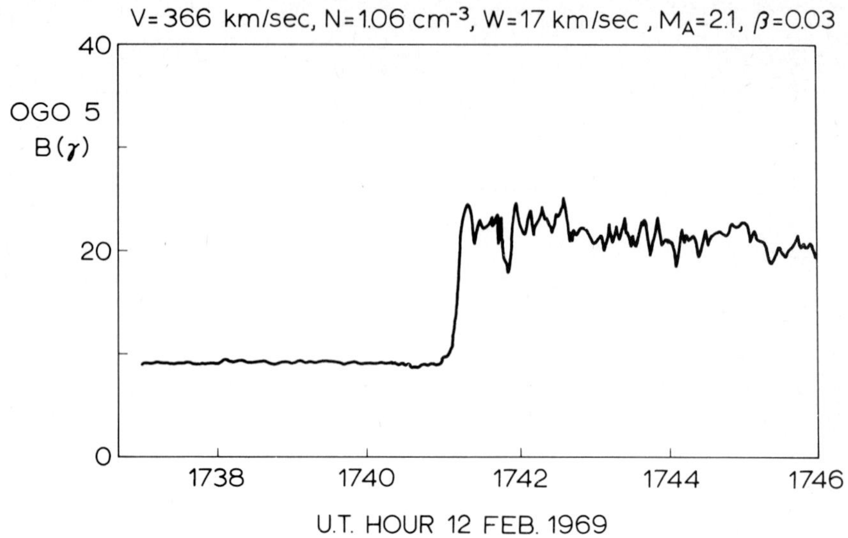

Fig. 16. Magnetic profile of a laminar shock observed by OGO 5. Plasma parameters, as measured by HEOS 1 at the same time are given on the top. The magnetic field intensity reaches the downstream value in 6 s.

respondingly, low frequency upstream waves, do not appear to be typical features of this structure in agreement with the angle $\widehat{Bn}$ being $>45°$ and the picture one has of particle reflection (see Section 10). Observation of electrostatic fluctuations have not been published for this kind of shock. The fine structure and its thickness have been studied by Formisano and Hedgecock for the case of Day 66, 1970 presented in Figure 18. Contrary to the laminar cases, downstream damped oscillations can now be seen which may be identified with solitary waves arising from dispersive effects. From Figure 18, we see that the time interval during which the damped waves are observed is about 24 s, while the first increase of the magnetic field (the main magnetic gradient) takes only 6s. The typical length for which downstream waves are seen is also of the same order and it amounts to about 400 km, i.e. $\sim 20\ R_i$ in this case.

8. Turbulent Structures

These structures are together with mixed structures the most commonly observed at the bow shock. The experimental information available is therefore very extensive. However, because of its complexity, our understanding of many of the phenomena revealed by the observations is still poor. The turbulent shock is characterized by $M_A > 3$, $\beta > 0.1$, $\widehat{Bn} > 45°$. The downstream proton velocity distribution shows a high energy tail corresponding to a $k = 2$ distribution. Going from the upstream to the downstream state, one observes a precursor in front of the steepening of the magnetic field intensity (see Figure 19). This precursor although observed only for a short, or very short, length of time shows the features of the low frequency waves

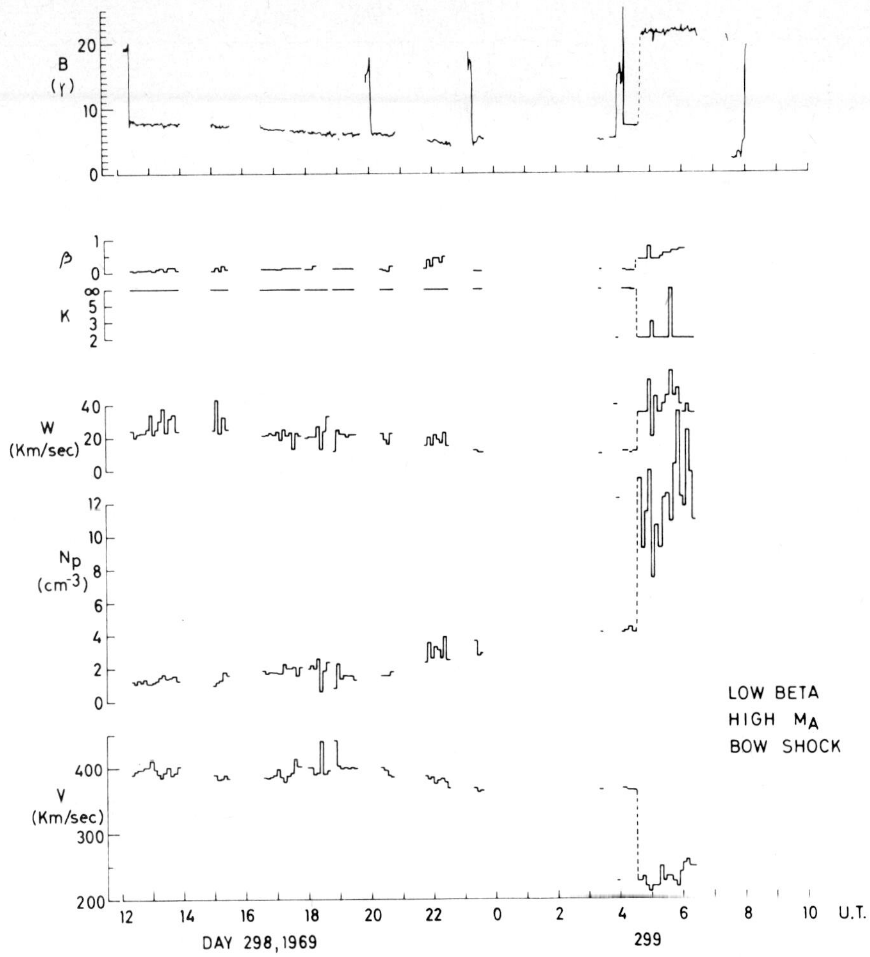

Fig. 17. Plasma and magnetic field data for a quasi-laminar bow shock crossing on Day 298–299, 1969.

observed by Fairfield (1969) at great distances from the shock in the upstream plasma (see Section 10). Positive ions appear to undergo initial deceleration in the precursor region (Neugebauer, 1970). It is possible because of their very low frequency (Formisano and Hedgecock, 1973a) ($\omega < \Omega_i$ since the observed period is 5–10 s) to exclude the theory that the precursor waves are generated at the shock and then propagate upstream. It appears also that these waves are not stationary with respect to the shock, and hence are probably generated upstream. The amplitude of these wave precursors is low and it is therefore likely that they do not essentially affect the main shock structure. Other higher frequency waves have been observed by several authors (Heppner *et al.*, 1967; Holtzer *et al.*, 1972; Olson, 1970; Olson *et al.*, 1969) in connection with this kind of shock (see Figure 20), upstream wave periods being about 1 s and downstream periods about 3 s. These waves, identified as whistlers, are also not stationary in the shock frame (Holtzer *et al.*, 1972) and are probably convected

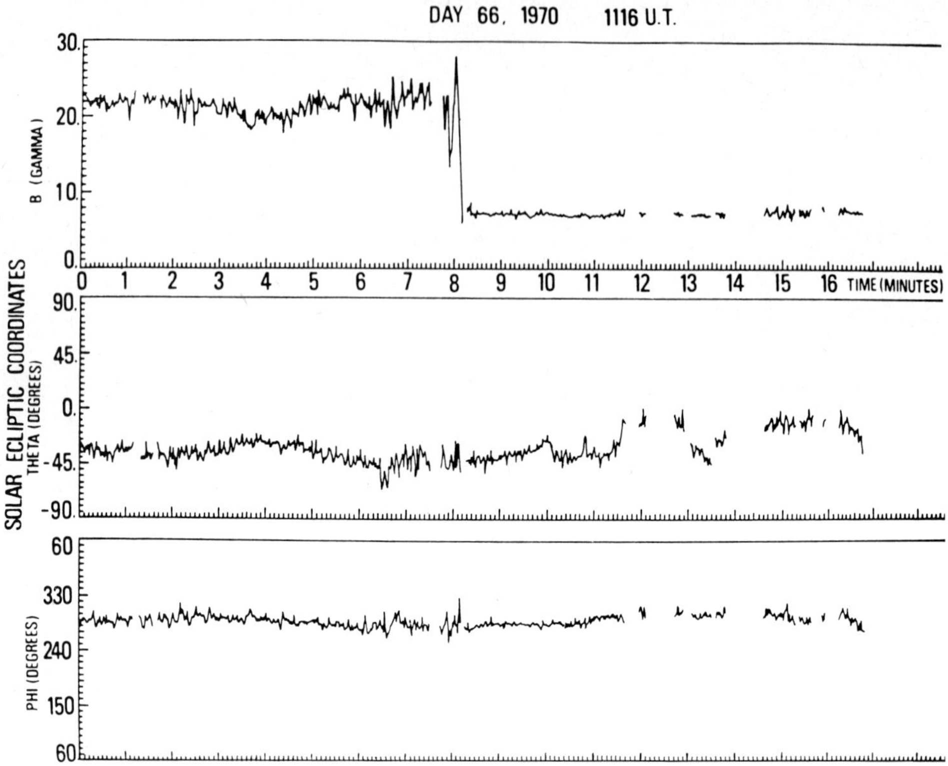

Fig. 18. Magnetic field data for a quasi-laminar shock observed by HEOS 1 on Day 66, 1970, 1124 UT – Plasma parameters were $V=414$ km s^{-1}, $N=3$ p cm^{-3}, $T=8\times10^3$ K, $B=7.5\ \gamma$, $M_A=4.3$, $\beta=0.05$, $\widehat{Bn}=78°$.

from upstream to downstream since the propagation vector is found to be the same in both regions.

As whistler waves can propagate at large angles to the magnetic field and have a group velocity larger than the solar wind speed, they can be observed in regions which are forbidden for the low frequency hydrodynamic waves, i.e. outside the knowing region (see Figure 1 and Section 10).

Acceleration of some incoming protons, within the magnetic gradient of the shock, to an energy from 2 to 4 times the cold solar wind energy, has been reported by Formisano and Hedgecock (1973b) for magnetic crossings where both types of waves mentioned above (periods 5 ÷ 10 s and 1 ÷ 2 s) were present. This acceleration is observed as a second peak (Montgomery *et al.*, 1970) in the energy spectrum (see Figure 21). When this second peak is observed, the main distribution fluctuates in energy loosing 100–200 eV without being thermalized and at the same time the electrons, which appear to be preheated at the beginning of the main magnetic gradient, fluctuate in energy gaining 100–200 eV. The electron preheating starts at the first enhancement of an electrostatic turbulence of 1–8 KHz, which is always observed in connection with the turbulent shock crossings (see Fredricks, 1969;

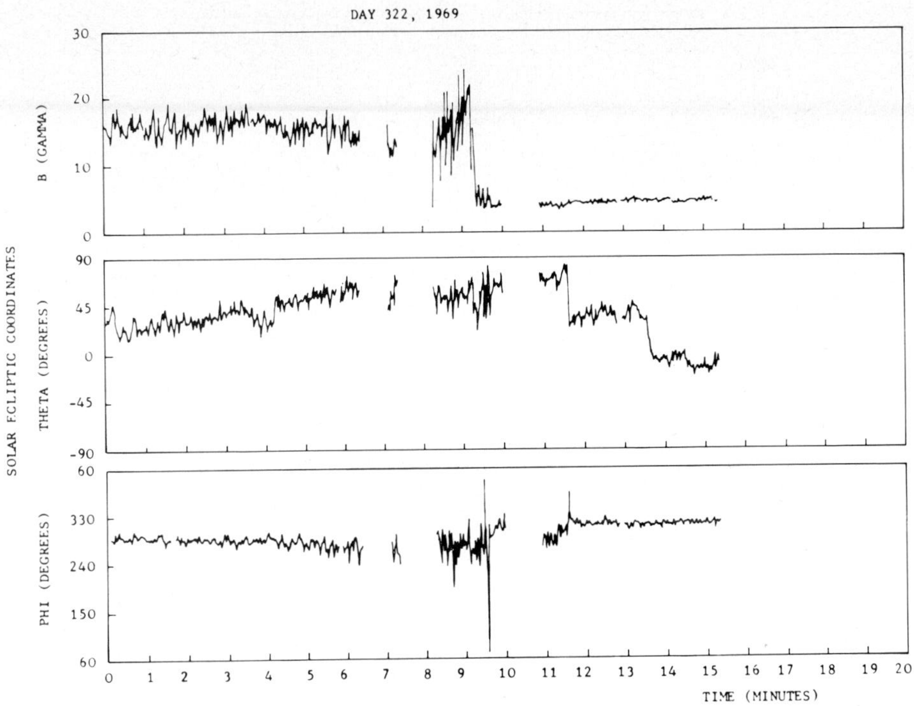

Fig. 19. Turbulent bow shock magnetic profile observed by HEOS 1 on Day 322, 1969. Note the short magnetic precursor before the large B increase.

Fredricks *et al.*, 1970; Fredricks *et al.*, 1968; Fredricks and Scarf, 1969; Scarf, 1971). It appears that, while the electrons are heated at the first appearance of the electrostatic noise, protons are heated only at (or after) the maximum in the plasma wave noise. This maximum is sometimes indicated by a spike in the electrostatic noise profile, the location of the spike being roughly in the middle of the magnetic gradient profile. The second peak in the proton energy spectrum has been observed also after ion thermalization, but with decreasing importance. Figure 22 shows the behaviour of electrostatic noise through the shock, the data (Fredricks *et al.*, 1970) being from OGO 5 and referring to Day 12, 1968. Figure 22a shows the magnetic field profile and positive ion fluxes. Nine arrows indicate the times at which the nine frequency spectra shown in Figure 22b were taken. Figure 22c shows magnetic field and ion flux profiles for another bow shock crossing on the same day. The electrostatic noise level for fluctuations at 1.3 KHz is also shown.

The shock thickness, as defined by the magnetic gradient, is of the order of a few ion gyroradii but the ion thermalization appears to be related to a much thinner region. The high energy tail of the downstream proton distribution is indicative of some acceleration process taking place at the shock.

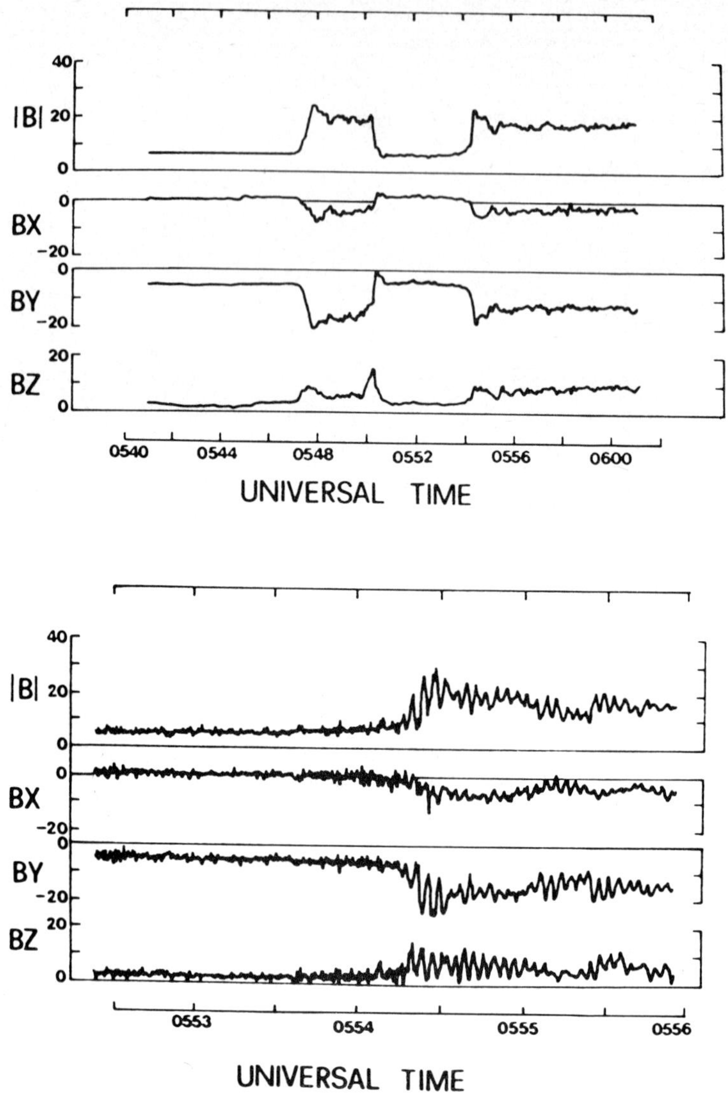

Fig. 20. (Top) Three bow shock crossings on OGO-5 orbit 2, inbound on March 9, 1968. The magnetic field in gammas is a 5-s average. (Bottom) Expanded time-scale (7 points s^{-1}) diagram of the third shock crossing shown in Figure 20 (top).

9. Quasi-Turbulent Structures

Only one shock structure with high β, low Mach number and large $\widehat{Bn}$ ($\beta > 0.1$, $M_A < 3$, $\widehat{Bn} > 45°$), has been observed (Neugebauer *et al.*, 1971) and this recorded by OGO 5 on March 5, 1969, 2321 UT, is shown in Figure 23 where the shock structure is clearly seen to be turbulent. The solar wind parameters are $V \simeq 480$ km s^{-1},

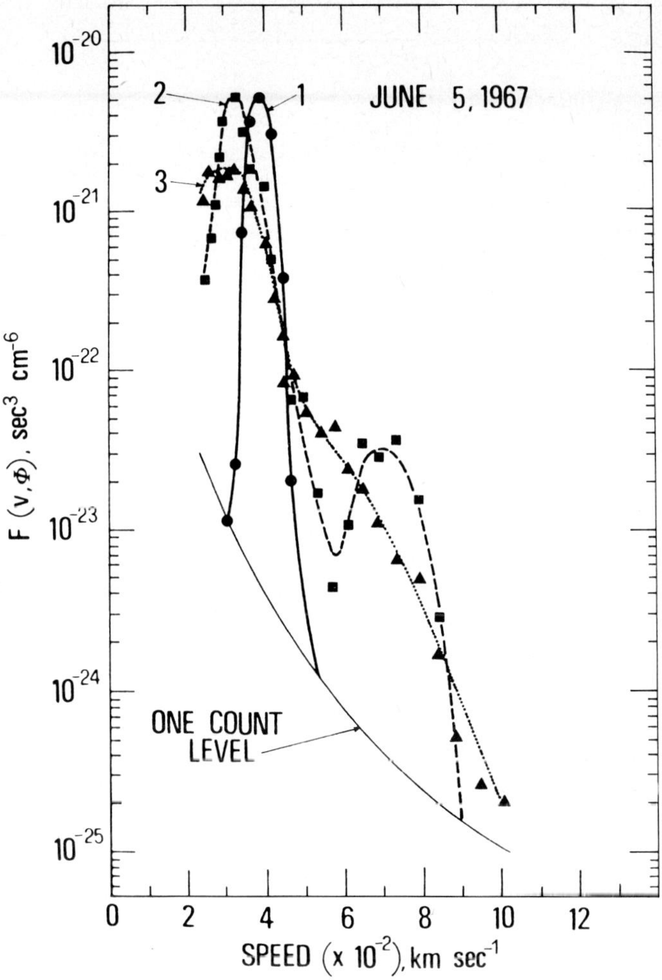

Fig. 21. Three proton velocity distributions observed by Vela 4 in connection with a bow shock crossing on June 5, 1967. Curve 1 shows cold solar wind protons, curve 2 shows a distribution taken within the shock layer where the protons have a second peak at 3–4 times the upstream energy. Curve 3 shows a distribution typical of the downstream state with a long high energy tail.

$T_p \simeq 2.2 \times 10^5$ K, $n \simeq 6$ cm^3, $B \simeq 9$, $\beta = 2$, $M = 3$, $\widehat{Bn} = 46°$. As $M = 3$, it is not certain that the shock was subcritical. A few differences are however noted by the authors, in comparison with Day March 12, 1968 0056 UT., when a turbulent shock was observed ($\beta = 0.6$, $M = 5$, $\widehat{Bn} = 45°$). The main difference is that the magnetic turbulence shows a large enhancement at 200 Hz. The electrostatic turbulence also peaks at ~ 200 Hz while in a turbulent shock the peak is at a much higher frequency. The noise appears therefore to be due to whistler mode electromagnetic radiation and the electrostatic plasma waves are absent. This could be related to absence of a

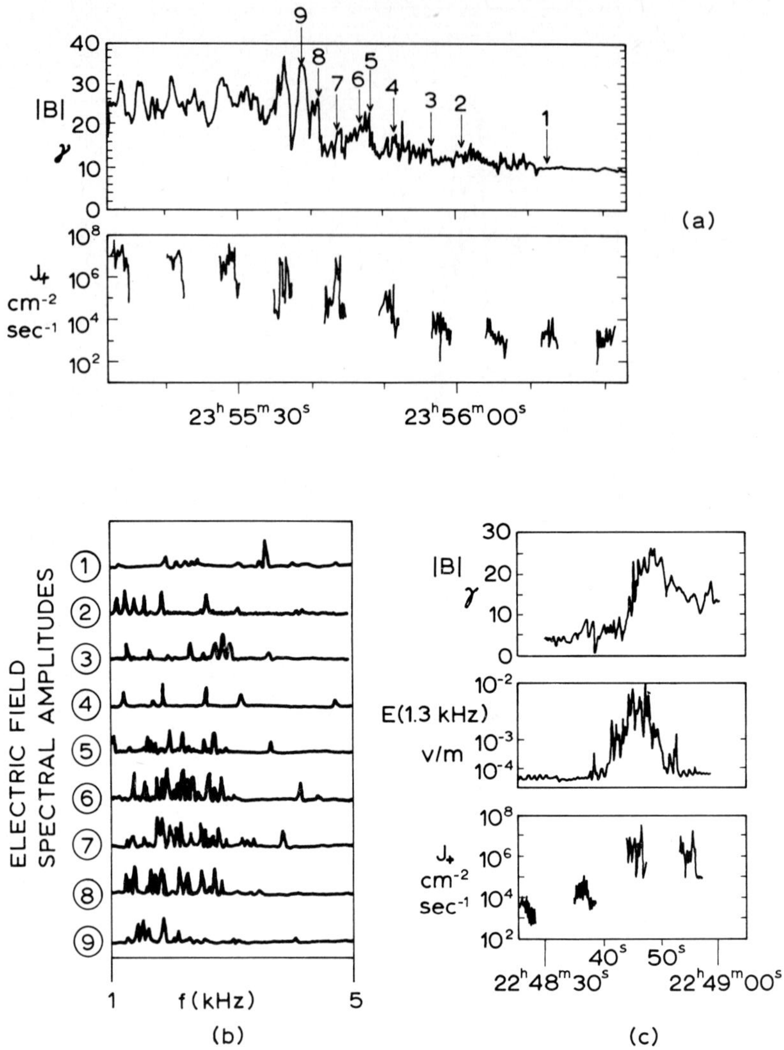

Fig. 22. Electrostatic noise observations in connection with the bow shock: OGO-5 data – (a) magnetic field intensity and positive ions total flux for the bow shock crossing of March 12, 1968. Arrows indicate the times when the electric field spectral amplitude shown in (b) were taken. (c) shows another bow shock crossing on the same day March 12, 1968. The amplitude profile across the shock of 1.3 kHz electrostatic noise is shown.

substructure in the shock. It appears also that the preheating of electrons observed for turbulent shocks is also absent.

10. Mixed Structures

When the solar wind β is comparable to unity and the Mach number is larger than

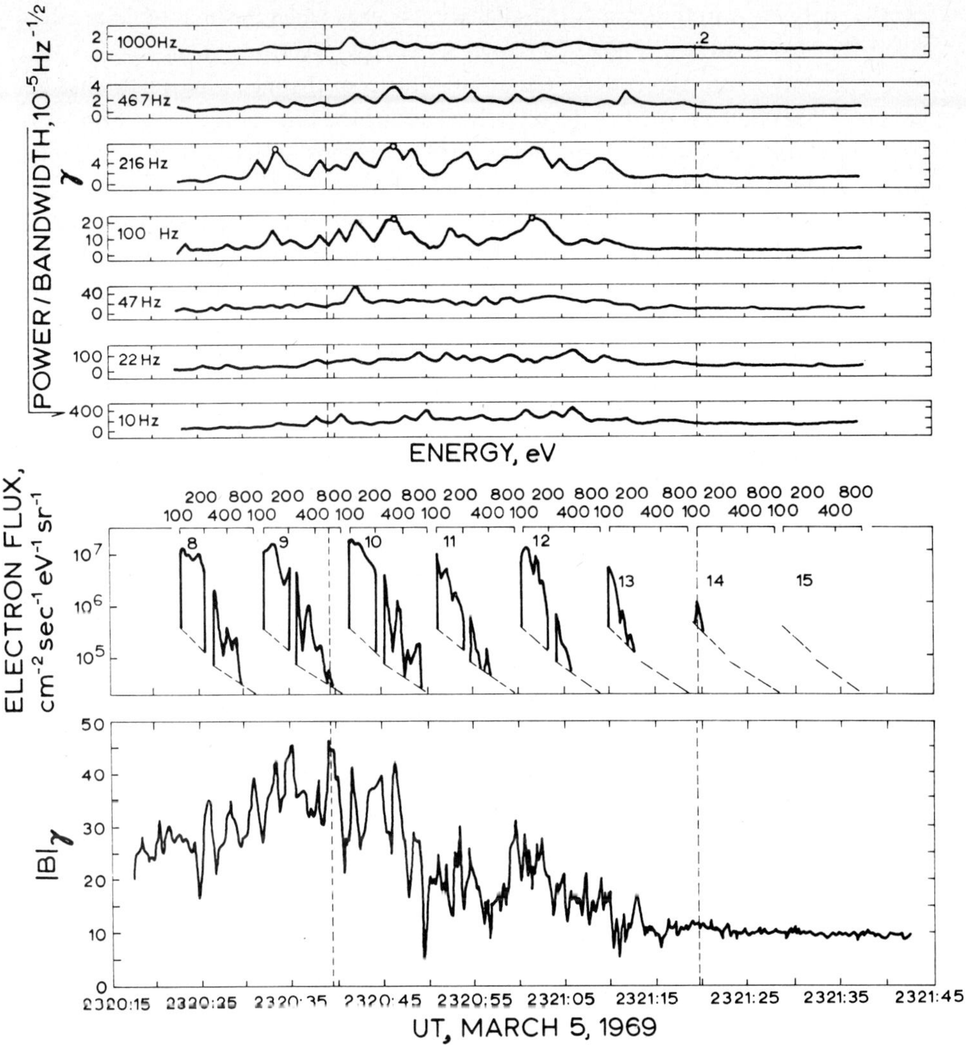

Fig. 23. Magnetic field oscillations, electron flux and magnitude of the magnetic field for the bow shock crossing observed on March 5, 1969. The electron data show the single energy spectra, the energy of each channel being indicated at the middle of the figure.

a critical Mach number, as well as the turbulent shock, (see Section 8), a different and more complicated situation always exists in some region of the bow shock (Greenstadt, 1972a). This situation (mixed structure bow shock), is characterized by the magnetic field lines of force being at a small angle to the local shock normal. In this case it is possible for some particles of the incoming solar wind to be reflected by the shock magnetic gradient {Hudson (1965)}, and accelerated by the solar wind electric field [$(E = -\bar{V} \times \bar{B}/c)$ Sonnerup, 1969]. Proton streams have in fact been observed by Asbridge *et al.* (1968) travelling upstream along the magnetic field direction with an energy two to three times the solar wind energy when the lines of force intersected the shock surface. Figure 24 shows an energy spectrum taken by Vela 4A, when reflected protons were observed at 30°–90° from the incoming solar wind direction. A heat flow away from the bow shock along the lines of force was also measured. This heat flux can be interpreted as due to electrons reflected at the bow shock and accelerated by the solar wind electric field. Figure 25 shows two simultaneous energy spectra taken by the OGO 5 Lepedea (Anderson, 1968; Fredricks *et al.*, 1971; Scarf *et al.*, 1971) experiment. Lepedea B, which pointed towards the bow shock, observed high energy particles coming along the field line, while Lepedea A (transverse to this direction) measured only solar wind electrons.

Both protons and electrons reflected at the bow shock appear to interact with the solar wind generating four kinds of phenomena: upstream low frequency magnetosonic waves, plasma waves, whistler wave packets and whistler waves. Figure 26 shows three of these features simultaneously, namely, magnetosonic low frequency waves, whistler wave packets and plasma waves (Scarf *et al.*, 1970b; Scarf *et al.*, 1970a; Russell *et al.*, 1971). The low frequency magnetosonic waves were first studied by Fairfield (1969). Almost sinusoidal waves are observed (see Figure 27a), being circularly polarized (see Figure 27b) and having frequencies around 0.02 Hz, i.e. periods of 30–50 s. (see Figure 27c). These waves found only along the lines of force intersecting the shock surface, have an amplitude amounting to 10–50% of the ambient field and are mainly transverse. Their amplitude decreases by roughly a factor 2 in

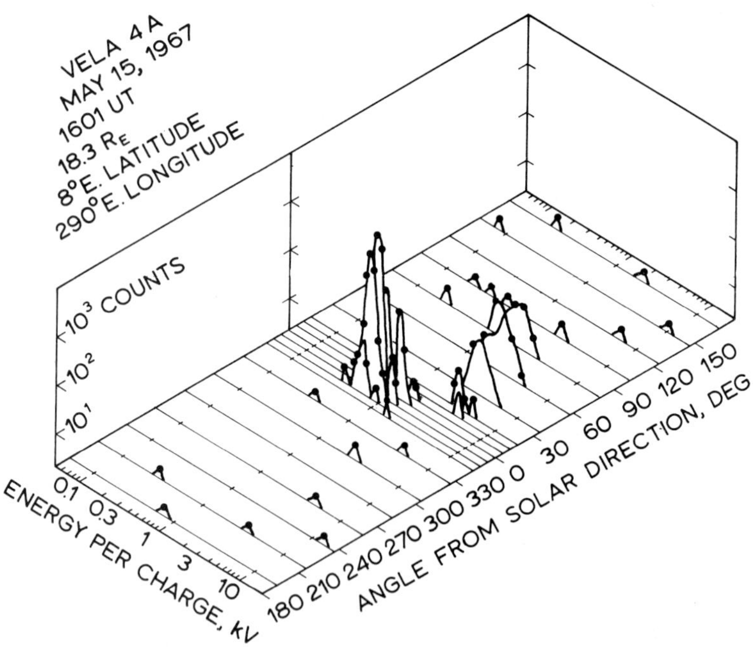

Fig. 24. An example of backstreaming protons from the bow shock observed by Vela 4a on May 15, 1967. The satellite was immediately outside the bow shock on the dawn site.

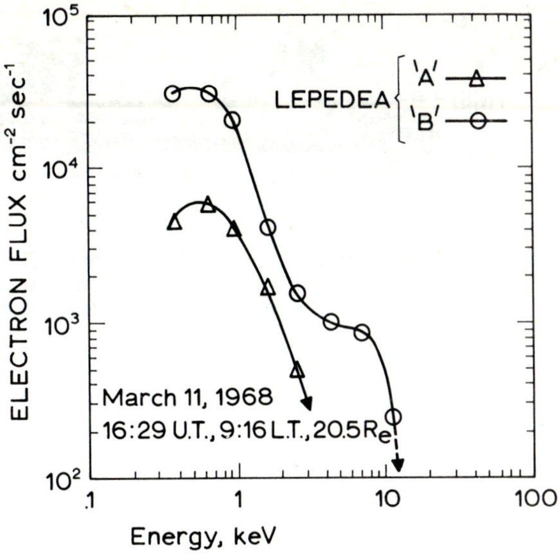

Fig. 25. An example of backstreaming electrons from the bow shock observed by OGO 5 on March 11, 1968 The sensors Lepedea B points back at the Earth, whereas Lepedea A is transverse to this direction.

10–15 R_E from the shock and their polarization is lefthand circular. OGO-5 data have shown (Olson, 1970; Russell *et al.*, 1971; Childers and Russell, 1971) that other waves are present when the magnetosonic low frequency waves are observed. Figure 28a shows a power spectrum with the low frequency peak ($F \sim 0.02$ Hz), and with a higher frequency peak at $F \sim 0.55$ Hz. This peak appears to be due to whistler wave packets which are often observed upstream (see Figures 26 and 28b), or to higher frequency waves, (~ 1 Hz) observed both with the low frequency magnetosonic waves (see Figure 28c) and without these waves (see Figure 20b). These whistler waves observed upstream of the bow shock have recently been extensively studied by Fairfield (1973). While the low frequency magnetosonic waves appear to be related to back streaming protons interacting with the incoming solar wind (see the solar wind flux oscillations in Figure 26), the origin of the high frequency whistler waves is less clear. A possibility is that they are related to back-streaming electrons (notice that the frequency ratio of low frequency and whistler waves which are observed is of the order of $(m_e/m_i)^{1/2}$). Very high energy electrons are in any case responsible for the very high frequency electron plasma waves observed by OGO 5 (Fredricks *et al.*, 1971; Scarf *et al.*, 1971; Fredricks *et al.*, 1972). Figure 29 shows 30 KHz electrostatic noise correlated with the integral flux of non-thermal electrons backstreaming from the bow shock. The local plasma frequency was $27 \sim 32$ KHz. While the electron plasma waves and the high frequency whistlers have a group velocity high enough to propagate upstream the low frequency magnetosonic waves and the wave packets do not propagate upstream. These waves have therefore a limited region of space where they can be observed. This region which is shown in Figure 1 where it is called knowing region, has been geometrically defined in Greenstadt *et al.* (1970a), Green-

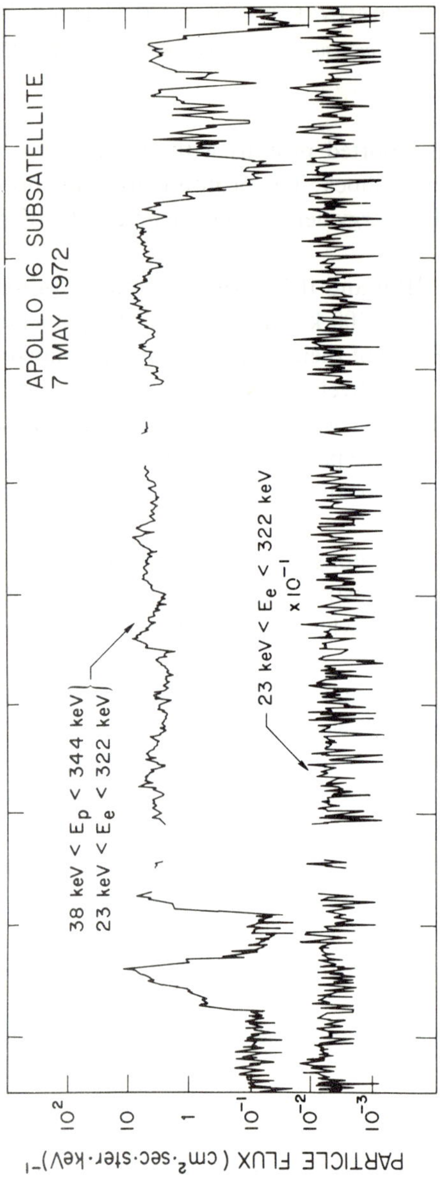

Fig. 1. Measurement showing the presence of upstream protons for many hours.

Acknowledgements

This work was supported by NASA contracts NAS 5-11038 and NAS 9-10509.

References

Anderson, K. A.: 1969, *J. Geophys.* **74**, 95.

Asbridge, J. R., Bame, S. J., and Strong, I. B.: 1968, *J. Geophys. Res.* **73**, 5777.

HEOS 2 IN RELATION TO EARLIER POLAR CUSP MEASUREMENTS

K.-P. WENZEL

Space Science Department (ESLAB), European Space Research and Technology Centre, Noordwijk, The Netherlands

Abstract. The purpose of this paper is to place the HEOS-2 polar cusp observations that will be described in more detail in the following two papers, in proper perspective with regard to earlier measurements. Previous satellite results are shortly reviewed. The orbit of HEOS 2 and its experiments contributing to the polar cusp observations are presented. The topology of the polar cusp as seen by HEOS 2 is compared to the results from IMP 5.

1. Introduction

The purpose of this paper is to place the HEOS-2 polar cusp observations that will be described in the following two papers (Bahnsen *et al.*, 1974; Paschmann *et al.*, 1974) in proper perspective with regard to measurements on previous satellites. No attempt is made to review all pertinent literature, only recent observational results concerning plasma entry into the magnetosphere and the topology of the polar cusp – both areas to which HEOS 2 may contribute – being shortly mentioned.

The interaction of the Earth's magnetic field with the solar wind causes the high latitude geomagnetic field lines to be blown back into an elongated tail. This morphology implies that a pair of neutral points (or lines) must exist on the dayside high-latitude magnetopause in the northern and southern hemispheres. In the noon-midnight meridional plane, all field lines located at latitudes higher than the neutral points are assumed to be swept back into the geomagnetic tail and field lines at latitudes below the neutral points pass through the sunlit hemisphere of the geomagnetic cavity.

Many speculations relating these neutral points to auroral and magnetospheric phenomena arose in the past. One main question was: Can solar or magnetosheath plasma flow directly into the magnetosphere through these polar neutral points?

2. Review of Earlier Results

Until recently only indirect evidence was available to support the idea that magnetosheath plasma had access to the dayside auroral oval. Most of this evidence came from various dayside ionospheric observations (see review by Heikkila and Winningham, 1971).

Within the past two years four satellite measurements of direct influxes of magnetosheath plasma have been reported: two at low altitudes and two at mid- to high altitudes (see Table I). Heikkila and Winningham (1971) have observed on the near Earth polar satellite ISIS 1 electron and proton intensities similar to those of the magnetosheath over a broad latitude band with a width of approximately several

D. E. Page (ed.), Correlated Interplanetary and Magnetospheric Observations, 229–235.

degrees in magnetic latitude in the dayside auroral zone during a magnetic storm. Frank and Ackerson (1971) concluded from observations on the low altitude near-polar satellite INJUN 5 that a narrow band located at the high latitude boundary of the dayside auroral precipitation pattern is the low-altitude signature of the polar neutral point. This band was observed over a large range of local times around the sunlit auroral zone.

TABLE I

Satellite observations of polar cusp

Satellite	Launch	Apogee (R_E)	Observer	Polar cusp observation
ISIS 1	30 Jan. 69	1.6	Heikkila and Winningham (1971)	Low-altitude
INJUN 5	8 Aug. 68	1.4	Frank and Ackerson (1971)	Low-altitude
OGO 5	4 Mar. 68	23.4	Russell *et al.* (1971) Scarf *et al.* (1972)	One mid-altitude observ. during large magnetic storm
IMP 5	21 Jun. 69	28.8	Frank (1971) Fairfield and Ness (1972)	Mid-to-high-altitude
HEOS 2	31 Jan. 72	38.7	See Table II	Mid-altitude

The first in situ observations of the character of the plasmas and the magnetic field in the vicinity of this neutral line at the high latitude dayside magnetopause were made by IMP 5, launched into a highly eccentric high-latitude orbit Frank (1971) reports that IMP 5 repeatedly encountered the direct influx of magnetosheath plasma. He concludes that this direct influx of protons and electrons is a usual feature of the magnetosphere and occurs in a band across the magnetosphere rather than in a single tube of magnetic flux. Frank has termed this region the polar cusp.

Figure 1 shows the geometry and location of the polar cusp in the noon meridional plane during periods of relative magnetic quiescence as derived by Frank (1971). The co-ordinates are geocentric radial distance R and magnetic latitude λ_m (magnetic latitude, the angle between a point of observation and the magnetic equator, should not be confused with invariant latitude, the magnetic latitude of a field line's intersection with the surface of the Earth). The polar cusp was observed at geocentric distances between about 4 and 9 R_E on inbound segments of the IMP-5 orbits. Several important features of the polar cusp are its relatively well-defined poleward and equatorward boundaries, its position poleward of and adjacent to the trapping boundary for energetic electrons with $E > 45$ keV, and the bulk flow of magnetosheath plasma in the high-altitude polar cusp within several R_E of the magnetopause.

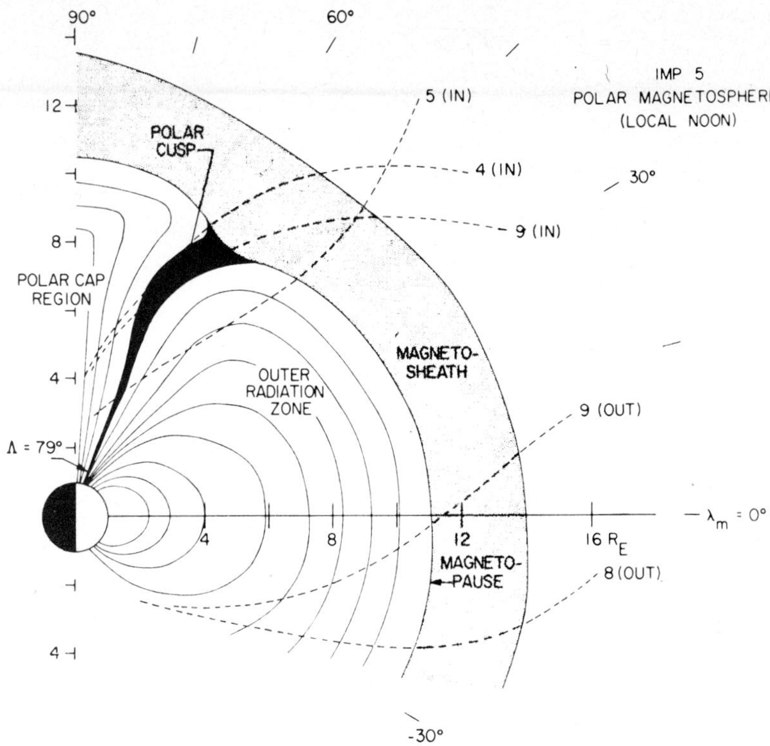

Fig. 1. Geometry and location of polar cusp within the polar magnetosphere in the noon meridional plane during periods of relative magnetic quiescence. The co-ordinates are geocentric radial distance R and magnetic latitude λ_m. The intersection of the polar cusp with the auroral zone is positioned at dipole invariant latitude $\Lambda = 79°$. Several sample trajectories of IMP 5 are also shown (after Frank, 1971).

Simultaneous measurements of the polar magnetic fields with IMP 5 have been presented by Fairfield and Ness (1972).

The mid-altitude polar cusp was encountered on only one other occasion, namely by OGO 5 (Russell *et al.*, 1971; Scarf *et al.*, 1972; D'Angelo, 1973; Kivelson *et al.*, 1973). During the large magnetic storm of 1 November 1968 the apparent direct penetration of magnetosheath plasma into the dayside magnetosphere and wave motion in the cusp were detected, when the cusp was severely displaced to a low geomagnetic latitude of 43°.

3. HEOS-2 Orbit and Experiments

HEOS 2 launched on 31 January 1972 into a highly eccentric polar orbit offers a good opportunity for measurements in the high-latitude magnetosphere in spite of its distant apogee and its low bit rate. Figure 2 shows two outbound trajectories of HEOS 2 projected into the dayside polar magnetosphere as sketched by Frank.

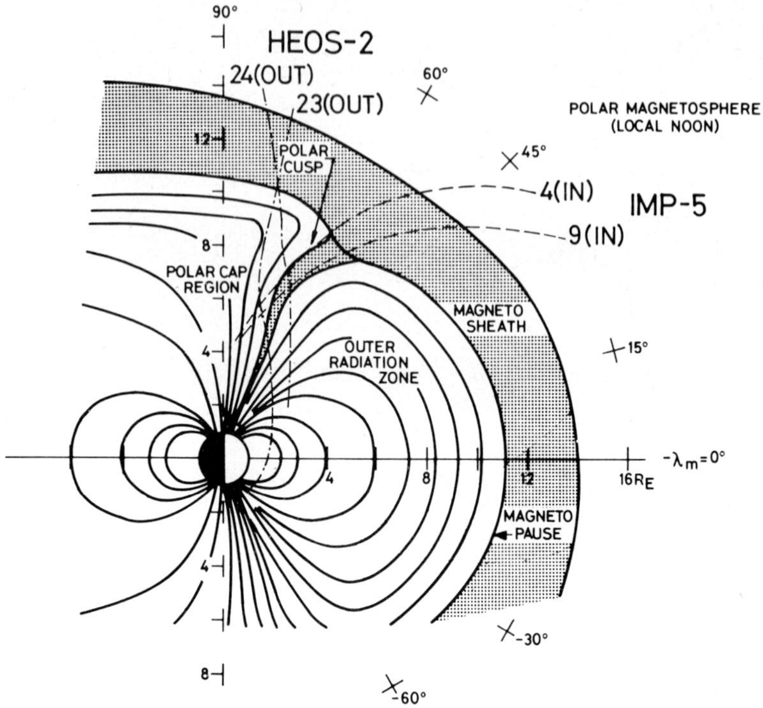

Fig. 2. Dayside polar magnetosphere with HEOS-2 and IMP-5 sample trajectories.

TABLE II

HEOS-2 experiments associated with polar cusp observations

Experiment	Group	Measurement	Measurement range
S 201	Imperial College, London	DC magnetic field	0 – ±144 Gammas
S 203	Danish Space Research Institute, Copenhagen	ULF magnetic fields	20–236 Hz
S 202	University of Rome	Plasma (electrons and protons)	10 eV–10 keV (per unit charge)
S 210	Max-Planck-Institut, Garching	Plasma (electrons and protons)	100 eV–40 keV (per unit charge)
S 204	Space Science Dep. ESTEC	Energetic electrons	>40 keV

Two IMP-5 orbits are also shown. The HEOS-2 polar cusp crossings during the summer of 1972 lie typically at 4–6 R_E, through the mid-altitude polar cusp. Due to the high satellite velocity of about 230 km min^{-1} in this region, the crossing times are from a minimum of a few minutes up to about half an hour only. It is to be noted that the re-orientation capability of HEOS 2 allows the spin axis to be placed into any direction perpendicular to the Sun–Earth line, permitting for example the study of the plasma flow in more than one plane during the spacecraft's lifetime.

The payload of HEOS 2 (de Waard *et al.*, 1970) comprises several experiments suitable to observe phenomena associated with the dayside polar cusp. These experiments are summarized in Table II. Observations of the ULF magnetic fluctuations at the polar cusp boundary as observed by the S203 instrument have been interpreted by Bahnsen *et al.* (1974). The observations of the plasma population around the polar cusp and in the tail region near the magnetopause as seen by the S210 instrument are described by Paschmann *et al.* (1974).

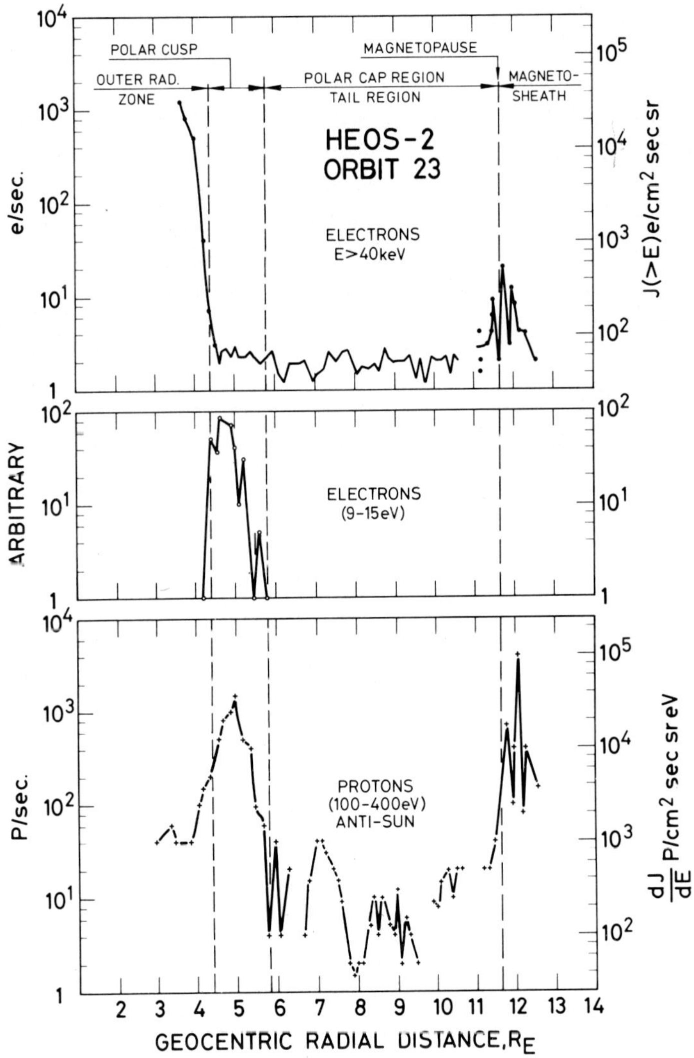

Fig. 3. Spatial distributions of energetic electrons >40 keV (S204), 9–15 eV electrons (S202) and 100–400 eV protons (S210) on outbound pass of HEOS 2 through polar magnetosphere.

4. HEOS-2 Results

Figure 3 summarizes characteristic observations from 3 experiments as HEOS 2 moves outbound through the various regions of the high-latitude magnetosphere. (See top of the figure). Three particle count rates are plotted versus geocentric radial distance. The top panel shows the electron rate >40 keV as observed by a Geiger counter of the S204 experiment of ESTEC, the centre panel the count rate of 9–15 eV electrons as observed by the S202 instrument of the University of Rome and the bottom plot the sum of 3 proton channels between about 100 and 400 eV from the S210 instrument of the Max-Planck-Institut, Garching.

The sharp drop-off of the energetic electron intensity near 4.4 R_E (orbit 23) marks the termination of the outer radiation zone. The position of this trapping boundary is interpreted as the high-latitude termination of closed field lines, i.e. it is coinciding with the equatorward boundary of the polar cusp. It is to be noted that on most orbits a short burst of low frequency magnetic noise is observed by the S203 experiment coinciding with this boundary. The polar cusp is observed as a region with intense fluxes of low energy electrons and protons, typically about 1 R_E wide at a distance of 5 R_E. The polar cap region (tail region) is characterized by very low intensities of energetic electrons and magnetosheath plasma. The high-latitude tail magnetopause is encountered around 11.7 R_E when HEOS 2 enters the magnetosheath.

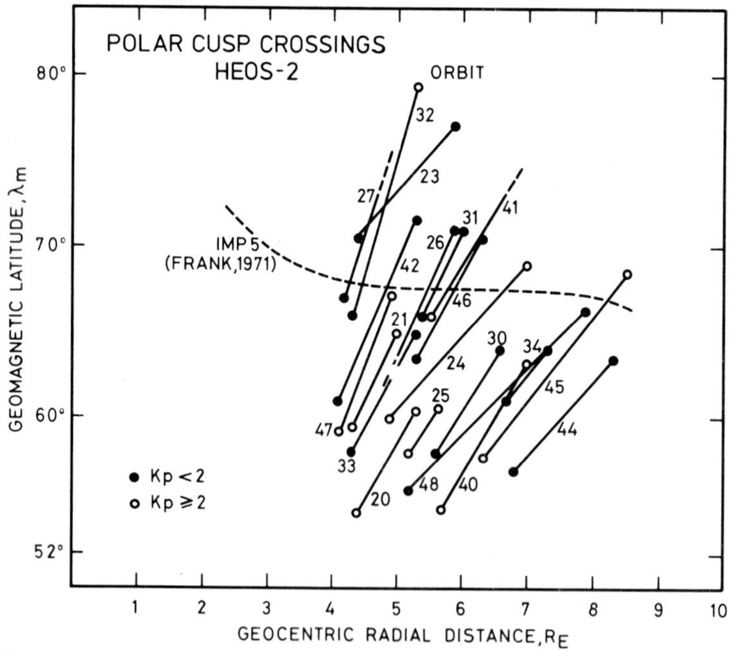

Fig. 4. Summary of positions of 20 polar cusp crossings at various longitudes in the dayside polar magnetosphere as function of geocentric radial distance R and geomagnetic latitude λ_m as observed by several instruments on HEOS 2.

Figure 4 shows the observed positions of the dayside polar cusp for a series of HEOS-2 orbits in 1972. The co-ordinates are geomagnetic latitude versus geocentric radial distance. The data are derived from the plasma, ULF and energetic particle experiments. Each line represents one crossing marked by the orbit number. The low-latitude termination of each passage corresponds to the last closed field line, the high latitude termination to the entry into the tail region. Crossings at magnetically quiet times ($Kp<2$) and disturbed times ($Kp \geqslant 2$) are shown. The HEOS-2 data cover all GSM longitudes between about 14.5 and 6 h local time (ϕGSM $\sim$ 38 to 270 deg). For comparison the average polar cusp location near the Sun-Earth line for magnetically quiet times as derived by Frank (1971) from the IMP-5 data is plotted.

One may conclude:

(1) The polar cusp is observed by HEOS 2 over a broad band of longitudes on the dayside magnetosphere.

(2) The HEOS-2 polar cusp passages near local noon (orbits 23–24) at magnetically quiet times are scattered around the average polar cusp location given by Frank (1971).

(3) The polar cusp when crossed at longitudes away from the Sun-Earth line (away from local noon) is found on the average at latitudes a few degrees lower than at local noon. This is in agreement with the known dependence of the electron trapping boundary on local time.

(4) The HEOS-2 data show indications as did IMP 5 that the polar cusp moves equatorwards at times of relative magnetic disturbance.

Acknowledgement

The HEOS-2 project scientist is grateful to the Garching, Rome and Copenhagen groups for allowing to show part of their data for comparison in this summary paper.

References

Bahnsen, A., D'Angelo, N., and Ungstrup, E.: 1974, this volume, p. 237.
D'Angelo, N.: 1973, *J. Geophys. Res.* **78**, 1206.
Fairfield, D. J. and Ness, N. F.: 1972, *J. Geophys. Res.* **77**, 611.
Frank, L. A.: 1971, *J. Geophys. Res.* **76**, 5202.
Frank, L. A. and Ackerson, K. L.: 1971, *J. Geophys. Res.* **76**, 3612.
Heikkila, W. J. and Winningham, J. D.: 1971, *J. Geophys. Res.* **76**, 883.
Kivelson, M. G., Russell, C. T., Neugebauer, M., Scarf, F. L., and Fredricks, R. W.: 1973, *J. Geophys. Res.* **78**, 3761.
Paschmann, G., Grünwaldt, H., Montgomery, M. D., Rosenbauer, H., and Sckopke, N.: 1974, this volume, p. 249.
Russell, C. T., Chappell, C. R., Montgomery, M. D., Neugebauer, M., and Scarf, F. L.: 1971, *J. Geophys. Res.* **76**, 6743.
Scarf, F. L., Fredricks, R. W., Green, I. M., and Russell, C. T.: 1972, *J. Geophys. Res.* **77**, 2274.
Waard, J. de, Kooter, C. J., and Summerhill, S.: 1970, HEOS-A2 Satellite Functional Description, ESRO TN-104 (ESTEC), June.

ULF MEASUREMENTS AT THE POLAR CUSP

A. BAHNSEN, N. D'ANGELO, and E. UNGSTRUP
Danish Space Research Institute, Lyngby, Denmark

Abstract. Measurements of ULF fluctuations at several HEOS-2 polar cusp crossings are described. A possible source of the fluctuations appears to be a Kelvin-Helmholtz instability exited at the cusp boundary by velocity shear in the ion flow. The influence of this instability on the gross properties of plasma flow in the cusp region is discussed.

1. Introduction

Several investigations have been reported in the last few years on the penetration of magnetosheath plasma down to ionospheric heights in the so-called polar cusp (Heikkila and Winningham, 1971; Frank and Ackerson, 1971; Frank and Gurnett, 1971; Frank, 1971). It is interesting that already in 1948, in a paper concerned with solar-diurnal variation in terrestrial magnetism, Johannes Olsen of the Danish Meteorological Institute came very close to predicting the existence of the polar cusp (Olsen, 1948).

The polar 'neutral points' (e.g. Spreiter *et al.*, 1968) of mathematical models for the high latitude dayside magnetosphere are to be replaced by 'bands', approximately 1 R_E in north-south extent, permanently present in a $\gtrsim 100°$ longitude sector centered around the noon meridian. From these 'bands' magneto-sheath plasma gains direct access to the ionosphere through the polar cusp region. During quiet periods the intersection of the polar cusp with the auroral zone is positioned, according to Frank (1971), at an invariant latitude $\Lambda = 79°$ ($\pm 1°$). The equatorward boundary of the polar cusp appears to coincide with the high-latitude boundary of the outer radiation zone energetic ($E > 40$ keV) electrons. In fact, the sharp variations in the counting rate of $E > 40$ keV electrons observed near the day-side auroral zone by polar orbiting satellites is generally taken as defining the location of the equatorward boundary of the polar cusp.

In the midaltitude cusp (~ 5 R_E geocentric radial distance) the angular distribution of protons is peaked at small pitch angles. According to Frank (1971), "the protons are literally flowing down the magnetic field toward the Earth's atmosphere". Measurements of directional, differential energy spectra of proton intensities at the magnetosheath (9.8 R_E), distant polar cusp (7.2 R_E) and midaltitude polar cusp (5.35 R_E) led Frank to conclude that the polar cusp acts on the magnetosheath plasma as a kind of 'velocity filter', by suppressing low-energy protons. This observation points to an acceleration mechanism acting on the magnetosheath plasma as it travels from the magnetosheath to midaltitudes.

Wave motion in the cusp region was encountered by OGO 5 during a large storm, when the cusp was strongly displaced to low latitudes (Russell *et al.*, 1971; Scarf *et al.*, 1972). High-frequency (~ 1 kHz to ~ 70 kHz) oscillations were detected by

D. E. Page (ed.), Correlated Interplanetary and Magnetospheric Observations, 237–247. All Rights Reserved.

the OGO-5 electric antenna. In addition, magnetic fluctuations were observed in the ULF range (frequencies from 0.07 Hz to $\sim$30 Hz), the oscillation amplitudes being typically of the order of a fraction of a gamma to a few gammas. The most intense waves were observed near the cusp boundary, as defined by the local observations of energetic electrons. It was established that "the strong magnetic noise with $f \lesssim f_c^+$ is associated with the proximity of the cusp, rather than with any general characterizations of the over-all magnetic storm" (Scarf *et al.*, 1972). These OGO-5 results are in agreement with the HEOS-2 data to be described in the present paper (particularly the association of the magnetic noise to the proximity of the cusp). A discussion of the OGO-5 magnetic noise in terms of a Kelvin-Helmholtz instability, excited at the cusp boundary by downward flowing protons, has already been given (D'Angelo, 1973).

The present paper contains data obtained with the DSRI (Danish Space Research Institute) magnetic antenna on the HEOS-2 satellite, for a total of 20 polar cusp passes. Where appropriate, these data are correlated with those obtained with low-energy particle detectors and a GM counter ($E > 40$ keV electrons) on board HEOS 2. The magnetic noise observations are again discussed in terms of a Kelvin-Helmholtz instability at the cusp boundary. Finally, a picture is proposed for the polar cusp in terms of a 'leaky Venturi tube'. It appears at present to account for the acceleration of the magnetosheath plasma observed by Frank (1971) and for the observed widening, transverse to the geomagnetic field lines, occurring as the plasma travels along the cusp toward the ionosphere.

2. Satellite Orbit

The HEOS-2 satellite was intended for the study of the polar magnetosphere and the solar wind, and was therefore launched in a highly elliptical orbit with apogee at high latitudes. The orbital period is 5.2 days. The apogee is at an altitude of approximately 40 R_E (Earth radii), the plane of the orbit being almost perpendicular to the ecliptic plane with the major axis tilted at an angle of 11.5° away from the Sun at summer solstice. The satellite was launched on January 31st, 1971 from Western Test Range, California.

The orbit, at a time close to summer solstice, is shown in Figure 1. The satellite passes through the north polar cusp at approximately 6 R_E geocentric distance. It crosses the south polar cusp at low altitudes in both summer and winter, but very little data are available close to perigee.

3. Observations

The DSRI experiment was designed to measure low-level fields. By means of a loop antenna, the magnetic field fluctuations are measured at five fixed frequencies: 20, 44, 84, 148 and 236 Hz. The results are presented in Figure 2 for a short time period of the orbit shown in Figure 1. Inside the magnetosphere the signal stays generally close to a steady background noise level. As the satellite enters the magnetosheath region, where the plasma is turbulent, large amplitudes are measured, especially at the lowest frequencies.

The same type of VLF signal as observed in the magnetosheath is also seen earlier in the orbit (Figure 2), around 0610 UT. The trapping boundary for $E > 40$ keV

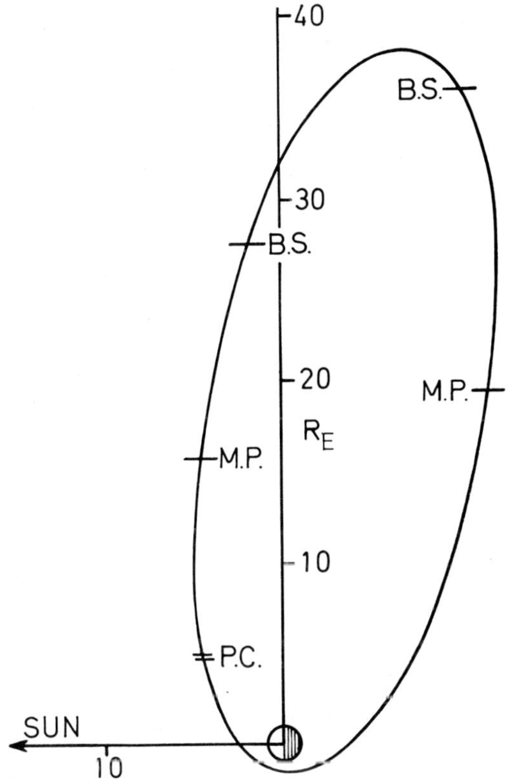

Fig. 1. HEOS-2 orbit near summer solstice. The crossings of polar cusp, magnetopause, and bow shock are also shown.

electrons is located at $\sim$UT 0610. We may thus assign the enhanced magnetic field noise to the cusp plasma.

More examples of cusp-crossings are shown on Figures 3a and 3b. Here, the VLF signal is plotted together with data from the low energy particle experiments. Figure 3a shows a burst of noise, close to the equatorward cusp boundary, with a full width at half maximum of $\sim$0.7 min. This duration corresponds to a distance of 160 km along the orbit, or approximately 100 km normal to the geomagnetic field lines. The noise persists at a lower level for a total of $\sim$15 min, i.e. 2000 km normal distance.

The count rates of low energy electrons (9 to 250 eV) are large immediately northward of the trapping boundary and decay gradually with increasing latitude. The

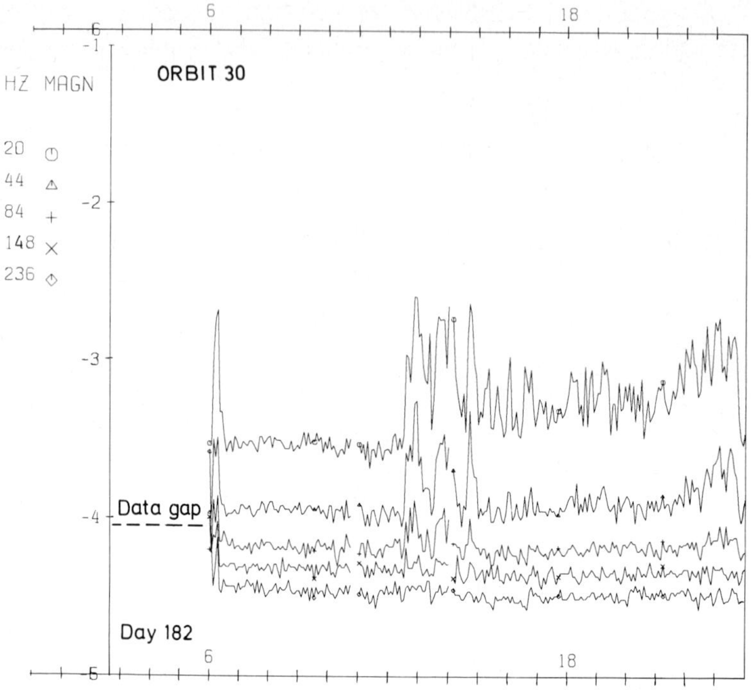

Fig. 2. Signals detected on orbit 30, day 182 (June 30) by the DSRI magnetic antenna, at $f = 20$, 44, 84, 148 and 236 Hz. The enhanced signal on all frequency channels at polar cusp crossing is seen near 06 UT. The magnetopause crossing occurs near 12:30 UT. The ordinate axis is logarithmic, going from 10^{-5} to 10^{-1} $\gamma \cdot Hz^{-1/2}$.

proton energy spectra, in sunward and anti-sunward directions, are given in Figure 3a, at 4-min intervals. The first spectra, (0207 UT), measured inside the trapping boundary, are deficient in low energy protons and show an isotropic angular distribution. The next measurement, (0212 UT), made right outside the boundary, shows a large flux of low energy (~ 1 keV) protons, with a very distinct angular asymmetry. The last energy spectra peak at still lower energies and indicate a nearly isotropic angular distribution.

The spectra at 0212 UT are quite similar to those measured by Frank (1971, Figure 11), thus confirming Frank's conclusion that "the protons are literally flowing down the magnetic field toward the Earth's atmosphere".

Likewise, we infer from these spectra a flow energy of the order of a few hundred eV.

Figure 3b shows a cusp crossing on a later orbit. The trapping boundary is detected at approximately the same position as in the previous orbit. Although the magnetic field fluctuations are somewhat enhanced in this region, they are not collocated with the trapping boundary. In addition, their amplitude is one order of magnitude smaller than seen in Figure 3a. The low energy electrons show an increase in flux similar to that in Figure 3a, occurring, however, not at the trapping boundary but at a distance of ~ 2000 km north of it.

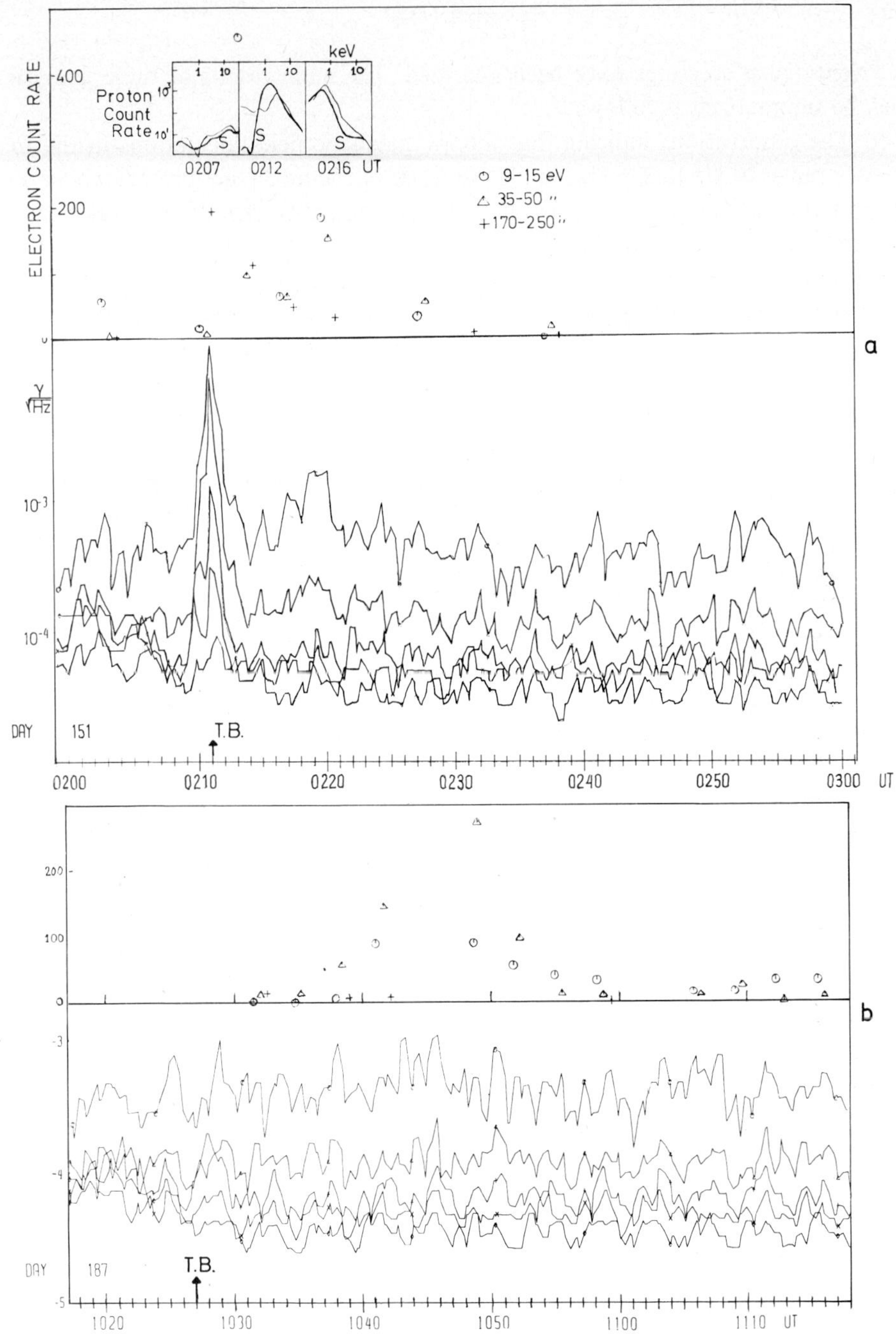

Fig. 3. (a) lower panel: intensity of the magnetic fluctuations, as a function of UT, detected at $f = 20$, 44, 84, 148 and 236 Hz. The trapping boundary for $E > 40$ keV electrons is indicated at 02:11 UT (courtesy of P. Wenzel). Upper panel: electron count rates vs. UT, in the energy intervals 9–15 eV 35–50 eV, and 170–250 eV (courtesy of A. Egidi). The insert shows at three different times proton spectra measured in the solar (S) and antisolar directions (courtesy of H. Grünwaldt). Notice the depletion in low energy protons at 02:12 UT. (Orbit 24, day 151, (May 30)). (b) Intensity of the magnetic fluctuations (lower panel) and electron count rates (upper panel) vs. UT, on orbit, 31, day 187 (July 5).

Twenty cusp crossings have been analyzed. The main results of these analysies may be summarized as follows:

(i) The observations indicate that intense magnetic field fluctuations in the frequency range 20 Hz to 236 Hz, similar to those in Figure 3a, are present whenever:

(a) the satellite is, in the polar cusp, in the immediate vicinity of a well defined trapping boundary for $E>40$ keV electrons, and

(b) the thermal plasma flux is substantial in that region.

(ii) The noise observed at the trapping boundary is very well localized, with a width of the noise peak a few hundred km measured transverse to the field lines. Note that this is considerably less than the width of the polar cusp.

(iii) To a very good approximation, the frequency spectra follow a power law over the entire measured frequency range. The amplitude of the signal varies with frequency as f^{-n}, with $n \sim 1.5$–2. Figure 4 shows in histogram form the distribution of n.

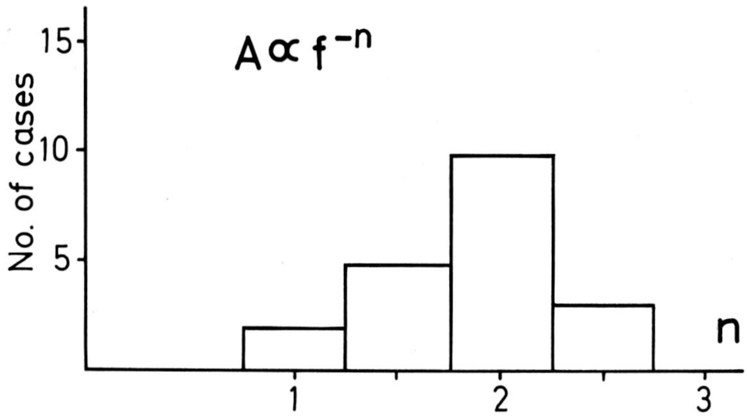

Fig. 4. The amplitude spectra of the magnetic fluctuations at polar cusp crossings are of the type $A \propto f^{-n}$. This figure shows in histogram form the distribution of n for a total of 20 cusp spectral measurements.

Intense fluctuations are also observed at cusp crossings in a channel centered at 3 Hz (Hedgecock, private communication). Although these measurements are not amplitude intercalibrated with ours, they do show a very good time correlation.

4. Discussion of Results

In this section we interpret the Section 3 results in terms of a Kelvin-Helmholtz instability produced by a shear in the ion velocity parallel to the magnetic field lines at the equatorward boundary of the cusp. The Kelvin-Helmholtz instability has already been invoked in order to explain various geophysical and cosmic plasma phenomena. For a review of the vast amount of previous work, the reader is referred to a paper by Gerwin (1968).

Several authors have proposed the Kelvin-Helmholtz instability on the magnetosphere boundary as the source of surface waves, in order to explain the origin and the characteristics of long-period micropulsations (e.g. Dungey, 1955; Atkinson and Watanabe, 1966; Dungey and Southwood, 1970; Southwood, 1968). Lanzerotti *et al.* (1972) have recently concluded, however, that a surface wave arising from the Kelvin-Helmholtz instability *on the magnetopause* cannot account for the observed polarizations of long period micropulsations. It may also be noted that this type of Kelvin-Helmholtz instability on the magnetopause, produced by shear in the flow normal to the geomagnetic field lines, should be absent in some longitude sector centered around the noon meridian, since the subsolar point on the magnetopause must be a stagnation point for the flow. A state of particular 'quiescence' near the noon meridian is, however, suggested neither by studies of geomagnetic micropulsations nor by the HEOS data of Section 3. In fact, micropulsation data by Gupta *et al.* (1971) indicate maxima in the Pc 3 and Pc 4 occurrence frequency near the noon meridian.

In addition, measurements of magnetic field fluctuations on IMP 5 by Fairfield and Ness (1972), show that spectra in the cusp contain "more power than in any of 70 magnetosheath spectra computed by Fairfield and Ness at lower latitudes". This last result is also not easily understandable on the basis of a Kelvin-Helmholtz instability on the magnetopause. It rather points to an instability excited in the cusp itself.

A Kelvin-Helmholtz instability excited at the cusp boundary by shear in the ion velocity parallel to the geomagnetic field lines was suggested (D'Angelo, 1973) as a possible source for the ULF magnetic fluctuations observed on OGO 5 by Russell *et al.* (1971). The more detailed HEOS-2 data give support to this suggestion. In what follows, we add a few remarks based upon and relevant to the HEOS-2 observations:

(a) The most prominent increases in magnetic noise are observed near the equatorward cusp boundary, when a velocity shear appears to occur (see Figure 3a).

(b) The interpretation of the enhanced noise peaks in terms of a Kelvin-Helmholtz instability is consistent with the fact that, when the bulk of the cusp plasma is observed not in the immediate vicinity of the cusp boundary, little or no enhanced noise is observed (see Figure 3b).

(c) The measured power spectra of the fluctuations are roughly of the type expected for ion-acoustic noise. They are in most cases in disagreement with an interpretation of the noise in terms of a drift-wave instability. This point has already been remarked upon when discussing the OGO-5 data (D'Angelo, 1973).

(d) There is substantial power in magnetic fluctuations at $f=3$ Hz. Since the ion-gyrofrequency $f_{ci} \approx 6$ Hz at 6 R_E, this observation appears to exclude an interpretation of the noise in terms of ion-cyclotron waves. These waves could be excited by sufficiently intense current flow along the geomagnetic field lines. The wave frequency, however, should always exceed f_{ci}. In addition, from explicit expressions for growth and damping rates of ion-cyclotron waves, as given for instance by Kai Fong Lee (1972), one should expect to observe these waves, if at all excited, over distances comparable with the entire latitudinal extent of the polar cusp. The HEOS-2 observations show, however, that the cusp noise is much more localized.

(e) The HEOS-2 frequency range partially overlaps with, and extends to the high-frequency side of, the OGO-5 range. The HEOS-2 data thus provide information at and beyond the short wavelength end of the fluctuation spectrum measured by OGO 5.

5. Model for Plasma Flow in the Polar Cusp

In this section we propose a tentative model for the plasma flow in the polar cusp region. This model should only be viewed as a rough picture, based on the plasma flow properties indicated by Frank's (1971) measurements and by preliminary results from the HEOS-2 observations. Undoubtedly modifications will be required as more data become available and further phenomena come to light (see, for instance, the observation of current sheets by Russell *et al.*, 1973).

The motion of solar wind particles in the neighbourhood of a neutral point has been discussed, for instance, by Ershkovich *et al.* (1967). They analyzed the trajectories of individual particles and the formation of a 'neutral region' through which solar wind particles may be injected into the magnetosphere. Our model is based on a fluid picture for both electrons and ions. A fluid picture, in spite of the fact that mean-free-paths for particle-particle collisions are larger than the characteristic dimensions of the system, is justified by its success in describing many features of the solar wind plasma, where a similar relation obtains between mean-free-paths and characteristic dimensions. The picture we propose here for the polar cusp is that of a 'magnetic Venturi tube'. The material walls of a normal Venturi tube are replaced by the converging magnetic flux tube forming the cusp. A similar technique has already been employed in describing the behavior of a magnetic Laval nozzle (Andersen *et al.*, 1969), in which a converging-diverging magnetic flux tube replaces the material walls of a normal nozzle. The model of Andersen *et al.* (1969) accounted for the experimental results obtained in an alkali-plasma device. We also treat the cusp plasma in an isothermal approximation for both ions and electrons, i.e. $\gamma_e = 1$ and $\gamma_i = 1$, where $\gamma = C_p/C_v$. Similar results are obtained if an adiabatic law is used ($\gamma = 5/3$).

The walls of our Venturi tube are not, however, perfectly inpenetrable to the plasma, since a considerable amount of diffusion probably occurs across the geomagnetic field lines at the highly turbulent cusp boundary. A diffusion rate roughly given by the Bohm diffusion coefficient is suggested by the 'anomalous' latitudinal width of the cusp at ionospheric heights (Heikkila and Winningham, 1971). We may then picture the polar cusp as a 'leaky' Venturi tube, the 'leak' being provided by turbulent diffusion. On the basis of the continuity and momentum equations for ions and electrons, neglecting electron inertia and restricting the analysis to a one-dimensional, steady-state problem, we obtain:

$$-(1 - M^2)\frac{1}{M}\frac{\mathrm{d}M}{\mathrm{d}x} = \frac{1}{A}\frac{\mathrm{d}A}{\mathrm{d}x} + \frac{2\pi\mu}{C_s}\frac{1}{M} \tag{1}$$

where x is the coordinate along the flow, M the flow Mach number, A the cross-sectional area of a cusp 'element', and C_s the ion-acoustic speed. The quantity μ is related to the diffusion coefficient through $\mu = D_\perp/A$ and is independent of x if, as we assume, $D_\perp \propto 1/B$ (Bohm diffusion) and $A \propto 1/B$ (conservation of magnetic flux). Equation (1) coincides with Equation (1) of Andersen *et al.* (1969) when $\mu = 0$, i.e. when the effect of turbulent diffusion is absent. Equation (1) applies to a cusp 'element' of circular cross-section.

Before we proceed to solving Equation (1) we must make some assumption as to the nature of the particular type of plasma instability responsible for the turbulence at the cusp boundary. Since in Sections 3 and 4 we have given evidence to the effect that a Kelvin-Helmholtz instability at the cusp boundary is likely to be the source of the observed turbulence, we make here the assumption that the Kelvin-Helmholtz instability is indeed responsible for plasma diffusion as inferred from laboratory experiments with alkali-plasmas (D'Angelo and von Goeler, 1966). The condition for excitation of the Kelvin-Helmholtz instability at the cusp boundary can be written as (D'Angelo, 1965):

$$M > M_c = \frac{l_v}{l_n}, \tag{2}$$

where the 'critical' Mach number is given by the ratio of the e-folding length of the velocity profile to the e-folding length of the plasma density profile. This ratio is estimated to be of order 0.1–0.3. For $M < M_c$ we then set $\mu = 0$ in Equation (1), whereas $\mu \neq 0$ if $M > M_c$. We have taken $\mu(M) = \mu^* f(M/M_c)$, with $\mu^* = \text{const}$ and $f(M/M_c)$ being a smoothly varying function of M/M_c, equal to 0.5 at $M/M_c = 1$, to 0 for $M/M_c < 0.95$ and to 1 for $M/M_c > 1.05$.

μ^* was chosen as $3 \times 10^{-3}\ \text{s}^{-1}$ consistent with the assumption $D_\perp \approx D_{\text{Bohm}}$; C_s as 500 km s^{-1}. The results of a computer integration of Equation (1) are shown in Figure 5, for various values of M_c and of M_0 (the Mach number of the flow at the cusp 'entrance'). The term $(1/A)(\mathrm{d}A/\mathrm{d}x)$ was computed on the basis of a dipole field model. This was done for convenience in the calculations. Not a very great difference in the results would occur were the dipole field replaced by a more realistic model. Model calculations by Mead (1964) show the assumption of a dipole field to be reasonably good up to 8 or 9 R_E geometric radial distance.

An upper bound to $M_0 \simeq 0.1$ is suggested both by theoretical arguments (e.g. Willis, 1969) and by plasma flow measurements in the magnetosheath (e.g. Olbert, 1968). The curves in Figure 5 show that the plasma is accelerated up to the point where $M = M_c$. Thereafter the flow velocity remains constant, and only close to the Earth an additional acceleration takes place. If we consider the curve with $M_0 = 0.05$ and $M_c = 0.3$, we find that, over a distance of 4–5 R_E from the magnetosheath, ions have gained a directed kinetic energy of the order of one tenth of their thermal energy, i.e. 100–200 eV. This energy increase is sufficient to account for the acceleration envisaged by Frank (1971) (see Section 1). Note that an important feature of the Kelvin-Helmholtz instability in our model of the cusp is that the instability is excited, and diffusion sets in, only beyond a certain distance from the magnetosheath (i.e. once M has reached the value M_c). It is thus possible to connect together several cusp observations, namely:

(a) the acceleration of the ion flow (directed kinetic energy of a few hundred eV) occurring predominantly at large distances from the Earth,

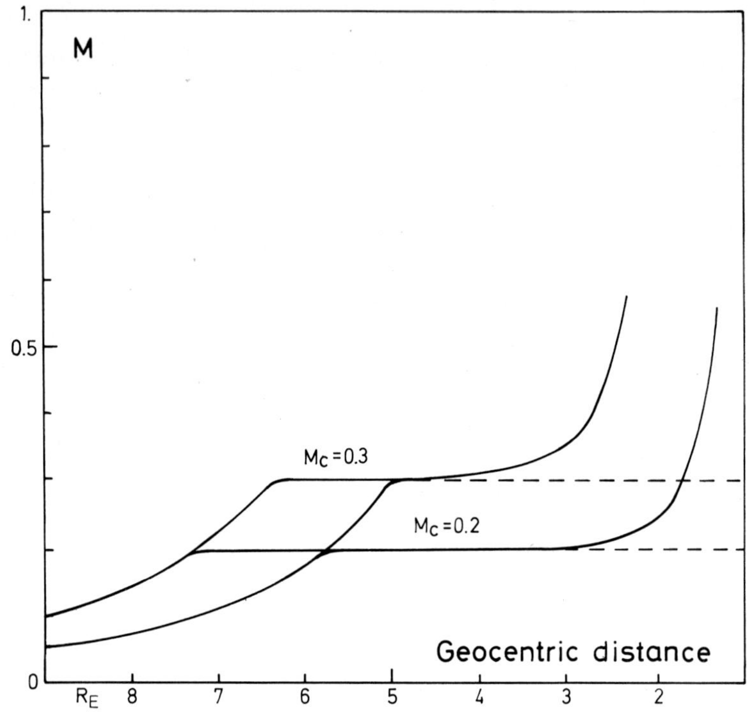

Fig. 5. Model of the plasma flow in the polar cusp, in terms of a 'leaky Venturi tube'. The Mach number, M, of the flow vs. distance along the cusp (in Earth radii). The full lines show the results of the integration of Equation (1) with $\mu^* = 3 \times 10^{-3}\ s^{-1}$, for values of $M_0 = 0.05$ and 0.1, and with $M_c = 0.2$ and 0.3.

(b) the apparent discrepancy between the large observed latitudinal extent of the polar cusp at ionospheric heights (Heikkila and Winningham, 1971) and the 'projected' width from distant cusp measurements (Frank, 1971),

(c) the high Mach number of the flow ($M \gtrsim 0.5$) at ionospheric heights, inferred by Rostoker *et al.* (1972) from the measurements of Heikkila and Winningham (1971).

Acknowledgements

We thank Drs A. Egidi, H. Grünwaldt, P. Hedgecock and P. Wenzel for making available to us part of their data prior to publication.

The support of many staff members of the Danish Space Research Institute in the design, construction, and data reduction phases is deeply appreciated.

We are indebted to Prof. B. Peters for encouraging, and being instrumental in the DSRI participation in the HEOS-2 project.

HEOS 2 is an ESRO satellite, launched by NASA on Jan. 31, 1971 from Western Test Range, California.

References

Andersen, S. A., Jensen, V. O., Nielsen, P., and D'Angelo, N.: 1969, *Phys. Fluids* **12**, 557.
Atkinson, G. and Watanabe, T.: 1966, *Earth Planetary Sci. Letters* **1**, 89.
D'Angelo, N. and von Goeler, S.: 1966, *Phys. Fluids* **9**, 309.
D'Angelo, N.: 1965, *Phys. Fluids* **8**, 1748.
D'Angelo, N.: 1973, *J. Geophys. Res.* **78**, 1206.
Dungey, J. W.: 1955, in *Physics of the Ionosphere*, Physical Society, London, p. 229.
Dungey, J. W. and Southwood, D. J.: 1970, *Space Sci. Rev.* **10**, 672.
Ershkovich, A. J., Pletnev, V. D., and Skuridin, G. A.: 1967, *J. Atmospheric Terrestr. Phys.* **29**, 367.
Fairfield, D. M. and Ness, N. F.: 1972, *J. Geophys. Res.* **77**, 611.
Frank, L. A.: 1971, *J. Geophys. Res.* **76**, 5202.
Frank, L. A. and Ackerson, K. L.: 1971, *J. Geophys. Res.* **76**, 3612.
Frank, L. A. and Gurnett, D. A.: 1971, *J. Geophys. Res.* **76**, 6829.
Gerwin, R. A.: 1968, *Rev. Mod. Phys.* **40**, 652.
Gupta, J. C., Stening, R. J., and van Beek, G. J.: 1971, *J. Geophys. Res.* **76**, 933.
Heikkila, W. J. and Winningham, J. D.: 1971, *J. Geophys. Res.* **76**, 883.
Lanzerotti, L. J., Hasegawa, A., and Tartaglia, N. A.: 1972, *J. Geophys. Res.* **77**, 6731.
Lee, K. F.: 1972, *J. Plasma Phys.* **8**, 379.
Mead, G. D.: 1964, *J. Geophys. Res.* **69**, 1181.
Olbert, S.: 1968, in R. L. Carovillano, J. F. McClay, and H. R. Radoski (eds.), *Physics of the Magnetosphere*, D. Reidel Publ. Co., Dordrecht-Holland, p. 641.
Olsen, J.: 1948, *Terrestr. Magn. Atmospheric Electr.* **53**, 123.
Rostoker, G., Samson, J. C., and Higuchi, Y.: 1972, *J. Geophys. Res.* **77**, 4700.
Russell, C. T., Chappell, C. R., Montgomery, M. D., Neugebauer, M., and Scharf, F. L.: 1971, *J. Geophys. Res.* **76**, 6743.
Russell, C. T., Fredricks, R. W., Kivelson, M. G., Neugebauer, M., and Scarf, F. L.: 1973, 'OGO-5 Observations of the Physical Processes Occurring in the Disturbed Polar Cusp and the Cusp-magnetosheath Interface', Paper b-9a, XVI Cospar meeting, Konstanz 23 May–5 June.
Scarf, F. L., Fredricks, R. W., Green, I. M., and Russell, C. T.: 1972, *J. Geophys. Res.* **77**, 2274.
Southwood, D. J.: 1968, *Planetary Space Sci.* **16**, 587.
Spreiter, J. R., Alksne, A. Y., and Summers, A. L.: 1968, in R. L. Carovillano, J. F. McClay, and H. R. Radoski (eds.), *Physics of the Magnetosphere*, D. Reidel Publ. Co., Dordrecht-Holland, p. 301.
Willis, D. M.: 1969, *Planetary Space Sci.* **17**, 339.

PLASMA OBSERVATIONS IN THE HIGH-LATITUDE MAGNETOSPHERE

G. PASCHMANN, H. GRÜNWALDT, M. D. MONTGOMERY, H. ROSENBAUER, and N. SCKOPKE

Max-Planck-Institut für Physik und Astrophysik, Institut für extraterrestrische Physik, 8046 Garching b. München, Germany

1. Introduction

This report is intended to give a brief phenomenological summary of the observations obtained with the MPI plasma experiment on HEOS 2 in the mid-altitude polar cusp and the distant polar magnetosphere.

The polar cusp is a region in the dayside high-latitude magnetosphere associated with the neutral line which separates magnetic field lines closed on the dayside and those swept back into the geomagnetic tail. Since magnetosheath plasma is apparently able to penetrate deep into the magnetosphere along magnetic field lines in the cusp, it has been considered to represent a major source of magnetospheric particles.

The first strong evidence for the direct entry of magnetosheath plasma through the polar cusp has been the similarity of the plasma characteristics observed at low altitudes over the dayside auroral oval with those typical for the magnetosheath (Heikkila and Winningham, 1971; Frank and Ackerson, 1971). This idea has received further verification by in situ measurements of Frank in the mid-altitude and distant polar cusp on IMP 5 (Frank, 1971), and by magnetic field observations on board the same spacecraft (Fairfield and Ness, 1972). Plasma wave measurements in the polar cusp have been reported for the magnetic storm of Nov. 1, 1968 when the cusp was so strongly displaced towards lower latitudes to be observable by OGO 5 (Russell *et al.*, 1971; Scarf *et al.*, 1972).

Although the polar cusp has been the subject of earlier investigation, the distant polar magnetosphere is being explored for the first time by HEOS 2. On the basis of such measurements it should be possible to observe the plasma behavior that results from the reconnection of terrestrial and interplanetary magnetic fields believed to take place at the front side of the magnetosphere.

2. Observations

The observations reported here were obtained with instrument 2 of the MPI plasma experiment. It consists of a pair of hemispherical electrostatic analyzers with channeltrons as detectors, and measures electrons and protons in the energy range between 100 eV and 40 keV. The instrument is mounted such that the normal to its aperture points perpendicular to the satellite spin axis and measures the particle fluxes at 8 and 4 angles about the spin axis for protons and electrons, respectively. The polar

D. E. Page (ed.), Correlated Interplanetary and Magnetospheric Observations, 249–253. *All Rights Reserved.*

acceptance angle of the instrument is approximately 30° FWHM. The measurements thus provide 2-dimensional cuts through the plasma velocity distribution functions. Since, however, the satellite spin axis has undergone several reorientations, 3-dimensional flow properties can be inferred from the measurements under favourable conditions.

3. The Polar Cusp

In presenting the observations let us first consider the mid-altitude polar cusp. In this report we will only address the question of the plasma flow. Frank has reported that the protons in the mid-altitude cusp have angular distributions corresponding to a flow down the magnetic field lines (Frank, 1971). In Figure 1 we present an example of our observations. The left panel shows the profile of the cusp as seen in the directional differential fluxes of electrons and protons of approximately 130 eV and 860 eV, respectively (top). The bottom graph shows the fluxes of electrons and

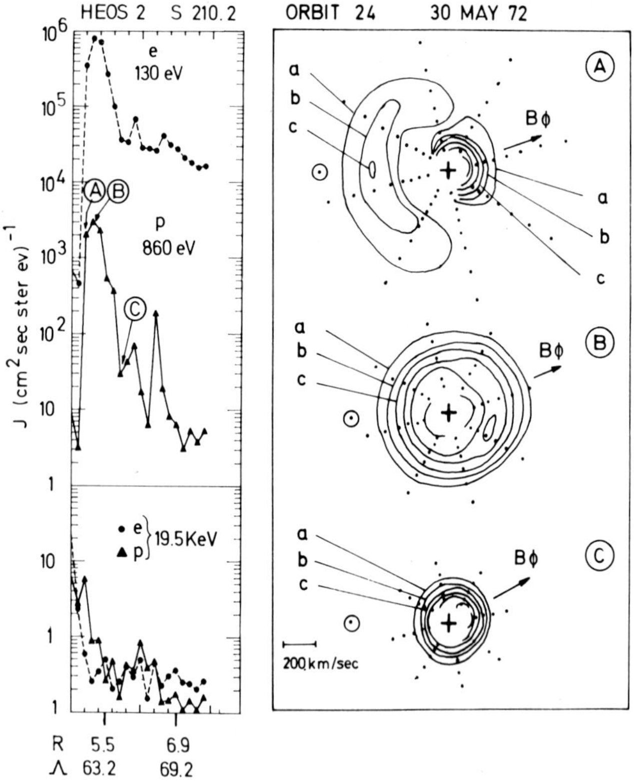

Fig. 1. Polar Cusp Observations on May 30, 1972. Left panel shows differential directional fluxes of protons and electrons for selected energies averaged over azimuthal angle. Geocentric radial distance and magnetic latitude are indicated at the bottom. Measurements are taken approximately 4 min apart. Right panel shows 2-dimensional proton velocity distributions for the times indicated by the letters on left panel. For a description of these presentations see text.

protons at 19.5 keV which indicate the high-latitude boundary of trapping. The panel on the right presents 2-dimensional representations of the shapes of the proton velocity distribution functions at positions A, B, and C as lettered on the curve in the left panel. The contours are lines of constant density in velocity space. A few selected contours have been labeled a, b, and c, in order to facilitate comparison of the various distributions. The points represent the positions in velocity space where the differential flux measurements are made. The cross in the middle of each picture represents the origin of the coordinate system in the spacecraft frame. Any shift of the 'center of gravity' of the distributions with respect to the origin is, of course, a measure of the plasma bulk flow velocity, projected onto the measurement plane. Note that the angles at which the data are plotted are the instrument look directions. Therefore the true flow directions are opposite to the apparent directions presented by the density contours. The symbol $\odot$ indicates the direction of the Sun, and the arrow B_ϕ shows the magnetic field direction projected onto the measurement plane. In the particular example shown in Figure 1, the measurement plane was oriented approximately parallel to the ecliptic plane.

Distribution A exhibits remarkable proton flow behaviour. At energies characteristic for the magnetosheath, the protons show a strong flow ($\sim$300 km s^{-1}) down the magnetic field, whereas at lower energies ($\sim$100 eV) the protons at the same time flow up the field lines at a velocity less than 100 km s^{-1}. (It should be noted that the example only shows that the proton flow is along the projection of the magnetic field into the measurement plane; however, putting together examples with varying spin axis orientations, it can be shown that the flow coincides indeed with the actual 3-dimensional magnetic field direction.) Distribution B, taken immediately (i.e. approximately 4 min) after distribution A, does not contain the downward flow, but only the upward flow, and the increased spread of the contours show a considerable increase in width of the distribution and therefore the internal energy or temperature. The relatively short time that distribution A was observed in comparison with the entire width of the cusp as defined by enhanced fluxes of low energy particles illustrates the fact that the type of distribution shown in A is normally either a very transient impulsive phenomenon or appears in only a relatively thin region. It is also possible that distribution A is likely to be somewhat distorted by time variations. We have, however, observed enough examples of distributions of the type A to feel confident that they represent a real phenomenon. Distribution C, taken near the high-latitude edge of the cusp, still shows the upward flow, but does not contain protons at higher energies. Our observations seem to indicate that the earthward flow of magnetosheath-like protons is generally restricted to a narrow region on the low-latitude side of the polar cusp whereas there is upward flow of colder protons throughout most of the region. We presently consider the upward flow to consist either of ionospheric particles that have been accelerated by field aligned electric fields and heated by wave turbulence, or particles of magnetosheath origin that have been reflected back from a region below the observing satellite after flowing down cusp field lines and being decelerated and scattered.

4. The Polar Magnetosphere and Magnetopause

We now present observations obtained in the polar magnetosphere and magnetopause. Figure 2 shows, in basically the same representation as Figure 1, a larger section of an outbound orbit, starting with the polar cusp and reaching to the magnetosheath. Inserted into the figure are the proton distribution functions which were measured at the times indicated by the arrows. (A description of this representation of the distribution functions is given in conjunction with the discussion of Figure 1.)

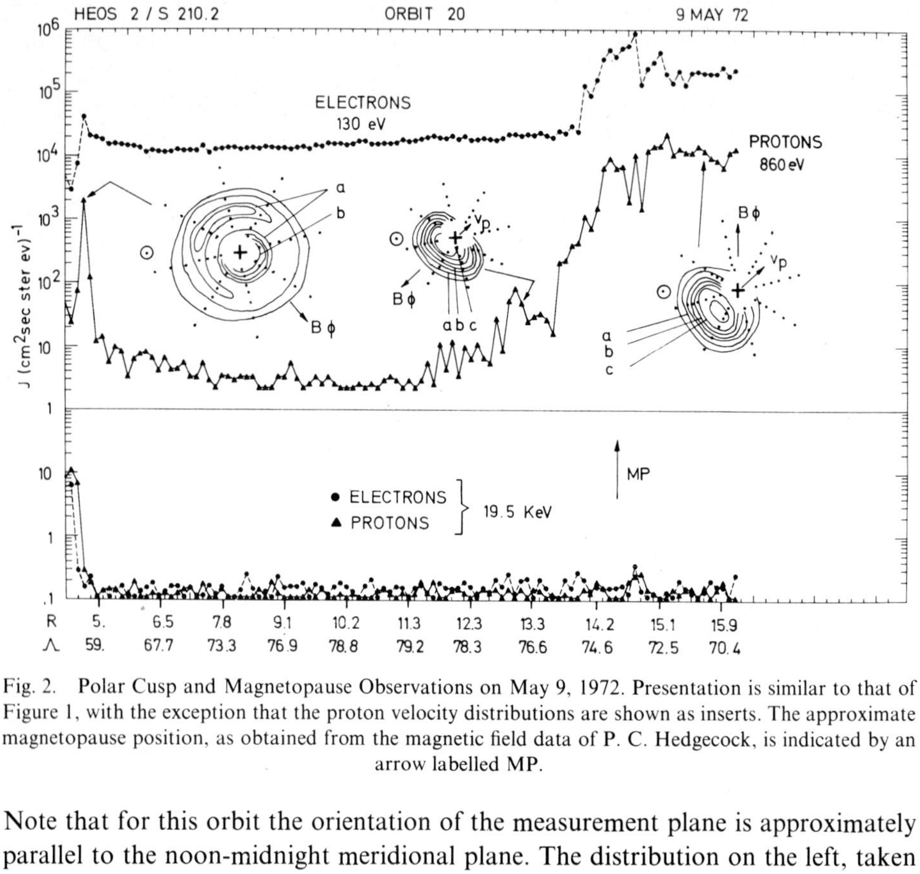

Fig. 2. Polar Cusp and Magnetopause Observations on May 9, 1972. Presentation is similar to that of Figure 1, with the exception that the proton velocity distributions are shown as inserts. The approximate magnetopause position, as obtained from the magnetic field data of P. C. Hedgecock, is indicated by an arrow labelled MP.

Note that for this orbit the orientation of the measurement plane is approximately parallel to the noon-midnight meridional plane. The distribution on the left, taken at the peak of the polar cusp, resembles the distribution A of Figure 1, again representing the simultaneous occurrence of upward and downward flow as discussed above.

The magnetopause boundary in this example is crossed at approximately the position indicated by the arrow. This conclusion is based upon the magnetic field data, kindly provided by P. C. Hedgecock, Imperial College London. The figure shows that for several earth radii inside the magnetopause, there is considerable flux of low-energy protons. This is in contrast to the appearance of the magnetopause at lower latitudes, where a sudden transition between very low plasma densities typical of the magnetosphere and the high densities of the magnetosheath is observed. The protons in this high-latitude region have magnetosheath-like energy spectra and flow characteristics, except that the flow direction is aligned with the magnetic field. We will refer to this region hereafter as the 'polar transition layer'. The main characteristics of the proton distribution in the transition layer are illustrated by the middle insert in Figure 2 and may be compared to those of the magnetosheath protons as shown in the right. In the transition layer the flow is directed along the projected magnetic field away from the Earth and has a velocity of approximately 130 km s^{-1}. In the magnetosheath the flow velocity is nearly twice as high and the magnetic field now forms a large angle with the velocity vector. Our observations of more than 30 magnetopause crossings at high latitudes show the thickness of the polar transition region to vary considerably. In more than 50% of the cases the thickness (as measured along the z-axis of a solar magnetospheric coordinate system) is greater than 1 R_E, and often is greater than 3 R_E. At times, however, the transition can be as abrupt as observed at low latitudes, and often the layer is highly structured. In all cases where the proton flow speed is reasonably well determined, its direction is along the magnetic field and away from the Earth.

From these observations (and others, not reported herein) we conclude that the polar transition layer is not simply a magnetopause boundary layer, but a part of the magnetosphere. Similar behavior of plasma near the magnetopause further back in the magnetotail at somewhat lower latitudes has been reported by Akasofu *et al.* (1973). It is our tentative interpretation that the polar transition region is the result of a combination of particle entry through the day-side polar cusp and field-line reconnection at the front of the magnetopause. It may be possible to estimate the rate at which field-line reconnection takes place, but this will be the subject of a later study.

Acknowledgement

We would like to thank Dr P. C. Hedgecock, Imperial College, London, for making available to us the magnetic field data.

References

Akasofu, S.-I., Hones, E. W., Bame, S. J., Asbridge, J. R., and Lui, A. T. Y.: 1973, 'Magnetotail and Boundary Layer Plasmas at a Geocentrical, Distance of 18 R_E: Vela 5 and 6 Observations', Los Alamos Scientific Laboratory Preprint LA-UR 73-547, submitted to the *J. Geophys. Res.*

Fairfield, D. H. and Ness, N. F.: 1972, *J. Geophys. Res.* **77**, 61.

Frank, L. A.: 1971, *J. Geophys. Res.* **76**, 5202.

Frank, L. A. and Ackerson, K. L.: 1971, *J. Geophys. Res.* **76**, 3612.

Heikkila, W. J. and Winningham, J. D.: 1971, *J. Geophys. Res.* **76**, 883.

Russell, C. T., Chappell, C. R., Montgomery, M. D., Neugebauer, M., and Scarf, F. L.: 1971, *J. Geophys. Res.* **76**, 6743.

Scarf, F. L., Fredericks, R. W., Green, I. M., and Russell, C. T.: 1972, *J. Geophys. Res.* **77**, 2274.

PART 3

MAGNETOSPHERIC PLASMA FLOW AND ELECTRIC FIELDS

THE MAGNETOSPHERIC ELECTRIC FIELD

JUAN G. ROEDERER
University of Denver, Denver, Colo. 80210, U.S.A.

Considerable progress has been made in recent years in the study of the electric field of the magnetosphere. This electric field is quite difficult to measure because of its weakness (millivolts per meter or less) and high variability in time and space. Direct measurements from satellites have begun only in recent years (Heppner, 1972a); indirect techniques based on the study of energetic particle motion (McDiarmid *et al.*, 1969; Roederer and Hones, Jr., 1970; DeForest and McIlwain, 1971) and drifts of natural (Carpenter *et al.*, 1972) for artificial (Haerendel, 1972) plasmas have been historically the first to provide information on the electric field. Balloon-borne measurements (Mozer, 1971) of the horizontal component of the stratospheric electric field – assumed to be roughly proportional to the horizontal electric field in the overlying ionosphere – are becoming an increasingly popular and relatively cheap technique, particularly useful if carried out simultaneously over periods of many hours at different geographic locations.

The general electric field configuration in the quiet magnetosphere is sketched in Figure 1, with electric field vectors shown in the northern twilight (dawn-dusk) meridian. Three main regions can be identified: (1) the region of open magnetic field

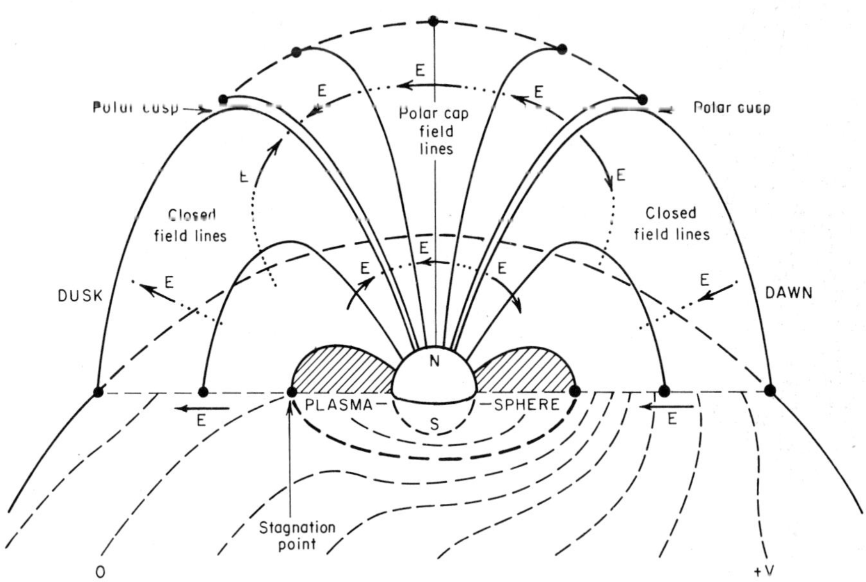

Fig. 1. A sketch of the steady-state electric field configuration on the dawn-dusk meridian of the magnetosphere. Solid lines: projections of the magnetic field lines on the dawn-dusk meridian; dotted lines: electric field lines; broken lines: electric equipotentials on the geomagnetic equator.

D. E. Page (ed.), Correlated Interplanetary and Magnetospheric Observations, 257–261. *All Rights Reserved.*

lines linked to the polar cap, carrying an electric field directed mainly from dawn to dusk; (2) the region of closed field lines crossing the magnetic equatorial surface, carrying an electric field that is directed poleward on the dusk side and equatorward on the dawn side; and (3) the region of the plasmasphere with an electric field (not shown in Figure 1) that on the equatorial plane is directed radially earthward, responsible for the Earth-locked corotation of the plasmasphere. Electric field lines are believed to be everywhere perpendicular to magnetic field lines, in absence of anomalous resistivity in regions of strong precipitation. Note in Figure 1 the sudden electric field reversal across the polar cusps on the dawn-dusk meridian. Typical electric potential drops across the polar cap are of the order of 50 kV. A stagnation point (gradual reversal of the radial electric field) is believed to occur on the dusk-side plasmapause. Also sketched in Figure 1 are typical (quiet-time) electric equipotential curves on the equatorial surface (Kavanaugh *et al.*, 1968). All field lines passing through a given curve define an equipotential surface. The plasmapause separates a family of open equipotentials from the closed equipotentials that encircle the earth inside the plasmasphere. A low energy plasma particle follows these equipotentials in its drift motion; the result is a general sunward flow from the tail along the open equipotentials, and a corotational east-west flow around the Earth in the plasmasphere.

The electric field behavior on the noon-midnight meridian is more complex and appears to be more irregular. Low altitude satellite measurements have detected multiple reversals of the electric field (Heppner, 1972a), probably due to motions of polar cusps; near the midnight meridian the pronounced electric field irregulatities observed in the auroral zone (which happen where the electric field switches from a poleward direction before midnight to equatorward after midnight) could be related to the relatively sharp separation of convection patterns to the east and to the west, respectively (Harang discontinuity) (Heppner, 1972b). Quite generally, the region around the midnight meridian is extremely sensitive to all major magnetospheric perturbations, and so is the electric field therein.

Some features of the electric field show distinct correlations with the interplanetary magnetic field. One of these correlations links the spatial dependence of the dawn-dusk polar cap electric field intensity with the azimuthal component of the interplanetary magnetic field (Heppner, 1972a): the electric field tends to be stronger on that side of the polar cap where interplanetary and geomagnetic field lines tend to be roughly of the same direction. As a result of this asymmetry, the whole ionospheric current system in and around the polar cap shifts toward the dawn or the dusk side, controlled by the azimuthal component of the interplanetary magnetic field. The effect of these current shifts is measurable on the Earth's surface with conventional magnetometers at appropriately located high latitude stations (Bassolo *et al.*, 1972; Friis-Christensen *et al.*, 1972). As an interesting spinoff from this recent finding, it was possible to infer the periods of time when the interplanetary magnetic field was directed toward or away from the Sun, for several solar cycles *before* the space age had begun (Svalgaard, 1972).

Another correlation, perhaps not so distinct as in the preceding case, exists between the overall intensity of the polar cap electric field and the north-south component of the interplanetary magnetic field. During periods of sustained southward direction of the latter, the polar electric field increases 2–3 times with respect to the periods in which the interplanetary field is directed northward (Gonzalez and Mozer, 1972). This process seems to be related to the growth phase of a substorm (Russell, 1974).

Both types of correlations are in support of the idea of magnetospheric field lines being connected with the interplanetary magnetic field across the magnetopause. As a matter of fact, the interplanetary field seems to be strongly related to the external 'driving mechanism' of the magnetospheric electric field. This can be pictured in the following way: interplanetary field lines embedded in the solar wind move with the velocity $\mathbf{v}$ of the latter; an observer at rest in the Earth's magnetosphere thus 'sees' an induced interplanetary electric field $\mathbf{E} = -\mathbf{v} \times \mathbf{B}$, where $\mathbf{B}$ is the interplanetary magnetic field vector. Assuming magnetic field lines to be electric equipotentials everywhere, it would be possible to map this 'external' $-\mathbf{v} \times \mathbf{B}$ electric field into the magnetosphere down to the polar ionosphere, if one knew exactly the manner in which the field lines are connected through the magnetopause (Stern, 1973). At the present time it is not at all clear whether the interplanetary magnetic field is the sole external source of the magnetospheric electric field. The existence of a dawn-dusk electric field and associated convection pattern at *all* times, regardless of the current characteristics of the interplanetary magnetic field, suggests the coexistence of another, steady-state, external driving mechanism for the electric field.

In addition to external driving mechanisms, dynamo effects in the ionosphere caused by neutral winds dragging plasma across magnetic field lines, represent an additional source of the electric field in the magnetosphere. The corotational electric field inside the plasmasphere is entirely of such origin.

Plasma conductivity in the ionosphere and energetic particle distributions trapped in the magnetic field control both the general configuration and local perturbations of the electric field. For instance, in regions of *localized* enhancements of the conductivity caused by auroral particle precipitation (e.g., auroral arcs), the electric field $\mathbf{E}$ adjusts to the irregularity of the conductivity $\boldsymbol{\sigma}$ (decreasing where $\boldsymbol{\sigma}$ increases) so as to keep the current density $\mathbf{j} = \boldsymbol{\sigma}\mathbf{E}$ a 'smooth' function. This phenomenon may also explain the fact that the electric field over the sunlit polar cap is usually more regular than that over the winter polar cap: electric irregularities are quickly shorted out in the enhanced conductivity of the dayside ionosphere. On the other hand, when the current intensities in the auroral zone increase orders of magnitude during magnetospheric substorms, the electric field is found to increase only a few times. This indicates that on these occasions the electric field is controlled mainly by an external driving mechanism and it is now the *current* which adjusts to the large increase in ionospheric (Hall) conductivity.

Energetic particles in the radiation belt influence the electric field configuration indirectly through the electric currents (perpendicular and parallel to the magnetic field lines) that are associated with these particles. In the closed field line region, the

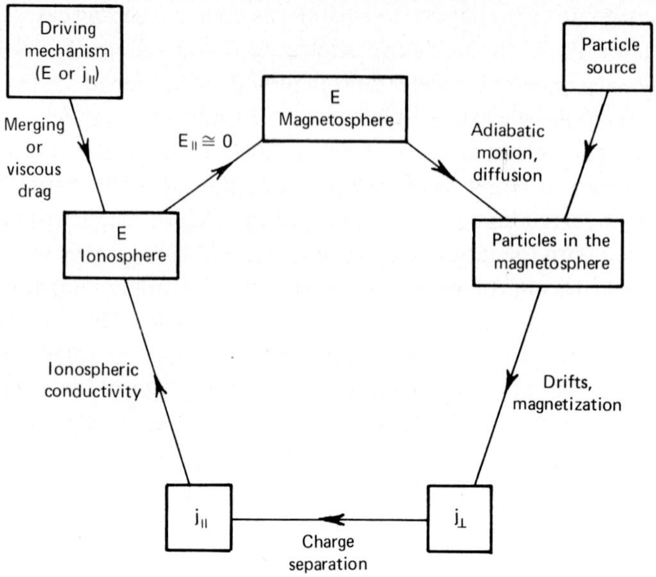

Fig. 2. Closed, self-consistent chain of cause-and-effect relationships for the electric field in the magnetosphere (Vasyliunas, 1972). j: electric current densities parallel and perpendicular to the magnetic field.

electric field is part of a closed, self-consistent chain of cause-and-effect relationships (Vasyliunas, 1972) (Figure 2), governing the strong feedback system of magnetosphere-ionosphere coupling. This requirement of self-consistency makes a quantitative theoretical study of the magnetospheric electric field (and plasma) particularly difficult.

References

Bassolo, V. S., Mansurov, S. M., and Shabansky, V. P.: 1972, *Issledovanija po Geomagnetismu, Aeronomii i Fizika Solntsa* **23**, 125.

Carpenter, D. L., Stone, K., Siren, J. C., and Crystal, T. L.: 1972, *J. Geophys. Res.* **77**, 2819.

DeForest, S. E., and McIlwain, C. E.: 1971, *J. Geophys. Res.* **76**, 3587.

Friis-Christensen, E., Lassen, K., Wilhjelm, J., Wilcox, J. M., Gonzalez, W., and Colburn, D. S.: 1972, *J. Geophys. Res.* **77**, 3371.

Gurnett, D. A.: 1972, in E. N. Dyer (ed.), *Critical Problems of Magnetospheric Research*, IUCSTP Secretariat, c/o National Academy of Sciences, Washington D.C.

Gonzalez, W. and Mozer, F. S.: 1972, High Latitude Electric Fields and the Three-Dimensional Interaction between the Interplanetary and Terrestrial Magnetic Fields, preprint, University of California at Berkeley.

Haerendel, G.: 1972, 'Results from Barium Cloud Releases in the Ionosphere and Magnetosphere', *Proceedings of the XV COSPAR Plenary Meeting*, Madrid.

Heppner, J. P.: 1972a, in E. N. Dyer (ed.), *Critical Problems of Magnetospheric Research*, IUCSTP Secretariat, c/o National Academy of Science, Washington D.C.

Heppner, J. P.: 1972b, *Geofysiske Publikasjoner* **29**, 105.

Hones, Jr., E. W.: 1970, *J. Geophys. Res.* **75**, 3923.

Kavanaugh, Jr., L. D., Freeman, Jr., J. W., and Chen, A. J.: 1968, *J. Geophys. Res.* **73**, 5511.

McDiarmid, I. B., Burrows, J. R., and Wilson, M. D.: 1969, *J. Geophys. Res.* **74**, 3554.

Mozer, F. S.: 1971, *Pure Appl. Geophys.* **84**, 32.
Roederer, J. G.: 1970, *Dynamics of Geomagnetically Trapped Radiation*, Springer-Verlag, New York, N.Y.
Russell, Christopher T.: 1974, this volume, p. 3.
Stern, D. P.: 1973, *J. Geophys. Res.* **78**, 7292.
Svalgaard, L.: 1972, *Bull. Amer. Astron. Soc.* **4**, 393.
Vasyliunas, V. M.: 1972, 'Mathematical Models of Magnetospheric Convection and Its Coupling to the Ionosphere', p. 60 of B. M. McCormac, ref. (2).

MAGNETOSPHERIC PLASMA FLOW AND THE NATURE OF THE MAGNETOSPHERIC BOUNDARY LAYER

EDWARD W. HONES, JR.

University of California, Los Alamos Scientific Laboratory, Los Alamos, N.M. 87544, U.S.A.

Abstract. Bulk flow of plasma in the outer magnetosphere has been studied using measurements of protons (30 eV $< E_p <$ 30 keV) made with electrostatic analyzers on the Vela satellites in circular orbits at geocentric distance, $r \sim 18\ R_E$. The flow is detected, and its velocity is measured by the spin modulation it causes in the response of the analyzers on the rotating (64-s period) satellites. Very rapid flow of plasma in the magnetotail plasma sheet occurs during magnetospheric substorms. As the plasma sheet thins early in a substorm expansion phase, bulk flow in the anti-Earthward direction at several hundred kilometers per second is sometimes detected, possibly indicating the formation of a neutral line at $r < 18\ R_E$. The plasma that re-inflates the plasma sheet late in a substorm is observed to be flowing Earthward at speeds as great as ~ 1000 km s^{-1}, suggesting that the neutral line is re-established at $r > 18\ R_E$ at that time. As the satellites pass from the magnetotail to the magnetosheath, they traverse a region, the boundary layer, whose thickness ranges up to a few thousand kilometers. The boundary layer plasma has a temperature similar to that of the magnetosheath plasma, but its number density and bulk flow speed are substantially less than those of the magnetosheath plasma. It is suggested that the boundary layer is the magnetic projection of the dayside cusp and is a site of energy and plasma transfer from the magnetosheath into the plasma sheet.

1. Introduction

The magnetosphere is the product of a dynamic interchange of energy and plasma between the solar wind on the one hand and the Earth's immediate magnetic environment on the other. That this dynamic interchange involves large-scale motions of plasma, particularly in the outer magnetosphere, has been widely recognized; the occurrence of such motions is a basic premise in nearly every theory of the magnetosphere. Until quite recently, however, the bulk motions of plasma in the outer magnetosphere had to be inferred from indirect evidence, e.g., measurements of ionospheric electric fields and observations of motions of auroral forms and of barium ion clouds.

Now these bulk motions of plasma have been more directly determined through the directional anisotropies they cause in the flux of plasma protons. Instruments on rotating satellites reveal the proton flux anisotropy as a spin modulation (with one peak per rotation) of their responses (Hones *et al.*, 1972; Prakash, 1972; Hones *et al.*, 1973b). Modulation of this type clearly cannot be due to anisotropic pitch angle distributions of the proton flux with respect to the magnetic field, for such distributions would be evidenced by modulations with an even number of peaks per rotation.

This paper is a report of the measurements of plasma flow made in this way by the Vela satellites in the magnetotail plasma sheet and in the magnetotail boundary layer. The reader is referred to Hones *et al.* (1972, 1973b) and to Akasofu *et al.* (1973) for more complete reports of these measurements. These more direct determinations of the bulk flow of plasma in the outer magnetosphere substantially strengthen the observational base upon which theoretical ideas concerning such problems as plasma

D. E. Page (ed.), Correlated Interplanetary and Magnetospheric Observations, 263–275.

entry into the magnetosphere and the nature of magnetospheric substorms can be founded.

2. Description of Experiment

The plasma measurements to be discussed here were made with hemispherical-plate electrostatic analyzers on the Vela satellites, in circular orbits at geocentric distance, $r \approx 18\ R_E$. We shall be concerned primarily with energy spectral measurements of proton fluxes in the energy range ~ 30 eV to ~ 30 keV made by these instruments. The analyzers have fan-shaped acceptance solid angles which are centered in the rotational equatorial planes of the rotating ($\sim$64-s period) satellites. The long dimension ($\sim 100°$) of the fan is parallel to the rotation axis and the narrow dimension ($\sim 3°$) is perpendicular to this. A complete energy spectrum (at 20 to 24 logarithmically-spaced energy levels) is measured every four seconds (every 22.5° of rotation) when the satellites are operated in the real-time (high bit-rate) mode. More detailed descriptions of these instruments have been given by Montgomery *et al.* (1970), by Bame *et al.* (1971) and by Hones *et al.* (1972).

The central topic of this paper is the bulk flow of protons (and thus, essentially, of plasma) measured by the spin-modulation that the flow causes in the response of the analyzers on the rotating satellites. Having only a single, equatorially situated plasma detector, each Vela satellite measures something approximating the component of the proton flow velocity perpendicular to its axis of rotation. (See Hones *et al.*, 1972, for a more detailed description of the method of interpretation of the observations.)

3. Observations

3.1. Flow of Plasma in the Plasma Sheet

The most striking feature of the flow of plasma in the plasma sheet is the very high velocity in the sunward (approximately Earthward) direction that is encountered as the plasma sheet thickens (or recovers) late in a substorm. Figure 1 shows magnetograms from auroral zone stations, from the low polar cap station, Baker Lake and from the low latitude stations, San Juan and M'Bour, recorded on July 2, 1968. Onset of the expansion phase of a substorm at $\sim$0130 UT is signified by the sharp deepening of auroral zone negative bays that started then and by the onset of a positive bay at the low-latitude, near-midnight station, M'Bour. The auroral zone bays started to recover at $\sim$0145 UT. At the same time a negative bay started to develop at the low polar-cap station, Baker Lake, signifying the occurrence of a substantial poleward shift of the auroral electrojet.

Vela 4A was near the dawn side of the magnetotail, very near the neutral sheet, whose position was estimated by the formula of Russell and Brody (1967). Data from that satellite are portrayed in Figure 2. A sharp drop of the energetic electron flux ($E_e > 30$ keV) and sudden cooling of the lower energy 'plasma' electrons occurred at the substorm's expansion phase onset. At $\sim$0145 UT the plasma electron temperature started to increase and the energetic electron flux increased toward new higher

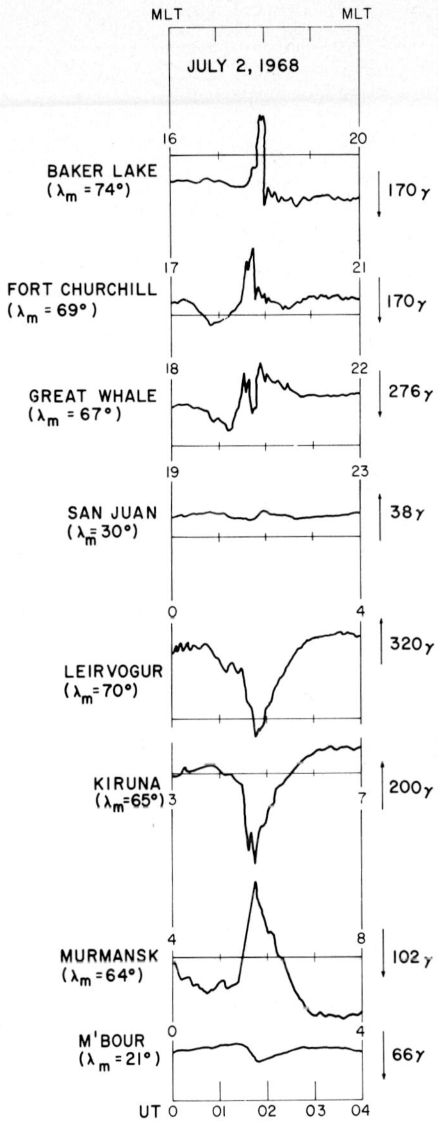

Fig. 1. H- or X-component magnetograms from several stations on July 2, 1968. The magnetic latitude (λ_m) of each station is given and the approximate magnetic local time (MLT) at 0000 UT and at 0400 UT is indicated at the ends of each trace.

levels. This was the onset of plasma sheet recovery, and, as is often observed, it coincided very closely with the sudden poleward shift of the auroral electrojet (Hones *et al.*, 1970; 1973a). Also at ~0145 a very strong spin modulation of the plasma proton flux was encountered and was of such phasing with the sun signals as to indicate plasma flow toward the Sun. (The rotational axis of this satellite was approximately perpendicular to the solar magnetospheric equatorial plane so that

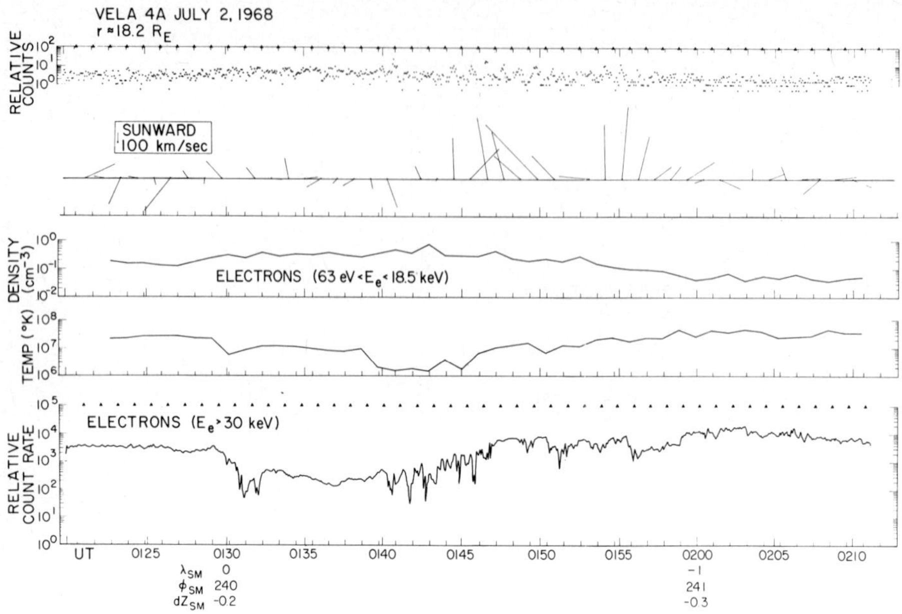

Fig. 2. Data from Vela 4A on July 2, 1968. At the top the average number of counts per energy channel of the electrostatic analyzer, recorded in each proton spectrum, is plotted. The black triangles along the line above the data points mark the times when the analyzer aperture crossed the subsolar meridian of the satellite. Flow velocity vectors derived from the spin modulations of the proton data points are drawn emanating from the second line. The Sunward direction is toward the top of the diagram, as illustrated by the small vector in the box indicating Sunward flow at 100 km s^{-1}. The next two graphs show the density and temperature of 'plasma' electrons (63 eV $< E_e <$ 18.5 keV) measured with the electrostatic analyzer. The count rate of a solid state detector sensitive to electrons ($E_e \gtrsim 30$ keV) is plotted at the bottom. Multiply by 20 to obtain absolute flux (electrons cm^{-2} s sr). The solar magnetospheric latitude and longitude (λ_{SM} and ϕ_{SM}) are shown at the bottom. dZ_{SM} is the estimated distance of the satellite from the neutral sheet.

the spin modulation provided a measure of the flow vector components in this plane. Note caption of Figure 2 and see Hones *et al.* (1972)). Flow velocities as great as 700 km s^{-1} were measured. This evident, rapid flow persisted for about 10 minutes.

At the expansion phase onset of substorms, when the plasma sheet thins (i.e., the plasma intensity is observed to decrease), a flow of plasma away from the Earth is sometimes encountered. An expansion phase onset occurred at ~1305 UT, March 5, 1968. This was marked by the sharp onset of a large negative bay at College and by the onset of a positive bay at Honolulu, both near local magnetic midnight (Figure 3). The College all-sky camera also recorded sudden brightening of auroras nearly overhead at that time. At Vela 4A, very near local midnight and ~1 R_E from the neutral sheet, decrease of energetic electron flux and of plasma electron density began at ~1305 UT (Figure 4). The plasma proton flux began to decrease at ~1309 UT and displayed a very strong spin modulation during the decrease. The phasing of the modulation suggested a flow in the anti-Sunward direction. The maximum speed of the flow was ~500 km s^{-1}. Unfortunately, operation of the satellite in the real-

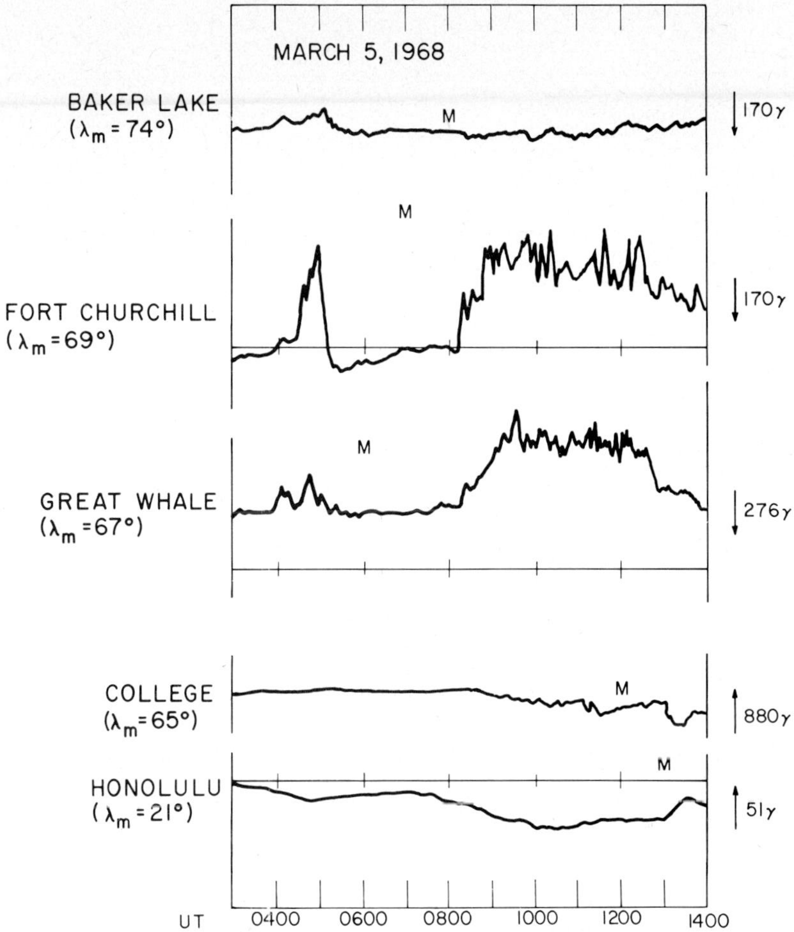

Fig. 3. H or X-component magnetograms from several stations. *M* marks the approximate time of magnetic local midnight.

time (high bit-rate) mode was interrupted at 1315, and further monitoring of the plasma flow became impossible. But available store-mode data show that the plasma intensity continued to decrease to background intensity levels and did not recover again until ~1330 UT when magnetic bay recovery began at the Earth.

The two examples of plasma flow in the plasma sheet which have just been described illustrate a feature that is found generally to be true, i.e., evident spin modulation having constant phasing with the Sun usually lasts only a few minutes at a time. This suggests that plasma flow in a given direction (at speeds above the 50–100 km s^{-1} threshold of these measurements) usually does not persist for more than a few minutes at a time.

Data obtained by Vela 4B were used to determine whether there is a long term 'average' pattern of plasma flow in the plasma sheet. The spin axis of Vela 4B was

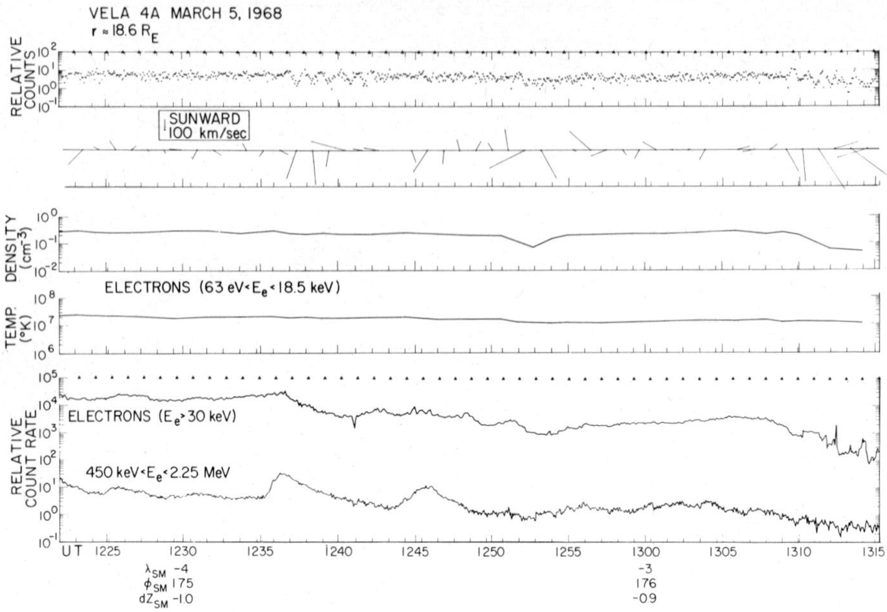

Fig. 4. Data from Vela 4A on March 5, 1968. Details are the same as in the caption of Figure 2 except here the count rate of a solid state detector of electrons (450 keV $< E_e <$ 2.25 MeV) has been added. Multiply this count rate by 20 to obtain absolute flux (electrons cm^{-2} s sr).

kept aligned along a radius vector from the Earth's center by an active orientation system. Thus, when that satellite was near the center of the magnetotail, the proton flux measurements did not provide a very sensitive means for detecting plasma flow in the Sunward and anti-Sunward directions. Figure 5 illustrates approximately the flow vectors that would be derived from measurements made by Vela 4B at various locations in the magnetotail if an actual flow of plasma toward the sun at 100 km s^{-1} existed throughout the tail. Note that measurements near the axis of the tail do not provide good indications of the velocity in this hypothetical flow distribution.

The vectors in the 'map' of Figure 6 portray flows derived by averaging all of the proton measurements made in each $10° \times 10°$ box during 6 months of operaton of Vela 4B. The number in each box, when multiplied by 9, gives the number of satellite rotations from which data were obtained in that box (when $K_p > 1_+$). (The absence of vectors near the center of the diagram is due to a few-week gap in data acquisition). No orderly pattern of plasma flow is evident in Figure 6. A similar map, compiled from data acquired with $K_p \leqslant 1_+$ also failed to show any systematic pattern of plasma flow. However, a map (Figure 7) compiled from flow vectors derived from groups of two or more successive satellite rotations during which the spin modulation of the proton counts was clearly evident *does* show a systematic flow pattern. Comparison with Figure 5 shows that the pattern of this very evident plasma flow, occurring primarily in conjunction with substorms, is consistent with flow in the general Sunward (approximately Earthward) direction.

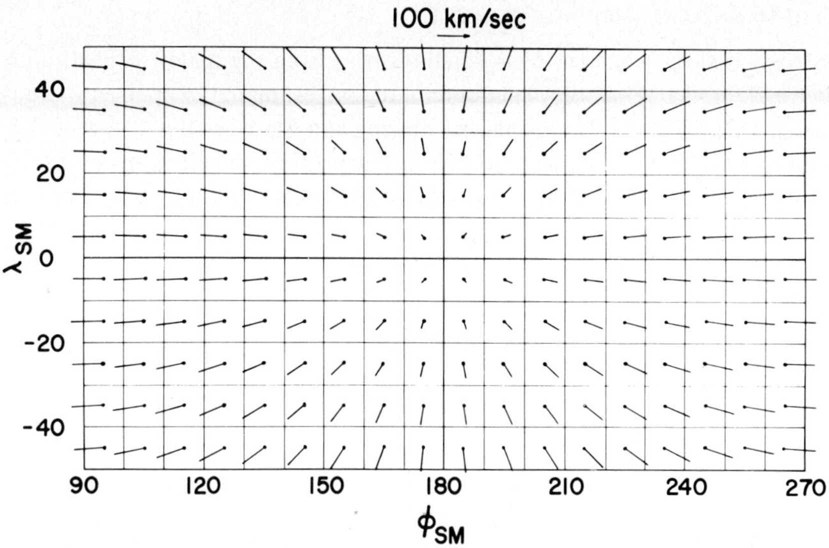

Fig. 5. A mercator map of the anti-Earthward portion of the ~18 R_E spherical surface on which the Vela satellites orbit. The coordinates are solar magnetospheric latitude and longitude. The map is divided into 10° by 10° sectors. The line radiating from the center of each sector represents the apparent magnitude and direction of plasma flow that would be derived from the Vela 4B analyzer's response to an actual bulk flow of protons toward the Sun (i.e., parallel to the solar magnetospheric X axis) at 100 km s^{-1}. The length of the arrow above the diagram represents 100 km s^{-1}.

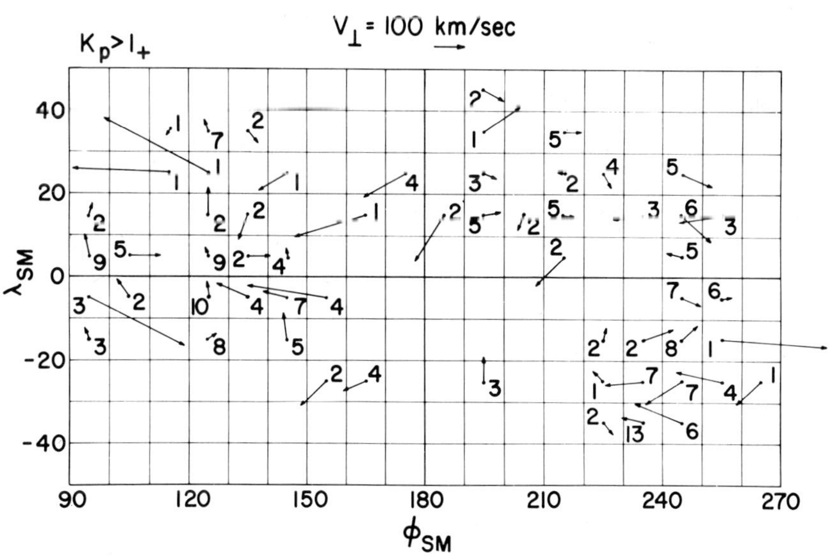

Fig. 6. Map, as in Figure 5, showing the flow vectors derived by averaging all flow measurements made by Vela 4B in each 10° by 10° range of solar magnetospheric latitude and longitude. Multiply the number in each box by 9 to determine the total number of satellite rotations from which the data were obtained to derive the corresponding flow vector.

3.2. The magnetotail boundary layer

On many occasions when the Vela satellites pass from the magnetosheath into the magnetotail, or vice versa, they encounter a region, estimated to be several thousand kilometers thick, which contains plasma flowing anti-Sunward but at speeds reduced from those measured in the magnetosheath. The protons and electrons in this region have temperatures comparable to those in magnetosheath plasma, but their number

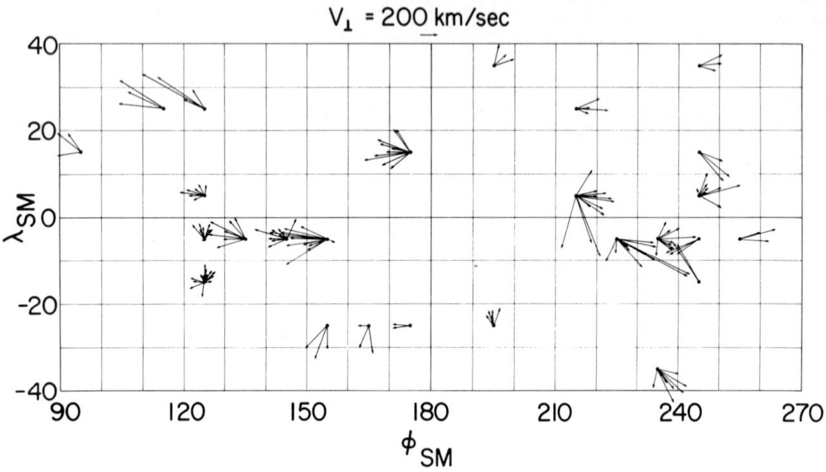

Fig. 7. Map, as in Figures 5 and 6. The flow vectors shown here were derived from single rotations of Vela 4B for which the spin modulation of the proton flux was very clear (see text).

densities are substantially lower than those in the magnetosheath. This region is the magnetotail boundary layer. It has been detected at other distances along the magnetotail by Intriligator and Wolfe (1972) and by Howe and Siscoe (1972).

Figure 8 shows data from Vela 4B, acquired during a passage through the boundary layer at the dawn side of the magnetotail on August 10, 1967. Three types of spin modulation of the plasma proton flux are evident in the figure. From ~0100 UT to ~0117 UT and again from ~0139 UT to ~0145 UT there was a moderately strong spin modulation of proton flux, with peaks seen in the Sunward direction, indicating quite rapid anti-Sunward flow. During these periods the satellite was in the boundary layer. From ~0120 UT to ~0139 UT the proton count rate was not distinctly modulated. The satellite was in the plasma sheet during that interval, a fact which is indicated also by the higher flux of energetic electrons that was encountered then. After ~0146 UT the proton count rate was very strongly modulated. The satellite was then in the magnetosheath.

The distribution of boundary layer-like proton fluxes measured during six months of operation of Vela satellites 5A, 5B, 6A and 6B is shown by the thin lines in Figure 9. The boundary layer is encountered most frequently at large values of $|dZ_{SM}|$ at the dawn and dusk edges of the magnetotail. The Vela satellites do not reach distances

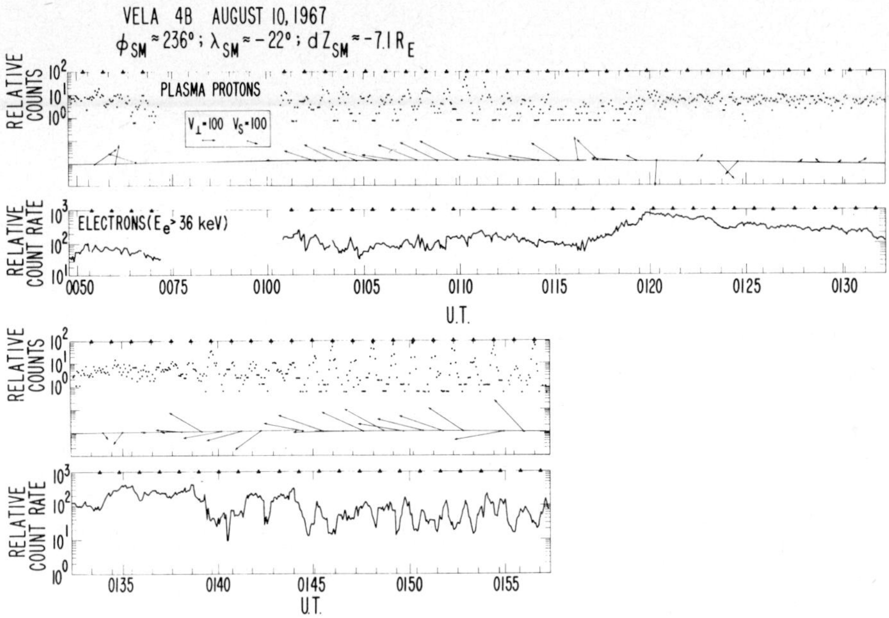

Fig. 8. Data from Vela 4B on August 10, 1967. At the top the average number of counts per energy channel of the electrostatic analyzer, recorded in each proton spectrum, is plotted. The back triangles along the line above the data points mark the times when the analyzer aperture crossed the subsolar meridian of the satellite. Flow velocity vectors derived from the spin modulations of the proton data points are drawn emanating from the line below the proton data. The small inserted box contains two arrows. One, labeled $V_\perp = 100$, designates the vector that would be derived from a proton flow at 100 km s^{-1} perpendicular to the satellite's spin axis. The other arrow, labeled $V_s = 100$, shows the vector that would be derived from a proton flow at 100 km s^{-1} *toward the Sun*. Many of the derived vectors point in a direction nearly opposite to that of the latter vector, indicating a proton flow approximately in the anti-Sunward direction. The response of a solid state detector of electrons, $E_e \gtrsim 36$ keV is shown at the bottom. Multiply the count rate by 20 to obtain electrons cm^{-2} s sr. The satellite's position is shown at the top of the figure.

$|dZ_{SM}|$ greater than $\sim 13\ R_E$. Thus they do not provide direct evidence as to whether the boundary layer extends completely around the top and bottom of the tail. But extrapolation to larger $|dZ_{SM}|$ of the observations in Figure 9 suggests that it may well do so. At small $|dZ_{SM}|$, near the tail's midplane, the boundary layer is very thin or perhaps non-existent. Many passes into the tail at small $|dZ_{SM}|$ reveal an essentially instantaneous transition ($\Delta t \leqslant 64$ s) from magnetosheath plasma to plasma sheet plasma. In fact there is often seen an instantaneous reversal of plasma flow from ~ 400 km s^{-1} anti-Sunward in the magnetosheath to perhaps 200 km s^{-1} Sunward in the immediately adjacent plasma sheet.

4. Discussion and Conclusions

4.1. Plasma Sheet Flow

The most evident and most rapid flow of plasma in the plasma sheet occurs in association with substorms. The character of this flow tends to support substorm models

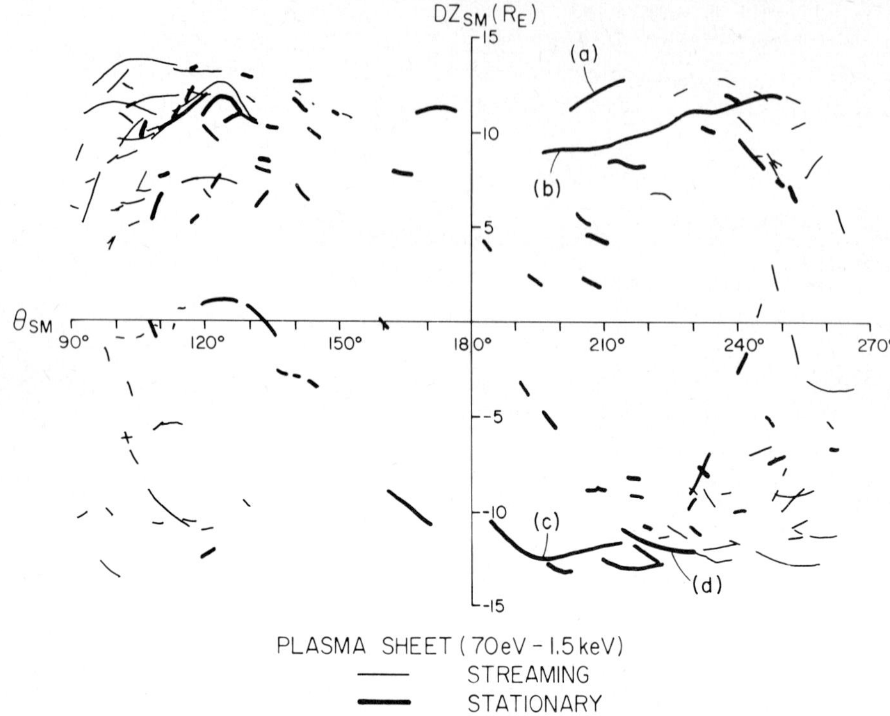

Fig. 9. Map showing segments of Vela orbits through the magnetotail along which were detected proton fluxes characteristic of the boundary layer (the thin lines). (The heavy lines in the figure are not relevant to the present discussion.) The coordinates are solar magnetospheric longitude (ϕ_{SM}). and distance from the neutral sheet (dZ_{SM}).

which incur reconnection of tail magnetic field lines in the near-Earth region of the tail as the mechanism responsible for substorm 'breakup' (i.e., the onset of the expansive phase). This is illustrated in Figure 10. Figure 10a depicts a gradual thinning of the plasma sheet such as has been observed to precede the expansive phase of some substorms. The figure implies that thinning begins in response to a southward turning of the interplanerary magnetic field (IMF). The escaping plasma may flow earthward or tailward or out the sides of the tail but must do so at relatively slow speed because the Vela satellites do not detect any consistent pattern of flow at such times. In (b) the plasma sheet has grown thin enough to allow magnetic reconnection to begin suddenly somewhere between the Vela satellite (V) and the Earth. Rapid tailward flow of plasma ejected from the reconnection region is observed by Vela for a short time before it is left outside the more rapidly thinning plasma sheet. Earthward of the reconnection region plasma flows rapidly earthward, causing that portion of the plasma sheet to expand rapidly (b) and (c). Reconnection continues in the near-Earth region for some time (c). Tailward of this, the plasma sheet remains thin, possibly disconnected magnetically from the near-Earth region which is supplying the auroral particle flux to the Earth.

Late in the substorm (Figure 10d) the reconnection region quite suddenly and rapidly moves much farther from the Earth, in association with a rapid poleward shift of the auroral electrojet (Hones *et al.*, 1973a), and Vela suddenly detects the rapid Earthward flow of plasma from the newly positioned reconnection region. Figure 10d

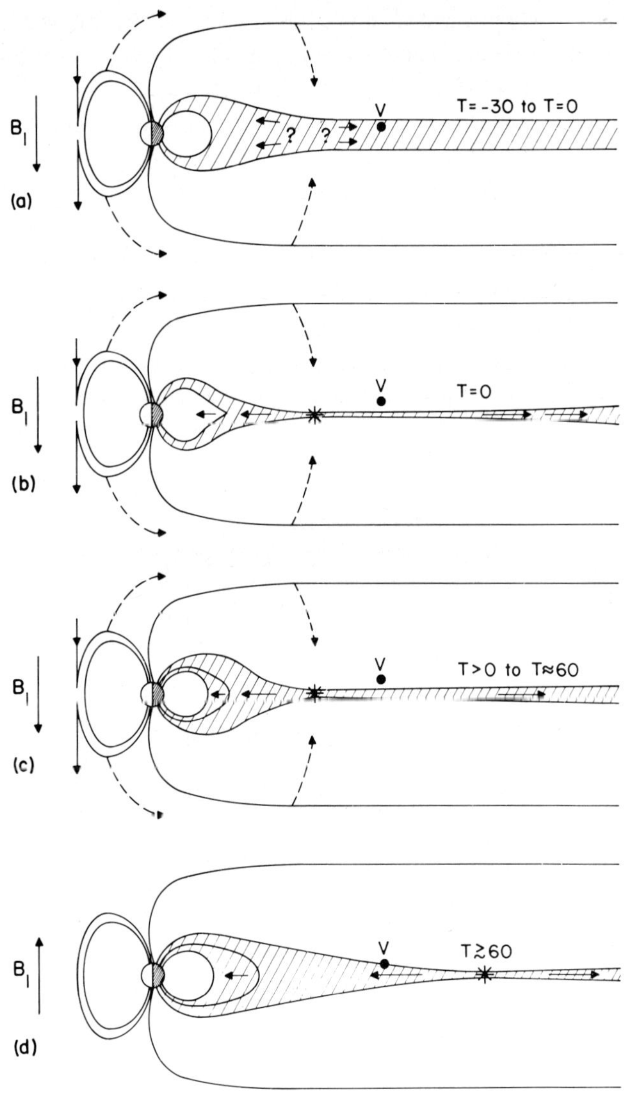

Fig. 10. Suggested variations of plasma sheet dimensions and of plasma flow during a substorm. The plasma sheet is the hatched portion of each diagram. Arrows indicate the direction and magnitude of plasma flow. The hypothetical position of a Vela satellite is shown by the dot labeled V. The times (T) refer to substorm time, with $T=0$ being the time of 'breakup' or beginning of the expansive phase of the substorm. The interplanetary field direction is shown by $\bar{B}_I$ and the dashed curves indicate flow of magnetic field lines. The star in (b), (c) and (d) suggests the site of magnetic reconnection in the tail.

implies that the tailward shift of the reconnection region might occur in response to a northward turning of the IMF.

4.2. THE BOUNDARY LAYER

It is reasonable to suppose that the boundary layer is an extension, along the tail surface of field lines emanating from the dayside cusp. That is, they may be magnetosheath field lines, newly merged with the Earth's field at the forward magnetopause, that have lost some of their plasma to the Earth. The plasma flow velocity is reduced because of the attachment of the field lines to the Earth. The boundary layer thus is the region where momentum of magnetosheath plasma is imparted to the tail, producing the tangential stress that draws out the tail. The boundary layer is also the site of the currents which confine the magnetotail to its cylindrical form.

The supposition that the boundary layer contains magnetic field lines connected to the dayside cusp is supported in a semi-quantitative way by magnetospheric model calculations portrayed in Figure 11. Magnetic field lines from the polar regions of the model were projected onto the spherical surface at $r = 18.5\ R_E$ where the Vela satellites orbit. The shaded region is the projection, onto this sphere, of the dayside cusp, as identified in particle measurements made with the polar orbiting ISIS 1

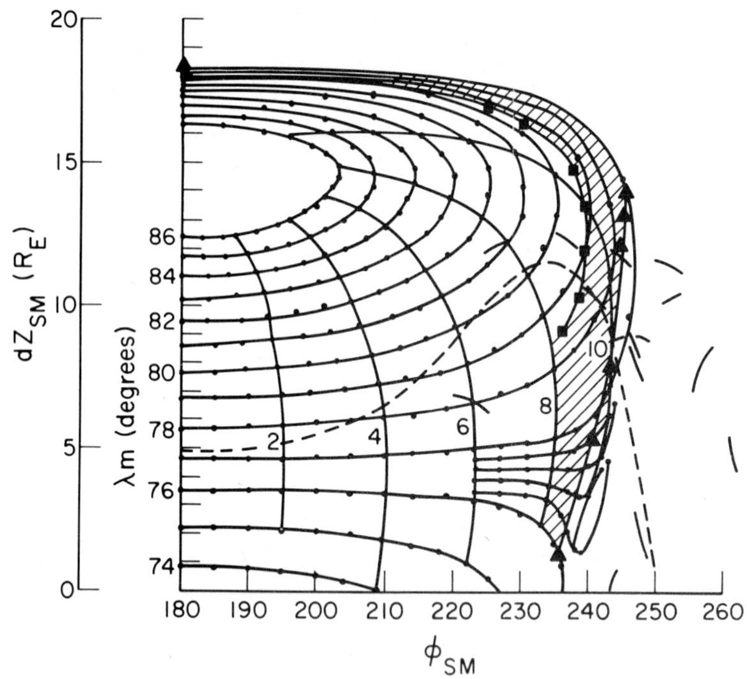

Fig. 11. A map of the 'Vela sphere' at $r = 18.5\ R_E$. The dots indicate intersections of magnetic field lines from the polar regions with the sphere. The field lines are identified by the latitude (λ_m) and the magnetic local time of their polar cap intersections. (The model used in this projection is briefly described in the text.) The hatched area contains dayside cusp field lines projected to the Vela sphere. See text for further discussion of details.

satellite (Winningham, 1972). The dashed curve is the approximate profile of the plasma sheet derived from Vela plasma measurements. The short line segments are copied from the upper right-hand quadrant of Figure 9 and are the orbit sections along which Vela satellites detected boundary layer plasma.

In the model magnetosphere used here, the last closed field line at the noon meridian is from 76° latitude at the Earth and crosses the equator at $r \approx 10\ R_E$. The radius of the tail is $\approx 20\ R_E$ and the tail field strength is 20 γ. It is notable that field lines in the thick dawn (and dusk) portions of the plasma sheet intersect the Earth two to four hours Sunward of the dawn (and dusk) meridians, in the region where the dayside cusp field lines also intersect the Earth. The relative positions, in Figure 11, of the cusp projection, the thick portion of the plasma sheet, and the boundary layer samples suggest that processes in the boundary layer (e.g., merging of boundary layer field lines along the flanks of the tail) may serve to populate the plasma sheet with the relatively hot plasma that is one of its identifying features.

Acknowledgements

This research was performed as part of the Vela Satellite Program, which is jointly sponsored by the U.S. Department of Defense and the U.S. Atomic Energy Commission. The program is managed by the U.S. Air Force and satellite operation activities are under the jurisdiction of the Air Force Satellite Control Facility, Sunnyvale, California.

Ground magnetograms used in this article were obtained from WDC-A for geomagnetism.

References

Akasofu, S.-I., Hones, E. W., Jr., Bame, S. J., Asbridge, J. R., and Lui, A. T. Y.: 1973, *J. Geophys. Res.* **78**, 7257.

Bame, S. J., Hones, E. W., Jr., Akasofu, S.-I., Montgomery, M. D., and Asbridge, J. R.: 1971, *J. Geophys. Res.* **76**, 7566.

Hones, E. W., Jr., Akasofu, S.-I., Perreault, P., Bame, S. J., and Singer, S.: 1970, *J. Geophys. Res.* **75**, 7060.

Hones, E. W., Jr., Asbridge, J. R., Bame, S. J., Montgomery, M. D., Singer, S., and Akasofu, S.-I.: 1972, *J. Geophys. Res.* **77**, 5503.

Hones, E. W., Jr., Asbridge, J. R., Bame, S. J., and Singer, S.: 1973a, *J. Geophys. Res.* **78**, 109

Hones, E. W., Jr., Asbridge, J. R., Bame, S. J., and Singer, S.: 1973b, *J. Geophys. Res.* **78**, 5463.

Howe, H. C. and Siscoe, G. L.: 1972, *J. Geophys. Res.* **77**, 6071.

Intriligator, D. S. and Wolfe, J. H.: 1972, *J. Geophys. Res.* **77**, 5480.

Montgomery, M. D., Asbridge, J. R., and Bame, S. J.: 1970, *J. Geophys. Res.* **75**, 1217.

Prakash, A.: 1972, *J. Geophys. Res.* **77**, 5633.

Russell, C. T. and Brody, K. I.: 1967, *J. Geophys. Res.* **72**, 6104.

Winningham, J. D.: 1972, in B. M. McCormac (ed.), *Earth's Magnetospheric Processes*, D. Reidel Publ. Co., Dordrecht, Holland, p. 68.

THE CONVERGENCE OF FACT AND THEORY ON MAGNETOSPHERIC CONVECTION

C. R. CHAPPELL

Lockheed Palo Alto Research Laboratory, Palo Alto, Calif. 94304, U.S.A.

Abstract. Since the early work of Dungey, Axford and Hines, and Piddington, the concept of magnetospheric convection has become more and more widely accepted as the fundamental driving force for magnetospheric processes. Although the early convection theories have been modified to include the effects of magnetospheric-ionospheric coupling, the original steady large-scale convection models remain largely intact. This paper discusses a variety of recent magnetospheric measurements which support the basic convection model. These measurements include the dynamics of thermal plasma, the drift patterns of medium-energy protons and electrons, and the direct measurement of magnetospheric electric fields by satellites and balloons. All of these measurements have added significantly to our understanding of convection processes, and taken collectively, they offer a fairly unified picture of magnetospheric convection.

1. Introduction

Magnetospheric convection is the basic driving process controlling the dynamics of the magnetosphere. The fundamental importance of the convection electric field has led to an increasing number of attempts to study the details of its spatial and temporal variations. The magnitude of this approximately dawn to dusk directed field is quite small, of the order of 1 mV m^{-1} or less in the equatorial plane, and it is consequently very difficult to measure by satellite electric field probes in the low density regions of the outer magnetosphere. Over the past three or four years several new direct and indirect measurements of the convection electric field have been accomplished and the general agreement among the measurements as well as the convergence of the convection picture as given by the data with that given by the theory is very encouraging. This review will discuss these recent measurements and demonstrate the way in which they support the basic convection theory.

2. The Theory of Magnetospheric Convection

The early convection models of the magnetosphere were proposed by Dungey (1961), and Axford and Hines (1961). These models recognized the interaction of the solar wind with the Earth's magnetosphere through merging of magnetic field lines in the case of Dungey and through a viscous interaction in the case of Axford and Hines. The anti-Sunward flow of the solar wind-magnetosphere interaction causes a Sunward return flow of plasma through the magnetosphere. Figure 1 from Axford (1969) shows this basic circulation pattern. The process begins with the merging of the interplanetary field with the Earth's dipole field at the magnetopause. The flow of the solar wind then sweeps the merged field lines across the polar caps and back into the magnetospheric tail. The tail field lines convect toward the center of the tail where they

D. E. Page (ed.), Correlated Interplanetary and Magnetospheric Observations, 277–295. All Rights Reserved.

reconnect and drift generally Sunward through the inner magnetosphere returning to the magnetopause and completing the circulation.

The motion of the field lines through the magnetosphere as viewed in the equatorial

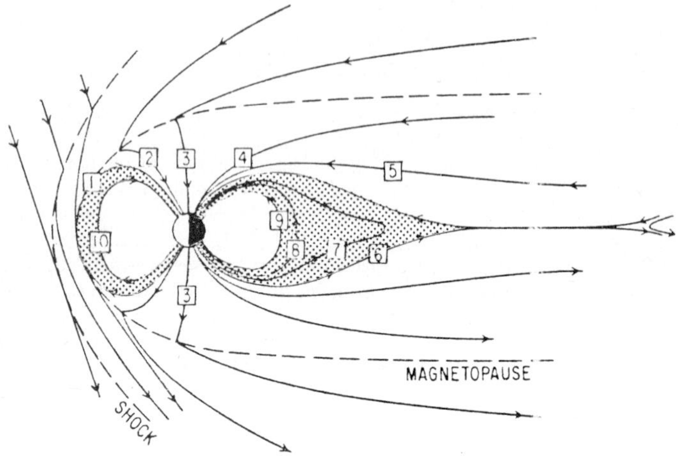

Fig. 1. Steady convection in the noon-midnight geomagnetic meridian. The numbers indicate the successive positions of geomagnetic field lines with reconnection occurring at points 1 and 6 (Axford, 1969).

plane is shown in Figure 2. This shows sketches by Axford and Hines (1961), Nishida (1966) and Brice (1967). The flow is Sunward in nature from the tail to the magnetopause with a diversion in direction around the Earth. In addition to the Sunward flow, these authors included the effect of a corotating Earth which causes a region of eddy flow turbulence in the flow pattern at the dusk side of the Earth.

Following the qualitative ideas of Axford and Hines (1961), Nishida (1966), and Brice (1967), Kavanagh et al., (1968) made some more quantitative calculations of the drift paths of particles in the magnetospheric convection field. These authors assumed different values for the strength of the convection electric field, and superimposed this field on the Earth's corotation electric field in a model magnetic field due to Mead (1964). An example of the resulting drift path for zero energy particles (or equivalently for the drift of magnetic flux tubes using the frozen in flux concept) is shown in Figure 3. The drift paths shown are also equipotentials since the drift of zero energy particles is in the $\mathbf{E} \times \mathbf{B}$ direction which is along equipotential surfaces. The zero energy particle drift paths can be varied by changing the magnitude of the convection electric field. A change in this convection field will not however change the shape of the flow pattern only the size of the 'teardrop' shape. For example, an increase in the convection field will decrease the teardrop region surrounding the closed flow paths. Note in this figure that there are two basic types of flow paths – those which roughly corotate on closed flow paths corresponding to magnetic field tubes which always remain closed, and those whose drift paths reach the magnetopause corresponding to magnetic field tubes which can become 'open' through magnetic merging. The boundary between the always closed flux tubes and the some-

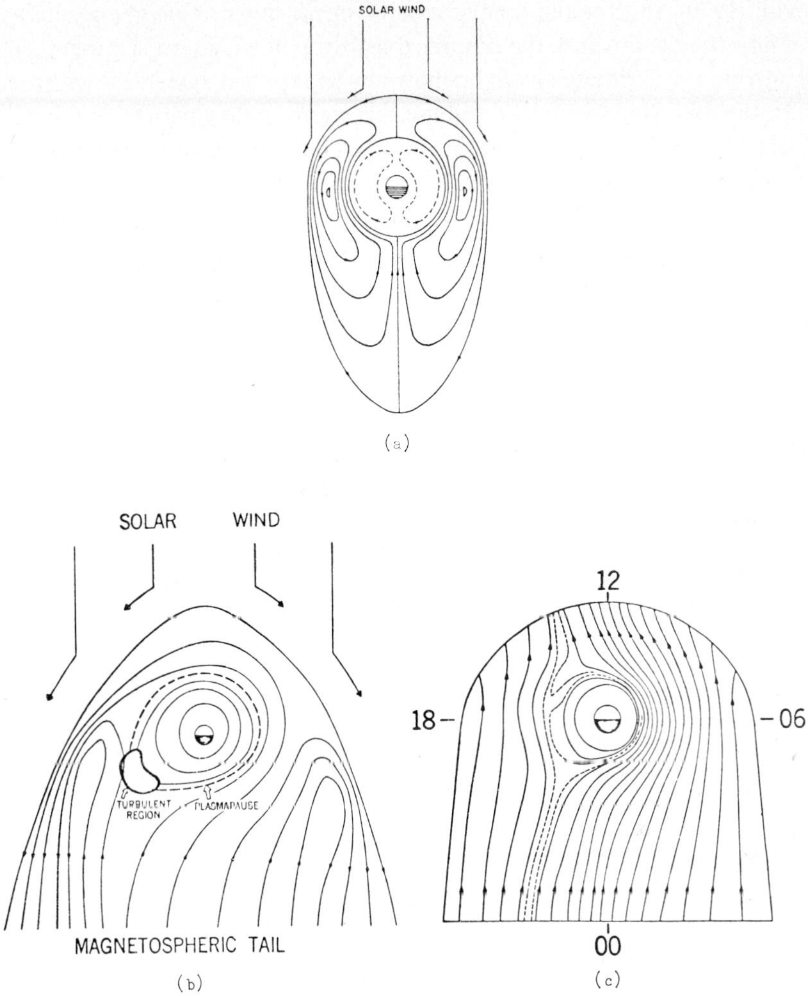

Fig. 2. The convective flow pattern of zero-energy plasma shown in the magnetospheric equatorial plane as drawn by (a) Axford and Hines, (1961), (b) Nishida (1966) and (c) Brice (1967).

times open flux tubes is thought to represent the plasmapause position (Carpenter, 1962; Nishida, 1966; Chappell *et al.*, 1970).

The teardrop flow pattern shown in Figure 3 represents a very simplified approach to magnetospheric convection. It does not include any of the effects of magnetospheric-ionospheric coupling processes or of shielding processes caused by the differential drift paths of energetic particles whose combined curvature, gradient and $\mathbf{E} \times \mathbf{B}$ drifts can cause the development of charge layers (Alfvén layers) which alter the driving convection field pattern. These effects are doubtlessly important in understanding detailed magnetospheric convection flow as pointed out by Block (1967), Falthämmer (1972), Vasyliunas (1972), Shield *et al.* (1969), Kavanagh *et al.* (1968), and Wolf (1970). But for the limited case of *steady quiet* to *moderate* activity convective flow (non substorm), the simplified teardrop flow pattern is a useful concept, and is surprisingly accurate as will be shown by its excellent agreement with measurement. In the following section we will use this teardrop flow pattern as a focal point from which to examine the agreement between different measurements of the convection electric fields.

3. Measurements of Steady Magnetospheric Convection Fields

The basic problem in comparing different types of data on convection electric fields stems from the wide variety of ways in which the data are normally presented. Because of the low magnitude of the outer magnetospheric electric fields, direct probe measurements are difficult, and consequently the equipotentials or equivalently the flow paths have been inferred mainly from indirect measurements. Convection electric fields have been measured directly in the ionosphere using balloons, radar, rockets and satellites. These measurements are possible because of the higher field

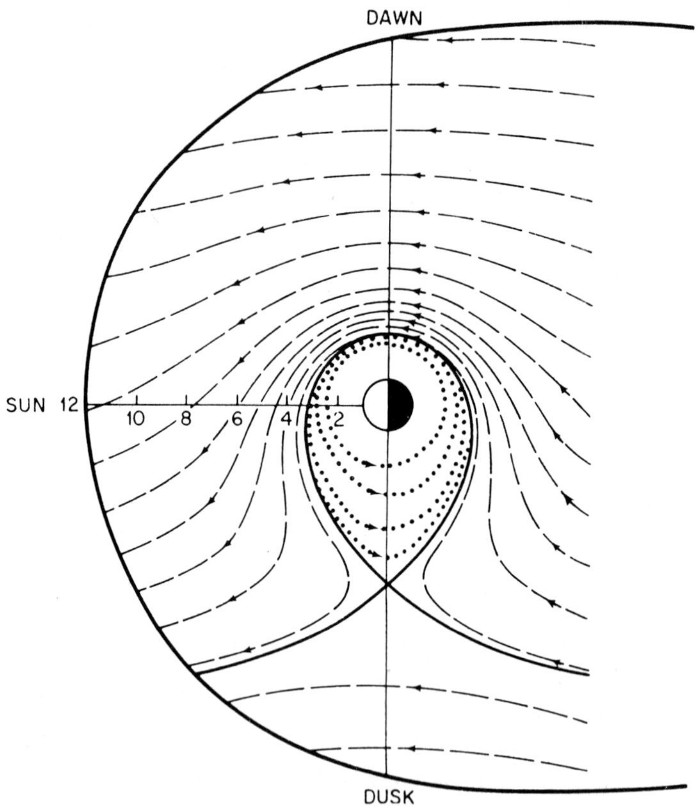

Fig. 3. The calculated drift of zero energy particles in the equatorial plane. This 'teardrop' model is derived from a superposition of the corotation and convection electric fields in a model magnetic field (Kavanagh *et al.*, 1968).

magnitudes ($\sim$20–30 mV m^{-1}, resulting from the 'focusing' caused by mapping along magnetic field lines as shown in Figure 4. Comparison of the indirect measurements in the equatorial plane with the direct probe measurements in the ionosphere requires a mapping of the measurements of one region into the other and this mapping necessitates several assumptions.

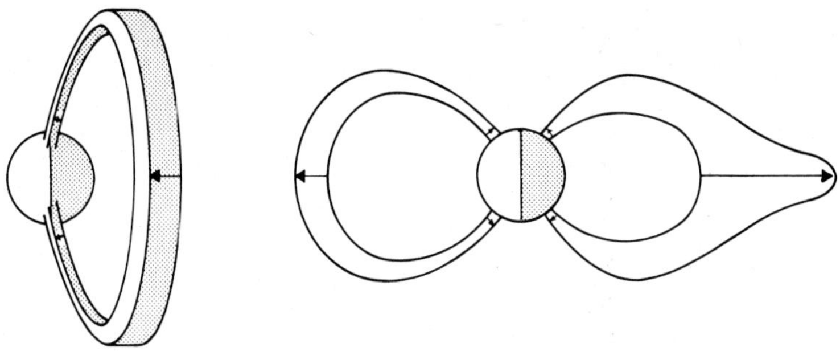

Fig. 4. The focusing effect of the dipole field lines in transforming a given potential drop in the equatorial plane into the ionosphere. The convergence of the magnetic field lines causes an increase in the electric field.

First, one must decide on a magnetic field model. This model can be extremely important in the case of higher L-shells in the nightside auroral zone where the field lines can be stretched out to very large distances in the magnetospheric tail. In some cases it is possible for a predominantly east-west directed convection field in the equatorial plane to become predominantly north-south when mapped into the ionosphere (Mozer, 1970). Even in a dipole field, the north-south component of $\mathbf{E}$ is approximately doubled relative to the east-west component when mapped into the ionosphere at higher L-shells. Figure 5 shows an example of mapping at $L=8.2$ at Fort Churchill (Mozer, 1970). Note that the north-south component of the electric field using a dipole magnetic field increased a factor of 40 in going from the equatorial plane to the ionosphere while the east-west field is only increased a factor of 20. Note also that if the same mappings are done using an empirical field model due to Fairfield (1968), the enhancement in the north-south component near midnight can be tremendous.

A second assumption required in mapping from ionosphere to equatorial plane is that the electric field parallel to the magnetic field lines is equal to zero. This requirement that the magnetic field lines are equipotentials may be difficult to satisfy during substorm times when large field aligned currents may be flowing. These large currents could drive instabilities which could give an anomalous resistivity and a resulting potential drop along the field lines. Because we are restricting ourselves to quiet to moderate magnetic activity with more or less *steady* convection situations, we will adopt the $E_{\|}=0$ assumption and treat the field lines as equipotentials.

4. Electric Field Measurements Inferred from Equatorial Particle Drift

The distribution of the steady convection electric field pattern has been inferred from observations of the drift of medium energy (50 eV to 50 KeV) protons and electrons (McIlwain, 1972) and low energy (~ 1 eV) ions and electrons (Carpenter *et al.*, 1972; Morgan *et al.*, 1973; and Chappell, 1972).

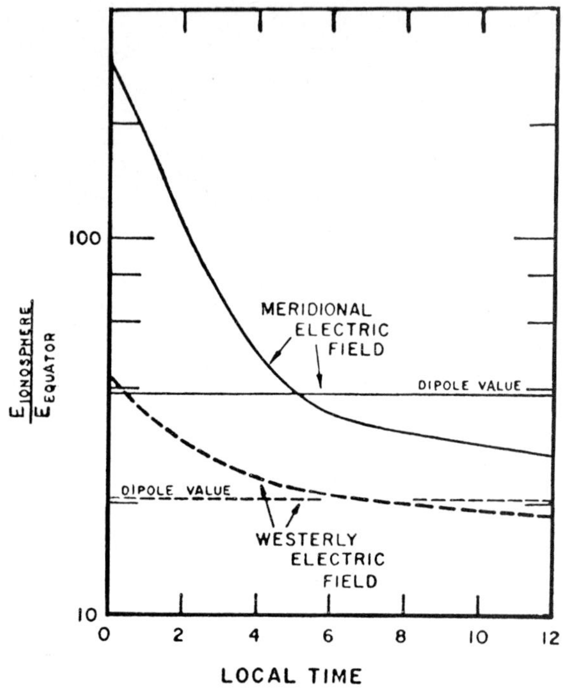

Fig. 5. The ratio of ionospheric to equatorial electric field components for a dipole magnetic field and for an empirical magnetic field model (Mozer, 1970).

Using medium energy plasma data from ATS 5, McIlwain (1972) studied the drift paths of equatorially mirroring plasma which intersected the synchronous orbit. By adopting a magnetic field model and including the corotation electric field of the Earth, and by assuming that each observed cloud of particles was initially injected near midnight, McIlwain was able to derive a static model of the equatorial electric field. The equipotential diagram for the ATS 5 measurements is shown in Figure 6. This figure represents the flow during very quiet times, $\Sigma K_p = 6$ to 10, with small isolated injections of 'tracer' particles. The ATS-5 data on electric field patterns is limited to the region of 5 to 7 R_E in the dawn side and 5 to 10 R_E on the dusk side because of the nature of the particle drift. Figure 6 has been compiled by combining ATS-5 data with some low altitude satellite electric field data. The similarity between the inferred ATS-5 pattern and the simple teardrop pattern of Figure 3 is very striking with the main difference being a more nearly radial drift near the dawn magnetopause. McIlwain reports that the maximum field is 1.2 mV m^{-1} near midnight with a minimum field of 0.04 mV m^{-1} near 1900 LT.

Data on the flow of low energy plasma in the equatorial plane has been inferred from density profiles measured through the plasmasphere by OGO 5 (Chappell, 1972). The characteristic changes in the density profiles with varying magnetic activity was

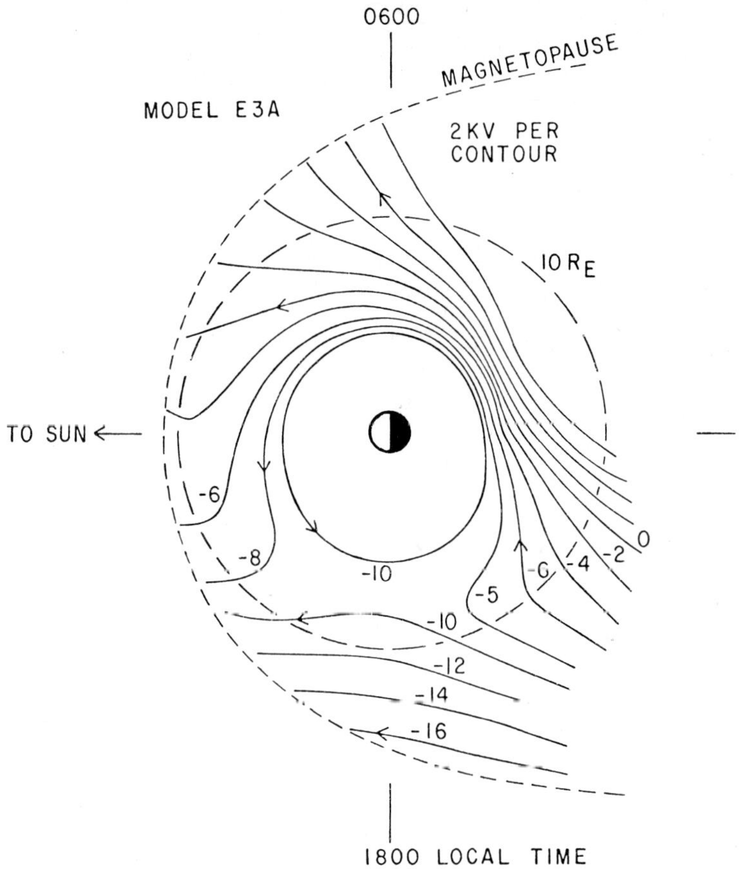

Fig. 6. The convection equipotential surfaces in the magnetospheric equatorial plane calculated from the drift of medium energy particles observed at synchronous orbit (McIlwain, 1972). (With October 1972 additions.)

used to give information on the shape of the equatorial flow paths and hence in the equipotentials of the convective electric field. Chappell (1972) found that the basic dynamic characteristics of the plasmasphere could be explained by the teardrop model of the convective flow. Specifically, the general shape of the plasmasphere as shown in Figure 7 agrees well with the expansive flow of plasma out into the bulge region given by the teardrop. Also the density profiles through the bulge region show the characteristic $1/R^4$ decrease in density expected if they are caused to drift to higher L-shells by the combined corotation and convection drifts as shown in Fig-

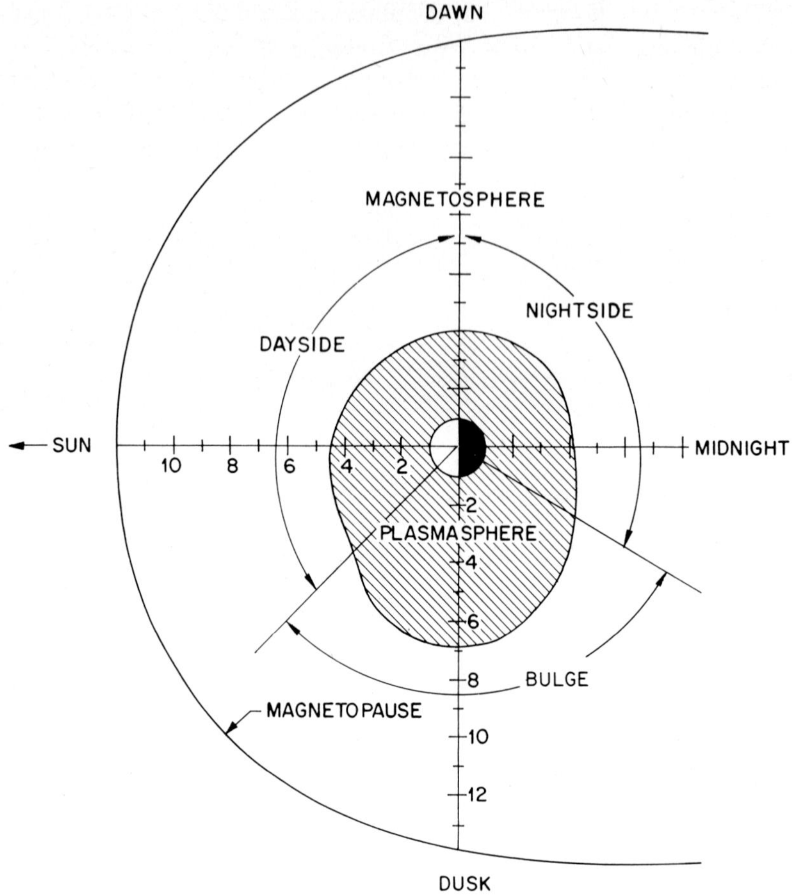

Fig. 7. The average position of the plasmapause shown in the equatorial plane of the magnetosphere from OGO-5 measurements.

ure 8. In addition, the reaction of the plasmasphere to increasing magnetic activity can be modeled nicely using the simplified teardrop model with an enhanced value of the convection field. The enhanced convection causes a compression of the plasmasphere on the nightside and a detachment and loss of plasma on the dayside as shown schematically in Figure 9 in agreement with data. Also, the observation of detached plasma regions in the afternoon-dusk sector of the magnetosphere adds support to the teardrop convection picture. In general then, measured plasmasphere characteristics are in excellent agreement with the steady teardrop model.

Because the OGO 5 data does not give direct information on ion or electron drifts, the expansive flow in the bulge region cannot be directly measured. An alternative picture of plasmasphere bulge flow is the idea of an eddy current flowing at the position of the bulge. This eddy flow would be in contrast with the simple steady picture given by the teardrop model. Recent whistler measurements by Morgan *et al.* (1973) have given new information on this picture. Figure 10 shows the drift paths of whistler ducts measured from Farewell, Alaska ($L=4.3$) over a steady quiet 24-h period. Note that in the dusk sector there is an expansive flow out into the bulge region before 1800 LT and a return flow between 1800 and 2400 LT. Accompanying this outward drift is a decrease in equatorial electron density from ~ 400 el cm^{-3} to ~ 250 el cm^{-3} exactly as would be expected from the R^4 increase in flux tube

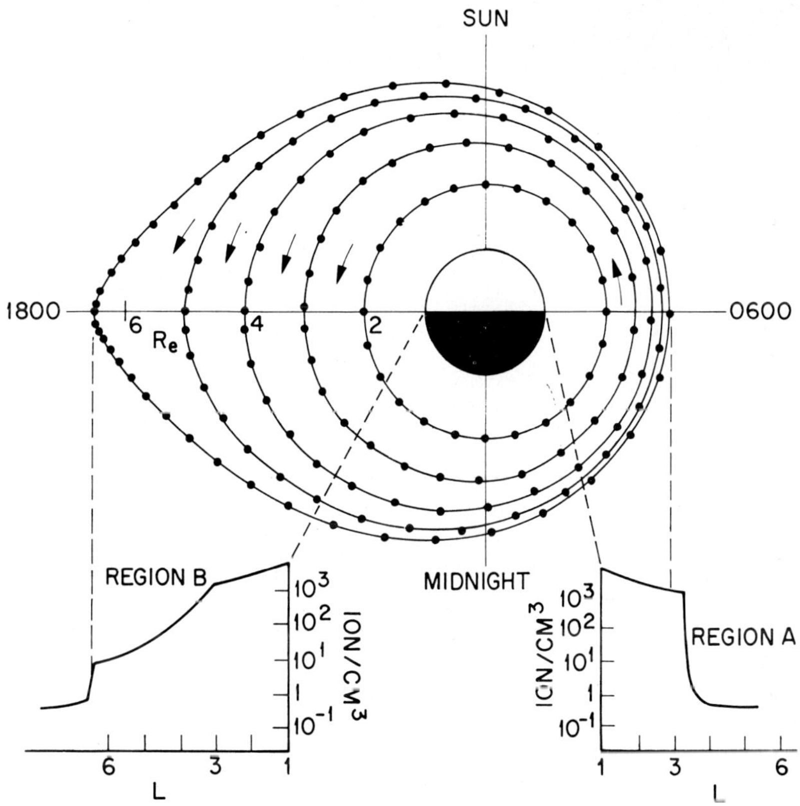

Fig. 8. The motion of the flux tubes inside the plasmasphere. The expansion in the dusk region transforms the sharp profile at $L=3.5$ shown in region A to the $1/R^4$ slope between $L=3$ and $L=6.5$ in region B (Chappell *et al.*, 1970).

volume. After dusk the inward flow is accompanied by an increasing density back up to pre-bulge values. These whistler data then act as excellent support for the steady teardrop picture of magnetospheric convection.

5. Electric Field Measurements Made from Low Altitudes

Low altitude electric fields have been measured from balloons, rockets, incoherent scatter radar, barium clouds, and satellites. Because of our concern here with the steady global convection field picture we will concentrate on the balloon data and

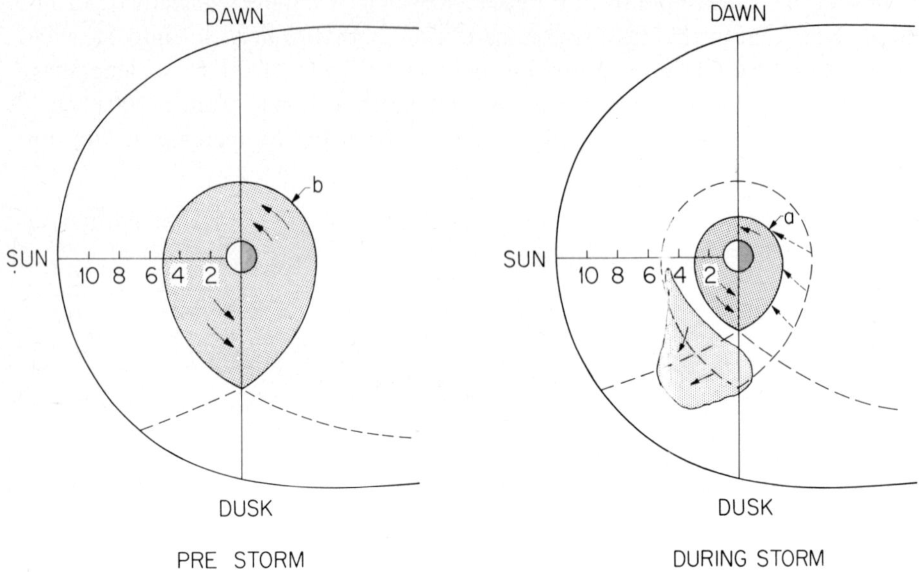

Fig. 9. A sketch of the effect of a magnetic storm on the size and shape of the plasmasphere. Following the onset of enhanced convection during the storm, the nightside of the plasmasphere is convected to lower L-shells; the outer edge of the dayside and bulge regions is peeled off and convected sunward (Chappell *et al.*, 1971).

the satellite data which have been interpreted by the different investigations in terms of global convection.

In a recent paper, Mozer and Lucht (1973) have presented measurements of electric fields at balloon altitudes averaged over some 400 h of data. These measurements were made in the auroral zone covering an L-range of 5.4 to 8.2. The averaged value of the north-south component and the east-west component are shown in Figure 11a. Using a model magnetic field for $\Lambda = 67°$ these components have been mapped to the equatorial plane in part b of Figure 11 and the vector directions are shown along with equipotential contours drawn through the vectors. The agreement between the flow paths shown here and those inferred from particle drift measurements is remarkable. The average convection field calculated by Mozer is about 1 mV m^{-1} which is in good agreement with the field measured by McIlwain. The balloon measurements show that the convection field is always present at some value. When K_p increases say from 1 up to 6, Mozer finds that the average convection field increases by roughly a factor of 2.

A second notable characteristic about the convection flow pattern as measured by the balloons is the change with increasing K_p of the position of the crossover point from southward to northward directed field. Figure 12a shows plots of the east-west and north-south average fields for $K_p > 3$ and $K_p < 3$. Note that for the higher K_p the crossover from south to north $E_\perp$ fields (which corresponds to a shift from predominantly eastward to westward drift) takes place at an earlier local time (1300 LT

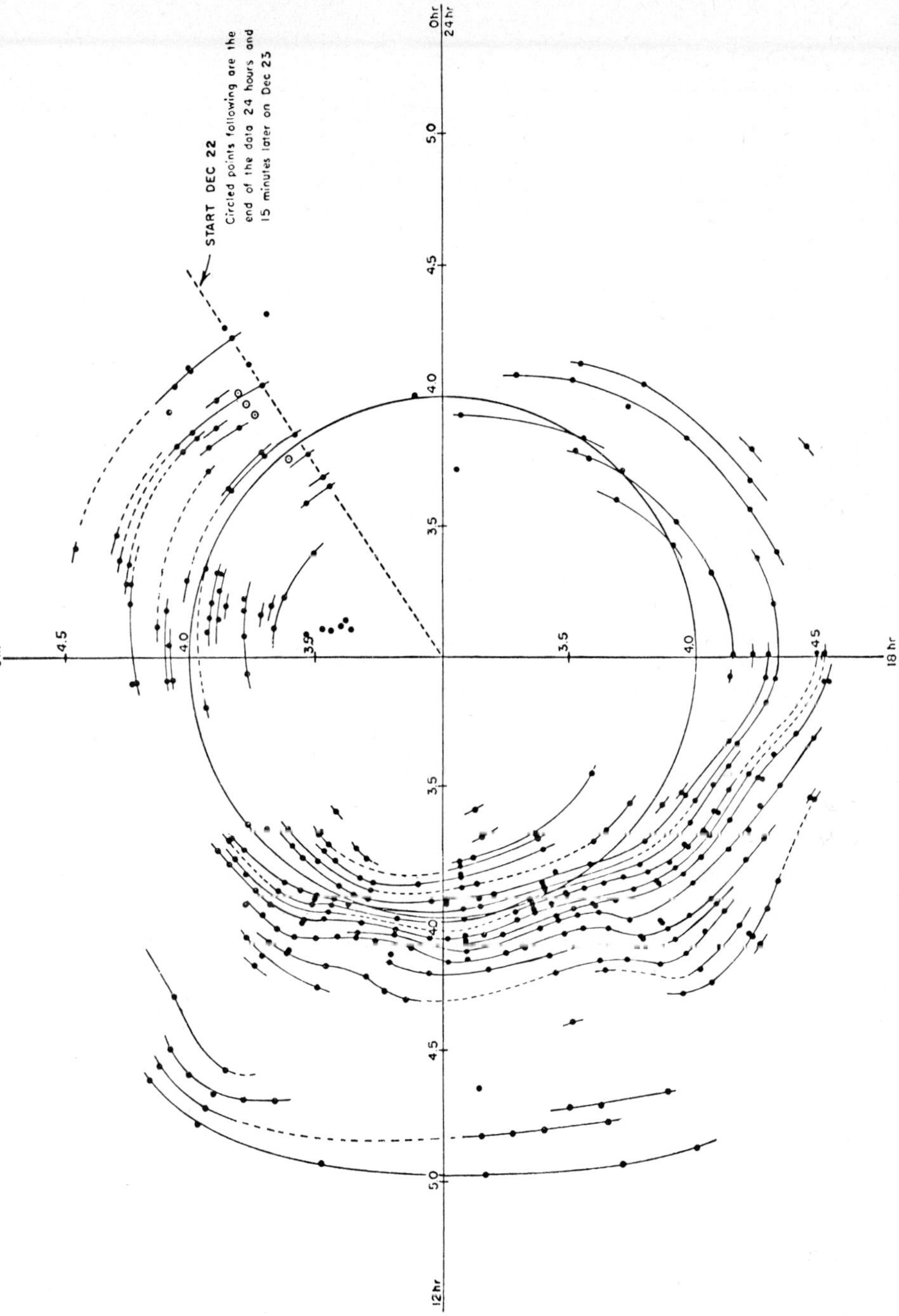

Fig. 10. The drift paths of whistler ducts in the equatorial plane over a 24 h period. These data show the smooth expansion of thermal plasma to higher L-shells in the bulge region near 1800 LT (Morgan *et al.*, 1973).

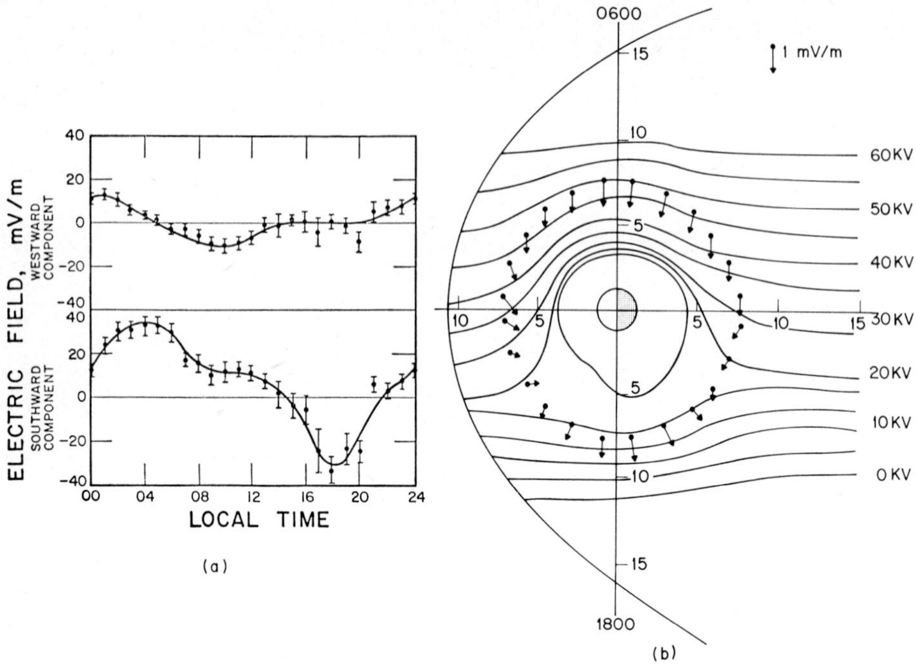

Fig. 11. The average direction of the convection electric field as measured by balloons (Mozer and Lucht, 1973). Part (a) shows the components of the field as measured at balloon altitudes and part (b) shows the projection of these fields into the magnetospheric equatorial plane.

as opposed to 1700 LT) for higher K_p. This effect can be understood in terms of the decreasing teardrop size as shown in Figure 12b. The enhanced K_p corresponds to an enhanced convection field which causes a reduced teardrop size shifting the east-west flow reversal region at point B to an earlier local time (point A). The corresponding predicted shift is not seen at the nightside flow reversal but this descrepancy may be caused by the uncertainty of the true magnetic field configuration – which has not been taken into account here.

Finally it should be noted that the balloon measurements do show an eastward electric field on the dayside (0600 to 1300 LT) which is enhanced with enhanced K_p. It is this field which would cause the detachment and sunward convection of the cold plasma of the plasmasphere and partially filled plasma trough as discussed by Chappell (1972).

The average balloon measurements offer strong support for the simplified steady convection picture. However, as Mozer consistently points out, the balloon electric field data is characterized as much by its large fluctuations as it is by its average value and these large fluctuations must of course also be understood if the complete convective flow process is to be modeled in detail.

Another set of ionospheric electric field measurements has been made on the Injun-5 satellite by Gurnett and Frank (1973). These authors find a dramatic reversal

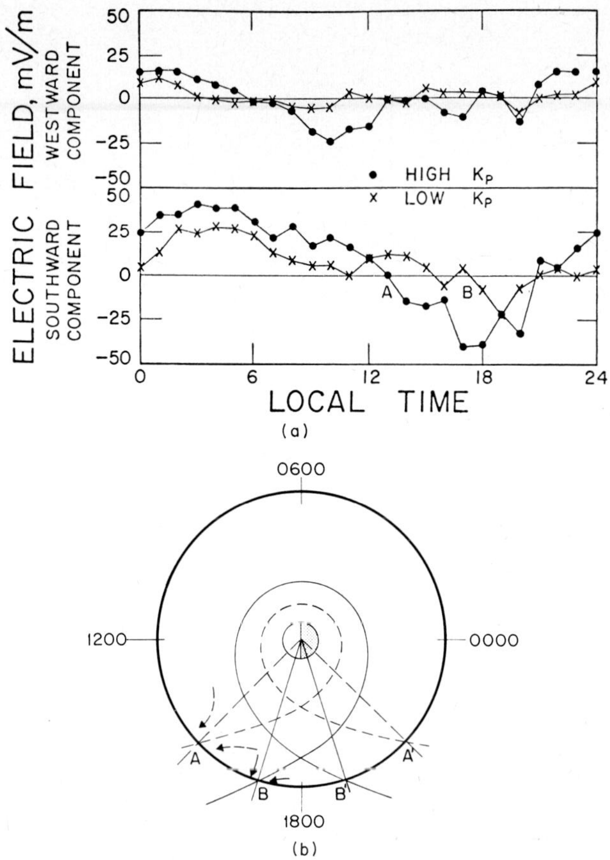

Fig. 12. Part (a) shows the average ionospheric electric field components for high and low K_p (Mozer and Lucht (1973). Part (b) gives the expected location of the zero points in the north-south components as predicted using a 'teardrop' model of the convection field.

in electric field direction at the position of the >50 keV electron trapping boundary and the polar cusp. This field reversal represents a shift from anti-Sunward flow poleward of the trapping boundary to Sunward just equatorward of the trapping boundary. This general trend is in agreement with the convection pattern discussed above but should not be confused with the reversal of flow direction found across the plasmapause in the bulge region. Figure 13 illustrates the Injun-5 measurements. The flow reversals observed on Injun-5 correspond to the flow reversals expected near the magnetopause boundary at points A and C (at field lines which change from closed to open through the cusp). The flow reversal which should exist across the bulge stagnation point at point B is too weak in magnitude to be observed by Injun-5.

Note that the Injun-5 electric fields die away as one goes to lower invariant latitudes. This effect is caused on the dusk side by the weakening of the actual equatorial field value as the satellite samples lower *L*-shells approaching the stagnation point and

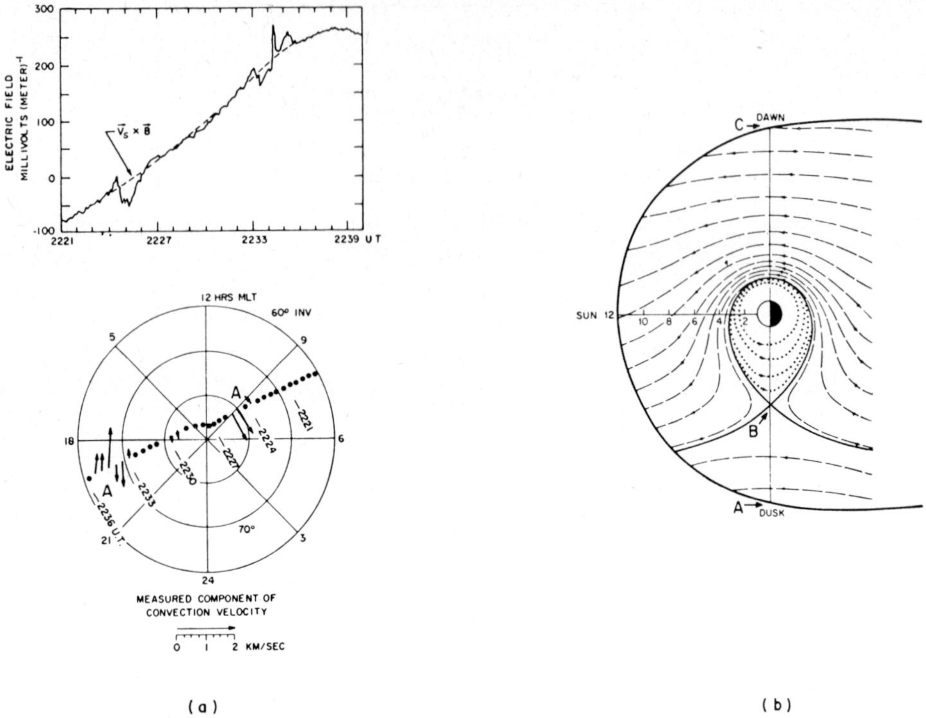

Fig. 13. The convection electric field as measured on Injun 5 (Gurnett and Frank, 1973). Part (a) shows the electric field measurements and a projection of the resulting drift direction on the polar cap. Part (b) shows the expected location of the field reversals at points A and C in the magnetospheric equatorial plane.

most importantly because as the satellite goes to lower L-shells on both the dawn and dusk sides the effective focusing of the equatorial field into the ionosphere by the magnetic field becomes less significant. This focusing effect is given by Mozer (1970) as:

$$\frac{E_{\text{ion}}}{E_{\text{eq}}} \quad \text{north-south} = 2L(L-\tfrac{3}{4})^{1/2}$$

$$\frac{E_{\text{ion}}}{E_{\text{eq}}} \quad \text{east-west} = L^{3/2},$$

where E_{ion} is the electric field in the ionosphere and E_{eq} is the electric field magnitude in the equatorial plane at the particular L-shell. Note that as the L value of the measurement decreases the magnitude of the focusing decreases by $L^{3/2}$. Therefore even if the convection field is uniform in the equatorial plane it becomes progressively harder and harder to measure at low altitudes at the lower invariant latitudes. This reduction in the magnitude of the measurable electric field at lower latitudes combined with the sensitivity threshold of the Injun-5 instrument of about 10 mV m^{-1} on the passes shown preclude any measurements of the convection field in the inner magnetosphere. The general Sunward flow inside the flanks magnetosphere and the anti-Sunward flow just outside the trapping boundary observed by Injun-5 certainly agree with the general convection picture in the outer part of the magnetosphere. Also, during one substorm event, Gurnett and Frank (1973) report an enhancement in the convection field by a factor of 4 in agreement with the enhanced fields reported by Mozer. In this substorm case the enhanced Sunward convecting zone moved to invariant latitudes of 59° Λ on the dusk side. This change in the convection pattern would certainly cause the detachment of plasma from the bulge region of the plasmasphere such as that reported by Chappell *et al.* (1970).

Very high sensitivity measurements of the convection electric field at ionospheric altitudes have been made by Heppner (1972) on the OGO-6 satellite. These measurements are presented in terms of the electric field along the satellite orbit which is in the dawn-dusk meridian plane. Figure 14 shows the plane of the orbit in the magnetosphere as viewed at the north pole (a) and also as viewed from the Sun (b). The data are shown in part (c) of the figure. In presenting his data Heppner has taken out the $\mathbf{V}_{sat} \times \mathbf{B}$ fields due to the satellite motion and also the $\mathbf{V}_{cor} \times \mathbf{B}$ field due to the corotation of the Earth which transforms the data into an Earth-centered rotating coordinate system. In order to be able to visualize the measured convection pattern in the frame of reference fixed with the magnetosphere we have to add back in the $\mathbf{V}_{cor} \times \mathbf{B}$ electric field due to the Earth's corotation. In Figure 14c this is equivalent to subtracting about 10 mV m^{-1} on the duskside and adding 10 mV m^{-1} on the dawn side. This change of coordinate system will then give a field reversal at the stagnation point ($\Lambda = 60°$) in the flow as predicted in the teardrop convection model. Heppner's high sensitivity data therefore give a direct measure of the combined convection and corotation field in the inner magnetosphere which is observed to penetrate well into the plasmasphere and thereby to alter particle motion in excellent agreement with the cold plasma drift data discussed above.

In the (d) part of Figure 14 the data from the June 11, 1969 pass has been mapped into the equatorial plane using the (E_{ion}/E_{eq}) north-south correction given above. The table shows the equivalent value of the dawn-dusk electric field when mapped to the equatorial plane. Note that the ionospheric field values are greatly decreased to values of a few mV m^{-1} in agreement with the fields estimated from particle flow in the equatorial plane. Some flow paths are drawn perpendicular to the projected electric field vectors. These flow paths illustrate well the flow reversal near the magnetopause as seen by Gurnett and Frank (1973) and the flow reversal at the bulge stagnation point as predicted by the teardrop model.

In addition to the different points of agreement discussed above, there are other points that are common to all the measurements. For example, the total potential drop across the polar cap and correspondingly across the magnetosphere is estimated by Gurnett and Frank (1973) at 50 kV, by Mozer (1973) at 60 kV, by Heppner (1972) at 40 to 70 kV, and by Chappell *et al.* (1970) from plasmapause measurements at 60 kV. All of these are in excellent agreement with the potentials estimated in the original models by Axford and Hines (1961), Brice (1967), and Nishida (1966). Also, the magnitude of the convection field is found to vary proportionally with K_p. Mozer,

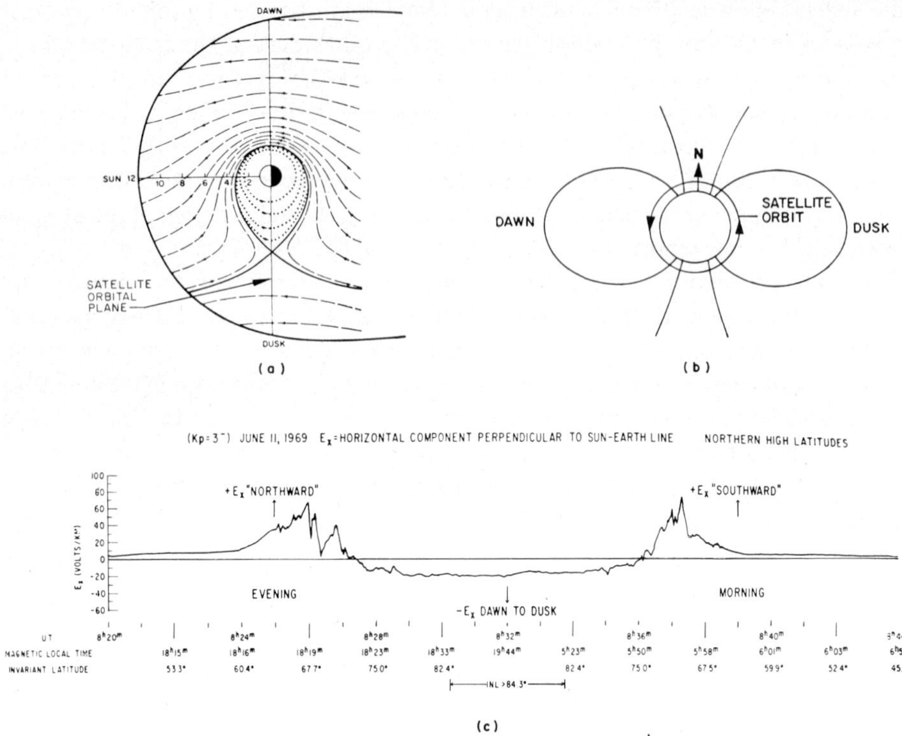

Λ	L	E_x(Equatorial)
53.3	2.82	-.65 mV/m
60.4	4.1	.11 mV/m
Northward 67.7	6.9	1.82 mV/m
75.0	15.	-.16 mV/m*
75.0	15.	-.03 mV/m*
67.5	6.85	.8 mV/m
Southward 59.9	3.98	1.1 mV/m
52.4	2.69	2.4 mV/m
45.0	2.0	3.7 mV/m

* The dipole correction of these values to the equatorial plane is probably in error.

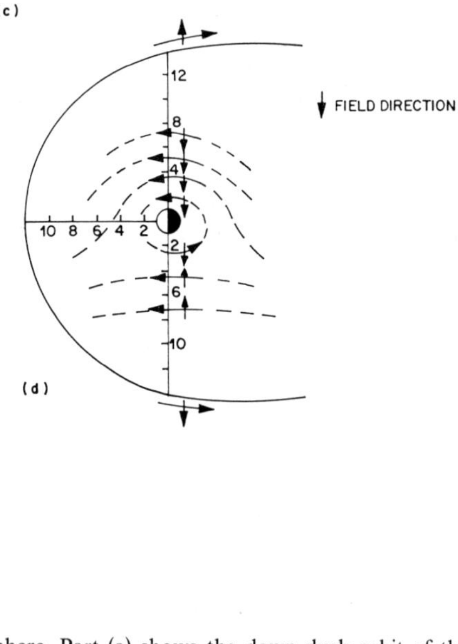

Fig. 14. A pass of OGO 6 over the northern hemisphere. Part (a) shows the dawn dusk orbit of the satellite. Part (b) shows the satellite orbit in the dawn dusk meridian plane. Part (c) shows the satellite electric field data. $+E_x$ is in the plane of the orbit northward-directed on the dusk side and southward on the dawn side. Part (d) shows a table of the measured field values transformed into the magnetospheric equatorial plane with the electric field of corotation included. The field direction and the flow paths are shown sketched in the equatorial plane.

Gurnett and Heppner all report an increase in the measured convection field with an increase in magnetic activity. In Mozer's measurements, the increased convection field could be associated with increased occurrences of a southward component of the interplanetary field. This gives support to the reconnection model of Dungey (1961).

Both Mozer and Heppner found that the details of the convection pattern were responsive to changes in the sector structure of the solar wind. The magnitude of the convection field was found to vary across the magnetosphere being maximum on the dawn side in the northern hemisphere when the interplanetary field was toward the Earth and maximum on the dusk side of the northern hemisphere when the field was away from the Earth. This most interesting observation may well lead to shifts in the convection pattern which might for example cause detachments of the bulge regions of the plasmasphere preferentially when the magnetosphere is engulfed in an away sector.

6. Summary

In summary, there is definitely a convergence of fact and theory concerning convective flow in the magnetosphere. If one considers averaged steady (non substorm) convection, there is excellent agreement among the different types of data as to the large scale flow pattern as is shown in Figure 15. One must be cautious however about over generalizing the agreement past the average steady picture. On short time scales (< 1 h) and in spatially confined regions (< 1 *L*-shell) there can be large fluctuations in the electric field as shown by the balloon, radar, barium cloud, and

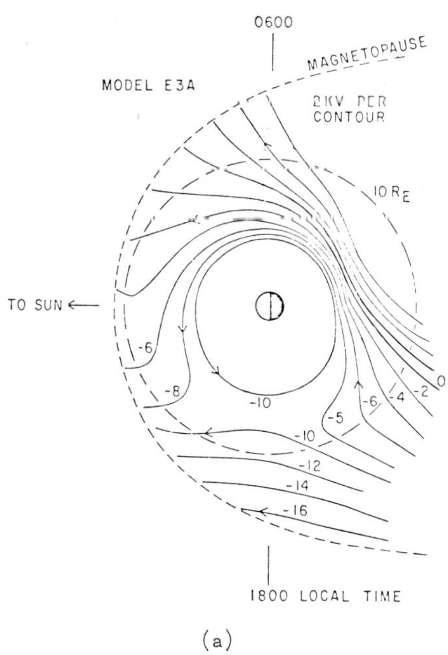

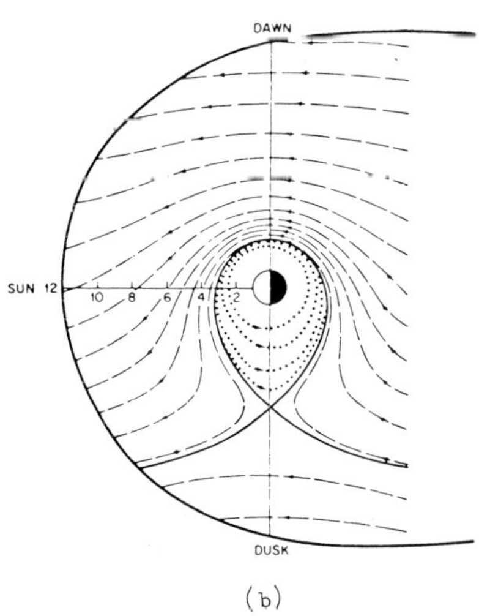

Fig. 15a–b.

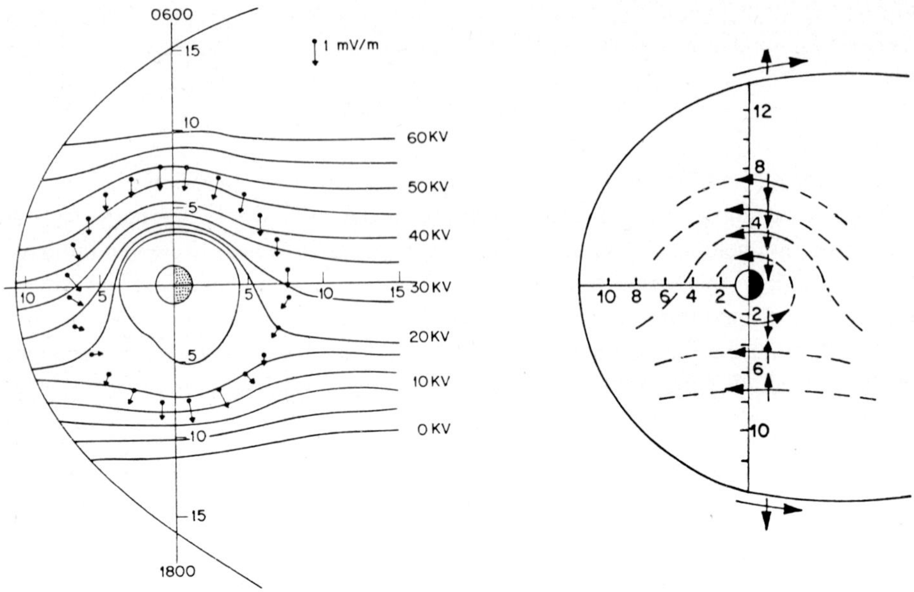

(c) (d)

Fig. 15a–d. A composite showing the equipotentials or drift patterns of zero-energy particles in the equatorial plane for a steady convection field configuration. The measurements are due to (a) McIlwain (1972), (b) Chappell (1972), (c) Mozer and Lucht (1973) and (d) Heppner.

whistler data which can vary significantly from the large scale picture. These effects probably reflect the influences of magnetosphere-ionosphere coupling processes such as parallel electric fields, shielding by charge layers, and neutral winds. These coupling processes will certainly have to be included before the detailed magnetospheric convection picture can be completed. The agreement in the steady convection picture is excellent and offers a first step toward understanding the dynamics of the magnetosphere.

References

Axford, W. I.: 1969, *Rev. Geophys. Space Phys.* **7**, 421.

Axford, W. I. and Hines, C. O.: 1961, *Can. J. Phys.* **39**, 1433.

Block, L. P.: 1967, *Space Sci. Rev.* **7**, 198.

Brice, N. M.: 1967, *J. Geophys. Res.* **72**, 5193.

Carpenter, D. L.: 1962, 'The Magnetosphere During Magnetic Storms', *Rep. SEL-62-059*, Radio Science Lab., Stanford Univ., Stanford, Calif.

Carpenter, D. L., Stone, K., Siren, J. C., and Crystal, T. L.: 1972, *J. Geophys. Res.* **77**, 2819.

Chappell, C. R.: 1972, *Rev. Geophys. Space Phys.* **10**, 951.

Chappell, C. R., Harris, K. K., and Sharp, G. W.: 1970, *J. Geophys. Res.* **75**, 3848.

Dungey, J. W.: 1961, *Phys. Rev. Letters* **6**, 47.

Fairfield, D. H.: 1968, *J. Geophys. Res.* **73**, 7329.

Falthämmer, C. G.: 1972, in B. M. McCormac (ed.), *Earth's Magnetospheric Processes*, D. Reidel Publ. Co., Dordrecht-Holland, p. 16.

Gurnett, D. A. and Frank, L. A.: 1973, *J. Geophys. Res.* **78**, 145.

Heppner, J. P.: 1972, 'Electric Fields in the Magnetosphere', Goddard Preprint X-645-72-154.

Kavanagh, L. D., Jr., Freeman, Jr., J. W., and Chen, A. J.: 1968, *J. Geophys. Res.* **73**, 5511.

McIlwain, C. E.: 1972, in B. M. McCormac (ed.), *Earth's Magnetospheric Processes*, D. Reidel Publ. Co., Dordrecht-Holland, p. 268.

Mead, G. D.: 1964, *J. Geophys. Res.* **69**, 1181.

Morgan, M. G., Brown, P. E., Johnson, W. C., and Taylor, Jr., H. A.: 1973, 'Electron Densities and Electric Fields in the Outer Equatorial Plasmasphere Deduced from Whistlers Observed on the Ground Compared to Ion-Mass, VLF, and Electric Field Observations on a Polar-Orbiting Spacecraft near the same Geomagnetic Meridian', Preprint.

Mozer, F. S.: 1970, *Planetary Space Sci.* **18**, 259.

Mozer, F. S. and Lucht, P.: 1973, 'The Average Auroral Zone Electric Field', Preprint, Univ. of Calif., *Spa. Sci. Lab. Series* **14**, Issue 21.

Nishida, A.: 1966, *J. Geophys. Res.* **71**, 5669.

Schield, M. A., Freeman, J. W., and Dessler, A. J.: 1969, *J. Geophys. Res.* **74**, 247.

Vasyliunas, V. M.: 1972, in B. M. McCormac (ed.), *Earth's Magnetospheric Processes*, D. Reidel Publ. Co., Dordrecht-Holland, p. 29.

Wolf, R. A.: 1970, *J. Geophys. Res.* **75**, 4677.

PARTICLES AND PLASMAS IN THE EARTH'S MAGNETOTAIL AT 60 R_E

K. A. ANDERSON

Physics Dept. and Space Sciences Laboratory, University of California, Berkeley, Calif., U.S.A.

L. M. CHASE, R. P. LIN, R. E. McGUIRE

Space Sciences Laboratory, University of California, Berkeley, Calif., U.S.A.

and

J. E. McCOY

Johnson Spaceflight Center, Houston, Tex., U.S.A.

1. Introduction

A small scientific satellite was launched from the Apollo 16 service module on April 26, 1972 into an approximately circular orbit about the moon of about 100 km altitude. The particles and field subsatellite passed through the magnetotail twice before mascon perturbations caused it to crash into the Moon on 29 May 1972. This satellite is nearly identical to the one launched by Apollo 15, which has been previously described (Anderson *et al.*, 1972).

The main purpose of this article is to describe the plasma and particle populations in the magnetotail near 60 R_E geocentric distance. We discuss both the plasma sheet and the high latitude portions of the magnetotail. Electron and proton spectra at energies above about 20 keV and electron spectra down to 0.5 keV have been obtained. Another topic of importance is the comparison of the deep magnetotail plasma sheet with the Vela satellite measurements at $\sim 20\ R_E$ geocentric distance.

2. Experimental Method

The plasma and particle detectors used in the experiment are described in Table I. The method used to separate high energy protons and electrons in the solid state telescopes has been described in some detail (Anderson *et al.*, 1972). The basic idea of the particle separation technique is to use two solid state telescopes that are identical except that one is covered with a thin foil of thickness 375 $\mu g\ cm^{-2}$ (Figure 1). This foil stops protons or else produces very large changes in their energy over the range of interest while changing the energy of electrons by only a small fraction of their total kinetic energy. An electron of initial energy 26 keV is decreased by 5 keV in passing through the foil.

Examples of the technique's effectiveness are shown in Figure 2. Following a solar flare on 25 May 1972 particles appear in both telescopes at energies of 20 keV up to about 100 keV where the counting rate becomes insignificant. There are no particles of any kind with higher energy. The lowest energy channel in the open and in the foil

D. E. Page (ed.), Correlated Interplanetary and Magnetospheric Observations, 297–316.

TABLE I

Summary of detector characteristics

Detector type	Energy range		Geometric factor, cm^2 sr	Angular aperture	Angle to spin axis, deg	Minimum detectable flux particles $(cm^2\ s\ sr)^{-1}$
	Protons, MeV	Electrons, keV				
Open telescope	0.004 to 2	20 to 300	0.045	40°, cone	0	0.01
Foil telescope	0.3 to 2	20 to 300	0.045	40°, cone	0	0.01
HESA-CEM	No response	0.52 to 0.58	5.3×10^{-5}	20° by 60° full width half max. (FWHM)	90	10^4
HESA-CEM	No response	1.9 to 2.1	2.13×10^{-4}	20° by 60° (FWHM)	90	10^4
HESA-CEM	No response	5.9 to 6.4	2.4×10^{-4}	15° by 60° (FWHM)	90	10^4
HESA-CEM	No response	5.8 to 6.5	0.08	18° by 60° (FWHM)	90	0.1
HESA-CEM	No response	13.5 to 15.0	0.17	13° by 60° (FWHM)	90	0.1

HESA = Hemispheric plate Electrostatic Analyzer
CEM = Continuum Electron Multiplier

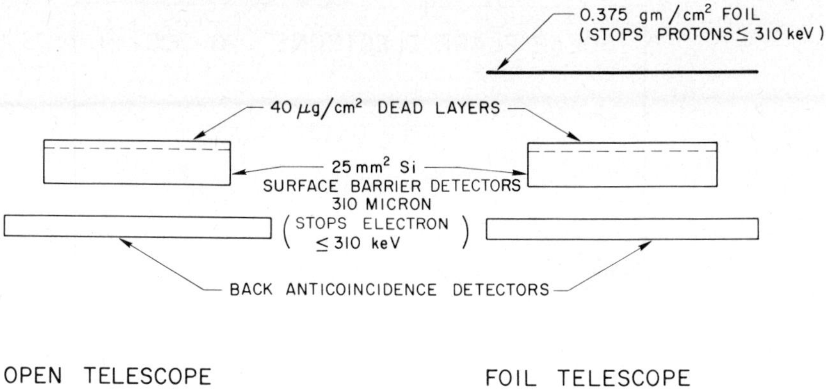

Fig. 1. Proton and electron energy spectra are measured by a pair of solid state detector telescopes, one covered by a thin foil which produces very large, known energy losses on incident protons in the energy range of interest. However, the foil produces only very small effects on electrons.

telescope show closely the same counting rates, as they should if the incident particles are electrons. A second example to illustrate the technique is given in Figure 2. Particles with energy from 30 keV up to about 100 keV appear in the open telescope only. There are no detectable effects in the foil-covered telescope. Thus, the particles are stopped by the foil. But since >22 keV electrons can be detected trough the foil, the particles must be ions, presumably mainly protons, although the technique gives little information about the charge composition of the particles. These particles are upstream protons almost continually present on interplanetary field lines that intersect the Earth's bow shock.

3. Results

3.1. Particles in the High Latitude Magnetotail

Whenever the subsatellite was in the magnetotail and above or below the plasma sheet measurable particle fluxes were present over the proton and electron energy ranges covered by the subsatellite detectors. These particle fluxes are stable compared to the plasma sheet particle fluxes. There is some variability from one hour to the next and one day to the next of roughly a factor of three and the variability is often energy dependent. On the basis of many energy spectra obtained in the solar wind and in the magnetosheath, we arrive at the following results:

(1) The HLM electron fluxes above 20 keV and the proton fluxes above 35 keV have the same intensity and spectral characteristics as those particle spectra observed in interplanetary space. Previously we have ascribed the origin of the latter particle fluxes to the Sun and/or the Galaxy. We therefore conclude that the higher energy protons and electrons in the HLM originate outside the Earth's magnetosphere and have access to the high latitude magnetotail with little effect on their energy and intensity. Little can be said about the mechanism of entry from these observations.

(2) The electron spectra from 0.5 to 20 keV in the HLM closely resemble those in

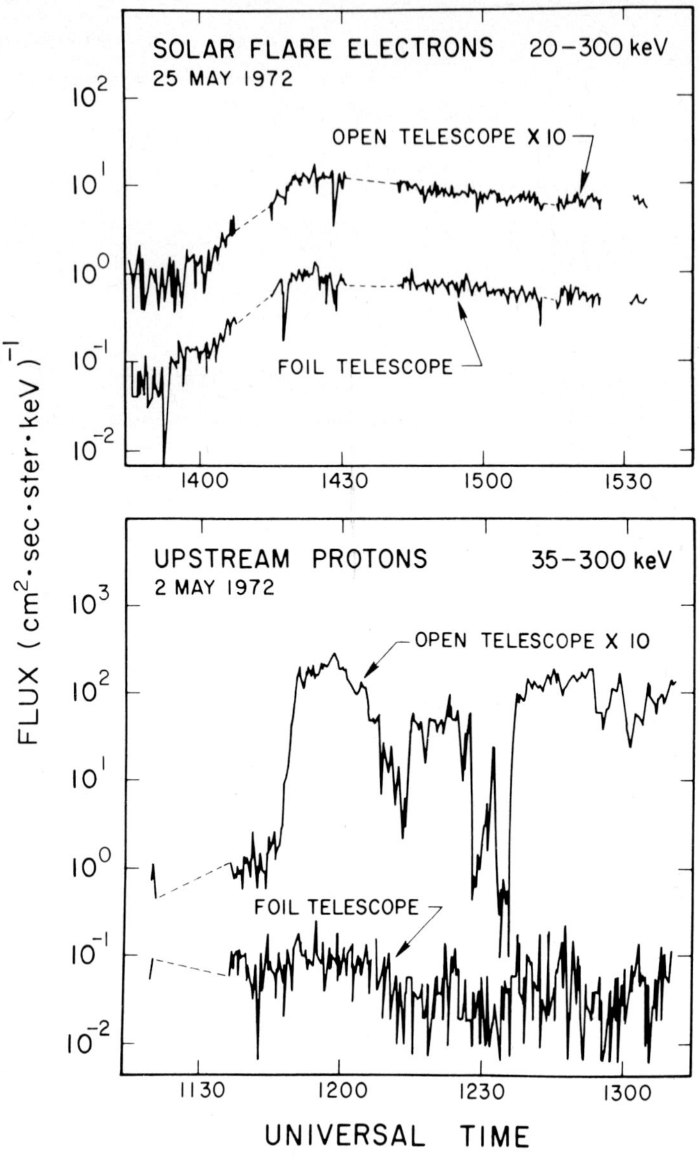

Fig. 2. Examples of how the foil separation technique works with a beam of pure electrons (upper panel) and pure protons (lower panel). The solar flare electrons produce equal counting rates in the open and the foil telescopes while the upstream protons are only able to enter the open telescope.

the magnetosheath. If there are differences, they are hidden in the present data by the variability of the particle fluxes. This means that the particle intensities do not vary by more than a factor of two from magnetosheath to HLM. Figure 3a shows that as the satellite moves from the magnetosheath into the magnetotail the 0.5 keV plasma electron flux does not change appreciably. The two regions are however easily recognized in our experiment: in the magnetosheath the shadows formed behind the Moon in the plasma flow resemble those formed in the solar wind but are less well defined and more shallow. The HLM is immediately recognizable by the formation of shadows in front of the Moon. It is appropriate here to show that the magnetosheath plasma as seen at the higher electron energies does not resemble the plasma sheet electron fluxes. The intensities at all electron energies from 0.5 keV are much higher in the plasma sheet than in the magnetosheath (Figure 3b). At 0.5 and 2 keV we believe that the electrons are still part of the solar wind-magnetosheath plasma and have the collective properties of these fully ionized plasmas (Anderson *et al.*, 1972). However, somewhere above 2 keV the electrons begin to behave as single, independent particles that do not participate in the collective phenomena of the plasma. One might therefore suspect that particle entry for the two different spectral regions might be completely different. The higher energy particles probably enter on con-

(a)

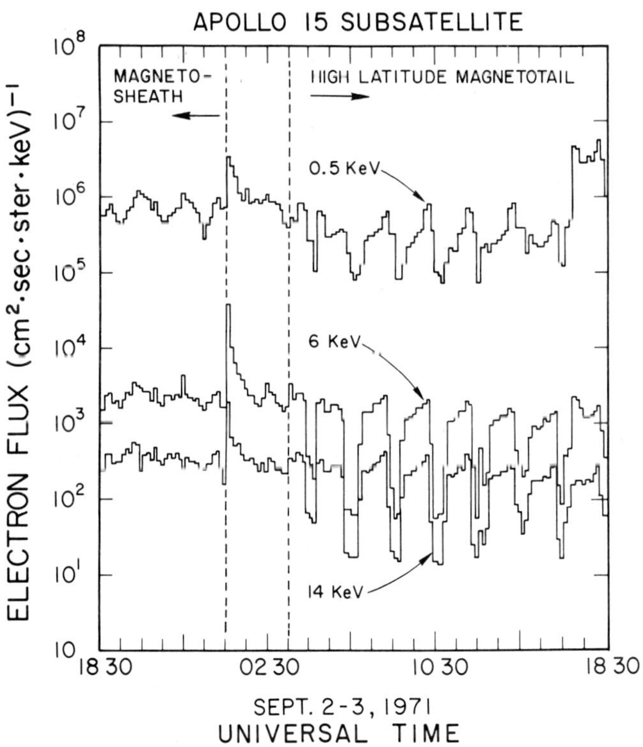

Figs. 3a–b. The upper panel shows the passage of the subsatellite from the magnetosheath into the high latitude magnetotail at a geocentric distance of 60 R_E. After entry into the tail deep shadows are seen on the Earthward side of the Moon reflecting the different topology of these field lines. The fluxes of 0.5, 2 and 14 keV electron change very little from the magnetosheath to the high latitude tail. The lower panel shows passage of the satellite from magnetosheath to plasma sheet. The flux of 0.5 keV electron is 1 to $1\frac{1}{2}$ orders of magnitude greater in the plasma sheet. This is true in all cases.

(b)

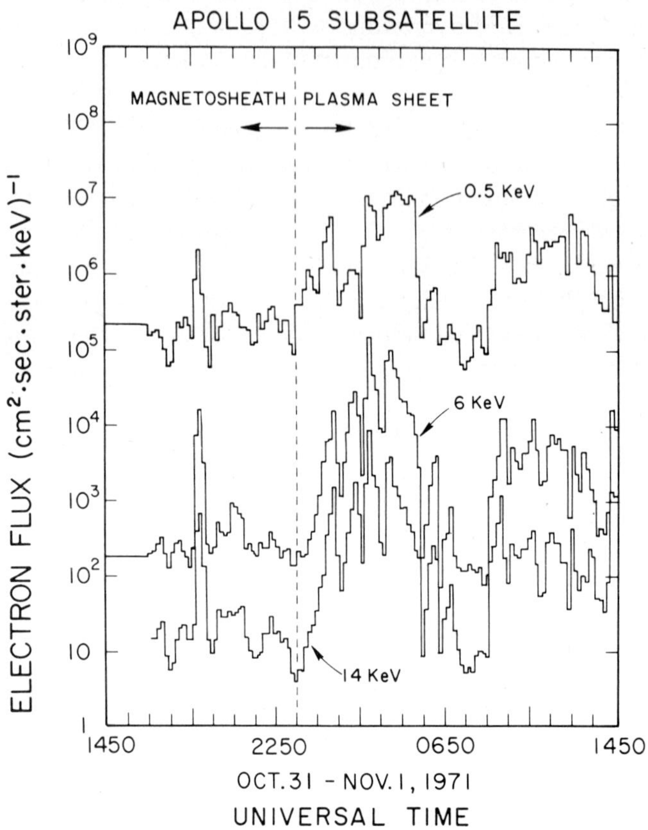

nected field lines in the manner suggested by Lin and Anderson (1966). The lower energy electrons that are still tied to the solar wind-magnetosheath plasma may not enter in this way, however. It is possible these particles may diffuse into the magnetotail although it must be said that to date there is no direct, physical evidence for particle transport by diffusion in the magnetotail regions. Table II shows that at times the 6 and 14 keV electron fluxes in the HLM and magnetosheath range to somewhat higher values than those encountered in the solar wind. At present we do not know if this is due to time variation that increases these fluxes everywhere or if there is consistently a higher flux in these regions than the solar wind.

(3) The shadowing of the electron fluxes above 2 keV by the Moon gives evidence that entry into the magnetotail occurs at geocentric distances greater than 60 R_E. Figure 4 shows that deep shadows occur on the Earthward side of the Moon while on the far side the fluxes are reduced only by a factor of two. This result confirms the earlier work of Anderson and Lin (1969) which clearly showed that 20 and 40 keV electrons must have entered the tail beyond the Moon's orbit. The argument is simply that on the Earthward side of the Moon these electrons were completely absent; on

TABLE II

Electron flux $(cm^2 \text{ sr s keV})^{-1}$

Late April – Mid May, 1972				
	0.5 keV	2 keV	6 keV	14 keV
Solar wind	$1–3 \times 10^6$	$1–4 \times 10^3$	40–150	8–15
Magnetosheath	5×10^5 to 10^6	$1–3 \times 10^4$	100–500	9–30
High latitude Magnetotail	3×10^5 to 2×10^6	4×10^3 to 3×10^4	$10^2–10^3$	12–35
Plasma sheet	Up to 3×10^7	Up to 4×10^6	Up to 3×10^5	Up to 10^4

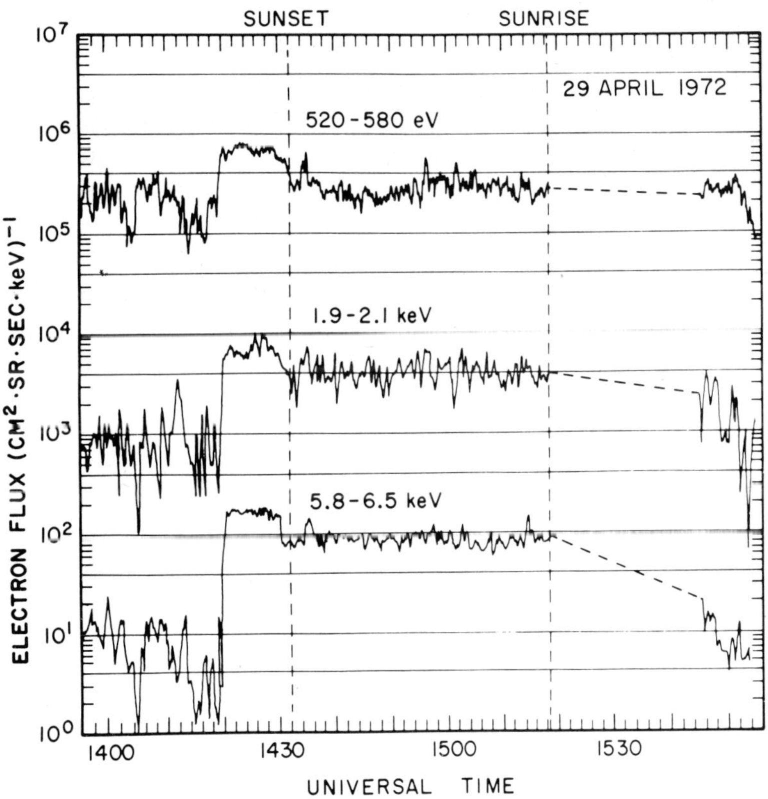

Fig. 4. Electron fluxes over one orbit of the subsatellite about the Moon while in the high latitude magnetotail. The highest fluxes occur when the satellite is moving on magnetic field lines that do not pass through the Moon. Thus the true flux of 0.5 keV electrons averaged over all directions is 7×10^5 $(cm^2 \text{ sr s keV})^{-1}$. These lowest energy electrons are present in nearly the same intensity on the Earthward side as they are on the far side of the Moon. Electric fields probably are responsible for moving these electrons into the Earthward shadow region where they otherwise would be excluded. These fields have less effect on higher energy electrons so that these particles do not appear in the Earthward shadow region.

the far side of the Moon the electron fluxes are present and hence can only come from the magnetotail beyond 60 R_E.

(4) Figure 4 shows that the 0.55 keV electrons appear on the Earthward side of the Moon with little reduction of intensity. Investigation of this effect has led McCoy *et al.* (1973) to the conclusion that these particles entered the HLM somewhere behind the Moon and were acted on by a large-scale, steady electric field that exists, at least at times, across the HLM. For a dawn-dusk electric field direction, particles in the northern HLM undergo a southward drift and those in the Southern HLM, a northward drift (Anderson, 1970; McGuire, 1972; McCoy *et al.*, 1973). Thus, if the electric field were sufficiently strong, the 0.55 keV electron fluxes could reach the subsatellite when on the Earthward side of the Moon. For example shown in Figure 4, a field of 3×10^{-4} V m^{-1} would be required to move 50% of an initially isotropic distribution of 0.55 keV electrons down near the Moon's equator. For other cases, fields ranging from 1×10^4 to 9×10^{-4} V m^{-1} would be required (McCoy *et al.*, 1972).

3.2. Particles in the Plasma Sheet at 60 R_E

The satellite entered the magnetotail late on 26 April 1972 and at once encountered the plasma sheet. Entry occurred on the evening side of the magnetopause at $Y_{SM} \cong 20$ R_E and $Z_{SM} = +5$ R_E. The satellite remained in or near the plasma sheet for about 1 day when it moved into the southern high latitude magnetotail. Figure 5 shows the great variability of the plasma sheet, some of which is accounted for by passage of its boundary back and forth across the satellite (0535 to 0600 UT).

The slowest variations, on the order of half an hour for factor of two changes, are either related to changes of the plasma sheet, properties as a function of time or else these properties are spatially dependent. The most rapid variations (a few minutes to produce factor of two changes) are attributed to MHD wave modulations of the electrons. From these data we see that the protons are modulated to a much lesser degree.

The plasma electron flux is $\sim 10^7$ (cm^2 sr s keV)$^{-1}$ at 0.55 keV, a rather typical value for the plasma sheet. The 13 keV electron flux reaches 4×10^2 e cm^{-2} sr s keV, a considerably enhanced value but not as large as on other plasma sheet encounters when the flux may be a factor of 10 higher.

The proton flux above 35 keV is as large as and usually somewhat larger than the electron flux above these energies. In other plasma sheet encounters, the proton intensity is often much higher than the electron intensity at the same energy.

The comparatively small energetic particle flux at these times may be due to the persistence of the dawn-dusk symmetry found at ~ 18 R_E (Montgomery *et al.*, 1965) back to ~ 60 R_E. On the other hand the high energy components simply may not have been particularly intense anywhere in the plasma sheet at these times.

The electron spectrum shows a feature at 6 ± 3 keV in many of the spectra taken at these times. This feature is better defined at some times than others as can be seen in the collection of spectra given in Figure 6.

The plasma sheet was encountered for about $1\frac{1}{2}$ h on 28 April 1971 for the last time

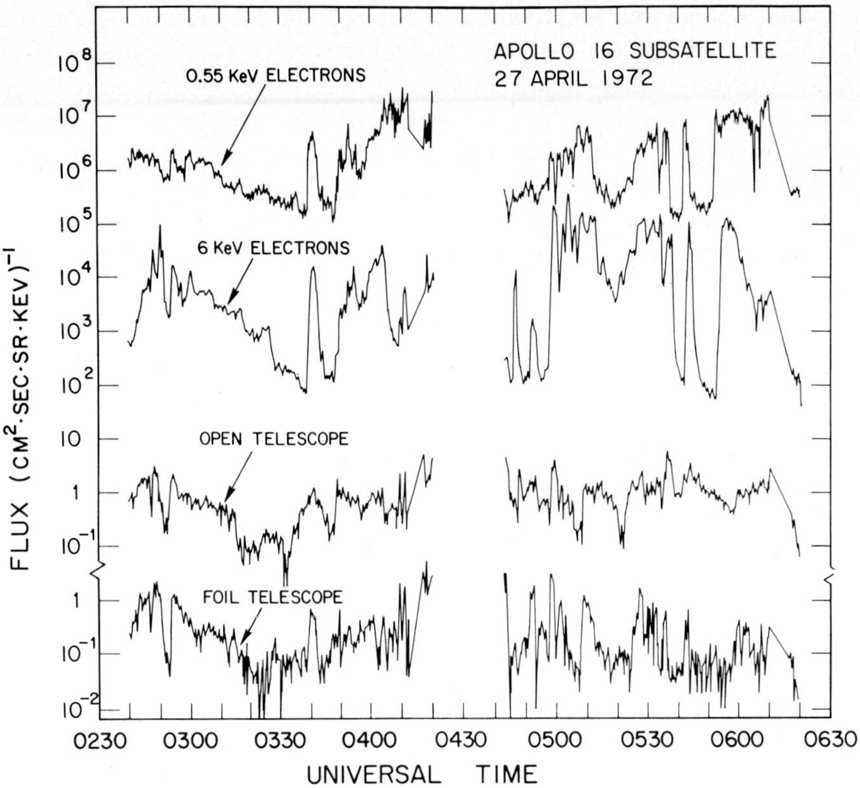

Fig. 5. This figure shows fluxes of particles in the dusk side of the plasma sheet at 60 R_E geocentric distance. The 0.5 keV electrons reach values above 10^7 (cm^2 sr s keV)$^{-1}$, at least 10 times their magnetosheath value. Six keV electrons are quite intense, showing values over 10^5 (cm^2 sr s keV)$^{-1}$. However, electrons above 20 keV are not particularly intense. Protons above 35 keV are present in intensities of about 2 (cm^2 sr s keV)$^{-1}$.

during the April 1972 passage of the Moon across the magnetotail (Figure 7). The brief encounter occurred when the Moon was located at $Y_{SM} \cong -7\ R_E$. The distance from the Murayama model neutral sheet was $-8\ R_E$. Previously there had been only one other brief encounter since early on 27 April. During these times the Moon was below the plasma sheet. The encounter was evidently due to a large motion of the plasma sheet, possibly an expansion. As might be expected in such a situation the particle fluxes were unusually high and large fluxes of high energy particles were present.

The 0.5 keV electrons exceeded 2×10^7 (cm^2 sr s keV)$^{-1}$, the 6 keV electron flux was 3×10^5, and the 11 keV flux reached 4×10^3. The proton flux reached the high value of 9×10^2 (cm^2 sr s keV)$^{-1}$ at ~45 keV. Figure 8 shows the high energy electron and proton spectra averaged over 10 min at the peak intensity. There is no well developed peak in the electron energy spectrum near 6 keV at this time, but the spectrum is markedly curved on the log-log plots in the $\frac{1}{2}$ to 10 keV region. Protons are present well above background in the energy interval 344 to 518 keV. Above 700 keV the proton flux is lost in the backgrounds. The proton spectrum from 38 to 518 keV is very well represented by a straight line on the log-log plots of Figure 8. The power law exponent, γ, is -5.1. The most remarkable aspect of the energy spectra is the much greater proton flux at the same kinetic energy. The proton flux is roughly 30 times as great as the electron flux at the same kinetic energy. The electron spectrum is also closely power law from 10 to 160 keV. However, the spectral slope is $\gamma = -4.5$ significantly less steep than the proton spectrum in the same energy range.

(a)

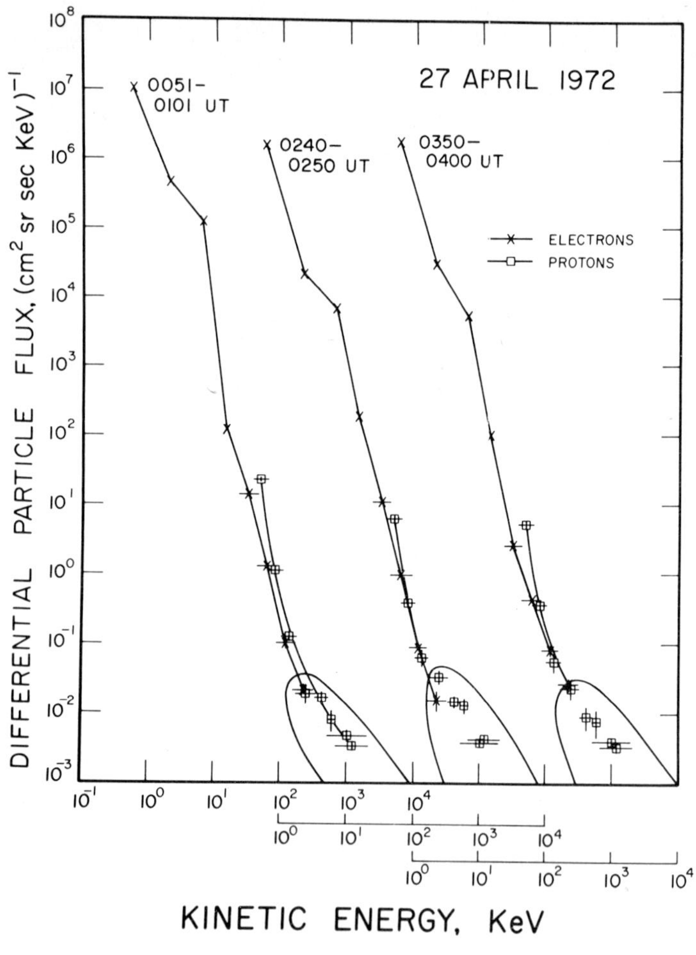

Figs. 6a–b. Electron and proton energy spectra constructed for times shown in Figure 5. There is a distinct break in the spectrum around 6 keV. At certain times the 6 keV flux exceeds the 2 keV electron flux. The last two spectra show a characteristic feature of the protons above 30 keV: their flux is much higher than that of electrons at the same energy. The loops enclose points that show no change above their interplanetary space levels.

(b)

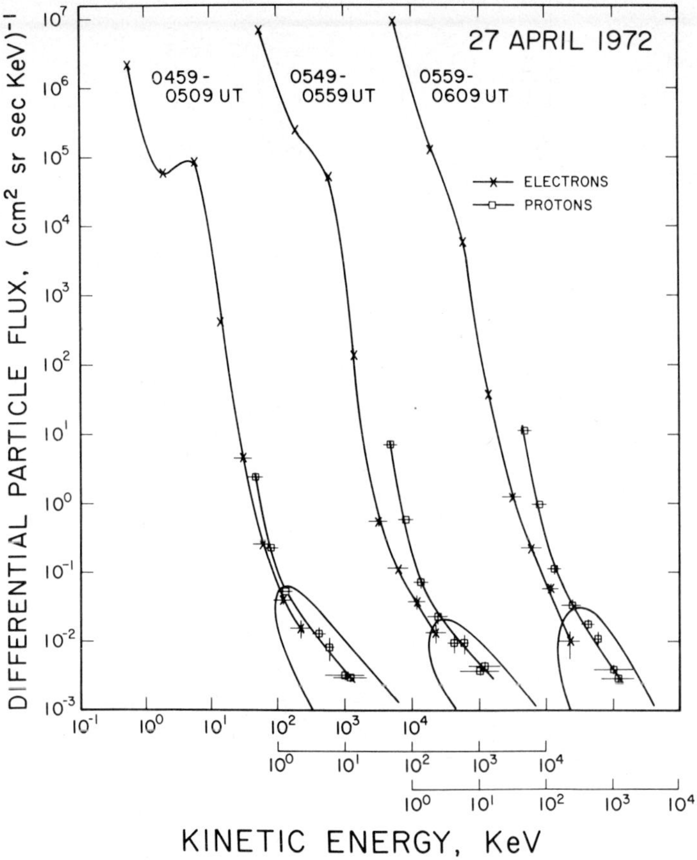

During late May of 1972 the subsatellite encountered the plasma sheet very close to the center of the tail. From 0111 to 1019 UT on May 28 particle fluxes consisting predominantly of energetic protons were detected (Figure 9). These decayed away and were followed by a spike of low energy electrons from 0117 to 0123 UT. These effects are evidently encountered at the edge of the plasma sheet but cannot be entirely accounted for by the fact that protons have larger gyroradius at the same energy. After 0124 UT, both protons and electrons are present with the proton flux higher than the electron flux by factors of 5 to 10. The electron spectra are shown in Figure 10. Much of the time the spectrum is somewhat curved from 1.2 keV to 20 keV and the degree of curvature changes with time. The well-defined feature at 6 keV which appeared in the spectra on 26 April is not present at these times.

The particle fluxes observed later on 28 May and shown in Figure 11 also occur in the plasma sheet close to the center of the magnetotail ($Z_{SM} \approx -5\ R_E$; $Y_{SM} \approx -5\ R_E$).

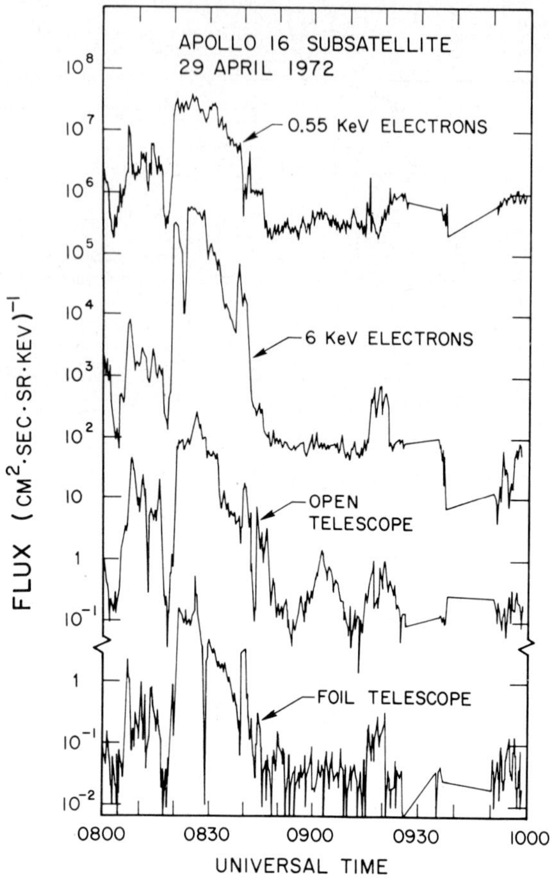

Fig. 7. The most intense plasma sheet fluxes during the April and May passes through the magnetotail are shown here. Shortly following 0900 UT there is an increase in the proton flux (open telescope) but no corresponding increase in the electrons.

The corresponding energy spectra are shown in Figure 12. These fluxes have the following characteristics:

(1) Electron spectra are somewhat steeper than proton spectra, except at end of sequence of plots where the slopes become nearly the same.

(2) The variability of the electron fluxes is greater than that of the protons. Response of the foil telescope shows that the highest energy electrons are the most variable.

(3) The proton time profile is generally like that of the electrons. However, late in the event the electrons decay away more rapidly, increasing the proton to electron ratio at a given energy.

We have searched the Apollo 15 and 16 subsatellite data in the plasma sheet for effects of particle shadowing by the Moon. We have not found any such effects except possibly factor of two decreases in the omnidirectional intensity behind and in front of the Moon compared to limb regions where particles can be received from all directions. In Figures 9 and 11 we have marked the times of sunrise and sunset on the spacecraft. At these times shadow effects would be expected if they indeed occur. However, no such effects can be seen in these examples. If the high energy electrons, frequently observed in the plasma sheet, were accelerated Earthward of the Moon's location we might expect a near absence of particles behind the Moon. Similarly, if

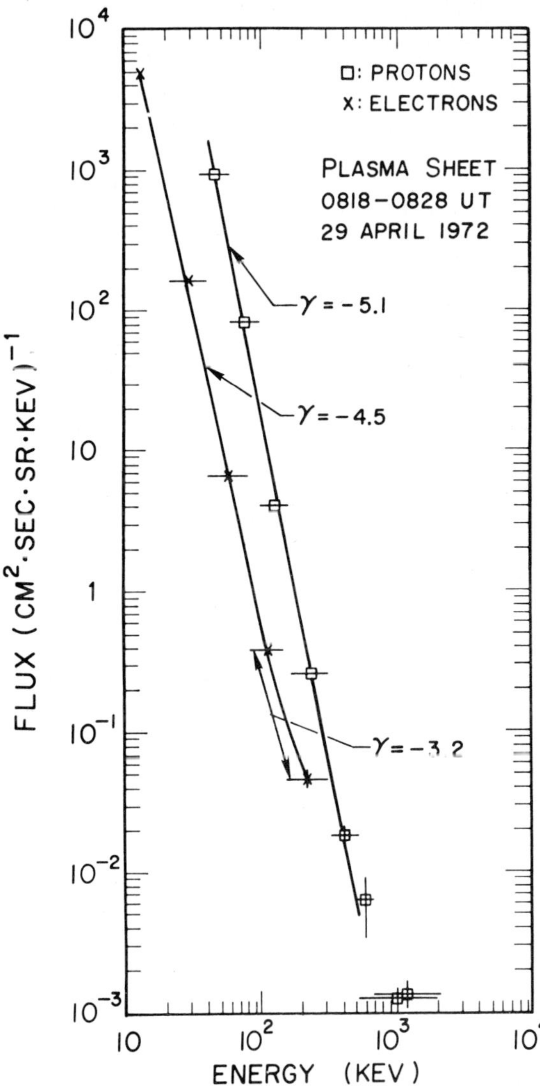

Fig. 8. High energy portion of the electron and proton spectra during the intense plasma sheet fluxes shown in Figure 7. Both the electrons and protons have power law spectra at these times, with the proton spectrum being somewhat steeper. At the same energy the proton flux is as much as 30 times higher than the electron flux.

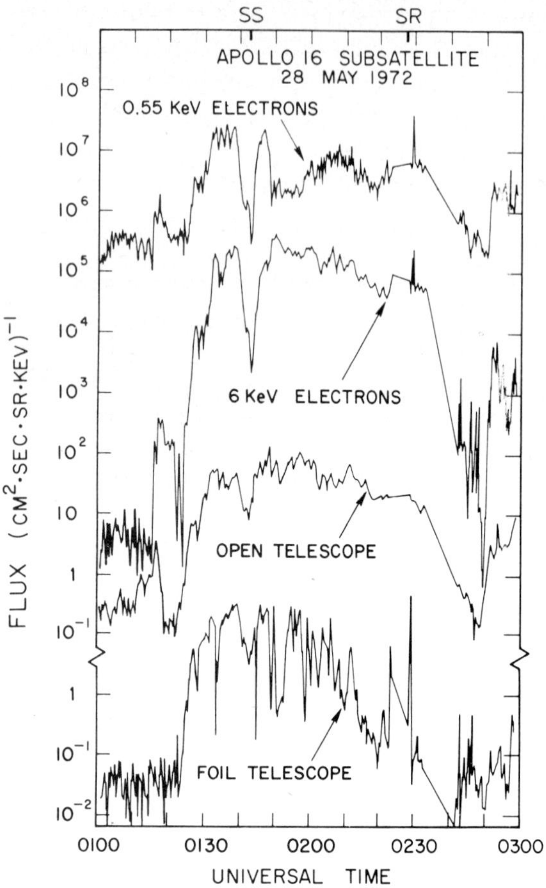

Fig. 9. Fluxes near the center of the plasma sheet. This example shows complicated effects at the edge of the plasma sheet: first protons appear for several minutes, followed by a spike of soft electrons. After 0200 UT the electrons decay more rapidly than the protons.

the particles are accelerated in the plasma sheet beyond the Moon's orbit we would expect large particle decreases on the Earthward side of the Moon.

4. Comparison with Vela Results

(1) In the plasma sheet the flux of electrons at 0.5 keV is $\sim 10^8$ (cm^2 sr s keV)$^{-1}$ at 18 R_E while at 60 R_E the flux at this energy is $\sim 10^7$. At 18 R_E the peak of the energy spectrum is near 0.8 keV when it can be determined (Montgomery *et al.*, 1968; Hones *et al.*, 1971). The Apollo experiment is not able to determine the peak energy in the electron spectrum but at no time has a peak occurred at or above 0.5 keV (except for the feature near 6 keV).

(2) There is occasionally a marked spectral feature in the energy spectrum near 6 keV at 60 R_E. This 'bump' is much like that reported in the Vela results at a some-

what higher energy (~15 keV). It thus appears that two spectral components characterize the plasma sheet electron spectrum at 60 R_E as well as at 18 R_E.

(3) The plasma sheet electron spectra at 60 R_E are definitely non-Maxwellian above 2 keV. Above about 15 keV the electron spectra are closely fit by power laws with exponents of −4.5 to −5. At 18 R_E the Vela measurements show that the energetic electron spectra (>10 keV) are also often non-Maxwellian. However, at that geocentric distance the spectra are somewhat flatter than at 60 R_E, having a power law exponent near −4 on the average.

5. Summary and Discussion

(1) The spectrum of electrons in the high latitude magnetotail is virtually indistinguishable from that in the magnetosheath. At the lowest energies (0.5 and 2 keV) this implies that the electron component of the magnetosheath and HLM plasmas

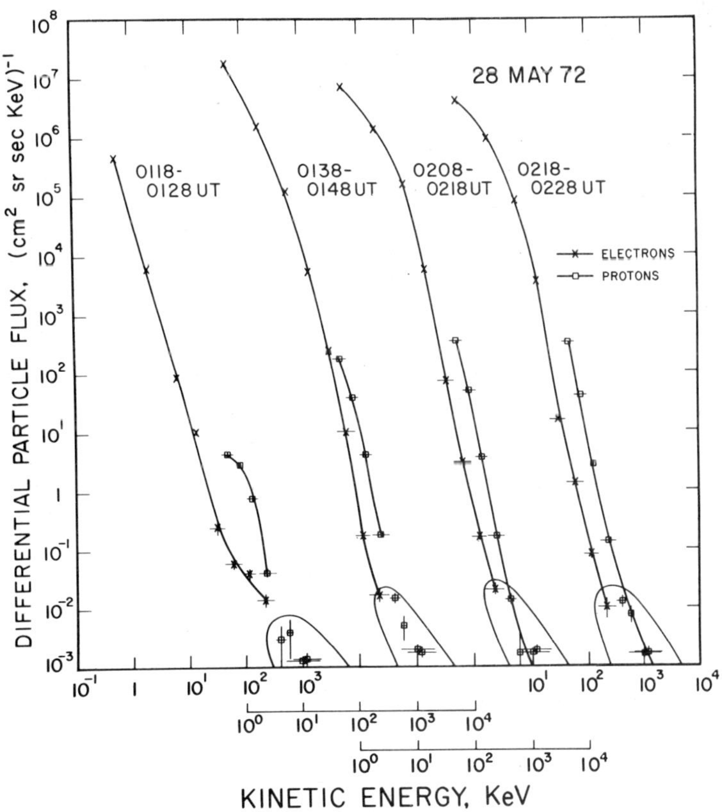

Fig. 10. Electron and proton spectra in the plasma sheet at times shown in Figure 10. The spectral feature at 6 keV seen in Figure 6 does not appear. The slope of these spectra changes rapidly over the energy range 0.5 to 14 keV. In most cases they become power law above 14 keV.

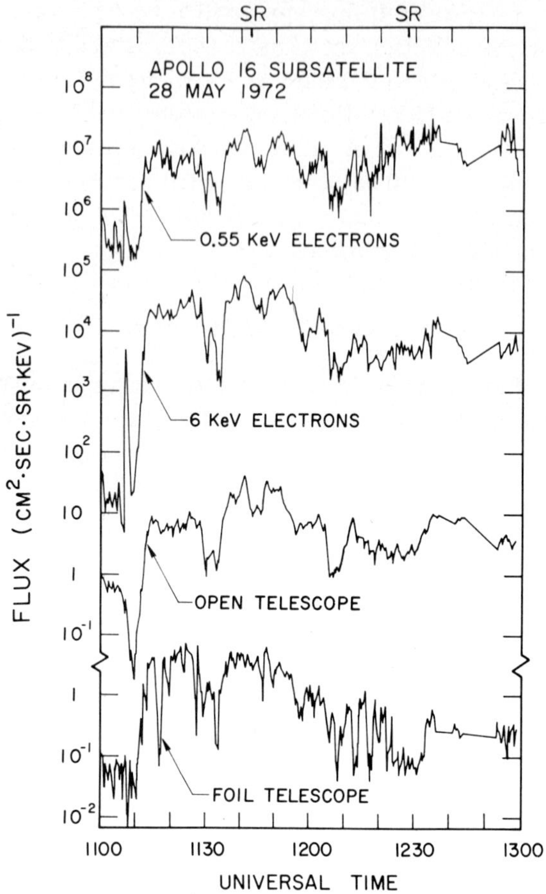

Fig. 11. Plasma sheet particle 10 hours after those shown in Figure 9. Again a spike of soft electrons is seen at the edge of the plasma sheet. It occurs at the same time as a large decrease in the proton fluxes.

are closely similar. Above an energy near 2 keV this implies that the interplanetary cosmic ray electrons enter the HLM with little impediment.

(2) The plasma sheet is very much different from the magnetosheath plasma as seen in electrons. At all energies 0.5 to 100 keV the plasma sheet intensities are higher by an order of magnitude or more.

(3) In the plasma sheet the flux of electrons at 2 keV is always less than the flux at 0.5 keV. We therefore conclude it is probable the peak of the plasma sheet electron spectrum at 60 R_E is always below 0.5 keV. The unidirectional, differential energy electron spectrum for a Maxwellian distribution of electrons would be given by

$$\frac{dJ}{dE} = CE\, e^{-E/kT} \frac{\text{electrons}}{\text{cm}^2 \text{ sr s keV}}.$$

The maximum intensity of this distribution occurs at $E = kT$. We have observed the maximum electron intensity never occurs above 0.5 keV. If 0.5 keV or a smaller

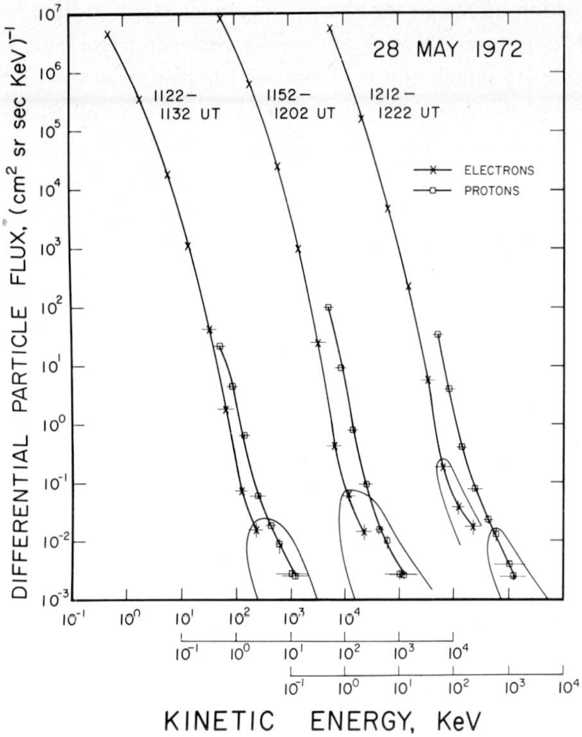

Fig. 12. Electron and proton spectra in the plasma sheet during times shown in Figure 11.

numerical value is used for kT in the expression for dJ/dE it is not possible to fit any of the electron spectra shown in Figures 6, 8, 10 and 12 above 2 keV.

(4) At times the energy spectrum of electrons in the plasma sheet shows a feature at 6 ± 3 keV. On one occasion the flux at 6 keV exceeded that at 2 keV by a factor of 2 (Figure 6). Most of the time, the 6 keV flux is less than the 2 keV flux but there will be a distinct break or at least a curvature of the spectrum over the interval 0.5 to 14 keV.

(5) The proton and electron energy spectra are closely fit by a power law from a few tens of keV to several hundred keV.

(6) The flux of protons above 20 keV usually is greater and often a factor of 10 to 30 times greater than electron fluxes compared at the same kinetic energy. The large proton to electron ratio is a common feature of the plasma sheet at 60 R_E. It is known from IMP-6 measurements that this property also holds in the range 20 to 30 R_E.

(7) Comparison of the Apollo subsatellite electron energy spectra with those spectra from Vela measurements we find the best agreement when the intensity scale is shifted by $\sim 10\,x$ and if the energy scale is shifted by a factor of about 2 (Figure 13). A transformation which would relate the plasmas at the two geocentric distance is

an energization which increases the energy of each electron in the plasma by a factor of 2 to 3 at 20 R_E compared to 60 R_E. If such a view is correct, this suggests that the basic physical quantity which controls the particle energy in the plasma sheet is the magnetic field energy density: at 20 R_E the field is $\sim 15\ \gamma$, at 60 R_E the field is $\sim 10\ \gamma$. Hence the ratio of magnetic field energy densities is ~ 2.5. Such a situation might be expected on general grounds in these plasmas where β is close to unity and where static equilibrium exists. It must be conceded that comparison of the Vela and Apollo Subsatellite data are subject to at least two uncertainties:

(a) The instruments have never been cross calibrated although both sets of detectors measure absolute fluxes.

(b) The spectral comparisons were made for different times. It is true, however, that at no time during the Subsatellite's operation do fluxes come closer than an

(a)

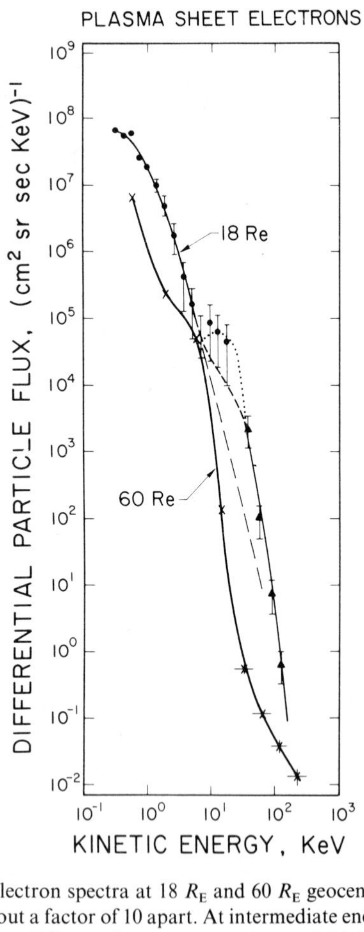

Figs. 13a–b. Comparison of electron spectra at 18 R_E and 60 R_E geocentric distance. At 0.5 keV and at higher energies the fluxes are about a factor of 10 apart. At intermediate energies the fluxes are comparable. The spectra were taken at different times so comparisons of this kind are not conclusive.

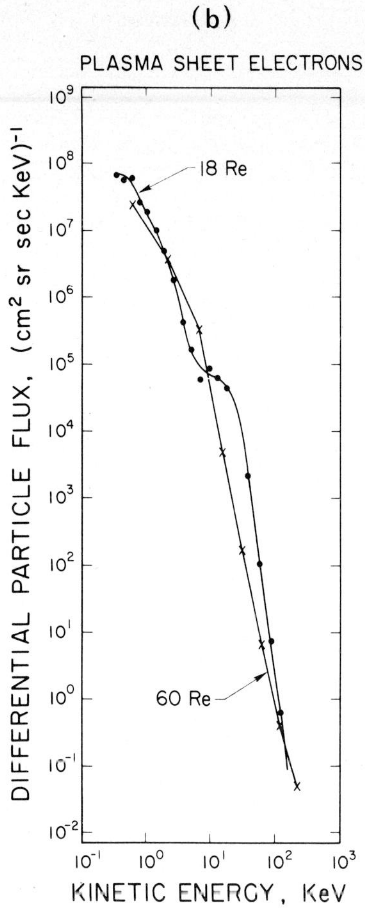

order of magnitude to those reported at 20 R_E at the lowest and highest energies.

(8) While there is a very large difference between the proton flux and electron flux at the same kinetic energy it is possible to reconcile the observed proton to electron ratios of up to 30 in the following way: If in the plasma sheet the protons can be preferentially energized by a factor k compared to electrons the proton spectrum would then move up by a factor of k^γ where γ is the power law exponent of the proton and electron spectra. For observed values of γ near 5, k only needs to be a factor of two. At present we have no suggestions for a mechanism which would preferentially energize protons compared to electrons by this amount.

Acknowledgements

This work was supported by the National Space and Aeronautics Administration under Contract NAS 9-10509.

References

Anderson, K. A., Chase, L. M., Lin, R. P., McCoy, J. E., and McGuire, R. E.: 1972, *J. Geophys. Res.* **77**, 4611.

Anderson, K. A.: 1970, *J. Geophys. Res.* **75**, 2591.

Anderson, K. A. and Lin, R. P.: 1969, *J. Geophys. Res.* **74**, 3953.

Anderson, K. A., Chase, L. M., Lin, R. P., McCoy, J. E., and McGuire, R. E.: 1972, *J. Geophys. Res.* **77**, 4611.

Hones, E. W., Jr., Asbridge, J. R., Bame, S. J., and Singer, Sidney: 1971, *J. Geophys. Res.* **76**, 63

Lin, R. P. and Anderson, K. A.: 1966, *J. Geophys. Res.* **71**, 4213.

McCoy, J. E., Anderson, K. A., Chase, L. M., Lin, R. P., and McGuire, R. E.: 1972, *EOS Trans.* **53**, 1101.

McCoy, J. E., Anderson, K. A., Chase, L. M., Lin, R. P., and McGuire, R. E.: 1973, *EOS Trans.* **54**, 427.

McGuire, R. E.: 1972, *Cosmic Electrodyn.* **3**, 208.

Montgomery, Michael D.: 1968, *J. Geophys. Res.* **73**, 871.

Montgomery, M. D., Singer, S., Conner, J. P., and Stogsdill, E. E.: 1965, *Phys. Rev. Letters* **14**, 209.

MAGNETOSPHERIC PLASMA FLOW AND ELECTRIC FIELDS DERIVED FROM WHISTLER OBSERVATIONS

M. J. RYCROFT
Dept. of Physics, The University, Southampton, England

Abstract. Whistlers contain information about the electron density distribution along their path of propagation, taken to be geomagnetic field lines. Their interpretation to give magnetic equatorial plane electron density values, at different L-values, is discussed. Changes in these plasma density profiles can be interpreted in terms of the cross-L drifting of magnetospheric plasma.

The magnetohydrodynamic treatment of magnetospheric plasma convection, under the action of either a constant dawn-dusk electric field ($n=1$) or the Grebowsky (1971) hyperbolic electric field model ($n=2$), is reviewed. The streamlines of flow are electric equipotentials of the general form $[L^{-1}(t)-k_n L^n(t) \sin n\phi]$. They represent the drift paths followed by ducts both within the plasmasphere (continuous for all values of ϕ, the local time angle, in the corotation dominated region) and beyond the plasmapause (discontinuous in ϕ, in the convection dominated region). Theory and observations are compared. Both the $n=1$ and $n=2$ models explain certain features of the observations.

It is proposed that networks of VLF goniometer receivers should be established to track the motions of whistler ducts in order to study magnetospheric convection and dynamics directly.

1. Introduction

Some audio-frequency electromagnetic radiation from a lightning discharge can penetrate the lower ionosphere and propagate through the dispersive magnetospheric plasma along geomagnetic field aligned ducts of enhanced electron density. Such signals can be received on the ground in the conjugate hemisphere as a set of one-hop multi-path whistlers. For each whistler trace, observations of the nose frequency and the minimum group delay, f_n and t_n respectively (see Figure 1), can be interpreted to give two parameters of the plasma distribution along the propagation path; these are usually the minimum magnetic field strength and the thermal plasma density at that point.

A dipole model of the geomagnetic field is generally used to derive the geocentric distance at which this field line crosses the magnetic equatorial plane (Angerami, 1966). For more accurate work, for example, when observations made in different geographical regions are to be compared, a spherical harmonic expansion of the geomagnetic field (Stassinopoulos, 1970) could be used. The L-value of the duct of each component in the set of multi-path whistlers, and the plasma density in the geomagnetic equatorial plane N_{eq}, are thus obtained. Thereby an instantaneous magnetospheric plasma density profile is constructed (see Figure 1).

The ring current, due to the westward longitudinal drift of radiation belt protons, reduces the geomagnetic field strength in the magnetosphere and distorts the geomagnetic field lines. Electron density profiles in the magnetospheric equatorial plane obtained from whistlers, making allowance for the ring current effects (see Figure 2), have been published recently by Mathur *et al.* (1972).

D. E. Page (ed.), Correlated Interplanetary and Magnetospheric Observations, 317–335. *All Rights Reserved.*

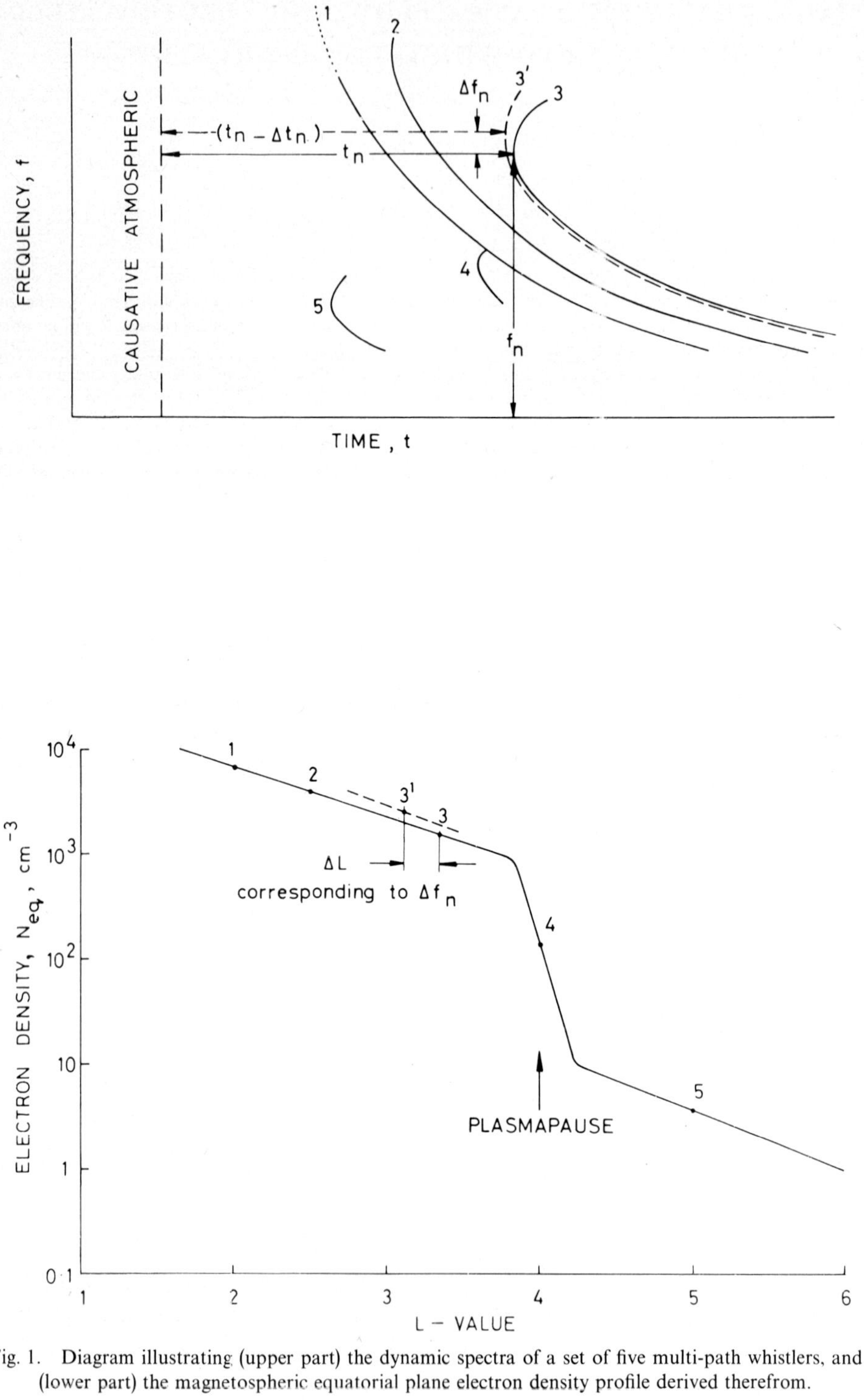

Fig. 1. Diagram illustrating (upper part) the dynamic spectra of a set of five multi-path whistlers, and (lower part) the magnetospheric equatorial plane electron density profile derived therefrom.

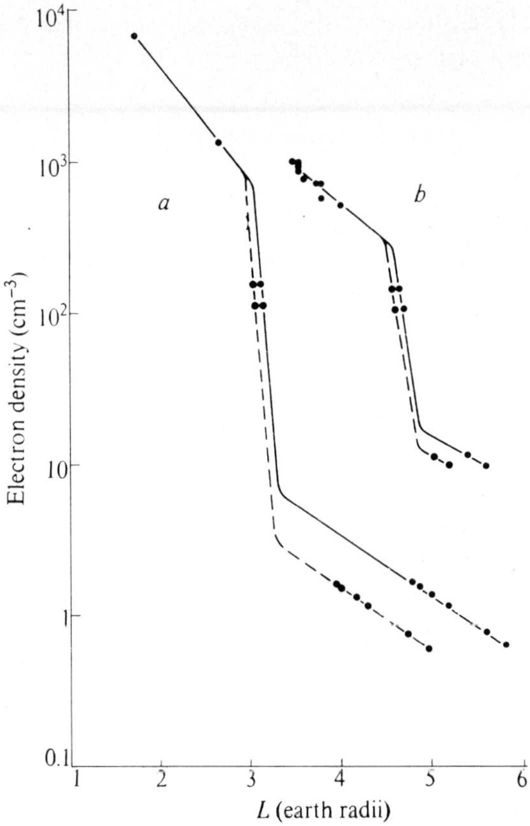

Fig. 2. Magnetospheric equatorial plane electron density profiles derived from whistlers; a, on March 8, 1970, with $K_p = 8o$, and b, on July 2, 1970, with $K_p = 1-$. The profiles are derived assuming (solid lines) a dipole geomagnetic field, and (dashed lines) a ring current model (Mathur *et al.*, 1972).

On occasions, thunderstorms are active in the correct positions and the magnetospheric duct structure persists for a few hours; then whistlers may propagate along the same ducts several times a minute for some hours. This happenstance allows a study to be made of changes of the magnetospheric plasma profile during this time. For a whistler trace whose nose frequency remains constant with time, and thus for a duct whose L-value remains constant, the net increase or decrease of plasma density may be studied. The filling or depletion of geomagnetic flux tubes may thus be investigated. The entire spectrographic pattern of whistler traces may, however, gradually change with time. A gradual increase of the nose frequency of each and every whistler trace, for example (see traces 3 and 3′ in Figure 1), would be interpreted as being due to the radially inward, cross-L drifting of each and every whistler duct and – presumably – of all magnetospheric plasma. The radial component of plasma drift velocity may thus be found quantitatively. Drifting of plasma in the equatorial plane perpendicular to the radial direction, that is at constant L and constant f_n, is detected as particular ducts drift out of the 'field of view' of the receiver

(Carpenter *et al.*, 1972); however, this component of plasma drift velocity cannot be measured directly – see also Section 4.

2. Theoretical Considerations

In the magnetohydrodynamic approximation, that is for magnetospheric changes which take place on a time scale much longer than the ion gyro-period, or longer than the time taken by an Alfvén wave to cross the magnetosphere, $\sim 10^2$ s, the thermal plasma moves in electric and magnetic fields **E** and **B** with velocity **v** such that

$$\mathbf{v} = \frac{\mathbf{E} \wedge \mathbf{B}}{\mathrm{B}^2} \tag{1}$$

Alternatively,

$$\mathbf{E} = -\mathbf{v} \wedge \mathbf{B}. \tag{2}$$

The plasma can be thought of as convecting, with velocity **v**, with the magnetic field lines. For steady conditions, that is for curl-free electric fields, both the streamlines of convective motion and the magnetic field lines are electric equipotentials (Axford, 1969).

It is within the framework of magnetohydrodynamic theory that the drift or convective flow of magnetospheric plasma in the equatorial plane is discussed here. The frame of reference is the solar magnetospheric frame. The magnetic field **B** is taken to be a dipole representation of the Earth's magnetic field with its axis along the Earth's rotational axis which, in the magnetic equatorial plane, is 3.1×10^{-5} T Northwards at the Earth's surface and which varies as L^{-3}. In the equatorial plane the scalar potential of this electric field is $91.5\ L^{-1}$ kV.

The interaction between the solar wind and the geomagnetic field also gives rise to an electric field in the magnetosphere. One way of considering this, hydromagnetically, is

$$\mathbf{E} = -\mathbf{v}_{\mathrm{sw}} \wedge \mathbf{B}_{\mathrm{sw}}, \tag{3}$$

where $\mathbf{v}_{\mathrm{sw}}$ and $\mathbf{B}_{\mathrm{sw}}$ are the velocity of the solar wind past the magnetopause and the magnetic field carried by it, respectively (see Driatsky and Shumilov, 1972).

The simplest case is with $\mathbf{v}_{\mathrm{sw}}$ constant and radially outwards from the Sun and $\mathbf{B}_{\mathrm{sw}}$ constant and antiparallel to the geomagnetic field. These are the conditions necessary for the connection of interplanetary magnetic field lines to geomagnetic field lines on the dayside of the magnetosphere and for steady magnetospheric convection, discussed by Dungey (1961), Levy *et al.* (1964) and Axford (1969) amongst others. Then this electric field is constant, and directed from dawn to dusk across the magnetosphere; a typical value would be $E_1 = (400\ \mathrm{km\ s^{-1}}) \times (5 \times 10^{-9}\ \mathrm{T}) = 2\ \mathrm{mV\ m^{-1}}$. For a typical value of the perpendicular-to-the-ecliptic component of the interplanetary magnetic field of $0.5\ \gamma$ (5×10^{-10} T), the dawn-dusk electric field E_1 would be $0.2\ \mathrm{mVm^{-1}}$ or about $1.3\ \mathrm{kV}\ R_E^{-1}$.

This situation is shown diagrammatically in Figure 3: it should be noted that the large + and – symbols are where charges at infinity would have to be positioned in order to produce this constant dawn-dusk electric field. The associated potential across the magnetosphere at a point in the equatorial plane whose geocentric polar coordinates are (L, ϕ), where ϕ is the local time angle measured from zero at local

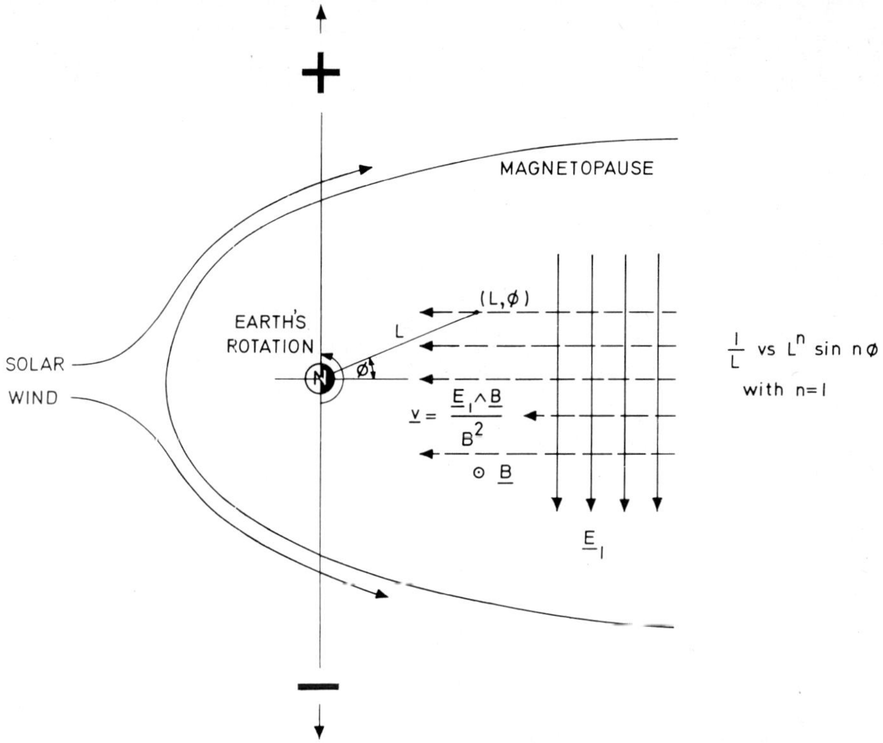

Fig. 3. Diagram illustrating the $n=1$ model for the magnetospheric electric field, which is a uniform dawn-dusk $\mathbf{E}_1$ (solid lines). This drives the inward convection of magnetospheric plasma at velocity $\mathbf{v}$ (dashed lines).

midnight to $+90°$ at 0600 LT, is $-E_1\, L \sin\phi$. Thus, at any time t, the total electric potential in which the charged particles find themselves is $[91.5\, L^{-1}(\text{t}) - E_1 L(t) \sin\phi]$ kV. The drift paths of thermal particles are calculated as the equipotentials,

$$91.5\, L^{-1}(t) - E_1 L(t) \sin\phi = 91.5\, L^{-1}(0000 \text{ LT}), \tag{4}$$

with E_1 in kV R_E^{-1}. Figure 4 shows typical results of such calculations, with $E_1 = 0.3$ mV m^{-1}, performed by Kavanagh *et al.* (1968). Similar results have also been presented by Schield (1969), Chen (1970), Grebowsky (1970), Rycroft (1971) and Chen and Wolf (1972).

There are two types of equipotential surface, open (down the geomagnetic tail) and closed (continuous for all values of ϕ). The former type represent the drift or flow paths of tubes of magnetic flux whose motion is dominated by convection. This convection is directed from the tail towards the Earth between 1800 LT and 0600 LT, during the local night, and out to the magnetopause during the local day. Since plasma is lost from flux tubes, the plasma density on them remains low. The latter type represent the drift or flow paths of tubes of magnetic flux whose motion is dominated by corotation with the Earth. Since these flux tubes are never broken with

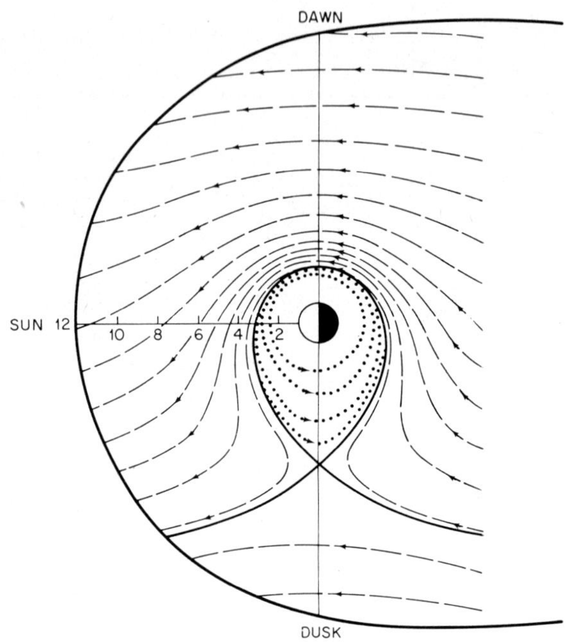

Fig. 4. Magnetospheric convective flow streamlines (or equipotentials) in the equatorial plane for the $n = 1$ model including corotation. The plasmapause (solid line) marks the boundary between corotating (dotted lines) and convecting (dashed lines) plasmas (Kavanagh *et al.*, 1968).

E_1 constant, the plasma density on them remains high. Nishida (1966) and Brice (1967) first pointed out that the special equipotential surface between these two types, the last closed equipotential which marks the boundary between corotating and non-corotating plasma, is the plasmapause, the marked knee in the equatorial plane plasma density profile apparent in Figure 1. A bulge of the plasmapause at 1800 LT is very evident.

Wolf (1970) and Grebowsky (1971) pointed out that Equation (4) and Figures 3 and 4 might be expected to be invalid on the dayside of the Earth, due to the 'shorting out' of magnetospheric electric fields by the highly conducting daytime ionosphere. Grebowsky (1971) proposed that the dayside electric field vanishes in the rest frame of the rotating ionosphere; this results in the dayside equipotentials, or streamlines, being circles concentric with the Earth. A nightside electric field which has the required properties is

$$\mathbf{E}_2 \propto -y\hat{\mathbf{x}} - x\hat{\mathbf{y}}, \tag{5}$$

where $x = L\cos\phi$, positive radially outwards at midnight, and $y = L\sin\phi$, positive radially outwards at dawn. The contours of $\mathbf{E}_2$ are rectangular hyperbolae, shown diagrammatically as the solid lines in Figure 5. The equipotentials, or streamlines, shown by the dashed lines in Figure 5, constitute an orthogonal set of rectangular hyperbolae.

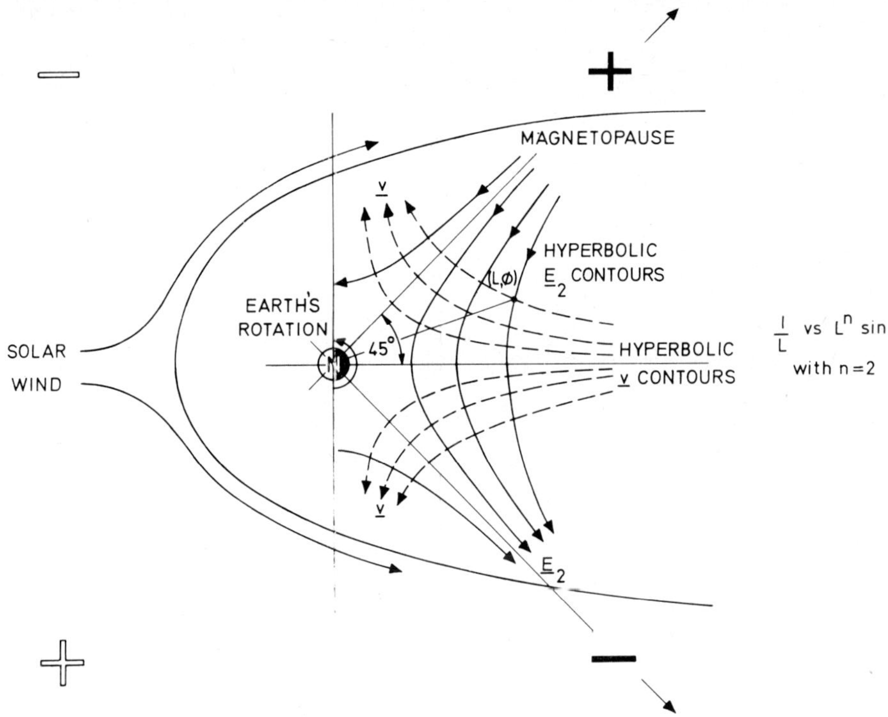

Fig. 5. Diagram illustrating the $n = 2$ model for the magnetospheric electric field $\mathbf{E}_2$ (solid lines), the contours of which are hyperbolae. The velocity contours are also hyperbolae.

Using the method of images in electrostatics, it is evident that $\mathbf{E}_2$ could be produced by a positive charge placed at infinity along the $\phi = 45°$ asymptote to the $\mathbf{E}_2$ hyperbolae, with infinite conducting planes in the noon-midnight and dawn-dusk meridian planes. Following rather tedious algebraic manipulations, it is possible to show that the electric equipotential is of the form $[L^{-1}(t) - k_2 L^2(t)\sin 2\phi]$.

Comparing the form of the equipotential for the constant cross-tail field $\mathbf{E}_1$ with this result, it is clear that both these expressions can be obtained as special cases, with $n = 1$ and $n = 2$ respectively, of a general result. This result is that the equipotentials are of the form $[L^{-1}(t) - k_n L^n(t)\sin n\phi]$.

If the dayside condition of Grebowsky (1971) is relaxed, a positive charge can also be positioned at infinity on the $\phi = 225°$ asymptote to the $\mathbf{E}_2$ hyperbolae, as shown by the outline + symbol in Figure 5. For the same imposed field on dayside and nightside, $n = 2$ positive charges at infinity and $n = 2$ infinite planes at angles of $\pi/(n = 2) = \pi/2$ radians apart enable $\mathbf{E}_2$ to be calculated anywhere. The general result is that n positive charges at infinity in n different directions and n infinite planes at angles of π/n apart enable $\mathbf{E}_n$ to be calculated at any value of L and ϕ.

Figure 6 shows diagrammatically the results of Grebowsky (1971) in a manner to allow comparison with Figure 4. Both convection-dominated flow paths and corotation-dominated flow paths are evident, as is the plasmapause. The most marked dif-

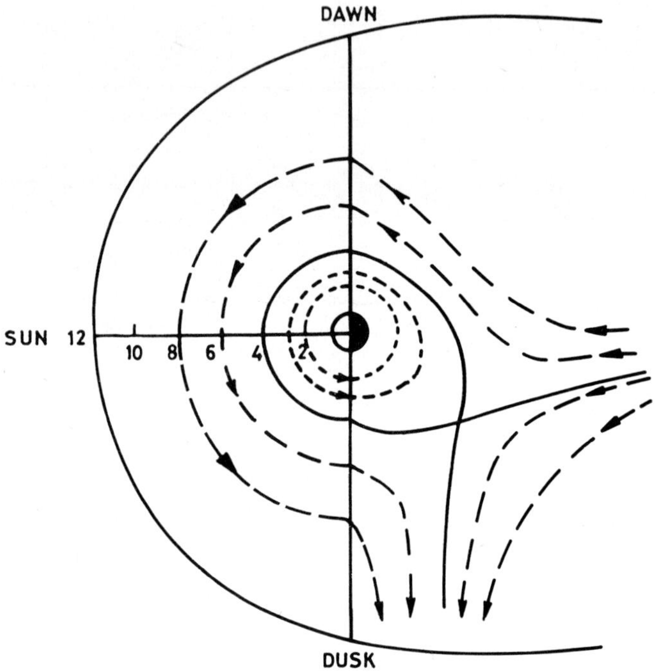

Fig. 6. Magnetospheric convective flow streamlines (or equipotentials) in the equatorial plane for the $n=2$ model on the nightside, with the dayside field 'shorted out', including corotation. The plasmapause (solid line) marks the boundary between corotating (short dashed lines) and convecting (long dashed lines) plasmas (Grebowsky, 1971).

ference from Figure 4 to Figure 6 is that the time of the extreme plasmapause bulge has changed from 1800 LT ($\phi=-\pi/2$) for $n=1$ to 2100 LT ($\phi=-\pi/(2\times 2)=-\pi/4$) for $n=2$. In both cases it occurs at a value of ϕ corresponding to that of the image negative charge. For high values of n, it changes to later times, to values of $\phi=-\pi/2n$.

3. Comparison of Observations with Theory

Carpenter and Stone (1967) presented detailed direct observations of the typically $0.4\ R_E\ h^{-1}$ inward motion of whistler ducts associated with a single substorm. Six ducts, shown by the different symbols in Figure 7 were tracked at $3.8<L<4.6$. The radially inward drift velocity measured was interpreted using Equation (2), to find the westward component of the electric field E_w (parallel to the $-y$ direction at local midnight and parallel to the $+x$ direction at local dawn) at the L value of the duct. Typical values were found to be 0.35 mV m^{-1} or 2.2 kV R_E^{-1}.

Carpenter *et al.* (1972) studied this event in more detail, and other similar case studies. Their analysis was based on the empirical result that, for a convecting whistler duct, $t_n \propto f_n^{-2/3}$. Combined with Equation (1), this leads to $E_w \propto d(t_n^{-1})/dt$.

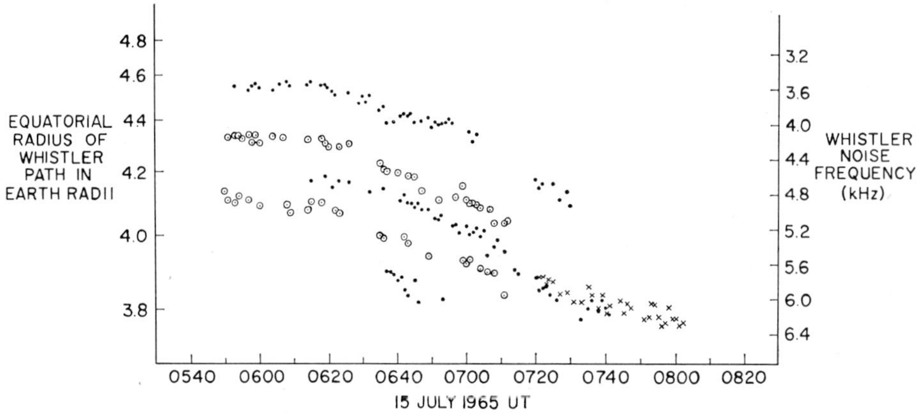

Fig. 7. Observations of whistler duct movements during the July, 15, 1965, substorm; the maximum depression of H at $L \sim 7$ occurred at $\sim$0655 UT (Carpenter and Stone, 1967).

Thus from the variation of t_n^{-1}, derived from the observations, with time (shown in Figure 8 with up to eight different ducts), E_w is derived by differentiation which is carried out numerically. Typical results are shown in Figure 9. The power spectral density of the fluctuations of E_w are presented in Figure 10; the fall off with increasing frequency roughly as $\propto f^{-2}$ agrees approximately with the results of Mozer (1971).

It is quite clear that E_w varies dramatically with time during a substorm. However the theoretical treatments discussed in Section 2 consider the electric field to be constant with time. Ways of reconciling these apparently contradictory views have to be investigated.

One way is to calculate the electric field $\mathbf{E}_1$ or $\mathbf{E}_2$ (or even $\mathbf{E}_n$) which causes the ducts and the plasmapause to be in their observed positions at some instant of time, say near local midnight, that is at the dots or small dashes and the solid lines at a certain value of ϕ in Figure 4 or Figure 6. At a later instant, the same ducts and plasmapause will only remain on the appropriate equipotentials if $\mathbf{E}$ has remained constant; they will, however, have suffered an inward motion. It should be remarked, in passing, that this inward plasma flow takes place naturally during this time interval. It can be understood in terms of the superposition of convection and corotation effects. It should not be ascribed just to a positive value of E_w. But it may be that, at the later instant, the same ducts and plasmapause may have suffered an even greater inward motion. In that case $\mathbf{E}$ has clearly increased. The new value of $\mathbf{E}$ can be cal-

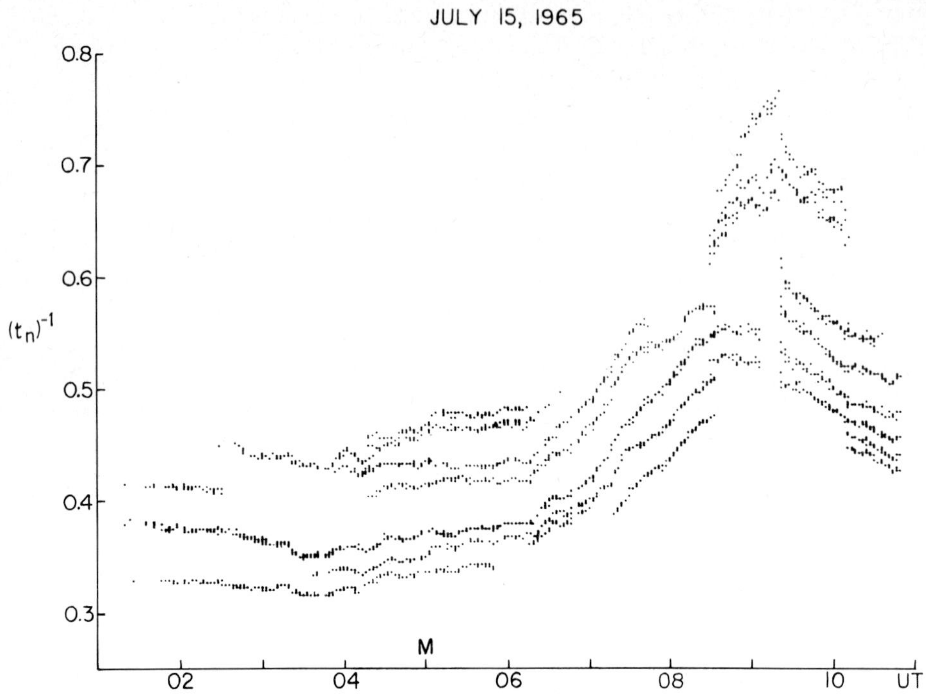

Fig. 8. $(t_n)^{-1}$ for several whistler paths at $3.7 < L < 4.5$ during the July 15, 1965, substorm; M shows local midnight (Carpenter *et al.*, 1972).

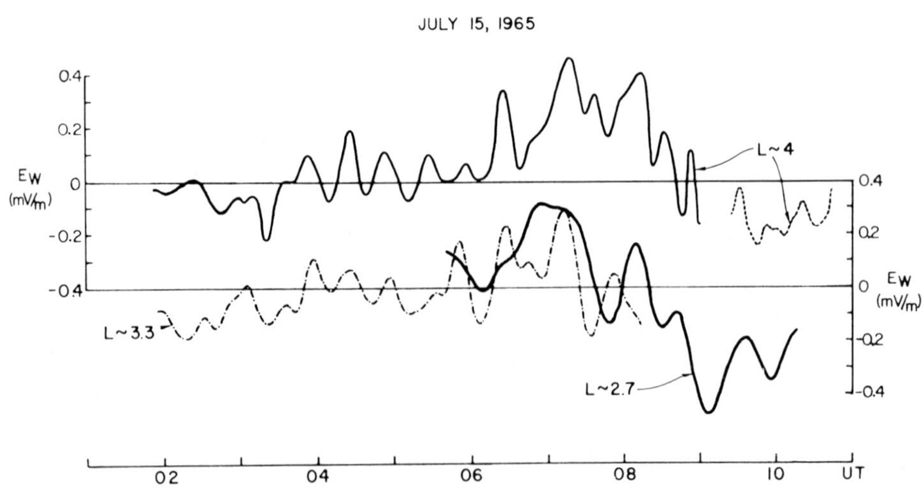

Fig. 9. Values of E_w, westward component of electric field, at three different L-values, derived from observations in Figure 8; note the scales with displaced zeros (Carpenter *et al.*, 1972).

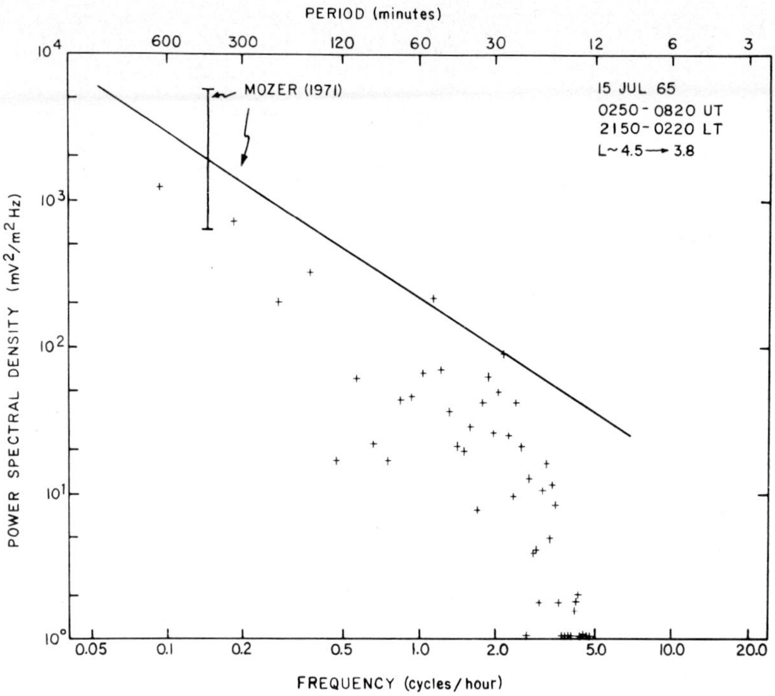

Fig. 10. Power spectral density of E_w fluctuations for the July 15, 1965, observations, upper solid curve in Figure 9. Also shown is the power spectral density of E measurements made aboard balloons (Mozer, 1971; Carpenter *et al.*, 1972).

culated ($\mathbf{E}_1$ by applying Equation (4)) if it is assumed that the inertia of the plasma is negligible and that the plasma fully responds to the changed $\mathbf{E}$ field within the time interval between the two instants. This method seems to merit detailed attention and application to observations such as those presented in Figure 7.

Another way is to examine the observations of particular ducts in a statistical manner. Then temporal changes of $\mathbf{E}$ will be averaged out, and the average quasi-static electric fields $\mathbf{E}_1$ or $\mathbf{E}_2$ (or even $\mathbf{E}_n$) may be studied. This method is applied here to hourly average values of the plasmapause position, $\overline{L_p}(t)$, derived from 102 observations of the plasmapause position made during July 1963 (Carpenter, 1966; Rycroft and Thomas, 1970), shown in Figure 11.

Considering first the uniform dawn-dusk electrostatic field across the magnetosphere $\mathbf{E}_1$, Chen and Wolf (1972) state that $L_p(1800 \text{ LT}) = 2L_p(0000 \text{ LT})$; this result is included in Figure 4. Using this result and Equation (4), $L_p(0000 \text{ LT}) = (91.5/4\, E_1)^{1/2}$. Inserting this back into Equation (4), solution of a quadratic equation reveals that $91.5\, E_1^{-1} = L^2 \sin^2 \phi\, [\sin \phi + 2 - 2\sqrt{1 + \sin \phi}]^{-1}$. The hourly average values of $\overline{L_p} = L$ are inserted into the right hand side of this expression, which is plotted against local time in Figure 12. Between 2200 LT and 1200 LT this expression is essentially constant as the least squares fit line shows. Thus the model of cons-

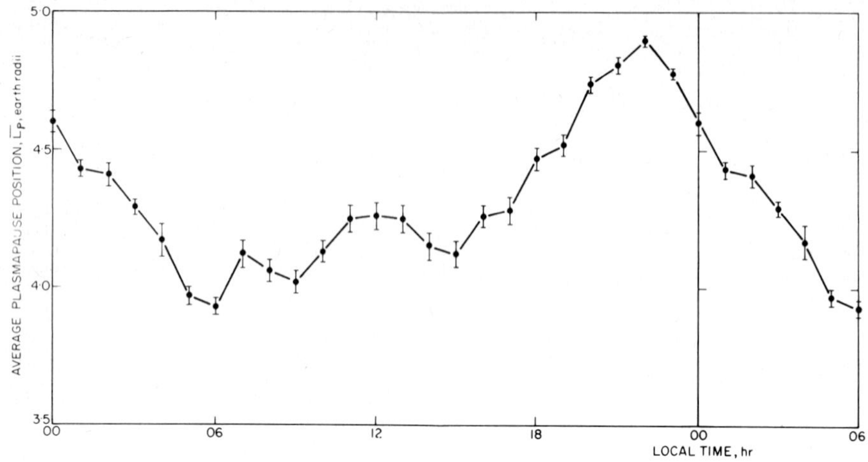

Fig. 11. Diurnal variation of average plasmapause position (showing standard deviations of mean values (Carpenter, 1966; Rycroft, 1971).

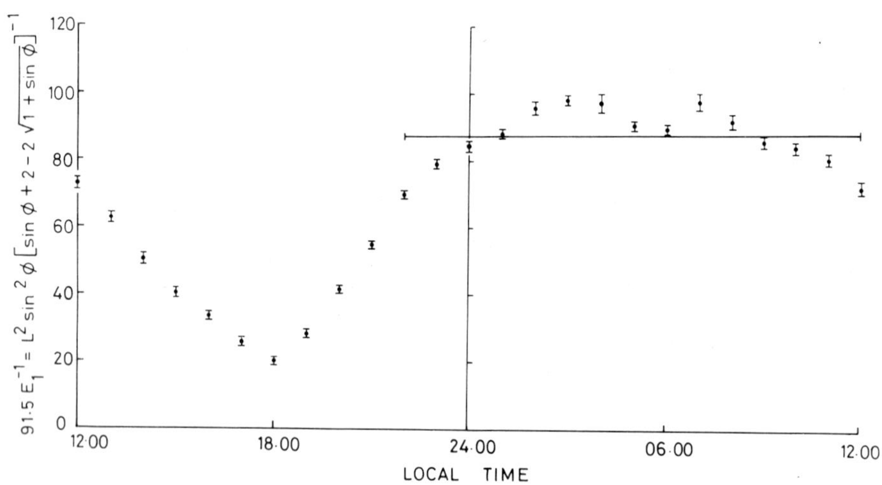

Fig. 12. Diurnal variation of $L^2 \sin^2 \phi[\sin \phi+2-2\sqrt{1+\sin \phi}]^{-1}$; this does not vary much between 2200 and 1200 LT signifying that the $n=1$ model, with E_1 constant, is quite a good model.

stant E_1 can be used with some confidence in this time range; E_1 is thus 0.17 mV m^{-1} or 1.05 kV R_E^{-1}. The potential difference across the magnetosphere is $\sim$30 kV, for a diameter of 30 R_E. At 1800 LT $\pm$4 h, the model is invalid. The observations of average plasmapause position are displayed in another way in Figure 13, in which L^{-1} is plotted against $L \sin \phi$ (Rycroft, 1971). Between 2200 and 0400 LT, and in fact also between 0900 and 1200 LT, there is a linear relationship between the variables, the gradient of which leads to an estimate of E_1 as $\sim$0.08 mV m^{-1} or $\sim$0.5 kV R_E^{-1}. Because this estimate is obtained by taking the difference between the two quantities, it is not as accurate as that obtained from Figure 12.

A similar procedure has been followed using Grebowsky's (1971) model with his dayside condition relaxed. The hourly average values of $\overline{L_p} = L$ are displayed in Figure 14, with L^{-1} plotted against $L^2 \sin^2 \phi$. Between about 1500 and 0500 LT there is a linear relationship between the variables with a positive gradient. It is very clear

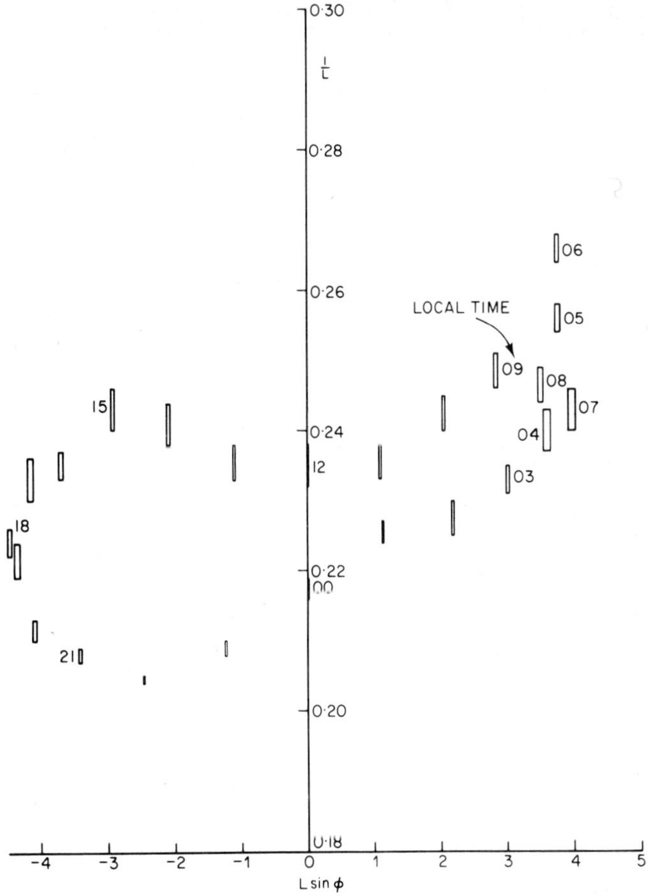

Fig. 13. Plot of L^{-1} vs. $L \sin \phi$; where this is linear, the $n = 1$ model is valid (Rycroft, 1971).

that a much better linear relationship is obtained if the observations are restricted to from 2200 to 0300 LT inclusive.

Between 0400 and 1500 LT, L^{-1} does not vary significantly with $L^2 \sin 2\phi$. It is concluded that, within this mainly daytime interval, Grewbowsky's (1971) original model with the equipotentials having constant L-values between 0600 and 1800 LT is better than the model with his dayside conditions relaxed.

It is necessary to investigate whether the nighttime average plasmapause position L-values are better accounted for by the uniform dawn-dusk electrostatic field model with $n = 1$ or by the Grebowsky (1971) model, with $n = 2$, or even by a model with a higher power of n. The general linear relationship that $L^{-1} = (m \pm \Delta m)\,(L^n \sin n\phi) + c$ was studied for integral values of n up to 6. The results of this analysis are presented in Figure 15, in which the fractional error of the gradient of the least squares fit straight line, $\Delta m/m$, is plotted against n. The criterion for the best model is that the most suitable value of n is that for which this fractional error is a minimum. Values of $n \geqslant 3$

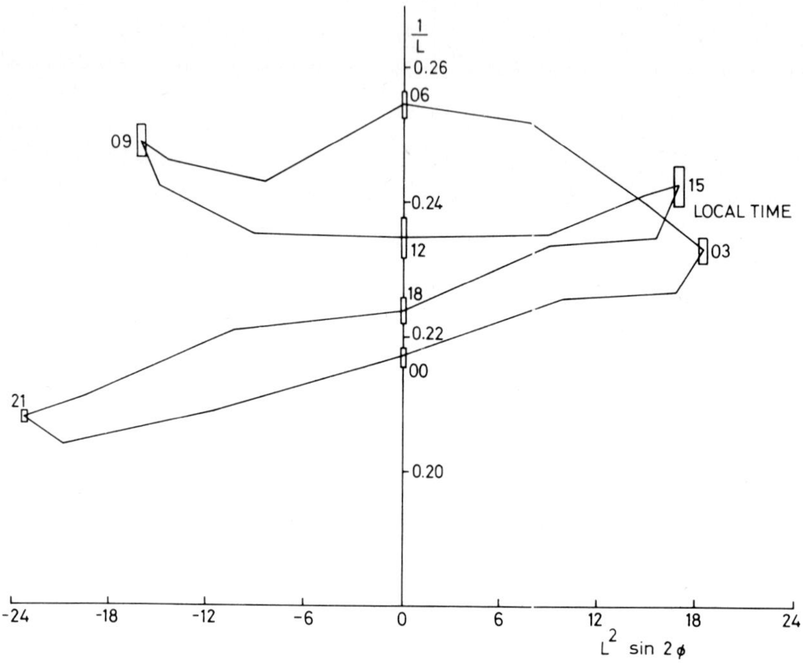

Fig. 14. Plot of L^{-1} vs. $L^2 \sin 2\phi$; where this is linear, the $n=2$ model is valid.

are ruled out. Unfortunately, however, there is no clear cut result between $n=1$ and $n=2$. This is because the better value of n depends on the exact range of local times, over which the observations were made, that is used in the analysis. It is evident from Figure 15 that both models are better for the 2200 to 0300 LT range of data than for the 1800 to 0400 LT range. For the shorter time interval, the $n=1$ model is slightly better than the $n=2$ model. For the longer time interval, the $n=2$ model is considerably better than the $n=1$ model.

To conclude this section, it is useful to compare these results with the electric equipotentials derived by a completely different technique. Figure 16 shows the updated results of McIlwain (1972) based on observations of the dynamic spectra of electrons and protons of <12 keV aboard a geostationary satellite. Since Figure 16 can be regarded as a cross between Figures 4 and 6, for the $n=1$ and $n=2$ models respectively, the agreement must be regarded as being very satisfactory.

4. New Experimental Techniques

Bullough and Sagredo (1970) have published observations of whistlers made at Halley Bay, Antarctica ($L \sim 4$) using a goniometer receiver. The signals induced in two orthogonal vertical loops are combined electronically to simulate the signal received by a rotating loop. By measuring the positions of the minima of this signal with respect to calibration pulses, the azimuthal bearing of the signal propagating in the Earth-ionosphere waveguide is found.

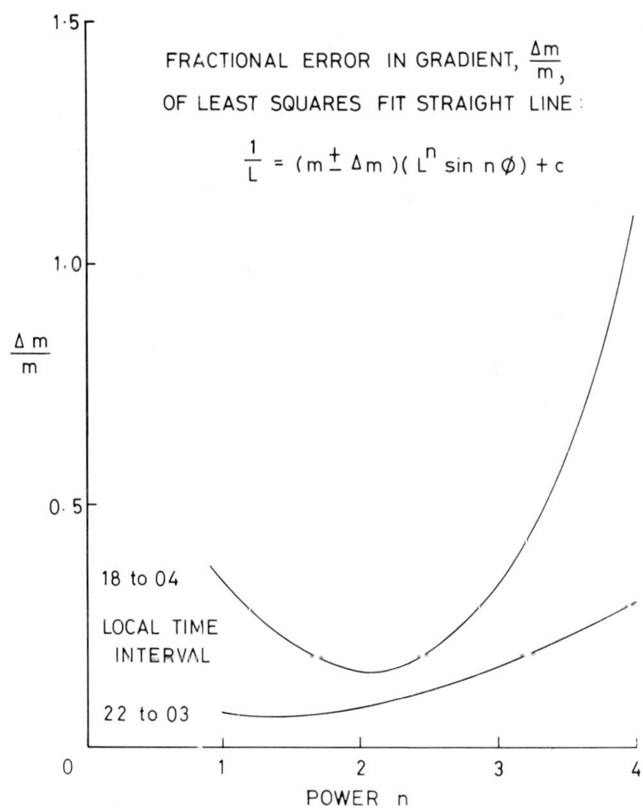

Fig. 15. Fractional error in gradient. $\Delta m/m$, of best fit straight line vs. n, order of model, for two different local time intervals.

The University of Southampton Radiophysics Group was instrumental in establishing a temporary network of four such goniometer receivers in Eastern Canada, and making observations there in connection with the total solar eclipse of July 10, 1972. Figure 17 shows a VLF spectrogram taken at this time at the field station near Antigonish, Nova Scotia ($L \sim 3.5$). Maxima and minima in the signals received from

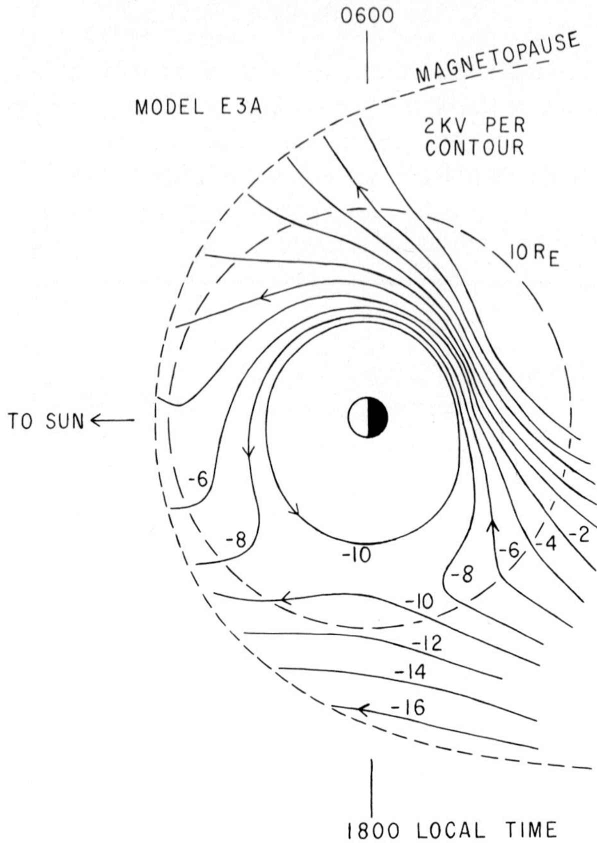

Fig. 16. Magnetospheric equipotentials derived by McIlwain (1972) (October 1972 version incorporating some ATS-5 data); compare with Figures 4 and 6.

two Omega transmitters at 12 kHz are very clear; the positions of the minima, in relation to the calibration pulses at 28 s^{-1} evident below 1 kHz, are different for the two transmitters. Their azimuthal bearings are therefore different. Also apparent is part of an intensity modulated whistler trace at 4 kHz, the azimuthal bearing of which can also be measured. Figure 18 shows the spectrogram of a set of multi-path whistlers, some of which have measurable nose frequencies. Each whistler trace has a different azimuthal bearing.

When the same set of whistlers, or of discrete VLF emissions, is received at two or three stations, the latitude and longitude of the points at which each and every component emerges from the ionosphere may be found by triangulation. If it is possible to perform this triangulation with sufficient accuracy, a search may be made for movements of each and every whistler mode signal exit point. These may then be related to movements of both ionospheric plasma and magnetospheric ducts. It may

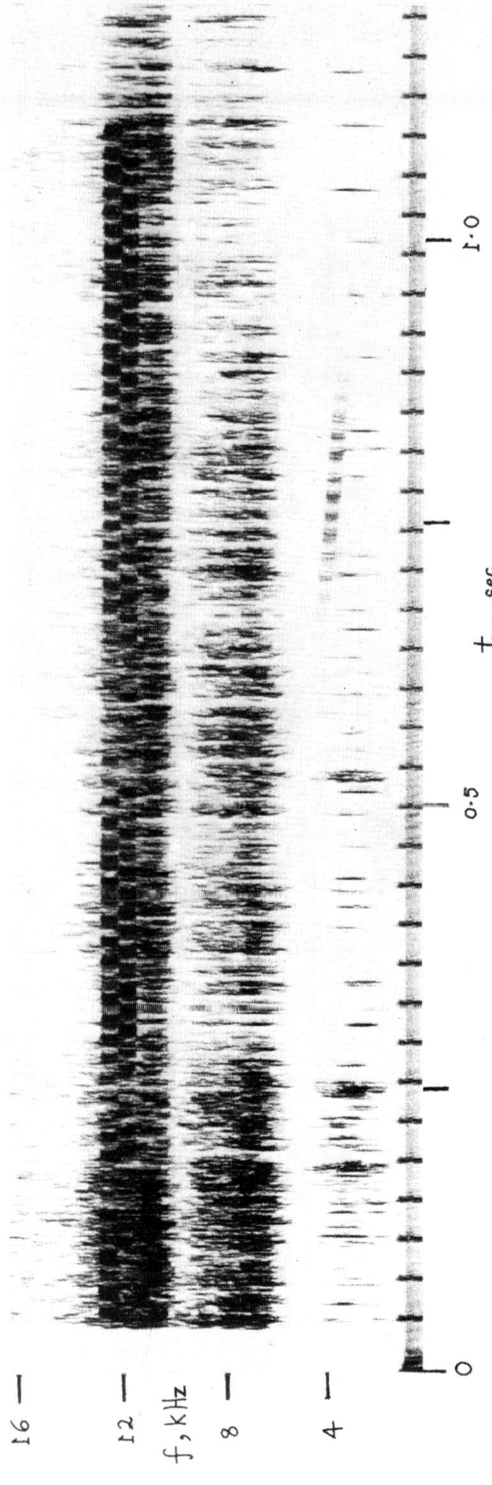

Fig. 17. Goniometer receiver spectrogram (wide bandwidth) showing Omega transmissions and a whistler.

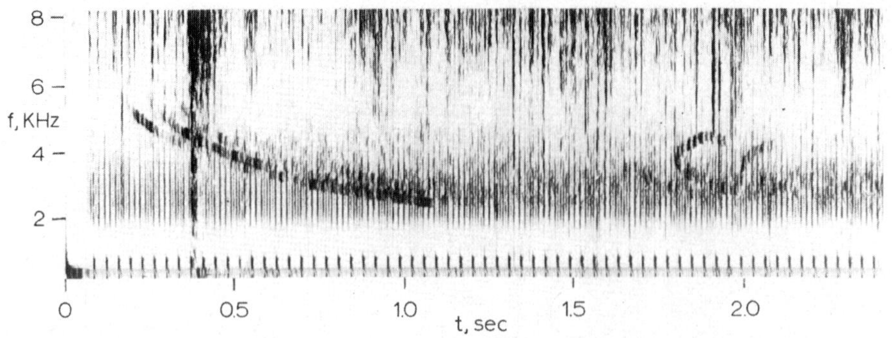

Fig. 18. Goniometer receiver spectrogram (wide bandwidth) showing several different whistler traces having different bearings, recorded at 1906 UT on July 10, 1972, near Antigonish, Nova Scotia ($L \sim 3.5$).

then be possible to study directly by VLF techniques both the latitudinal (cross $-L$) and longitudinal (at constant L) drifting of different ducts of magnetospheric plasma and of the plasmapause itself. The observations may be interpreted in terms of magnetospheric electric fields as discussed here.

Acknowledgments

The author is grateful to Prof. W. I. Axford, Dr K. Bullough, Dr D. L. Carpenter, Dr G. J. Daniell, M. J. Jarvis and S. Powell for useful discussions.

References

Angerami, J. J.: 1966, 'A Whistler Study of the Distributions of Thermal Electrons in the Magnetosphere', SEL-66-017, Radioscience Lab., Stanford University, California, May.
Axford, W. I.: 1969, *Rev. Geophys.* **7**, 421.
Brice, N. M.: 1967, *J. Geophys. Res.* **72**, 5193.
Bullough, K. and Sagredo, J. L.: 1970, *Nature* **225**, 1038.
Carpenter, D. L.: 1966, *J. Geophys. Res.* **71**, 693.
Carpenter, D. L. and Stone, Keppler: 1967, *Planetary Space Sci.* **15**, 395.
Carpenter, D. L., Stone, Keppler, Siren, Jan C., and Crystal, T. L.: 1972, *J. Geophys. Res.* **77**, 2819.
Chen, A. J.: 1970, 'Thermal and Quasi-Energetic Plasma Flow in the Magnetosphere – A Theoretical Study', Ph.D. thesis, Rice University, Texas.
Chen, A. J. and Wolf, R. A.: 1972, *Planetary Space Sci.* **20**, 483.
Driatsky, V. M. and Shumilov, O. I.: 1972, *Planetary Space Sci.* **20**, 1375.
Dungey, J. W.: 1961, *Phys. Rev. Letters* **6**, 47.
Grebowsky, J. M.: 1970, *J. Geophys. Res.* **75**, 3848.
Grebowsky, J. M.: 1971, *J. Geophys. Res.* **76**, 6193.
Kavanagh, L. D., Jr., Freeman, J. W., Jr., and Chen, A. J.: 1968, *J. Geophys. Res.* **73**, 5511.
Levy, R. H., Petschek, H. E., and Siscoe, G. L.: 1964, *AIAA J.* **2**, 2065.
Mathur, A., Rycroft, M. J., and Sagredo, J. L.: 1972, *Nature* **237**, 508.
McIlwain, C. E.: 1972, in B. M. McCormac (ed.), *Earth's Magnetospheric Processes*, D. Reidel Publ. Co., Dordrecht-Holland, p. 268.
Mozer, F. S.: 1971, *J. Geophys. Res.* **76**, 3651.
Nishida, A.: 1966, *J. Geophys. Res.* **71**, 5669.

Rycroft, M. J.: 1971, 'Dawn-Dusk Electric Fields Across the Magnetosphere Derived from Plasmapause Observations', ESRO SP-72, Wave-Particle Interactions in the Magnetosphere, p. 163.

Rycroft, M. J. and Thomas, J. O.: 1970, *Planetary Space Sci.* **18**, 65.

Schield, M. A.: 1969, 'Drift of Non-Interacting Charged Particles in a Simple Geomagnetic Field', University of Iowa Preprint No. 69–54.

Stassinopoulos, E. G.: 1970, 'World Maps of Constant B, L, and Flux Contours', NASA SP-3054.

Wolf, R. A.: 1970, *J. Geophys. Res.* **75**, 4677.

ELECTRIC FIELD MEASUREMENTS AT THE GEOSTATIONARY POSITION

A. PEDERSEN

Space Science Dept. (ESLAB), ESTEC, Noordwijk, The Netherlands

Abstract. During the last years several models have been made of the large scale electric field of the magnetosphere, and recently these models have been refined by data from plasma experiments flown on geostationary satellites, and by measurements of ionospheric electric fields. Superimposed on the average large scale electric field there are changes in time and space related to substorms.

This paper will outline how direct measurements of electric fields at the geostationary position with the GEOS satellite, possibly supplemented with similar measurements on the future International Magnetospheric Explorer (IME), can contribute to magnetospheric physics.

1. Introduction

In expectation of electric field measurements on GEOS and IME with long double probes it is interesting to sum up the present knowledge of electric fields within the magnetosphere and to indicate what new areas can be surveyed by such measurements.

In the past it has been stated repeatedly that it is not possible to measure electric fields when the Debye length, λ_D, is larger than the double probe. The correct statement is that when the Debye length is very small compared to the double probe system, electric field experiments are relatively easy because the plasma is effectively shielding the probes from the spacecraft in the middle; however, with proper care of symmetry problems electric fields can also be measured when λ_D is very large, such as in the magnetosphere where λ_D can be several hundred metres. The large scale magnetospheric electric field will be superimposed on the symmetric electric field from the satellite and can be measured by a spinning probe.

In this presentation experimental problems will not be discussed further; a detailed study of double probe techniques in the magnetosphere has been published by Grard *et al.* (1973). Furthermore, this discussion will be limited to DC and quasi DC electric fields; however, it should be kept in mind that electric fields up to several hundreds of kHz can be measured with double probe techniques.

Some aspects of ULF waves will be mentioned because the experimental problems are similar to those experienced in measuring quasi-static electric fields.

Finally a possible coordination of electric field measurements on GEOS and IME during the forthcoming International Magnetospheric Study (IMS) will be outlined.

2. Magnetospheric Electric Fields

Very few direct measurements of electric fields have been carried out in the magnetosphere so far and our knowledge about electric fields has, in most cases, been

D. E. Page (ed.), Correlated Interplanetary and Magnetospheric Observations, 337–342. *All Rights Reserved.*

derived in an indirect way. The large scale convection in the magnetosphere is today well described both by theory and observations. The model described by Axford and Hines (1961) was inspired by ground observations of magnetic fields and auroral phenomena, and assumed that a visceous interaction between the solar wind and the magnetosphere was the driving force behind the convection. Several observations carried out during the last years have demonstrated that there is close interrelation between the solar wind magnetic field and processes in the magnetosphere, and it is today generally accepted that field line reconnection is the most important mechanism that makes it possible to derive energy from the solar wind and support an internal convection in the magnetosphere.

When we speak about a model convection pattern we can only refer to steady state conditions and we have to limit ourselves to the convection in the magnetic equatorial plane.

The electric field $\bar{E}$ can be expressed in terms of a scalar potential, ϕ, and a vector magnetic potential, $\bar{A}$:

$$\bar{E} = -\Delta\phi - \partial\bar{A}/\partial t. \tag{1}$$

The convection velocity of magnetospheric plasma and the electric field are related by the following equation:

$$\bar{E} + \bar{v} \times \bar{B} = 0 \tag{2}$$

In steady state, i.e. no variation of the magnetic field, $\bar{E} = -\nabla\phi$ and plasma movements in the magnetosphere can be determined by knowledge of $\bar{E}$ and the stationary magnetic field, $\bar{B}$.

Equation (2) is sufficient to describe the motion of low energy plasma; however, for higher energies above 1 keV gradient drift and curvature drift in opposite directions for electrons and protons must be considered.

Two recent observations of the connection pattern in the magnetospheric equatorial plane have been given by McIlwain (1972) and Mozer and Lucht (1973). (Both observations are shown in the presentation of C. R. Chapell in this volume.)

The potential pattern reported by McIlwain (1972) has been calculated on the basis of measurements of electron and proton spectra during quiet periods on ATS 5. This quiet time electric field pattern is the final model resulting from an iterative process where this electric field model and gradient and radial drifts determine the electron and particle trajectories. The iterative process then consists of bringing the electric field model in the best agreement with the particle observations.

The magnetospheric electric field model reported by Mozer and Lucht (1973) has been derived from a large number of electric field measurements on balloons at L-values between 5,4 and 8,2. Altogether 478 h of balloon observations for K_p values between 0 and 6 have been used to produce average ionospheric electric field variations, which in turn have been projected into the equatorial plane. The equipotential lines have been calculated on the basis of a uniform electric field across the magnetospheric tail at 15 earth radii.

The order of magnitude of the potential across the magnetosphere agrees with McIlwains model, considering that an increase with a factor 2 was observed in electric fields in the ionosphere when K_p varied from 0 to 6. However, there are differences between the two patterns, which is not surprising considering that both are based on methods where the magnetospheric convection is determined indirectly by different simplifying assumptions.

Magnetospheric electric fields of the order 1 mV m^{-1}, which is typical for the convection patterns referred to above, are well within the limits of a double probe experiment on a satellite, particularly as the important electric field component across the magnetosphere is less disturbed by photoemission than the component parallel to the Earth-Sun line (see Grard *et al.*, 1973). Such measurements will therefore be very important for determination of the magnetospheric convection and its dependence on solar wind and magnetospheric processes.

3. Magnetospheric Electric Fields During Substorms

The real value of a direct measurement is under conditions when the magnetosphere is disturbed. Under these conditions $\partial B/\partial t \neq 0$ and information about the magnetic field must be fed into Equation (1) to relate the electric field to variations of the magnetic field. From typical variations of the magnetic field observed on a number of satellites, it can be estimated that during disturbed periods the induced electric field is of the same order of magnitude as the convection electric field resulting from the steady average motion of magnetospheric plasma. It is very unlikely that magnetic field lines in this case are equipotential and it is difficult to derive information about magnetospheric plasma behaviour from observation of electric fields in the ionosphere. Similarly, under disturbed conditions it will be impossible to infer an electric field model from particle observations as it was done on ATS 5.

Our knowledge of ionospheric electric fields during substorms has increased during the last years due to extensive balloon measurements, Mozer (1971), Mozer *et al.* (1973), by measurements on polar orbiting satellites Cauffman and Gurnett (1972), Heppner (1972), and by rocket measurements with double probes and Ba releases. The entry of high energy plasma clouds near the nighttime geostationary orbit in connection with substorms has been reported by DeForest and McIlwain (1971) and constitutes the most detailed study of plasma behaviour in the magnetosphere at a location where we must expect precipitating auroral electrons to be accelerated. All these observations have provided many details and have made it possible to agree on characteristic features of substorms. Both ionospheric observations and ATS 5 observations show that during the initial phase (growth phase) of a substorm, an increased sunward flow of hot plasma takes place from the magnetospheric tail towards the nighttime geostationary orbit. The particle precipitation and aurora take place during the so-called expansion phase when a violent rearrangement of magnetic field lines, implying large values of $\partial B/\partial t$, have been observed on earlier ATS satellites. However, the basic question of how the solar wind plasma is injected and accelerated is still not solved. Data from the Injun 5 satellite, Gurnett and Frank (1973) indicate that the solar wind plasma entry in the dayside polar cusp region is important, but other mechanisms, which allow for entry of solar plasma in the tail region, are not excluded.

4. Electric Fields of Magnetospheric Oscillations and Alfvén Waves

Micropulsations of the magnetic field have been observed on the ground and by magnetometers on satellites. In connection with future measurements on GEOS, it is of particular interest to note the magnetic measurements on earlier ATS satellites Barfield and Coleman (1970), Cummings *et al.* (1968). Transverse waves with amplitudes of 10–20 γ and with periods of the order of a hundred, or a few hundred seconds have been observed mainly during quiet periods from early morning to early afternoon local time. During substorms compressional waves observed near the midnight sector seem to dominate; however, transverse waves with characteristic amplitudes of a few γ and periods of 10–20 s have been observed. It is of special interest that on several occasions there is a clear correlation between waves and particles, see Figure 1. The induced electric field of these waves is of the order 1 mV m^{-1} and a spinning probe will easily detect the electric field of these waves in cases when the electric field has a component in the equatorial plane. In fact, it should be easier to detect the electric component of such waves with amplitudes near 1 mV m^{-1} than a DC electric field of the same magnitude. The spin period of the double probe on GEOS will be of the order 6 s, which yields an effective time resolution of 3 s for such measurements. The first measurement of this type was done by a double probe on Imp 6, Aggson (1973). During the large August 1972 event large electric waves

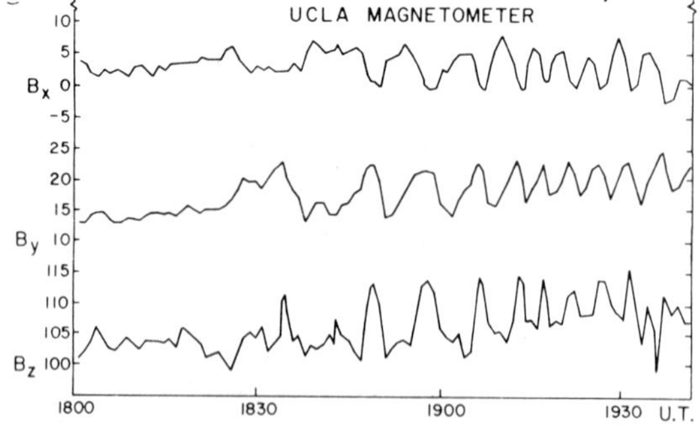

Fig. 1. The magnetic components of an ULF wave and electrons observed on ATS 1. After Freeman and Young (1971).

with a period of several minutes were observed on the background of a 5 mV m^{-1} dawn-dusk electric field. It is obvious that knowledge about the electric component of these waves can give information about their origin and propagation characteristics to a much larger extent than what is possible from the magnetometer observations alone. It could very well be that these waves play a very important part in the solar wind magnetosphere system and can therefore be valuable diagnostic tools in further studies of the magnetosphere.

It is also interesting to note that with knowledge of the electric and magnetic components of an Alfvén wave it will be possible to determine the local plasma density.

5. Possible Coordination of Measurements on GEOS and the International Magnetospheric Explorer (IME)

One can argue that measurements of particles and fields on a geostationary satellite still cannot dissolve temporal from spatial variation, and undoubtedly such a study will benefit from simultaneous measurements on a geostationary and a highly eccentric orbit satellite, such as IME. The orbit of IME for different times of the year and the orbit of GEOS are sketched in Figure 2. It is obvious that simultaneous measurements on GEOS and IME could be very valuable for studies of electric fields during injections of plasma during substorms and the study of Alfvén waves at different locations can probably yield new information about the dynamic behaviour of the magnetosphere.

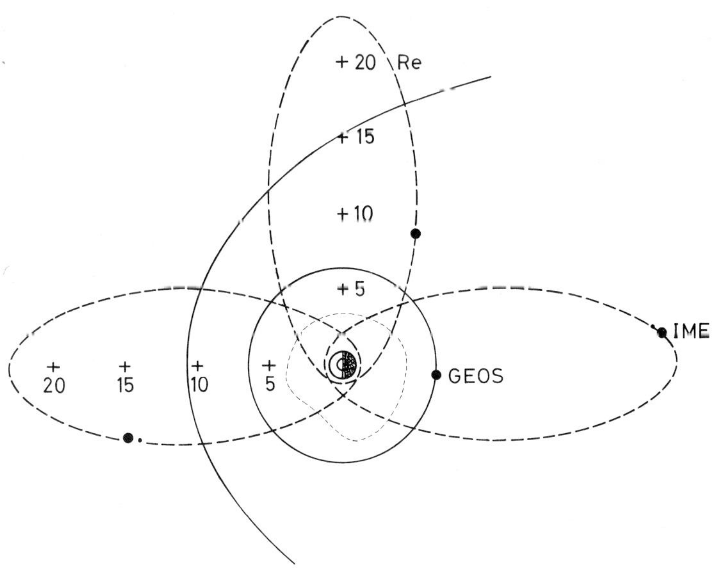

Fig. 2. Orbits of GEOS and IME.

Another valuable contribution to a substorm study would be simultaneous electric field measurements carried out on a polar orbiting satellite; this would make it possible to see the link between ionospheric and magnetospheric convection. However, it is at present uncertain whether a polar orbiting satellite will be operating during the lifetime of GEOS and IME.

References

Aggson, T. L.: 1973, *Trans. Amer. Geophys. Union* **54**, 417.

Axford, W. I. and Hines, C. O.: 1961, *Can. J. Phys.* **39**, 1433.

Barfield, J. N. and Coleman, P. J.: 1970, *J. Geophys. Res.* **75**, 1863.

Cauffman, D. P. and Gurnett, D. A.: 1972, *Space Sci. Rev.* **13**, 369.

Cummings, W. D., Barfield, J. N. and Coleman, P. J.: 1968, *J. Geophys. Res.* **73**, 6687.

De Forest, D. E. and McIlwain, C. E.: 1971, *J. Geophys. Res.* **76**, 3587.

Freeman, J. W. and Young, D. T.: 1971, 'Magnetospheric Plasma Phenomena at the Geostationary Orbit, "The Esro geostationary satellite"', ESRO SP-60.

Grard, R. J. L., Knott, K. and Pedersen, A.: 1973, in R. J. L. Grard (ed.), *Photon and Particle Interactions with Surfaces in Space*, D. Reidel Publ. Co., Dordrecht-Holland, p. 163.

Gurnett, D. A. and Frank, L. A.: 1973, *J. Geophys. Res.* **78**, 145.

Heppner, J. P.: 1972, *Planetary Space Sci.* **20**, 1475.

McIlwain, C. E.: 1972, in B. M. McCormac (ed.), *Earth Magnetospheric Processes*, D. Reidel Publ. Co., Dordrecht-Holland, p. 268.

Mozer, F. S.: 1971, *J. Geophys. Res.* **76**, 7595.

Mozer, F. S. and Lucht, P.: 1973, 'The Average Auroral Zone Electric Field', Internal report Space Science Laboratory, Un. of California, Berkeley, Series 14, Issue 21.

Mozer, F. S., Bogott, F. H. and Tsurutani, B.: 1973, *J. Geophys. Res.* **78**, 630.

PART 4

SOLAR PARTICLE PROPAGATION IN THE INTERPLANETARY MEDIUM

OBSERVATIONS OF SOLAR PARTICLE PROPAGATION

L. J. LANZEROTTI
Bell Laboratories, Murray Hill, N.J., U.S.A.

Abstract. A review is given of the observations of the propagation characteristics of solar particles from impulsive flare events that principally have been reported since the Third ESLAB Symposium in 1969. The review is limited to current interpretations of the observations in terms of the complete particle transport equation and the recent attempts at reconciling values of the elements of the diffusion tensor as obtained from particle observations and from measurements of interplantetary magnetic field fluctuations. The importance of understanding and modeling the effects of flare-produced shock waves on the diffusive-convective low energy particle time-intensity profiles is stressed. The present observational data on particle modulations by shock waves is reviewed.

1. Introduction

At the time of the Third ESLAB Symposium in Noordwijk in the Fall of 1969 (Manno and Page, 1970) four distinct types of solar particle events were apparently evident in satellite data. These types of events, as classified as that time by McDonald (1970), were (a) flare-associated events, (b) recurrent events, (c) events associated with solar active regions, and (d) energetic storm-particle events. In addition, solid evidence was becoming available that the Sun was a 'continuous source' of low energy ($E \lesssim 20$ MeV) protons and alpha particles (e.g., McDonald, 1970; Kinsey, 1970).

Since the Noordwijk symposium, solar particle observations have made advances on a number of fronts. One of the more interesting of these, and one that will not be discussed here, is the problem of solar particle composition at low energies (e.g., Price *et al.*, 1971; Armstrong and Krimigis, 1971; Mogro-Campero and Simpson, 1972 a, b; Lanzerotti *et al.*, 1972, Braddy *et al.*, 1973; Crawford *et al.*, 1972; Teegarden *et al.*, 1973; Krimigis and Armstrong, 1973). Advances in solar particle propagation studies have been made not so much in the area of uncovering new morphological characteristics of solar particle events, but rather in trying to understand the abundant morphology that is known to exist. In particular, much effort is being devoted to understanding low-energy propagation phenomena, mainly from non-flare events, in terms of propagation regions controlled by large-scale coronal structures (Roelof, 1973a, b; Roelof and Krimigis, 1973).

This present paper will confine itself to a discussion of current approaches being made in the study of observations of impulsive solar flare events in the theoretical context of the complete particle transport equation, including the convection and energy loss terms. These approaches consist not only of studying the solar particles, but also of attempting to extract diffusion parameters from the interplanetary field observations. During the period since the Noordwijk conference, it has become increasingly clear to the experimentalists that an understanding of low energy impulsive events requires a theoretical treatment including all terms in the cosmic ray transport equation. However, it has also been obvious that the intensity-time profiles

D. E. Page (ed.), Correlated Interplanetary and Magnetospheric Observations, 345–379.

of low energy events are often not of a 'diffusive' type. Thus, the nature of energetic storm particles and the over-all relationship of a propagating flare-produced shock in modifying the interplanetary cosmic ray population assume additional importance. Accordingly, current observational information on energetic storm particle events is also discussed in this review with the hope and expectation that it will ultimately be possible to merge diffusion-convection-energy loss theory with a theory of shock effects on solar particles to understand the propagation characteristic of an entire impulsive event.

Because of limitations of space, the status of the observational situation with respect to recurrent events and active region events, and the implications therefrom for particle storage and generation on the Sun, are largely omitted from this paper. Some will claim that these topics are of paramount importance in interpreting impulsive events as well, and that therefore the theoretical formalism of the transport equation is somehow deficient. If this be so, the deficiency should ultimately manifest itself in the failure of the theory to exhibit consistency between interpretations from different (i.e., particle and interplanetary field) sets of observations. This point is briefly discussed in the final section.

2. Impulsive Solar Events

That impulsive solar flare proton events can exhibit a 'diffusive' time-intensity profile for protons of energy as low as ~0.5 keV is demonstrated by the data plotted in Figure 1. These particle fluxes resulted from a flare (McMath region 9740) beginning at ~0520 UT on, or just beyond, the west limit of the Sun (and at S15°). Although it is true that such 'well-behaved' time-intensity profiles are an unusual occurrence for such low energy particles, it is interesting to note that the shock wave from the originating flare was not observed at the Earth. The existence of a diffusive time-intensity profile in this impulsive event may well be a resultant of this fact. This point will be discussed at more length in the last section of the review. The event in Figure 1 is rather short-lived, with a 'decay-time', after the peak intensity, increasing somewhat with decreasing energy.

Several years ago the data of Figure 1 would have been discussed in theoretical terms of the simple diffusion of particles in an infinite, static, spherically symmetric interplanetary medium (Parker, 1963) with a diffusion coefficient that perhaps depended upon radial distance (e.g., Krimigis, 1965) or that was anisotropic (Axford, 1965; Burlaga, 1967). However, McCracken *et al.* (1967) had presented evidence that the equilibrium anisotropy of ~10 MeV particles in the decay phase of a solar event was ~10%, in a direction radially outward from the Sun. Also, although fits of low energy ($\lesssim$10 MeV) data to simple theoretical pictures gave reasonable-appearing values of the diffusion coefficient, at the lowest energies the particle decay times appeared to be essentially independent of energy (e.g., Lanzerotti, 1969).

That the particle decay from a solar flare event can be energy-independent over a wide energy range is shown in Figure 2. In the inset to Figure 2 is plotted the particle

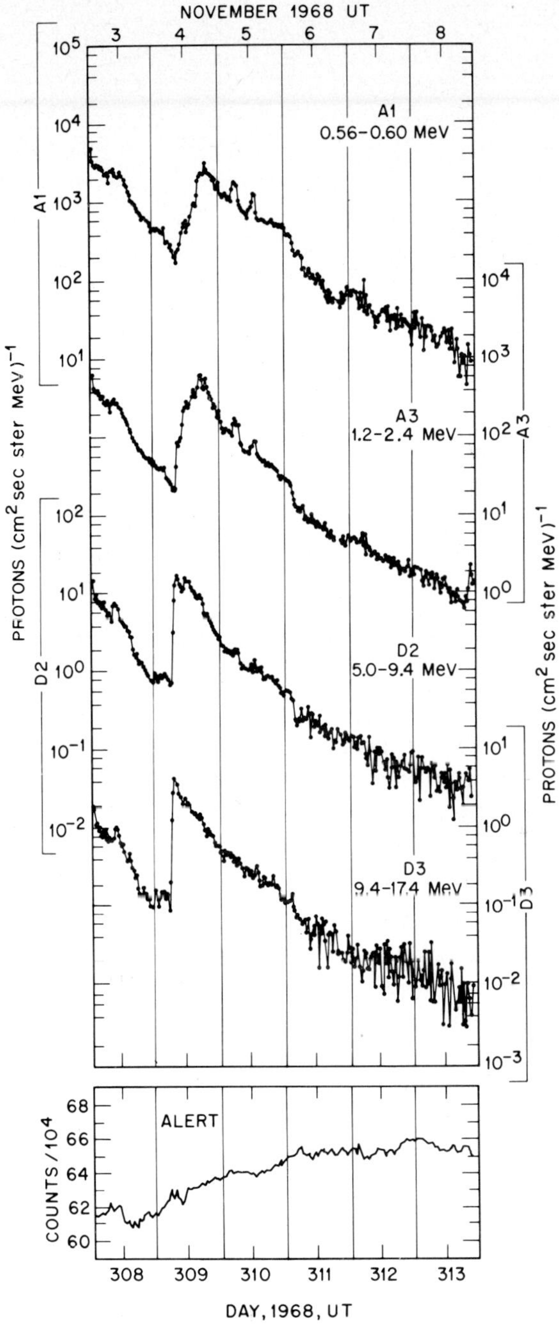

Fig. 1. Proton fluxes measured by Explorer 34 during the solar particle event of 4 November 1968. The counting rate of the Alert neutron monitor shows the recovery from an earlier Forbush decrease.

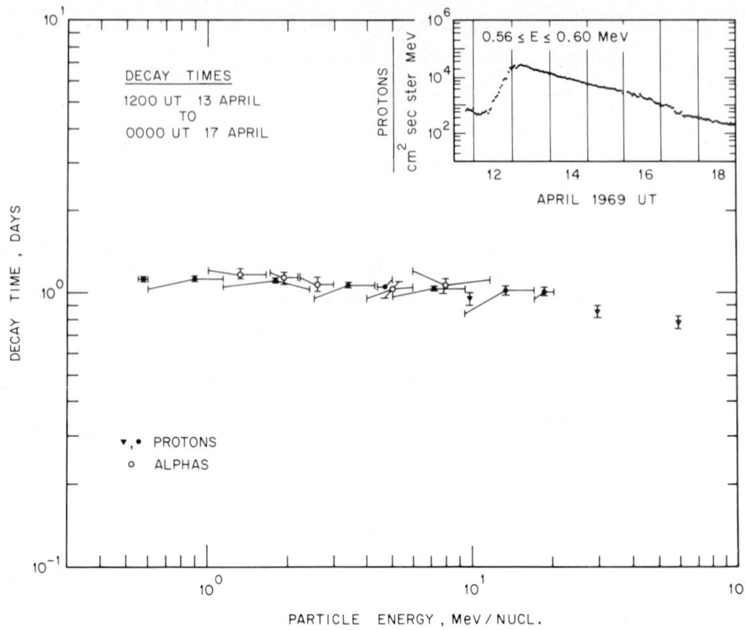

Fig. 2. Inset: Intensity-time profile of 0.56–0.60 MeV solar fluxes during the event of 10 April 1969. Plotted in the main portion of the figure are the exponential decay times measured for protons and alpha particles during the period 12 UT 13 April to 00 UT 17 April.

flux-profile for protons of energy 0.56–0.60 MeV resulting from an east limb event at $\sim$0410 UT, 10 April 1969. The main section of the figure contains the event decay times as a function of particle energy obtained from linear fits to the log intensities for particle fluxes measured between 12 UT, 13 April and 00 UT, 17 April. For particles with energies as high as $\sim$15 MeV, the time to decay by a factor of e^{-1} is $\gtrsim$1 day, independent of energy.

On the basis of the essentially radial residual anisotropy results and the energy-independent decay-time results, Forman (1970, 1971) and Axford (1970) pointed out that treating low energy events with propagation models in which convection and energy change effects are ignored can lead to serious errors for times late in the event.* Fisk and Axford (1968) had obtained a specialized solution to the complete transport equation (Parker, 1965), but the solution was generally ignored by the experimental community. However, the observations of low energy solar particles from discreet events have been increasingly discussed and analyzed in the past several years in terms of solutions to the complete transport equation including diffusion, energy losses (from adiabatic deceleration), and convection,

$$\frac{\partial U}{\partial t}+\nabla\cdot\left[\mathbf{V}\left(U-\frac{1}{3}\frac{\partial}{\partial T}(\alpha TU)-\boldsymbol{\kappa}\cdot\nabla U\right)\right]=-\frac{V}{3}\frac{\partial}{\partial r}\frac{\partial}{\partial T}(\alpha TU), \tag{1}$$

* The energy-independent decay observations in Figure 2 cannot be attributed purely to convective effects as the rotation of the Sun over the five or six day period of observation brings the active-region field lines closer to the connection area of the Sun-Earth spiral field lines.

where $\boldsymbol{\kappa}$ is the diffusion tensor (e.g., Parker, 1965, 1966; Fisk and Axford, 1968; Gleeson and Axford, 1967), $\mathbf{V}$ is the solar wind velocity, T is the particle kinetic energy and $\alpha=(T+2m_0c^2)/(T+m_0c^2)\approx 2$ for non-relativistic particles ($m_0c^2 \gg T$). In (1) U is the particle density and equals Φv^{-1} where v is the particle velocity and Φ is the experimentally measured particle flux $(\text{cm}^2 \text{ s sr})^{-1}$.

The transport Equation (1) is frequently written in terms of the particle current density $\mathbf{S}$

$$\mathbf{S}=C\mathbf{V}U-\boldsymbol{\kappa}\cdot\nabla U, \tag{2}$$

where C is the Compton-Getting factor

$$C=1-\frac{1}{3}\frac{1}{U}\frac{\partial}{\partial T}(\alpha TU). \tag{3}$$

Using (2), (1) can be re-written as (Ng and Gleeson, 1971; Gleeson, 1973)

$$\frac{\partial U}{\partial t}+\nabla\cdot\mathbf{S}=-\frac{V}{3}\frac{\partial^2}{\partial r\partial T}(\alpha TU). \tag{4}$$

The derivation and discussion of these equations from first principles had been given by a number of authors (Parker, 1965, 1966; Gleeson and Axford, 1967; Gleeson, 1969; Jokipii and Parker, 1967, 1970) substantially before their usefulness and importance became necessary for most experimental observations of solar flare particles; a general review of transport phenomena has been presented by Jokipii (1971). The terms involving α in (1) are those associated with particle energy change. It is interesting to note that if $(\partial/\partial T)(\alpha TU)=0$, the equations reduce to the standard transport equations where energy changes are neglected (Parker, 1963). However, in reality, this factor could never equal zero because, for low energy events ($\alpha\approx 2$), it would imply that the differential energy spectrum of the particle flux varied as $T^{-1/2}$, which is a spectral variation much harder than any observed to date.

The validity of (1) in describing cosmic ray transport phenomena, particularly for low energy particles, has been questioned by Klimas and Sandri (1971) who have introduced a new (integral) transport equation (Klimas and Sandri, 1973a, b). These theoretical questions are currently being debated vigorously in the literature (Jokipii, 1972; Jones *et al.*, 1973).

Suggestions for the use of multi-spacecraft measurements to study the problem of solar cosmic ray energy loss as implied by (1) were made by Gleeson and Palmer (1971); their initial results, using sparse data, were inconclusive. Additional studies of the energy-loss problem have been reported by Murray *et al.* (1971), Gleeson *et al.* (1971), Lezniak and Webber (1972), and Palmer (1973).

The 'equilibrium' anisotropies measured during discrete flare events are found not to be limited to just the radial direction in the decay phrase. McCracken and his co-workers have also presented evidence for the existence, late in an event, of an anisotropy stage oriented approximately at right angles to the nominal interplanetary field direction. Examples of the intensity-time and anisotropy observations from such an event are shown in Figure 3. Plotted on the left-hand side are the fluxes of protons (7.4–10.5 MeV) measured at Pioneer 8 (24° E of the Earth), following a flare at ~0248 UT at ~110° W on 30 March 1969. Plotted on the right-hand side of Figure 3 are the observed anisotropies as a function of time for some ten days following the event. During the first two days, 30 and 31 March, the anisotropies were

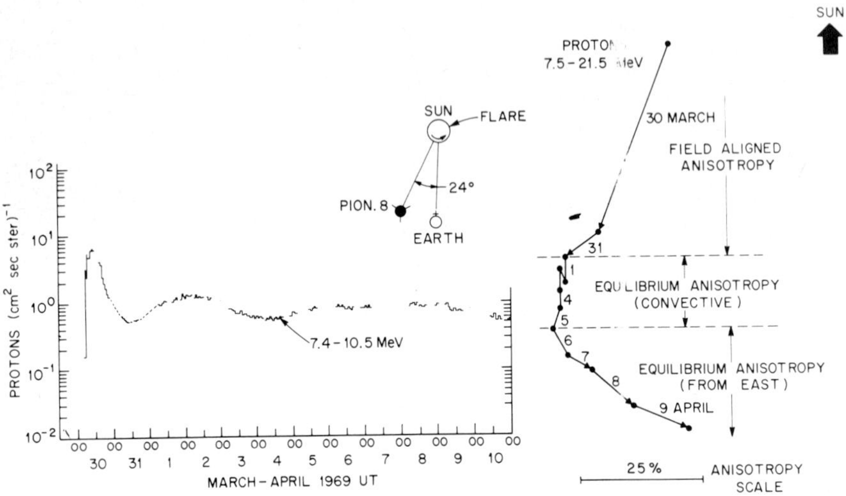

Fig. 3. Left side: time-intensity profile of 7.4–10.5 MeV protons measured on Pioneer 8 following the flare at W110° (denoted by arrow) on 30 March 1969. Right side: diagram indicating magnitude and direction of proton anisotropies measured following the 30 March events (adapted from McCracken *et al.*, 1971).

directed along the nominal interplanetary field direction. A small, ~10%, radial anisotropy was measured during the next five days while a somewhat larger anisotropy was measured in a direction ~45° E of the Earth-Sun line during the period 6–9 April.*

The two different equilibrium anisotropy phases can be understood in a phenomenological way by referring to the expression for the cosmic ray particle current in (2). This is in essence what McCracken *et al.* (1971) have done in the sketches of Figure 4 (see also Ng and Gleeson, 1971). In Figure 4, the streaming due to azimuthal gradients is neglected so that only streaming in the radial direction is considered. Thus (2) can be considered to consist of convective and diffusive scalar terms. At very early times in the impulsive event, the particle density gradient $\partial U/\partial r$ is negative; the two components in the streaming expression (2) add vectorially to produce a net anisotropy approximately along the nominal spiral direction. When the peak of the injected particle distribution passes the observer, the second term in (2) is approximately zero so that the current becomes purely convective. At late stages in the event

* Rigorous understanding of particle anisotropies can only result when the particle fluxes are related to the measured, actual directions of the interplanetary field. The work of McCracken *et al.* (1968) and Kirsch and Münch (1973) point up the importance of simultaneous particle/field observations.

the gradient of the solar particle density is positive, giving a diffusive current *toward* the Sun along the nominal spiral field. Added to the convective current, the net streaming is from a direction approximately perpendicular to the interplanetary field.

In order to understand solar events in a more quantitative fashion, solutions to (1)

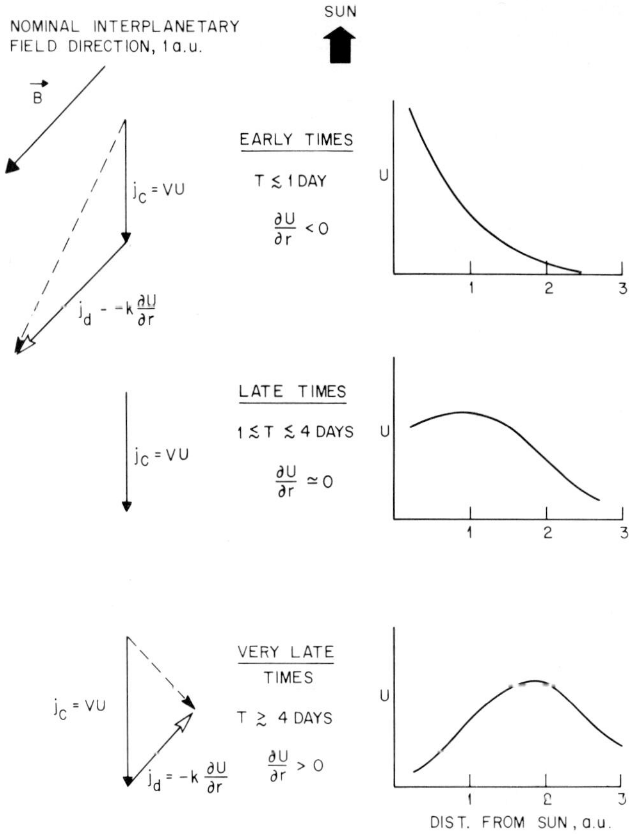

Fig. 4. Schematic illustration of solar particle anisotropies at different times during an hypothetical event assuming negligible azimuthal propagation (adapted from McCracken *et al.*, 1971).

must be fit to both the observed anisotropies and time-intensity flux profiles in order to determine the value of the transport tensor $\boldsymbol{\kappa}$ as a function of particle energy. Solutions to (1) have been obtained by several different authors on the basis of different assumptions as to the analytic form of $\boldsymbol{\kappa}$ and the outer solar system boundary conditions (Fisk and Axford, 1968; Forman, 1971; Ng and Gleeson, 1971). Recently, another analytical solution to (1) was found by Lupton and Stone (1973), who then used their solution to obtain $\boldsymbol{\kappa}$ as a function of proton energy for three well-behaved impulsive flare events.

The analytical solution of Lupton and Stone was obtained on the basis of assumptions that $\boldsymbol{\kappa}$ is diagonal in a reference frame aligned with the radial direction and that

$\kappa_\theta = \kappa_\varphi = \kappa_1 r^2$ and κ_r = constant, with all κ independent of energy within a specified, small energy domain. They also assumed a perfectly absorbing, 'escape', boundary at a distance $r = L$. (Forman (1971) found a solution to (1) for the decay phase of an event with much the same set of assumptions as Lupton and Stone with the major exception that she used $\kappa_r \propto r$.)

The best visual fits of Lupton and Stone's analytical solutions to four channels of proton data are shown in Figure 5. These proton fluxes resulted from a 3B flare at

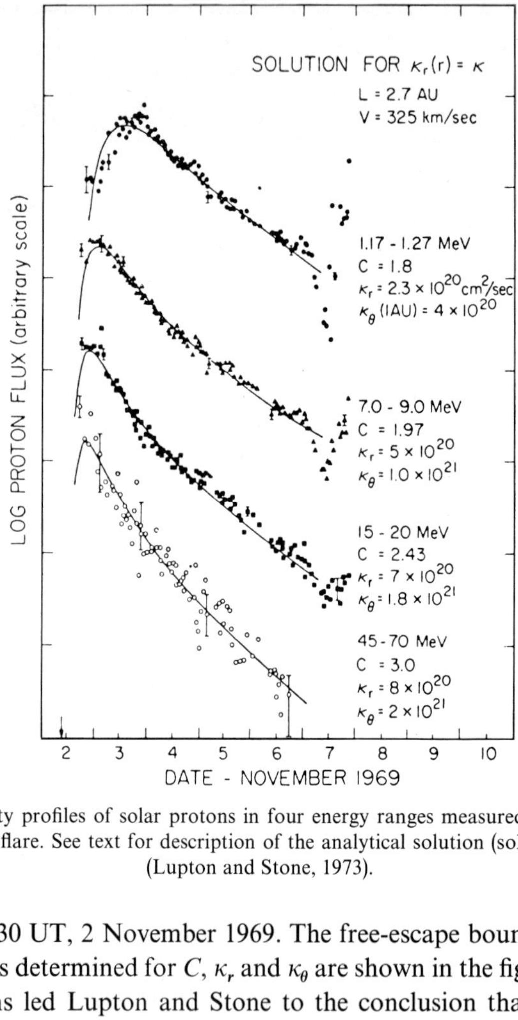

Fig. 5. Time-intensity profiles of solar protons in four energy ranges measured on OGO 5 following the 2 November 1969 flare. See text for description of the analytical solution (solid line) fits to the data (Lupton and Stone, 1973).

N15° W90° at 0930 UT, 2 November 1969. The free-escape boundary was taken at 2.7 AU; the values determined for C, κ_r and κ_θ are shown in the figure. Note that the initial assumptions led Lupton and Stone to the conclusion that $\kappa_r \approx \kappa_\theta$ at 1 AU. This conclusion on the magnitude of the transverse diffusion coefficient is rather doubtful for 1 MeV protons in the light of the work of, for example, Krimigis *et al.* (1971), which supports the general idea of the maintenance of steep longitudinal co-rotating flux gradients about flux tubes corresponding to the position of favorable solar injection.

Values of the radial diffusion coefficient κ_r obtained by Lupton and Stone as a function of proton energy for the 2 November 1969 event observations are shown in Figure 6. There is a slight increase in κ_r, from $\sim 3 \times 10^{20}$ cm^2 s^{-1} to $\sim 10^{21}$ cm^2 s^{-1} over the proton energy range ~ 1 MeV to ~ 50 MeV.

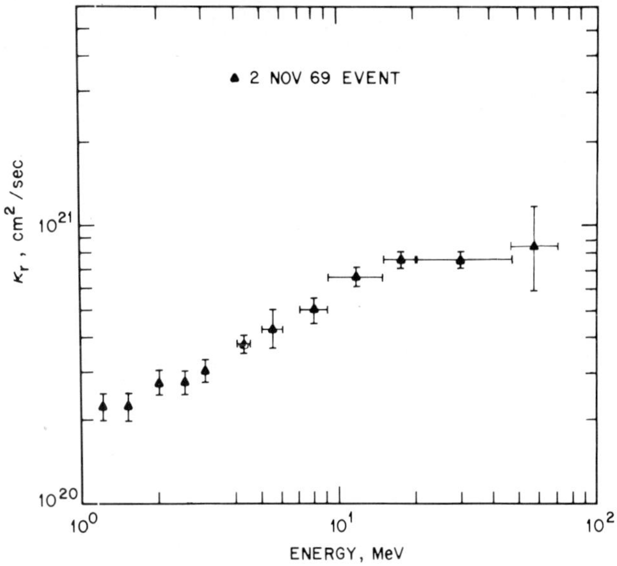

Fig. 6. Radial diffusion coefficient as a function of proton energy resulting from analytical solution fits to the time-intensity particle profiles measured following the 2 November 1969 flare (Lupton and Stone, 1973).

3. Field and Particle Observations

Several recent investigations have attempted to relate the values of the components of the interplanetary transport tensor, obtained from fits of various solutions of (1) to the observed data, to measurements of variations in the interplanetary field. Extensive work, reviewed by Jokipii (1971), has been carried out on the derivation of the diffusion tensor from the power spectrum of the interplanetary field fluctuations (Jokipii, 1966, 1967; Roelof, 1966, 1968; Hasselmann and Wibberenz, 1968). Because of the limitations of satellite observations, the majority of the studies of the interplanetary field fluctuations are limited to a spatial region close to 1 AU; thus, the diffusion tensor derived from measurements of these fluctuations is a quantity describing propagation in the local environment.

General approaches to the computation of the diffusion tensor using interplanetary field data have been made by Jokipii and Coleman (1968), Sari and Ness (1970), and Wibberenz *et al.* (1970). More recently, several studies have been made of simultaneous interplanetary particle and field observations in order to investigate the transport tensor from as complete a data set as possible (e.g., Quenby and Sear, 1971; Webb *et al.*, 1973). Such studies appear to be the optimum way in which to ultimately understand propagation phenomena and will undoubtedly become more common in the future.

Plotted in Figure 7 are the interplanetary magnetic field data (field magnitude, solar-ecliptic angles) measured on HEOS 1 during 23 January–3 February 1969 (Balogh *et al.*, 1971). At the bottom of Figure 7 are the proton data (1–13 MeV) ob-

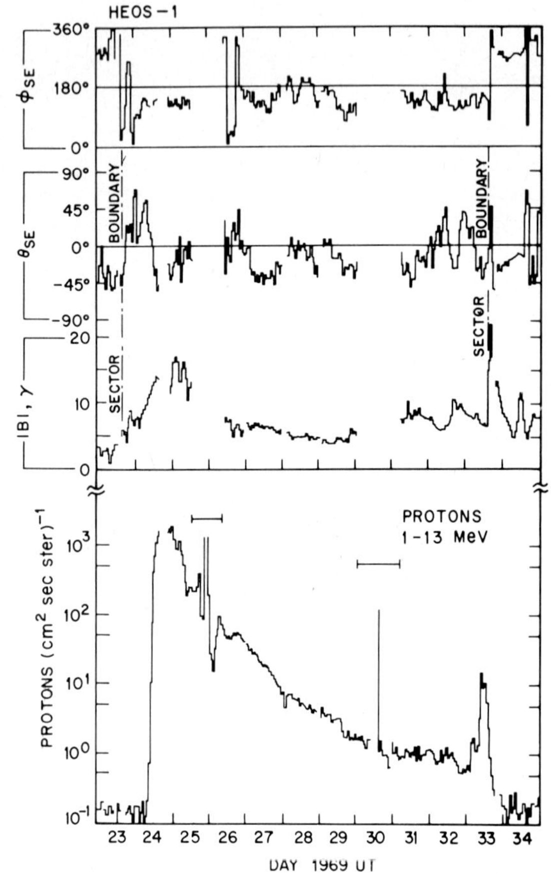

Fig. 7. Solar proton fluxes and interplanetary field parameters measured on HEOS 1 following the 23 January 1969 flare. The horizontal 'error' bars over the proton data indicate periods when the satellite was inside the magnetosphere. An interplanetary field sector boundary on 23 January and 2 February (Day 33) is indicated in the field data (adapted from Balogh *et al.*, 1971).

served on HEOS 1 following a 2B flare at N20° W08° at ~0700 UT on 24 January (Balogh *et al.*, 1971). The two periods of HEOS-1 passage through the magnetosphere are denoted in the plot of particle data.

Webb and Quenby (1973) and Webb *et al.* (1973) have studied the January 1969 event shown in Figure 7 in order to compare theoretical and experimental estimates of the solar proton diffusion coefficient. Numerical solutions of the cosmic ray transport equation were obtained by Webb and Quenby (1973) using the assumptions that azimuthal diffusion is negligible, a free escape boundary exists at $r > 1$ AU, and $\kappa_r = \kappa_0 r^b$, $-3 \leqslant b \leqslant 1$. The assumption of negligible azimuthal diffusion differs substantially from the numerical work of Englade (1971a) who assumed $\kappa_\perp \propto r^2$. (It also differs, of course, from the analytical results of Lupton and Stone (1973) discussed above in connection with Figure 5).

Sketched in Figure 8, from Webb and Quenby (1973), is a time history of the

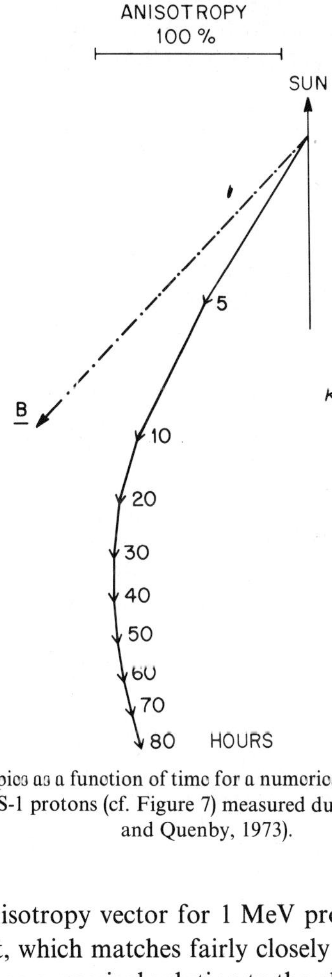

Fig. 8. Calculated anisotropies as a function of time for a numerical solution fit of the transport equation with $\kappa_\parallel \alpha r^{-1}$ $\kappa_\perp = 0$ to HEOS-1 protons (cf. Figure 7) measured during the 23 January 1969 event (Webb and Quenby, 1973).

development of the anisotropy vector for 1 MeV protons in the 24 January event. This time development, which matches fairly closely that reported by Balogh *et al.* (1971), was obtained for a numerical solution to the observed data using the assumptions outlined in the previous paragraph, with $\kappa_r = \kappa_0 r^{-1}$. Such a radial dependence of κ has not been used in previous analytical or numerical solutions of the transport equation. This radial dependence implies an increasingly large κ closer to the Sun.

Webb *et al.* (1973) have examined in detail relationships between the diffusion coefficient obtained from numerical solutions to the 24 January event (Figure 7) and the diffusion coefficient implied by the interplanetary field and plasma parameters measured at the earth. For the 24 January event, Webb *et al.* found it necessary to divide the event into four separate propagation 'regimes', which they asserted corresponded to changes in the intensity-time slope during the decay phase of the event. The results of Webb *et al.* for their analysis of 1–13 MeV protons are not straightforward; if they assume that the diffusion proceeds via magnetic fluctuation power residing in small-amplitude transverse waves, they find $\kappa = \kappa_0 r^{-3}$. Such a large negative value of b implies either very rapid wave growth with r or purely convective transport (which is not consistent with the observations; cf., Figure 8). They seem to find more 'reasonable' results ($b \approx 0$) with an assumption of diffusion proceeding via discontinuous changes (rotational discontinuities) in the field magnitude along the propagation path.

Two other results of important value were emphasized by the numerical work of Webb and Quenby (1973). The first of these is that the time to peak particle flux at the Earth is insensitive to the nature of either the inner (solar) or outer boundary conditions (see also Ng and Gleeson, 1971; and Englade, 1971b, for similar conclusions). Hence, for times to peak flux (after the flare) that are short compared to when convection effects become important* the result for the time to peak flux

$$t_{max} = \frac{r^{2-b}}{3(2-b)\,\kappa_0} \tag{5}$$

derived from the pure diffusion picture for propagation in an infinite scattering medium is valid for $0 \leqslant b < 2$. For $b = -1$, (5) differs from the numerical solution of Webb and Quenby by about a factor of two. Thus, the timing of particle arrival at the Earth can be used to investigate conditions at $r \lesssim 1$ AU. The second result emphasized by Webb and Quenby is that an exponential decrease of intensity with time in the decay phase of an event does not have to result from a free escape boundary at a few AU distance (Axford, private communication, 1969; Ng and Gleeson, 1971). Ng and Gleeson (1971) mention that a parallel diffusion coefficient with $\kappa \propto (1 + r^3)$ will produce an effect similar to an exponential decay.

The most detailed and extensive study to date of the power spectra of interplanetary field observations has been carried out by Sari (1972). He analyzed magnetic field data from Pioneer 6 for four solar rotations in 1965–1966 in terms of the spectra produced by the discontinuities as well as the fluctuations between discontinuities. Sari's spectra covered the frequency range 10^{-4}–10^{-2} Hz. Sari then computed the daily values of the local interplanetary proton diffusion coefficients for 30 MeV, 50 MeV, and 70 MeV protons that would be implied by the spectra 'between' discontinuities and the total interplanetary field spectra (including discontinuities). A correlation study between observations of 50 MeV and 70 MeV proton modulations on the same space craft and the diffusion coefficients obtained from the two different spectra indicated that the between spectra gave the best representa-

* One way in which this can be determined is to use (2) and determine $\Re = CVU/\boldsymbol{\kappa} \cdot \nabla U$ for a specific model for $U(\mathbf{r}, t)$. For example, for the isotropic diffusion model treated by Krimigis (1965), where $\kappa = \kappa_0 r^b$, convection becomes important for non-relativistic particles ($\alpha = 2$) for $b \sim 1$ at $\sim$20–40 h after the flare, depending upon the exponent of the power-law energy spectrum (Axford, 1972).

tion of the measured particle modulations. Plotted in Figure 9 are histograms (Lanzerotti *et al.*, 1973), for 50 MeV and 70 MeV protons, of the percentage daily occurrence of a given value of the interplanetary diffusion coefficient as tabulated by Sari (1972) for four solar rotations. The two separate histograms in each half of the figure correspond to the transport coefficient obtained from the between spectra and the 'real' (or total) spectra.

Sari also gave the results from a statistical study of the power-law slopes of the interplanetary spectra. He found that during the four-month period of study the average value of the power-law slope of the discontinuity spectrum was 1.95, the

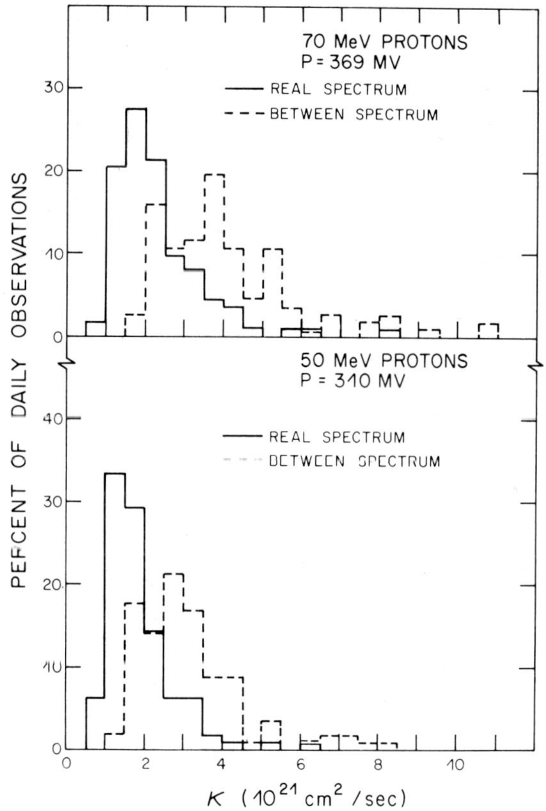

Fig. 9. Histogram of the percentage daily occurrences of a given value of the diffusion coefficient as determined by Sari (1972) for 50 MeV and 70 MeV protons. The distributions are shown for the diffusion coefficient calculated for both the total ('real') power spectra of the interplanetary field and the spectral power measured between discontinuities (Lanzerotti *et al.*, 1973).

average value of the slope of the spectra between discontinuities was 1.65, and the average value of the slope of the total power was 1.71. Furthermore, on days of quiet interplanetary field conditions (which he characterized as days of low terrestrial K_p values) the spectral slopes of the total power tended to be steep, characteristic of the discontinuity spectra. During disturbed times, more over-all power was measured at all frequencies and the spectra tended to have flatter slopes characteristic of the spectra between discontinuities.

The statistical study of interplanetary diffusion coefficients represented by Sari's results as plotted in Figure 9 have been used by Lanzerotti *et al.* (1973) to study the well-known variability in the time to maximum flux of solar particle events. In particular, Lanzerotti *et al.* proposed that by using the times to maximum flux for solar electron events from the western hemisphere, the variations in the high-frequency component of the interplanetary power spectrum could be studied. The energies of solar electrons that they used (>300, >600, >1000 keV) undergo resonance scattering with interplanetary field fluctuations (at 1 AU) with frequencies $\sim 10^{-1}$ Hz. Thus, the statistical study of Sari can be extended to higher frequencies using particle observations.

Plotted as open and closed circles in Figure 10 are the values of the mean-free path ($\lambda_r = 3\,\kappa_r/\beta c$) obtained from the 90%-limit points on the histograms (Figure 9) for the diffusion coefficients for 50 MeV and 70 MeV protons. The limit points are plotted for both the real and the between spectra. The lines with slopes corresponding to the rigidities $P^{0.05}$ and $P^{0.35}$ correspond to extrapolations of Sari's spectra for quiet and disturbed times. Sari noted that for disturbed times the field between discontinuities (as well as the total spectrum) had an average slope of ~ -1.65. Thus, $\lambda_r(P)$ has been extrapolated to higher frequencies from the lower points using

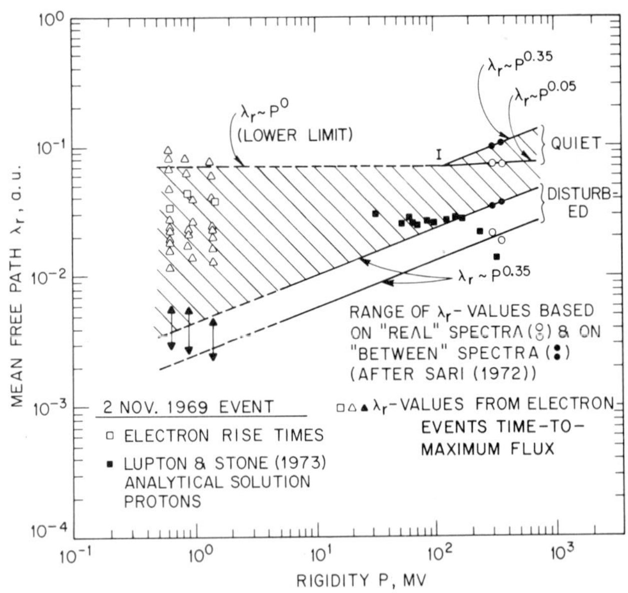

Fig. 10. Solar electron mean free paths determined from the rise to maximum of the electron events plotted as a function of the particle rigidity. Also shown are the ranges (90% level) of λ_r from the histograms of Figure 9 as well as extrapolations of these λ_r using average slopes of the power spectra (see text; Sari, 1972). The values of λ_r derived from the Lupton and Stone results in Figure 6 are shown as solid square points. Shown as open squares are the λ_r values from the electron rise times measured during the 2 November 1969 event (adapted from Lanzerotti *et al.*, 1973).

$\lambda_r(P) \sim P^{0.35}$*. The λ_r-values should in general lie above this line, in the shaded area.

For the quiet conditions, the real spectrum is steeper ($q \sim -1.95$) than the between spectrum, so that at a certain frequency, the two spectra intersect at point 'I'. At rigidities below I (at higher frequencies) the spectra have been extrapolated as $\lambda_r \sim P^0$ in Figure 10. This is consistent with the quiet time spectral dependence ($\sim f^{-2}$) reported at higher frequencies by Childers and Russell (1972).

Plotted as open and closed triangles in Figure 10 are the mean-free paths determined by Lanzerotti *et al.* (1973) from the time to maximum flux of a number of solar electron events originating from flares on the western hemisphere in 1967–1972. The expression (5) was used to derive κ_0 from the time to maximum flux with the assumption that $b=0$ (*i.e.*, a radially-independent diffusion coefficient). For the one or two electron events with the smallest inferred values of λ_r, convection effects are important. Consideration of the numerical work of Webb and Quenby (1973) indicates that the lowest deduced value of λ_r should be moved downward by the length of the arrows, as indicated in Figure 10.

The values of λ_r deduced from the time to maximum electron flux, assuming instantaneous particle injection (no storage) at the beginning of the flare, are quite consistent with the statistical variation expected from an extrapolation of Sari's results to higher frequencies. Points lying outside the shaded area might be expected from the high κ_r 'tail' on the distributions in Figure 9. Further, spectra over time periods shorter than one day exhibit more variability than the daily spectra used to compute the results of Figure 9 (Wibberenz, 1973).

Denoted by the open squares in Figure 10 are the mean-free-paths derived from the times to maximum flux for the electrons from the event of 2 November 1969. Plotted as closed squares in Figure 10 are values of λ_r implied by the diffusion coefficient results for the proton observations (Figure 6) during the 2 November event (Lupton and Stone, 1973). Except for the two highest rigidities, the diffusion coefficient results for protons from the Lupton and Stone analytical model lie within the statistical range found by Sari. The time to maximum electron flux results for the 2 November event, compared to the proton results, would suggest an increase in mean-free path (increase in the diffusion coefficient) at lower rigidities.

Electron (or proton) events that have a mean-free path at the upper edge (or above) the extrapolated Sari quiet-time spectrum could be considered to be approximately 'scatter-free' events. Indeed, Lin (1970) reported what appear to be a large number of scatter-free electron events resulting from McMath plage region 8905. Figure 11 contains the time-intensity profiles of two 'scatter-free' events observed from region 8905 on 30 July 1967 (Wang *et al.*, 1971). Sharp, short-lived enhancements are seen for electrons with energies as high as 500–1100 keV. The scatter-free nature of these events is based primarily upon the timing of the event onset following the identification of a flare in the active region.

The anisotropies, averaged over the duration of the events of Figure 11, are

* This follows since $\lambda_r \sim P^{2-q}$ if the interplanetary field power spectrum has a power-law dependence upon wave number; i.e., $P(k) \sim k^{-q}$.

$\lesssim 50\%$, in contradiction to the highly anisotropic flow expected of scatter-free events. It was suggested (Wang *et al.*, 1971) that upstream waves (Fairfield, 1969) from a greatly extended bow shock ($>20\ R_E$ during this period of observation) could produce the isotropy observed. The anisotropy characteristics of some eighteen solar electron events (electrons of energy >70 keV) observed in 1967–1968 have been

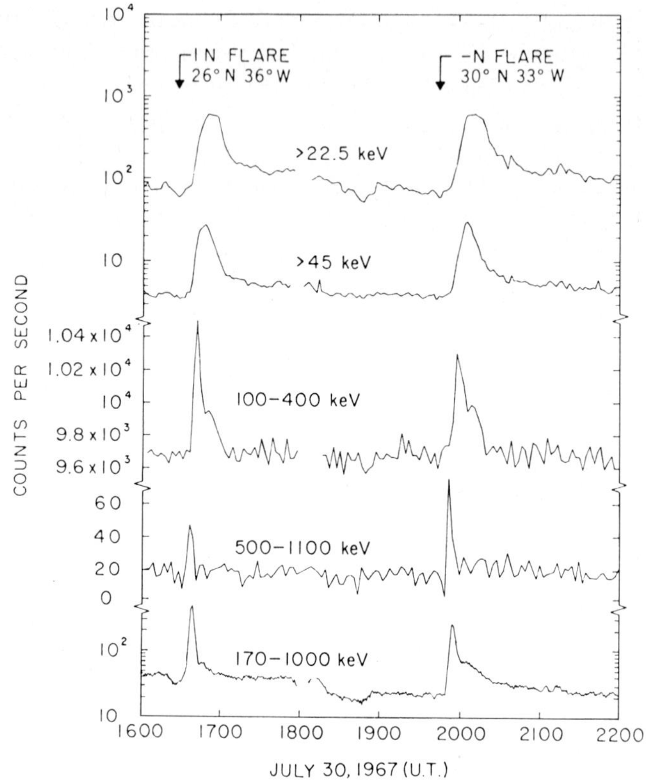

Fig. 11. Two 'scatter-free' solar particle propagation events indentified by Wang *et al.* (1971) in electron data measured on Explorer 34.

discussed by Allum *et al.* (1971). They found that although the electron anisotropy was quite large during the onset stage of most impulsive flare events (as would be expected for the initial particles) the anisotropy decayed to $\lesssim 10\%$ within an hour or so after the onset. Such a rapid decrease in the anisotropy is not immediately compatible with the very frequent occurrence of scatter-free events.

One event ($\sim$0520 UT, 30 July 1967) termed scatter-free by Lin (1970) was reported by Allum *et al.* (1971) to have an anisotropy of $\sim 85\%$ in the onset stage. No mention is made by Allum *et al.* of the anisotropy during the remainder of the event. Simnett (1971) has proposed that the events of Lin (1970) are not scatter-free but rather that the electrons arise from the release, at the time of a flare or disturbance, of electrons previously accelerated and trapped in the corona.

4. Longitudinal Extent of Solar Events

As discussed above, the assumptions that have been made by different investigators concerning the azimuthal propagation of impulsive events differ among themselves rather drastically. For example, in studying analytically the decay phase of impulsive events Forman (1971) assumed $\kappa_\perp = \kappa_1 r^2$ (as was assumed in the non-convective anisotropic diffusion model of Burlaga (1967)). The study of Ng (1972) assumed $\kappa_\perp = 0$, as did that of Webb and Quenby (1973) in their numerical work. The results of Lupton and Stone (1973; cf. Figure 6) found $\kappa_\perp \approx \kappa_\parallel$ at 1 AU as a result of the initial assumption that $\kappa_\perp \propto r^2$.*

The azimuthal propagation characteristics of solar cosmic rays is obviously of great importance in interpreting impulsive events. That significant fluxes of protons are seen at large distances from the nominal injection field-line bundle has been known for a long while (e.g., O'Gallagher and Simpson, 1966). The mechanisms by which these particles arrive at large azimuthal distances from a flare site are closely related to the questions of field-line 'wandering' (Jokipii and Parker, 1969; O'Gallagher, 1970) and to the more general problems of particle storage and emission at the Sun (e.g., Anderson, 1969; Fan *et al.*, 1968; McDonald and Desai, 1971; Simnett and Holt, 1971; Anderson, 1972) and coronal propagation of solar particles (e.g. Reid, 1964; Axford, 1965; Fisk and Schatten, 1972). The work of Krimigis *et al.* (1971), using data from multi-spacecraft, indicate that low energy particles ($\lesssim 1$ MeV) are closely tied to a given flux tube.** On the problem of storage, Krimigis and Verzariu (1971) and Verzariu and Krimigis (1972) claim that their measurements of spectra of solar protons from impulsive events do not support the thesis of significant particle storage near the Sun. Specifically, they found no evidence for a flattening, or turn over, of the proton spectra at low energies that could not be readily explained by transit-time considerations.

Bukata *et al.* (1972) have examined the azimuthal extent near 1 AU of the distribution of solar particles following the 30 March 1969 event (cf. Figure 3). Using the data from four Pioneer space craft positioned about the solar system they could make plots of the azimuthal distribution of particle fluxes at various times during the event. Two such azimuthal plots for 31 March and 1 April are shown in Figure 12. The simple straight-line connections suggest that a peak particle flux on 31 March would probably have been measured by an instrument on a satellite located at W55° (with respect to the Sun-Earth line at 1 AU on that day). Taking into account the spiral field direction, this would imply a source near W110°, quite consistent with the radio burst observations of the event by Smerd (1970).

An analysis of the decay phase of the 10 April 1969 event (cf. Figure 2) by

* Lupton and Stone actually used the two azimuthal polar coordinates $\kappa_\theta = \kappa_\varphi = \kappa_1 r^2$ in their analytical solution.

** This is an important low-energy observational result which gives additional support for the $\kappa_\perp = 0$ assumption in the theoretical works (cited earlier) of Ng and Gleeson (1971), Ng (1972), and Webb and Quenby (1973).

McCracken *et al.* (1971) showed that the azimuthal gradient did not change appreciably during the decay; they concluded that, for this event, the processes that initially distributed particles longitudinally in an event were no longer operative several days following injection. McCracken *et al.* found that an *e*-folding angle of $\sim 30°$ for the flux intensity decrease from the injection site for ~ 10 MeV particles

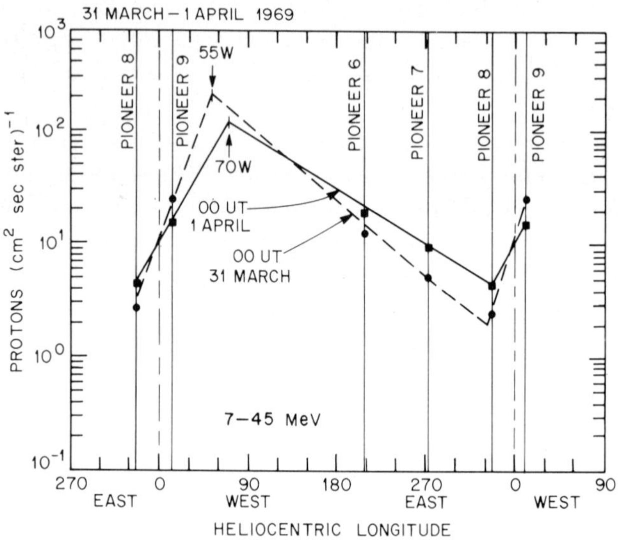

Fig. 12. Solar-azimuthal distribution of proton fluxes measured on four Pioneer spacecraft by Bukata *et al.* (1972) on two successive days following the 30 March 1969 solar flare.

described the observations rather well (see also discussion above concerning the time to maximum flux of impulsive events).*

McKibben (1972) examined several proton events as measured with instruments on three satellites distributed in solar azimuth. In several events that exhibited two distinct decay phases, McKibben found large azimuthal gradients during the first phase and some evidence of almost no gradient during the second. The work of Roelof (1973a, b) and his collaborators indicates that large-scale interplanetary structures, controlled by coronal conditions, must be considered in interpreting the azimuthal extent of solar particle events.

One possible way for particles from impulsive events to distribute themselves in longitude around the Sun is via coronal propagation (as discussed theoretically by Reid, 1964; Axford, 1965; and Fisk and Schatten, 1972). Some information about the importance of coronal propagation, at least early in an event, can perhaps be gleaned from studies of the initial time of arrival of the first particles following a flare. Simnett (1972) determined the time difference between the initial observations of 0.3–0.9 MeV electrons and the time of the flash phase of the identified solar flare.

* O'Gallagher (1970), in an analysis of the 5 February 1965 event, found an half-width *e*-folding angle of $\sim 15°$ for the first thirty-two hours following the flare.

These data, plotted in Figure 13 as a function of the longitude of the originating flare, indicate a solar azimuthal dependence to the onset time. Even for the events at W30°–W60° (at the foot of the nominal interplanetary field bundle) there is an average delay of ~20–30 min beyond that expected for instantaneous particle release from a flare at the foot of the field line.

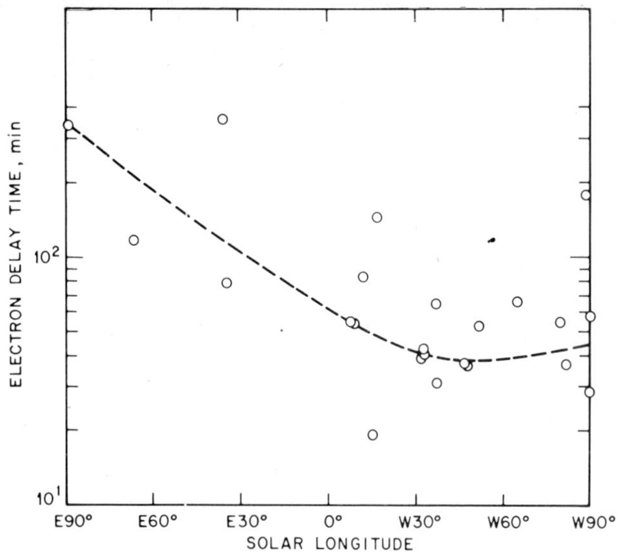

Fig. 13. Time delay between the first observations of solar electrons at 1 AU and the flash phase of the identified flare, plotted as a function of the longitude of the originating flare (Simnett, 1972).

Under the assumptions (a) that the initial particles measured at the Earth undergo diffusion in the corona in making their way to the foot of the field line connecting with the Earth and (b) that these first particles then undergo no interplanetary scattering, data such as those in Figure 13 can be used to estimate the coronal diffusion coefficient. This has been done by Lanzerotti (1973a) for solar protons of various energies resulting from a number of solar flares in 1967–1969. The results are plotted in Figure 14. Similar to the results of Simnett (cf., Figure 13), there seems to be a finite time delay, over that expected for travel along the spiral field, for particles originating from flares near the foot of the interplanetary field lines connecting to the Earth. The onset times tend to have a (longitude)^2 dependence; assuming diffusion in the corona at an altitude of five solar radii, a diffusion coefficient $\approx 3 \times 10^{18}$ cm^2 s^{-1}, comparable to that estimated by Fisk and Schatten (1972), is obtained. Fisk and Schatten, in attempting to explain the 30° azimuthal *e*-folding distribution (see above) in the decay phase of the events of McCracken *et al.* (1971), proposed coronal diffusion along thin current sheets that divide the solar granular sectors.*

* In the future, in addition to the onset times, the anisotropies of the initial low-energy particles from eastern hemisphere events need to be examined more closely than they have been in the past. Strong positive evidence that the initial anisotropies from eastern-hemisphere events were large and field-aligned would place severe limits on the magnitude of $\kappa_\perp$ and/or on interplanetary field irregularity models that had an isotropic diffusion shell at $r > 1$ AU.

The azimuthal distribution of solar particles, and its development during the initial and decay phases of impulsive events, remains a topic of particular importance for understanding solar particle propagation. That the current picture is far from complete was pointed out in a paper by Keath *et al.* (1971), wherein the authors

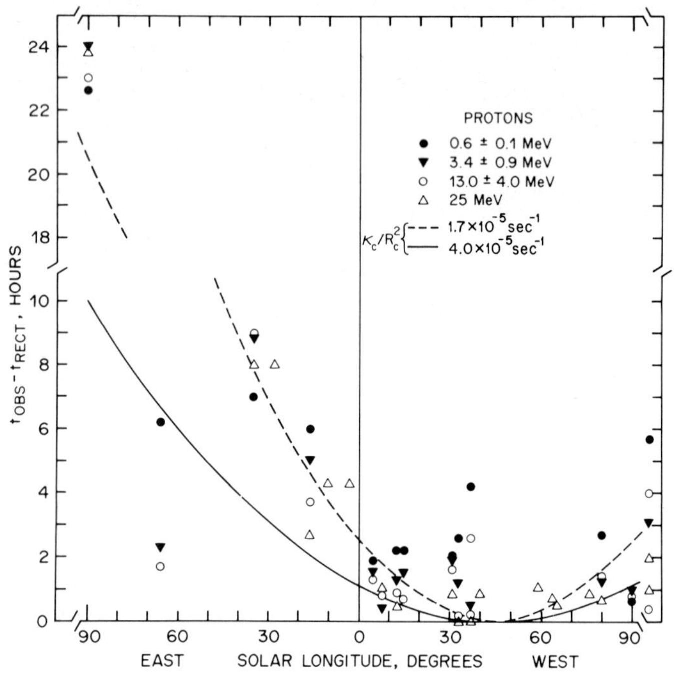

Fig. 14. Time delay between the first observations of solar protons at 1 AU and the rectilinear transit time following the Hα maximum of the originating flare, plotted as a function of the longitude of the originating flare (Lanzerotti, 1973a). The delay time is plotted for four different proton energies. The 25 MeV proton data are from Barouch *et al.* (1971).

discuss an event (12 March 1969) observed on multiple spacecraft in which the particle flux rise time to a maximum, the flux time-intensity profile, and the flux anisotropy all indicate that the favored field line for propagation was ~40° east of the nominal spiral field line. This would be an ideal case for examination and study of the local diffusion coefficients using the magnetic field data measured on the individual spacecraft.

5. Energetic Storm Particles and Shock Spikes

The discussions in the previous sections concerned with modeling energetic particle events using (1) have been concerned principally with events that 'looked' diffusive. However, for lower energy particles ($\lesssim$10 MeV), the majority of impulsive events deviate strongly from a diffusive profile within a day or two after the event. The deviation does not result from the increasing importance of convective flow, but rather appears to be related to the presence of the flare-produced blast wave traveling out through the interplanetary medium toward the observer.

It has been recognized for quite some time that the arrival of the flare-produced shock at the Earth is accompanied by large enhancements of low energy solar particles (e.g., Axford and Reid, 1963; Bryant *et al.*, 1962; Rao *et al.*, 1967). The presence of the shock at the location of the observations of the low energy particles can greatly alter the time-intensity profiles of a solar particle event. This effect can be seen in comparing the diffusive profiles of the 4 November 1968 (Figure 1) and 2 November 1969 (Figure 5) events, when no shock from the originating flares were observed at the Earth, to the profile of the 18 November 1968 event (Figure 15, Lanzerotti, 1970a) when the shock produced by the flare, together with an energetic storm particle enhancement (in this case rather symmetrical about the shock) was measured at the observing spacecraft. All three of these flares occurred at ~W90° longitude.

The phenomena of energetic storm particle enhancements have been attributed

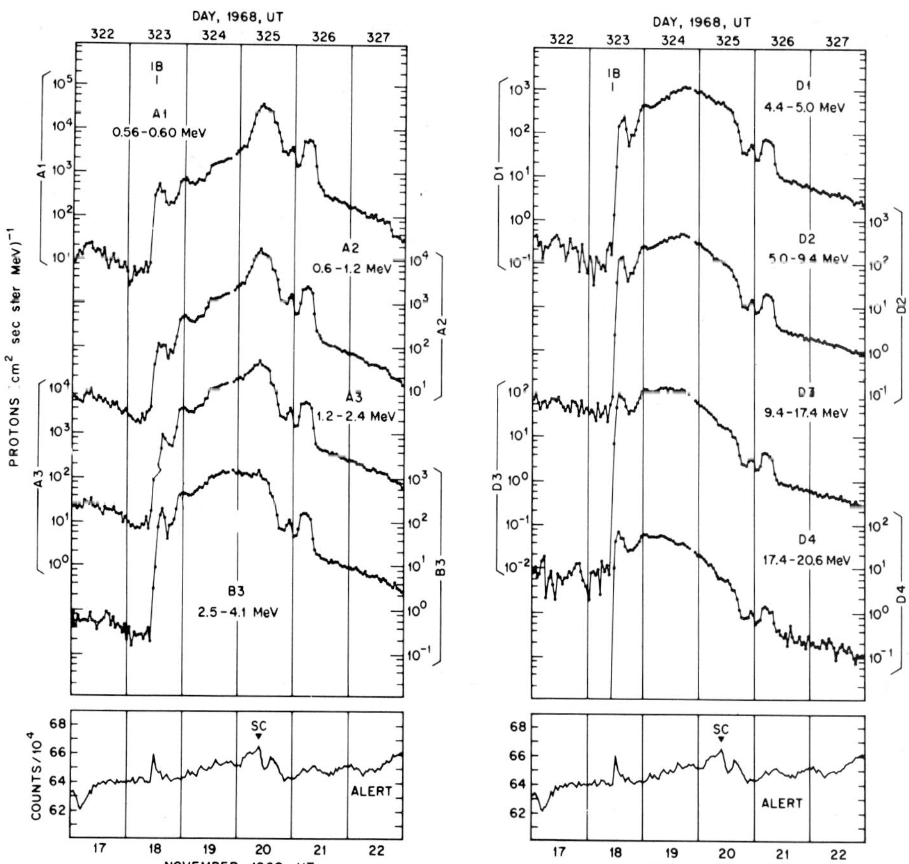

Fig. 15. Solar proton fluxes measured in eight narrow differential energy channels on Explorer 34 following the flare of 18 November 1968. Plotted at the bottom are the counting rates of the Alert neutron monitor showing the ground-level event on 18 November and the Forbush decrease on 20 November.

to acceleration of particles by the outward-moving shock (Rao *et al.*, 1967, 1968) and to a manifestation of the observation of the 'core-halo' (Lin *et al.*, 1968) injection and co-rotation process (Kahler, 1969). This last interpretation was challanged for a number of the cited events (Lanzerotti, 1970b). Kunstmann and Wibberenz (1973) have shown that in a number of the events treated by Rao *et al.* (1967), the energetic storm particle enhancements appear to be more closely connected with the tangential discontinuity following the shock than with the shock itself.

A possible way in which to approach the study of the time development of impulsive events that are followed within a couple of days by a flare shock wave is to first study and model the impulsive event at both high and low energies using (1). The effect of the shock could then be modelled in terms of interplanetary acceleration and convection at, or close to, the shock front. Indeed, significant low-energy particle convection may well occur in the vicinity of the propagating shock. Possible evidence for this is observed in the anisotropy data published by Rao *et al.* (1971) for the 18 November event and shown in Figure 16.

The interpretation of the first stage of the event anisotropy in Figure 16 (the non-

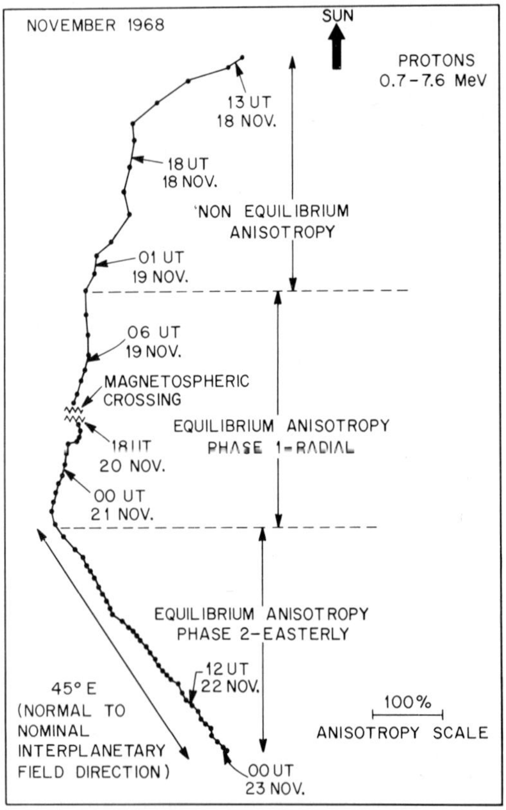

Fig. 16. Low-energy proton anisotropies measured on Explorer 34 as a function of time following the 18 November 1968 event (Rao *et al.*, 1971).

equilibrium stage) is not completely unambiguous because of the lack of the corresponding interplanetary field data (see also Kirsch and Münch, 1973). Nevertheless, at the time of the dip in the particle fluxes at ~1700 UT, 18 November (Figure 15; see also Wibberenz *et al.*, 1973) the anisotropy becomes quite radial at the low energies. Except for an hour of anisotropy approximately along the nominal spiral field direction at ~2200 UT, 18 November, the anisotropy is radial until after the passage of the shock and energetic storm particle event at ~0600 UT, 21 November (cf. Figure 15).

Although not completely understood as yet, this predominantly radial anisotropy early in the event could represent some convective flow of the low-energy particles. This suggestion is given possible further support by the development with time during the event of the particle energy spectra. Plotted in Figure 17 are proton spectra taken at several different times during the event. These plots show that the spectra were almost flat at the lower energies for a considerable period of time (>1 day) after the beginning of the event. The flatness of the spectra did not arise from purely travel-time considerations. As the intensity of particles associated with the energetic storm particle enhancement began to increase, the spectrum began to resemble a power-law shape more closely. It could be argued that the flat spectra represent particles that are trapped close to the Sun and are only slowly being released. However, the almost radial anisotropy shown in Figure 16 would tend to argue against such a storage interpretation.

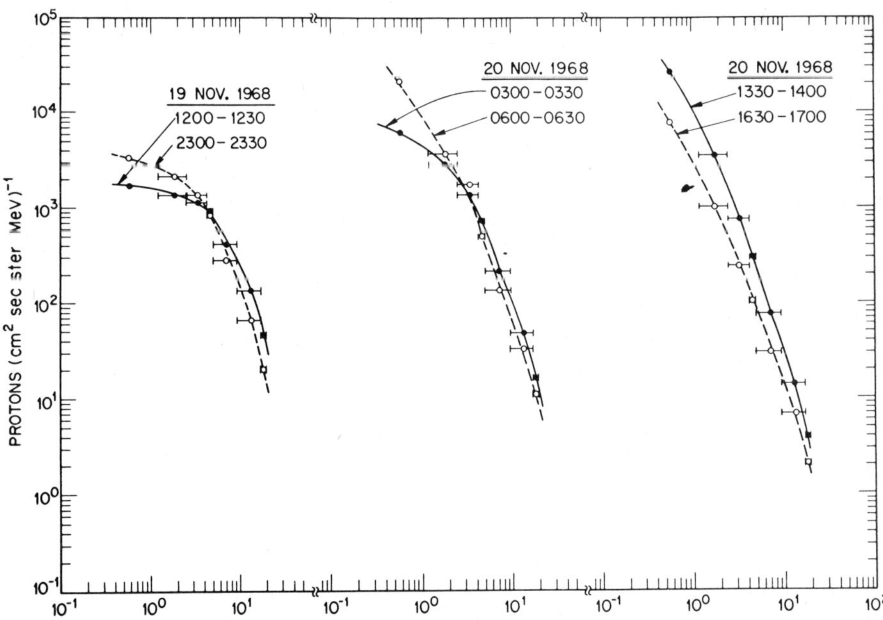

Fig. 17. Solar proton spectra measured on Explorer 34 at six selected times during the 18 November 1968 particle event. The two spectra at the extreme right were measured following the passage of the flare-produced shock beyond the Earth on 20 November (see Figure 15).

The importance of a shock in altering the decay profile of an impulsive event is perhaps evident in the comparison of McCracken *et al.* (1971) of the time-intensity profiles (7.5–45 MeV protons) of the 18 November 1968 event as observed at Pioneer 9 ($\sim$450 R_E east of the Earth) and at Pioneer 8 ($\sim$22° east of the Earth). Apart from co-rotation effects, the Pioneer 9 profile is similar to that seen at the Earth by an instrument on Explorer 34 (Figure 15; Wibberenz *et al.*, 1973). However, the Pioneer 8 data show no evidence of an energetic storm particle enhancement and no evidence of a shock being observed at the spacecraft (McCracken *et al.*, 1971; Wibberenz *et al.*, 1973).

Further evidence of the effect of the flare shock wave in altering the profiles of low-energy particles can be seen in the rather complicated 7–8 March 1970 event (UAG Report 12). During this event, the particle flux changes (in particular the shock spike event and the alpha to proton ratios), occurring with the measurement of the shock wave at the earth are typical of those changes observed with the passage of the shock from other flares. Plotted in Figure 18 are proton fluxes measured on Explorer 34 in several different channels. The two integral-channel proton data are taken from the solar proton monitoring experiment (SPME); the low-energy differential-channel data are from the Bell Laboratories solid state telescope. A large number of flares were observed on the Sun prior to, and during, the period included in Figure 18. A 2N flare (S12°, E50°) at 1318 UT 6 March in McMath region 10618 could have been the source of the particles observed on 6 March (Sakurai, 1971). On the other hand, the 6 March particles may have resulted from a flare at N15°, W99° in McMath region 10595 (McIntosh, 1971; Scholer *et al.*, 1972). The 7 March particles perhaps came from region 10614 (S11°, E09°) at 0143 UT (Sakurai, 1971; Scholer *et al.*, 1972).

The high-energy ($E>10$ MeV, $E>30$ MeV) protons on 7 March rise slowly to a peak flux and then decay away. The 1.1–2.5 MeV proton fluxes also rise to a peak intensity early on 8 March and then decrease in intensity. However, at this energy, evidence of energetic storm particles appears near the time of the sudden commencement, about 1417 UT, 8 March. At the lowest measured proton energy, the fluxes rise to a peak intensity at the time of the sudden commencement. This intensity-time behavior of the lowest energy protons is similar to that seen in the 18 November 1968 west limb event (cf. Figure 15) and is quite different from the particle time-intensity profiles observed with no accompanying shock (cf. Figures 1, 2, and 5).

Alpha particle fluxes from the 6–9 March time period are plotted in Figure 19. Evidence of an energetic storm particle enhancement is also observed at the lower energies.

A new development associated with interplanetary shock waves and energetic storm particle events that was only recognized just before (Lanzerotti, 1969) and at the time of (Singer, 1970) the Third ESLAB Symposium is the short lived 'shock spike' associated with the passage of the shock disturbance (Ogilvie and Arens, 1971; Armstrong *et al.*, 1970; Singer and Montgomery, 1971; Palmeira *et al.*, 1971). The nature of shock-spike events can be seen in the data of Figure 20, where 9.28 s aver-

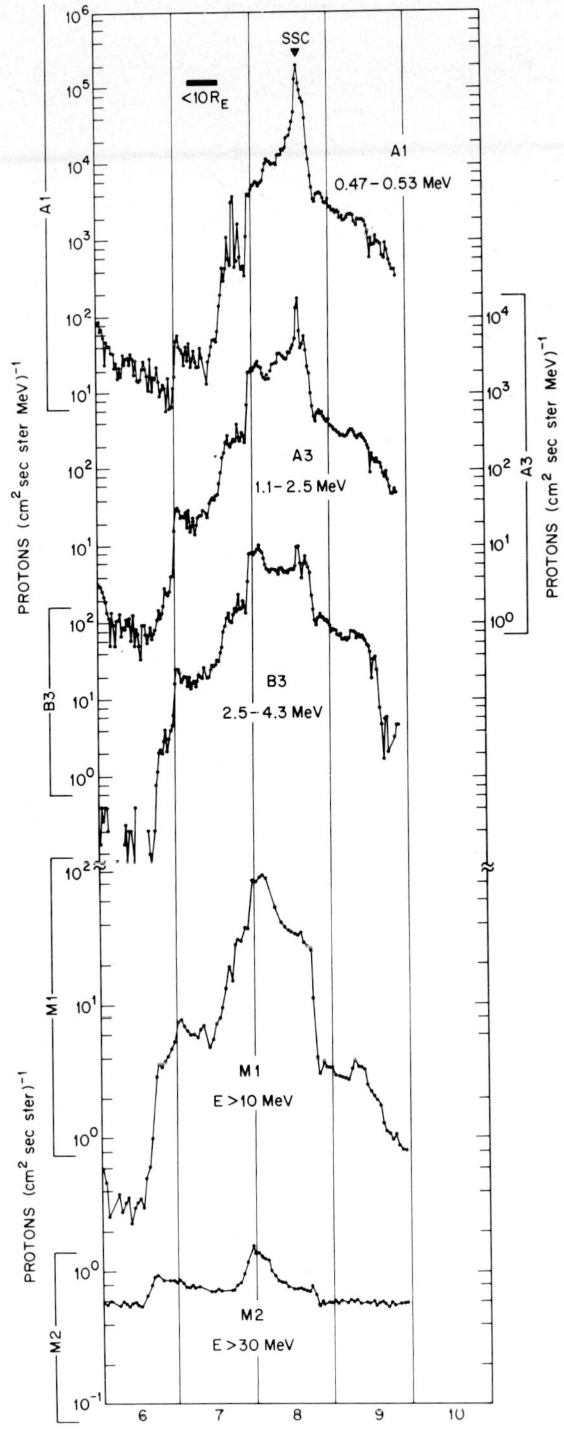

Fig. 18. Solar proton fluxes measured on Explorer 41 during the solar-active period 6–9 March 1970. The solid bar at the top of the figure indicates a period when the spacecraft was within the magnetosphere.

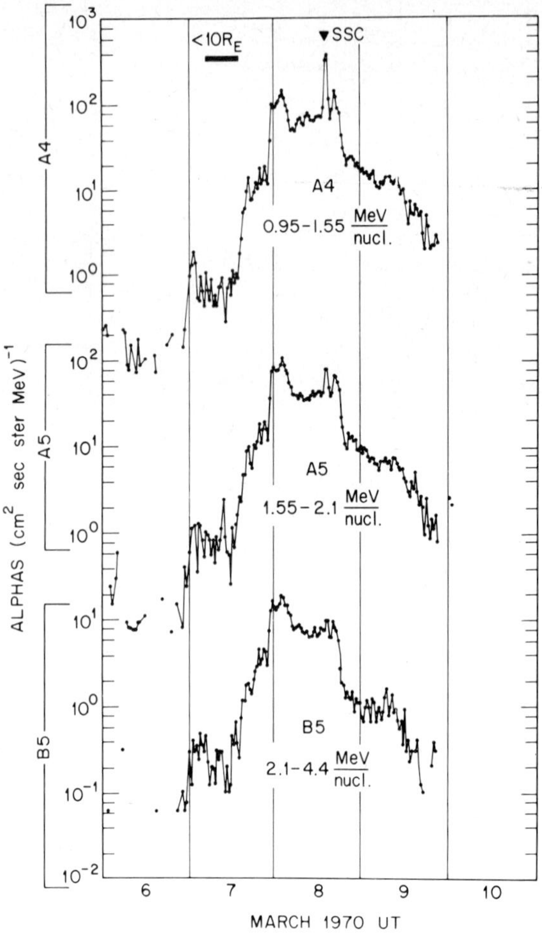

Fig. 19. Solar alpha particle fluxes measured on Explorer 41 during the period 6–9 March 1970.

ages of proton, alpha particle, and electron fluxes are plotted during a five hour period around the time of the 8 March 1970 sudden commencement (cf. Figures 18 and 19). Momentary (few minute) increases in both the low-energy proton and alpha fluxes are observed immediately following the sudden commencement. No effect of the shock is apparent in the electron fluxes (Armstrong *et al.*, 1970).

Singer and Montgomery (1971) have examined the nature of the anisotropies (summarized in the following paragraph) of the protons before and after the observation of shock spikes. Two of the events they examined are shown in Figure 21. In the data at the top of the figure, the shock spike is observed following the shock arrival; in the data at the bottom of the figure, the shock spike occurs simultaneous with the shock arrival. The small arrows indicate the times when the detector is measuring particles from the sunward direction. Energy spectra of the protons before, during, and after the shock spikes shown in Figure 21 are plotted in Figure 22. The particles in the spikes are always very soft (see also Lanzerotti, 1969).

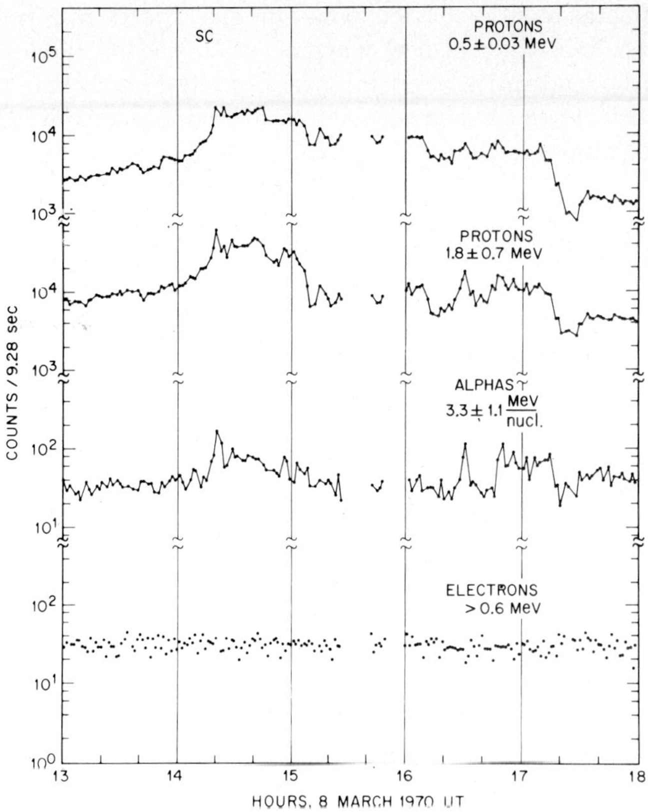

Fig. 20. Proton, alpha particle, and electron counting rates (9.28 s spin-averaged rates) measured on Explorer 41 during the time interval of the shock arrival at the Earth on 8 March 1970.

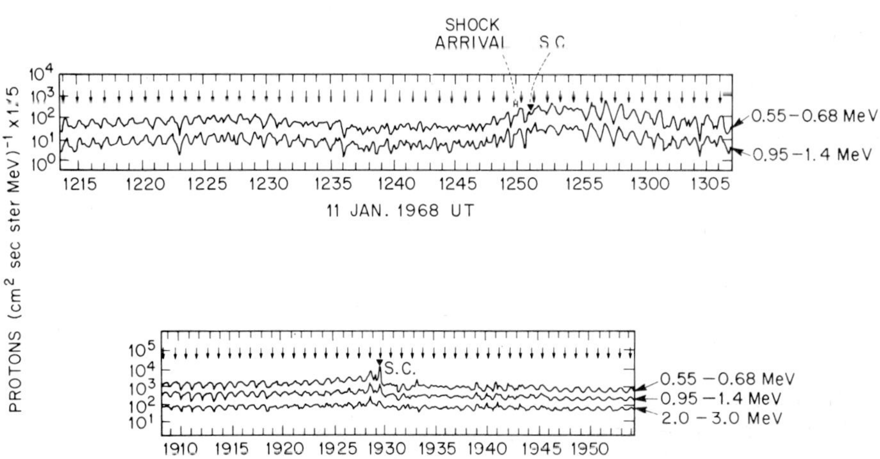

Fig. 21. Proton fluxes measured in several energy channels on Vela 4A during two shock-spike events (Singer and Montgomery, 1971).

Summarizing the results published to date on shock spikes (of which the data in Figures 21 and 22 are representative), it appears that (a) the spike need not be coincident with the shock arrival; however, the spike is usually located within a few minutes of the shock time; (b) shock spikes do not have to be associated with energetic storm particle events; the presence of a shock spike is not clearly related to the magnitude of the background particle fluxes; (c) prior to the spike, the particles tend to flow in a direction away from the oncoming shock; after the shock passage the flow direction is from the Sunward direction; (d) the pitch-angle distribution in the region of the shock spike is nearly isotropic.

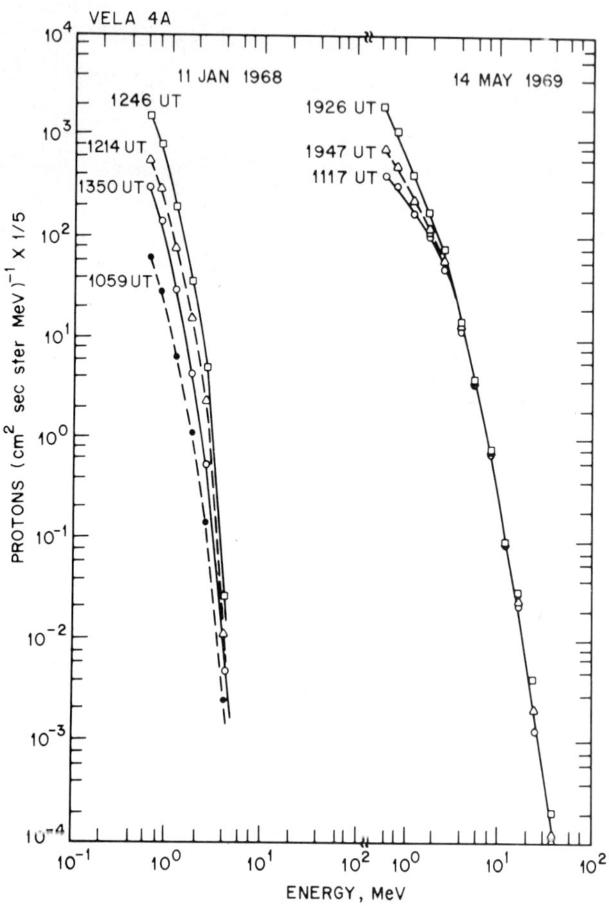

Fig. 22. Proton spectra at selected intervals during the two shock-spike events shown in Figure 21 (Singer and Montgomery, 1971).

The shock-spike observations do not seem to be explained by either the Axford-Reid model for energetic storm particle enhancements (Axford and Reid, 1963) or the Fisk (1971) model of particle energy gain by collisions with the moving shock front.

Singer and Montgomery (1971) proposed a model of single-encounter particle acceleration at the shock.

Another observational development associated with energetic storm particles and shock-spike events is the relative effects of the shock on the different particle species. As noted above, Armstrong *et al.* (1970) indicated that shock spikes are observed for protons, alpha particles, and $Z>3$ nuclei, but not for electrons (cf. Figure 20). Energetic storm particle events appear to be principally observed in proton fluxes (e.g. Bryant *et al.*, 1962; Rao *et al.*, 1967) and alpha particles (see e.g., Lanzerotti, 1969, 1972). However, the relative abundances of lower energy alpha particles and protons appear to change across the shock; only when these particles are compared as to equal energy per charge (Lanzerotti and Robbins, 1969; Lanzerotti, 1970a, 1972, 1973b) do the abundances on both sides of the shock appear equal. This effect is evident in the proton-to-alpha flux ratios published by Scholer *et al.* (1972) for the 8 March event and included here as Figure 23. Changes in the flux ratios at the time of the sudden commencement are observed for particles of equal velocity v (equal energy per nucleon) and equal rigidity P (equal energy); changes are not seen for particles of equal $P \cdot v$ (equal energy per charge).

A model for the alpha to proton ratio changes that invokes electric fields (Lanzerotti and Robbins, 1969) is probably unreasonable (Scholer *et al.*, 1972). Other models invoking some type of particle acceleration at the shock have been proposed by Scholer *et al.* (1972), Lanzerotti (1973b), and Van Allen *et al.* (1974).

Also associated with energetic storm particle events are the Forbush decreases that occur for higher energy solar and galactic cosmic ray particles at the time of, or just after, the shock or sector boundary passage (e.g., McCracken and Palmer, 1968; Vernov and Lyubimov, 1970). Figure 24, from Vernov and Lyubimov (1970), contains proton data from Venus 6 taken during the time interval of an interplanetary disturbance (identified as a sector boundary by Balogh *et al.*, 1971; see Figure 7) on 2 February 1969. The 1–5 MeV solar proton anisotropy during the energetic storm particle event indicates large streaming from the solar direction. Coincident with the end of the energetic storm particle event is a decrease in the 30 MeV proton fluxes and a small Forbush decrease in the Alert and Deep River neutron monitor data.

6. Discussion

Although the basic theoretical formulation (1) for the treatment of impulsive events now exists and is being exploited, much more work needs to be carried out in the areas of (a) simultaneously fitting, in a consistent fashion, the observed particle time-intensity profiles as well as the observed anisotropies, (b) comparing the values of the elements of the diffusion tensor obtained from such fits to local (and, hopefully, distant) values of the transport tensor as calculated from in-situ measurements of field fluctuations; and (c) using the present knowledge about energetic storm particle events and shock spikes, to model the solar particle events in terms of particle convection and possible acceleration (as well as any co-rotational effects) close to the

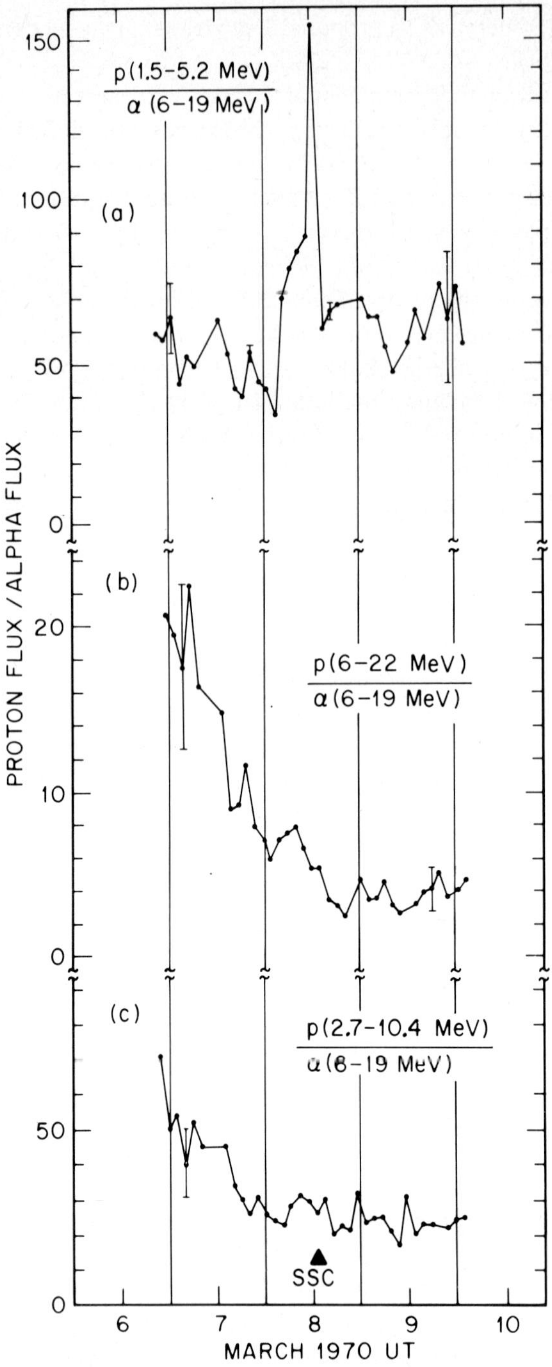

Fig. 23. Proton-to-alpha particle flux ratios measured on HEOS 1 during the period 6–9 March (Scholer *et al.*, 1972). The ratios at the top are plotted for particles of equal energy per nucleon, the ratios in the center are for particles of equal total energy, and the ratios at the bottom are for particles of equal energy per charge.

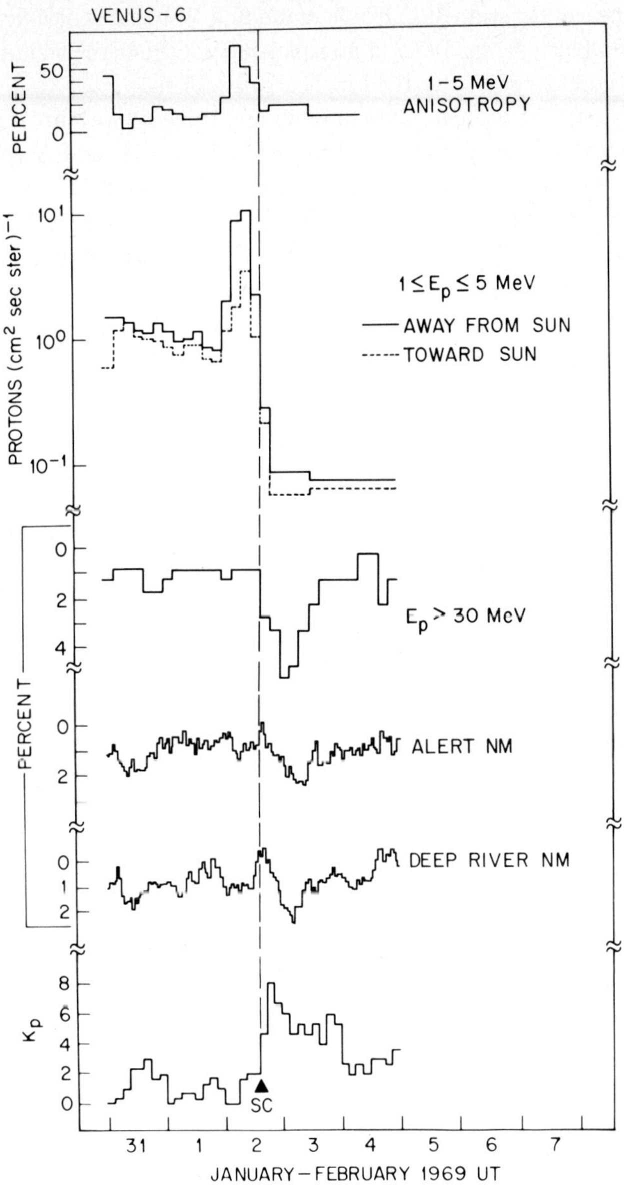

Fig. 24. Proton data (1–5 MeV, >30 MeV) measured on Venus 6 during the energetic storm particle event of 2 February 1969 (see also data in Figure 7). Also plotted are the anisotropies of the lower energy proton fluxes and neutron monitor data from Alert and Deep River (Vernov and Lyubimov, 1970).

propagating shock in order to understand the low energy particle profiles and anisotropies when a shock passes the observing location. Some modeling as mentioned in (c) has been initially attempted (Barouch and Raguideau, 1970; Palmer, 1972). Models of the interplanetary shock waves (e.g., De Young and Hundhausen, 1971, 1973) undoubtedly also must be considered when treating area (c) above. Considerations of Fermi acceleration (e.g. Hasselmann and Wibberenz, 1968; Jokipii, 1971; Wibberenz and Beuermann, 1972) in interplanetary space in the vicinity of the shock front may be necessary to solve the low-energy problem.

A major assumption in the treatment of all impulsive-type events is the invocation of an essentially delta-function (in time) injection of the particles from the flare region into interplanetary space. This assumption may well not hold for a number of events discussed in the literature in the last several years (e.g., Feit, 1973). For example, the initial phase of the 18 November 1968 event (Figure 15) appears interpretable in terms of instantaneous particle injection whereas the second phase may be solar controlled (Wibberenz *et al.*, 1973). This latter conclusion was arrived at by considering the event in terms of pure interplanetary transport after instantaneous injection and then finding that consistency of interpretation was not maintained. Such a procedure may well be valuable and necessary for the quantitative study of other events.

The observational time-intensity situation with respect to impulsive events can be pictured schematically as in Figure 25. Part A of Figure 25 describes an idealized impulsive injection event, observed at 1 AU, produced by a western-hemisphere flare where the shock wave from the flare is seen at the location of the Earth. The understanding of the propagation characteristics of this event will result from a

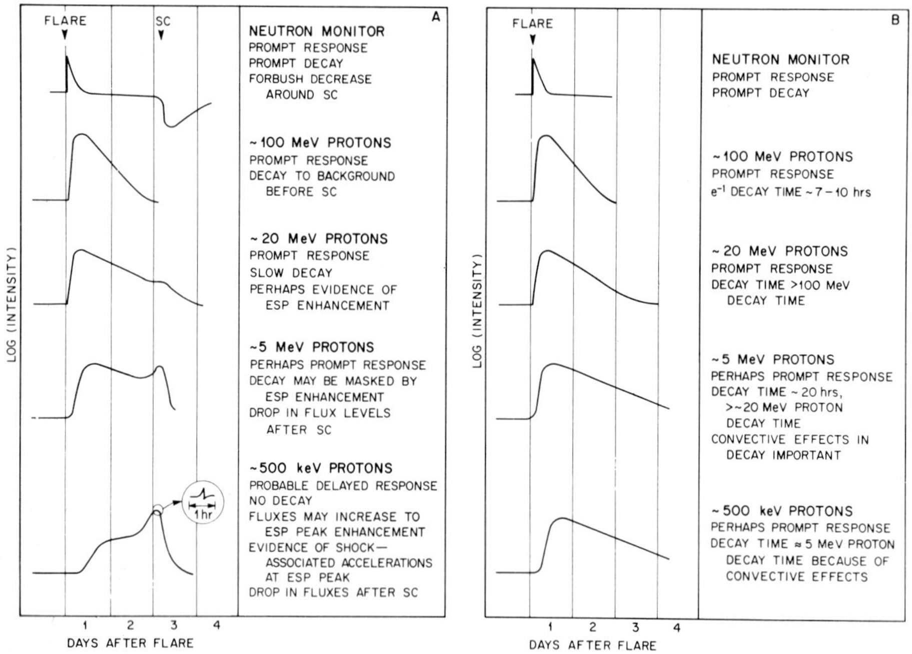

Fig. 25. A: Idealized, schematic illustration of proton fluxes measured at 1 AU following a western-hemisphere flare when the flare-produced shock is also observed at the Earth. – B: Idealized, schematic illustration of proton fluxes measured at 1 AU following a western-hemisphere flare if the flare-produced shock were not observed at the Earth.

superposition of (a) the analysis, using (1), of the idealized western-hemisphere-produced event without an accompanying shock, sketched in part B of Figure 25, together with (b) a convection-acceleration model for the energetic storm particle/shock spike event at the lower energies.

Acknowledgments

I sincerely thank a number of my colleagues for their valuable comments at various times in the past on several aspects of these observational problems: T. P. Armstrong, W. I. Axford, E. Barouch, L. A. Fisk, M. A. Forman, L. J. Gleeson, J. R. Jokipii, S. M. Krimigis, R. P. Lin, E. C. Roelof, S. Singer, D. Venkatesan and G. Wibberenz.

References

Allum, F. R., Palmeira, R. A. R., Rao, U. R., McCracken, K. G., Harries, J. R., and Palmer, I.: 1971, *Solar Phys.* **17**, 241.
Anderson, K. A.: 1969, *Solar Phys.* **6**, 111.
Anderson, K. A.: 1972, *Solar Phys.* **27**, 442.
Armstrong, T. P. and Krimigis, S. M.: 1971, *J. Geophys. Res.* **76**, 4230.
Armstrong, T. P., Krimigis, S. M., and Behanon, K. W.: 1970, *J. Geophys. Res.* **75**, 5980.
Axford, W. I.: 1965, *Planetary Space Sci.* **13**, 1301.
Axford, W. I.: 1970, 'A Review of Theoretical Work on the Effects of Solar Wind Transport on Energetic Solar Particles', USSR Academy of Sciences Seminar on Cosmic Ray Generation on the Sun, Leningrad, December.
Axford, W. I.: 1972, in E. R. Dyer (ed.), *Solar-Terrestrial Physics/1970, III*, D. Reidel Publ. Co., Dordrecht-Holland, p. 110.
Axford, W. I. and Reid, G. C.: 1963, *J. Geophys. Res.* **68**, 1793.
Balogh, A., Hedgecock, P. C., Hynds, R. J., and Sear, J.: 1971, *Solar Phys.* **20**, 150.
Barouch, E. and Raguideau, J.: 1970, *Astron. Astrophys.* **9**, 146.
Barouch, E., Gros, M., and Masse, P.: 1971, *Solar Phys.* **19**, 483.
Braddy, D., Chan, J., and Price, P. B.: 1973, *Phys. Rev. Letters* **30**, 669.
Bryant, D. A., Cline, T. L., Desai, U. D., and McDonald, F. B.: 1962, *J. Geophys. Res.* **67**, 4983.
Bukata, R. P., Rao, U. R., McCracken, K. G., and Keith, E. P.: 1972, *Solar Phys.* **26**, 229.
Burlaga, L. F.: 1967, *J. Geophys. Res.* **72**, 4449.
Childers, D. D. and Russell, C. T.: 1972, in C. P. Sonett, P. J. Coleman, and J. M. Wilcox (eds.), *Solar Wind*, NASA SP-308, 375.
Crawford, H. J., Price, P. B., and Sullivan, J. D.: 1972, *Astrophys. J.* **175**, L149.
De Young, D. S. and Hundhausen, A. J.: 1971, *J. Geophys. Res.* **76**, 2245.
De Young, D. S. and Hundhausen, A. J.: 1973, *J. Geophys. Res.* **78**, 3633.
Englade, R. C.: 1971a, *J. Geophys. Res.* **76**, 768.
Englade, R. C.: 1971b, *J. Geophys. Res.* **76**, 6190.
Fairfield, D. H.: 1969, *J. Geophys. Res.* **74**, 3541.
Fan, C. Y., Pick, M., Pyle, K. R., Simpson, J. A., and Smith, D. R.: 1968, *J. Geophys. Res.* **73**, 1555.
Feit, J.: 1973, *Solar Phys.* **29**, 211.
Fisk, L. A.: 1971, *J. Geophys. Res.* **76**, 1662.
Fisk, L. A. and Axford, W. I.: 1968, *J. Geophys. Res.* **73**, 4396.
Fisk, L. A. and Schatten, K. H.: 1972, *Solar Phys.* **23**, 204.
Forman, M. A.: 1970, *J. Geophys. Res.* **75**, 3147.
Forman, M. A.: 1971, *J. Geophys. Res.* **76**, 759.
Gleeson, L. J.: 1969, *Planetary Space Sci.* **17**, 31.
Gleeson, L. J.: 1973, in D. Venkatesan (ed.), *Proc. Solar-Terrestrial Relations Conference*, Univ. of Calgary, p. 79.
Gleeson, L. J. and Axford, W. I.: 1967, *Astrophys. J.* **149**, L115.

Gleeson, L. J. and Palmer, I. D.: 1971, *Astrophys. Space Sci.* **12**, 137.
Gleeson, L. J., Krimigis, S. M., and Axford, W. I.: 1971, *J. Geophys. Res.* **76**, 2228.
Hasselmann, K. and Wibberenz, G.: 1968, *Z. Geophys.* **34**, 353.
Jokipii, J. R.: 1966, *Astrophys. J.* **146**, 480.
Jokipii, J. R.: 1967, *Astrophys. J.* **149**, 405.
Jokipii, J. R.: 1971, *Rev. Geophys. Space Phys.* **9**, 27.
Jokipii, J. R.: 1972, *Astrophys. J.* **172**, 319.
Jokipii, J. R. and Parker, E. N.: 1967, *Planetary Space Sci.* **15**, 1375.
Jokipii, J. R. and Coleman, Jr., P. J.: 1968, *J. Geophys. Res.* **73**, 5495.
Jokipii, J. R. and Parker, E. N.: 1969, *Astrophys. J.* **155**, 777.
Jokipii, J. R. and Parker, E. N.: 1970, *Astrophys. J.* **160**, 735.
Jones, F. C., Birmingham, T. J., and Kaiser, T. B.: 1973, *Astrophys. J.* **181**, L139.
Keath, E. P., Bukata, R. P., McCracken, K. G., and Rao, U. R.: 1971, *Solar Phys.* **18**, 503.
Kinsey, J. H.: 1970, *Phys. Rev. Letters* **24**, 246.
Kirsch, E. and Münch, J. W.: 1973, 'The Transport Process of Low-Energy Solar Protons During the Decay Phase of the Events', preprint, Max-Planck-Institut für Aeronomie.
Kahler, S. W.: 1969, *Solar Phys.* **8**. 166.
Klimas, A. J. and Sandri, G.: 1971, *Astrophys. J.* **169**, 41.
Klimas, A. J. and Sandri, G.: 1973a, *Astrophys. J.* **180**, 937.
Klimas, A. J. and Sandri, G.: 1973b, *A.R.A.P. Report No. 192*, February.
Krimigis, S. M.: 1965, *J. Geophys. Res.* **70**, 2943.
Krimigis, S. M. and Armstrong, T. P.: 1973, *EOS. Trans. Am. Geophys. Union* **54**, 410.
Krimigis, S. M. and Verzariu, P.: 1971, *J. Geophys. Res.* **76**, 792.
Kunstmann, J. and Wibberenz, G.: 1973, *Proc. of 13th International Cosmic Ray Conf.*, paper 301, Denver.
Lanzerotti, L. J.: 1969, *J. Geophys. Res.* **74**, 2851.
Lanzerotti, L. J.: 1970a, *World Data Center A*, Report UAG-9, April, p. 34.
Lanzerotti, L. J.: 1970b, *Solar Phys.* **11**, 145.
Lanzerotti, L. J.: 1972, in E. J. Warman (ed.), *Proc. National Symp. Natural and Manmade Radiation in Space*, NASA TM-X-2440, 193.
Lanzerotti, L. J.: 1973a, *J. Geophys. Res.* **78**, 3942.
Lanzerotti, L. J.: 1973b, in D. Venkatesan (ed.), *Proc. Solar-Terrestrial Relations Conference*, Univ. of Calgary, p. 283.
Lanzerotti, L. J. and Robbins, M. F.: 1969, *Solar Phys.* **10**, 212.
Lanzerotti, L. J., Maclennan, C. G., and Graedel, T. E.: 1972, *Astrophys. J.* **173**, L39.
Lanzerotti, L. J., Venkatesan, D., and Wibberenz, G.: 1973, *J. Geophys.* **78**, 7986.
Lezniak, J. A. and Webber, W. R.: 1972, *Solar Phys.* **26**, 474.
Lin, R. P.: 1970, *J. Geophys. Res.* **75**, 2583.
Lin, R. P., Kahler, S. W., and Roelof, E. C.: 1968, *Solar Phys.* **4**, 338.
Lupton, J. R. and Stone, E. C.: 1973, *J. Geophys. Res.* **78**, 1007.
Manno, V. and Page, D. E. (eds.): 1970, *Intercorrelated Satellite Observations Related to Solar Events*, D. Reidel Publ. Co., Dordrecht-Holland.
McCracken, K. G. and Palmer, I.: 1968, *Proc. Astron. Soc. Australia* **1**, 145.
McCracken, K. G., Rao, U. R., and Bukata, R. P.: 1967, *J. Geophys. Res.* **72**, 4293.
McCracken, K. G., Rao, U. R., and Ness, N. F.: 1968, *J. Geophys. Res.* **73**, 4159.
McCracken, K. R., Rao, U. R., Bukata, R. P., and Keath, E. P.: 1971, *Solar Phys.* **18**, 100.
McDonald, F. B.: 1970, in V. Manno and D. E. Page (eds.), *Intercorrelated Satellite Observations Related to Solar Events*, D. Reidel Publ. Co., Dordrecht-Holland, p. 34.
McDonald, F. B. and Desai, U. D.: 1971, *J. Geophys. Res.* **76**, 808.
McKibben, R. B.: 1972, *J. Geophys. Res.* **77**, 3957.
McIntosh, P. S.: 1971, *World Data Center A*, Report UAG-12, p. 10.
Mogro-Campero, A. and Simpson, J. A.: 1972a, *Astrophys. J.* **171**, L5.
Mogro-Campero, A. and Simpson, J. A.: 1972b, *Astrophys. J.* **177**, L37.
Murray, S. S., Stone, E. C., and Vogt, R. E.: 1971, *Phys. Rev. Letters* **26**, 663.
Ng, C. K.: 1972, 'Propagation of Solar-Flare Cosmic Rays', Thesis, Monash University.
Ng. C. K. and Gleeson, L. J.: 1971, *Solar Phys.* **20**, 166.
O'Gallagher, J. J.: 1970, *J. Geophys. Res.* **75**, 1163.

O'Gallagher, J. J. and Simpson, J. A.: 1966, *Phys. Rev. Letters* **16**, 1212.

Ogilvie, K. W. and Arens, J. F.: 1971, *J. Geophys. Res.* **76**, 13.

Palmer, I.: 1972, *Solar Phys.* **27**, 466.

Palmer, I. D.: 1973, *Solar Phys.* (in press).

Palmeira, R. A. R., Allum, F. R., and Rao, U. R.: 1971, *Solar Phys.* **21**, 204.

Parker, E. N.: 1963, *Interplanetary Dynamical Processes*, Interscience Publishers, New York.

Parker, E. N.: 1965, *Planetary Space Sci.* **13**, 9.

Parker, E. N.: 1966, *Planetary Space Sci.* **14**, 371.

Price, P. B., Hutcheon, I., Cowisk, R., and Barber, D. J.: 1971, *Phys. Rev. Letters* **26**, 916.

Quenby, J. J. and Sear, J. F.: 1971, *Planetary Space Sci.* **19**, 95.

Rao, U. R., McCracken, K. G., and Bukata, R. P.: 1967, *J. Geophys. Res.* **72**, 4325.

Rao, U. R., McCracken, K. G., and Bukata, R. P.: 1968, *Can. J. Phys.* **46**, S844.

Rao, U. R., McCracken, K. G., Allum, F. R., Palmeira, R. A. R., Bartley, W. C., and Palmer, I.: 1971, *Solar Phys.* **19**, 209.

Reid, G. C.: 1964, *J. Geophys. Res.* **69**, 2659.

Roelof, E. C.: 1966, 'Statistical Theory of Charged Particle Transport in Disordered Magnetic Fields', Thesis, Univ. Calif., Berkeley.

Roelof, E. C.: 1968, *Can. J. Phys.* **46**, S990.

Roelof, E. C.: 1973a, in D. Venkatesan (ed.), *Proc. Solar-Terrestrial Relations Conf.*, Univ. of Calgary, p. 341.

Roelof, E. C.: 1973b, in D. Venkatesan (ed.), *Proc. Solar-Terrestrial Relations Conf.*, Univ. of Calgary, p. 411.

Roelof, E. C. and Krimigis, S. M.: 1973, *J. Geophys. Res.* **78**, 5375.

Sakurai, K.: 1971, *World Data Center A*, Report UAG-12, p. 52.

Sari, J. W.: 1972, 'Modulation of Low-Energy Cosmic Rays', Thesis, Univ. of Maryland, NASA X-692-72-309, August.

Sari, J. W. and Ness, N. F.: 1970, *Acta. Phys. Acad. Sci. Hungaricae* **29**, 373.

Scholer, M., Hovestadt, D., and Häusler, B.: 1972, *Solar Phys.* **24**, 475.

Simnett, G. M.: 1971, *Solar Phys.* **20**, 448.

Simnett, G. M.: 1972, 'Relativistic Solar Electrons in Space', XV Plenary Meeting of COSPAR, Madrid, May.

Simnett, G. M. and Holt, S. S.: 1971, *Solar Phys.* **16**, 208.

Singer, S.: 1970, in V. Manno and D. E. Page (eds.), *Intercorrelated Satellite Observations Related to Solar Events*, D. Reidel Publ. Co., Dordrecht-Holland, p. 571.

Singer, S. and Montgomery, M. D.: 1971, *J. Geophys. Res.* **76**, 6628.

Smerd, S. F.: 1970, *Proc. Astron. Soc. Australia* **1**, 305.

Teegarden, B. J., von Rosenvinge, T. T., and McDonald, F. B.: 1973, *Astrophys. J.* **180**, 571.

Van Allen, J. A., Venkatarangan, P., and Venkatesan, D.: 1974, *J. Geophys. Res.* **79**, 1.

Vernov, S. N. and Lyubimov, G. P.: 1970, 'Low-Energy Cosmic Rays in Interplanetary Space', Moscow State University report.

Verzariu, P. and Krimigis, S. M.: 1972, *J. Geophys. Res.* **77**, 3985.

Wang, J. R.: 1972, 'The Coronal Transport of the Flare-Associated Scatter-Free Electrons', NASA X-661-72-76, March.

Wang, J. R., Fisk, L. A., and Lin, R. P.: 1971, in *Proc. 12th Int. Cosmic Ray Conference*, p. 438.

Webb, S. and Quenby, J. J.: 1973, *Planetary Space Sci.* **21**, 23.

Webb, S., Balogh, A., Quenby, J. J., and Sear, J. F.: 1973, *Solar Phys.* **29**, 477.

Wibberenz, G.: 1973, in H. Pilkuhn (ed.), *Summer School on Space Physics*, vol. II, Black Forest (to be published).

Wibberenz, G. and Beuermann, K. P.: 1972, in K. Schindler (ed.), *Cosmic Plasma Physics*, Plenum Publ. Co., New York, p. 339.

Wibberenz, G., Hasselmann, K., and Hasselmann, D.: 1970, *Acta. Phys. Acad. Sci. Hungaricae* **29**, 37.

Wibberenz, G., Lanzerotti, L. J., and Venkatesan, D.: 1973, *Proc. 13th Intern. Cosmic Ray Conf.*, paper 148, Denver.

THE VALUE OF MULTI-SPACECRAFT MEASUREMENTS

M. A. SHEA and D. F. SMART

Air Force Cambridge Research Laboratories, Bedford, Mass. 01730, U.S.A.

Abstract. The utilization of multi-spacecraft and multi-detector measurements has long been recognized as a valuable tool in unraveling the intricacies and complexities of solar particle propagation through the interplanetary medium and in the magnetosphere. To provide as complete a solar particle picture as possible for various analyses, a catalog of solar particle events during the first part of the 20th solar cycle is being prepared. This requires examining the data from detectors on various satellites and space probes and identifying the onset times for various solar particle energies, the times of maximum intensities and the fluxes. Examples of problems encountered in this type of compilation, some of which can be resolved by multi-spacecraft detection are:

(1) The resolution of multiple particle injection from successive flares in a short time period or on different regions of the Sun;

(2) The possibility of sensor contamination by significant fluxes of higher energy particles or different particle species, and

(3) The identification and separation of enhanced low energy intensities associated with active region transits across the solar disk from discrete low energy solar-flare associated particle increases.

1. Introduction

During the early years of satellite experiments each significant particle event was enthusiastically reported in the literature with the identification of onset times, spectra, maximum intensity, association with solar flare and a few deductions on particle propagation in the interplanetary medium and/or the magnetosphere. As more satellites were flown there were significant technological advances and sophistication in satellite experiments such that considerably more detail could be obtained about each particle event. The number of satellites and experiments increased with the increasing solar cycle and experimenters, in analyzing their data, rapidly became aware that detailed analyses of several detectors preferably on different spacecraft were essential in unraveling the intricacies and complexities of solar particle propagation through the interplanetary medium and in the magnetosphere.

Nevertheless, scientists still continued, for the most part, to report the large, significant and interesting events of the 20th solar cycle, with most of the events discussed in the literature being those detected by relatively high energy detectors; very little effort was expended into the analyses of the relatively smaller and minor solar particle events. Although some experimenters did attempt to identify particle events detected by their instrumentation during specific periods of time, a systematic comprehensive survey of solar particle events has not been undertaken. The reasons for this deficiency are perhaps the following: (1) such a survey is not glamorous compared to the fascinating analyses of specific events, and (2) it requires a significant amount of time and effort.

For the past three years a systematic compilation of solar particle events has been undertaken by Working Group 2 of the Inter-Union Commission on Solar-Terres-

D. E. Page (ed.), Correlated Interplanetary and Magnetospheric Observations, 381–396. *All Rights Reserved.*

trial Physics. When completed this survey will include the identification of proton and electron events during the 19th and 20th solar cycles, the flare association whenever possible, the associated X-ray and radio emissions and the optical and magnetic characteristics of the identified solar active region.

It is not our intent to discuss this catalog in detail, but to confine ourselves to the problems associated with the identification of the particle events detected by satellite instrumentation during the 20th solar cycle. Initially it was believed that the identification of these events would be a relatively simple matter of a survey of published literature supplemented, if necessary, by inspection of original data records. We anticipated that perhaps approximately 200 discrete events would be identified between 1965–1969.

Starting with the published lists of Allum *et al.* (1971), Bryant *et al.* (1965), Krimigis and Van Allen (1966), Lin and Anderson (1967), Lin (1970) and McCracken *et al.* (1969), as well as publications discussing individual events, we compiled a list of events which included whenever possible onset times, times of maximum intensities, maximum fluxes and duration of event. The detection of these events at different energies and on different spacecraft were all included as separate entries. In examining the initial listings it became apparent that (1) a uniform, comprehensive, systematic survey could not be obtained from the literature, (2) many of the desired parameters could not be ascertained from the literature, and (3) inconsistencies in the literature necessitated inspection of the original satellite data records whenever possible. With the aid of the satellite data in the U.S. Space Science Data Center and the welcomed assistance of many of the experimenters, we proceeded to examine microfilm and computer records from several spacecraft – principally the IMP and Pioneer series. To the lists compiled from these records, we added the published and privately provided data from the ESRO and Soviet spacecraft. At the present time over 500 separate particle enhancements have been identified between 1965–1969 as detailed in Table I.

TABLE I

Number of particle events identified during 1965–1969

Year	Number of events
1965	43
1966	100
1967	162
1968	120
1969	112

The large well documented events presented us with few problems; however, as events containing primarily low energy particles were included, several annoying and persistent problems were all too frequently encountered. In this paper we will present and discuss some of the most annoying problems, most of which fortunately can be resolved by multi-sensor or multi-spacecraft measurements.

2. Sensor Contamination

One of the most frequent problems encountered in the identification of particle events in the 20th solar cycle is the contamination of proton sensors by the presence of relativistic electrons. This problem is not readily apparent in the available data from the 19th solar cycle because the differences between solar electrons and solar protons could not be distinguished by the relatively insensitive instruments by contemporary standards used at that time. In scanning various proton energy channels, it was found necessary to refer to the published electron data to ascertain whether the apparent proton events were real and, if so, what onset times could be selected from the available data.

An example of relativistic electrons being detected by a proton sensor is shown in Figure 1 where the response of two comparable energy channels on IMP 4 is given

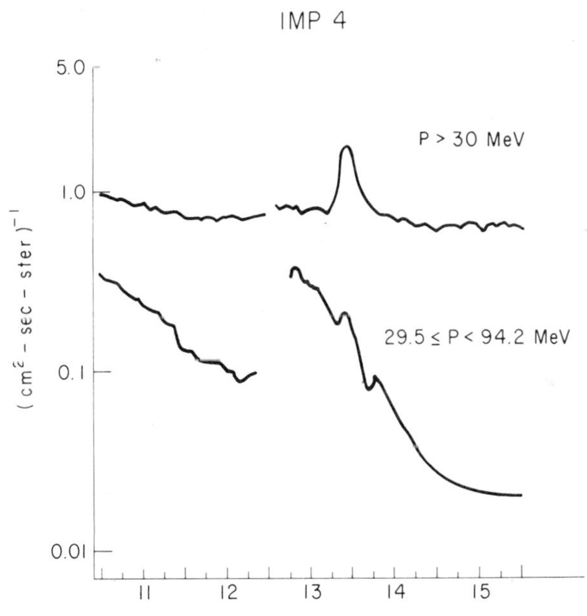

Fig. 1. An example of relativistic electrons detected by the $P > 30$ MeV sensor on IMP 4.

for 11–14 July 1968. Inspection of the >30 MeV Solar Proton Monitoring Experiment (SPME) data would indicate an apparent enhancement on 13 July; inspection of the 29.5–94.2 MeV University of Chicago (UC) data would indicate only a decay from a previous event. Since the background intensity of the UC detector is much lower than that of the SPME detector, this apparent event, if real, should have been detected by both experiments.

To resolve this type of discrepancy it was necessary to consider the possible presence of electrons and also to consider the configuration of the actual sensor. Simnett (1970) has discussed this event in detail describing a sudden increase in the

0.3–0.9 MeV electron flux at about 16 UT on 13 July. In this event the electron flux had a three-fold increase within an eight minute interval that continued until a maximum flux of 150 electrons $cm^{-2} s^{-1} sr^{-1}$ occurred at 22 UT. Relativistic electrons (12–45 MeV) were reported by Datlowe (1971) with an onset of ~18 UT, maximum at 22 UT and maximum flux of 0.02 electron $cm^{-2} s^{-1} sr^{-1}$.

In considering the response of the actual instruments we find that the UC 29.5–94.2 MeV channel requires a triple coincidence (200 keV energy deposition in the first two elements and 15 MeV energy deposition in a third detector element) and further requires the guard counter surrounding the sides of the detector assembly (having a 200 keV threshold) not be actuated. In contrast, the >30 MeV channel of the SPEM will respond to any event capable of penetrating the detector shielding of 16 mm of copper and depositing more than 150 keV in the thick detector sensitive volume. When considering the flux of relativistic electrons present at this time and the relative response of the detectors to electron contamination, it becomes apparent that almost the entire increase on the >30 MeV SPME channel is due to electron contamination, and that this event does not contain a significant flux of 30 MeV protons.

Another example of electron contamination is shown in Figure 2 where the 0.8–9.6 MeV proton energy channel of the UC experiment on IMP 4 has an apparent

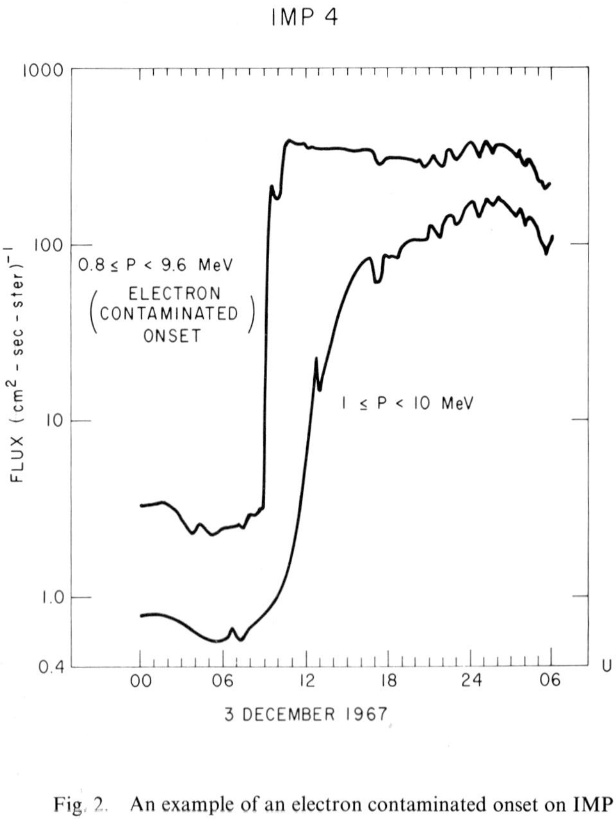

Fig. 2. An example of an electron contaminated onset on IMP 4.

onset time of 0920 UT on 3 December 1967 whereas the 1–10 MeV SPME channel has an onset time of approximately 1030 UT. The onsets of the electrons at a number of energies have been reported by Lin (1970) with onset at 0918 UT $\pm$02 min for >40 keV electrons, by Allum *et al.* (1971) with onset at 0910 UT for >70 keV electrons, and by Simnett (1972) with onset at 0920 UT for electrons in the interval between 0.3 to 0.9 MeV and also for electrons >2.7 MeV. Since the UC 0.8–9.6 MeV channel on IMP 4 is sensitive to electrons >200 keV (Hsieh, private communication), the onset recorded by this channel could be due to electron contamination.

The 1–10 MeV SPME channel is reported to be relatively insensitive to electrons (Bostrom *et al.*, 1972). The instrument design has a very thin sensitive region in the detector, consequently the probability that an electron can deposit the 200 keV of energy required to trigger the counter within the sensor active volume is small. The fact that this sensor does not indicate an enhanced 1–10 MeV proton flux until about 1030 UT is convincing evidence that the 0920 onset on the UC 0.8–9.6 MeV channel is due to electron contamination.

3. Sensitivity Levels

An example of sensitivity levels is illustrated in Figure 3. In this case the 29.5–94.2 MeV UC proton detector on IMP 4 records a factor of three increase between 16–18 UT on 27 October 1968; however, this event is not discernible on the >30 MeV SPME detector on the same satellite. This detector has a much higher background

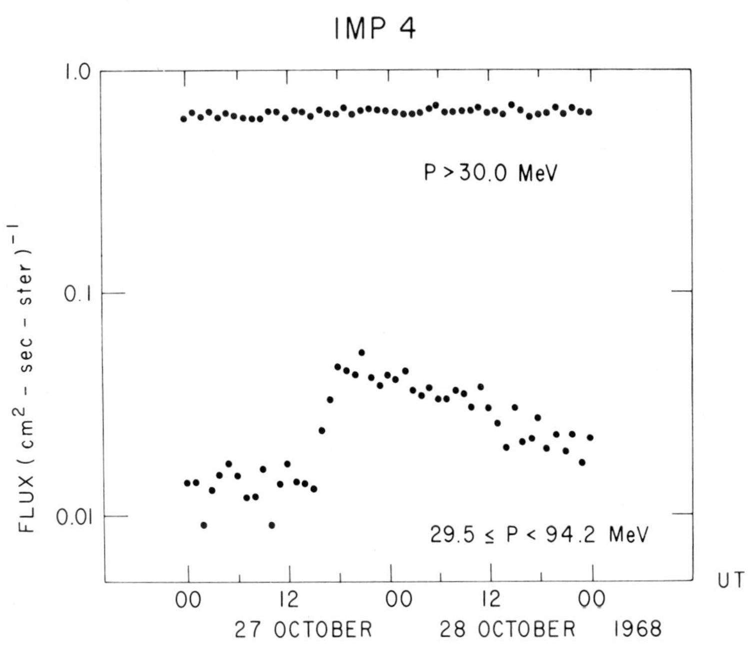

Fig. 3. An example of relative sensitivity levels of two sensors on IMP 4. The background of the $P > 30$ MeV detector is about 0.7 particle cm^{-2} s^{-1} sr^{-1}

level and consequently small fluxes of 30 MeV protons cannot be recognized until there is sufficient flux to increase above the cosmic-ray background.

This example indicates the difficulty of trying to place a limit on the maximum energy of particles associated with a specific enhancement. An observer utilizing data from the >30 MeV and the >10 MeV channels of the SPME could conclude that although this event did produce a proton flux of 0.15 particle $cm^{-2}\ s^{-1}\ sr^{-1}$ >10 MeV there were no particles >30 MeV associated with this event. Inspection of the comparable University of Chicago data illustrates that a relatively small flux of >30 MeV protons was associated with this event. In this case the proper nomenclature with respect to the possibility of protons being detected on the >30 MeV SPME detector would have been to state that it recorded no measurable proton flux in excess of a background of 0.7 particle $cm^{-2}\ s^{-1}\ sr^{-1}$.

Additional discussion, giving an example of circumstances where an event is observed on the Pioneer-6 low energy channel but not discernible on the IMP-3 low energy channel because there is not a sufficient particle flux to exceed the IMP-3 background, is contained in Section 5.

4. Perigee Interference

Another problem which often requires multi-sensor information from an Earth-orbiting spacecraft or multi-spacecraft analysis for resolution is the determination of the onset time of a solar particle enhancement when the spacecraft is undergoing a perigee passage and a sensor is contaminated by the magnetospheric energetic particle population. Figure 4 illustrates this problem where there is an obvious increase on four independent sensors after the completion of a perigee passage of IMP 4 on 1 August 1967. This event is during a period when a number of solar particle outbursts occurred and early on 1 August the low energy (10 MeV) particle population is enhanced above the quiet time background. In this case the UC detectors for the 0.8–9.6 MeV and the 9.6–18.8 MeV energy range show an apparent increase at approximately 10 UT on 1 August. There is obvious perigee contamination on all UC channels after 15 UT when the spacecraft altitude is less than 10 R_E. When the spacecraft completes perigee passage and is again at a distance greater than 10 R_E, a factor of two enhancement in the 29.5–94.2 MeV channel and correspondingly higher flux values on the lower energy UC channels are noted. Inspection of the data from the two low energy UC channels alone would lead to the temptation to ascribe an onset time of approximately 10 UT because the spacecraft altitude is greater than 10 R_E when the initial apparent particle increase appears. However, inspection of possible candidate solar flares indicates no prominent solar activity before 10 UT and the most likely flare to produce particles occurs after this time at 1721 UT.

An examination of the SPME 1–10 MeV channel data (relatively insensitive to electrons), after elimination of the most obvious trapped particle contamination when the satellite is less than 4 R_E, indicates an apparent onset at approximately 19 UT. A completely independent spacecraft, Venus 4, located at 0.93 AU with an Earth-Sun

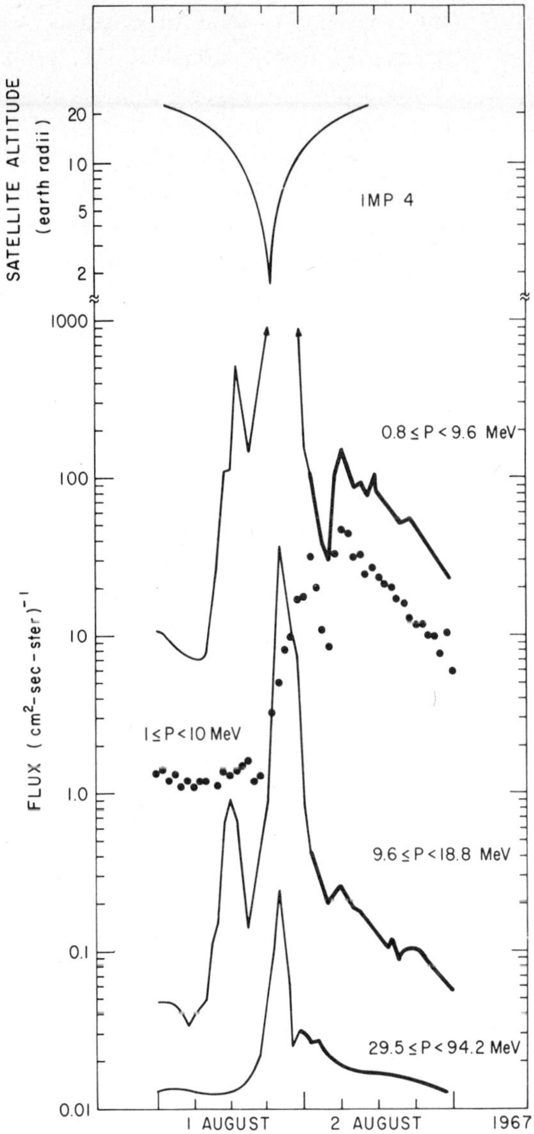

Fig. 4. An example of the response of several detectors to a particle event which occurs during a perigee passage of IMP 4.

probe (ESP) angle of -1 degree, measures an increase in the 1–5 MeV energy range at about 19 H (Vernov *et al.*, 1970) confirming the 1–10 MeV onset of IMP 4. In addition Lin (1970) identifies a prompt electron event at 1734 UT $\pm$ 02 min and a prompt proton event with energy >15 MeV at 1743 UT.

While the flare at 1721 UT satisfactorily accounts for the major proton increase after this time, the source of the 10 UT increase has not been identified. Graedel and Lanzerotti (1971) confirm this earlier event by reporting a 0.56–0.6 MeV proton onset at about 11 UT. Our evaluation of these various data is that there are two independent proton increases, the first at about 10 UT, probably a convective structure that has only an accidental coincidence with the satellite crossing inside the magnetosheath, the second a flare initiated particle event where the energetic protons arrived promptly and the lower energy protons arrived later.

Additional confusion occurs at 0515 UT ± 15 min on 2 August when an impulsive increase superimposed on the decay of the previous event is observable on the two low energy channels of the UC experiment and the SPME 1–10 MeV channel. This occurs after the satellite has passed beyond the magnetosheath boundary on its outbound orbit and is probably a flare initiated event since it is in approximate time coincidence with small solar flares on the Western Hemisphere of the sun (Dodson and Hedeman, private communication, 1973).

5. Multiple Events

The majority of solar particle experimentalists welcome the occurrence of a significant and interesting event clearly associated with a specific solar flare. However, it is well known that an active proton producing region on the sun may produce a multiplicity of solar proton events. When these events occur over a short period of time, it becomes a case of 'too much of a good thing' and the utilization of several detectors and spacecraft may be required to answer the relatively simple question of 'how many discrete events occurred?' While the classic events lend themselves to relatively easy analyses, these multiple events with the associated problems of sensor contamination, sensitivity levels and saturation, can become an 'interesting time' for analysts. In this section we will discuss two periods where the sun was most generous in producing particle events and the problems associated with the identification of each event. An event (recognizable discrete increase or significant change in slope of an increase) may have a number of possible sources (flare initiation, geomagnetic storm modulation, or convective enhancement) and, in general, we do not attempt to identify the source of each 'event' we have identified in this paper.

5.1. The period 17–27 March 1966

Figure 5 illustrates the proton flux between 17–27 March 1966 as observed on the UC 0.9–190 MeV detector on IMP 3 and the 0.6–13.9 MeV and the 13.9–73.2 MeV channels on Pioneer 6. The Pioneer 6 low energy measurements of this period were discussed briefly by Fan *et al.* (1968). The position of the Pioneer 6 spacecraft during this time is about 0.9 AU with an ESP angle of 4 degrees. Table II lists the individual proton events detected on these three sensors and Table III lists the electron events for >40 keV electrons (Lin and Anderson, 1967).

The first enhancement (event number 1 in Table II) of this period is observed by the low energy Pioneer 6 detector at 13 UT on 17 March with an additional increase (decided change in slope) at about 20 UT on 18 March. There is no apparent increase on the IMP 3 detector until ~0.6 UT on 19 March when a small increase is ob-

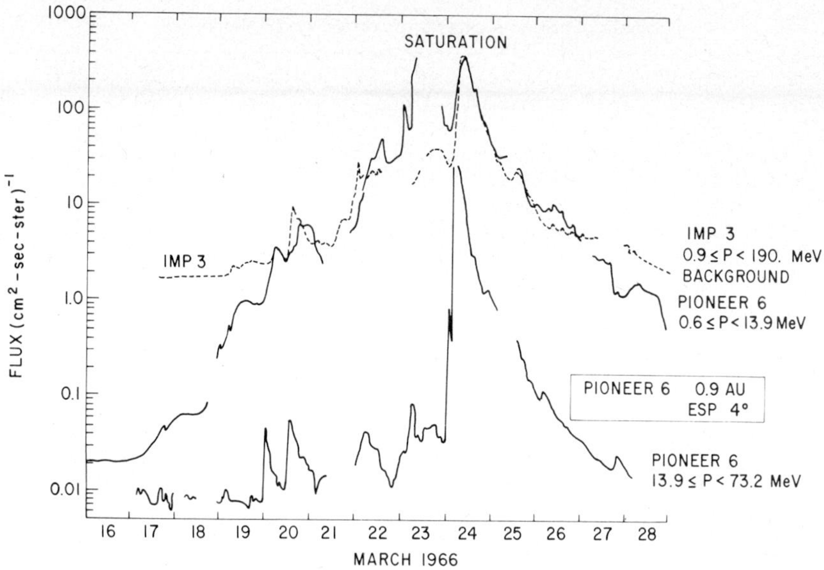

Fig. 5. The proton flux as detected by one sensor on IMP 3 and two sensors on Pioneer 6 for the period 17–27 March 1966.

served above the background of 1.6 particles $cm^{-2} s^{-1} sr^{-1}$. The next identifiable increase (event number 4 in Table II) occurs around midnight on 19 March and is observed on both spacecraft. Time association with an electron event reported by Lin and Anderson (1967) at 2223 UT$\pm$2 min would imply that both electrons and protons were released by the source flare on the Sun.

The next event (event number 5 in Table II) occurring at about noon on 20 March is indicative of another flare accelerating both electrons and protons. Inspection of the individual response between the low energy IMP and low energy Pioneer detectors illustrates the problem of electron contamination on the low energy IMP channel. The difference in the time-intensity profile between the low energy IMP 3 and Pioneer 6 detectors coupled with the essentially simultaneous onsets of the electrons (Lin and Anderson, 1967) and the low energy IMP 3 channel (Simpson, 1972a) strongly implies electron contamination. Furthermore, the maximum electron flux >40 keV reported by Lin and Anderson of 500 particles $cm^{-2} s^{-1} sr^{-1}$ extrapolated to the 150 keV energy deposition threshold of the 0.9–190 MeV detector on IMP 3, multiplied by the detection efficiency of this sensor to electrons, roughly corresponds to the observed counting rate.

The next increase observable on IMP 3 at ~12 UT on 21 March occurs in a period when there was no tracking of the Pioneer 6 spacecraft. Using the criteria that every identifiable increase (significant change in slope) is an 'event', then an apparent event is observed on both the Pioneer 6 and IMP 3 spacecraft at ~20 UT on 21 March (event number 7 in Table II). During this increase there is another change in slope around 23 UT observable on both IMP 3 and Pioneer 6. We note that the 'spike' at

TABLE II

Details of proton events for the period 17–27 March 1966

Number	Date and time of onset (UT)	Date and time of maximum flux (UT)	Maximum flux $(cm^2\ s\ sr)^{-1}$	Energy range (MeV)	Satellite
1	17 March 13^{h}	(Start of enhancement)	—	0.6– 13.9	Pioneer 6
2	18 March 20^{h}	19 March 12^{h}	1.03	0.6– 13.9	Pioneer 6
3	19 March 06^{h}	19 March 15^{h}	1.2	0.9–190.0	IMP 3
4	20 March 00^{h}	20 March 07^{h}	2.0	0.9–190.0	IMP 3
	19 March 23^{h} 26^{m} ± 5	20 March 00^{h} 13^{m}	0.026	13.9– 73.2	Pioneer 6
	20 March 00^{h} 03^{m} ± 5	20 March 06^{h} 08^{m}	3.45	0.6– 13.9	Pioneer 6
5	20 March 10^{h} 45^{m} ± 15	20 March 12^{h}	10.63	0.9–190.0	IMP 3
	20 March 11^{h} 53^{m} ± 10	20 March 12^{h} 28^{m}	0.03	13.9– 73.2	Pioneer 6
	20 March 13^{h} 28^{m} ± 20	20 March 16^{h} 48^{m}	6.2	0.6– 13.9	Pioneer 6
6	21 March 12^{h}	21 March 16^{h}	4.0	0.9–190.0	IMP 3
7	21 March 19^{h} 30^{m} ± 30	21 March 22^{h}	11.25	0.9–190.0	IMP 3
	21 March 20^{h}	21 March 21^{h}	0.017	13.9– 73.2	Pioneer 6
	21 March 20^{h}	22 March 00^{h}	10.3	0.6– 13.9	Pioneer 6
8	21 March 23^{h} 00^{m} ± 20	22 March 02^{h}	27.5	0.9–190.0	IMP 3
	22 March 00^{h} 23^{m} ± 15	22 March 03^{h} 30^{m}	0.03	13.9– 73.2	Pioneer 6
9	23 March 05^{h}	23 March 18^{h}	48.25	0.9–190.0	IMP 3
	22 March 22^{h}	23 March 07^{h}	0.05	13.9– 73.2	Pioneer 6
[a]	22 March 23^{h}	23 March 00^{h}	112	0.6– 13.9	Pioneer 6
	23 March 05^{h}	23 March 07^{h}	345	0.6– 13.9	Pioneer 6
10	24 March 00^{h}	24 March 01^{h}	26.2	0.9–190.0	IMP 3
	24 March 00^{h} 18^{m} ± 5	24 March 01^{h} 06^{m}	0.6	13.9– 73.2	Pioneer 6
11	24 March 02^{h} 45^{m} ± 15	24 March 06^{h}	>373 (Saturated)	0.9–190.0	IMP 3
	24 March 03^{h} 01^{m} ± 5	24 March 03^{h} 44^{m}	43.0	13.9– 73.2	Pioneer 6
	24 March 04^{h} 58^{m} ± 2	24 March 06^{h} 08^{m}	379.0	0.6– 13.9	Pioneer 6
12	25 March 10^{h}	25 March 11^{h} 30^{m}	20.0	0.9–190.0	IMP 3
13	27 March 17^{h}	27 March 20^{h}	0.017	13.9– 73.2	Pioneer 6

[a] These two entries were included to illustrate coincidence between the increases observed on the IMP 3 and Pioneer satellites.

Data sources: – IMP 3: Simpson (1972a).
Pioneer 6: Fan *et al.* (1967) and Simpson (1972b).

TABLE III

Details of electron events ($E > 40$ keV) for the period 17–27 March 1966 as observed by IMP 3

Number	Date and time of onset (UT)	Date and time of maximum flux (UT)	Maximum flux $(cm^2\ s\ sr)^{-1}$
1	19 March 01^{h} 04^{m} ± 4	19 March 01^{h} 30^{m}	12
2	19 March 14^{h} 51^{m} ± 2	19 March 15^{h} 20^{m}	100
3[a]	19 March 22^{h} 23^{m} ± 2	19 March 22^{h} 40^{m}	35
4[a]	20 March $<12^{h}$ 24^{m}	(undetermined)	500
5[a]	22 March $<02^{h}$ 20^{m}	(undetermined)	200
6[a]	23 March 12^{h}	(undetermined)	500
7[a]	24 March $<02^{h}$ 41^{m}	24 March 03^{h} 00^{m}	5700
8	24 March 19^{h} 30^{m} ± 2	24 March 19^{h} 35^{m}	640
9	26 March 19^{h} 16^{m} ± 1	26 March 19^{h} 30^{m}	280

[a] These events are believed associated with proton events.

Data source: Lin and Anderson (1967).

~02 UT on 22 March observable on the IMP 3 low energy channel corresponds to an electron event (Lin and Anderson, 1967).

On 22 March another apparent increase is observed initially on Pioneer 6 (event number 9 in Table II). The IMP-3 satellite is undergoing perigee passage at this time and so cannot contribute to the identification of a unique proton event onset. The interesting structure observed after midnight does not appear to be directly correlated with any unique solar circumstance (Dodson and Hedeman, private communication, 1972). After completion of perigee passage, when IMP 3 is again outside the magnetosphere boundary, the low energy particle increase is observed on IMP 3.

The next major proton increase, observed on the 13.9–73.2 MeV proton channel on Pioneer 6 at 0018 UT $\pm$5 min on 24 March (Fan *et al.*, 1967) is attributed to a flare at 2248 UT (23 March) by Dodson and Hedeman (private communication, 1972). Lin and Anderson (1967) specifically state the absence of >40 keV electrons with a flux greater than 10 particles cm^{-2} s^{-1} sr^{-1} (the lower threshold of the electron detector at this time) which strongly argues against an electron contaminated onset on this channel. The next increase (event number 11 in Table II), containing a significant flux of protons greater than 13.9 MeV, has an 03 UT onset and is observed on both the Pioneer 6 and IMP 3 spacecraft. This major event, from McMath Plage 8207, contains a large flux (5200) of >40 keV electrons (Lin and Anderson, 1967) and also a flux of 0.01 particle cm^{-2} s^{-1} sr^{-1} of >3.0 MeV electrons (Cline, 1968). The presence of high energy protons is measured on both the IMP 3 and Pioneer 6 spacecraft with the highest energy channel on Pioneer 6 (>175 MeV) recording a flux of approximately 0.3 particle cm^{-2} s^{-1} sr^{-1}, (Simpson, 1972b). This dual injection of protons, best discernible from the 13.0–73.2 MeV data from Pioneer 6, can be associated with two independent solar events (Dodson and Hedeman, private communication, 1972).

A significant electron event of 640 particles cm^{-2} s^{-1} sr^{-1} above 40 keV is reported by Lin and Anderson (1967) at 1930 UT $\pm$2 min on 24 March. There is no convincing evidence of a proton increase at this time due to the fact that the flux being measured in the various proton sensors is at a relatively high level and undergoing rapid decay. The modulation of the decay observable on IMP 3 on 25 March occurs during the time when Pioneer 6 was not being tracked. It is not possible to determine from the data presented in Figure 4 whether this modulation is structure on the decay of the large particle event on 24 March or perhaps another discrete proton injection; however, additional information is provided by the fact that when the Pioneer 6 data coverage resumed at ~15 UT, McCracken (1969) indicates a higher flux on both his 44.0–77.1 MeV and his 7.4–44.0 MeV channels on Pioneer 6.

The magnetic storm modulation is readily apparent after 25 March. The apparent increase on 27 March at 17 UT on the 13.9–73.2 MeV channel on Pioneer 6 corresponds roughly to a sudden decrease on the Pioneer 6 low energy channel. It is tempting to ascribe this to a sudden commencement effect attributed to a geomagnetic storm beginning at ~19 UT as reported by several stations (Lincoln, 1966). Again, at this particular time IMP 3 was undergoing perigee passage and the low energy IMP 3 data cannot be used to confirm the observations of the Pioneer 6 low energy channel.

5.2. The Period 22 October–9 November 1968

The period late in October to early November 1968 shown in Figure 6 constitutes another time of multiple injections principally from solar phenomena associated

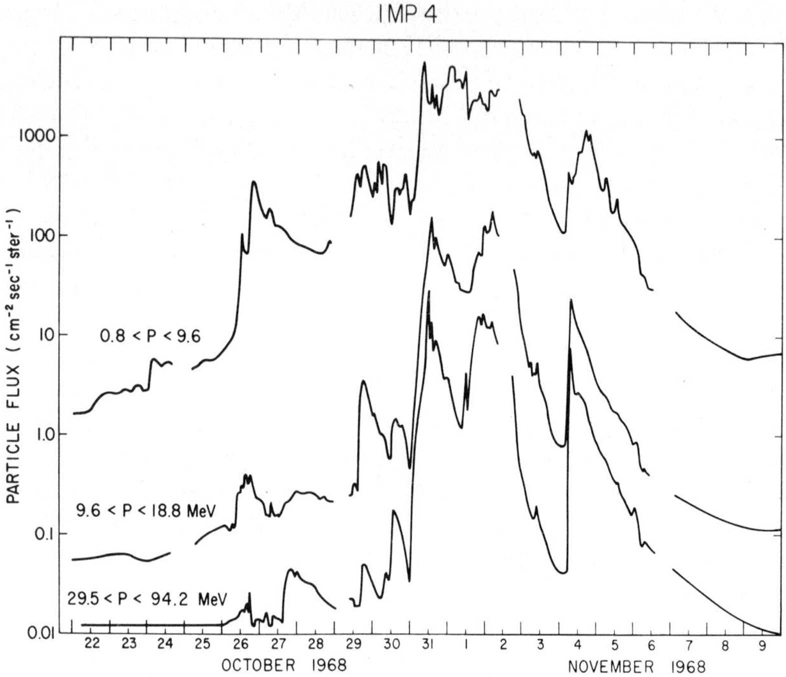

Fig. 6. The proton flux as observed by three detectors on IMP 4 during the period 22 October to 9 November 1968.

with McMath Plage 9740. The phenomena that occurred during this time have been the subject of a special report (UAG-8, 1970). The events we have identified are detailed in Table IV. Starting on 17 October there commences a long, slow gradual enhancement (not shown in Figure 6 or listed in Table IV) which is observable on the 0.8–9.6 MeV and the 9.6–18.8 MeV channels on IMP 4. The particle flux gradually increases until an apparent small impulsive increase is observable early on the 24th on the low energy channel with a corresponding change in the rate of increase on the 9.6–18.8 MeV channel.

The next apparent discrete increase (event number 2 in Table IV) occurs on 26 October and is discernible on all three channels. Since the low energy proton channel is quite susceptible to contamination by electrons >200 keV and the 9.6–18.8 MeV channel has some sensitivity to relativistic electrons, we will confine further event identification in this period to those observed on the 29.5 94.2 MeV channel. The

TABLE IV

Details of proton events for the period 23 October–6 November 1968 as observed on three energy channels on IMP 4

Number	Date and time of onset (UT)	Date and time of maximum flux (UT)	Maximum flux $(cm^2 \text{ s sr})^{-1}$	Energy range (MeV)
1	24 Oct. 10^h	26 Oct. 06^h	1.25	9.6–18.8
	24 Oct. 02^h	26 Oct. 02^h	80	0.8– 9.6
2	26 Oct. 08^h	26 Oct. 16^h	0.006	29.5–94.2
	26 Oct. $09^h\,30^m \pm 30$	26 Oct. 15^h	0.40	9.6–18.8
	26 Oct. 09^h	26 Oct. 20^h	359	0.8– 9.6
3	27 Oct. 16^h	27 Oct. 21^h	0.04	29.5–94.2
	27 Oct. 14^h	28 Oct. 02^h	0.25	9.6–18.8
4	29 Oct. $15^h\,30^m$	29 Oct. 19^h	0.03	29.5–94.2
	29 Oct. $16^h\,00^m \pm 30$	29 Oct. $18^h\,30^m$	3.6	9.6–18.8
	29 Oct. 12^h	30 Oct. 04^h	602	0.8– 9.6
5	30 Oct. $12^h\,30^m$	30 Oct. 14^h	0.15	29.5–94.2
	30 Oct. $12^h\,50^m \pm 10$	30 Oct. 15^h	1.46	9.6–18.8
	31 Oct. 00^h	31 Oct. 09^h	5936	0.8– 9.6
6	31 Oct. 02^h	31 Oct. $14^h\,00^m$	14.2	29.5–94.2
	31 Oct. $01^h\,00^m \pm 30$	31 Oct. $14^h\,00^m$	169	9.6–18.8
	31 Oct. (undetermined)	1 Nov. 02^h	5368	0.8– 9.6
7	1 Nov. 13^h	1 Nov. 23^h	17.7	29.5–94.2
	1 Nov. $14^h\,00^m \pm 40$	2 Nov. $00^h\,30^m$	136	9.6–18.8
	2 Nov. 02^h	2 Nov. 19^h	5473	0.8– 9.6
8	4 Nov. 06^h	4 Nov. 07^h	7.8	29.5–94.2
	4 Nov. $06^h\,00^m \pm 10$	4 Nov. $08^h\,20^m$	25.3	9.6–18.8
	4 Nov. $07^h\,00^m \pm 70$	4 Nov. 18^h	1289	0.8– 9.6

Data source: Simpson (1972a).

relative insensitivity of this channel to electrons was discussed in Section 1 and the presence of the guard counter surrounding the sides of this detector enables the sensor to respond to extremely low solar proton fluxes.

On 27 October after 1530 there is an apparent flare initiated particle increase time associated with a double maximum flare on the Sun at S17, E17 initially starting at 1232 UT. While this event is clearly discernible with an impulsive profile on the 29.5–94.2 MeV channel, it has a distinctly slower time intensity profile on the 9.6–18.8 MeV channel. It is not even discernible as an increase above the decay of the previous enhancement on the 0.8–9.6 MeV channel until the next day, and this enhancement is likely associated with an SC magnetic storm. The next impulsive increase (event number 4 in Table IV) occurs on 29 October at 1530 UT in good time coincidence with a relativistic electron event reported by Simnett (1972). This increase, clearly discernible on the 29.5–94.2 MeV channel (which is very insensitive to electrons) and also discernible on the 9.6–18.8 MeV channel, is difficult to recognize as a separate event on the 0.8–9.6 MeV channel due to magnetic storm modulation. The next event on the afternoon of 30 October is again more clearly discernible on the higher energy channel.

At ~02 UT on 31 October a major flare initiated event (event number 6 in Table

IV) occurs that is clearly discernible on all channels. It is difficult to select an accurate onset time for the 0.8–9.6 MeV channel for this event because the apparent increase could be due to either electron contamination or low energy particles from the previous flare or a combination of both. The next increase occurs on 1 November which is again clearly discernible on the upper two energy channels. However, the presence of an unique event is not as readily apparent on the low energy channel because of the complex magnetic storm modulation structure. Although the time of maximum intensity for the event on 1 November appears to be relatively easy to identify on the upper two energy channels, the maximum intensity on the low energy channel may well be obscured by a perigee passage on IMP 4 with the satellite at L-values less than 10 R_E from 09 UT to 18 UT on 2 November. The intensity on all three channels appears to undergo a relatively simple decay from the multiple injections on 31 October and 1 November until on 4 November another major increase (event number 8 in Table IV) occurs in good time association with a flare at 0524 UT. Once again the onset time of this event is easily identified on the 29.5–94.2 MeV channel and again the presence of relativistic electrons (Simnett, 1972), in addition to the enhanced proton intensity from the previous events, makes identification of an onset time difficult on the lower energy channels. After this event on 4 November, the flux on all channels slowly decays.

6. Differentiation of Enhancement Type

In scanning the individual data records from various satellites, each obvious increase in intensity, or a sudden change of slope in a period of increasing intensity, was identified as an 'event'. Many of these events, particularly those with a rapid rise in intensity were easily associated with a flare on the Sun. However, as the detection sensitivity increases and lower energy particle enhancements are identified, the association of some of the increases with solar phenomena becomes more difficult. One problem of this type is the differentiation between a discrete low energy solar particle increase and an enhancement associated with the rotation of active regions across the solar disk. The probability of these accidental coincidences is enhanced by the tendency of the co-rotating type of events to be associated with the central meridian passage of an active region, and the tendency of the detection of small discrete injections to maximize west of central meridian. McDonald and Desai (1971) have previously noted that there is a correlation between central meridian passage of some active regions and the enhancement of the low energy particle flux. These small co-rotating type of events become more numerous and difficult to identify with enhanced solar activity and, therefore, the danger exists of erroneously identifying a low energy proton event when it is really a co-rotating type of event.

An example of such a problem is illustrated in the identification of the low energy (0.8–9.6 MeV) proton events observed by the UC detector on IMP 4 during the period 25 September–6 October 1967. As shown in Figure 7 three distinct enhancements are obvious on the 0.8–9.6 MeV channel during this period, none of which are discernible on the 9.6–18.8 MeV channel. The first increase, with onset at ~08 UT on 26 September, may be associated with a flare late on 25 September. Graedel and Lanzerotti (1971) report an associated enhancement for 0.56–0.6 MeV protons with an onset time of 10 UT.

Graedel and Lanzerotti report another proton event at 07 UT on 27 September.

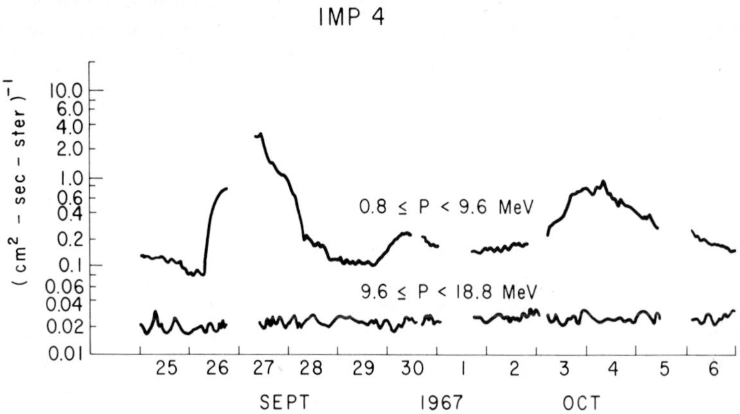

Fig. 7. An example of low energy proton enhancements detected on IMP 4 between 25 September and 6 October 1967.

This is a time when IMP 4 has just finished perigee passage. It cannot be determined from these data alone whether this is from a new injection of solar particles or is a continuation of the previous event. Certainly, any low energy proton event that occurred prior to 07 UT on 27 September cannot be associated with the weak electron event at 0708 UT ± 2 min reported by Lin (1970).

The next enhancement has an onset at ~19 UT on 29 September and does not appear to be associated with any specific solar phenomena. This increase is not included in the list given by Graedel and Lanzerotti (1971). The last enhancement of this period, on 3 October, has an onset at 0330 UT on both the 0.8–9.6 MeV and the 0.56–0.6 MeV channels on IMP 4 (Graedel and Lanzerotti, 1971). Again there does not appear to be any specific solar flare associated with this enhancement. It is our opinion that these last two enhancements are a convective phenomena of the type described by Roelof and Krimigis (1973) and are not directly related to a unique flare phenomena. These convective type structures happen to be observable because of the relatively quiet low level of solar activity at this time.

7. Summary

In this paper we have presented examples of the following problem areas related to the analyses of spacecraft data: sensitivity levels of various detectors; electron contamination; the resolution of multiple events; perigee interference; and the identification of type of enhancement. Certainly an understanding of these problems and their relationship with the sensor and type of data being studied is essential for a meaningful analysis of solar particle data. Fortunately, the majority of these problems can be resolved either by examining data from other sensors on the same spacecraft, or in cases where the problem relates to the entire satellite such as perigee passage, by examining data from other spacecraft.

In the past few years our satellite measurements and techniques have vastly improved. However, as our sensitivity levels continue to increase and our detection techniques continue to improve, we will essentially exchange the problems discussed herein with other more complicated and puzzling problems. In this way our basic knowledge of solar particle acceleration and propagation through the interplanetary medium and the magnetosphere will be increased.

Acknowledgements

We wish to thank the principal investigators (especially Drs J. A. Simpson and C. O. Bostrom) and the U.S. National Space Science Data Center for providing the satellite data used in this analysis.

References

Allum, F. R., Palmeira, R. A. R., Rao, U. R., McCracken, K. G., Harrier, J. R., and Palmer, I.: 1971, *Solar Phys.* **17**, 241.

Bryant, D. A., Cline, T. L., Desai, U. D., and McDonald, F. B.: 1965, *Phys. Rev. Letters* **14**, 481.

Bostrom, C. O., Williams, D. J., and Arens, J. F.: 1972, Unpublished data, Experiment description, Solar Proton Monitoring Experiment, U.S. National Space Science Data Center.

Cline, T. L. and McDonald, F. B.: 1968, *Solar Phys.* **5**, 507.

Datlowe, D.: 1971, *Solar Phys.* **17**, 436.

Fan, C. Y., Pick, M., Pyle, R., Simpson, J. A., and Smith, D. R.: 1968, *J. Geophys. Res.* **73**, 1555.

Graedel, T. E. and Lanzerotti, L. J.: 1971, *J. Geophys. Res.* **76**, 6932.

Krimigis, S. M. and Van Allen, J. A.: 1966, *Phys. Rev. Letters* **16**, 419.

Lin, R. P.: 1970, *Solar Phys.* **12**, 266.

Lin, R. P. and Anderson, K. A.: 1967, *Solar Phys.* **1**, 466.

Lincoln, J. V.: 1966, *J. Geophys. Res.* **71**, 4219.

McCracken, K. G., Rao, U. R., and Bukata, R. P.: 1969, *J. Geophys. Res.* **72**, 4293.

McDonald, F. B. and Desai, U. D.: 1971, *J. Geophys. Res.* **76**, 808.

Roelof, E. C. and Krimigis, S. M.: 1973, *J. Geophys. Res.* **78**, 5375.

Simnett, G. M., Cline, T. L., Holt, S. S., and McDonald, F. B.: 1970, *Acta Physica Academiae Scientiarum Hungaricae* **29**, Suppl. 2, 649.

Simnett, G. M.: 1972, *Solar Phys.* **22**, 189.

Simpson, J. A.: 1972a, Unpublished data from IMP Satellites, deposited in the U.S. National Space Science Data Center.

Simpson, J. A.: 1972b, Unpublished data from Pioneer Spacecrafts, deposited in the U.S. National Space Science Data Center.

UAG-8: 1970, 'Data on Solar-Geophysical Activity October 24–November 6, 1968', compiled by J. Virginia Lincoln.

Vernov, S. N., Chudakov, A. E., Vakulov, P. V., Gorchakov, E. V., Kontor, N. N., Logachev, Yu. I., Lyubimov, G. P., Pereslegina, N. V., and Timofeev, G. A.: 1970, in V. Manno and D. E. Page (eds.), *Intercorrelated Satellite Observations Related to Solar Events*, D. Reidel Publ. Co., Dordrecht-Holland, p. 53.

PART 5

SOLAR PARTICLE ENTRY TO THE EARTH

SOLAR PARTICLE ACCESS INTO THE INNER MAGNETOSPHERE

G. MORFILL

Max-Planck-Institut für Physik und Astrophysik, Institut für extraterrestrische Physik, 8046 Garching bei München, Germany

Abstract. There are four modes of access of energetic particles into the inner magnetosphere, defined here as the region of 'closed' field lines, i.e. regions of latitude below, for instance, the 40 keV electron boundary. These are direct entry to all latitudes above the cut-off, longitudinal drift from the nightside to the dayside pseudotrapping region, radial diffusion through the dayside magnetopause and a combination of radial and pitch angle diffusion. The first two modes of access are reversible processes, and the access boundaries can, in principle at least, be determined using numerical calculations in static model magnetospheric fields.

Apart from this static description of particle access, it is also possible to use energetic solar particle measurements on closed field lines as diagnostics for magnetospheric motions and wave distribution as function of time and location within the magnetosphere, and information can be obtained on the motion of boundaries and the strength of pitch angle scattering as function of position within the magnetosphere and as function of geomagnetic activity. Observations and theoretical treatments of these problems are discussed.

1. Introduction

Four modes of access of energetic cosmic particles into the inner magnetosphere are discussed in the literature. These are direct impact, longitudinal drift into the pseudotrapping region in an asymmetric magnetosphere, radial diffusion through the dayside magnetopause and a combination of radial and pitch angle diffusion. The first two modes of access are determined by the static topology of the geomagnetic field, whereas the last two rely on temporal and spatial fluctuations in the magnetospheric field.

Using calculations of particle cut-offs in model geomagnetic fields, as well as determinations of gradient and curvature drifts, it is possible to investigate the first two modes of particle access into the inner magnetosphere. The results of such an investigation then provide the basis to which wave-particle interactions must be added.

Energetic particle pitch angle distributions yield information about the strength of possible scattering mechanisms, and correlation of types of pitch angle distributions with magnetic activity can provide hints as to the nature of the activities concerned. Temporal and spatial variations provide additional information.

Low altitude polar orbiting satellites can be very useful for determining the particle pitch angle distribution near the atmospheric loss cone, and even omnidirectional detectors can differentiate between regions of isotropy and double loss cone distributions, i.e. regions where atmospheric loss cones from both hemispheres exist on closed field lines.

Scattering theory has been very successful in explaining some of the observations made of the energetic particle distribution within the magnetosphere, such as the stability of the inner radiation zone, the location of the outer edge of the electron

D. E. Page (ed.), Correlated Interplanetary and Magnetospheric Observations, 399–417. *All Rights Reserved.*

radiation belt as a function of energy and the removal of electrons to levels near zero throughout the quiet-time electron slot which divides the radiation belt electrons into an inner and outer zone (e.g. Kennel and Petschek, 1966; Cornwall, 1964; Dungey, 1963; Lyons *et al.*, 1972; Roberts, 1969).

The same scattering theory of wave-particle interaction with varying types of magnetic fluctuations (e.g. Alfvén waves, ion cyclotron or electron cyclotron waves) should be applicable in the inner magnetosphere, which might then yield possible clues as to the nature and spatial inhomogeneity of geomagnetic activity.

2. Diagnostic Uses of Solar Particles

The information obtained from solar particle fluxes is the intensity and pitch angle distribution as a function of time and space (invariant latitude and longitude).

Comparisons of interplanetary and magnetospheric particle fluxes have shown a close temporal tracking of intensities in the latitude range between the cut-off and the location of the last closed field line, whereas at higher latitudes delays relative to interplanetary intensity changes are often seen. This aspect is particularly obvious when interplanetary particle intensity changes, bounded by tangential magnetic discontinuities, are convected past the Earth (see Figure 1). Trajectory calculations in model geomagnetic fields (e.g. Morfill and Quenby, 1971; Engelmann *et al.* 1971; Gall *et al.*, 1968, 1972) indicate that particle access to the latitudes where close temporal tracking occurs is via the dawn magnetopause close to the Earth, or possibly the neutral sheet on the nightside of the magnetosphere, and through subsequent longitudinal drift in the curved, dipole-like geomagnetic field line regime.

The observed fact that solar particles penetrate the inner magnetosphere to latitudes well below the theoretically computed cut-offs is usually interpreted as being due to radial diffusion caused by suitable temporal and spatial fluctuations of the Earth's electromagnetic field.

Observation of the particle pitch angle distributions yields information about the strength of scattering processes in the following way.

(1) If a single loss cone distribution is observed on closed field lines, particle scattering is strong enough to fill the second loss cone within $\frac{1}{2}\tau_b$, where τ_b is the bounce period.

(2) If a double loss cone distribution is observed, and the loss cone width corresponds to the geometric atmospheric loss cone, pitch angle scattering is negligible over the time period which the particle distribution has spent in the closed field line regime, i.e. $\beta \cdot \tau_d$, where τ_d is the longitudinal drift period and β is a factor giving the fraction of drift periods since the particle injection into the closed field line regime.

(3) If a double loss cone distribution with wide loss cones is observed, this is an indication for weak pitch angle scattering (pitch angle change $\Delta\alpha \sim$ few degrees during a bounce period) and consequently for preferential loss of particles close to the geometric atmospheric loss cone which leads to a broadening.

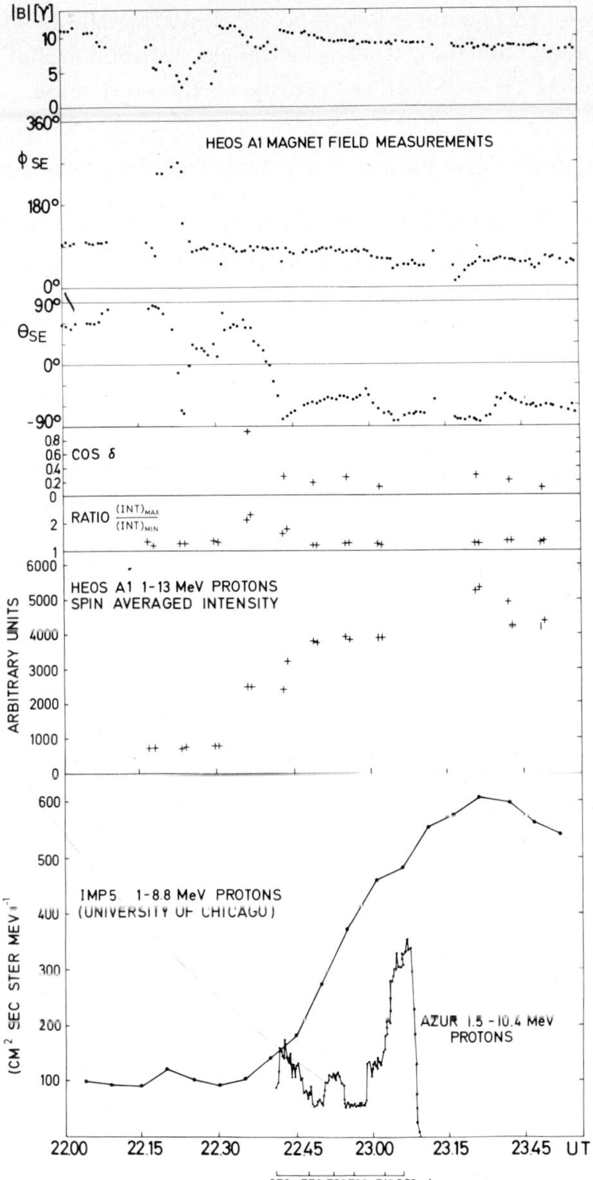

Fig. 1. Simultaneous observations of the interplanetary magnetic field and solar protons by HEOS A1, IMP 5 and AZUR. The upper three panels show the magnitude and directions of the interplanetary magnetic field measured by HEOS A1. The next three panels represent the angle δ between the particle detector scanning plane and the magnetic field vector, the anisotropy, represented as a ratio of maximum/minimum intensity, and the spin averaged change in 1–13 MeV proton intensity as function of time. The lowest panel shows the absolute 1–8.8 MeV proton fluxes measured by the University of Chicago telescope on IMP 5, and the AZUR 1.5 10.4 MeV polar cap fluxes. The corresponding invariant latitudes of the AZUR polar pass are indicated. The close tracking of polar cap intensities near cut-off ($\sim 66°$) with the near isotropic interplanetary IMP 5 fluxes is clearly demonstrated, whereas the high latitude intensities ($\Lambda \gtrsim 74°$) do not show this close temporal correspondence (From Scholer, 1972).

It is clear from this heuristic description of scattering effects on the appearance of loss cones that both the particle pitch angle distribution and its longitudinal development along the drift shell can provide useful information.

3. Low Altitude Solar Particle Pitch Angle Distributions in a Static Field

In determining the static behaviour, to which the effects of wave-particle interactions have to be added, we shall use the same method as that employed by Morfill (1973). This approach rests on the assumptions that particle access to the inner magnetosphere is governed only by the static geomagnetic field that the pitch angle distribution at low altitudes is, in general, isotropic outside the atmospheric loss cones, and that nonadiabatic motion, at any time during the particle bounce, results in an aperiodic drift motion.

The first assumption yields the zero order particle orbits, $\mathbf{r}(t)$, from the numerical integration of $\ddot{\mathbf{r}}=(Ze/mc)\,\dot{\mathbf{r}}\times\mathbf{B}(\mathbf{r})$ in the geomagnetic field, which is approximated by a model field $\mathbf{B}(\mathbf{r})$, for particles of rigidity $m\,|\dot{\mathbf{r}}|\,c/Ze$. The second assumption is realistic because the angular extent of the low altitude particle distribution, projected to the equator using the first invariant condition

$$\sin^2\alpha_e=\frac{B_e(r=L)}{B(r\approx 1)}\approx\frac{1}{L^3},$$

is given by the angle α_e which is small at high latitudes ($L\gtrsim 6$). The third assumption is certainly correct in an asymmetric magnetosphere, and implies that particles which undergo nonadiabatic motion, even if this is confined to a particular longitude region during their drifts around the Earth, cannot be on truly trapped orbits. Life times then are dependent on the degree and spatial extent of the nonadiabatic motion. Application of a particle intensity outside the magnetosphere (e.g. through a solar particle event) can lead to a magnetosphere penetration as far as the last adiabatic drift shell, and a single loss cone distribution as far as the last adiabatic field line. Calculations made with the Williams and Mead (1965) geomagnetic field model are shown in Figure 2 in an invariant latitude magnetic local time representation.

The use of the adiabatic motion criterion in trying to determine the possible extent of particle penetration into the inner magnetosphere is not so rigorous as the cut-off calculations made with trajectory integrations in a static model field. Since the latter calculations are limited, technically, by available computer time, such that particle orbits which complete several drifts around the Earth cannot be followed far enough, the two approaches complement each other. The close agreement on the dayside between the last adiabatic field line and the vertical cut-off emphasises this, as well as some orbit calculations made by Maseley, Olson and Pfitzer (private communication), where particles which had arrived at the Earth just above the location of the last adiabatic drift shell were found to complete several drifts around the Earth before being traced back out of the magnetosphere.

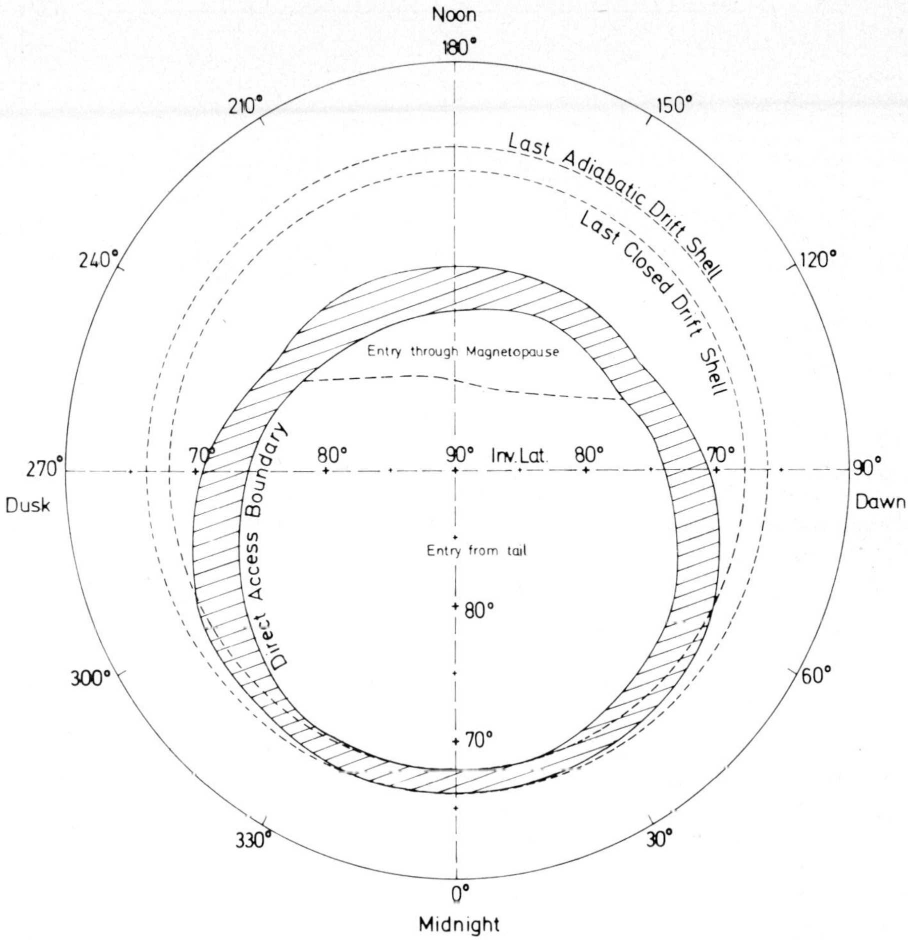

Fig. 2. Position of the regions of nonadiabaticity. The direct access boundary and the breakdown of the guiding centre approximation boundary are drawn in solid lines, whereas the last closed drift shell and the last adiabatic drift shell are dotted lines. The hatched area represents that part of the pseudo-trapping region where nonadiabatic motion in a spatially varying field does not allow double loss cone formation (from Morfill, 1973).

Observational support for the usefulness of the adiabatic motion criterion is depicted in Figure 3, where the location of the intensity 'edges' observed by Flindt (1970) with an omnidirectional proton detector are compared with the position of the adiabaticity contours. The 'edges' in Flindt's measurements are due to the observation of single and double loss cone distributions, yielding high and low particle intensities respectively, when measured with an omnidirectional detector. These edges should occur at the location of the last adiabatic field line, and as can be seen, substantial agreement is indeed found.

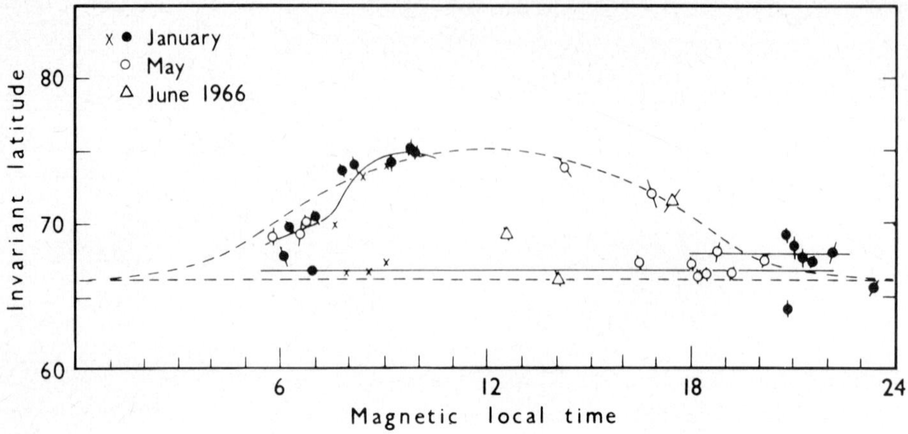

Fig. 3. Data taken from Figure 5 of Flindt (1970) show the location of edges in counting rate profiles of protons in the energy range 0.52 to 4 MeV. The solid lines are the best empirical fit given by Flindt, the upper dashed line is the contour of breakdown of adiabaticity, and the lower line is the last adiabatic drift shell (from Morfill, 1973).

4. Loss Times Due to Atmospheric Absorption

The loss cone is given, at the equator, from the first invariant

$$\sin^2 \alpha = \frac{B(r=L)}{B(r\approx 1)}$$

$$\sin^2 \alpha = \frac{1}{L^3}. \tag{1}$$

This loss cone occupies a solid angle Ω given by

$$\Omega = \int_0^\alpha 2\pi \sin\theta \, \mathrm{d}\theta$$

$$\Omega = 2\pi(1-\cos\alpha)$$

which for small values of α becomes

$$\Omega = \pi \sin^2 \alpha. \tag{2}$$

Thus we have

$$\Omega = \frac{\pi}{L^3} \tag{3}$$

(1) *Strong pitch angle scattering*. – Under the assumption of strong pitch angle scattering, the particle intensity at the equator is isotropic. The two atmospheric

loss cones, which occupy a volume $2\pi/L^3$ steradian at the equator are each filled and emptied twice during a bounce period, τ_b. Thus the equatorial flux which is lost in time τ_b, is

$$\mathrm{d}J=\frac{J}{L^3}, \tag{4}$$

where J is the omnidirectional equatorial flux. The rate of change of flux is given by

$$-\frac{\mathrm{d}J}{\mathrm{d}t}=\frac{J}{L^3\tau_b} \tag{5}$$

which yields under the boundary condition that $J=J_0$ at $t=0$

$$J=J_0\,e^{-t/L^3\tau_b}.$$

The longitudinal drift per bounce is

$$\Delta\phi=0.128\sqrt{E}\,L^2 \text{ degrees}, \tag{6}$$

where E is the kinetic energy of the particle in MeV. Hence the longitudinal development of particle intensity in the strong scattering limit, under the assumption of particle injection occurring at one specific longitude ϕ_0 is given by

$$J=J_0\,e^{-(\phi-\phi_0)/L^5\,0.128\,\sqrt{E}}. \tag{7}$$

Equation (7) shows that in our region of interest ($L\gtrsim 6$) and at the energies considered ($E\sim 1$ MeV) only little flux loss occurs during the longitudinal drift so that the intensity is approximately constant and of single loss cone type when pitch angle scattering is strong. The characteristic loss time is $L^3\tau_b$.

(2) *No pitch angle scattering.* – In the case of no pitch angle scattering it will take just one bounce period, or a drift in longitude of $\Delta\phi$ given by Equation (6) to establish the two loss cones after a particle distribution has been injected into a closed field line regime or has drifted out of a locally bounded region of strong scattering. This double loss cone then remains until different conditions are again encountered.

(3) *Weak pitch angle scattering.* – The situation of weak scattering is most difficult to describe quantitatively, since it represents the intermediate case between (1) and (2). Clearly, the omnidirectional equatorial intensity remains approximately constant, but the exact form of the pitch angle distribution must depend on the angular dependence of the pitch angle diffusion coefficient, the time phase relative to the time needed to set up an equilibrium distribution, the form and angular dependence of energy diffusion etc. (See e.g. Haerendel, 1971; Gendrin, 1968, 1972; Kennel and Engelmann, 1966.)

Qualitatively, however, one can say that the redistribution from an injected, very likely isotropic flux, must take the form of somewhat enlarged loss cones similar to those of the radiation belt particles, with timescales for the approach to equilibrium

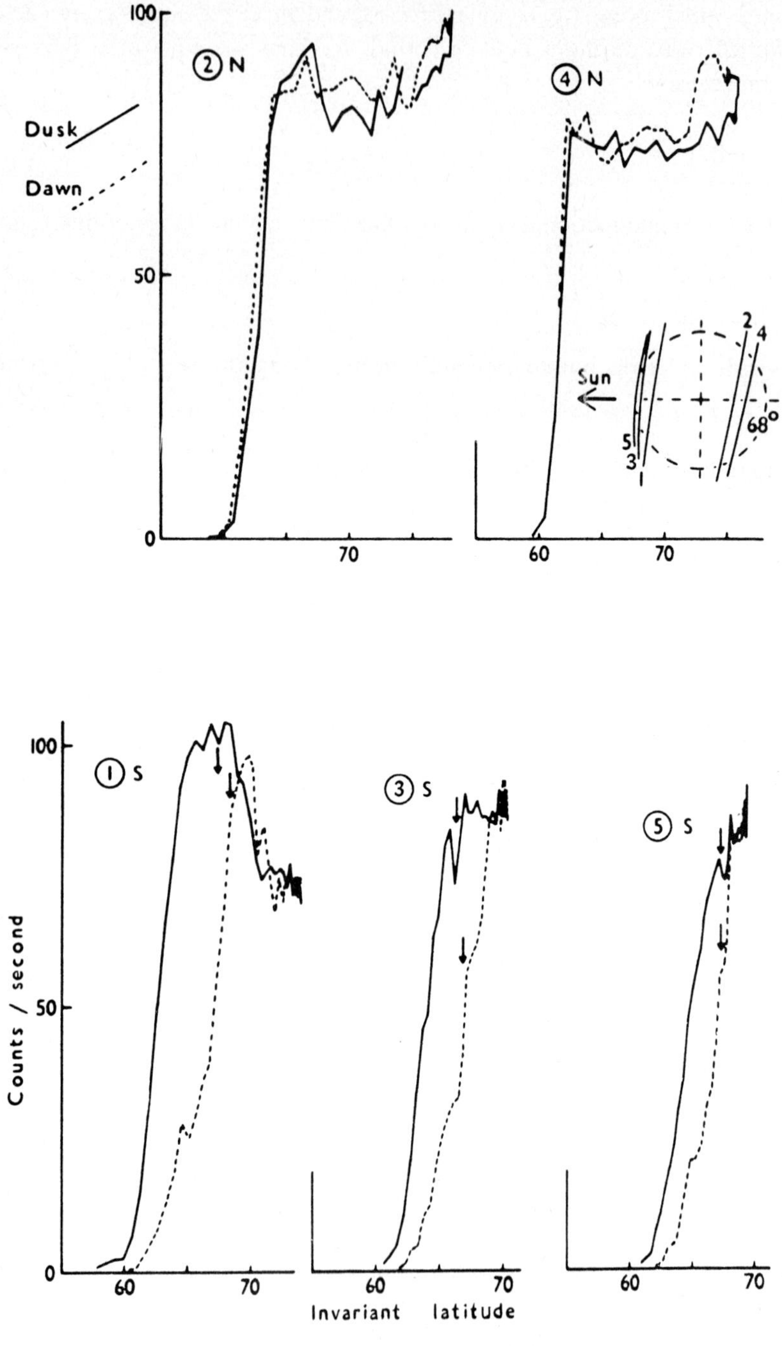

Fig. 4. Five consecutive passes, alternately from the north and south polar caps, showing the loss of flux between dusk and dawn on the dayside. The arrows mark the boundary of trapped energetic electrons. The insert is a plot of invariant latitude versus magnetic local time and shows the path of the satellite for each of the passes (from Bewick *et al.*, 1973).

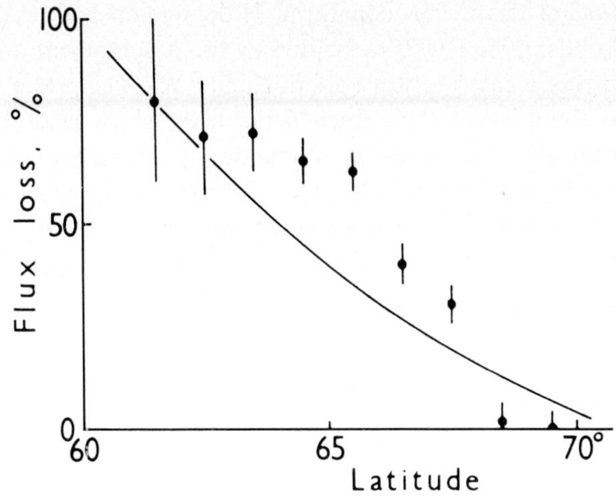

Fig. 5. The points show the percentage 'loss' of flux between dusk and dawn at given latitudes on pass 3 of Figure 4. The curve shows, on an arbitrary scale, the relative number of bounces between measurements, calculated for a dipole field (from Bewick *et al.*, 1973).

being of the same order of magnitude as $L^3\tau_b$. Observations of weak pitch angle scattering with omnidirectional detectors at low altitudes are often interpreted as anomalous asymmetric cut-offs if the cut-off is defined as the intensity edge from the polar plateau level rather than some low percentage intensity contour. A comment on this was made in Bewick *et al.* (1973). Figure 4 shows observations made with the ESRO 2 satellite of large flux 'losses' at low altitudes near cut-off, which are very likely due to a pitch angle redistribution caused by weak pitch angle scattering.

Figure 5 shows the percentage flux 'loss' between the dawn and dusk passages of the ESRO 2 detector (Figure 4), as a function of invariant latitude. The solid line shows the relative variation (with latitude) of the number of bounces between the satellite intersections at dawn and dusk of the respective drift shells. In the weak pitch angle scattering situation, there are some good reasons to believe that the removal of particles near the atmospheric loss cone must be correlated to the number of bounces executed by the particle distribution, and Figure 5 seems to lend support to these.

5. Systematic Investigation of Disturbed Magnetospheric Conditions

5.1. Geomagnetic storms

Geomagnetic storms are usually triggered by the passage of an interplanetary shock front over the magnetospheric cavity. They are identified by a s.s.c. and subsequent depression in the equatorial H-component of the geomagnetic field, which reaches a maximum within the order of hours and then slowly decays back to the pre-storm value over a period of days. This equatorial H-depression has been shown (e.g. Frank, 1967; Sckopke, 1966, 1972) to be due to the development and decay of a symmetric ring current located between about 2 and 7 R_E. The presence of this ring current alters the topology of the geomagnetic field, which leads to changes in energetic particle cut-offs. The increased plasma density at the equator, where most of the ring current particles are found, gives rise to wave growth, and alters the phase velocity of the waves so that wave-particle interactions will be modified drastically. Williams and Heuring (1973) have found that the pitch angle distribution of solar protons becomes isotropised on the dayside, when ring current plasma drifts round the earth to form a symmetric ring current. This is strong evidence for increased wave activity which can cause pitch angle scattering.

Fairly realistic ring current models are available (Sckopke, 1972) and it has become possible to repeat the adiabaticity analysis of Morfill (1973), but now applied to the purely geometrical changes in the magnetospheric field caused by the presence of symmetric ring currents. For this purpose we have superimposed the ring current models of Sckopke (1972) onto the Williams and Mead (1965) magnetic field model, and calculated $\mathbf{B}\times\nabla B$ and the particle rigidity at which the Alfvén criterion $X_r = \rho(\mathbf{B}\times\nabla B/B^2)=0.1$. Here ρ is the particle gyroradius.

In order to obtain a realistic field model two points have to be considered; firstly, the increased interplanetary plasma pressure behind the shock tends to push the magnetopause inwards, and secondly, the ring current field outside the ring current location is dipole-like and adds to the geomagnetic dipole field, which would tend to push the magnetopause outwards. Both effects return to normal gradually after the shock passage. We have chosen, as sample calculations, the situation where these two effects just cancel, so that care must be taken to not apply these calculations universally. Figure 6 shows the radial distance from the geocentre to the computed iso-rigidity contours of adiabatic breakdown in the equatorial plane for a few selected rigidities as a function of Dst. The upper curve refers to the noon meridian and the lower curve to the midnight meridian. As can be seen, the addition of a ring current does not affect the position of the radial iso-rigidity contours too much, except to add a rigidity 'plateau' at $5R_E$, indicated by the two 0.1 GV contours. However, due to the ring current inflation of the magnetosphere, field lines originating at the same point on the Earth reach further out, so that the latitude iso-rigidity contours are affected considerably more.

In Figure 7, actual measured positions of particle cut-offs are compared with the contours of adiabatic breakdown. Data were obtained on board the GRS–AZUR satellite during the geomagnetic storm on March 1970. Measured values of interplanetary plasma pressures (Palmiotto, 1971) were inserted into the field model parameters and thus the contours of breakdown of adiabatic motion shown for comparison were computed. Apart from two time periods, 000 UT–400 UT and 1500 UT–1730 UT, when strong substorm activity distorted the measurements, good correlation can be seen between the location of experimental cut-off and the non-adiabatic motion contour, as well as their temporal variations.

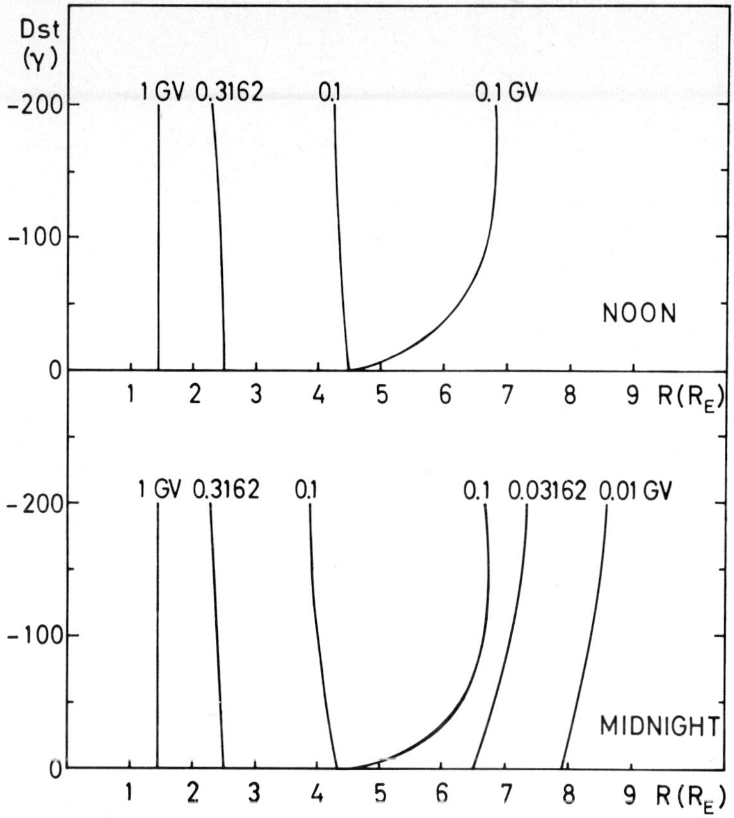

Fig. 6. Computed iso-rigidity contours of breakdown of the guiding centre approximation in a magnetosphere distorted by the solar wind and by the presence of a symmetric ring current. Contours are drawn in the equatorial plane and represent the radial distance from the geocentre as a function of Dst, both for noon and midnight meridional cross sections.

5.2. Geomagnetic substorms

The trigger mechanism for substorms is not known, although substorm activity appears to be correlated with southerly interplanetary magnetic fields (e.g. Arnoldy, 1971; Foster *et al.*, 1971 and Tsurutani and Meng, 1972). Substorms are identified by increasing tail fields and thinning plasma sheet during the 'growth phase' and then a fast expansion of the field into a more dipole-like configuration, associated with weak tail fields, thick plasma sheet and a depression in the H-component of midnight auroral zone magnetograms.

Since the whole process lasts no more than perhaps 2 to 4 hours before the field topology is 'normal' again it is necessary to develop a statistical approach in order to investigate temporal and spatial phenomena associated with this disturbance. One such attempt has been made (Scholer *et al.*, 1974) and will be described here. The method used was to define a parameter $\alpha = C_{90}/C_{45}$, i.e. the ratio of 90° pitch angle intensity to 45° pitch angle intensity and to use this ratio as a description of the pitch

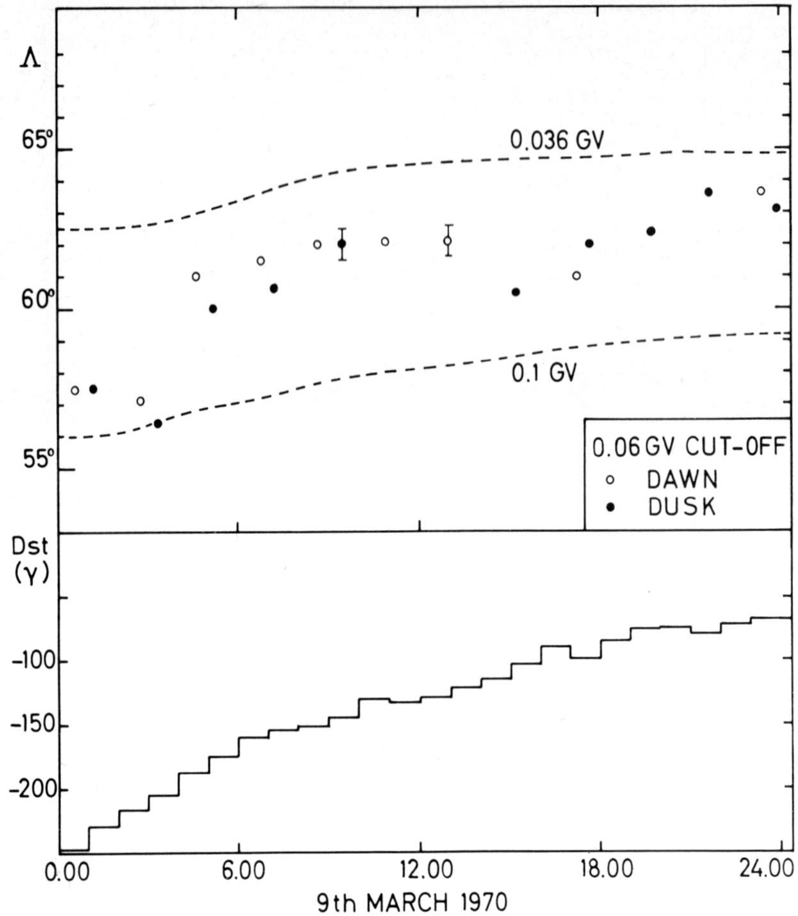

Fig. 7. Solar proton penetration cut-off latitudes (Λ) during the geomagnetic storm of March 9th, 1970. Cut-off latitudes for ~0.06 GV protons are shown, as well as the theoretically calculated adiabaticity contours for 0.036 GV and 0.1 GV protons for comparison. The lower panel shows the hourly averaged Dst values during the decay of the ring current.

angle distribution at low altitudes. The 45° detector was looking upwards, and due to the finite half opening angle (22°) of the instruments used (Experiment 88/1 and 88/2 on the magnetically stabilised GRS-A Azur satellite) the 45° detector should observe the atmospheric loss cone due to particle absorption in the opposite hemisphere, which is typically ~37° for the satellite altitudes concerned.

If $\alpha = C_{90}/C_{45} > 1$ we have a double loss cone distribution, if $\alpha = 1$ the flux is isotropic with the upper loss cone filled up (≡strong pitch angle scattering) and if $\alpha < 1$ there is particle precipitation along the field lines. In a statistical analysis, the data are ordered with respect to invariant latitude Λ, magnetic local time MLT, time T relative to the substorm expansion onset and B_0 the maximum depression of the midnight auroral zone H-magnetograms. The range of α is subdivided into 12 intervals of width $\Delta\alpha=0.2$, from $\alpha=0$ to $\alpha=2.4$ and the measurements are represented as weighted histograms of N_0 versus α, where $N_0(\Lambda, \text{MLT}, T, B_0, \alpha)$ is given by

$$N_0=\sum_{l=1}^{k} 1-\kappa(l, \Lambda, \text{MLT}, T, B_0, \alpha) \tag{8}$$

$$\kappa=\frac{1}{\sqrt{C_{90}}}+\frac{1}{\sqrt{C_{45}}} \tag{9}$$

and only mean counting rates >4 counts/frame are considered. In order to decide whether the histograms obtained in this way correspond to a trapping, isotropic or precipitating distribution, they must be compared with the theoretical isotropic case, which is determined in the following way.

From the two counting rates C_{90} and C_{45} a mean value $\bar{C}$ is obtained. The probability of measuring a counting rate 'm' in one of the detectors is given by the Poisson distribution

$$P(m)=\frac{\bar{C}^m e^{-\bar{C}}}{m} \tag{10}$$

The probability $P(\alpha)$ of obtaining a ratio α where $\alpha_0 \leqslant \alpha < \alpha_1$ is then

$$P(\alpha_0 \leqslant \alpha < \alpha_1)=\sum_m \sum_n P(m)\, P(n), \tag{11}$$

where the summation is taken over all counting rates such that $\alpha_0 \leqslant m/n < \alpha_1$. Multiplication of $P(\alpha)$ with the weighting factor $(1-\kappa)$ and summation over all k events gives the theoretical histogram N_T for the case of an isotropic pitch angle distribution, with which the observed distribution can be compared.

Figure 8 shows histograms of the probability $P(\alpha)$ for obtaining a given ratio $\alpha=C_{90}/C_{45}$ in the case of an isotropic pitch angle distribution with mean counting rates $\bar{C}$.

From the statistical analysis performed by Scholer *et al.* (1974) the following picture seems to emerge:

In the latitude region straddling the last closed field line precipitation is observed (see Figure 9) at the beginning of the substorm growth phase.

In the latitude region just below the last closed field line, an isotropic pitch angle distribution (strong pitch angle scattering) is observed (see Figure 10) at the beginning of the substorm expansion phase.

In this same latitude region, a double loss cone distribution (no or very weak pitch angle scattering) is observed (see Figure 11) approximately 2 h after the substorm expansion, indicating very quiet field conditions.

These observations are in qualitative agreement with measurements of McPherron and Coleman (1970) that enhanced wave-activity occurs just after substorm expansions, and that this period is followed by a magnetically quiet period.

Using the pitch angle scattering theory (e.g. Kennel and Engelmann, 1966) applied to different types of wave phenomena, under the condition that Δt, the scat-

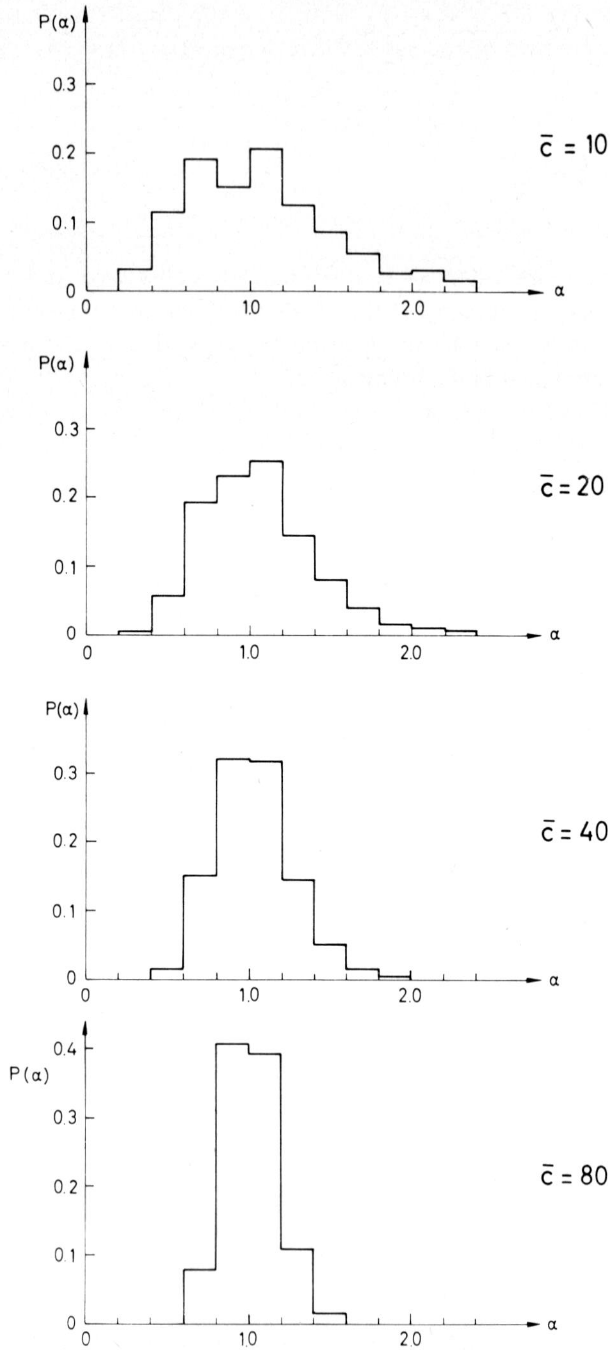

Fig. 8. Theoretical histogram of the probability $P(\alpha)$ of obtaining a ratio in counting rates $C_{90}/C_{45}=\alpha$. Probabilities are added in parameter blocks $\Delta\alpha=0.2$. $P(\alpha)=1$ corresponds to isotropic pitch angle distributions. Probability distributions are shown for four different mean counting rates $\bar{C}$ (from Scholer *et al.*, 1974).

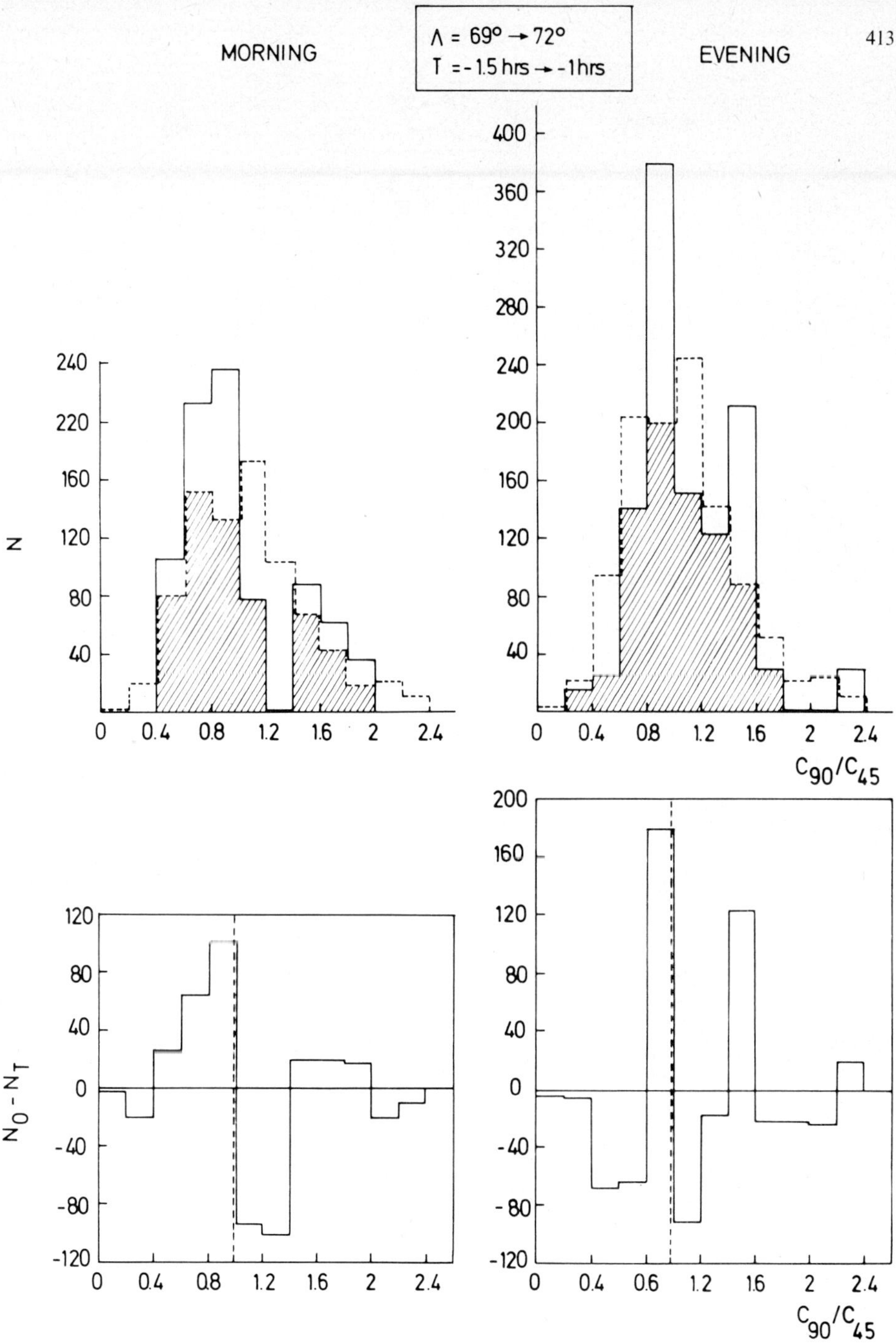

Fig. 9. Histograms of the weighted number of events having $\alpha_0 \leqslant C_{90}/C_{45} < \alpha_1$ where $\alpha_1 - \alpha_0 = 0.2$, as function of C_{90}/C_{45}. Solid lines represent the observed distribution, N_0, dotted lines the equivalent theoretical distribution, N_T, if the particle flux is assumed isotropic. The region of overlap is shaded. In the lower panel the difference $N_0 - N_T$ is plotted. Positive values at $C_{90}/C_{45} < 1$ are indicative of a precipitating particle flux. Data are shown for morning and evening passes at invariant latitude, $\Lambda = 69°–72°$ and at $T = 1.5$ h to 1 h before the substorm expansion phase (from Scholer *et al.*, 1974).

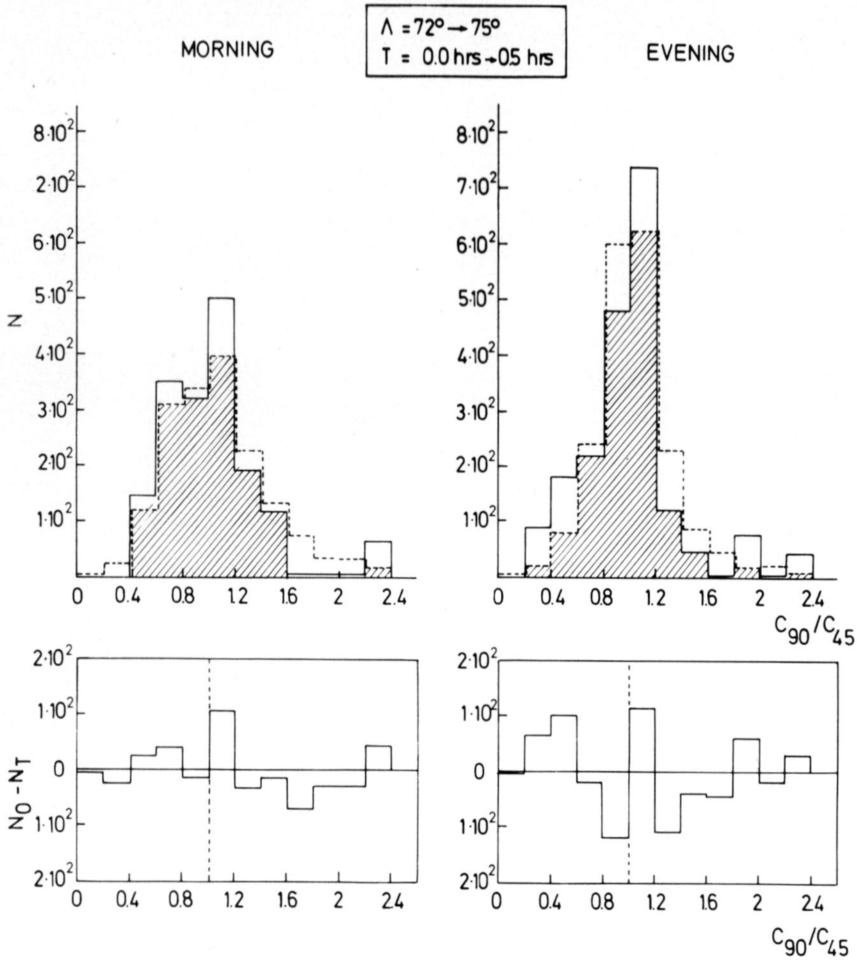

Fig. 10. As for Figure 9, except that the particle flux is isotropic, $\Lambda = 72°–75°$ and $T = 0$ h to 0.5 h after the substorm expansion (from Scholer *et al.*, 1974).

tering time, is less than $\frac{1}{2}\tau_b$ (in other words the strong pitch angle scattering criterion) yields the power necessary in the magnetic fluctuations to account for the observation of an isotropic particle distribution. Comparison of these power levels with observations should yield further insight into the type of fluctuations, the mechanics of wave-particle interactions and the spatial distribution of waves in the magnetosphere.

5.3. K_p – GENERAL ACTIVITY

The relation of particle cut-offs to K_p the general index of magnetic activity has been investigated by for instance Imhof *et al.* (1971) who found that these cut-offs for ~2 MeV particles, even extrapolated back to quiet geomagnetic times, occur at

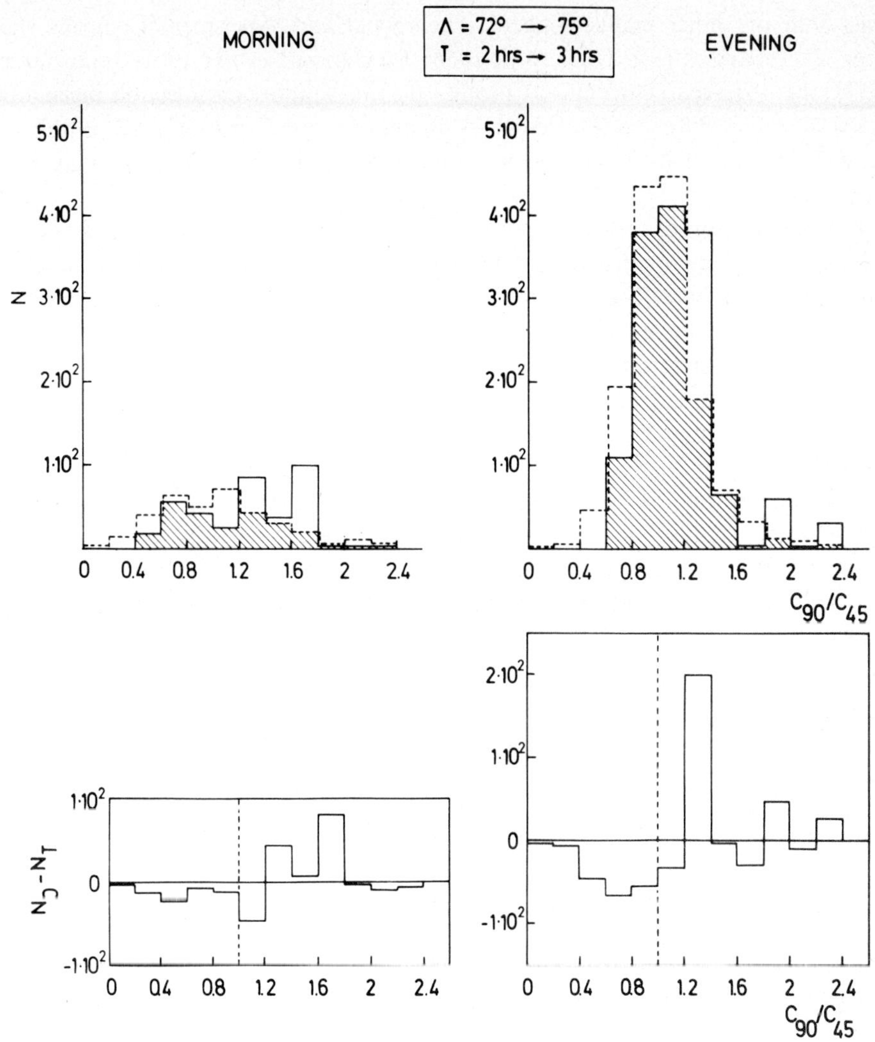

Fig. 11. As for Figure 9, except that the particle flux exhibits a double loss cone distribution (negative values of $N_0 - N_T$ at $C_{90}/C_{45} < 1$ and positive values at $C_{90}/C_{45} > 1$), that $\Lambda = 72°–75°$ and $T = 2$ h to 3 h after the substorm expansion (from Scholer *et al.*, 1974).

lower invariant latitudes than those calculated by Smart *et al.* (1969), but that they exhibited the same local time variation. Their cut-off, defined as the latitude where the intensity has decreased to $1/e$ of the polar cap value, exhibits a marked and almost linear dependence on K_p, such that for instance $\Lambda_c = -K_p + 66.3$ for 2.16 MeV protons at close to local midnight.

6. Conclusions

Clearly there are a number of unexplained facets as regards the entry of solar particles into the inner magnetosphere. Theoretical and experimental cut-offs show large discrepancies (e.g. Lanzerotti, 1968; McDiarmid *et al.*, 1971; Imhof *et al.*, 1971), the observed sharp cut-off profiles do not support an injection mechanism based on radial diffusion (Fanselow, 1970) and the topological changes in the geomagnetic field and its associated wave phenomena are only just beginning to be quantifiable.

The observation of low altitude solar particle pitch angle distributions and cut-offs as a diagnostic method to aid the understanding of magnetospheric processes has been treated here. The basis of this approach was firstly to determine, in a semi-quantitative way, the effects of a static but distorted magnetosphere on the general access characteristics, and secondly to investigate the contribution which solar particle observations can make to our understanding of wave-particle processes and the spatial distribution of magnetospheric fluctuations during geomagnetically disturbed times.

It was found that the iso-rigidity contours marking the breakdown of the guiding centre approximation yield close quantitative agreement with dayside observations of the counting rate 'edges' observed by Flindt (1970) and with the location and temporal variation of the cut-off latitude during the decay phase of geomagnetic storms.

References

Arnoldy, R. L.: 1971, *J. Geophys. Res.* **76**, 5189.
Bewick, A., Haskell, G. P., Hynds, R. J., and Morfill, G.: 1973, *J. Geophys. Res.* **78**, 597.
Cornwall, J. M.: 1964, *J. Geophys. Res.* **69**, 1251.
Dungey, J. W.: 1963, *Planetary Space Sci.* **11**, 591.
Engelmann, J., Hynds, R. J., Morfill, G. E., Axisa, F., Bewick, A., Durney, A., and Koch, L.: 1971, *J. Geophys. Res.* **76**, 4245.
Fanselow, J. L.: 1970, *Trans. AGU*, **51**, 800.
Flindt, H. R.: 1970, *J. Geophys. Res.* **75**, 39.
Foster, J. C., Fairfield, D. H., Ogilvie, K. W., and Rosenberg, T. J.: 1971, *J. Geophys. Res.* **76**, 6971.
Frank, L. A.: 1967, *J. Geophys. Res.* **67**, 3753.
Gall, R., Jimenez, J., and Comacho, L.: 1968, *J. Geophys. Res.* **73**, 1593.
Gall, R., Bravo, S., and Orozco, A.: 1972, *J. Geophys. Res.* **77**, 5360.
Gendrin, R.: 1968, *J. Atmospheric Terrestr. Phys.* **30**, 1313.
Gendrin, R.: 1972, in B. M. McCormac (ed.), *Earth's Magnetospheric Processes*, D. Reidel Publ. Co. Dordrecht-Holland, p. 311.
Haerendel, G.: 1971, 'Proceedings of ESRO Colloquium on Wave and Particle Interactions', ESRO/S.P. 72, p. 63.
Imhof, W. L., Reagan, J. B., and Gaines, E. E.: 1971, *J. Geophys. Res.* **76**, 4276.
Kennel, C. F. and Petschek, H. E.: 1966, *J. Geophys. Res.* **71**, 1.
Kennel, C. F. and Engelmann, F.: 1966, *Phys. Fluids* **9**, 2377.
Lanzerotti, L. J.: 1968, *Phys. Rev. Letters* **21**, 929.
Lyons, L. R., Thorne, R. M., and Kennel, C. F.: 1972, *J. Geophys. Res.* **77**, 3455.
McDiarmid, I. B., Burrows, J. R., and Wilson, M. D.: 1971, *J. Geophys. Res.* **76**, 227.
McPherron, R. L. and Coleman, Jr., P. J.: 1970, *J. Geophys. Res.* **75**, 3927.
Morfill, G.: 1973, *J. Geophys. Res.* **78**, 588.
Morfill, G. and Quenby, J. J.: 1971, *Planetary Space Sci.* **19**, 1541.
Palmiotto, F.: 1971, Techn. Report LPS-71-15, University of Rome.
Roberts, C. S.: 1969, *Rev. Geophys.* **7**, 305.
Scholer, M.: 1972, *J. Geophys. Res.* **77**, 2762.
Scholer, M. Morfill, G., and Hovestadt, D.: 1974, *J. Geophys. Res.* (to be published).

Sckopke, N.: 1966, *J. Geophys. Res.* **71**, 3125.
Sckopke, N.: 1972, *Cosmic Electrodyn.* **3**, 330.
Smart, D. F., Shea, M. A., and Gall, R.: 1969, *J. Geophys. Res.* **74**, 4731.
Tsurutani, B. T. and Meng, C.-I.: 1972, *J. Geophys. Res.* **77**, 2964.
Williams, D. J. and Heuring, F. T.: 1973, *J. Geophys. Res.* **78**, 37.
Williams, D. J. and Mead, G. D.: 1965, *J. Geophys. Res.* **70**, 3017.

ENERGETIC SOLAR PARTICLE ACCESS INTO THE GEOMAGNETIC TAIL

MANFRED SCHOLER
Max-Planck-Institut für Physik und Astrophysik, Institut für extraterrestrische Physik, 8046 Garching bei München, Germany

1. Introduction

The magnetospheric models can essentially be divided into open and closed magnetospheric models. In the open model first proposed by Dungey (1961) the polar field lines are connected to the interplanetary magnetic field (Figure 1a). These field lines are convected downstream in the antisolar direction with the solar wind to form the geomagnetic tail. Solar particles can reach the polar cap by following these connected open field lines adiabatically.

Dessler (1964) and Michel and Dessler (1965), on the other hand, propose a closed magnetosphere. In their model the geomagnetic tail extends from the Earth for 1000–2000 R_E with no field line interconnection. The tail field lines are parallel to the magnetopause (Figure 1b), and only a diffusive mode of particle entry into the tail and from there into the polar cap is possible.

Figure 1c shows the nonadiabatic reconnection model, examined in detail by Morfill and Scholer (1972a). In this model some field line draping exists near the front of the magnetopause, but the transition from interplanetary-like to tail-like field lines occurs as a sharp rotational discontinuity. In addition, a turbulent boundary layer is proposed to exist, which may give rise to particle scattering at ~ 0.04 GV, but which should be negligible at $\gtrsim 0.2$ GV.

Finally Michel and Dessler (1970) have modified the closed magnetospheric model to fit contrary observational evidence into a diffusion model. They describe the tail

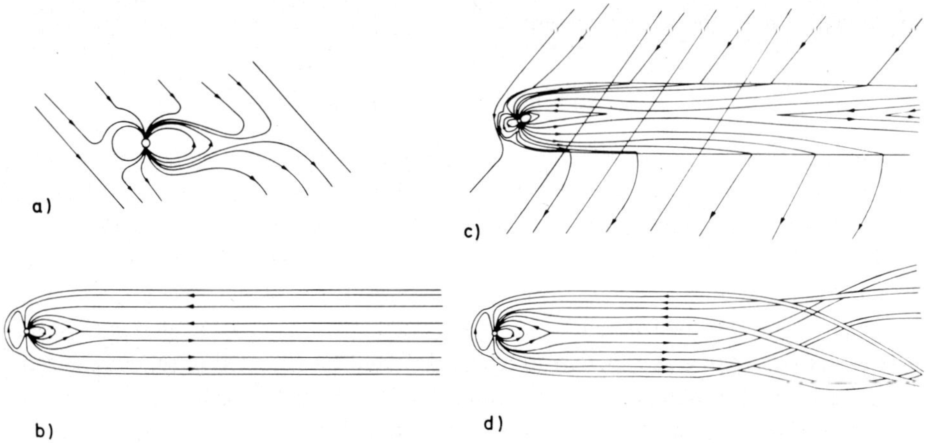

Fig. 1. Schematic diagram of different magnetospheric models. (see text)

D. E. Page (ed.), *Correlated Interplanetary and Magnetospheric Observations*, 419–431.

as being composed of filaments with slightly different plasma and field properties (Figure 1d). Beyond about 300 R_E the tail becomes flattened, and beyond about 0.1 AU the filaments intermingle with the interplanetary field lines. In this tail model, access of solar protons is diffusive and by gradient drift.

Since Michel and Dessler in 1965 first proposed to use low energy solar protons as test particles in order to study the topology of the magnetosphere a great number of polar cap and magnetotail measurements of such particles were reported in the literature. These measurements, however, revealed a rather complex spatial and/or temporal particle distribution in the magnetosphere, so that by emphasizing only a few and neglecting the others these observations could be interpreted to support the open magnetospheric model as well as the closed magnetospheric model. It became apparent that the complexity of the polar cap observations was mainly a result of different interplanetary particle conditions. Therefore, before interpreting polar cap observations in terms of different models one had to know the interplanetary conditions, i.e. only multisatellite observations could resolve the controversy which has arisen over such measurements. Once the interplanetary conditions are known they have to be linked with the polar cap measurements by some method which takes account of different magnetospheric models. This can be done with trajectory integrations, i.e. to predict the polar cap directional intensity from the known interplanetary particle distribution it is necessary to follow a typical particle trajectory back to the boundary of the magnetosphere or down the tail (e.g. Morfill and Quenby, 1971). Different magnetospheric models will result, in principle, in different predicted polar cap structures and comparison with the observed structure should give an answer to the question of the magnetospheric topology.

In this report we will discuss the effect of particle anisotropies aligned along interplanetary magnetic field lines and of local intensity gradients in interplanetary space on the magnetospheric particle structure in terms of different entry models. We will confine ourself here to low energy protons, i.e. $\sim$1 MeV, since they can reveal due to their smaller gyroradius a more detailed picture of the magnetospheric topology as the higher energy protons.

2. Interplanetary Anisotropies

During the early stages of solar particle events, interplanetary fluxes are usually anisotropic, especially in the case of west limb flares. This anisotropy should lead in the case of the open magnetosphere to a north-south asymmetry: Particles moving outward from the Sun will reach the north polar cap or the south polar cap, respectively, depending on the interplanetary sector structure, whereas those moving inward towards the Sun will reach the opposite polar cap. This is depicted in Figure 2. The upper part shows the particle intensity distribution expected in the magnetotail during a north-south anisotropy; the interplanetary magnetic field was assumed to have a southern component and a positive sector. The intensity change across the neutral sheet is not expected to be sharp since the solar proton mean free path in

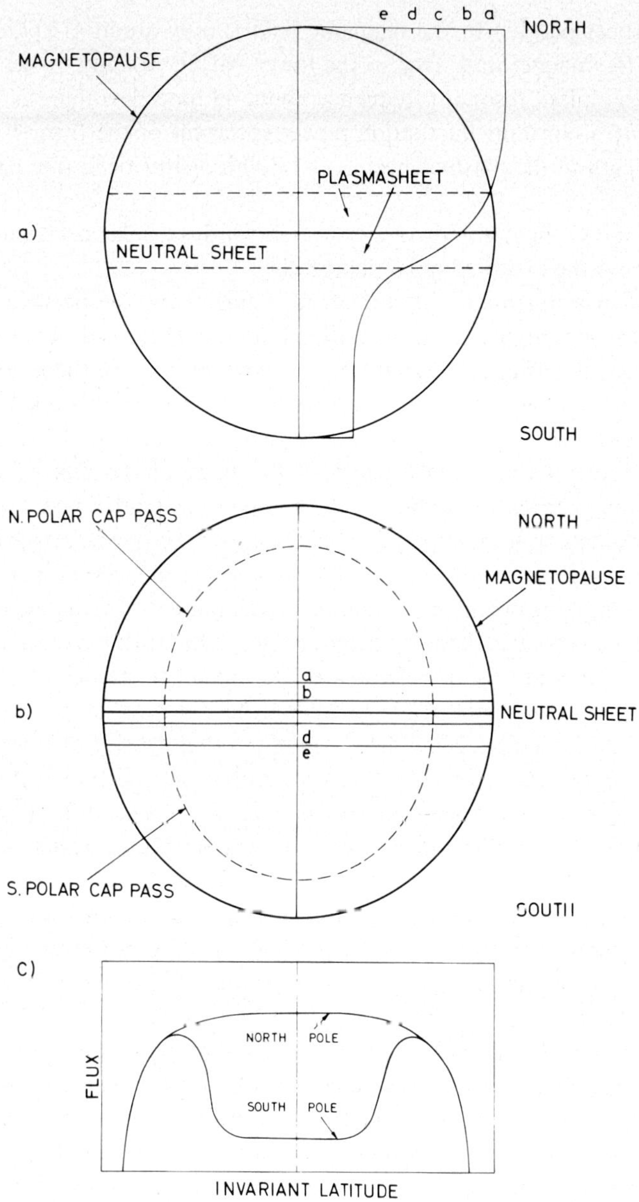

Fig. 2. Qualitative intensity structure in the magnetotail, if an 'open' tail topology exists. (a) shows the cross section of the magnetotail, including the approximate extent of the plasmasheet. Also indicated is the particle intensity variation in the x-z plane (perpendicular to the neutral sheet), assuming some diffusion in the plasma sheet. (b) represents qualitative isoflux contours in the magnetotail. Also shown is the projection into the tail of a dawn-dusk polar satellite orbit. (c) gives the intensity variation seen over the polar cap for the dawn-dusk orbiting satellite pass of Figure 2b. The figure shows the particle flux plotted as a function of invariant latitude. Intensity cut-off corresponds to $\sim 65°$, the polar 'trough' extends from $\sim 90°$ down to $\sim 75°$ (Scholer and Morfill, 1972).

the plasma sheet parallel to the magnetic field is only about 1/100th of the mean free path in the magnetotail. Due to the lower Alfvén velocity in the plasma sheet is the Doppler shifted resonant frequency about 40 times less than in the rest of the tail. Assuming a constant fluctuation power spectrum of the form $P \sim \omega^{-2}$ we get a mean free path in the plasma sheet of $\sim 1/100$th of the mean free path in the rest of the tail.

Figure 2b shows the mapping of a typical dawn-dusk polar orbit into the tail and Figure 2c shows the expected polar cap profile.

In the case of a north-south interplanetary anisotropy the particle intensity distribution in the magnetotail is identical for both adiabatic and nonadiabatic access modes, i.e. high intensity in one tail lobe and low intensity in the other. In the case of a dawn-dusk anisotropy the adiabatic entry mode will also result in this same intensity distribution. However, in the nonadiabatic access model isotropisation in a boundary layer at the magnetopause of that part of the flux which is moving towards the magnetopause should lead to a particle gradient in the tail. In this model one expects a somewhat sharp change-over between forward and reverse particle fluxes at a plane through the tail axis which is inclined at right angles to the interplanetary anisotropy direction. A correspondent flux change over the polar cap is expected whenever the projection of the satellite orbit crosses this plane.

In order to compare the predictions of the open tail model with the diffusion model Scholer and Morfill (1972) calculated the diffusion of particles into a tail consisting of two cylinders which have initial internal density of zero and in which diffusion is purely radial. The boundary conditions depend on the direction of the anisotropy in the plane perpendicular to the Sun-Earth line and are represented by a first or second order harmonic expansion, a description, which is quite appropriate for the interplanetary particle distribution (e.g. Engel *et al.*, 1970). Solving the two dimensional time-dependent diffusion equation Scholer and Morfill (1972) produced iso-intensity contours of the particle intensity in the two cylindrical tail lobes, which can be mapped afterwards via trajectory integrations onto the polar cap.

Van Allen *et al.* (1971) made a comprehensive comparison of interplanetary anisotropy measurements with simultaneously observed polar cap structures. The interplanetary proton fluxes incident from the north and south ecliptic directions onto the ecliptic plane are shown in Figure 3 together with the high latitude north and south polar cap average fluxes. As can be seen the north-south polar asymmetry tracks well with the north-south anisotropy, which was interpreted by van Allen *et al.* (1971) as an indication for direct interconnection between interplanetary and tail magnetic field lines. Scholer and Morfill (1972) compared individual polar cap measurements of van Allen *et al.* (1971) with their diffusion calculation. Figure 4 shows the calculated polar cap structure at the north and south pole for a dawn-dusk orbit passing through 90° invariant latitude together with data points of van Allen *et al.* (1971), where the polar satellite orbits were approximately dawn-dusk. The diffusion coefficient was chosen so as to give a good fit to the data. From the

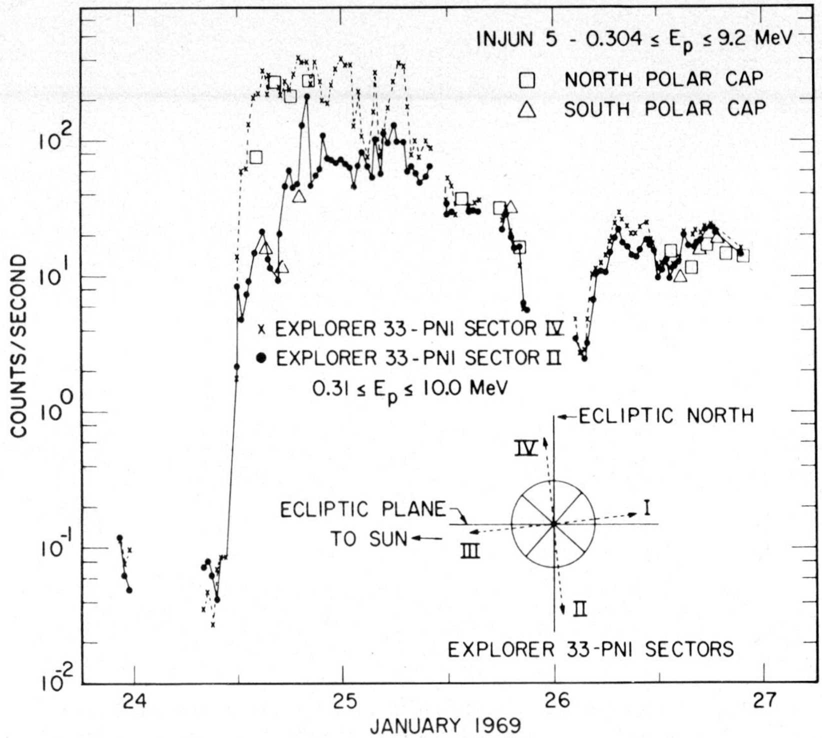

Fig. 3. Graphical comparison of the normalised north and south polar cap averaged counting rates from Injun 5 with the half hourly averaged counting rates from Explorer 33's sector 4 and 2 (southward and northward-directed intensities). The arrows in the auxiliary diagram indicate the mean directions of the collimator axis in the four sectors (Van Allen *et al.*, 1971).

good agreement of the measured polar cap structure with the calculations of the diffusion model as well as with the predictions of the open tail model, Scholer and Morfill (1972) suggest that it is too difficult to differentiate between magnetospheric models on the basis of north-south asymmetries alone.

In general, if the characteristic diffusion time τ_D is less than a characteristic interplanetary time τ_i given by the particle intensity or anisotropy changes, the diffusion case becomes indistinguishable from the open model predictions, so that the polar cap asymmetry reflects the instantaneous interplanetary north-south anisotropy. However, if rapid variations in the interplanetary anisotropy ratios occur the speed with which the polar cap structures follow these variations will be different for the two tail models. Therefore despite of the fact that individual polar cap measurements of van Allen *et al.* (1971) can be interpreted as being due to diffusive access, the close temporal correlation between the interplanetary anisotropy and the polar cap asymmetry favours an open tail interpretation.

Polar cap observations during an interplanetary east-west anisotropy were first reported by Haskell and Hynds (1972). Figure 5 shows the counting rate of two consecutive northern polar passes vs. invariant latitude. From the systematic dif-

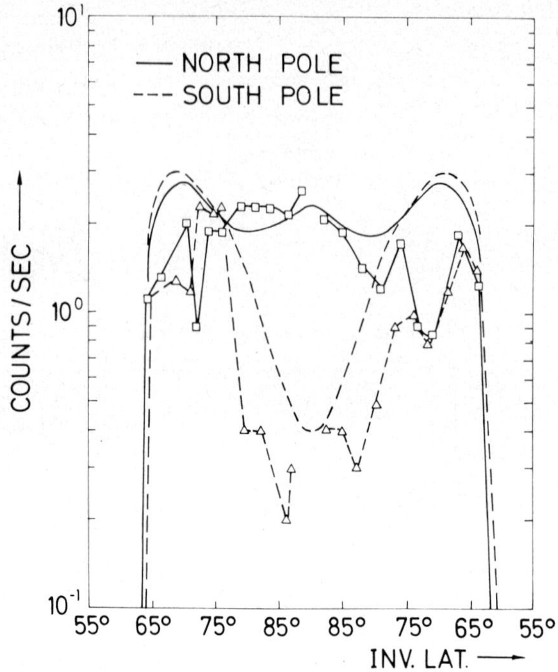

Fig. 4. Polar cap intensity structure as a function of invariant latitude for a dawn-dusk orbiting satellite passing through 90°. The theoretical curve uses an anisotropy ratio of 13 at a value for $T = \kappa t / R^2 = 0.15$, and the measurements are taken from Van Allen *et al.* (1971) Figure 7 (Scholer and Morfill, 1972).

ferences in counting rates in the open field line region over the same polar cap it can be concluded that the dawn half of the northern tail lobe holds a higher particle intensity than the dusk half, which cannot be explained by the adiabatic reconnection model. Also shown in Figure 5 are the results obtained from diffusion calculations for three different times in the case of an interplanetary east-west anisotropy. Since diffusive interpretation yields as little agreement with the measurements as the adiabatic reconnection model, it can be concluded that the easiest interpretation of the measurements of Haskell and Hynds (1972) is in terms of the nonadiabatic access mode.

Rampling (1972) has recently investigated the effect of a change in the interplanetary anisotropy direction on the polar cap structure which results. The bottom panel in Figure 6 shows the polar cap particle flux averaged in the open field line region in the dawn and dusk halves of the polar cap respectively as a function of time during an anisotropic solar particle event. The upper three panels show the magnitude and direction of the interplanetary magnetic field together with the ϕ component of the particle anisotropy. At ~1630 UT a tangential discontinuity swept past the Earth, and whilst previously the anisotropy maximum came from the dawn side, along the Parker spiral direction it came from the dusk side after the passage of the discontinuity, although the anisotropy ratio was then significantly

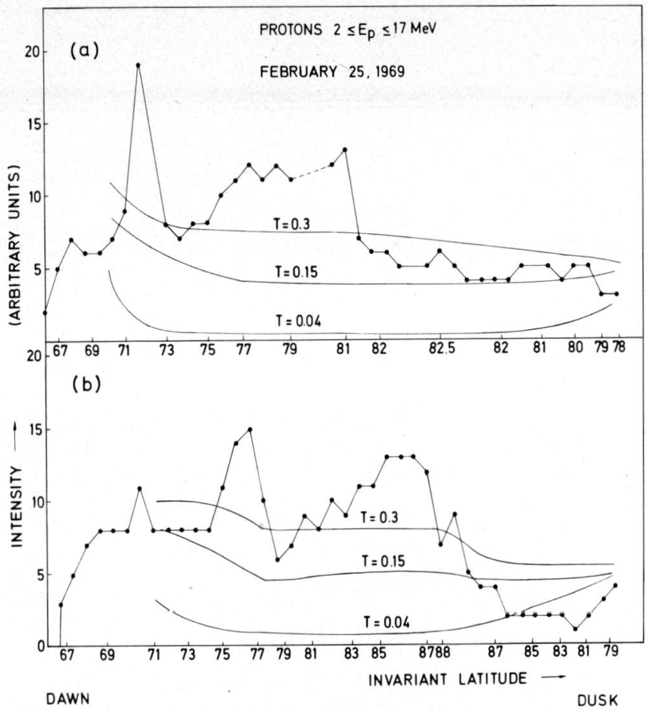

Fig. 5. Intensity vs invariant latitude for two satellite passes of the ESRO 2 satellite (2033 to 2047 UT and 2347 to 2402 UT on February 25, 1969) together with the results of the diffusion calculation for an east-west anisotropy (Morfill and Scholer, 1972a).

reduced. Figure 6 shows that if maximum interplanetary flux impinges on the dawn flank of the tail and the minimum flux on the dusk flank, the measured particle intensity in the dawn side of the polar cap and hence in the dawn half of the tail exceeds that in the dusk half by an amount similar to the anisotropy ratio. Furthermore, there is hardly any time delay between the direction changes in the interplanetary anisotropy and the corresponding establishment of the asymmetry over the polar cap. This result argues strongly for the nonadiabatic reconnection model.

3. Interplanetary Discontinuities

Tangential discontinuities in interplanetary space can separate regions of different proton flux. As such distinct flux variations are convected by the solar wind along the magnetotail, they provide an ideal means to study the proton access from different regions of the magnetotail and the time delay between the intensity changes outside and inside the magnetosphere can be used to estimate the length of a coherent magnetotail.

Figure 7 shows a schematic cross section through one tail lobe. A discontinuity has reached a distance L down the tail and particles entering at a point M down

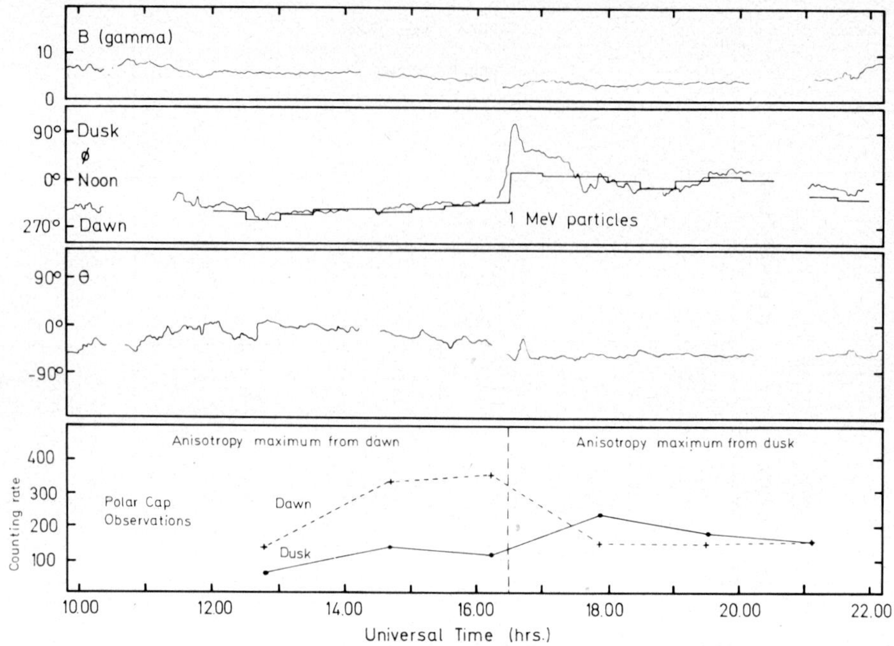

Fig. 6. The upper three panels show the interplanetary magnetic field (B, ϕ and θ components) measured during November 18th 1968 by Explorer 35. Superimposed on the ϕ component is the anisotropy direction of 1 MeV protons measured in the ecliptic plane by Venkatarangan *et al.* (1970). The lower panel shows the ESRO 2 polar cap observations in the dawn and dusk halves at 2–17 MeV proton energies, which demonstrates the correlation between the polar cap dawn-dusk asymmetry and the interplanetary anisotropy direction (Rampling, 1972).

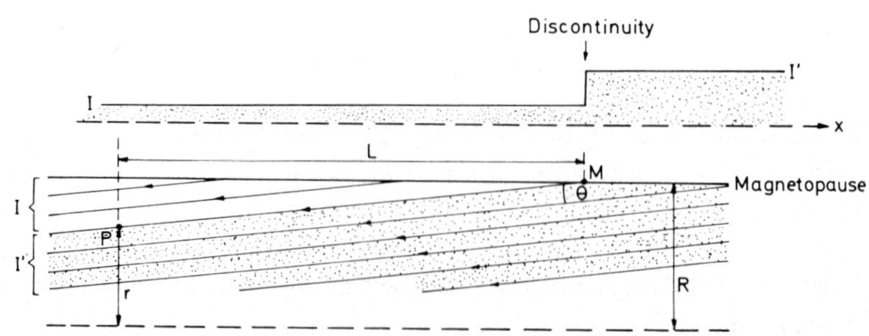

Fig. 7. Schematic cross section through one tail lobe in order to demonstrate the effect of a discontinuity in the particle intensity caused by interplanetary magnetic field structure.

the tail can be observed at a point P laying at a radial distance r. The magnetic field is assumed to make a small angle θ with the magnetopause due to reconnection with the interplanetary magnetic field. The normal component of the magnetic field at the magnetopause is $B_N = B_0(R-r)\cos\theta/L$. If the interplanetary discontinuity travels the distance L in time τ one gets with the solar wind speed V_S and $\cos\theta \approx 1$:

$$\langle B_N \rangle \approx B_0 \frac{R-r}{V_S \tau}.$$

In order to maintain a stationary reconnection between interplanetary and tail field, magnetic flux has to reconnect across the neutral sheet and field lines must move inwards to the earth. This inward motion leads to a $\mathbf{V} \times \mathbf{B}$ cross tail electric field

$$E_y = \frac{1}{c} \langle V_{TN} B_{TN} \rangle,$$

where V_{TN} is the velocity of the field lines moving to the Earth and B_{TN} is the perpendicular component of the magnetic field in the neutral sheet. Since in a steady state the rate of flux loss out of the tail must be equal to the flow into the neutral sheet one gets one the average:

$$\langle V_{TN} B_{TN} \rangle \propto \langle B_N \rangle \; V_S,$$

so that the cross tail electric field is

$$E_y \propto B_0 \frac{R-r}{\tau}.$$

The time τ can be determined by multisatellite observations, the distance r can be obtained by projecting the observed polar cap structure of the discontinuity via trajectory calculations on a cross section of the tail. The quantity $B_0(R-r)$ gives the amount of magnetic flux (or field lines) between the measuring position and the magnetopause and has to be replaced in a more realistic magnetospheric model by the 'magnetic depth' $D - \int \mathbf{B} \times \mathbf{dl}$. The integral is taken from a point inside the tail which connects via a trajectory with the corresponding measurement position over the polar cap to the magnetopause along a path which is perpendicular to B everywhere (Morfill, 1972).

By using intensity variations over the polar cap which are produced by interplanetary discontinuities, it is possible to calculate the mean free path of the particles in the magnetotail: After the passage of a discontinuity past the Earth a polar orbiting satellite will observe an intensity change during some stage of its orbit and depending on the amount of scattering this change is abrupt or smeared out. Morfill (1972) assumes that particle scattering is due to gyroresonance interaction with small amplitude Alfvén waves. The mean free path parallel to the magnetic field is then given by

$$\lambda_{\parallel} = \frac{4R_i^2 \mathbf{V}_S \tau}{3D_e^2},$$

where R_i is the particle rigidity and D_e is the scale length of the measured spread in the particle distribution measured as magnetic depth. From the mean free path for motion along the magnetic field lines one gets the power in the transverse fluctuation spectrum at the particle resonant frequency.

Detailed multisatellite observations of such discontinuities were reported by Scholer (1972) and Rampling (1972). Scholer (1972) identified an interplanetary filament by two sudden changes in the interplanetary magnetic field θ_{SE} component and simultaneous changes in the interplanetary 1 MeV solar proton flux and a following single discontinuity that separates two regions of different proton flux. These discontinuities led to corresponding structures over the polar cap. For an observed solar wind velocity of 500 km s^{-1} the measured time delays gave a tail length of $\sim 790\ R_E$ for the filament and $\sim 550\ R_E$ for the discontinuity and corresponding cross tail electric fields of 2.2×10^{-4} V m^{-1} and 2.6×10^{-4} V m^{-1}, respectively. The sharpness of the polar cap profile due to the filament resulted in a mean free path parallel to the magnetic field of $\sim 1000\ R_E$ (Morfill and Scholer, 1972b). The differences in tail lengths were attributed to different directions of the interplanetary field during the passage of the filament and the discontinuity, thus resulting in different reconnection rates.

Rampling (1972) has also made detailed investigations of polar cap particle structures during a time when an interplanetary discontinuity was convected past the Earth. During that time polar cap observations were made almost simultaneously by two spacecraft, ESRO 1 in a noon-midnight orbit and ESRO 2 in a dawn-dusk orbit. Figure 8 shows the measured polar cap intensity structure during the northern passage of the two satellites. The orbits of the two satellites were projected onto a cross section of the tail and the measured particle flux is represented as the length of the lines perpendicular to the orbit projection. The higher flux was observed in front of the tangential discontinuity. As can be seen from Figure 8, the lower intensity observed on the Sunward side of the discontinuity has already filled the outer rim of the tail at the time of the passage of the two polar satellites, so that they observed the higher flux (shaded area) in the centre of the tail lobe.

From the observations of Rampling (1972) a cross tail electric field of $\sim 1.4 \times 10^{-4}$ V m^{-1} can be deduced. Despite the fact that these observations were made during times with a southern interplanetary magnetic field component (see Figure 6) is this value smaller than that obtained by Morfill and Scholer (1972b) during a time with a northern interplanetary magnetic field and illustrates the variability of the magnetospheric topology.

Fennell (1973) compared the time histories of proton fluxes over the polar cap with interplanetary fluxes at times when interplanetary particle gradients existed. Assuming that these gradients were bounded by tangential discontinuities he obtained a measure of the access distance down the tail by timing the delay required to match the interplanetary and polar cap intensities. The results of this statistical study are subdivided into open field line region and pseudo-trapping region at positive or negative interplanetary sector. The statistical distribution of access points

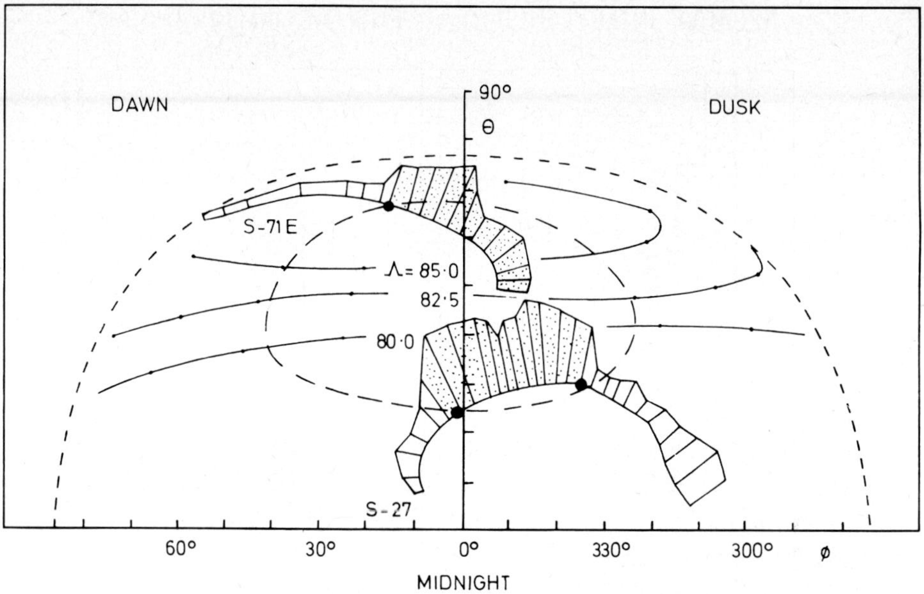

Fig. 8. Projections of satellite orbits (ESRO 1 with experiment S-71E and ESRO 2 with experiment S-27) along 0.08 GV trajectories to 15 R_E in the tail during the November 1968 flare, after the passage of an interplanetary tangential discontinuity past the geomagnetic cavity. The measured particle intensities are represented as the lengths of the lines drawn normal to the orbit projections. Also shown is the magnetopause and some trajectory contours at $\Lambda = 85°$, 82.5° and 80° (Rampling, 1972).

down the tail is shown in Figure 9. According to Fennell (1973) no significant relation between access distance and magnetic sector structure nor any dependence on the inclination of the interplanetary field with respect to the ecliptic could be found. The mean access distances are 95 to 130 R_E for the low polar region and about 260 R_E for the open field line region. However, such a mean access distance is difficult to interpret and might depend on the event sampling technique employed. For instance averaging counting rates on the open field line region irrespective of magnetic local time and invariant latitude should lead to a statistically more or less uniform sampling of the whole tail lobe. The probability of a sampling point being on a field line, distance r from the centre of the tail lobe is r^2. Since in the open tail model field lines at the largest values of r undergo reconnection with interplanetary field lines closer to the Earth than those with small r-values, one gets more frequent measurements of access positions close to the Earth than at larger distances (Morfill, private communication). The normalized probability of a particular sampling point being on a field line distance r from the centre of the tail lobe is

$$P(r) = \frac{3r^2}{R^3}.$$

The maximum tail length can be defined by the distance L_{max} over which the field line originally at the centre of the tail lobe reaches the magnetopause. Since accord-

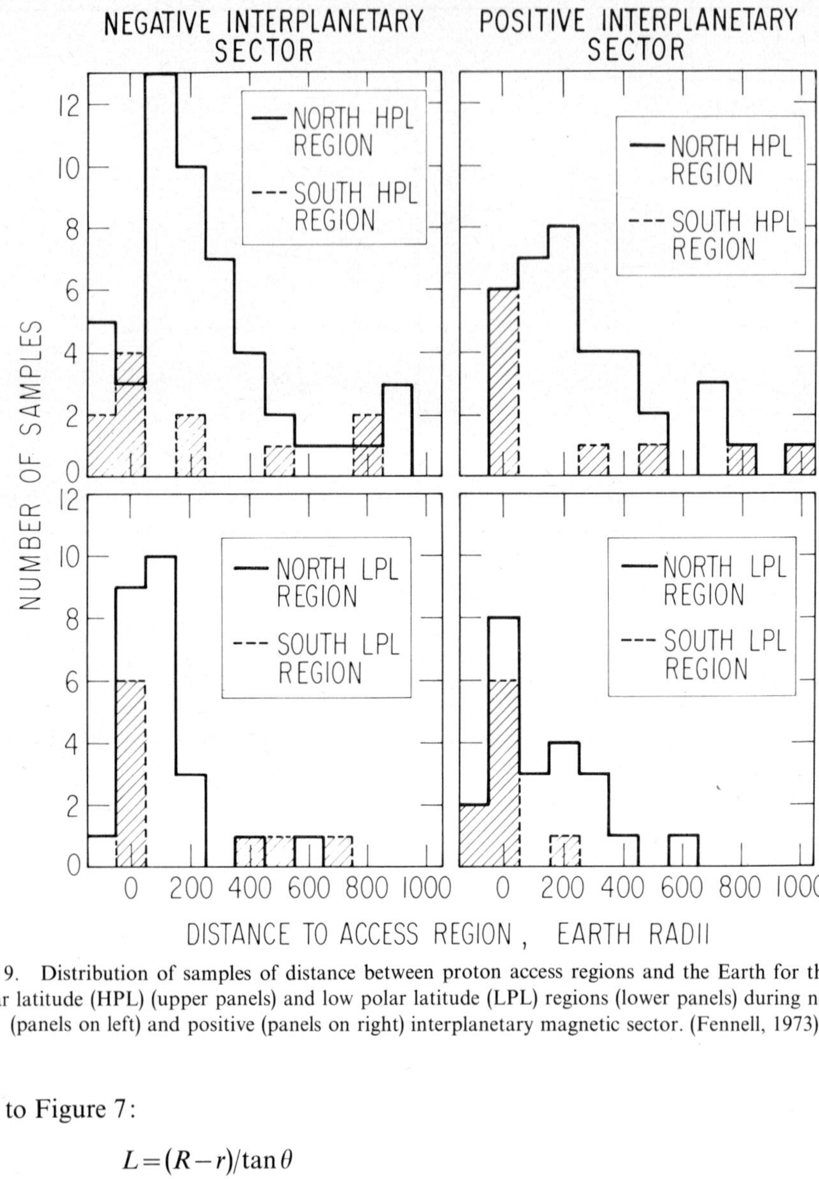

Fig. 9. Distribution of samples of distance between proton access regions and the Earth for the high polar latitude (HPL) (upper panels) and low polar latitude (LPL) regions (lower panels) during negative (panels on left) and positive (panels on right) interplanetary magnetic sector. (Fennell, 1973).

ing to Figure 7:

$$L=(R-r)/\tan\theta$$

$$L_{\max}=R/\tan\theta$$

we get

$$r=(L_{\max}-L)\frac{R}{L_{\max}}.$$

Hence the probability of observing a 'tail length' L is given by

$$P(L)=\frac{3}{RL_{\max}^2}(L_{\max}-L)^2.$$

It can be shown that using $R \sim 20\ R_E$ and $L_{max} \sim 1000\ R_E$ gives close agreement with the statistical distribution of observed access points given by Fennell (1973). Thus the experimental data could be interpreted as favouring a 'mean tail length' of $\sim 1000\ R_E$, with a probable error of $\sim 300\ R_E$, since the form of $P(L)$ is not too sensitive to $\sim 30\%$ changes in L_{max}.

4. Summary

Multisatellite measurements of solar particles at times when large anisotropies or intensity gradients exist in interplanetary space are a promising method to study the magnetospheric topology. The evidence of such measurements is in favour of the reconnection model first proposed by Dungey (1961) with a turbulent boundary layer at the magnetopause. The average convection electric field appears to lie between 2×10^{-4} and 5×10^{-4} V m^{-1}. The length of the coherent geomagnetic tail lies between 500 and 1000 R_E. Scattering in the magnetotail seems to be weak; mean free paths in the tail are of the order of $\sim 1000\ R_E$ for 1 MeV protons. The variations of the convection electric fields and corresponding tail length with the interplanetary magnetic field direction and solar wind parameters is unclear and remain to be investigated.

References

Dessler, A. J.: 1964, *J. Geophys. Res.* **73**, 3913.
Dungey, J. W.: 1961, *Phys. Rev. Letters* **6**, 47.
Engel, A., Balogh, A., Elliot, H., Hynds, R. J., and Quenby, J. J.: 1970, *Acta Phys. Hungaricae* **29**, Suppl. 2, 439.
Fennell, J. F.: 1973, *J. Geophys. Res.* **78**, 1036.
Haskell, G. P. and Hynds, R. J.: 1972, in B. M. McCormac (ed.), *Earth's Magnetospheric Processes*, D. Reidel Publ. Co., Dordrecht-Holland, p. 81.
Michel, F. C. and Dessler, A. J.: 1965, *J. Geophys. Res.* **70**, 4305.
Michel, F. C. and Dessler, A. J.: 1970, *J. Geophys. Res.* **75**, 6061.
Morfill, G.: 1972, *J. Geophys. Res.* **77**, 4010.
Morfill, G. and Quenby, J. J.: 1971, *Planetary Space Sci.* **19**, 1541.
Morfill, G. and Scholer, M.: 1972a, *Planetary Space Sci.* **20**, 2113.
Morfill, G. and Scholer, M.: 1972b, *J. Geophys. Res.* **77**, 4021.
Rampling, R.: 1972, Ph. D. thesis, Imperial College, London.
Scholer, M.: 1972, *J. Geophys. Res.* **77**, 2762.
Scholer, M. and Morfill, G.: 1972, *Planetary Space Sci.* **20**, 1051.
Van Allen, J. A., Fennell, J. F., and Ness, N. F.: 1971, *J. Geophys. Res.* **76**, 4262.
Venkatarangan, P., Venkatesan, D., and Van Allen, J. A.: 1970, *Acta Phys. Hungaricae* **29**, Suppl. 2, 421.

ON A MODEL OF PROPAGATION OF 5–300 MeV PROTONS IN THE EARTH MAGNETIC CAVITY

The Role of Neutral and Plasma Sheets

RUTH GALL* and SILVIA BRAVO

Instituto de Geofisica, Universidad Nacional Autonoma de Mexico, Mexico, D.F. Mexico

Abstract. The general pattern of access and propagation of nonrelativistic (5–300 MeV) protons in the Earth cavity is derived by simulation of motion with trajectory techniques. The neutral and plasma sheets are shown to play a preferential role in the access to auroral latitudes. The windows of access lie along the dawn-night boundary of the sheets at $R < 60\ R_E$ and are energy and pitch angle dependent. Based on the general pattern a model for auroral peak-valley flux profile during the anisotropic phase of solar event is described and compared with other models. The limitations of the available theoretical models are discussed and a method for comparison with observed flux profiles proposed.

1. Introduction

Due to their rigidity range, 5–300 MeV, the non-relativistic protons that we shall be concerned with are sensitive to the major structural characteristics and dynamics of the Earth magnetic domain. Consequently one can expect the following relations between some structural and propagation characteristics, listed in Table I. Due to the geomagnetic cut-off these particles find access at low altitudes over polar caps where they have been observed by detectors onboard satellites and by balloons flown over auroral and polar stations.

To relate the observed flux with the interplanetary flux and magnetospheric

TABLE I

Cavity structural characteristics and C.R. propagation characteristics

	Cut-off
Tail (neutral and plasma sheet)	Cut-off below internal field value
Axial-asymmetry	Daily cut-off variation
Partial tail collapse	Sudden increase of cut off
	Anisotropic incident flux
Magnetosheat and Magnetopause	Flux modulation Isotropization (~ 1 MeV)
	Illumination of polar caps by anisotropic flux
Neutral and plasma sheet	Auroral flux peaks: at all local times, at both polar caps
Lobes of tail	Illumination of high polar latitudes
Reconnection of geom. lines across sheets (B_z)	Turning mechanism (in the neutral and plasma sheets)
Merging on $\bar{B}_{\text{geom}}$ and $\bar{B}_{\text{interpl}}$	Turning mechanism (in tail lobes)
Difference in N, S merging rate	N/S illumination asymmetry (delay)

* Also at Instituto Nacional de Energia Nuclear, Mexico, D.F. Mexico.

D. E. Page (ed.), Correlated Interplanetary and Magnetospheric Observations, 433–447. *All Rights Reserved.*

structure the following fundamental question must be answered: Through which region has the particle entered into the geomagnetic cavity and which region of the cavity has it transversed on its way to the detector? A theoretical model on non-relativistic proton propagation should offer a satisfactory answer to this fundamental question. We shall first discuss some general aspects of theoretical models obtained by trajectory tracing technique and emphasize the role played by the neutral and plasma sheet. We shall finally elaborate a model for the auroral flux peaks observed during the anisotropic phase of solar events and discuss it in the light of other models proposed.

2. The Making of a Propagation Model

As the particle undergoes no scattering and its motion is non-adiabatic, except in the Earth immediate vicinity, the motion can best be simulated by numerical integration of trajectories in a magnetospheric model. The magnetospheric models presently available are sufficiently good to allow the prediction of the general characteristics of access and propagation. However, due to the fact that the models deviate from the 'real' field and due to the field variability one cannot expect precision in the quantitative values obtained and comparison with experimental results should be made with care. This is discussed in detail in the second part of this paper. Our study on propagation is done, as illustrated in Figure 1, in three successive steps,

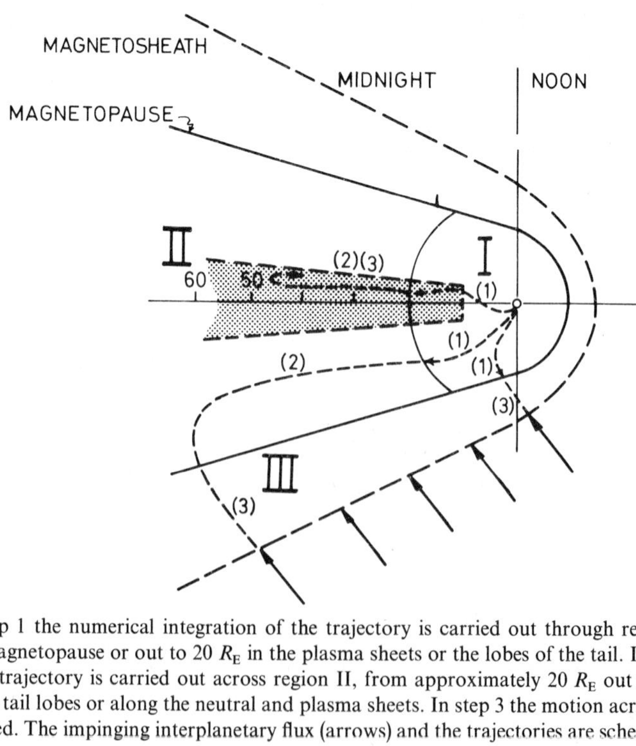

Fig. 1. In step 1 the numerical integration of the trajectory is carried out through region I, from the Earth to the magnetopause or out to 20 R_E in the plasma sheets or the lobes of the tail. In step 2 the integration of the trajectory is carried out across region II, from approximately 20 R_E out to the boundary lying along the tail lobes or along the neutral and plasma sheets. In step 3 the motion across the magnetosheat is analysed. The impinging interplanetary flux (arrows) and the trajectories are schematically drawn.

retracing the motion outward from the Earth across three different regions up to the interplanetary medium. The separation in steps is due to the fact that models which represent adequately the field both in the inner and outer magnetosphere as well as in the near and distant tail and in the magnetosheat are not available.

In the 1st step the motion is simulated starting from the position of the detector out to distances not larger than 20 R_E. Several hundreds trajectories along which 5 to 300 MeV protons reach all invariant latitudes of southern and northern polar caps at four different local times, namely noon, midnight, dawn and dusk were computed (Gall *et al.*, 1968, 1969, 1972). The magnetospheric model used in the computation is a combination of high simulation of the internal field with IGFR coefficients to which the Williams and Mead expressions for the external field $\bar{B}_{cs}$ and $\bar{B}_s$ were added, taking into account the angle between the geomagnetic and geographic axes. The parameters used for B_{cs} are: $R_1 = 10\ R_E$, $R_2 = 40\ R_E$, $B_{cs} = 40\gamma$. For B_s corresponding to currents in the magnetopause the 2-term expansion was used; the computations were further checked with a 6-term expansion that lead to practically the same results (Gall and Orozco, 1973). In the 1st step the motion is traced up to the magnetopause, where the point of entry and the direction of approach to the magnetopause are computed or if the trajectory stretches out into the tail the integration is stopped at 20 R_E in the lobes or in the neutral and plasma sheets. This is equivalent to mapping the point of observation with the point of access of flux into region I (see Figure 1). One should bear in mind that the computed direction of approach to the boundary are not necessarily the same as the direction of approach from the interplanetary medium since the flux might have undergone changes when traversing the magnetosheath.

The main characteristics of propagation to northern and southern polar caps derived from the 1st step can be summarised as follows:

(1) cut-off energies reach any latitude at all local times via the neutral and plasma sheets.

(2) Most of the radiation of energies above the cut-off will reach the auroral latitudes at all local times via the neutral and plasma sheets.

(3) In the hypothetical case in which radiation of any energy would move exclusively in the neutral and plasma sheets, peaks of flux will appear over the auroral latitudes, at any local time (see Figure 2).

(4) Interplanetary anisotropic flux, in the narrow range of pitch angles that finds access into the cavity across the dawn side of the nose of the magnetopause will appear in form of flux peaks only at specific local times, at specific latitudes (see Figure 3).

(5) All protons with $\varepsilon_p \lesssim 80$ MeV will reach high latitudes of N(S) cap at any local time via the N(S) lobe of the tail; protons of $\varepsilon_p \gtrsim 80$ MeV that reach high N(S) latitudes find access across to the northern (southern) tail magnetopause at $R < 20\ R_E$.

The exact upper limit to the energies arriving via lobes of the tail depends on the magnetospheric model used. An experimental proof of the theoretical prediction on

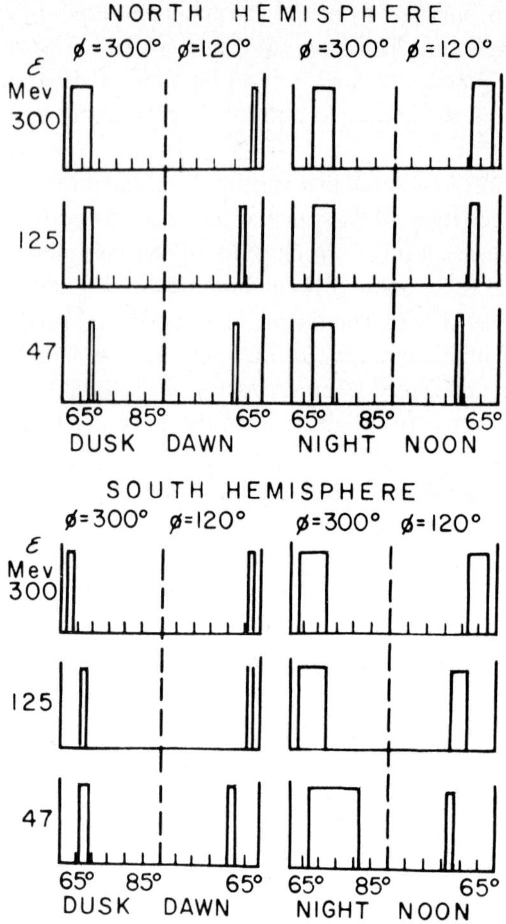

Fig. 2. Schematic representation of computed auroral peaks illuminated by monoenergetic fluxes of protons propagating exclusively via the neutral and plasma sheet of the tail. Simultaneously auroral latitudes of both hemispheres are illuminated at all local times. For simplicity equal fluxes are considered for all energies. The invariant latitudes from 60° to 90° are located in the 300°–120° geographic meridian. In the real case in which an anisotropic flux finds access into the entire cavity, the peaks will be modulated by the energy spectrum and the pitch angle distribution and valleys between the peaks will appear. (Reproduced from Gall *et al.*, *J. Geophys. Res.* **77** (1972), 5360.

propagation to high latitudes was given by Greenhill *et al.* (1972). The neutral and plasma sheet obviously play an important role in propagation since they are extensive regions of the Earth's magnetic cavity of weak magnetic field of the order of the interplanetary field and several fold weaker than the field in the lobes of the tail. Consequently they constitute a preferential region of access for propagation of protons to the polar caps at any local time. This preferential role is illustrated in Figure 3.

Extensive analysis of the first step of proton propagation was made by Morfill and Quenby (1971) using a similar model with a six-term expansion of B_s. The computed access points do not differ significantly from those obtained by the authors.

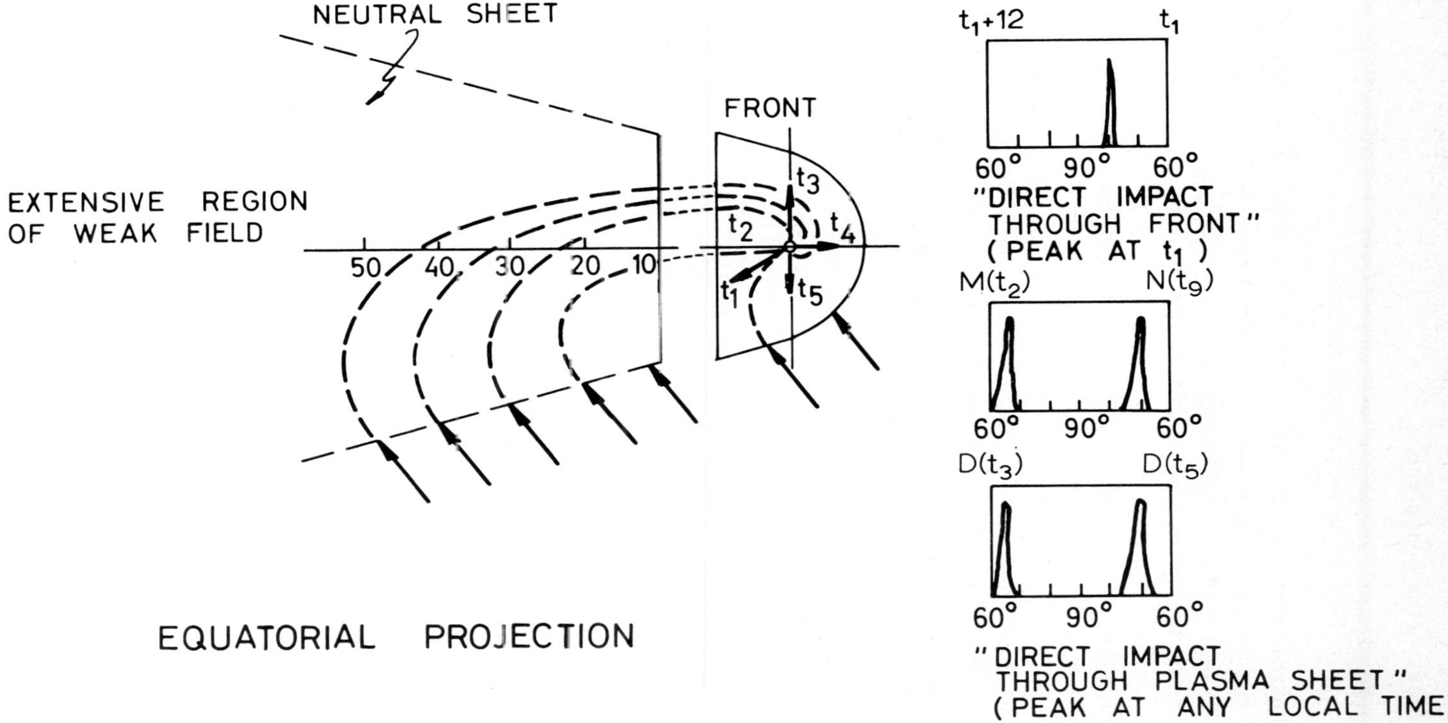

Fig. 3. Schematic representation of the propagation of the interplanetary anisotropic flux across the cavity to the Earth. The preferential role of the plasma sheet is illustrated. Flux impinging along the sheet will reach the auroral latitudes at all local times, while the flux moving across the dawn side of the magnetopause will reach a given latitude at a special local time. The peaks formed by the flux moving across the front or across the sheet are schematically illustrated in the flux vs invariant latitude plots.

Let us now apply the obtained results on propagation to the problem of auroral flux peaks observed during the solar events.

3. Flux Peaks over Auroral Latitudes

The fact that during the solar event the polar caps are not evenly illuminated by solar protons but instead present flux latitudinal profiles have been known since 1965. The complexity and variability of these profiles can probably only be matched by the complexity of interpretations that one finds in numerous papers written since on the subject.

One feature that is common to the majority of latitudinal profiles observed during the anisotropic phase by orbiting polar satellites are the auroral flux peaks (see for example Evans, 1972). These peaks observed in the orbital plane of the satellite at any local time are joined over the higher latitudes by a valley. We shall be concerned here only with this fundamental structure and will not discuss other aspects such as polar peaks or N/S asymmetry.

Several interpretations of the formation of auroral peaks based on computed trajectories have recently been proposed (Engelmann *et al.*, 1971; Gall *et al.*, 1971, 1972; Morfill and Quenby, 1971; Morfill and Scholer, 1973). We shall first develop further the model proposed by the authors and later review different models and discuss adequate methods of their application to experimental results.

The results on the 1st step of motion simulation discussed in the previous section indicate that the auroral peaks are engendered by radiation moving via the neutral and plasma sheets.

4. Windows of Access and the Turning Mechanism

To complete the model of Auroral Peak-Valley structure we proceed with the 2nd step of analysis and study the access at the boundary of the anisotropic flux into the neutral and plasma sheets and the mechanism responsible for turning the impinging flux into an Earthbound flux. The Mead and Williams expressions cannot be used in region II. Moreover at $R > 10\ R_E$ in region I they lack the B_z component of the field in the sheets whose presence introduces a strong dawnward shift of the computed points of access (See Figure 4). In the models of Willis and Pratt (1972) and Olson and Pfitzer (1973), which are valid up to large distances in the tail, the separation of 1st and 2nd steps are unnecessary.

The fact that the elongated geomagnetic field lines 'reconnect' across the neutral and plasma sheet was confirmed by the detection of the transversed B_z component (Behannon and Ness, 1966) and discussed by Bird and Beard (1972); B_z decreases with distance from 5 γ at 20 R_E to 1 γ at 60 R_E. The B_z component provides the effect of turning towards the Earth the radiation impinging on the dawn-night side of the boundary along the neutral and plasma sheet. The operation of this turning is illustrated in Figure 4 where trajectories computed with the help of a simple tail field

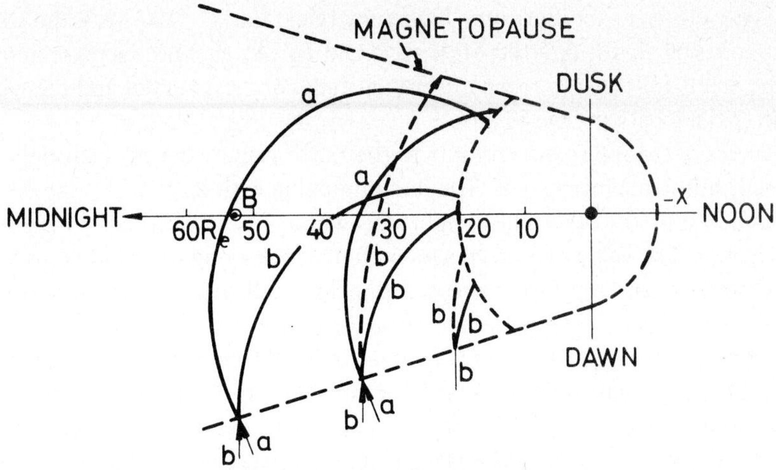

Fig. 4. The computed trajectories followed by 5 MeV (solid lines) and 50 MeV (pointed lines) protons in the neutral sheet. The protons approach the boundary with pitch angles of $\alpha = 15$ and $\alpha = 40$. (α is the angle between the particle velocity vector and the interplanetary field vector.) The trajectories illustrate the turning mechanism due to the B_z component. (Reproduced from Gall and Bravo, *J. Geophys. Res.* **78** (1973),

including a B_z component are plotted. About 100 trajectories of 5–300 MeV protons impinging with pitch angles of $0° < \alpha < 60°$ at different points located between 20 to 60 R_E along the night-dawn side of the boundary were computed and the windows of access derived (Gall and Bravo, 1973). Trajectories at $R < 17$ R_E was further studied with the help of the Mead-Fairfield (1972) model. A typical window of access is illustrated in Figure 5. The windows are wide open for low energies and narrow

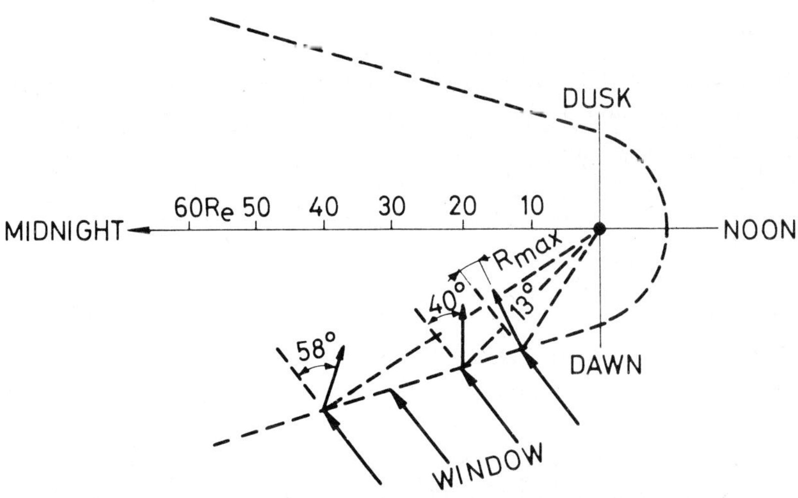

Fig. 5. Window of access for an anisotropic flux of 47 MeV protons along the boundary of the neutral sheet. Thick arrows indicate the interplanetary field, the thin arrows the particle velocity vector corresponding to the minimum pitch angle under which the particle that impinges at a given point of access will rotate towards the Earth. (Reproduced from Gall and Bravo, *J. Geophys. Res.* **78** (1973), 6773)

down, lying closer to the Earth for higher energies. For $\varepsilon_p < 100$ MeV the windows lie at $20 < R \lesssim 60\ R_E$ for $\varepsilon_p \gtrsim 100$ MeV at $R < 40\ R_E$. As energies increase above 300 MeV the orbits followed to reach auroral latitudes are less curved and the windows lie close to the local time meridians.

The access of the interplanetary flux to the Earth vicinity depends strongly on the pitch angle at the magnetopause. The flux impinging with $\alpha < \alpha_l$ will cross the cavity and abandon it on the other, dusk-night side, while the flux with $\alpha > \alpha_1$ will reach the Earth vicinity. The value of α_l decreases with distance along the window so that the maximum interplanetary flux corresponding to small pitch angles will reach the Earth coming from the nearest points of access of a given window.

The windows discussed here were computed for flux moving in the plane of the ecliptic. They represent only the first approximation to windows of access for the illumination of auroral peaks. The latter will necessarily extend in width up to $\pm 8\ R_E$ to include the boundary along the plasma sheets as well.

A tail model that includes B_z component was also extensively used by Morfill and Quenby (1971) and applied to the study of the dispersion of low energy trajectories and the dawnward bending of trajectories linked to the dusk side of the auroral regions. The dawnward displacement of the points of access induced by the presence of B_z is clearly exhibited by Thomas *et al.* (1973).

5. A Model for Auroral Peaks

By combining the results of the 1st and 2nd step of motion simulation with the computations on the daily variation of cut-off (Smart *et al.*, 1969), the following theoretical model for auroral peak-valley profile – based entirely on trajectory simulation – is derived. The model applies to the anisotropic phase of the event. In the absence of scattering, by Liouiville theorem the interplanetary flux of a given pitch angle is conserved all along the dynamical path of the particle within the cavity. The following are the characteristics of the model:

(1) *Access:* the interplanetary flux that illuminates the auroral peaks of the northern and southern polar caps, at any local time finds access into the cavity across the dawn-night side along the boundary of the neutral and plasma sheets. The windows of access are energy dependent and lie at $R \lesssim 60\ R_E$.

(2) *Turning*: the transverse B_z component in the neutral and plasma sheets provides the turning mechanism.

(3) The rising slope of the peak is a function of the energy spectrum of the anisotropic interplanetary flux and is modulated by its pitch angle distribution. It starts at the cut-off latitude of the highest energy of the channel of the detector and rises until it reaches first the latitudinal cut-off of the lowest energy of the channel.

(4) The shape and the height of the top as well as the falling slope of the peak is a function mainly of the pitch angle distribution of the interplanetary flux and is modulated by the spectrum.

(5) The outer edge of the dawn auroral peak at higher latitudes may be illuminated

– during certain events – by radiation that has access across the dawn side of the nose of the magnetopause.

(6) The valleys lie at latitudes far above the latitudinal cut-off and are illuminated mainly by the flux that crosses the magnetopause at $10 < R < 20\ R_E$ under large pitch angles.

(7) As an isotropic interplanetary flux illuminates evenly all latitudes of the polar caps above the cut-off, the evolution of the structure with time into a plateau is due to flux isotropisation.

A schematic representation of the model for the auroral peak-valley profile is shown in Figure 6. The main part of the peak is illuminated via the neutral and plasma sheets (N.S.). The flux illuminating latitudes very near the cut-off (C.O.) is for same events of very low intensity as for these latitudes the points of access and directions of approach may vary by as much as 90° in longitude for 1° change in the latitude of arrival (Fig. 36, Engelmann *et al.*, 1971). Some points of access may indeed be located at the dawn front side of the magnetopause. The computed values of the location and width of auroral peaks for 50 MeV at N. hemisphere are listed in Table II. The regions of access along the magnetopause for the flux that reaches the valleys adjacent to auroral peaks are listed in Table III. Due to the daily variation of the cut-off and the daily variation of the direction of approach to the magnetopause the location and latitudinal width of the auroral peaks is local time dependent.

Experimental proof that auroral peaks are associated with the flux moving along the neutral sheets was recently presented by Fennell (1973).

6. Application of the Theoretical Models to the Observed Structures

The theoretical models of latitudinal structures are very useful guides for the interpretation of the experimental results but should not be applied directly to the ob-

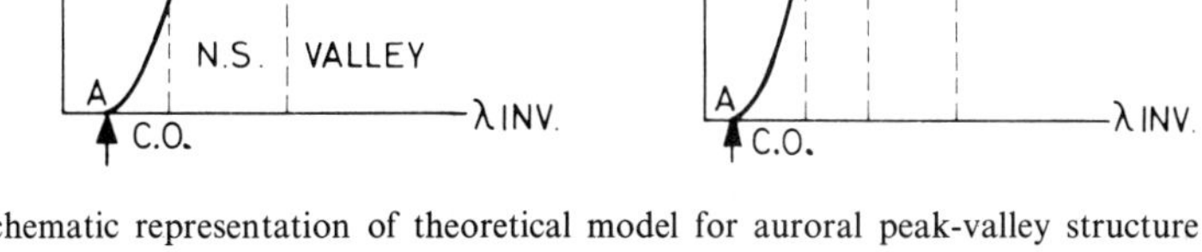

Fig. 6. Schematic representation of theoretical model for auroral peak-valley structure for a mono-energetic anisotropic flux of protons from 5 to 300 MeV. The regions of access at the magnetopause to peaks and valleys over the polar cap are indicated: (C.O.) cut off (N.S.) access at $R < 60\ R_E$ along the neutral and plasma sheets, (D) access at dawn side of the front magnetopause $10 < R < 14\ R_E$; the arrows indicate the cut-offs; *AB* and *BC* are the rising and falling slopes modulated by the pitch angle distribution of the interplanetary flux; the valleys are illuminated by the flux with large pitch angles finding access across different regions of the magnetopause listed in Table III.

TABLE II

Model for auroral peaks and adjacent valleys

Computed invariant latitudes corresponding to cut-offs, to the near cut-off regions (C.O.) to auroral peaks and to valleys are listed for 50 MeV protons, for different local times. The regions illuminated via the neutral and plasma sheets (N.S.) and the front side of the dawn magnetopause (D) are also shown. The listed corrected values ($\lambda_{obs} = \lambda_{theor} - 5°$) can be applied directly to the observed latitudinal flux structures.

Local Time	Energy	λ Cut-off	Auroral peak ←→ ├(C.O.)┤(N.S.)┤	Valley	Corrected values: λ Cut-off	Corrected values: Auroral peak ←→ ├(C.O.)┤(N.S.)┤	Corrected values: Valley
Midnight	50 MeV	65.0°	65.0°–66.5°–73.5°	$\lambda > 73.5°$	60.0°	60.0°–61.5°–68.5°	$\lambda > 68.5°$
Noon	50 MeV	66.5°	66.5°–68.0°–71.5°	$\lambda > 71.5°$	61.5°	61.5°–63.0°–66.5°	$\lambda > 66.5°$
Dusk	50 MeV	65.5°	65.5°–66.5°–69.5°	$\lambda > 69.5°$	60.5°	60.5°–61.5°–64.5°	$\lambda > 64.5°$
			├—(D)—┤			├—(D)—┤	
Dawn	50 MeV	66.0°	66.0°–66.5°–69.5°–74.0°	$\lambda > 74.0°$	61.0°	61.0°–61.5°–64.5°–69.0°	$\lambda > 69.0°$

TABLE III

Illumination of valleys during the anisotropic phase of solar proton event

Valley	Access along the boundary
Noon valley	dusk-noon side $R < 14\ R_E$
Midnight valley	Lobe of the tail $R > 20\ R_E$
Dawn valley	noon-dawn side $R < 14\ R_E$
Dusk valley	dusk-midnight side $14 < R < 20\ R_E$

Regions of access along the magnetopause for fluxes of protons that illuminate the valleys adjacent to auroral peaks. The fluxes at polar caps are of low intensity corresponding to interplanetary fluxes of large pitch angles.

served profiles before a certain correction term is introduced. The direct application is bound to lead to erroneous estimation of the points of entry and consequently to an erroneous interpretation of the mechanisms responsible for the appearance of the latitudinal profiles. Theoretical profiles must be displaced along latitudinal axis as the computed cut-offs are too high due to the limitations of the magnetospheric models used. A correction term is obtained by subtracting the observed cut-off of the latitudinal flux profile (λ_{obs}) from the computed cut-off of the highest energy of the detector channel (λ_{h}). This term (Δ°) corresponds obviously to a linear correction.

$$\Delta^{\circ} = \lambda_{\text{h}}^{\circ} - \lambda_{\text{obs}}^{\circ} .$$

After displacing the invariant latitude axis of the theoretical model by Δ°, the cut-offs will coincide and the experimental flux vs invariant latitude curve can be compared with the model. The correction term is in many cases as large as half of the width of the peaks. Moreover the observed flux profiles should be compared with computations for all the energies of the channel and not only for the mean response energy. The latitudinal cut-offs for polar caps computed with models using Williams and Mead expressions for the external field (Smart *et al.*, 1969) are more than 5° above the observed cut-off. The corrected values, using $\Delta^{\circ} = 5^{\circ}$, of the location and width of auroral peaks and of valleys are listed in Table II.

We have applied the corrected theoretical profile to the dawn and dusk auroral peaks observed by ESRO 2 at 1115 UT during the November 18 th, 1968 event. This is shown in Figures 7a and b. There seems to be a satisfactory agreement between the theory and the observation: The main shape of the peak, its location and width, as well as some fine features of the latitudinal profile, can be explained by the model. The observed abrupt changes in the rising slope seem to match the theoretically predicted changes from the cut-off (C.O.) region to the (N.S.) auroral regions computed for the extreme energies of the channel.

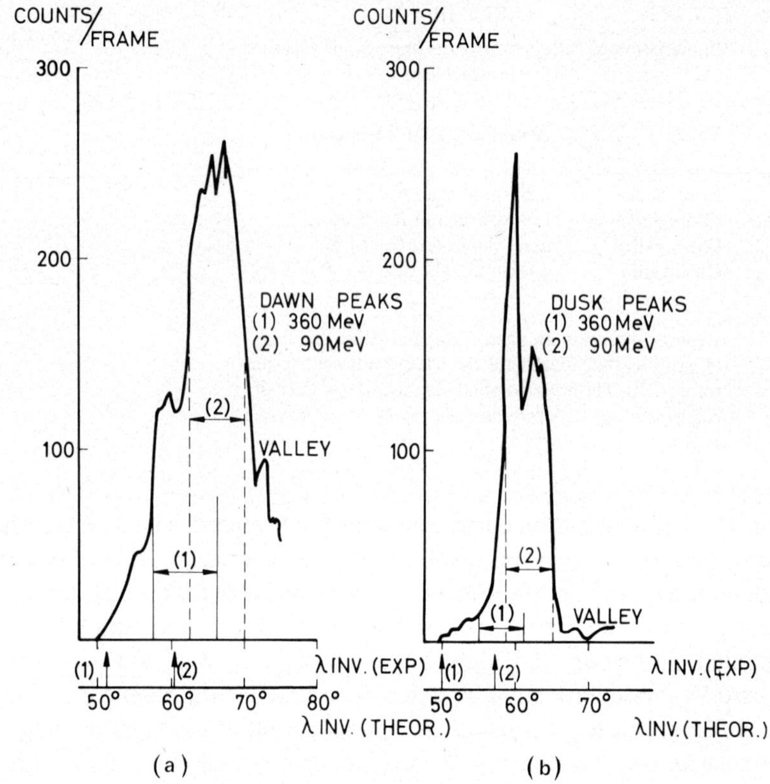

Figs. 7a and b. Theoretical analysis of the observed auroral peaks of 90–360 MeV solar-proton fluxes. (corrected values) The computed cut-offs indicated by arrows as well as the computed auroral peaks for 360 (solid lines) and 90 MeV (pointed lines) are drawn (see Table III). The rising and falling slope of the peaks are modulated by the energy spectrum and pitch angle distribution of the anisotropic interplanetary flux. The dawn auroral peak (7a) and the dusk auroral peak (7b) were reproduced separately from Morfill and Quenby, *Planetary Sqace Sci.* **77**, 5360, Figure 1a where the latitudinal flux profile observed by ESRO 2 at 1115 UT of the February 18th, 1968 event was plotted.

7. Other Models of Auroral Peaks Based on Trajectory Computations

Engelmann *et al.* (1971) have used the trajectory simulation to analyse the auroral peaks observed by the ESRO 2 satellite during the February 25, 1969 event along the 1700 to 0100 local time meridians. They suggest that the dusk auroral peak is due to the anisotropy of the interplanetary flux, the radiation arriving by direct impact through the dawn side of the front of the magnetopause. By applying the latitudinal correction to their results the region of access to the peaks falls within the (N.S.) region, in agreement with this model, and moves out of the (C.O.) region where high flux is not expected (the computed cut-off for 130 MeV is $\lambda_{\text{inv}} = 63.5°$).

Morfill and Quenby (1971) made extensive computations of points of entry using magnetospheric models with Williams and Mead expressions and 6 term B_s expansion. They propose multiple crossing of the boundary as the turning mechanism.

When applying directly the computed points of entry to the auroral peaks observed during the November 18th, 1968 event, Morfill and Quenby (1971) find that the dawn and dusk auroral peaks are illuminated by direct impact mostly through the front dawn magnetopause. Morfill and Scholer (1973) using these and some additional results, propose a model that is illustrated in Figure 8, in which the dawn peak and part of the dusk is illuminated by direct dawn impact through the nose of the magnetopause while the dawn valley is illuminated by radiation moving via the neutral sheet. This last model differs considerably from the one proposed here or from the simpler model proposed previously by the authors. The difference cannot be due – as argued by Morfill and Scholer (1973) – to the different B_s expansions since both the 2 and 6 term expansions (Gall and Orozco, 1973) lead to the model described here.

A significant indication that the use of different magnetospheric models leads to similar points of access was given by Thomas *et al.* (1973).

The difference between Morfill and Scholer's (1973) and the model presented here is mainly due to different matching of the computed points of entry with the observed flux profiles. The matching is done for the mean response energy and without the cut-off corrections (Morfill and Quenby, 1971, Figure 9). The important thing to stress is that in this model in contrast with the model presented here, the neutral sheet does not play a preferential role in the access and propagation of the flux to the auroral peaks.

To end the discussion on the propagation of non-relativistic cosmic rays in the Earth's magnetic domain we shall mention briefly one aspect of the motion of the interplanetary flux across the magnetosheath. There is experimental evidence backed by extensive theoretical work (Spreiter *et al.*, 1966) that the interplanetary field lines are

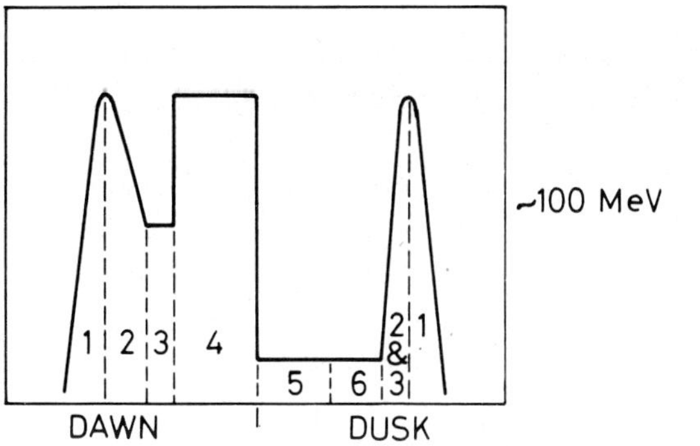

Fig. 8. Model of Morfill and Scholer for the auroral dawn and dusk peaks and the polar peak flux structure. The numbers indicate regions of access to different latitudes: (1) region below nominal derived cut-off; (2) direct impact from dawn magnetopause; (3) access from neutral sheet; and (4) access from dawn half of the magnetotail. (Reproduced from G. Morfill and M. Scholer. Uneven illumination of the polar caps by solar protons. A comparison of different particle entry models. Figure 4, *J. Geophys. Res.*, 1973, in publication).

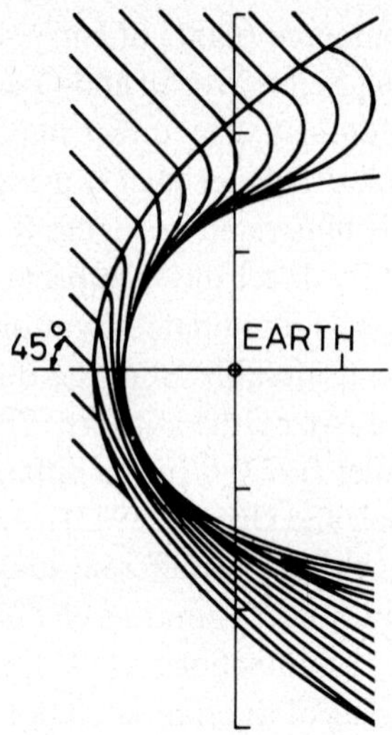

Fig. 9. Compression of magnetic field lines at the 'nose' of the magnetosphere (Spreiter *et al.*, 1966).

compressed around the nose of the magnetosphere (see Figure 9). A highly anisotropic flux with very small pitch angles may indeed follow the field lines, by-pass the nose of the Earth's cavity and never reach the detectors over the polar caps. This would no longer be true for the flux of larger pitch angles for which the radius of gyration is of the order of the width of the magnetosheath. The described effect or scattering of the flux in the magnetosheath by field irregularities might inhibit partly the direct access of anisotropic flux across the front of the magnetopause, narrowing the dawn auroral peak.

Acknowledgements

The authors wish to thank the Direccion de Computo y Electronica of the Nuclear Center of Salazar of the National Nuclear Institute where all the computations were performed.

One of us (R.G.) would like to thank the Cosmic Rays and Space Physics Group of Imperial College for their hospitality and helpful discussions on this paper. Special thanks are also due to Corri van de Stege and Mrs de Rosinski.

References

Bird, M. and Beard, D. B.: 1972, *Planetary Space Sci.* **20**, 2072.
Behannon, K. W. and Ness, N. F.: 1966, *J. Geophys. Res.* **71**, 2327.
Engelmann, J., Hynds, R. J., Morfill, G., Axisa, F., Bewick, A., Durney, A. C., and Koch, L.: 1971, *J. Geophys. Res.* **76**, 4245.
Evans, L. C.: 1972, Doctoral Thesis, Calif. Inst. of Technology.
Fennell, J. F.: 1973, *J. Geophys. Res.* **78**, 1036.
Gall, R. and Orozco, A.: 1973, *J. Geophys. Res.* **78** (in publication).
Gall, R., Jimenez, J., and Camacho, L.: 1968, *J. Geophys. Res.* **73**, 1593.

Gall, R. and Bravo, S.: 1973, *J. Geophys. Res.* **78** (in publication).
Gall, R., Jimenez, J., and Orozco, A.: 1969, *J. Geophys. Res.* **74**, 3529.
Gall, R., Bravo, S., and Orozco, A.: 1971, *Proc. 12th Int. Conf. Cosmic Rays, Hobart, Australia*, p. 1822.
Gall, R., Bravo, S., and Orozco, A.: 1972, *J. Geophys. Res.* **77**, 5360.
Greenhill, J. G., Fenton, K. B., Fenton, A. G., and White, K. S.: 1972, *J. Geophys. Res.* **77**, 6656.
Mead, G. D. and Fairfield, D. H.: 1972, *Trans. Amer. Geophys.* **5**,
Morfill, G.: 1971, *Planetary Space Sci.* **19**, 1061.
Morfill, G. and Quenby, J. J.: 1971, *Planetary Space Sci.* **19**, 1541.
Morfill, G. and Scholer, M.: 1973, *J. Geophys. Res.* **78** (in publication).
Olson, W. P. and Pfitzer, K. A.: 1973, 7th ESLAB Symposium, Saulgau, Germany.
Spreiter, J. R., Summers, A. L., and Alksne, A. Y.: 1966, *Planetary Space Sci.* **14**, 223.
Smart, D. F., Shea, M. A., and Gall, R.: 1969, *J. Geophys. Res.* **74**, 4730.
Thomas, G. R., Willis, D. M., and Pratt, R. J.: 1973, ESLAB Symposium, Saulgau.
Williams, D. J. and Mead, G. D.: 1965, *J. Geophys. Res.* **70**, 3017.
Willis, D. M. and Pratt, R. J.: 1972, *J. Atmospheric Terrestr. Phys.* **34**, 1955.

COMPARISONS OF SOLAR PROTON TRAJECTORY COMPUTATIONS USING MODEL MAGNETOSPHERES

G. R. THOMAS, D. M. WILLIS, and R. J. PRATT
S.R.C., Appleton Laboratory Ditton Park, Slough SL3 9JX, England*

Abstract. Solar proton trajectories are computed for three model magnetospheres in current use and the results are compared. The models are those of Morfill and Quenby (1971), Gall *et al.* (1972) and Willis and Pratt (1972). The first two model magnetospheres are based on the Mead-Williams model and the third model is an extension of the Hones-Taylor model. Generally similar results are obtained for the three models if the trajectories are terminated at 20 or 25 R_E in the tail. The results are interpreted in terms of direct access of solar protons to the Earth. Results obtained using the third model for trajectories extending to 100 R_E in the tail suggest that the entry of solar protons to the geomagnetic tail occurs close to the neutral sheet.

1. Introduction

In several recent papers (Morfill and Quenby, 1971; Durney *et al.*, 1972; Gall *et al.*, 1972; Thomas *et al.*, 1974a; Willis *et al.*, 1974) computations of proton trajectories in model magnetospheres have been used to investigate the access of solar protons to the Earth. By means of such computations the proton distribution at the magnetopause has been derived from the distribution measured over the Earth's polar caps by low-altitude, polar-orbiting satellites.

Information on the mode of access of the solar protons to the magnetosphere has then been obtained by comparing the proton distribution at the magnetopause with the corresponding interplanetary proton distribution and with the interplanetary magnetic field configuration.

Since results obtained using this procedure may depend on the model magnetosphere that is used, it is important to compare results obtained with different models. In the present paper we compare results for three models, and show that generally similar results are obtained for trajectories extending to a distance of 20 or 25 R_E (Earth radii) in the magnetotail. The three models, which are described in Section 2, are those of Morfill and Quenby (1971), Gall *et al.* (1972) and Willis and Pratt (1972). Results obtained with the Morfill-Quenby model and the Willis-Pratt model are compared in Section 3, and results for the Gall *et al.* model and the Willis-Pratt model are compared in Section 4. The conclusions are presented in Section 5.

2. The Model Magnetospheres

2.1. The Morfill-Quenby Model

In the Morfill and Quenby (1971) model, which is based on the Mead-Williams model (Mead, 1964; Williams and Mead, 1965), the Earth's main (internal) magnetic field is represented by a sixth-order Gaussian expansion and the external contribution to the field comprises two parts. The first part is a six-term harmonic expansion to

* formerly Radio and Space Research Station.

D. E. Page (ed.). Correlated Interplanetary and Magnetospheric Observations, 449–461. *All Rights Reserved.*

represent the magnetopause currents, assuming zero dipole tilt (Mead, 1964), and the second part is an infinitely thin current sheet of constant current density to simulate the neutral sheet in the geomagnetic tail. To allow for dipole tilt (Speiser and Ness, 1967; Bame *et al.*, 1967), the inner edge of this current sheet is 'hinged' at the magnetic equator at a distance of 10 R_E and the sheet is aligned with the solar wind flow. The outer edge is at 40 R_E and the characteristic magnetic intensity associated with the current sheet is 40 γ.

2.2. The Gall et al. Model

The Gall *et al.* (1972) model is also based on the Mead-Williams model, and is therefore similar to the Morfill-Quenby model. However, the magnetopause currents are represented by a two-term harmonic expansion (instead of a six-term expansion), and the inner edge of the current sheet is at 8 R_E in the Gall *et al.* model. To account for dipole tilt, the external field is transformed from geomagnetic to geographic coordinates. Morfill (1971) and Gall and Orozco (1973) have commented on the relative merits of the Morfill-Quenby model and the Gall *et al.* model with regard to trajectory computations.

2.3. The Willis-Pratt Model

In the Willis and Pratt (1972) model, which is an extension of the Hones-Taylor model (Hones, 1963; Taylor and Hones, 1965), the internal field is represented by a dipole, and the magnetopause currents are simulated by a larger 'image' dipole situated at 38 R_E on the sunward side of the terrestrial dipole. The magnetic moment of the image dipole is 40 times greater than that of the terrestrial dipole, and both dipoles are assumed to be perpendicular to the solar wind velocity vector; i.e. there is no allowance for dipole tilt. The geomagnetic tail is represented by 36 contiguous current sheets of width 2 R_E within the range 8–80 R_E together with a final current sheet extending from 80 to 200 R_E (included for the purposes of continuity). Each current sheet has a semi-thickness of 0.1 R_E. The current density is assumed to be uniform in each sheet but to vary from one sheet to another in such a way as to give a best fit to satellite measurements of magnetic intensity in the tail under relatively quiet magnetic conditions. The characteristic magnetic intensities associated with the Willis-Pratt current sheets in the interval 10–40 R_E vary between 10 γ and 15 γ, as compared with the 40 γ intensity of the Gall *et al.* model and the Morfill-Quenby model in the same interval.

The field line configuration of the Willis-Pratt model in the noon-midnight meridian plane is illustrated in Figure 1, the numbers being the latitudes at the feet of the field lines. It may be noted that the field is directed northwards across the equatorial plane everywhere in the midnight meridian. Thus there are no neutral lines that cross the midnight meridian in this model. In this respect the Willis-Pratt model differs from the Morfill-Quenby model and the Gall *et al.* model, which both have an equatorial neutral line crossing the midnight meridian near the inner edge of the single current sheet.

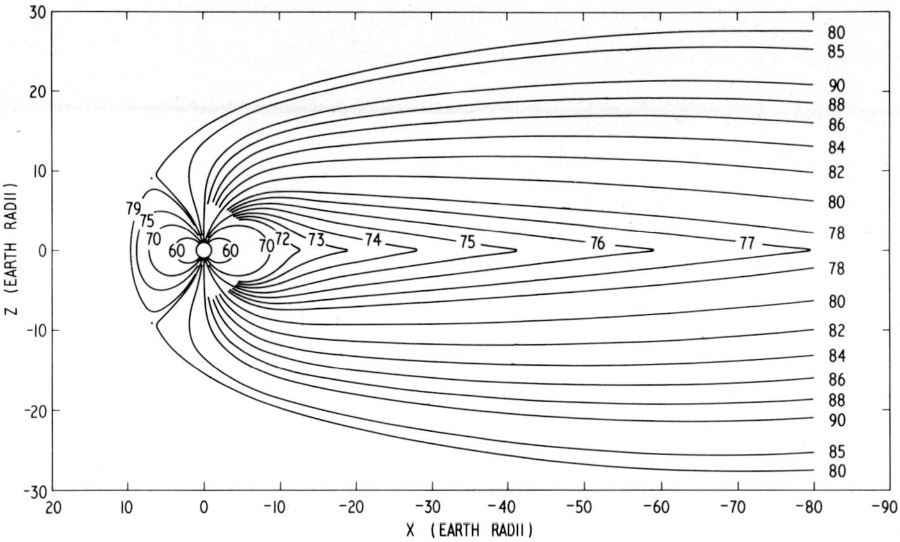

Fig. 1. The configuration of the magnetic lines of force in the noon-midnight meridian plane for the Willis and Pratt (1972) magnetic field model. The numbers refer to the geomagnetic latitudes of the field lines at the Earth's surface and the dots indicate the positions of the day-side neutral points.

2.4. Representation of the Magnetopause in the Three Models

Since the trajectory computations are performed in the 'reversed' sense, i.e. the trajectories are traced outwards from the Earth (with particles of opposite charge to those being studied), it is necessary to define a model 'magnetopause' at which the trajectories are terminated. The Morfill-Quenby model and the Gall *et al.* model both incorporate a magnetopause similar to that described by Gall *et al.* (1968). The magnetopause is represented by a spherical cap of radius 13.9 R_E on the day side of the Earth, centred at a distance of 3.5 R_E on the night side, and extended into a truncated cone of semi-angle 15.3° in the tail. In the Gall *et al.* (1972) model, the magnetopause is completed on the night side by a spherical cap of geocentric radius 20 R_E, shown as cross-section *a* in Figure 2. In the Morfill-Quenby model it is completed by a plane perpendicular to the tail axis at 25 R_E, shown as cross-section *b* in Figure 2.

The Willis-Pratt model magnetopause is defined by those field lines that diverge from the southern day-side neutral point or converge to the northern day-side neutral point (cf. Figure 1). Figure 3 illustrates the cross-sections of the Willis-Pratt model magnetopause by planes perpendicular to the tail axis at specified distances down the tail. The magnetopause is incompletely defined, in fact, by field lines passing through the neutral points, and the dotted lines in Figure 3 indicate how it is completed (somewhat arbitrarily) across the equatorial plane. A simplified representation of this magnetopause is discussed by Thomas *et al.* (1974b), and its maximum radial distance is shown as curve *c* in Figure 2.

Willis (1972) has demonstrated how the Willis-Pratt magnetopause agrees with the

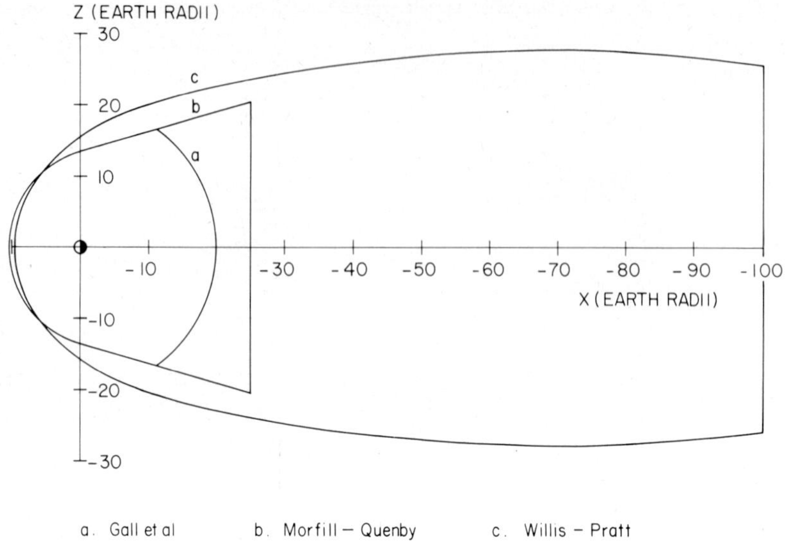

Fig. 2. Representations of the magnetopause for the three models discussed in this paper: (a) Cross-section of the Gall *et al.* (1972) magnetopause. (b) Cross-section of the Morfill and Quenby (1971) magnetopause. (c) Maximum radial distance of the Willis and Pratt (1972) magnetopause.

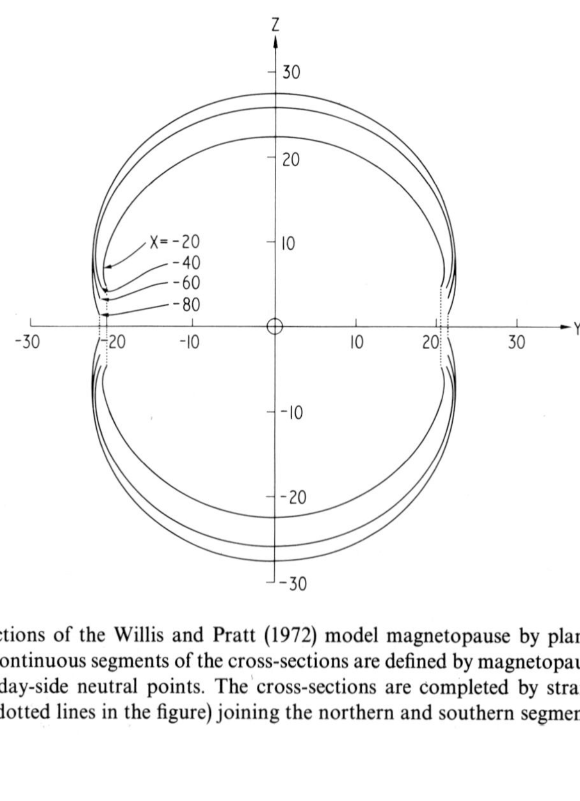

Fig. 3. Cross-sections of the Willis and Pratt (1972) model magnetopause by planes perpendicular to the tail axis. The continuous segments of the cross-sections are defined by magnetopause field lines passing through the two day-side neutral points. The cross-sections are completed by straight lines (shown as dotted lines in the figure) joining the northern and southern segments.

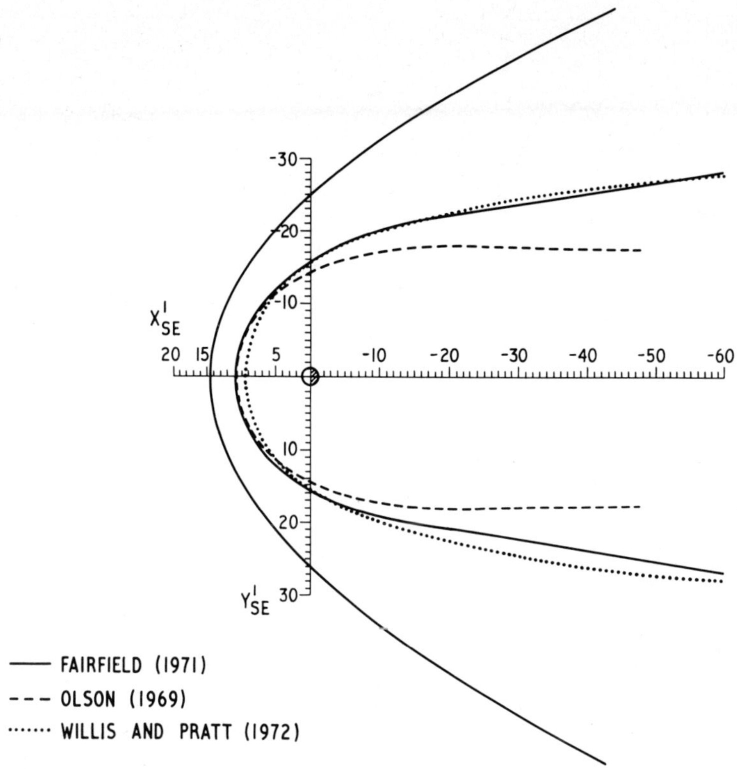

Fig. 4. The solid curve is the best-fit ellipse to the magnetopause crossings observed by six IMP spacecraft (Fairfield, 1971). The dotted curve is the Willis and Pratt (1972) model magnetopause, and the broken curve is the Olson (1969) model magnetopause. Also shown in the figure is the best-fit hyperbola to the observed bow shock crossings.

observed magnetopause. Figure 4, which is reproduced from his paper, shows (solid line) the best-fit ellipse derived by Fairfield (1971) to magnetopause crossings observed by six IMP spacecraft. The Willis-Pratt magnetopause is shown by the dotted line, and the theoretical estimate of the magnetopause position derived by Olson (1969) in the absence of a tail current sheet is indicated by the broken line. It can be seen that the Willis-Pratt magnetopause is in good agreement with the observed magnetopause, particularly in the geomagnetic tail.

3. Comparison of Results for the Morfill-Quenby Model and the Willis-Pratt Model

Morfill and Quenby (1971) and Durney *et al.* (1972) have performed trajectory computations using the Morfill-Quenby model to analyse 130 MeV proton data recorded by the polar-orbiting ESRO-2 satellite during the initial phase of the 18 November 1968 solar proton event. The satellite data, which are illustrated in Figure 5, show two lower-latitude maxima and a central polar cap maximum. Corresponding computations using the Willis-Pratt model have been described by Willis *et al.* (1974) and the results are reproduced in Figure 6. The upper panel gives the Morfill-Quenby re-

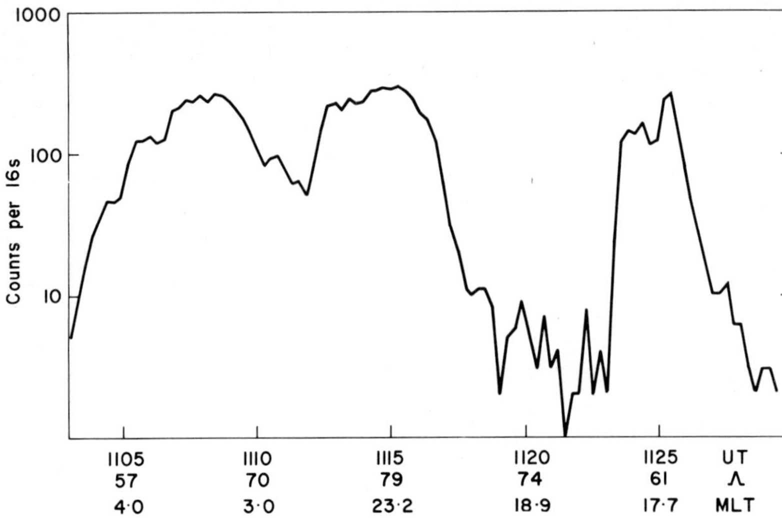

Fig. 5. Count rate of solar protons in the energy range 90–350 MeV measured by the ESRO-2 satellite over the northern polar cap on 18 November 1968. Counts are plotted on a logarithmic scale as a function of universal time UT with corresponding invariant latitude Λ and magnetic local time MLT also indicated.

sults, the middle panel gives the Willis-Pratt results with the trajectories terminated at 25 R_E in the tail, and the lower panel gives the Willis-Pratt results with the trajectories terminated at 100 R_E. In each panel the latitude and longitude of the magnetopause crossing points are plotted for a number of trajectories with starting points at specified positions along the satellite pass. The numbers attached to the points are the proton count rates measured by the satellite over the polar cap. The circles in the upper and middle panels depict the magnetopause at 25 R_E in the tail, and in the lower panel the magnetopause at 100 R_E is shown by the 'figure-of-eight' cross-section. The fact that the cross-section at 100 R_E appears to be smaller than at 25 R_E is a consequence of the latitude-longitude form of presentation. In the upper panel the broken line depicts the Morfill-Quenby model current sheet at 25 R_E in the tail; in the middle and lower panels the corresponding line is at 0° latitude since there is no allowance for dipole tilt in the Willis-Pratt model.

Comparing the upper and middle panels it is clear that, although there are differences in detail between the results for the two models to a distance of 25 R_E in the tail, in general they show basic similarities. In particular, both sets of results show that some of the higher count rates are associated with particle entry from the dawn

→

Fig. 6. Results of trajectory computations for the ESRO-2 satellite data shown in Figure 5. Upper panel: Results for Morfill and Quenby (1971) model (trajectories terminated at 25 R_E in the tail). – Middle panel: Results for Willis and Pratt (1972) model (trajectories terminated at 25 R_E in the tail). – Lower panel: Results for Willis and Pratt (1972) model (trajectories terminated at 100 R_E in the tail).

In each panel the points represent the magnetopause crossing points, and the number attached to each point denotes the proton count rate (see Figure 5); the curves are broken where few data points are available. The positions of the magnetopause and the current sheet at either 25 or 100 R_E in the tail are also indicated.

Morfill – Quenby
($25 R_E$)

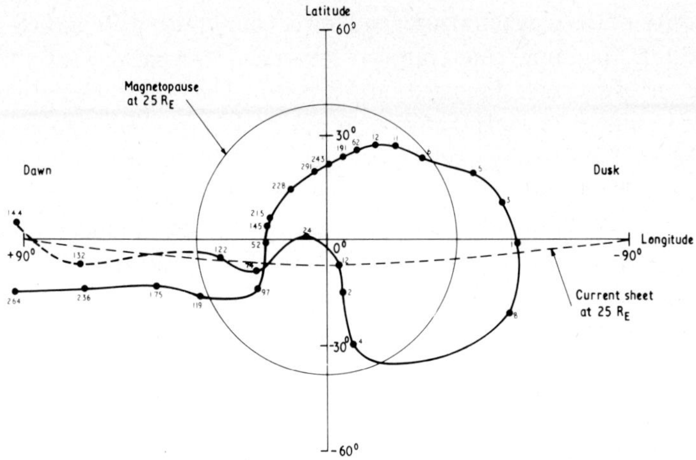

Willis – Pratt
($25 R_E$)

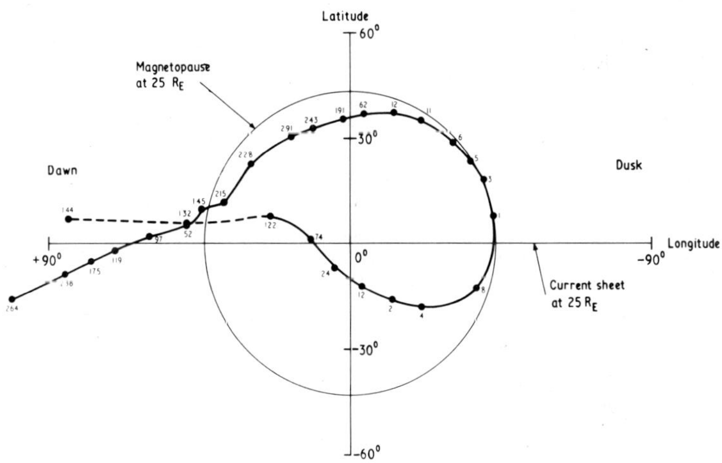

Willis – Pratt
($100 R_E$)

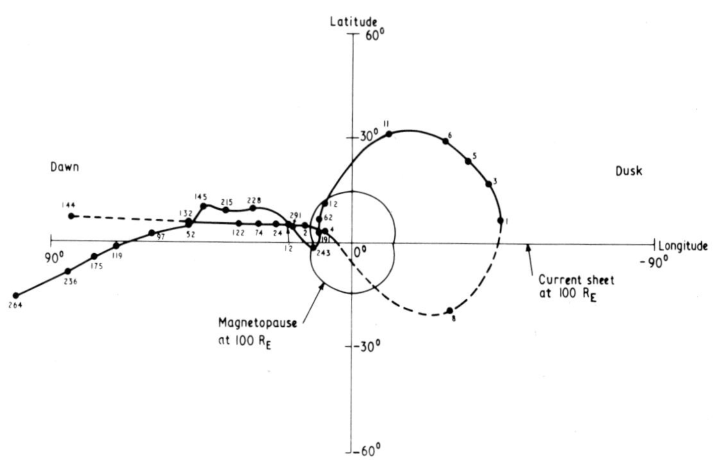

sector of the magnetopause, and lower count rates with particle entry from the dusk sector. Maximum count rates are associated with particle entry from the tail, and this mode of access is discussed further below. The main difference between the two sets of results to 25 R_E is that some of the dusk-sector magnetopause crossing-points in the Morfill-Quenby model (upper panel) are shifted by about 10–15° in longitude in the Willis-Pratt model (middle panel). As discussed by Willis *et al.* (1974), the protons have a greater total eastward drift in the Willis-Pratt model, which is consistent with the northward directed tail field of the Willis-Pratt model (since the proton trajectories are traced in the 'reversed' sense).

The interpretation of these results in terms of particle access to the Earth has been discussed by Morfill and Quenby (1971), Durney *et al.* (1972) and Willis *et al.* (1974). Thus, the lower-latitude maxima in the proton count rates measured by ESRO 2 over the polar caps (Figure 5) are associated with particle entry from the dawn magnetopause, and the minima observed at higher latitudes are associated with entry from the dusk magnetopause. These results are consistent with the field-aligned particle anisotropies measured outside the magnetosphere during the initial phase of the 18 November 1968 event, with maximum fluxes arriving from the Sun along the 'garden-hose' interplanetary magnetic field direction, and with minimum fluxes in the opposite direction. The results therefore imply direct access of the solar protons to the magnetosphere.

As stated above, both sets of results to a distance of 25 R_E in the tail show that the central polar cap maximum in the ESRO-2 proton count rates (Figure 5) is associated with particle entry from the tail. However, the results obtained with the Willis-Pratt model to a distance of 100 R_E in the tail (Figure 6, lower panel) show significant differences from those terminated at 25 R_E. Those trajectories that are still in the tail at 100 R_E (i.e. those points within the 'figure-of-eight') are in fact closer to the equatorial plane at 100 R_E than at 25 R_E. This indicates that the protons tend to follow the tail field lines in the Willis-Pratt model (Figure 1). There is also a greater eastward drift for the particles that reach 100 R_E than for those whose trajectories are terminated at 25 R_E; this is consistent with the northward directed tail field extending to 100 R_E in the Willis-Pratt model.

The Willis-Pratt results to 100 R_E do not support the interpretation of the central polar cap maximum proposed by Morfill and Quenby (1971) and Durney *et al.* (1972). These latter authors concluded that particle entry occurred only from the northern lobe of the tail and that, on the basis of a simple 'open' magnetosphere, this was inconsistent with the interplanetary magnetic field being directed towards the Sun in the 18 November 1968 event. The Willis-Pratt results to 100 R_E indicate that particle entry occurs close to the neutral sheet in the tail, and, as discussed by Willis *et al.* (1974), there is also observational evidence (Duggal *et al.*, 1971; Aarsnes and Amundsen, 1972) to suggest that maximum fluxes did not arrive from the northern lobe of the tail. Thus, contrary to Morfill and Quenby (1971) and Durney *et al.* (1972), we contend that a simple 'open' magnetosphere model cannot be rejected for the 18 November 1968 event.

4. Comparison of Results for the Gall *et al.* Model and the Willis-Pratt Model

Gall *et al.* (1972) have computed proton trajectories using their model, not for particular satellite passes but for four specific local times as the starting points of the computations. Their results are illustrated in the upper panel of Figure 7, where *A*, *B*, *C*, *D* refer to the four local times 1200, 1800, 2400, 0600, respectively. There are two curves for each local time because separate computations were performed for the northern and southern hemispheres. The numbers on the curves are the invariant latitudes of the starting points of the computations, and the results refer to protons of 125 MeV and an initial height of 4000 km. It should be noted that, although the latitude-longitude form of presentation of the magnetopause crossing points in Figure 7 is similar to that in Figure 6, the longitude scales of the two figures are in opposite senses. The circle in the upper panel depicts the magnetopause cross-section defined by the intersection of the spherical cap of radius 20 R_E and the truncated cone of semi angle 15.3° (Figure 2).

Corresponding results for the Willis-Pratt model are shown in the middle panel of Figure 7, with only one curve for each local time since the Willis-Pratt model assumes symmetry between the northern and southern hemispheres. The results again refer to protons of 125 MeV and an initial height of 4000 km, and the latitudes indicated on the curves have been computed for zero height using a dipole approximation; they are comparable, therefore, with the invariant latitudes in the upper panel. The magnetopause cross-section depicted by the circle corresponds to the intersection of the spherical cap of radius 20 R_E with the Willis-Pratt magnetopause (Figure 2).

The two sets of results in the upper and middle panels of Figure 7 show generally good agreement, the main difference being that some of the Willis-Pratt trajectories have an eastward displacement with respect to the Gall *et al.* trajectories. This displacement is again consistent with the northward directed tail field of the Willis-Pratt model. It is clear from Figure 7 that the tail field of the Willis-Pratt model exerts a strong influence on those trajectories that start at lower latitudes. As discussed by Gall *et al.* (1972), such trajectories are associated with the lower-latitude ('auroral') maxima in the proton count rates measured over the polar caps. Figure 7 illustrates that the magnetopause crossing points of these trajectories tend to occur in the dawn or tail sectors. The abrupt transition that occurs for curve *A*, with the crossing points moving from the day sector for the higher-latitude trajectories to the tail sector for the lower-latitude trajectories, is consistent with previous results discussed by Thomas *et al.* (1974a; cf. their Figures 7 and 9).

Gall *et al.* (1972) concluded that the lower-latitude count-rate maxima were due mainly to solar protons arriving via the plasma sheet of the tail. They rejected the idea that direct access of solar protons from the 'garden-hose' interplanetary magnetic field direction could produce the lower-latitude maxima at all local times. However, Gall and Bravo (1973) have extended the Gall *et al.* (1972) model to include a simple representation of the tail field that incorporates a component directed northwards across the equatorial plane. The results of Gall and Bravo (1973) show that

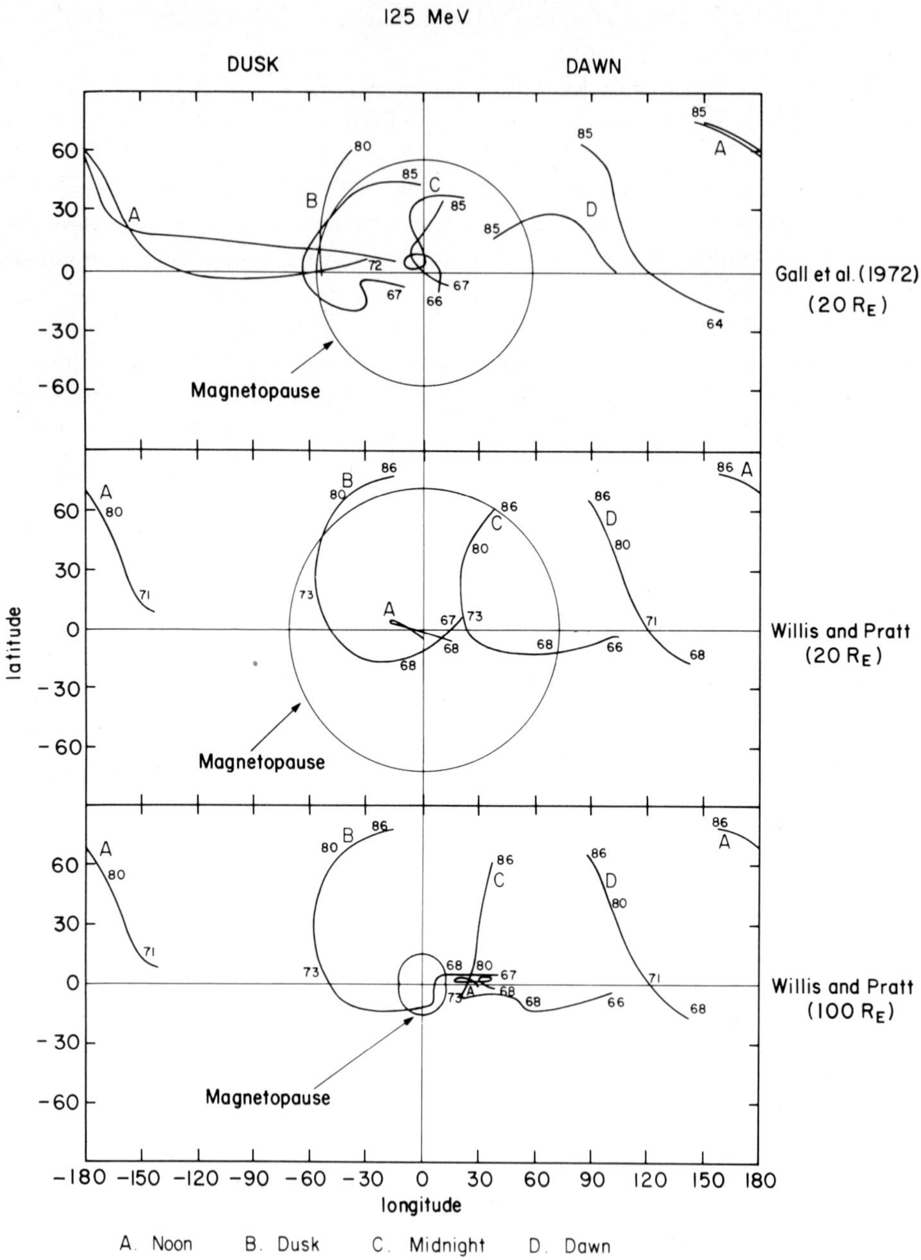

Fig. 7. Results of trajectory computations for starting points at four local times; *A* – noon, *B* – dusk, *C* – midnight, *D* – dawn. Upper panel: Results for Gall *et al.* (1972) model (trajectories terminated at 20 R_E in the tail). – Middle panel: Results for Willis and Pratt (1972) model (trajectories terminated at 20 R_E in the tail). – Lower panel: Results for Willis and Pratt (1972) model (trajectories terminated at 100 R_E in the tail).

In each panel the curves give the positions of the magnetopause crossing points as a function of the invariant latitude, as indicated; the position of the magnetopause is also indicated.

protons entering the dawn flank of the tail are turned towards the Earth by the northward-directed component of the tail field. The importance of the northward component of the tail field is similarly illustrated by the 100 R_E results in the lower panel of Figure 7, which relate to the more realistic magnetotail representation of the Willis-Pratt model. These results show that the lower-latitude maxima are associated with the entry of solar protons through the dawn flank of the tail at low latitudes. The results shown in Figure 7 can also be used to interpret the central polar cap maxima in the proton count rates measured over the polar caps. As discussed by Thomas *et al.* (1974a), these maxima are observed for satellite passes that reach invariant latitudes of about 80° near the midnight meridian, such as the ESRO 2 pass analysed in Section 3. It is pertinent, therefore, to examine curves *C* in Figure 7 and, from the results in the upper and middle panels, it is inferred that the central polar cap maxima (invariant latitude $\gtrsim 80°$) correspond to particle entry from the tail at relatively high latitudes ($\gtrsim 30°$). However, Gall *et al.* (1972) stated that this mode of access via the tail could not apply to protons of energy $\gtrsim 125$ MeV (cf. dusk-dawn diagrams of their Figure 8). Their conclusion is not supported by the ESRO-2 proton data presented by Morfill and Quenby (1971), Durney *et al.* (1972), and shown in Figure 5 of this paper, since a pronounced central polar cap maximum occurs for the energy channel 90–350 MeV. The schematic night-noon diagrams of Figure 8 of Gall *et al.* (1972), which show a semi-plateau on the night side, are also inconsistent with other observational evidence. Thus, for the ESRO-1 satellite data illustrated in Figure 4 of Thomas and Dalziel (1970), a broad maximum in the 1 30 MeV proton count rate is observed on the day side, while a narrow maximum is seen on the night side.

The results obtained for the Willis-Pratt model to a distance of 100 R_E in the tail (Figure 7, lower panel) provide additional information on particle entry from the tail. The trajectories for curve *C* that cross the spherical cap of radius 20 R_E do not reach the plane at 100 R_E, but have magnetopause crossing points at distances less than 60 R_E on the dawn flank of the tail. Those for invariant latitudes of about 80°, which correspond to the central polar cap maxima, have crossing points that are closer to the equatorial plane further down the tail ($\lesssim 60$ R_E) than they are at 20 R_E. This result is similar to that given in Section 3, and it suggests that the conclusion of Gall *et al.* (1972) that the central polar regions are illuminated by particles entering the tail at relatively high latitudes is a consequence of terminating the trajectories at 20 R_E.

5. Conclusions

Solar proton trajectories computed for the model magnetospheres of Morfill and Quenby (1971), Gall *et al.* (1972), and Willis and Pratt (1972) have been compared. The first two model magnetospheres are based on the Mead-Williams model, and the third is based on the Hones-Taylor model. ESRO-2 satellite data for protons of energy 130 MeV have been interpreted using the Morfill-Quenby and Willis-Pratt models. Similar results have been obtained for these two models when the trajectories are terminated at 25 R_E in the tail. The proton trajectories show a greater longitudinal drift in the Willis-Pratt model than in the Morfill-Quenby model. The results have been interpreted in terms of direct access of the solar protons through the dawn sector of the magnetopause. Results obtained with the Willis-Pratt model for trajectories that are extended to 100 R_E in the tail indicate that the particle entry points are close to the neutral sheet in the tail. These results are in conflict with the conclusion of Morfill and Quenby (1971) and Durney *et al.* (1972) that, for the 18 November 1968 solar event, particle entry occurred entirely through the northern lobe of the tail. Consequently, on the basis of these results, we do not exclude a simple 'open' magnetosphere model for this event.

Results for the Gall *et al.* model and the Willis-Pratt model have been compared for four specific local times and protons of energy 125 MeV. Similar results have been obtained for trajectories that are terminated at a radial distance of 20 R_E on the night side, but the trajectories have a greater longitudinal drift in the Willis-Pratt model than in the Gall *et al.* model. Results obtained with the Willis-Pratt model for trajectories extending beyond 20 R_E show that the lower-latitude count-rate maxima are associated with the entry of solar protons through the dawn flank of the tail. Similar results have been obtained by Gall and Bravo (1973) but, contrary to Gall *et al.* (1972), we contend that the results are consistent with the direct access of solar protons from the 'garden-hose' interplanetary magnetic field direction. Results obtained with the Willis-Pratt model for trajectories that are extended further down the tail also show that particles with entry points close to the neutral sheet in the tail are able to reach the Earth at high latitudes near the midnight meridian.

Acknowledgements

Figures 1 and 3 are reproduced from the paper by Willis and Pratt (*Journal of Atmospheric and Terrestrial Physics* **34** (1972), 1955–1976) by kind permission of Pergamon Press Ltd. The ESRO-2 satellite data shown in Figure 5 were supplied by kind permission of Drs A. C. Durney and J. J. Quenby of Imperial College, London; we are indebted to Dr R. J. Hynds, also of Imperial College, for helpful comments on the data. We are also indebted to Dr G. E. Morfill, formerly of Imperial College, for providing us with a copy of the computer program used to compute particle trajectories in the Morfill-Quenby model. The work described in this paper was carried out at the Radio and Space Research Station of the Science Research Council, and is published with the permission of the Director.

References

Aarsnes, K. and Amundsen, R.: 1972, *Planetary Space Sci.* **20**, 1835.
Bame, S. J., Asbridge, J. R., Felthauser, H. E., Hones, E. W., and Strong, I. B.: 1967, *J. Geophys. Res.* **72**, 113.
Duggal, S. P., Guidi, I., and Pomerantz, M. A.: 1971, *Solar Phys.* **19**, 234.
Durney, A. C., Morfill, G. E., and Quenby, J. J.: 1972, *J. Geophys. Res.* **77**, 3345.
Fairfield, D. H.: 1971, *J. Geophys. Res.* **76**, 6700.

Gall, R. and Bravo, S.: 1973, *J. Geophys. Res.* **78**, 6773.
Gall, R. and Orozco, A.: 1973, *J. Geophys. Res.* (in press).
Gall, R., Jimenez, J., and Comacho, L.: 1968, *J. Geophys. Res.* **73**, 1593.
Gall, R., Bravo, S., and Orozco, A.: 1972, *J. Geophys. Res.* **77**, 5360.
Hones, E. W. Jr.: 1963, *J. Geophys. Res.* **68**, 1209.
Mead, G. D.: 1964, *J. Geophys. Res.* **69**, 1181.
Morfill, G. E.: 1971, *Planetary Space Sci.* **19**, 1016.
Morfill, G. E. and Quenby, J. J.: 1971, *Planetary Space Sci.* **19**, 1541.
Olson, W. P.: 1969, *J. Geophys. Res.* **74**, 5642.
Speiser, T. W. and Ness, N. F.: 1967, *J. Geophys. Res.* **72**, 131.
Taylor, H. E. and Hones, E. W. Jr.: 1965, *J. Geophys. Res.* **70**, 3605.
Thomas, G. R. and Dalziel, R.: 1970, in T. M. Donahue, P. A. Smith, and L. Thomas (eds.), *Space Research* **10**, 837, North-Holland Publishing Co., Amsterdam.
Thomas, G. R., Willis, D. M., and Pratt, R. J.: 1974a, *J. Atmospheric Terrestr. Phys.* (in press).
Thomas, G. R., Willis, D. M., and Pratt, R. J.: 1974b, *J. Atmospheric Terrestr. Phys.* (in press).
Williams, D. J. and Mead, G. D.: 1965, *J. Geophys. Res.* **70**, 3017.
Willis, D. M.: 1972, in E. R. Dyer (ed.), 'Critical Problems of Magnetospheric Physics', IUCSTP Secretariat, Washington, D.C., U.S.A., p. 17.
Willis, D. M. and Pratt, R. J.: 1972, *J. Atmospheric Terrestr. Phys.* **34**, 1955.
Willis, D. M., Thomas, G. R., and Pratt, R. J.: 1974, *J. Atmospheric Terrestr. Phys.* (in press).

SOLAR ELECTRON ACCESS TO THE MAGNETOSPHERE

A. L. VAMPOLA

Space Physics Laboratory, The Aerospace Corporation Los Angeles, Calif. 90009, U.S.A.

1. Introduction

Solar flare particles, as suggested by Michel and Dessler (1970), can be used as probes of the geomagnetic topography. Low energy solar flare electrons, in particular, should be ideal natural test particles since their low magnetic rigidity permits them to be used to examine very small scale magnetic structure. The current status of magnetospheric topology (i.e., the lack of a clear consensus) and the fact that many groups are still making measurements of solar particles or analyzing data from previous events in an effort to establish a 'correct' magnetospheric model, indicate that our 'probes' have not been the ideal tools they were expected to be. As with any tool, quick success depends both upon the quality of the tool and the ability of the user. Here we shall examine the utility of the solar electron as a tool, its successes and failures, the use to which it has been put, and finally its areas of unexploited utility. We shall not dwell upon the solar proton access problem for several reasons: (a) The scale of geometry involved in solar proton access is such that apparently a complexity of access mechanisms is required for a complete solution; (b) The body of observational data indicate a marked dissimilarity in the electron and proton access phenomena (which may be due to the large difference in the rigidities of the particles involved); and, primarily, (c) The large volume of proton data tends to be contradictory. It is not clear that any theory, even an extreme or grotesque one, will ever explain *all* the proton observations due to the large number of conflicts that exist within the body of data. On the other hand, the electron data is much better behaved (due, probably, to the much smaller volume of observations and the small number of distinct research groups making the measurements). Hence, we shall limit ourselves to the small and pleasant task of organizing the electron observations.

2. Early Observations

Early studies of radio noise bursts from the Sun (pre-1960) had shown that electrons were accelerated by solar flares; however, no electron component had been identified in the particle radiation which impings upon the Earth's atmosphere at high latitude after some solar flares. It was thought that the intense magnetic fields on the Sun prevented the escape of the electrons during a flare. Some early evidence for interplanetary electrons was obtained with Pioneer 5 during the March–April 1960 solar proton events by Arnoldy *et al.* (1960), but it was inconclusive. Other reported electron events (Hoffman *et al.*, 1962; Frank and Van Allen, 1964; Frank, 1965) were observed near the magnetosphere and were probably magnetospheric in origin.

D. E. Page (ed.), Correlated Interplanetary and Magnetospheric Observations, 463–478.

Cosmic ray electrons were also measured (Cline *et al.*, 1964). The first unequivocal measurements, Figure 1, were made by Van Allen and Krimigis (1965) on Mariner 4 during the May 25–28, 1965 solar particle event, using a set of three Geiger-Muller tubes and a silicon surface-barrier detector. The set of detectors permitted positive identification of an electron component in the particle flux, the energy being between 45 keV and 170 keV. The angular distribution was reported as being isotropic. On the basis of the X-ray emission from the solar flare and the electron flux observed at 1.4 AU, it was estimated that 1 to 10% of the energetic electrons responsible for the X-ray burst escaped from the Sun.

Other measurements of solar flare electrons outside of the magnetosphere were soon reported (Anderson and Lin, 1966; Lin and Anderson, 1967) and one (Lin and Anderson, 1966) in the tail (Figure 2). No energy spectra were available from these

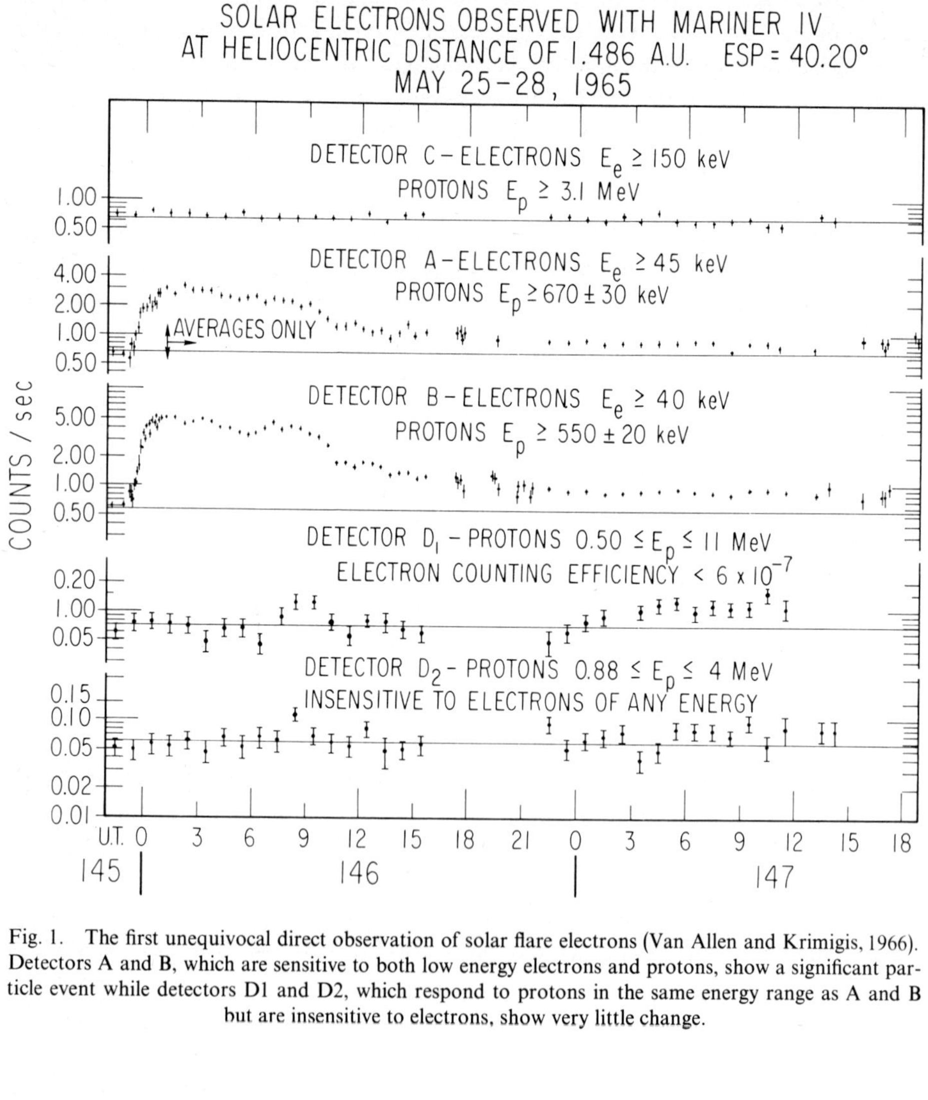

Fig. 1. The first unequivocal direct observation of solar flare electrons (Van Allen and Krimigis, 1966). Detectors A and B, which are sensitive to both low energy electrons and protons, show a significant particle event while detectors D1 and D2, which respond to protons in the same energy range as A and B but are insensitive to electrons, show very little change.

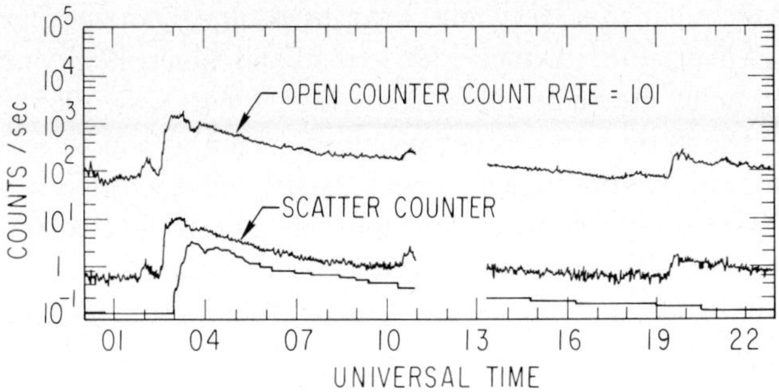

Fig. 2. The first reported solar flare electron observations in the geomagnetic tail (Lin and Anderson, 1966). The data show identical responses in an open counter (responds to protons and electrons) and in a counter with scatter geometry (responds to electrons) on IMP 3. The satellite was about 25 R_E downstream in the tail and about 5 R_E above the neutral sheet.

measurements. No definitive angular distributions could be derived, although some inferences could be drawn from the data. Unfortunately, the two sets of data available led the respective researchers to opposite conclusions. Lin and Anderson concluded the electrons propagate anisotropically, whereas Van Allen and Krimigis concluded they propagate isotropically.

The problem common to both of these experiments (which utilize geiger tubes as sensors) was an inability to make good electron measurements in the presence of a penetrating proton background. Good measurements were usually limited to the period between the arrival of the first electrons and the arrival of the first protons. Subsequent measurements were then contaminated by the protons. A second problem was the lack of energy discrimination in the experiments. A knowledge of the energy spectra is essential for applications involving the acceleration mechanisms on the Sun. Energy discrimination plus time-of-flight calculations (provided angular distributions are available) would allow the use of the electrons as a probe of fine structure in the magnetic field between the Sun and the point of observation. This had been done with protons, but the large cyclotron radius of a flare proton in the interplanetary field restricts their use to the observation of very large inhomogeneities. Measurements of electron spectra between 3 and 12 MeV were made using a Cerenkov counter (Cline, 1968) designed primarily for measuring galactic cosmic rays. The low intensity of solar flare electrons in this energy range precludes their use in a time-of-flight experiment. ('Time jitter' due to poor statistics wipes out onset effects.)

In addition to the use of electron spectra, flux anisotropies, and intensity variations in interplanetary space to investigate solar processes and particle propagation in the interplanetary medium, electron data from spacecraft could provide valuable data on the configuration of the geomagnetic field and its interaction with the solar magnetic field. Simultaneous measurements inside and outside the magnetosphere with good energy resolution are necessary to provide the data needed to make a stringent test of various geomagnetic field models. For solar flare electrons to be observed over the polar zone, they must have access due to an open magnetospheric configuration (Alfvén, 1963; Dungey, 1963; Reid and Sauer, 1967) or entry through the tail directly or by means of diffusion (Dessler, 1964; Ness, 1965; Axford *et al.*, 1965). Figure 3 shows the various configurations, including a more recent model designed to permit rapid diffusion (Michel and Dessler, 1970) which is based on the observations of filamentary structure in the magnetospheric tail (Fairfield, 1968; Mariani and Ness, 1969). A sufficiently detailed history of solar electron appearance and distribution over the polar zones in conjunction with simultaneous observations outside the magnetosphere would permit the determination of the actual field configuration.

3. Observations of Solar Flare Electrons in the Magnetosphere

Lin and Anderson (1966), in their study of several solar flare events, used X-ray and noise bursts from the Sun for the purpose of estimating the start time for the electrons to travel from the Sun to the location of observation in the tail. They observed an average delay time for arrival in the tail that was shorter than the average delay time outside the tail. Using two particular events, they estimated an upper limit of 0.2 AU for the point at which tail field lines connect to the interplanetary field. Because cross-

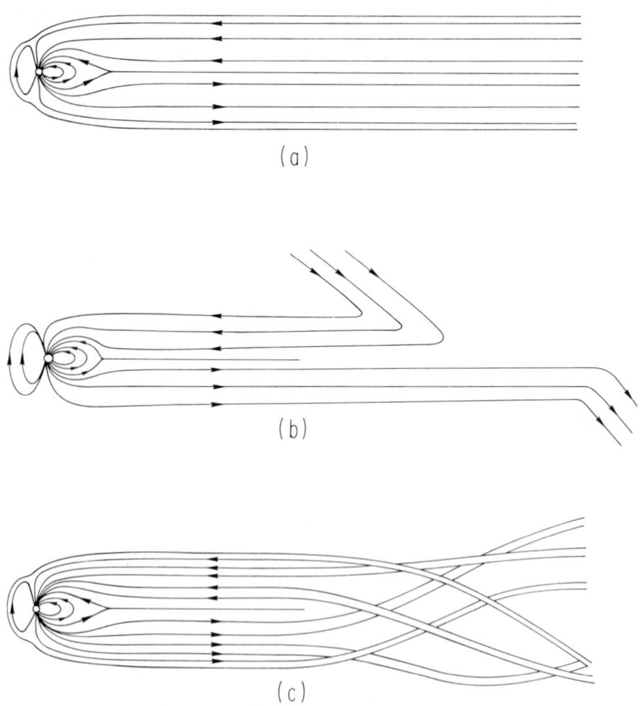

Fig. 3. Schematic representations of magnetospheric models: (a) Closed magnetosphere with a long tail; (b) Open magnetosphere with reconnection to the interplanetary solar field; (c) Closed magnetosphere with a filamentary tail.

field diffusion was expected to be very small (based on the generally quiet character of the tail field) and the electrons were observed far inside the tail many cyclotron radii away from the neutral sheet, they ruled out any method of access other than through direct connection. They also reported anisotropies in the intensity distributions. Although low-altitude polar proton measurements pre-date the satellite era, no report of solar electrons in the polar regions was made until 1967. Vampola (1967) using a multichannel electron spectrometer with magnetic analysis and a very high proton rejection efficiency on the polar-orbiting satellite OV 3-3, observed the August 25–September 2, 1966 solar event. Figure 4, from Vampola (1969), shows the polar flux of 300 ± 75 keV electrons due to the 3B solar flare on September 2, 1966. On the basis of both low and high altitude north- and south-polar passes, he concluded that the flux observed over the poles was independent of latitude, B (magnetic field strength), and altitude. The instrumentation was on a spinning satellite which permitted acquisition of pitch-angle distributions. All fluxes observed were isotropic except for the downward atmospheric loss cone. Since Anderson and Lin (1966) had interpreted their data as indicating the solar electron fluxes in interplanetary space

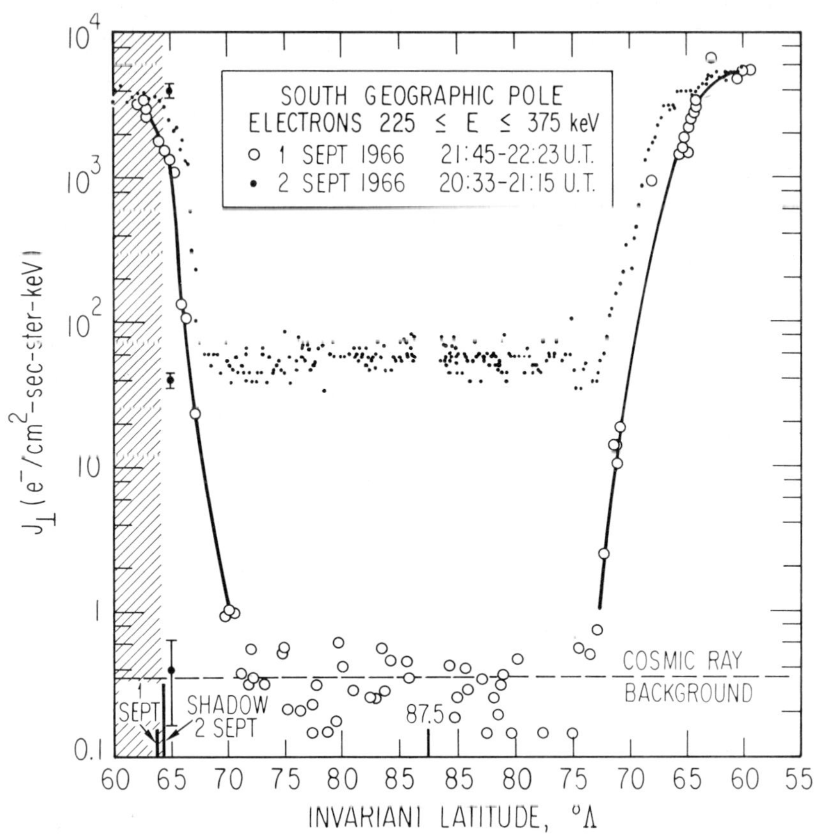

Fig. 4. First reported solar flare electron flux observations over the polar caps (Vampola, 1969). Two south-polar passes 10 h prior to and 15 h after the September 2, 1966 3B flare are shown.

are anisotropic and are preferentially along the field line, Vampola concluded that diffusion played a role in the access to low altitude polar field lines. Without diffusion, an anisotropic interplanetary flux will lead to different flux intensities over the two polar regions.

The launch of Explorer 35, the lunar-anchored member of the IMP series of satellites, provided an excellent opportunity for measurements in the tail of the magnetosphere. Lin (1968), Van Allen and Ness (1969), Van Allen (1969, 1970), Anderson and Lin (1969), and Anderson (1969) reported lunar shadowing of solar electrons in the magnetotail (Figure 5). (See McGuire (1972), for a detailed theoretical treatment of lunar shadowing.) The measurements show that solar electrons stream in from the magnetotail from beyond the orbit of the Moon at 64 R_E, are mirrored near the Earth, and stream back downstream past the Moon. On the basis of sharpness of occultation of the fluxes by the Moon, Van Allen and Ness show that the cross-B diffusion inside 64 R_E is less than 100 km s^{-1}. If the cross-B diffusion rate down the tail is of this same order of magnitude, diffusion is far too slow to account for the rapid appearance of solar electrons in the tail during a solar particle event. Anderson and Lin came to the same conclusion through somewhat different arguments.

In their analysis of lunar shadowing, Anderson and Lin (1969) observed one case of half-shadowing in front of the Moon. Their analysis of shadowing leads to the conclusion that the electrons gained access to the tail field lines *between* the Earth and the Moon; hence, they concluded that at least on this occasion the tail field-interplanetary field reconnection point was less than 64 R_E from the Earth. Figure 6 shows the tail topography used in their analysis.

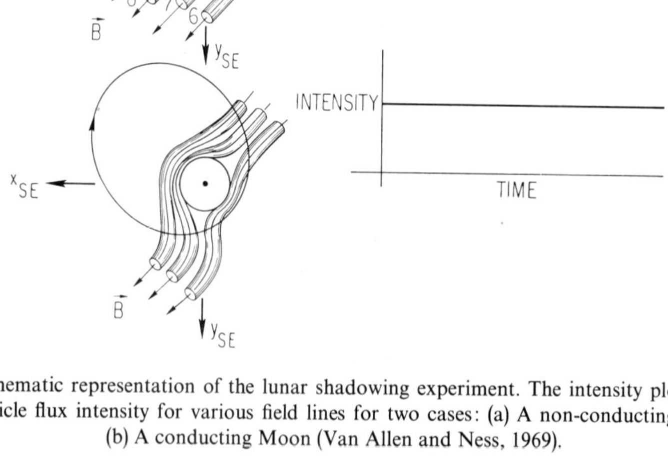

Fig. 5. A schematic representation of the lunar shadowing experiment. The intensity plots present an idealized particle flux intensity for various field lines for two cases: (a) A non-conducting Moon; and, (b) A conducting Moon (Van Allen and Ness, 1969).

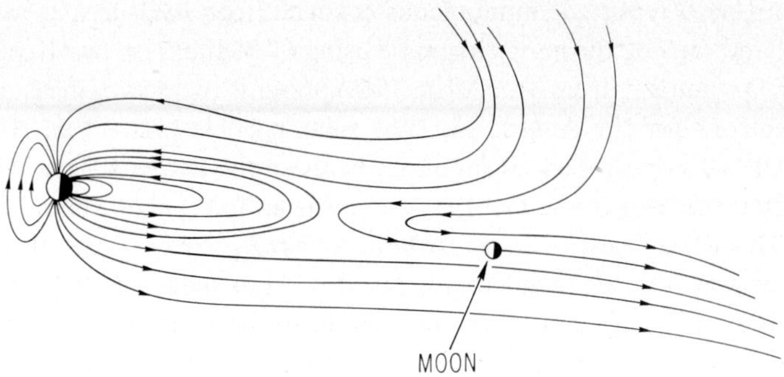

Fig. 6. A schematic representation of the lunar shadowing experiment in which half-shadows are observed on tail field lines between the Moon and the Earth. The interpretation utilizes an interconnection point that is within the orbit of the Moon (Anderson and Lin, 1969).

Since these Explorer 35 measurements had shown conclusively that, with one exception, electrons gained access to the tail beyond 64 R_E, efforts were made to determine the delay time between the appearance of the particles in the vicinity of the Earth outside the magnetosphere and within the magnetosphere. Anderson and Lin (1966) had made estimates based on the time difference between X-ray events on the Sun and the first arrival of electrons near the Earth. However, their measurements within the magnetosphere and outside were for different events and thus only a statistical analysis could be used as an estimate. This led to unsatisfactory results, but did show that the delay time was of the order of minutes. (Delays of up to several hours have been observed for low energy protons (Montgomery and Singer, 1969).)

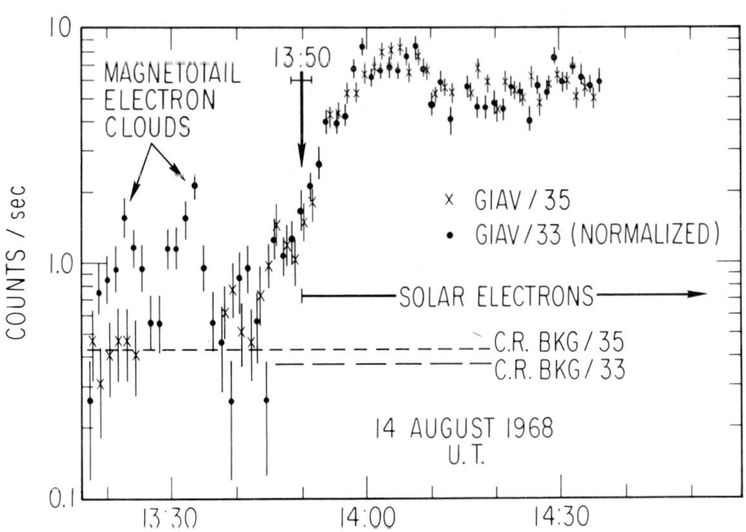

Fig. 7. A comparison of solar electron flux intensities observed outside the magnetosphere by Explorer 33 and within the tail by Explorer 35 (from Van Allen, 1969). These data set an upper limit of 100 s for the delay time in entering the tail.

Van Allen (1969) reported simultaneous solar electron measurements within the magnetotail and outside the magnetosphere using G-M tubes on two Explorer satellites, 33 and 35. Figure 7 (from Van Allen, 1969) presents two sets of measurements of >50keV electrons for the August 14, 1968 solar particle event. Van Allen set an upper limit of 100 s on the lack of similarity between the two sets. To a large extent, this upper limit was due to the fact that there was an interval of 81.8 s between observations. This delay leads to an upper limit of 900 R_E for the point at which solar electrons gain access to the magnetotail. The flux levels measured by the two detectors were within 10% of each other. The similarity between flux levels at the two locations ruled out large electric fields in the magnetotail as a source of cross-*B* diffusion (the kinetic energy of the diffusing electrons would be changed) and thus additional support for the interconnection model of the magnetotail was provided.

Since the point of access of electrons is intimately connected to the delay time between their appearance outside the magnetosphere and inside, other groups have also attacked the problem. Turtle *et al.* (1972) made use of ground-based VLF transmission measurements for determining the time of arrival of solar electrons at ionospheric heights over the polar caps. These were compared with particle observations in the interplanetary medium by satellite OV 5-6 and in the magnetotail by Explorer 41. Figure 8 shows the relationship of the particle measurements and VLF phase perturbation measurements. An upper limit of one minute delay, corresponding to a tail access point 600 R_E downstream, was set. This upper limit was due primarily to the time interval at which VLF phase perturbation data points were obtained. The

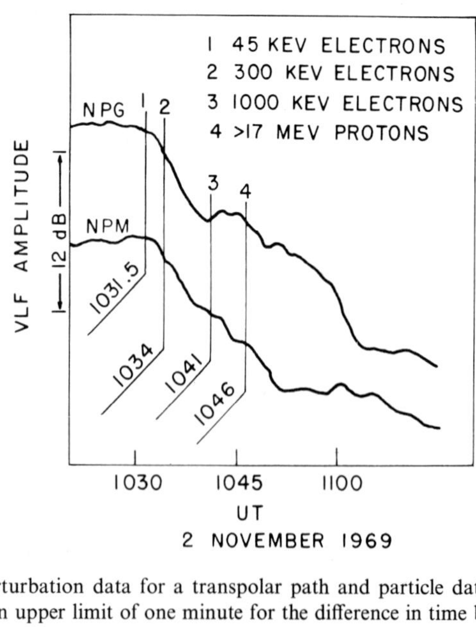

Fig. 8. VLF phase perturbation data for a transpolar path and particle data from the interplanetary region. These data set an upper limit of one minute for the difference in time between the appearance of the electrons in the interplanetary near-Earth region and their ionization effects in the polar atmosphere (Turtle *et al.*, 1972).

observations are consistent with no delay for the northern polar cap access time, and imply possible direct entry of the electrons other than by downstream entry in the tail.

Domingo and Page (1972) reported simultaneous small ($\sim \pm 10\%$) variations in flux intensity of solar particle fluxes observed over the north polar region by two widely separated satellites. They suggested that these variations, which were temporal rather than spatial, were due to magnetotail activity rather than fluctuations in the interplanetary intensity. Their measurements had several drawbacks, the most serious being proton contamination of the electron data; integral measurements; and low sample rate (12.8 s). Since electron measurements outside the magnetosphere are available for this time period from OGO 5, it is useful to obtain a correlation plot of the two sets of measurements. Figure 9 is a smoothed correlation function of the two sets of data. A Gaussian procedure was used in the smoothing which was necessary since a relatively small set of data points were available in Domingo and Page (1972) and the interval between those data points was larger than the interval between data points available from OGO 5 (H. I. West, Jr., private communication). Figure 9 shows that Domingo and Page were probably right in ascribing the fluctuations to magnetotail activity. If one speculates about the slight peak centered at -15 s, however, one would obtain an access region of $\approx 150\ R_E$ downstream. This speculative result is similar to the results obtained for protons recently by Fennell (1973). To make a definitive test of electron access, measurements over the polar regions are needed at the onset of an event. Since there is much evidence that solar protons have asymmetric access, it is plausible that electrons also have asymmetric access. Thus measurements over both poles are required. A difference in intensity could be ob-

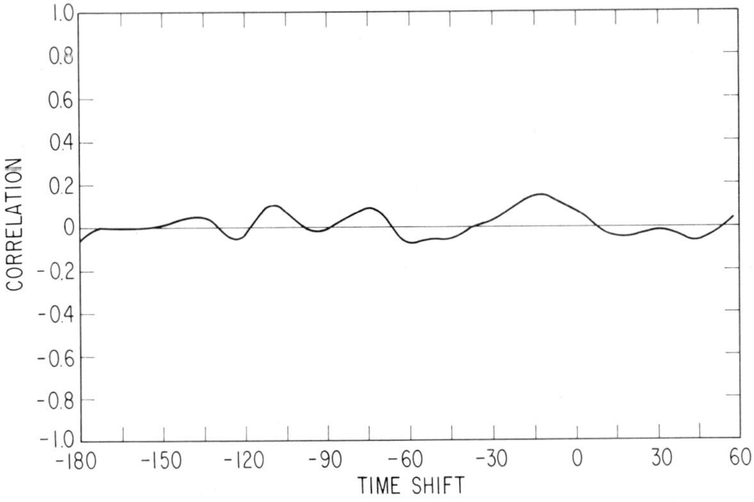

Fig. 9. A smoothed correlation function of the polar cap data from Domingo and Page (1972) and interplanetary electron data from OGO 5 (H. I. West, Jr., private communication). The plot supports the thesis that the fluctuations in the polar flux intensities observed by Domingo and Page were magnetospheric in origin.

served only while the intensity distributions are anisotropic in the interplanetary region. Allum *et al.* (1969) indicate that such a condition persists for less than an hour after the onset of an event. This short time makes a definitive bi-polar correlation very difficult and mostly a matter of luck.

West and Vampola (1971) made interplanetary and polar comparisons of solar flare electron spectra during the April 13–19, 1969 event. Figure 10 shows their comparisons for a south polar pass just prior to the peak flux intensity in the interplanetary region and a north polar pass just after. The two instruments (which were both 180° magnetic electron spectrometers) measured identical fluxes within the limits of accuracy of the measurements ($\simeq 10\%$) and showed the same decrease in flux between the two time periods. Attempts at obtaining a delay time by correlating the data from both spacecraft were unsuccessful due to the fact that no temporal structure with a short time constant was observed by either spacecraft. No systematic difference in absolute flux was observed, although individual measurements varied

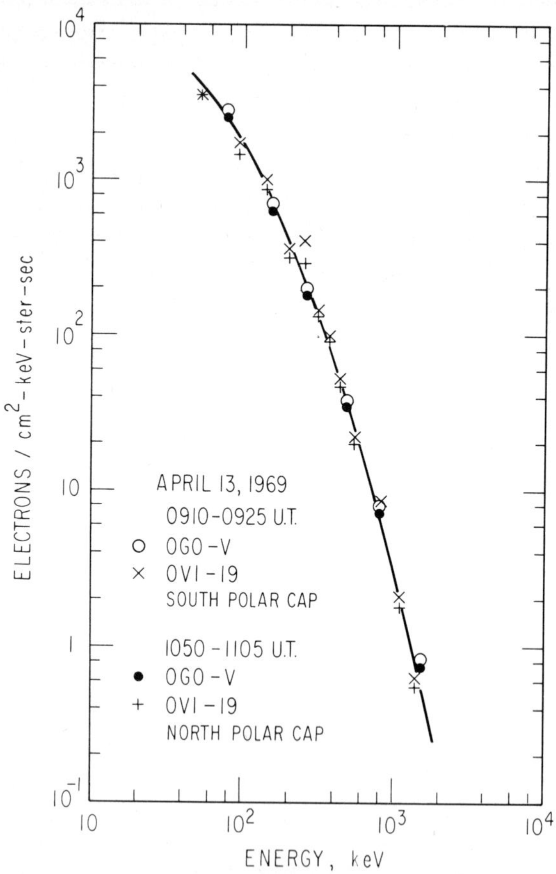

Fig. 10. A comparison of interplanetary data (OGO 5) and polar cap data (OV 1-19) for solar electrons. The same spectrum and absolute intensity was observed at both locations. The same change in intensity between the two time periods was also observed (West and Vampola, 1971).

by up to 10%. Energy spectra were identical within the limits of error of the individual measurements at all times. West and Vampora explained their results on the basis of particles starting out in diffusive equilibrium in the interplanetary medium and moving adiabatically along field lines to regions over the poles (adiabatically because the energy spectral measurements showed energy was being conserved in the access process).

The equality of the polar fluxes, independent of pole, with the interplanetary fluxes at all times during the event ruled out convection by the solar wind as a transport mode to the point of access in the magnetotail. Previous measurements in the magnetosphere examined only one tail lobe during an event and did not rule out the possibility that entry to the other lobe was delayed. The exclusion of convection as an access mechanism for the electrons can be contrasted with the proton observations which do require solar wind convection of structures (Fennell, 1973).

West and Vampola pointed out that an interplanetary magnetic field sector boundary crossing occurred during this event. If appreciable direct connection were involved between the interplanetary field and the geomagnetic tail, the rearrangement of the tail field topography should have produced effects in the particle observations. None were seen (Vampola, 1972). They concluded that the filamentary model of the tail structure (Michel and Dessler, 1970) best explained their observations *and* the long proton delays observed by Montgomery and Singer. This model permitted adiabatic motion of the particles and did not require significant cross-B diffusion. For the electron data alone, however, an interconnection model provided the simplest explanation. The lack of an effect due to the sector boundary crossing can be explained by assuming complete isotropy in the interplanetary electron fluxes coupled with a time constant in the flux intensity variations that is long compared to the difference in access time to the favored and unfavored polar regions.

4. Access to Closed Field Lines

Solar protons are often observed by synchronous orbit satellites on field lines which are closed; that is, bounce motion is permitted. In the case of electrons, synchronous orbit satellites have not been able to make measurements of solar electrons since the normal electron environment is orders of magnitude larger than the typical solar electron event. Low altitude satellites traversing the polar region, however, may observe solar electrons in the quasi-trapping region. This is a region of strong field in which an electron executing bounce motion is stably trapped for most of the drift period but in which it cannot execute a complete drift cycle. Using data from the polar orbiting satellite OV 1-19 during the April 13–19, 1969 solar particle event, Vampola (1971) showed that electrons gain access to the quasi-trapping region, drift eastward, and return to open field lines. By following the intensity changes as the electrons drift eastward on a given shell and obtaining profiles of flux intensity versus latitude in the quasi-trapping zone at various local times, Vampola concluded that although significant losses into the atmosphere at the end of the field line were occurring, diffusion to interior field lines was not significant. Figures 11 and 12 (from Vampola, 1971) show the quasi-trapping zone and the flux intensity profile along a drift shell. The method of access to the quasi-trapping region was explained as being simple gradient drift coupled with L-shell splitting.

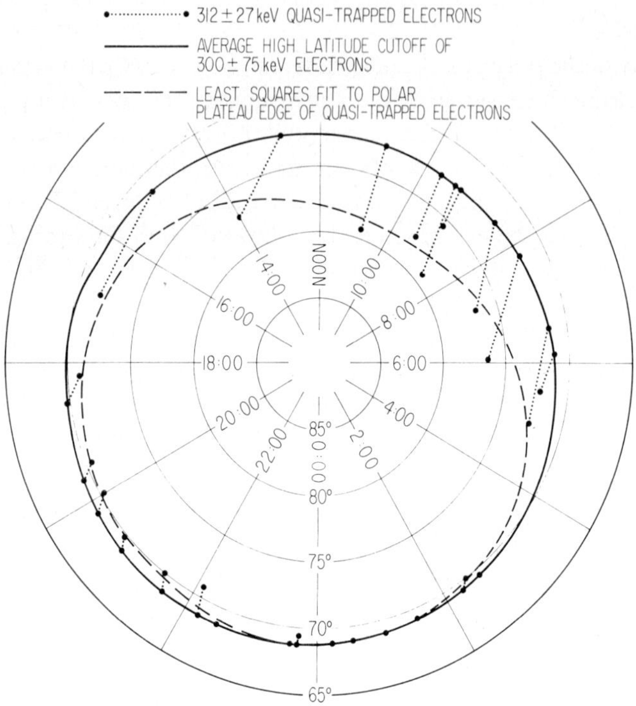

Fig. 11. A plot of the quasi-trapping zone for electrons as determined using solar-flare electrons (from Vampola, 1971).

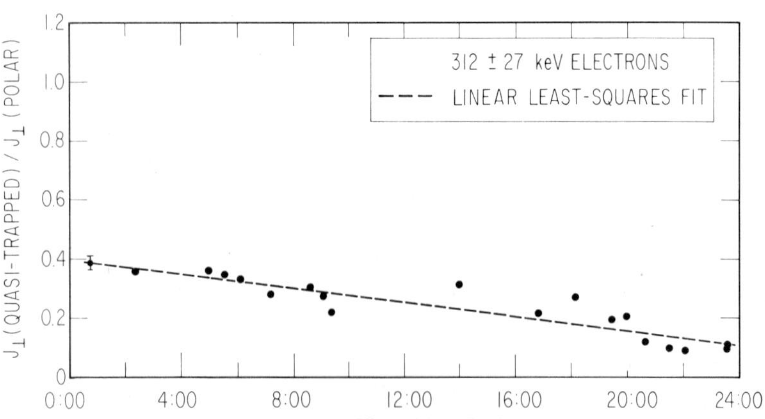

Fig. 12. A plot of the ratio of solar flare electron flux intensity in the quasi-trapping region to the flux intensity in the polar plateau as a function of magnetic local time (Vampola, 1971).

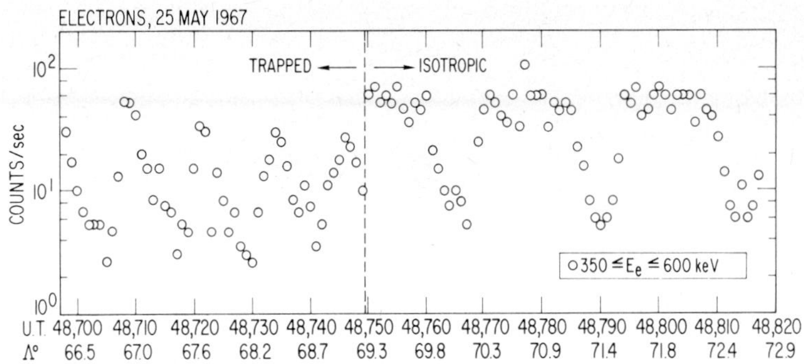

Fig. 13. The boundary between polar plateau solar electron flux and the quasi-trapped flux (from Vampola, 1969). The abrupt change in pitch-angle distribution from trapped (modulated twice per spin period due to two atmospheric loss cones) to isotropic (one loss cone) occurs within one sample period (1 s). The spinning satellite fortuitously scanned the detector aperture upward along the field line at the moment the boundary was crossed.

If one looks at the boundary between 'open' (bounce motion not permitted at both ends of the field line) and 'closed' (sustained bounce motion permitted) field lines in the polar region, one finds that it is exceedingly sharp. Vampola (1969) had a fortuitous measurement of the boundary on the noon side of the polar plateau during the May, 1967 solar particle event. Using a magnetic spectrometer on the spinning polar orbiting satellite OV 3-3, data was obtained on a boundary crossing during which the aperture of the instrument was aligned along the field line in the upward direction at the time the boundary was encountered. Figure 13 shows the change in one sample period (1 s) from a typical isotropic polar flux to a loss-cone trapped-flux profile, both obtained while the aperture was still aligned along the field line. The transition occurred simultaneously in all energy channels.

Other boundary measurements have been made (Evans and Stone, 1972; McDiarmid and Burrows, 1970; McDiarmid *et al.*, 1972). The most intriguing aspect of these is in the report of Evans and Stone in which a large number of polar electron boundary measurements resulted in a peculiar and unexplained 'zone' or 'band' structure. When the boundary measurements are plotted as invariant latitude vs magnetic local time, three distinct bands or zones are apparent (Figure 14). Evans and Stone were unable to account for these bands, but did an extensive study of their data base to exclude biases. They concluded that the bands were real and were not an artifact of their data acquisition or analysis procedures.

5. Summary

A fairly comprehensive body of data has been collected concerning solar electron phenomena in the interplanetary region and within the magnetosphere. Because of their high velocity and low magnetic rigidity, solar flare electrons should be excellent probes of the magnetospheric topology. Their failure to provide a clear picture of the geomagnetic field configuration in the tail and the method and location of inter-

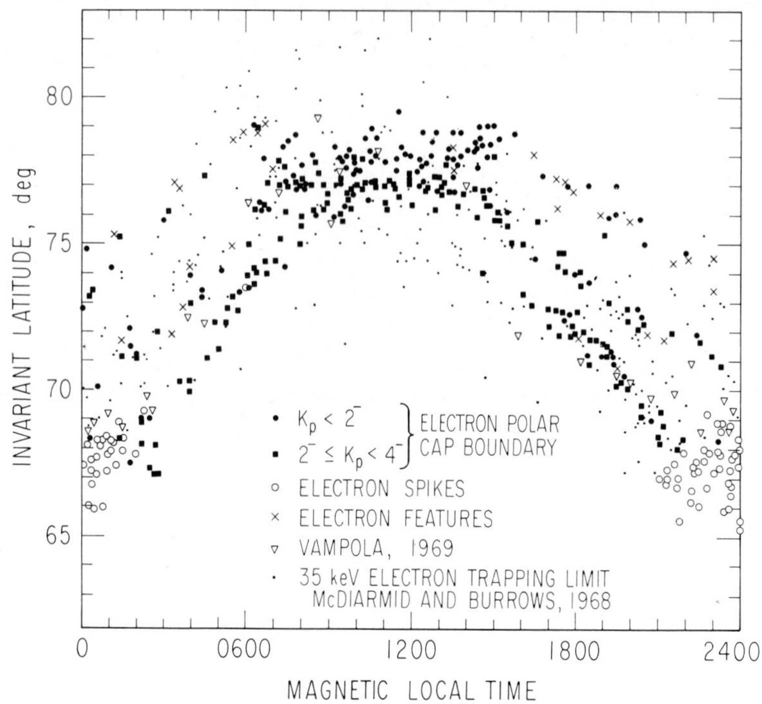

Fig. 14. A summary of electron polar-cap boundaries observed by OGO 4 (from Evans and Stone, 1972). The boundary observations group into three distinct (and unexplained) bands. Data from other satellites are shown for comparison.

connection, or lack thereof, is due not to their properties, but rather to the inadequacy of the measurements made. High resolution pitch-angle measurements are vital. These have not been available in the interplanetary region. During all except the initial onset of most events, and for the entire duration of the rest, only isotropic distributions are observed within the resolution of the instrumentation used. No detailed pitch-angle distributions outside the magnetosphere have been published, although several investigations reported gross anisotropies in the flux. Relatively few pitch-angle measurements are available in the polar regions, but all of these have shown complete isotropy at all latitudes above the regions where low-energy particles normally are precipitating. OGO 5 has made measurements of solar flare electron pitch-angle distributions in the tail, but these have all been made while the flux in the interplanetary region was isotropic (H. I. West, Jr., private communication). Hence it is not surprising that only isotropic fluxes were observed.

To make a valid pitch-angle comparison between polar fluxes and interplanetary fluxes, resolution of $\approx 1°$ along the field line in the polar region is necessary. Secondly, most interplanetary measurements have been made by instruments which have a very long interval between samples (10 to 100 s) and temporal and pitch-angle effects are obscured. The long sample duration reduces the value of the observations for cor-

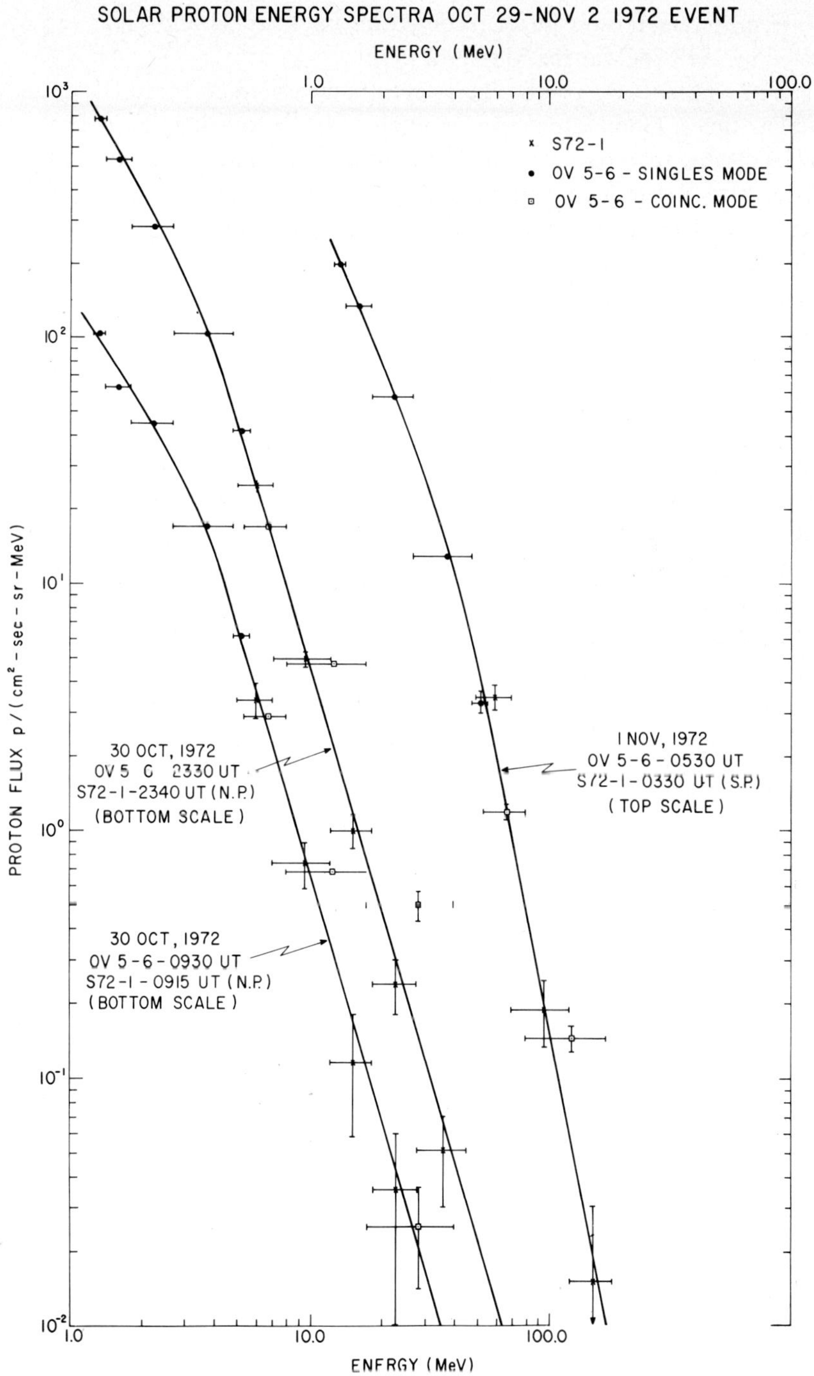

Fig. 2. Combined solar proton energy spectra.

the interplanetary proton fluxes recorded on Explorer 34 and polar cap averages detected by 1963-38C for the May 1967 event.

Figure 2 combines the proton energy spectra measured on board OV5-6 and S72-1 at similar times. Good agreement is obtained in the overlapping energy channels which indicates there is no normalization problem between the two instruments. Utilizing both detectors we were able to construct solar proton energy spectra between 1.33 and 45 MeV. Figure 2 shows that these spectra turn over at about 4 MeV and become softer with time.

Figure 3 shows a time-history of the alpha/proton ratio as measured on board both satellites.

The value of 6.8 MeV nucleon^{-1} was chosen as giving the closest overlap between the two instruments. The interesting feature of this graph is the progressive increase of the alpha/proton ratio from 0.1% on 30 Oct (0000–0600 h UT) to 1.5% on 31 Oct (0300–0800 h UT). During a period of four hours on Oct 31 (0800–1200 h UT), the ratio dropped to 0.4% from which, despite the poor statistics, it appears to remain constant or increase slightly. A close inspection of Figure 1 reveals that the 5–7 MeV proton flux increased by a factor of ~2 during the period in which the alpha/proton ratio dropped. A similar inspection of the alpha flux shows a decrease of about two during the same time.

Figure 4 shows the alpha/proton ratio as a function of energy in MeV nucleon^{-1}. Over the interval 5.1–14.6 MeV nucleon^{-1} we see the alpha/proton ratio increasing slightly with energy. This increase with energy is opposite to that observed by Lanzerotti (1969) on May 29, 1967 during the May 28, 1967 solar event.

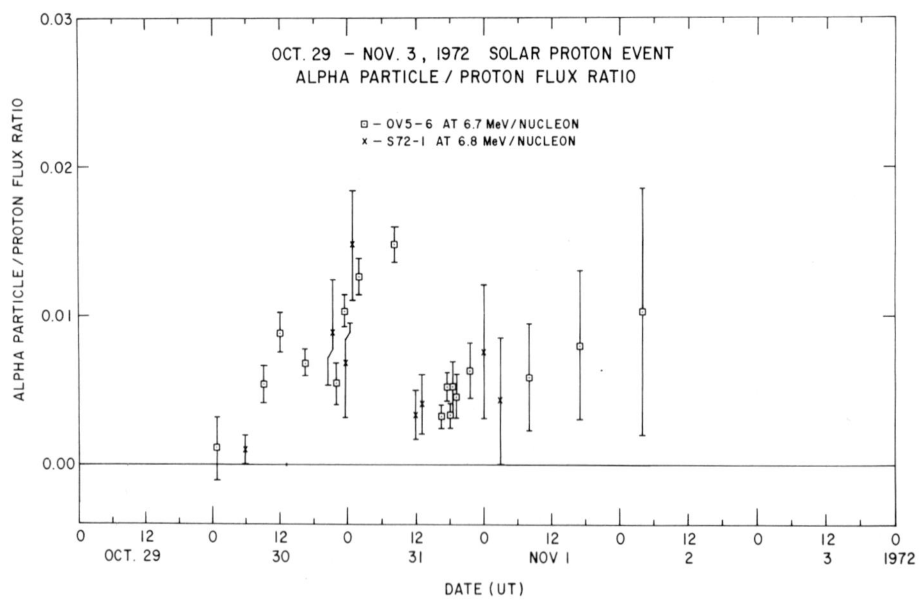

Fig. 3. Time history of alpha/proton flux ratio at 6.7–6.8 MeV nucleon^{-1}.

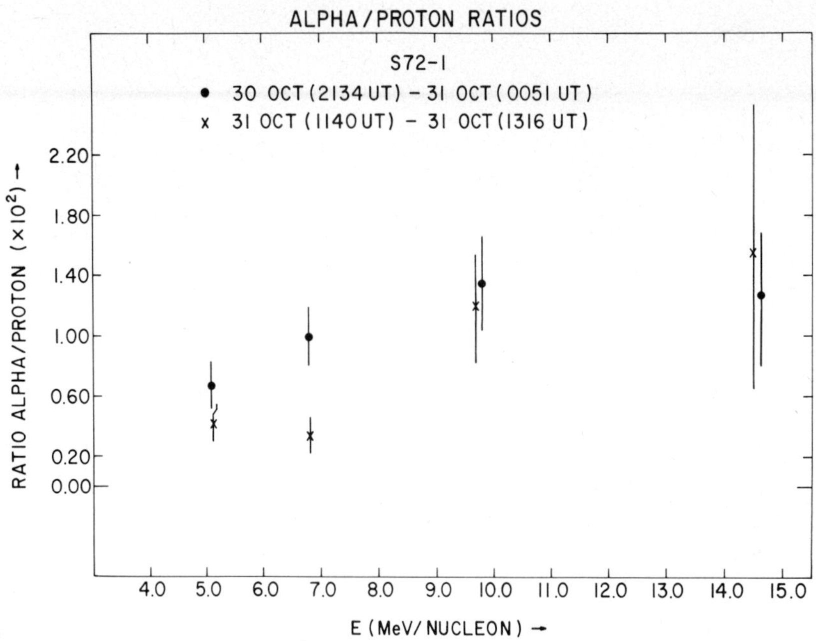

Fig. 4. Energy spectrum of alpha/proton ratio at two different times.

4. Conclusions

Simultaneous measurements of 5–45 MeV solar protons outside the magnetosphere and at low altitudes over the poles show good agreement. The 5–7 MeV proton flux recorded on the low-orbiting satellite S72-1, is at times 50% higher than similar fluxes recorded on the extra-magnetospheric satellite OV5-6. The solar proton energy spectra from the two satellites compare well showing softer spectra with time. The alpha/proton ratio shows a marked variability in time, varying from 0.1% to 1.5%. The alpha/proton ratio also increases from 0.4% to 1.2% for energies between 5.1 and 9.8 MeV nucleon^{-1} at a fixed time.

References

Bostrom, C. O.: 1970, in V. Manno and D. E. Page (eds.), *Intercorrelated Satellite Observations Related to Solar Events*, D. Reidel Publ. Co., Dordrecht-Holland, p. 229.

Hanser, Frederick A., Sellers, Bach, and Morel, Paul, R.: 1973, 'A Proton-Alpha Particle Detector for Use in Satellites', AFCRL-TR-73-0144.

Lanzerotti, L. J.: 1969, *J. Geophys. Res.* **74**, 2851.

Yates, G. K., Katz, L., Kelley, J. G., Sellers, B., Hanser, F. A., and Morel, P. R.: 1971, 'Proton, Alpha and Bremsstrahlung Fluxes Measured Aboard OV5-6', World Data Center A, Upper Atmosphere Geophysics Report UAG-12, Part I, 139.

Yates, G. K., Katz, Ludwig, Sellers, Bach, and Hanser, Frederick, A.: 1973, this volume, p. 597.

INITIAL OBSERVATIONS OF MAGNETOSPHERIC BOUNDARIES BY EXPLORER 45 (S^3)

T. A. FRITZ*

Space Environment Laboratory, NOAA/ERL, Boulder, Colo., U.S.A.

P. H. SMITH

NASA/Goddard Space Flight Center, Greenbelt, Md., U.S.A.

D. J. WILLIAMS

Space Environment Laboratory, NOAA/ERL, Boulder, Colo., U.S.A.

R. A. HOFFMAN

NASA/Goddard Space Flight Center, Greenbelt, Md., U.S.A.

and

L. J. CAHILL, JR.

Space Science Center, University of Minnesota, Minneapolis, Minn., U.S.A.

Abstract. We present a review of initial results obtained from the NASA Small Scientific Satellite (Explorer 45) concerning the nature of three magnetospheric boundaries observed by the satellite near the equatorial plane out to radial distances of 5.5 R_E (earth radii). We discuss features of the interaction region of the proton ring current and the plasmasphere during the developmental and decay phases of ring current enhancements. We have observed an energy and pitch angle dependent inner boundary of the proton ring current distribution which appears to occur near 1800 h local time during the developmental phase of the ring current build-up. Following the establishment of the ring current we have observed effects caused by an instability similar to (if not) the ion cyclotron instability acting in this interaction region which causes a decay of the storm-time ring current spectral distribution back into the quiet-time distribution. In addition we also report the observation of the magnetopause at a radial distance of less than 5.2 R_E at 14.6 h local time in connection with the large solar event of August 4, 1972. We further report the observation of an electron distribution consistent with all of the features of the inner edge of the electron plasma sheet at a position of 5–5.5 R_E and 2100 h local time in connection with a magnetic substorm. Quantitative calculation using models for the static electric and magnetic fields indicate that these electrons could not have reached the satellite as a result of simple convective flow from the geomagnetic tail. These S^3 results support the idea that a discontinuous change occurs at the initiation of a substorm which injects particles deep into the night side trapping region in a fashion consistent with the plasma cloud concept of DeForrest and McIlwain (1971).

1. Introduction

One of the topics under study at the 1973 ESLAB symposium was the nature of magnetospheric boundaries. We wish to present here a review and short compilation of the initial results of various studies being conducted using data obtained by the NASA Small Scientific Satellite, S^3-A (Explorer 45), near the equatorial plane which are concerned with the nature of some of these magnetospheric boundaries. In this review we will discuss the observations by S^3 of the nature of the inner edge of the proton ring current and its relationship to the plasmapause, the observation of the

* Presently spending a year in residence at the Max-Planck-Institut für Aeronomie, Institut für Stratosphären-Physik, Lindau/Harz, Federal Republic of Germany.

D. E. Page (ed.), Correlated Interplanetary and Magnetospheric Observations, 485–506. *All Rights Reserved.*

magnetopause during the large event of August 4, 1972, and the observation of the electron plasma sheet at the S^3 apogee of 5.2 R_E. The observations concerned with each of these three magnetospheric boundaries are presented in Sections 3, 4 and 5 respectively.

2. Description of Satellite, Instrumentation, and Program Concept

S^3-A was launched on November 15, 1971 into an elliptical orbit having an apogee of 5.24 R_E, a perigee of 220 km, a period of 7.82 h, and an inclination of 3.5°. The initial local time of apogee was approximately 2200 h and apogee moves to earlier local times at about 0.8 h per month. Proton intensities are measured from 0.73 keV to 2800 keV in 31 differential channels and electron intensities are measured from 0.73 keV to 560 keV in 20 differential channels. Three axis magnetic field measurements and single axis DC electric field measurements are made as well as electric and magnetic noise measurements (Longanecker and Hoffman, 1973). The location of the plasmapause can be identified by using the saturation characteristics of the DC electric field experiment (Maynard and Cauffman, 1973). The satellite contains an operational data processing system which permits complete format changes in flight in order to concentrate on specific problems. This feature has been used extensively.

The S^3 concept is unique. A single data tape is distributed to all experimenters in the program. The satellite is considered as a single experiment with each experimenter free to use any and all of the data regardless of his particular connection to the instrument making the measurements. The experimenter team consists of eight scientists representing five separate institutions. There are also a number of associate members on this team. With this operational situation and the strong cooperation of the team members, magnetospheric problems can be viewed and approached with a complete set of simultaneous data bearing on the problem and the compilation of results can be assembled into review papers such as the present one early in the data analysis program of the satellite.

3. The Inner Edge of the Proton Ring Current

Prior to the launch of S^3, the protons associated with the ring current had been observed directly on only one occasion (Frank, 1967). The S^3 satellite is providing the first total proton energy density measurements in the storm-time ring current region. We have shown that the storm-time ring current is produced by the injection of protons with energies from 1 keV to 140 keV with particles from 20 keV to 100 keV producing the largest increase from the quiet time distribution. The gross features of the ring current have been discussed by Smith and Hoffman (1972, 1973), Hoffman (1973), and Smith *et al.* (1973), and reviewed by Fritz *et al.* (1973). We now discuss work presented by Smith and Hoffman (1972) and by Smith *et al.* (1973) which deals with the inner edge of the proton ring current distribution.

3.1. Development Phase

In Figure 1 we present ground magnetograms during the period February 23–25, 1972, in which a small main phase magnetic storm occurred. The storm was initiated with a sudden commencement at 0642 UT on February 24 and fairly quickly developed a moderate main phase. Substantial substorm activity near midnight can be seen in the College magnetogram. At the bottom of Figure 1 the orbit of S^3 is shown for this period with apogee near a magnetic local time of 1900 h. The time periods for three S^3 orbits, 314, 316, 318, are indicated above the magnetograms. The three orbits are during the development, early recovery, and later recovery phase of the storm, respectively.

The energy density radial profiles were constructed by integrating the proton number flux over both pitch angle and energy from 1 keV to 138 keV during the three phases of the storm and the quiet-time period before the storm. These profiles are presented in Figure 2. The profile for orbit 312 shows the energy density distribution approximately 15 h before the sudden commencement during which time the geomagnetic conditions were quiet. The energy density reached a maximum at $L \sim 4$ and remained at about that level out to apogee. The sudden commencement occurred prior to the beginning of orbit 314. The profile for this orbit shows a large asymmetry between the inbound and outbound legs. Note the sharp inner boundary in which the energy density decreased by approximately two orders magnitude in less than one earth radii. It is this developmental aspect of the ring current distribution that we want to discuss in this section. The profiles for orbits 316 and 318 assumed the more standard form although the profile for orbit 316 remained somewhat asymmetric.

In Figure 3 we present the details associated with the energy density build-up which occurred on the outbound leg of orbit 314. In the center panel of Figure 3a the intensity of protons as a function of time and energy is shown in the form of a spectrogram. The particle energies range from 1 keV to 1000 keV and the grey shading is a measure of the flux for the 90° pitch angle protons in each energy band. The darker the shading the larger the flux. In the middle of the spectrogram around 1000 UT there is a prominent feature which has been called a 'nose' event by Smith *et al.* (1973) since the spectrogram appears to contain a projection or 'nose'. At the top of Figure 3a we demonstrate the timing features associated with this 'nose' event in more detail. Flux profiles are presented for the intensity measured at pitch angles of 90° for four energy passbands with energies centered at 6.0, 13.5, 25.6, and 42.8 keV. This panel shows that the flux in the 13.5 channel was the first to increase followed by an increase in the 25.6 keV channel and then respectively the channels at 6.0 keV and 42.8 keV. The differential energy spectra are shown in Figure 3b during four time periods in this event region. The spectrum at $L = 4.3$ appears to be the typically observed quiet-time spectrum (Smith and Hoffman, 1973; Fritz *et al.*, 1973) showing a relative minimum between 10 and 30 keV and a relative maximum near 100 keV. The subsequent spectra shown in this panel indicate that there was a rapid build up in the energy density in a very narrow energy range.

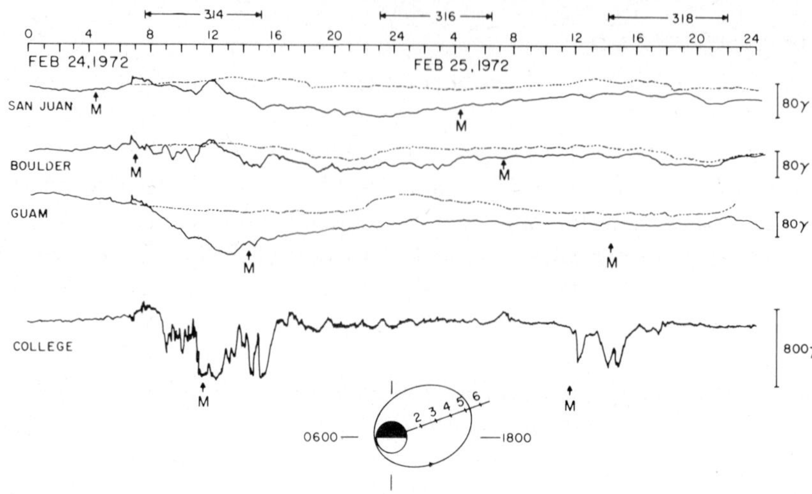

Fig. 1. Ground magnetograms for the period February 23–25, 1972. The orbit of S^3 is indicated as a function of L and magnetic local time for this period at the bottom of the figure (Smith and Hoffman, 1972).

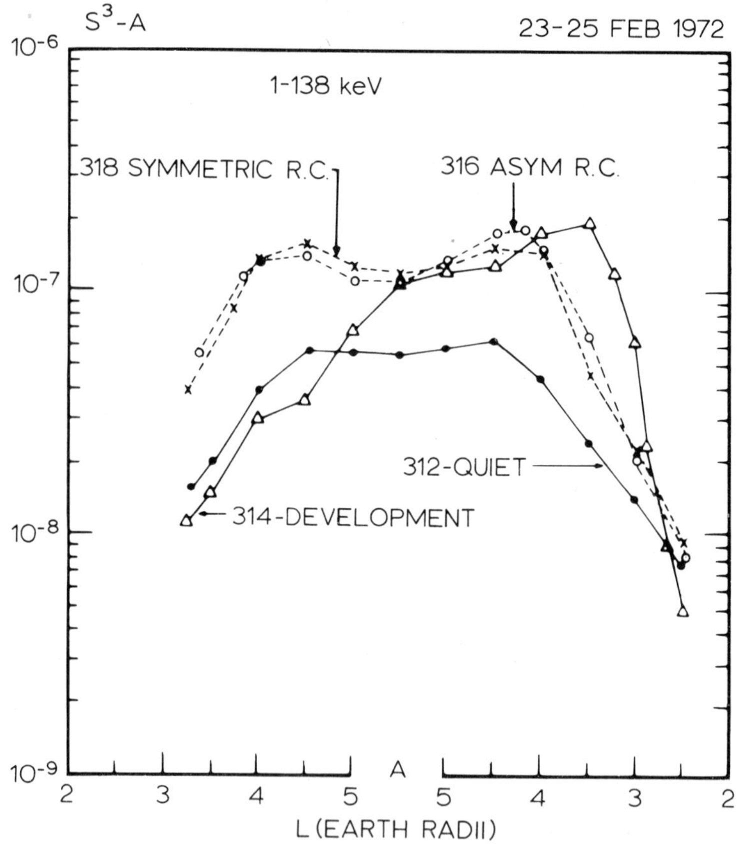

Fig. 2. Energy density radial profiles for the period of time associated with the February 72 storm of Figure 1 (Smith and Hoffman, 1972).

In the remaining panel in Figure 3a, pitch angle distributions for the four energy passbands used in the top panel are plotted for the five time periods indicated. The distributions are plotted from 20° to 90° and back to 20° with the flux scales being off-set one decade for each of the energies. The general trend in these pitch angle distributions was an initial large increase in the 90° pitch angle intensities followed by an increase in the intensity of the lower pitch angles. The entire distribution in-

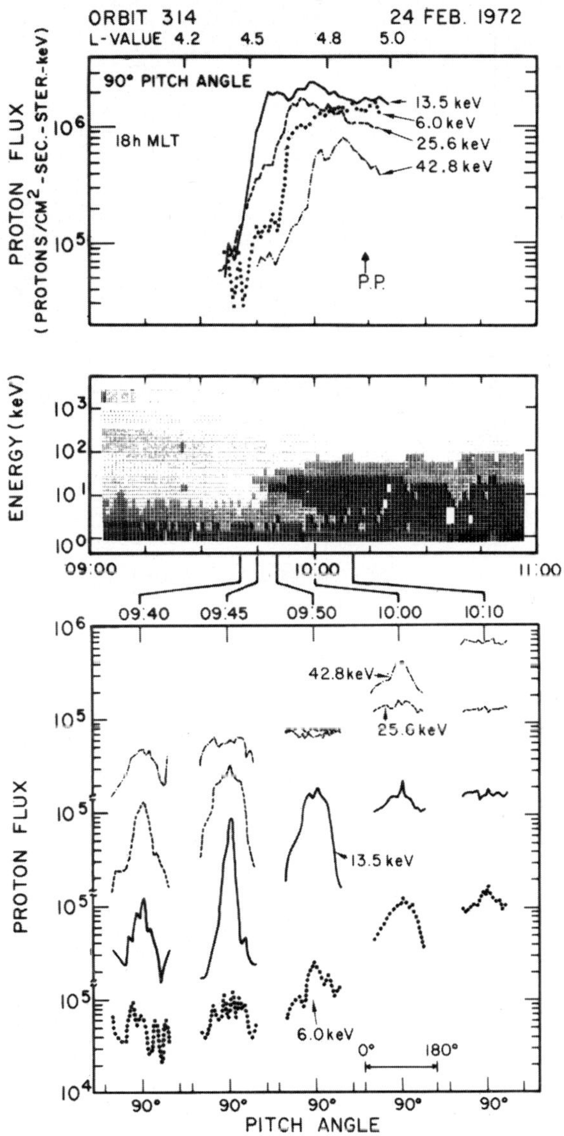

Figs. 3a–b. Details associated with the energy build up which occurred on orbit 314 associated with the February 72 storm. See text for details (Smith *et al.*, 1973).

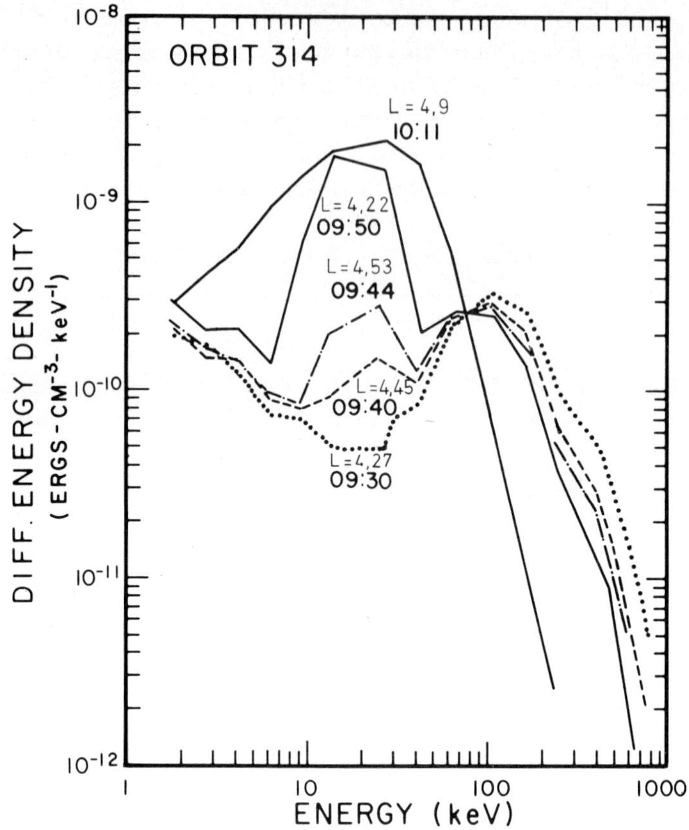

Fig. 3b.

creased until a nearly isotropic distribution was reached at considerably higher flux levels. This is especially evident in the pitch angle distributions shown for the 13.5 and 25.6 keV energy channels.

The behaviour of the protons shown in Figure 3 can not be explained by energy dispersion resulting from an injection near midnight nor can the pitch angle distributions of any one of the energy channels be explained by the effects of pitch angle dispersion resulting from a similiar injection and assuming the protons were gradient drifting around the Earth in the absence of an electric field (Smith *et al.*, 1973).

Data surveyed from S^3 thus far has tended to be concentrated in four storm-time periods. The associated main phase magnetic storms during these periods began on the following dates: December 17, 1971; February 24, 1972; March 6, 1972; and June 18, 1972. During the initial development phase of each of these storms an energy spectrogram was observed which displayed the 'nose event' characteristics described above. In Figure 4 the orbits in which these events were observed is shown in a magnetic local time – L coordinate presentation. The position of the satellite for each of these events is indicated along with its position at the time of the sudden commencement (SC) and the main phase decrease (MP) which were determined by ground magnetic records. The plasmapause crossings (PP) are also indicated and these were determined by the methods described by Maynard and Cauffman (1973). The time between the start of the main phase and the observation of the event is given by the quantity Δt.

From Figure 4 it is seen that the events (1) occur always just inside the plasmapause, (2) thus far, occur only in the local time interval near dusk, (3) are not time related to the beginning of the main phase or to the sudden commencement, and (4) are independent of the direction of the spacecraft since it is observed on the inbound as well as the outbound portion of the orbit.

Therefore we conclude that we are observing a feature of the magnetosphere (possibly a result of the interaction of the proton ring current with the plasmasphere) which permits a selected range of proton energies (10 to 30 keV) and particle pitch angles (90°) to approach closest to the Earth at 1800 h local time initially during the developmental phase of the proton ring current in main phase magnetic storms.

3.2. Decay Phase

After the ring current is established, the steep gradient which characterizes the inner edge of the ring current is in close proximity to the plasmapause. We have investigated the details of the interaction region between these two boundaries during the December 71 storm (Williams *et al.*, 1973a). In Figure 5 we present the radial profiles of seven energy channels recorded during an inbound pass of S^3. For the four lowest energy channels which are displayed, the profiles remain essentially constant into a given L-value which depends on energy and then they undergo a sharp decrease. A sharp decrease can also be observed in the profile associated with the energy channel from 78 to 138 keV, but the profiles of the two highest energy channels shown in Figure 5 display no sharp decrease. We have presented a discussion arguing that these data are consistent with the *initiation* of the ion-cyclotron instability (Williams *et al.*, 1973a). The conditions under which exponential wave growth can occur involve restrictions on the proton pitch angle anisotropy, A, and a minimum flux greater than a given particle energy, E_R, which must be exceeded. The resonant energy, E_R, is further defined by the following formula:

$$E_R \geqslant \frac{B^2}{8\pi N} A^{-2}(1+A)^{-1},$$

where B is the total magnetic field scaler intensity and N is the total plasma density (Kennel and Petschek, 1966; Cornwall *et al.*, 1970). In Figure 5 we can interpret the L-value where the sharp decrease occurred as the position where the resonant energy entered the particular energy passband of that profile. At this point the value of B and A can be determined from the recorded data. By this procedure we can determine all parameters in the above equation except the total plasma density, N. We therefore can extract the total plasma density as a function of radial distance. The results are shown in Figure 6. The vertical bars are the values of N extracted and reflect the

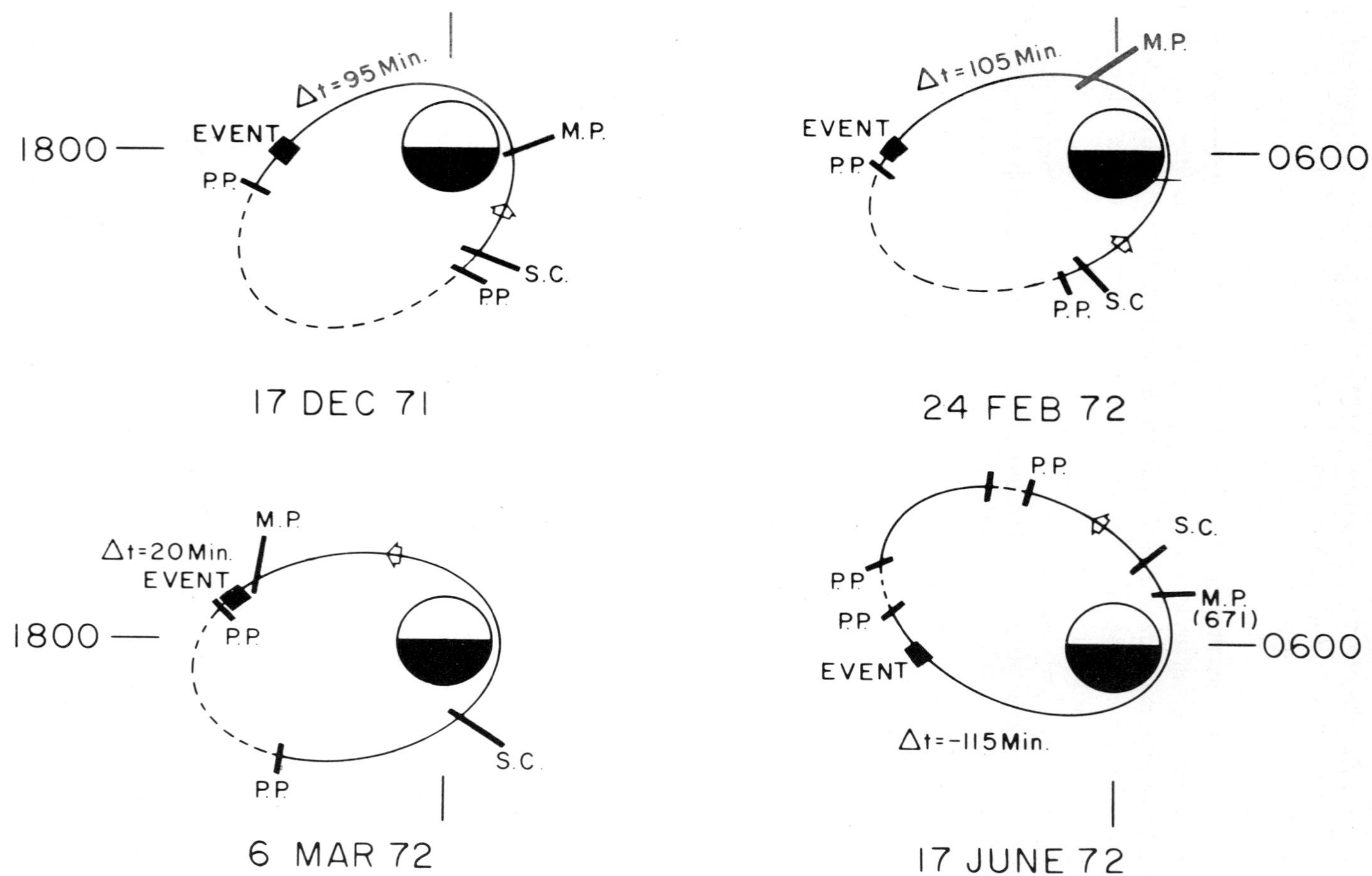

Fig. 4. Summary of the position of the S^3 satellite at the time of the occurrence of features associated with the 'nose' events observed during four main phase magnetic storms (Smith *et al.*, 1973).

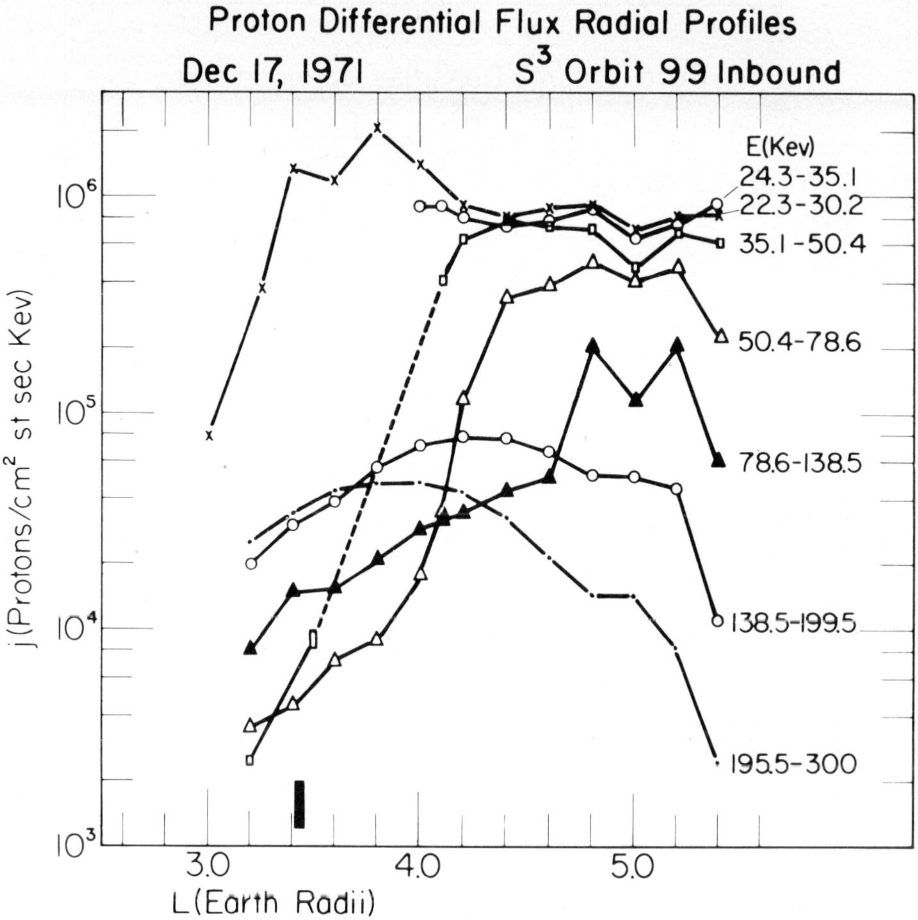

Fig. 5. Proton differential flux versus altitude for seven energy passbands during S^3 orbit 99. Vertical bar at $L = 3.45$ is where DC electric field experiment comes out of saturation thereby providing an indication of the plasmapause (Maynard and Cauffman, 1973; Williams *et al.*, 1973a)

fact that the resonant energy was measured as a differential energy band. A dashed line is drawn through the vertical bars to indicate that a normal plasmapause structure can be obtained from this interpretation. The plasmapause position as determined by the DC electric field experiment for this pass was $L \sim 3.45$. Recent attempts to calibrate the value of N associated with the DC E-field plasmapause position using simultaneous ground based measurements of whistlers have yielded values of N less than 100 particles cm^{-3} (Morgan and Maynard, 1973) rather than 1000 particles cm^{-3} as the ion-cyclotron instability analysis discussed above would indicate. The ion-cyclotron instability as a candidate to explain the observed behaviour has another problem. When the integral flux profiles are constructed from the response presented in Figure 5, it is observed that each profile continues to decrease indefinitely as a function of radial distance when according to the predictions of the theory, this in-

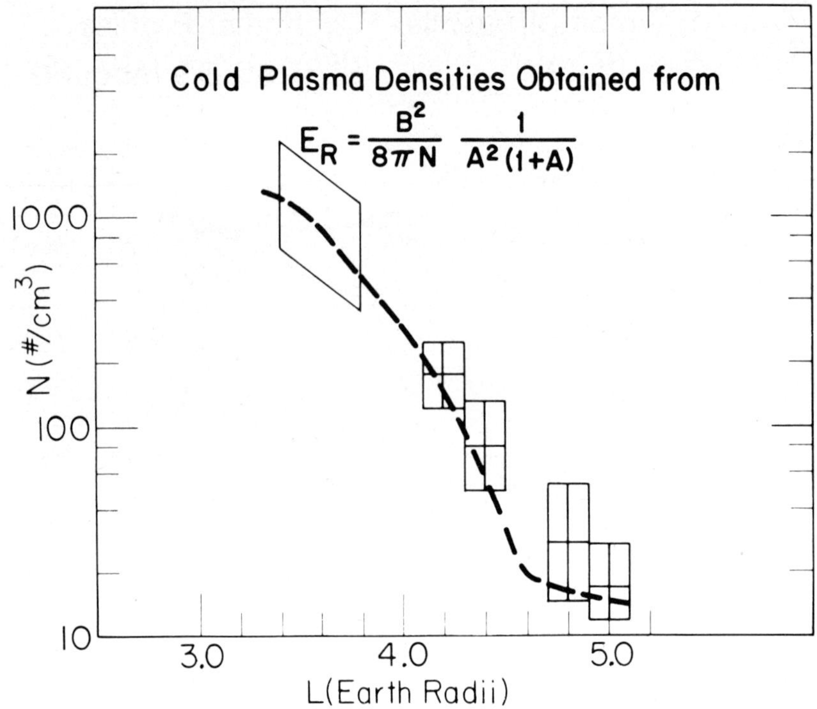

Fig. 6. Plasma densities obtained from resonant energy equation with E_R, B, and A measured at or just before the point of intensity decrease in Figure 5. Vertical range of N is due to the observation of E_R in a particular energy interval (Williams *et al.*, 1973a).

stability should cease as soon as the particle flux drops below the value established by the critical flux condition. Regardless of the exact instability or nature of the instability involved, one does apparently occur in the interaction region between the inner edge of the ring current and the plasmapause. This instability leads to a decay of the storm-time ring current spectral distribution back into the quiet-time distribution. The evolution of the quiet-time spectral minimum around 30 keV and relative maximum above 100 keV is observable in the series of spectra presented in Figure 7 for this pass we have been discussing.

Therefore we conclude that we have observed effects caused by an instability similar to (if not) the ion cyclotron instability acting in the interaction region between the inner edge of the ring current and the plasmapause which causes a decay of the storm-time ring current spectral distribution back into the quiet-time distribution.

4. The Magnetopause

The position of the magnetopause has been observed to move inward in conjunction with large solar events and has been observed as close to the Earth as the geostationary orbit ($L=6.6$) on at least three occasions (Opp, 1968; Cummings and Coleman,

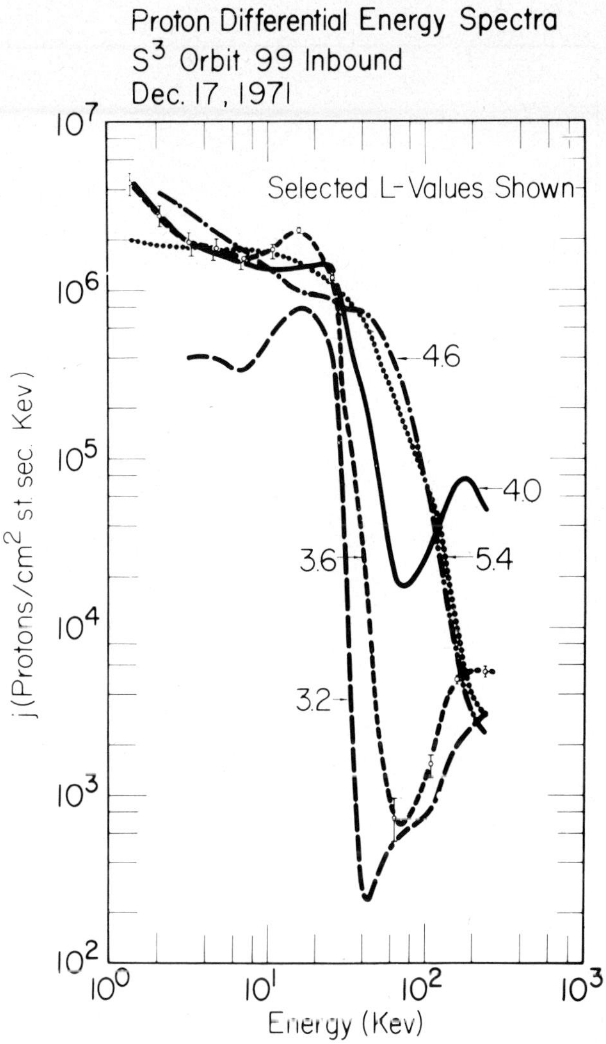

Fig. 7. Evolution of the proton differential energy spectra through the region of the steep gradients in Figure 5. See Williams *et al.* (1973a) for qualification to the data.

1968; Freeman *et al.*, 1968; Lezniak and Winckler, 1968; Paulikas *et al.*, 1968; Lanzerotti *et al.*, 1968; Skillman and Sugiura, 1971; Bogott and Mozer, 1971). On one of these occasions DeForrest (1973) has recently reported that the unshocked solar wind was observed at the geostationary orbit. We now review work presented by Cahill (1972), Hoffman and Smith (1972), and Cahill and Skillman (1973) concerning measurements taken by S^3 on August 4, 1972, which indicate that the magnetopause moved inward to at least 5.2 R_E for at least 25 min in connection with the very large solar event occurring on that date.

In Figure 8 we present the magnetic field measurements for a two hours period at the time of the suspected magnetopause crossing. The difference ΔB in gammas be-

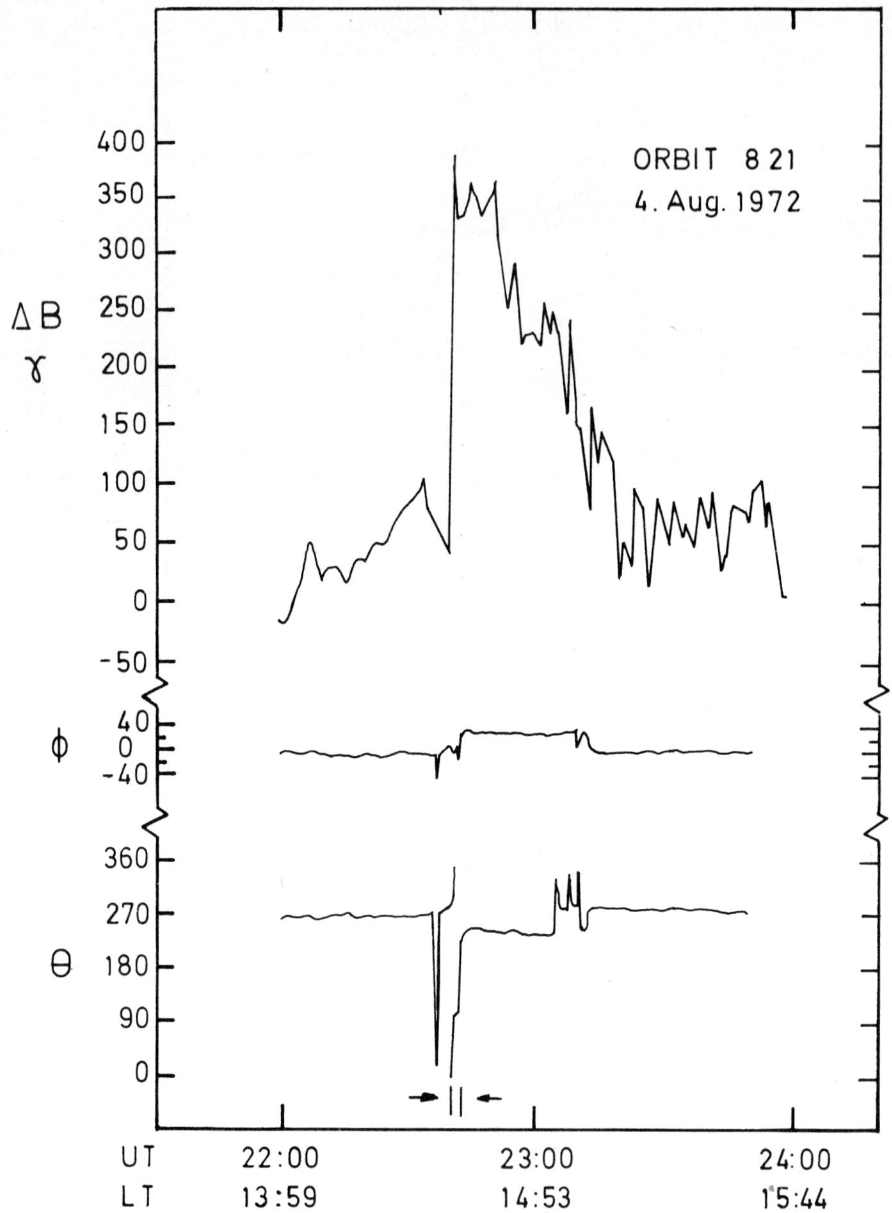

Fig. 8. Preliminary calculations of the change in the magnetic field magnitude and direction associated with the August 4, 1972 solar event. The angles are measured in terms of a spacecraft coordinate system and are defined in the text. (Cahill, 1972; Cahill and Skillman, 1973.)

tween the field magnitude measured and calculated using a reference field, $B_{\text{ref.}}$ is plotted at the top of the figure. (A reference field, $B_{\text{ref.}}$, is constructed from the August 1969 OGO-6 magnetic field model of Cain and Sweeney (1970).) The orientation of the local B-vector is given in terms of two angles associated with the spacecraft coördinate system. The spin axis ($\hat{\omega}$) of the satellite is nominally *in* the plane of the satellite orbit (the Earth's equatorial plane) and maintained within specified limits with respect to the Sun-Earth line ($\hat{s}$). At the time of this event $\hat{\omega}\cdot\hat{s}=\cos\alpha\approx\cos 60°$. The two angles plotted in Figure 8 are defined as

$$\Phi=\arcsin\frac{\mathbf{B}\cdot\hat{\omega}}{B}$$

$$\theta=\arccos\frac{\mathbf{B}\cdot(\hat{\omega}\times\hat{s})}{(B^2-(\mathbf{B}\cdot\hat{\omega})^2)^{1/2}\sin\alpha}$$

The angle φ is the angle between the B-vector and the plane perpendicular to the spin axis. It is positive when B has a component in the same direction as the spin axis (toward the Sun). The angle θ is the angle between the B-vector and the spin-axis-Sun plane, which in effect is the phase of the satellite spin. A value of $\theta\sim 270°$ is the normal magnetospheric northward pointing field direction. In Figure 8 it can be seen that the field was steadily compressed for the first 40 min up to the point where it executed a quick double reversal, i.e., the B-vector flipped from the northward pointing dipole direction to a southward pointing direction and then flipped back to the dipole direction again. This double reversal was accompanied by a small relaxation of the field. Then at 2240:36 UT the magnetic field again suddenly reversed direction in less than two seconds and a very strong compression of the field occurred. The observation of this second reversal and the strong compression has been identified as the magnetopause crossing the satellite (Cahill, 1972; Hoffman and Smith, 1972; Cahill and Skillman, 1973). S^3 was nearly stationary at its apogee of 5.24 R_E at about 9° magnetic latitude in the afternoon sector of the magnetosphere near 14.6 h local time at this time.

In Figure 9 we again present the magnetic field measurements but this time they are expanded to cover the period indicated by the two arrows in Figure 8. In addition the top panel of Figure 9 contains proton measurements made by the cylindrical electrostatic analyzer-channeltron experiment from 26 keV down to one keV. The energy step was changed each revolution of the spinning satellite and the curves are plotted alternately dashed and solid for ease in distinguishing each energy step. The second and third panels contain data from two channels of the solid state detector proton experiment. The blanks in the curves for these two channels are due to a different mode of instrument operation which occurred every fourth spin period.

Although the presence of the magnetopause in the vicinity of the satellite affected the proton intensities in a complex manner, it was possible to discern in much of the data the usual case in the magnetosphere of having two intensity maxima per spin period right up to the boundary crossing. Two intensity maxima per spin period can be observed in the higher energy protons plotted in Figure 9 prior to 2240:36.

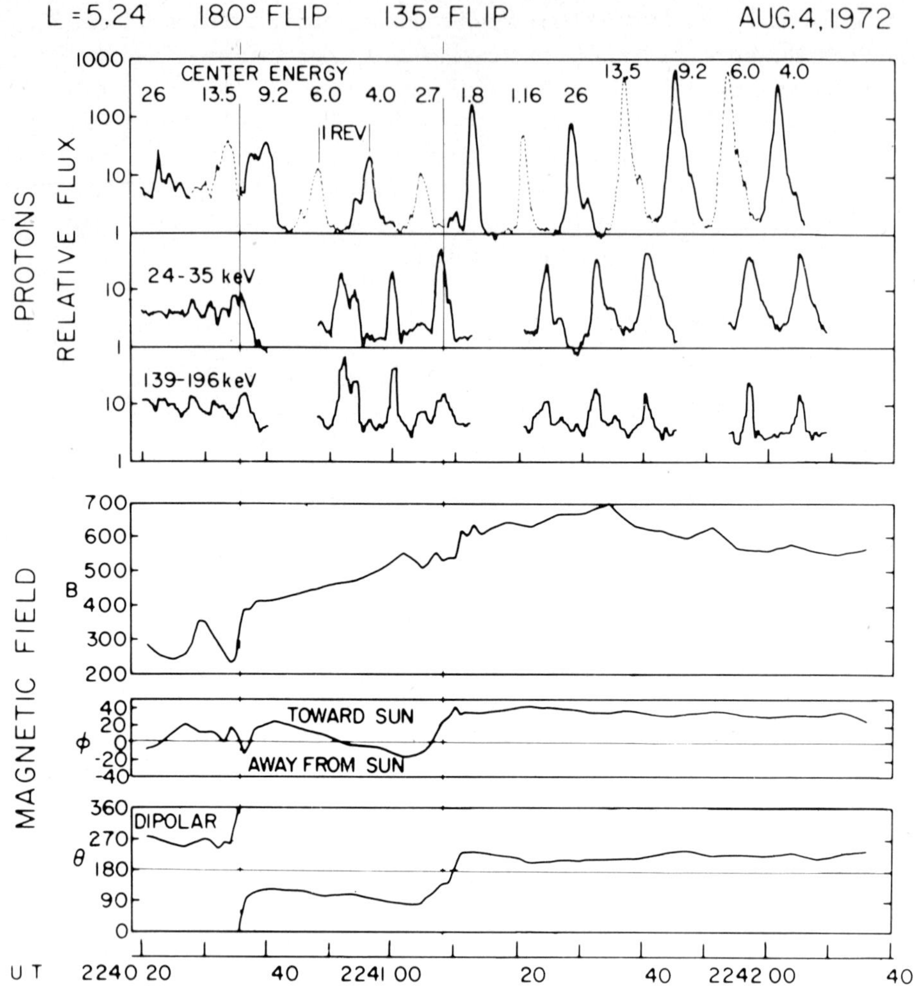

Fig. 9. Expanded presentation of the two minute period of time indicated in Figure 8 by the double arrows for the magnetic field variations. Also included are proton rates as a function of time for the energy channels indicated. The satellite spin period is indicated by '1 REV' (Hoffman and Smith, 1972).

As was noted in Figure 8 at 2240:36 the magnetic field suddenly reversed direction in less than two seconds so that it pointed southward. At the same time all proton measurements showed a streaming of protons with one large maximum per spin period. These maxima are 180° out of phase for the two detectors because the instruments are mounted on opposite sides of the spinning spacecraft. The strong compression noted in Figure 8 occurred in a steady manner over a period of about one minute. During this steady increase of the field strength the direction rotated in a matter of about 5 s to the northern hemisphere returning to a nearly dipolar configuration. The streaming protons came generally from the direction of the Sun and it is only because of the continued streaming of the protons that we claim that the satellite remained outside the magnetopause. S^3 apparently encountered the boundary for brief periods both before and after the one major crossing discussed here. Although the details associated with the streaming protons are complex, we have compared the differential energy spectrum just inside the boundary with that at the peak of the streaming protons. This comparison is presented in Figure 10. Fluxes greater than 10^9 protons cm^{-2} s^{-1} sr^{-1} keV^{-1} were encountered at around 4 keV outside of the magnetopause. The data have been fitted to a maxwellian with a temperature ($=kT$) of 3.9 keV (Hoffman and Smith, 1972).

Therefore we conclude that the magnetopause has been observed to approach the Earth to a radial distance of 5.2 R_E at the equatorial plane in the afternoon sector (14.6 h local time) in connection with the large solar event of August, 1972.

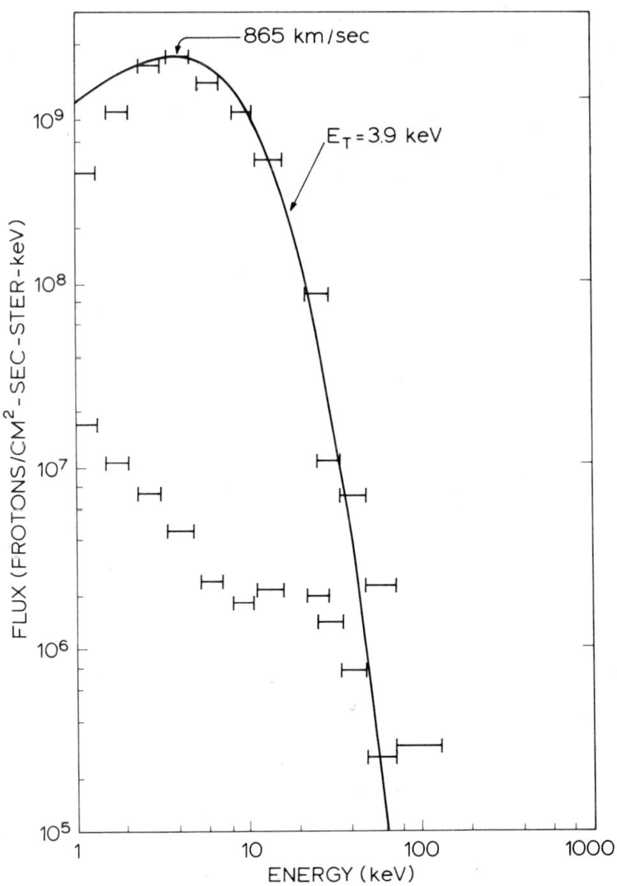

Fig. 10. Differential energy spectra recorded just before the magnetopause crossing and at the peak of the streaming proton distributions. A modified maxwellian distribution for dj/dE proportional to $E\,e^{-E/3.9\,\mathrm{keV}}$ has been fit to the data (Hoffman and Smith, 1972).

5. The Electron Plasma Sheet

The study of the characteristics of the injection of particles in association with substorms using S^3 has uncovered the observation that the gross features of the substorm signature persist deep into the nightside trapping region. We will now discuss an observation reported by Williams *et al.* (1973b) of one such signature at 5.0–5.5 R_E which has the characteristics associated with the electron plasma sheet usually observed at much greater radial distances.

In Figures 11a and 11b we present the electron intensity profiles for nine energy channels from 1.5 to 560 keV recorded on December 12–13, 1971. According to the ground magnetometers a substorm began at 2200 ± 10 UT on December 12, 1971.

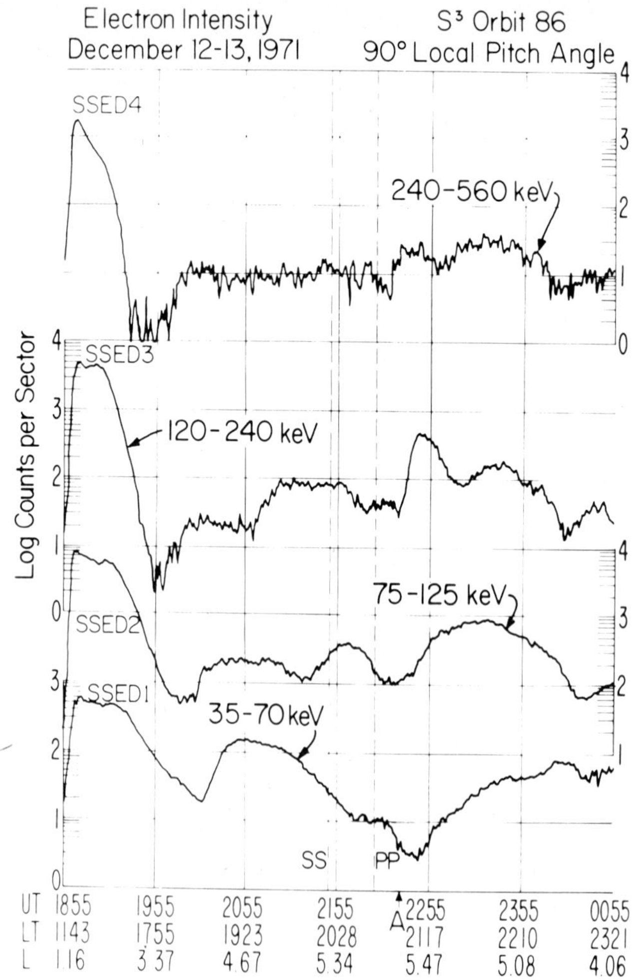

Figs. 11a–b. Response of S^3 solid state detector electron channels (a) and electrostatic analyzer-channeltron electron channels (b) for orbit 86. The substorm start (SS), plasmapause (PP) and apogee (A) are shown. Local pitch angles of $\pi/2$ are plotted. The spin period is divided into 32 sectors with a sector being 0.264 s (Williams *et al.*, 1973b).

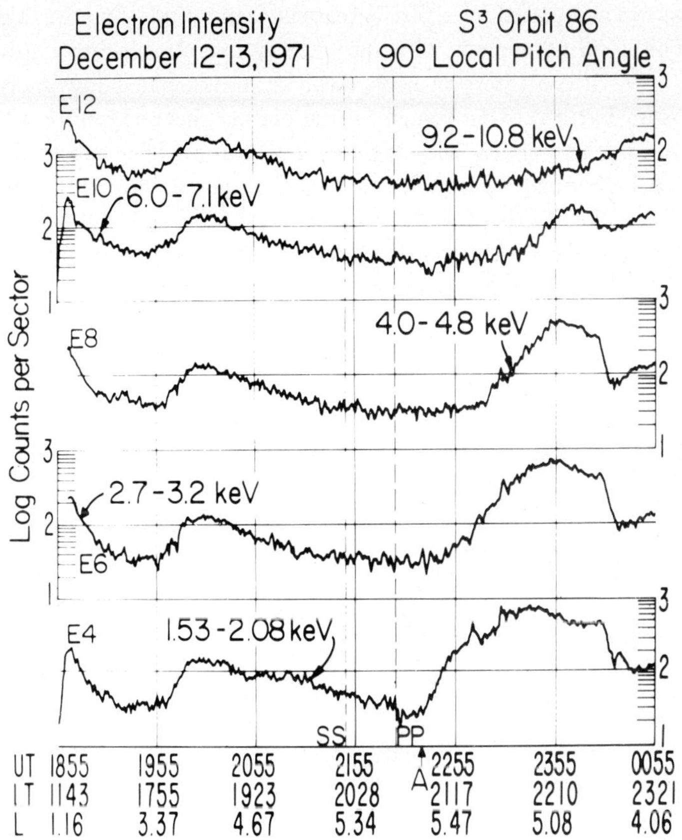

Fig. 11b.

At this time S^3 was outbound near apogee at a local time of ~2100 h but did not exit the plasmasphere until 2221 UT. The position of the plasmapause was determined by the method described by Maynard and Cauffman (1973). The substorm associated intensity increases for protons occurred while the satellite was still in the plasmasphere and any attempt to associate arrivals times could be affected by processes which we have discussed in Section 3. All electron responses that we have associated with the substorm noted above occurred after the protons responded and were outside of the plasmasphere. Time dispersion effects thus should be uneffected by effects of the plasmasphere.

In Figure 11a the time dispersion associated with the response of the higher energy electrons is qualitatively consistent with localized injection of these electrons at a position remote from the satellite and subsequent drift to the vicinity of S^3. Similar drift time dispersion effects have been observed at synchronous altitudes (Arnoldy and Chan, 1969; Lezniak and Winckler, 1970; DeForrest and McIlwain, 1971).

In Figure 11b the time dispersion associated with the response of the lower energy electrons (1.53–10.8 keV) display the opposite behaviour. The lower energy channels show a substorm associated intensity increase at earlier times than the higher energy channels. This is in agreement with the observations at synchronous altitude of Shelly *et al.* (1971) who interpret such a substorm associated low energy electron behaviour as an Earthward movement of the plasma sheet engulfing ATS 5.

Using the times associated with the first observation of a substorm associated increase in each channel, we obtain the dispersion curve shown in Figure 12. The vertical bars are the width of the energy channel and the horizontal bars are an estimate of timing errors. A dispersion curve due to gradient drift in a dipole field based on injection at midnight is shown for comparison. The dispersion character-

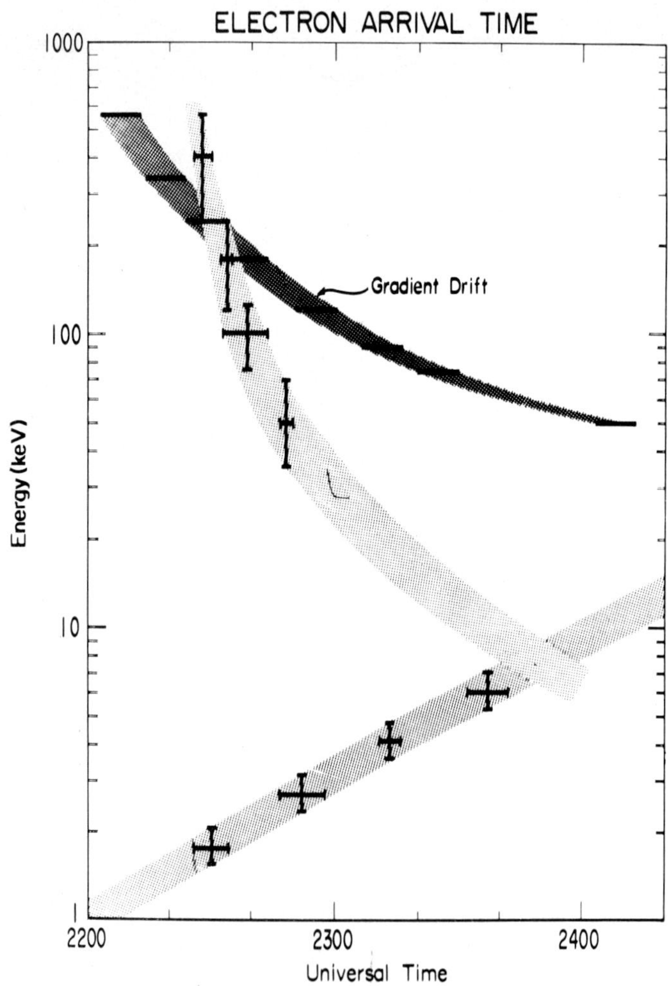

Fig. 12. Electron energy dispersion. The channel energy is plotted versus the time of the substorm associated response for that channel. Vertical bars give the energy width of channel and the horizontal bars are estimates of timing errors. The curve marked 'gradient drift' shows the expected dispersion profile resulting from the gradient drift of electrons in a dipole field which were injected initially near local midnight (Williams *et al.*, 1973b).

istics for the high energy electrons is consistent with a gradient drift through only four hours of local time following the initial enhancement. This indicates that the initial energetic electron enhancements may have extended well into the dayside region for this substorm.

In attempting to understand the appearence of the low energy electrons at S^3, two cases were considered: (a) S^3 was engulfed by the plasma sheet as it was convected inward or (b) a plasma cloud was formed at some position away from S^3 and was then detected at the satellite due to motion of the electrons in the combined electric and magnetic fields present during the substorm period.

The appearence of protons at S^3 prior to the electrons is qualitatively consistent with the inward convection of a high altitude plasma sheet consisting of a proton and electron component. In its ambient position such a plasma sheet in the geomagnetic tail is expected to have the positive ion component extend inward to lower altitudes than the negative electron component (Kennel, 1969).

In Figure 13 we present the variation of energy density as a function of time for both the electrons and protons during the period of the substorm. Note that the electron energy density is never more than 1% of the local proton energy density. The energy density of the equatorially mirroring protons through the substorm period and its spatial relationship to the plasmapause was similar to the spatial relation of

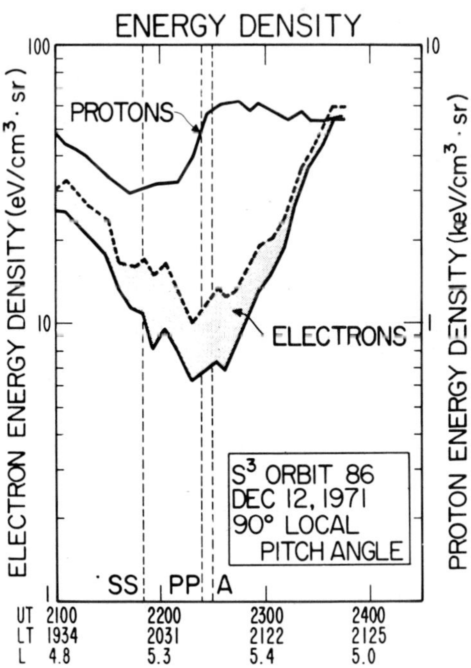

Fig. 13. Electron and proton energy density versus time. The substorm start (SS), plasmapause crossing (PP), and apogee (A) are indicated. The electron energy density profile show both a linear and a log-log interpolation used to estimate the energy density across the energy gap between 10 and 35 keV (Williams *et al.*, 1973b).

the ring current protons and the plasmapause reported by Frank (1971). This implied that these substorm associated protons are simply the inward flow of the ring current protons in the combined magnetic and electric fields.

Vasyliunas (1968) has reported an exponential spatial structure in the electron energy density at the Earthward edge of the plasma sheet. Frank (1971) has used this feature as a definition of the Earthward edge of the plasma sheet. If such a structure were simply transported inward past a quasi-stationary satellite an exponential time variation should be observed in the electron energy density and in Figure 13 such a variation is observed.

Using assumptions about the propagation of the inner edge of the plasma sheet inward, it is possible to determine a value of the convection electric field by a couple of methods. These values range from 0.11 mV m^{-1} to 0.34 mV m^{-1} which are of the right order of magnitude for the usual values of the convection electric field quoted.

Frank (1971) has reported that the inner boundary of the Earthward edge of the plasma sheet is usually 1 to 3 R_E beyond the plasmapause position in the pre-midnight sector. An electron 'trough' characterized by relatively low, constant electron densities lies between the plasmapause and the inner boundary of the Earthward edge of the plasma sheet in this sector. During enhanced substorm activity the electron 'trough' disappears and the inner boundary of the Earthward edge of the plasma sheet moves into the position of the plasmapause. This is consistent with the S^3 observations discussed here.

Therefore we find that all of the features usually ascribed to the inner edge of the plasma sheet are consistent with the S^3 observations. However, we run into serious difficulty when we attempt to become more quantitative. We have compared these results with flow patterns expected from a model geomagnetic-geoelectric field constructed specifically to explain particle signatures observed at ATS 5 in geostationary orbit (McIlwain, 1972).

Proton and electron trajectories in the McIlwain model were calculated for all values of the magnetic moment of interest. Protons in a wide energy band (~5 to 100 keV) were predicted in the region where the substorm proton responses were observed. They would flow in from the geomagnetic tail and would pass through the geostationary orbit near midnight. Thus, the proton observations made during the substorm are quantitatively consistent with the McIlwain model. On the other hand, the entire S^3 orbit was within the region defined by closed electron trajectories. *No* electrons originating in the geomagnetic tail could reach the S^3 position in this model field via a continuous flow process.

Therefore we feel that existing static electric field models are unable to explain the injection of both substorm associated protons and electrons into the evening sector at equatorial altitudes of 5 to 5.5 R_E. The apparent observation of the electron plasma sheet *cannot* be associated with a simple inward convection of these electrons from the geomagnetic tail in a steady electric field. A *discontinuous* change at the time of the substorm initiation is apparently required to explain the sudden appearence of these particles at altitudes well within the geostationary orbit. We therefore favor alternative *B* noted above as the preferred explanation to the S^3 observations of the substorm associated particle responses discussed here. This agrees with the plasma cloud concepts advanced by DeForrest and McIlwain (1971). After the plasma cloud has been injected to low altitudes, drift in the static electric field model does seem to explain subsequent S^3 observations (Williams *et al.*, 1973b).

6. Conclusions

We have reviewed work pretaining to the nature of three magnetospheric boundaries by S^3 (Explorer 45). We emphasize that this is a review of work still in progress. Some of the details of the interpretations may change as the work in each particular area of study is brought to completion. Although all of the work presented herein has been presented or published in some fashion elsewhere, almost all of it has yet to go through a critical review prior to journal publication.

Therefore with the above qualification we conclude that:

(1) We have observed two features in the interaction region between the inner edge of the proton ring current and the plasmapause.

(a) We have observed a spatial feature which permits a selected range of proton energies (10 to 30 keV) and particle pitch angles (90°) to approach closest to the Earth at 1800 h local time during the developmental phase of main phase magnetic storms.

(b) We have observed effects caused by an instability similar to (if not) the ion cyclotron instability which leads to a decay of the proton storm-time spectral distribution back into the quiet-time distribution.

(2) We have observed the presence of the magnetopause at a radial distance closer to the Earth than 5.2 R_E at the equatorial plane in the afternoon sector near 14.6 h local time.

(3) We have observed an electron distribution at 5–5.5 R_E which has all of the characteristics of the inner edge of the electron plasma sheet but was observed at a position into which it was not possible to transport electrons from the geomagnetic tail via a steady convection electric field.

Acknowledgements

We would like to acknowledge the participation of Dr N. C. Maynard of NASA/GSFC, Dr D. A. Gurnett of the University of Iowa, and Dr A. Konradi of NASA/JSC as members of the S^3 experimenter team.

One of us (T.A.F.) would like to thank Professor G. Pfotzer and Dr D. J. Williams for the special arrangements which have made the year of residence at MPI/Lindau and this review work a possibility.

References

Arnoldy, R. L. and Chan, K. W.: 1969, *J. Geophys. Res.* **74**, 5019.
Bogott, F. H. and Mozer, F. S.: 1971, *J. Geophys. Res.* **76**, 892.
Cahill, L. J., Jr.: 1972, *Trans. Amer. Geophys. Union* **53**, 1059.
Cahill, L. J., Jr. and Skillman, T. L.: 1973, *Trans. Amer. Geophys. Union* **54**, 433.
Cain, J. C. and Sweeney, R. E.: 1970, *J. Geophys. Res.* **75**, 4360.
Cornwall, J. M., Coroniti, F. W., and Thorne, R. M.: 1970, *J. Geophys. Res.* **75**, 4699.
Cummings, W. D. and Coleman, Jr., P. J.: 1968, *J. Geophys. Res.* **73**, 5699.
DeForrest, S. E.: 1973, *J. Geophys. Res.* **78**, 1195.
DeForrest, S. E. and McIlwain, C. E.: 1971, *J. Geophys. Res.* **76**, 3587.
Frank, L. A.: 1967, *J. Geophys. Res.* **72**, 3753.
Frank, L. A.: 1971, *J. Geophys. Res.* **76**, 2265.
Freeman, J. W., Jr., Warren, C. S., and Maguire, J. J.: 1968, *J. Geophys. Res.* **73**, 5719.
Fritz, T. A., Williams, D. J., Smith, P. H., Hoffman, R. A., Barfield, J. N., and Konradi, A.: 1973, 'Particle Entry into the Ring Current Region of the Equatorial Magnetosphere', *Space Res.* **14**, to be published (presented as paper b. 28 at the 16th meeting of COSPAR, Konstanz, FRG, June).
Hoffman, R. A.: 1973, *J. Geophys. Res.* **78**, 4771.
Hoffman, R. A. and Smith, P. H.: 1972, *Trans. Amer. Geophys. Union* **53**, 1059.
Kennel, C. F.: 1969, *Rev. Geophys.* **7**, 379.
Kennel, C. F. and Petschek, H. E.: 1966, *J. Geophys. Res.* **71**, 1.
Lanzerotti, L. J., Brown, W. L., and Roberts, C. S.: 1968, *J. Geophys. Res.* **73**, 5751.
Lezniak, T. W. and Winckler, J. R.: 1968, *J. Geophys. Res.* **73**, 5733.
Lezniak, T. W. and Winckler, J. R.: 1970, *J. Geophys. Res.* **75**, 7075.
Longanecker, G. W. and Hoffman, R. A.: 1973, *J. Geophys. Res.* **78**, 4711.
Maynard, N. C. and Cauffman, D. P.: 1973, *J. Geophys. Res.* **78**, 4745.
McIlwain, C. E.: 1972, in B. M. McCormac (ed.), *Earth Magnetospheric Processes*, D. Reidel Publ. Co., Dordrecht-Holland, p. 268.
Morgan, M. G. and Maynard, N. C.: 1973, 'Correlation of the Plasmapause Detection by the S^3-A Double Floating Probe with Whistler-Deduced Electron Densities', *Space Res.* **14** (to be published) (presented as paper b. 33 at the 16th meeting of COSPAR, Konstanz, FRG, June, 1973).
Opp, A. G.: 1968, *J. Geophys. Res.* **73**, 5697.
Paulikas, G. A., Blake, J. B., Freden, S. C., and Imamoto, S. S.: 1968, *J. Geophys. Res.* **73**, 5743.
Shelley, E. G., Johnson, R. G., and Sharp, R. D.: 1971, *Radio Sci.* **6**, 305.
Skillman, T. L. and Sugiura, M.: 1971, *J. Geophys. Res.* **76**, 44.
Smith, P. H. and Hoffman, R. A.: 1972, *Trans. Amer. Geophys. Union* **53**, 1088.
Smith, P. H. and Hoffman, R. A.: 1973, *J. Geophys. Res.* **78**, 4731.
Smith, P. H., Hoffman, R. A., and Fritz, T. A.: 1973, *Trans. Amer. Geophys. Union* **54**, 415.
Vasyliunas, V. M.: 1968, *J. Geophys. Res.* **73**, 2839.
Williams, D. J., Fritz, T. A., and Konradi, A.: 1973a, *J. Geophys. Res.* **78**, 4751.
Williams, D. J., Barfield, J. N., and Fritz, T. A.: 1973b, 'Initial Explorer 45 Substorm Observations and Electric Field Considerations', NOAA-ERL-SEL preprint.

MEASUREMENTS OF SOLAR PROTONS IN THE NEAR EARTH MAGNETOTAIL

V. DOMINGO, D. E. PAGE, and K.-P. WENZEL

Space Science Department (ESLAB), European Space Research and Technology Centre, Noordwijk, The Netherlands

1. Introduction

The penetration of solar protons into the magnetosphere has been studied mainly by measurements over the polar caps by near Earth polar orbiting satellites, by measurements made down in the tail of the magnetosphere at antisolar distances of 15 R_E or more, and in other equatorial magnetospheric regions (Lanzerotti, 1972; Fennell, 1973). All present theoretical studies are based on such measurements and on ground based measurements (Gall *et al.*, 1968; Morfill and Quenby, 1971).

The orbit of HEOS 2 offers the opportunity of observing a region not previously studied through which at least a part of the solar particles penetrate into the polar caps i.e. the high latitude magnetosphere. In this paper we present preliminary results of the measurement of protons in this region.

2. Instrumentation

Our detector placed at 90° to the spin axis of the satellite was formed by two solid state detectors and a shielding scintillator cup (Köhn *et al.*, 1972).

For the present study only data obtained from the front solid state detector in anticoincidence with the back detector and the scintillator are used. This channel measures protons of 1–9 MeV and electrons of roughly 0.4–0.6 MeV. Particle anisotropy measurements are obtained in the equatorial plane of the satellite, in 8 angular sectors of 90°, each one overlapping the next by 45°, when the experiment is operated in angular mode. The spin period of the satellite is about 6 s.

3. Studied Region

HEOS 2 is a spin stabilized spacecraft, launched on 31 January, 1972, in a highly eccentric polar orbit (apogee about 40 R_E, perigee 1.05 R_E) with its apogee at very high northern latitude. The spacecraft in the outbound pass of each orbit, crosses the near Earth magnetotail through the region lying just above the northern polar cap.

Seven outbound passes between February and September 1972 have been considered. The criterium used to select them, was that the detector was operating in angular mode and that the 1–9 MeV proton flux was suitable for the sensitivity of the instrument.

Figure 1 shows in geocentric solar magnetospheric co-ordinates a projection, as far as the magnetopause, of the trajectory of the satellite for the orbits considered.

D. E. Page (ed.), Correlated Interplanetary and Magnetospheric Observations, 507–516.

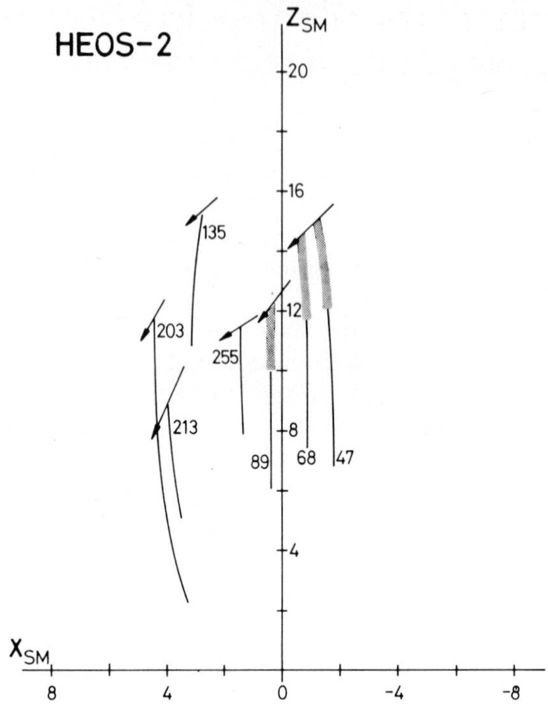

Fig. 1. Trajectory of HEOS 2 through the upper part of the near Earth magnetotail during the studied outbound passes. The co-ordinates are geocentric solar magnetospheric in Earth radii, the numbers indicating the day of the year in 1972. The arrow indicates the direction of magnetic field 10 min before crossing the magnetopause. The shadowed regions are those where an anisotropy is observed at 90° to the magnetic field (see text).

At the end of each trajectory the orientation of the magnetic field measured inside the magnetotail just before the crossing of the magnetopause is indicated to give an idea of the orientation of the surface of the magnetosphere.

If one takes into account the direction of the magnetic field and the relative position of the Earth dipole, one can assume that on 5 of the orbits (days 47, 68, 89, 135 and 255) the trajectory was crossing the magnetopause in the region B of Figure 2, while on 2 orbits (days 203 and 213) the magnetopause was crossed in region A.

4. Measurements

Figures 3 through 7 show the results of the measurements for some of the orbits. In all figures we show from top to bottom:

(1) The 10-min spin averaged count rate measured by the detector.

(2) The anisotropy coefficient (a) computed by a least squares fit of the angular values to the function

$$I = I_0 \{1 + a \cos(\phi - \phi_0)\},$$

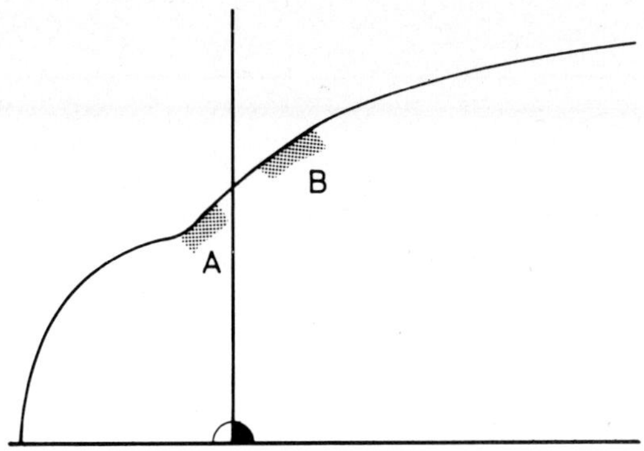

Fig. 2. Schematic view of the regions that were crossed by HEOS 2 during the days 203, 213 (Region A) and 47, 68, 89, 135 and 255 (Region B).

where I is the count rate measured at angle ϕ and I_0 and ϕ_0 are the results of the fit. The angle ϕ is measured in the equatorial plane of the spacecraft.

(3) a. The azimuth of the magnetic field vector measured in the same satellite by Dr P. C. Hedgecock (Imperial College) in spacecraft co-ordinates (lines).
b. The direction ϕ of the particle anisotropy in the same co-ordinates as the magnetic field (crosses).

(4) The dip of the magnetic field with respect to the spacecraft equator.

(5) The intensity of the magnetic field.

In all figures the magnetopause crossing and its distance are marked.

The angular measurements were made in the plane of the ecliptic on days 47, 68, 89, 203 and 213, and in a plane perpendicular to the ecliptic on days 135 and 255.

5. Discussion

Out of the seven chosen orbits a common feature appears on days 47, 68 and 89. A persistent anisotropy is observed inside the magnetotail, at near 90° to the magnetic field, and particularly on days 68 and 89 this anisotropy is coincident with an increase in the measured flux as one approaches the magnetopause. In Figure 1 shading indicates the region where the anisotropy occurs.

We therefore appear to have encountered a particle drift that can be produced only by an electric field perpendicular to the magnetic field and directed from Sun to anti-Sun, or by a particle density gradient, such that the particle density increases as one approaches the magnetopause from inside.

In order to set rough quantitative limits to help explain the origin of the drift we have used the expression (Forman, 1970):

$$a = \frac{(2 + \alpha\gamma)\, V}{v},$$

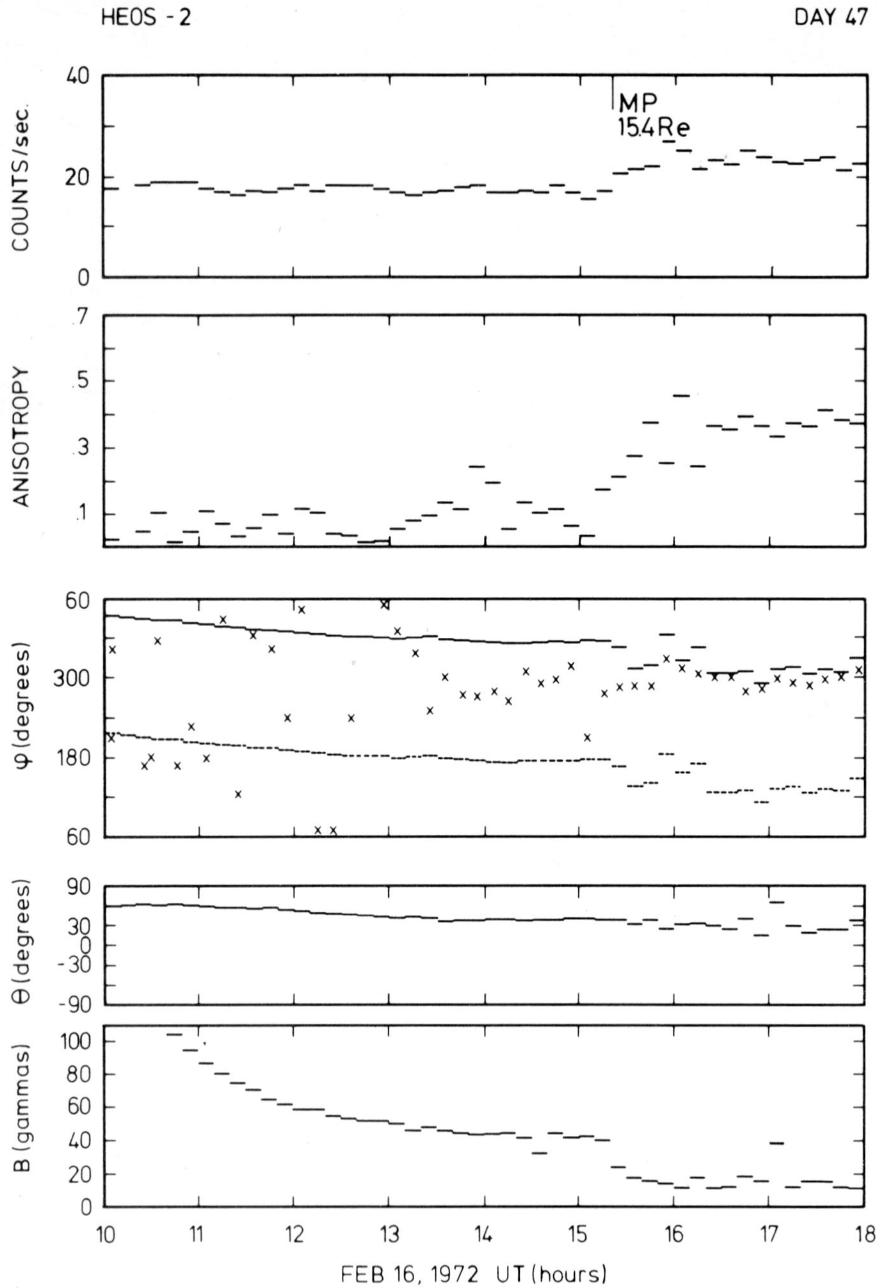

Fig. 3. From top to bottom, versus Universal Time (in hours), (a) the count rate of 1–9 MeV protons; (b) anisotropy coefficient (defined in text); (c) magnetic field azimuth (continuous line) direction and its opposite i.e. true azimuth plus 180° (dashed line) measured by the Imperial College magnetometer experiment (crosses are the direction of the anisotropy of 1–9 MeV protons); (d) latitude of the magnetic field vector in spacecraft co-ordinates; and (e) intensity of the magnetic field vector in gammas for February 16, 1972. The crossing of the magnetopause has been indicated and the distance at which it was observed.

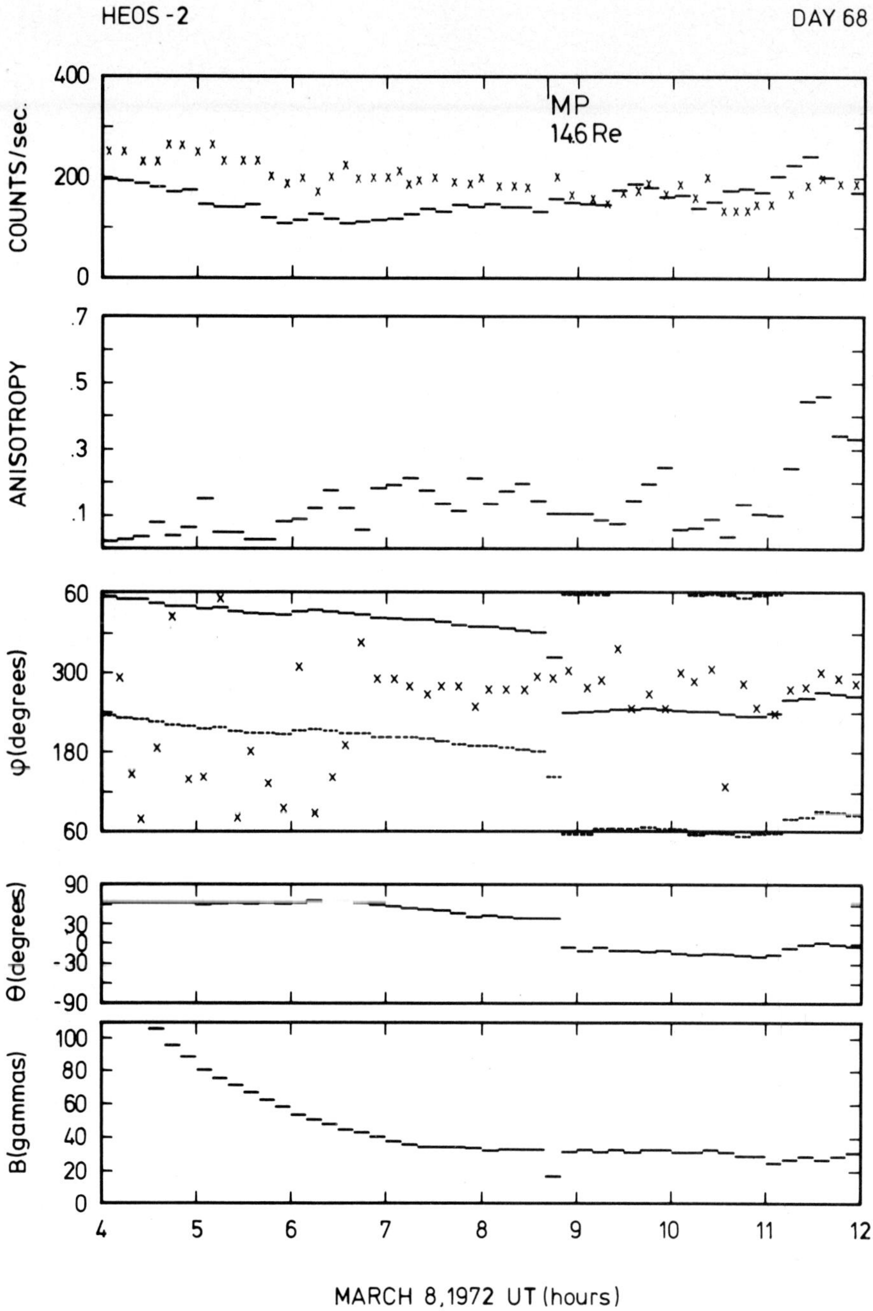

Fig. 4. Same as Figure 3 for March 8, 1972. On the top graph has been included the count rate observed in the 1–10 MeV channel of the NOAA laboratory, Boulder experiment onboard Explorer 41 in arbitrary units, normalized to the HEOS 2 count rate outside the magnetosphere.

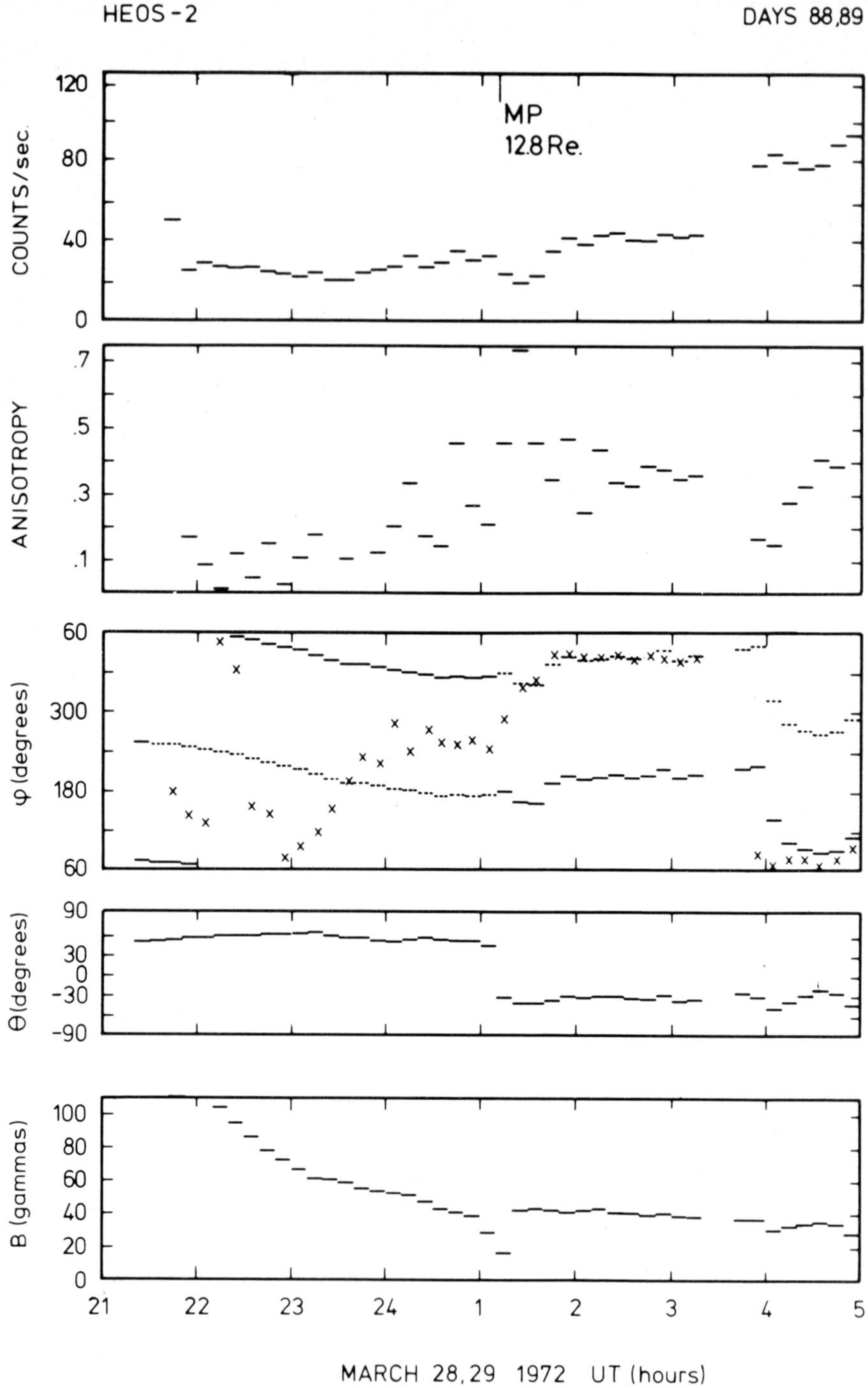

Fig. 5. Same as Figure 3 for March 28, 29, 1972.

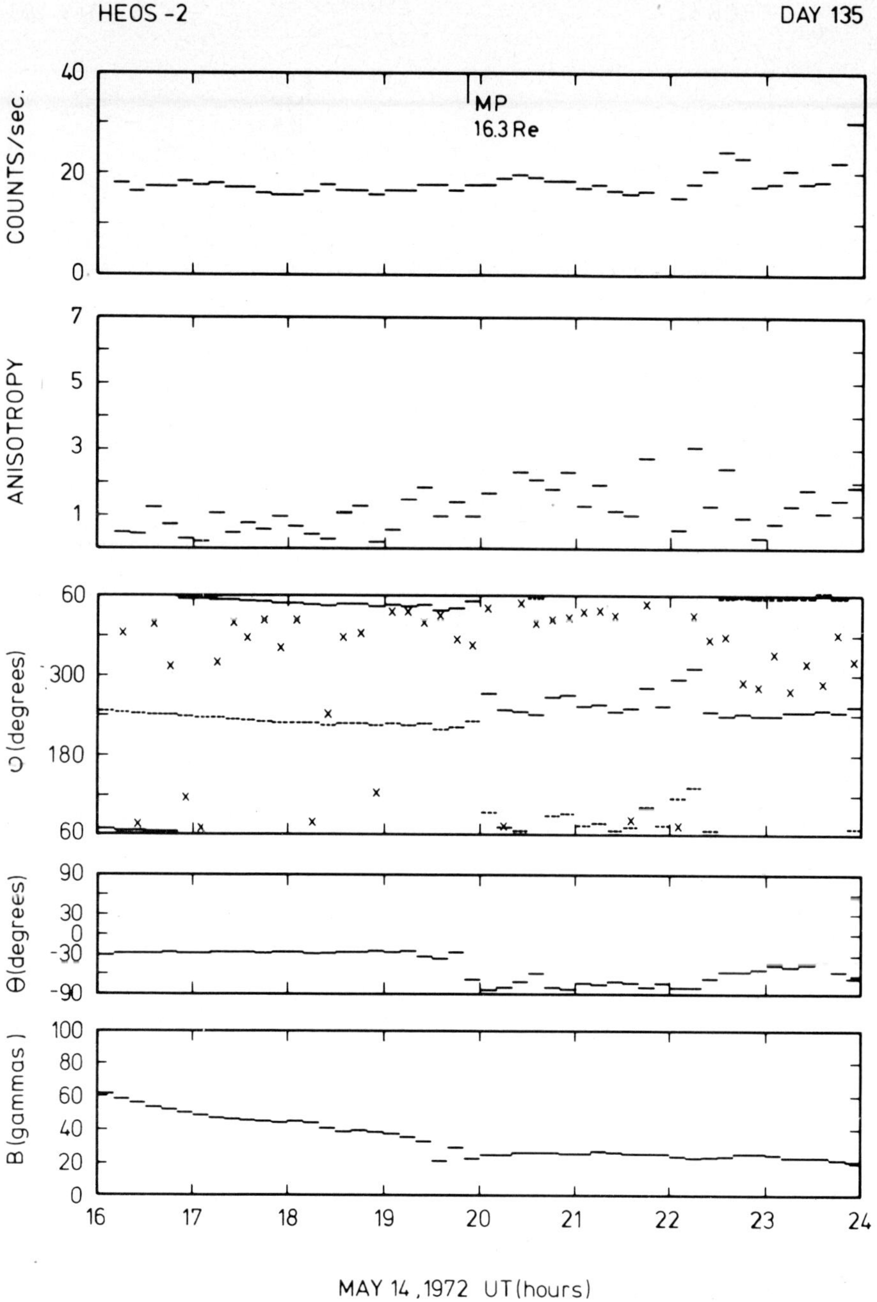

Fig. 6. Same as Figure 3 for May 14, 1972.

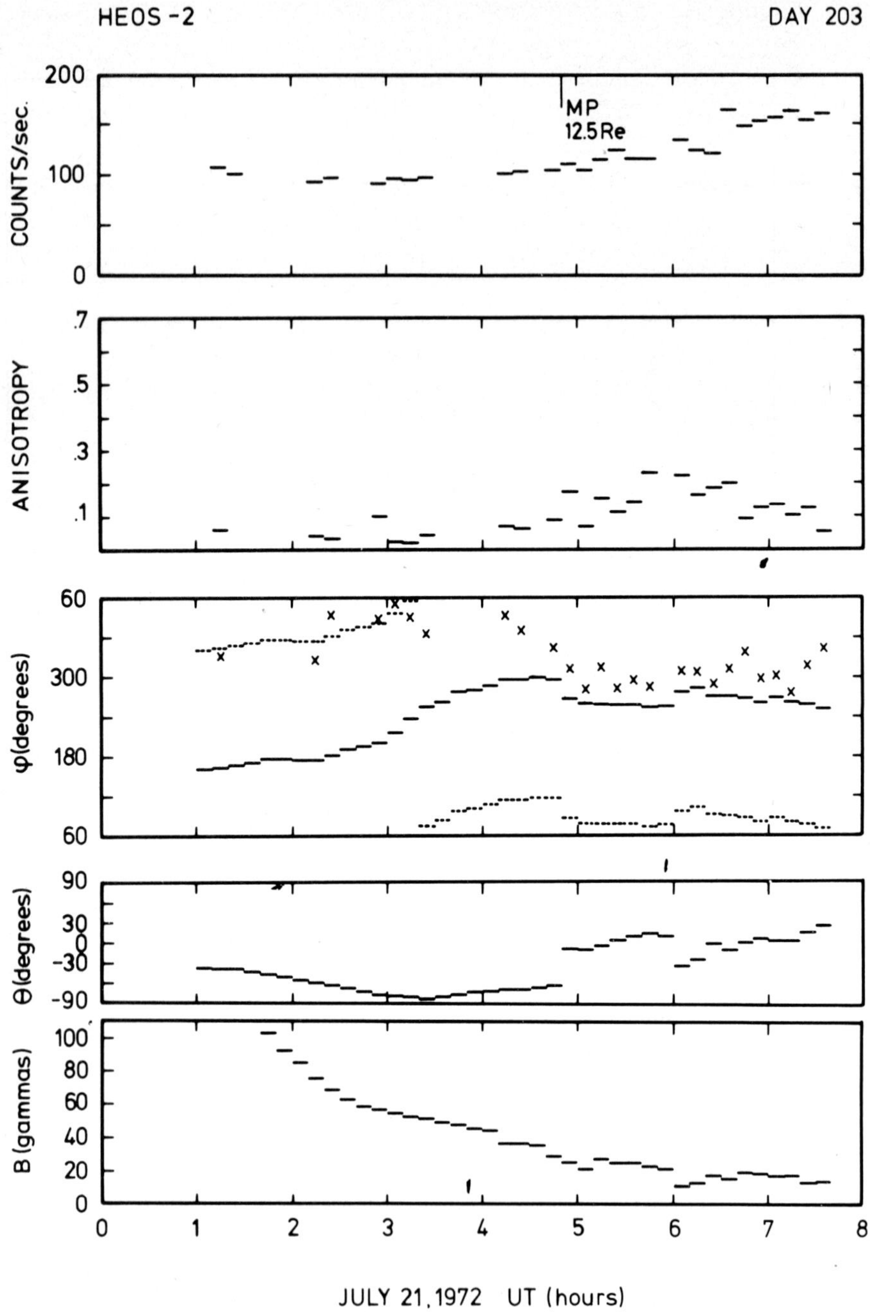

Fig. 7. Same as Figure 3 for July 21, 1972.

where a is the anisotropy coefficient defined above,

$$\alpha=(2m_0c^2+T)/(m_0c^2+T),$$

T is the kinetic energy of the particle, m_0 its rest mass, v its linear velocity, V the bulk (or external) velocity, and γ the spectral coefficient ($I=I_0E^{-\gamma}$). We have then computed the electric field that would produce the required bulk velocity and the particle density gradients that would produce the observed anisotropy (Lenhert, 1964). Since we do not know the spectral index in that energy region we assume that $\gamma=1$ and $\gamma=3$ are extreme values and obtain the results shown in Table I.

There are no direct measurements of the electric fields in this region of the magnetosphere, but all theoretical predictions and some measurements made in the ionosphere indicate more moderate electric fields directed at about 90° to the electric field needed to produce the present effect.

TABLE I

Day 1972	Anisotropy a (%)	Electric field ($mV\ m^{-1}$)		Particle density gradient ($\%/R_E$)	
		$\gamma=1$	$\gamma=3$	$\gamma=1$	$\gamma=3$
47	10	30	10	19	28
68	15	35	12	22	32
89	25	69	24	45	64

However, in one of the passes (day 68, 1972) we have been able to compare the fluxes measured inside the tail by our instrument with 1–10 MeV protons measured simultaneously outside the magnetosphere by Explorer 41 (courtesy Dr Williams). This measurement appears in Figure 4 after normalization to our measurements outside the magnetopause and indicates that there is a flux gradient of the right order of magnitude ($\simeq 20\%/R_E$), in the period in which we observe the anisotropy normal to the magnetic field.

On day 89 we do not have interplanetary data for reference but even with our own flux measurements we can observe the same pattern as on day 68 and 47: the anisotropy is enhanced during the period just before the encounter of the magnetopause, when there is indication of a particle density gradient increase. Moreover on both days 68 and 89 we see that deeper into the tail the flux gradient is in the opposite direction i.e. the satellite finds decreasing fluxes as it goes out, and then the anisotropy appears to be rotated 180° with respect to the one observed near the magnetopause, as should be expected if the reason for the measured anisotropy is the flux gradient. Its absolute value is expected to be smaller due to the higher magnetic field intensity.

We therefore conclude that the measurement of the anisotropy in the near Earth magnetotail for protons of about 1 MeV is a good indicator of the gradients of fluxes present in it, and can be used to map the penetration of protons into the polar cap.

6. Acknowledgements

We should like to thank Dr P. C. Hedgecock, Imperial College, London, and Dr D. J. Williams. Space Environment Laboratory, NOAA, Boulder, Colorado, for making available their data.

References

Fennell, J. F.: 1973, *J. Geophys. Res.* **78**, 1036.
Forman, M. A.: 1970, *J. Geophys. Res.* **75**, 3147.
Gall, R., Jimenez, J., and Camacho, L.: 1968, *J. Geophys. Res.* **73**, 1593.
Köhn, D., Page, D. E., Taylor, B. G., and Wenzel, K.-P.: 1972, *ELDO/ESRO Sci. Techn. Rev.* **4**, 19.
Lanzerotti, L. J.: 1972, *Rev. Geophys. Space Phys.* **10**, 8491
Lenhert, B.: 1964, *Dynamics of Charged Particles*, North-Holland Publ. Co., Amsterdam.
Morfill, G. and Quenby, J. J.: 1971, *Planetary Space Sci.* **19**, 1541.

PART 6

THE AUGUST 1972 SOLAR EVENTS

GAMMA RAY OBSERVATIONS DURING THE AUGUST 1972 SOLAR ACTIVITY

E. L. CHUPP*, D. J. FORREST, and A. N. SURI

University of New Hampshire, Durham, N.H. 03824, U.S.A.

Abstract. During the solar activity period in August 1972, the UNH gamma ray monitor on the OSO-7 satellite measured gamma ray line and continuum fluxes of solar origin on at least two occasions; August 4 and 7. In this discussion we will review the experiment and the gamma ray line observations with a few brief remarks on interpretation. A more complete discussion on the relation of these observations to the solar flare problem will be presented at the COSPAR-Working Group on Solar Flares in Konstanz, Germany – May 30, 1973. The specific observations to be discussed here are measurements of gamma ray lines at 0.51 and 2.2 MeV during two of the largest optical solar flares. Evidence for lines at other energies will also be given.

1. Introduction

The primary purpose of this paper is to review the observations of solar gamma rays made from the OSO-7 satellite in early August 1972. The results were first formally presented at the NASA Symposium on *High Energy Phenomena on the Sun* in September 1972 (Chupp *et al.*, 1973a) and a more detailed preliminary report has been published (Chupp *et al.*, 1973b). The results presented here are not in final form, however, a final report is in preparation.

We will describe here briefly the experiment itself and in more detail the observations made on August 4 and August 7, 1972.

At the outset it should be understood that we are discussing primarily here only gamma rays of discrete energy and not gamma ray radiation with a continuum of energies, often referred to as hard X-rays. As will be discussed, such continuum radiation was observed during the August flares with energies up to at least 3 MeV; however, a separate analysis is being carried out to analyze this data.

2. Experiment

Figure 1 shows a schematic of the detector and its arrangement in the rotating wheel compartment of the OSO-7 spacecraft. The gamma ray monitor (Higbie *et al.*, 1972), designed to measure the gamma ray spectrum from 0.3 to 10 MeV, consists of a $3'' \times 3''$ NaI (Tl) crystal surrounded by a cup-shaped anticoincidence shield of CsI (Na). This arrangement also enhances the relative sensitivity for lines by suppressing the Compton continuum. The instrument is automatically calibrated regularly each orbit at day/night and night/day transitions with a Co^{60} source giving photopeaks at 1.17, 1.33 and 2.50 MeV. The calibration source (Forrest *et al.*, 1972) consists of a neutron-activated Cobalt compound in a plastic scintillator button,

* Alexander von Humboldt Senior Fellow and Honorary Fulbright Senior Fellow, at the Max-Planck-Institut für extraterrestrische Physik, 8046 Garching, W. Germany.

D. E. Page (ed.), Correlated Interplanetary and Magnetospheric Observations, 519–531.

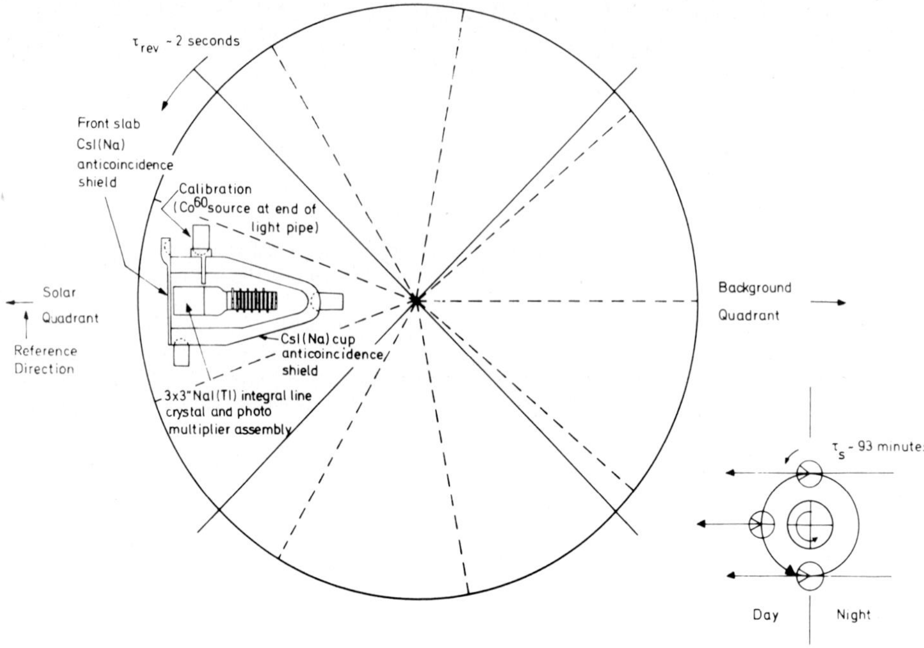

Fig. 1. A schematic drawing of the gamma ray spectrometer is shown with the OSO wheel in position for viewing the Sun. The different components of the detector discussed in the text are identified. The solar and background quadrant identifications are also shown. The small inset at the right shows satellite aspect at Sunrise and at noon and at Sunset.

which is viewed by a photomultiplier. The beta decay of the Co^{60} provides a gating pulse which is used to prevent the coincident gamma rays from Ni^{60} from being recorded during a normal gamma ray spectrum accumulation. At each day/night or night/day transition of the spacecraft, a signal automatically puts the Co^{60} beta pulse in coincidence with a signal in the main gamma ray detector, which is due, in this case, primarily to an absorption of one of the gamma rays at 1.17 MeV or 1.33 MeV. Occasionally, both gamma rays are simultaneously absorbed, giving a sum peak at 2.50 MeV. During the normal operation, when the calibration lines are gated out of the spectrum, there is about a 5% leakage of Co^{60} lines into their respective channels. This scheme allows the energy of a well defined peak to be determined to about 2% and verifies that the energy resolution is maintained at $\lesssim 8\%$ (referred to 662 keV). The energy resolution improves with energy approximately as $E^{-1/2}$ The gamma ray monitor has an auxiliary X-ray detector (not shown in Figure 1), consisting of a NaI (Tl) crystal of diameter $1\frac{1}{4}''$ and thickness $\frac{1}{4}''$, covered by a 10 mil thick Be foil. The X-ray energy spectrum from 7.5 to 120 keV is analyzed by a four channel analyzer.

During the normal mode of operation (see Figure 1) the gamma ray pulse height spectra are accumulated over ~90 wheel rotations (taking approximately 3 min) in both solar and background quadrants. The gamma ray spectrum is analyzed with a

377-channel quadratic pulse height analyzer and stored in a buffer until readout. A complete X-ray spectra is taken every thirty seconds.

3. Flare Observations

The 2B Hα on August 4 began at ~0621 UT reaching a maximum at 0638 UT and ending at ~0852 UT (Lincoln and Leighton, 1972). Before the satellite went into eclipse at ~0633 UT, excess gamma ray line and X-ray continuum radiation was observed. Additional evidence for gamma rays was obtained in the 3B flare of 7 August 1972. This flare began in the Hα at ~1500 UT during satellite nighttime. As the satellite emerged into the Sunlight, excess gamma ray line radiation was observed. Figure 1 (insert at the right) shows the approximate aspect of the satellite during the time intervals when excess solar associated gamma ray counts were recorded. We will discuss the basic data obtained in the two events in turn and defer interpretation.

3.1. August 4, 1972

Figure 2 shows the time integrated solar and background gamma ray spectrum taken during the rise of the optical flare and just before the commencement of the satellite night. The background quadrant spectrum shown is well below the corresponding solar quadrant flare spectrum. The solar spectrum is the sum of three normal, separate, consecutive three minute full scans with the total live time for the summed spectrum being 91.4 s. In the figure the left ordinate gives the counts observed in the indicated live time during the UT time interval shown. The abscissa shows the channel number in which the counts are accumulated. Also, above channel 200, the ordinate scale is shifted as shown on the right, and the total number of counts in five consecutive channels is shown, during the live time, to improve statistical accuracy. The numbers under the histogram give the typical number of background counts observed in the five channels. On the far right ordinate is shown the square root of the energy, $\sqrt{E}$, corresponding to a given channel according to the relation $E = k(n+n_0)^2$ and shown by the straight line in the figure. The constants k and n_0 are found from in-flight calibration data. The calibration curve is our best fit from the two calibrations which took place before and just after the event.

It is evident that only the solar quadrant gamma ray flux is significantly enhanced in both the ~0.5 and ~2.2 MeV spectral regions during the rise time of the optical flare. In addition, the general continuum below the line features is enhanced over the background quadrant thus indicating a continuum enhancement to above 2 MeV. Several comparison spectra showing solar and background rates before and after the flare were given by Chupp *et al.* (1973b). Above channel 200, where the counts are summed in successive groups of five channels, there are three or more statistically significant 'peaks'. Because of the necessity of summing channels to improve statistics, it is not possible to specify precisely the photo peaks, but they may be associated with features at energies 3.4→3.8 MeV, 4.2→4.6 MeV, and 5.9→6.3 MeV, re-

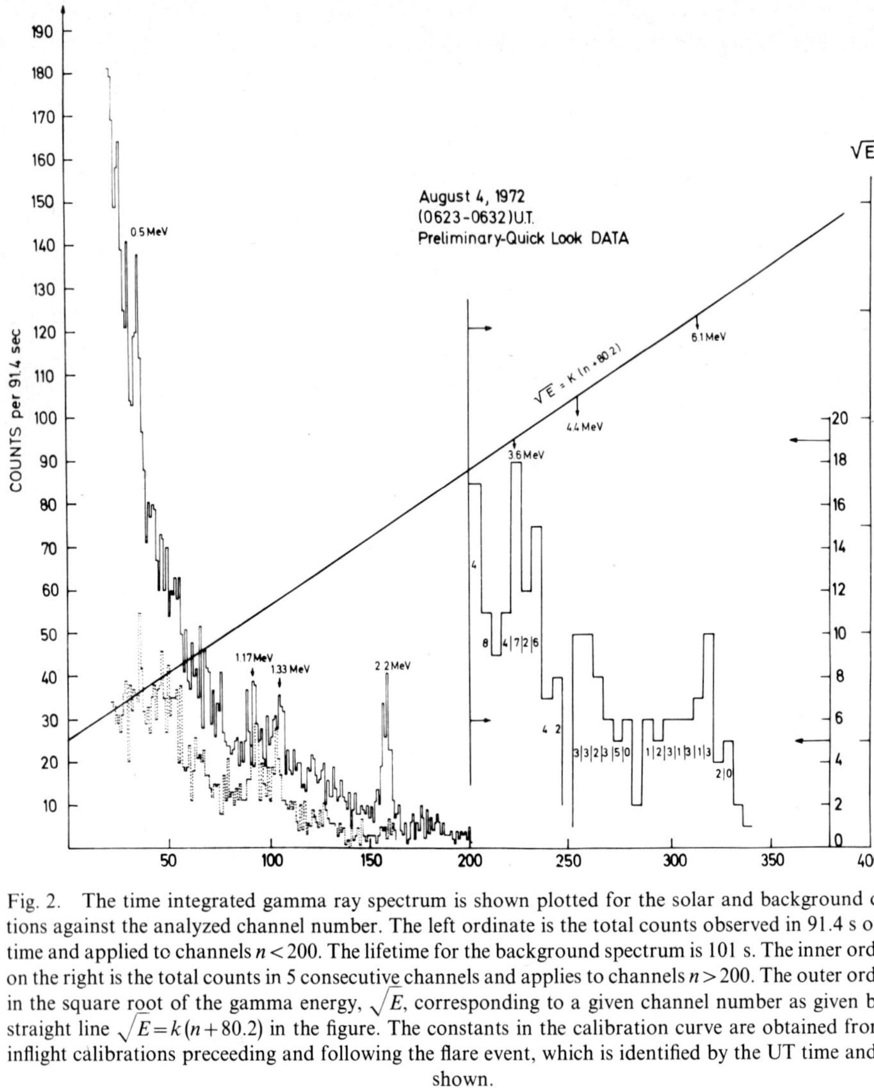

Fig. 2. The time integrated gamma ray spectrum is shown plotted for the solar and background directions against the analyzed channel number. The left ordinate is the total counts observed in 91.4 s of lifetime and applied to channels $n < 200$. The lifetime for the background spectrum is 101 s. The inner ordinate on the right is the total counts in 5 consecutive channels and applies to channels $n > 200$. The outer ordinate in the square root of the gamma energy, $\sqrt{E}$, corresponding to a given channel number as given by the straight line $\sqrt{E} = k(n + 80.2)$ in the figure. The constants in the calibration curve are obtained from the inflight calibrations preceeding and following the flare event, which is identified by the UT time and date shown.

spectively. The leakage Co^{60} calibration lines at 1.17 and 1.33 MeV are also clearly evident in both the solar and the background spectra. Other features or possible photo peaks may be evident in the flare spectrum shown in Figure 2; however we note with caution that it is necessary to separate out the hard X-ray burst spectrum before treating small features. This problem is the subject of a separate paper (Suri *et al.*, 1973). As will be emphasized later, gamma ray line emissions are expected from the Sun at all of these energies. In the case of the 4 and 6 MeV features the association with photopeaks at 4.4 and 6.1 MeV is circumstantial, but gamma ray lines at these energies are expected at about the intensity shown.

The four channel X-ray spectrometer covering the energy range 7.5→120 keV also recorded the event in coincidence with the gamma ray spectrometer. Figure 3 (center) shows the counting rates in each X-ray channel versus time from ~0500 UT on 8-4-72 until the eclipse of the satellite at ~0633 UT and then for three more successive orbits through the decay of the event. The rise of the event is recorded by each channel, with 30 s time resolution, until saturation occurs in the first three channels. The time of

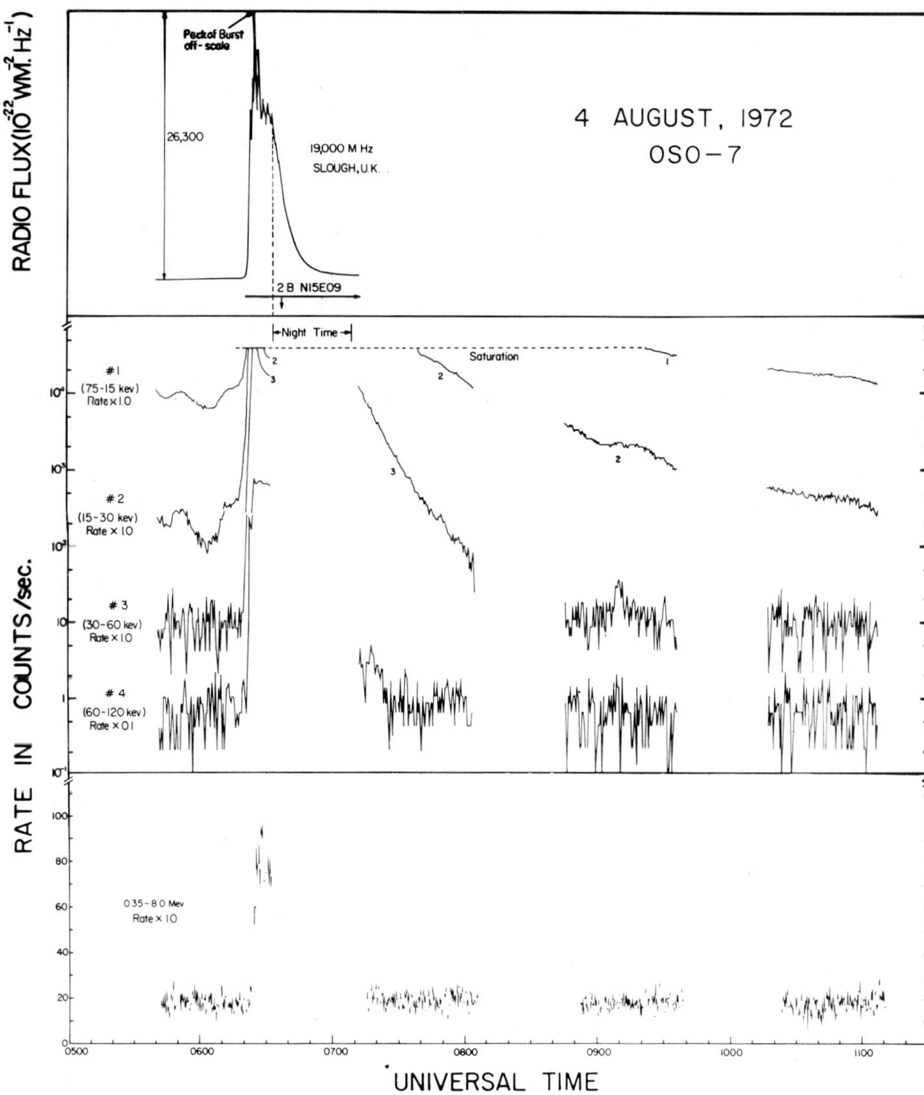

Fig. 3. In the center portion of the figure, the uncorrected counting rates in each of the four X-ray channels is shown plotted against Universal Time on August 4, 1972. The lower portion of the figure shows the uncorrected total counting rate in the gamma ray detector between the lowest and uppermost channels of the analyzer. The corresponding energy interval for this rate is 0.35–8.0 MeV. In the upper part of the figure the radio flux at 19000 MHz is shown during the event.

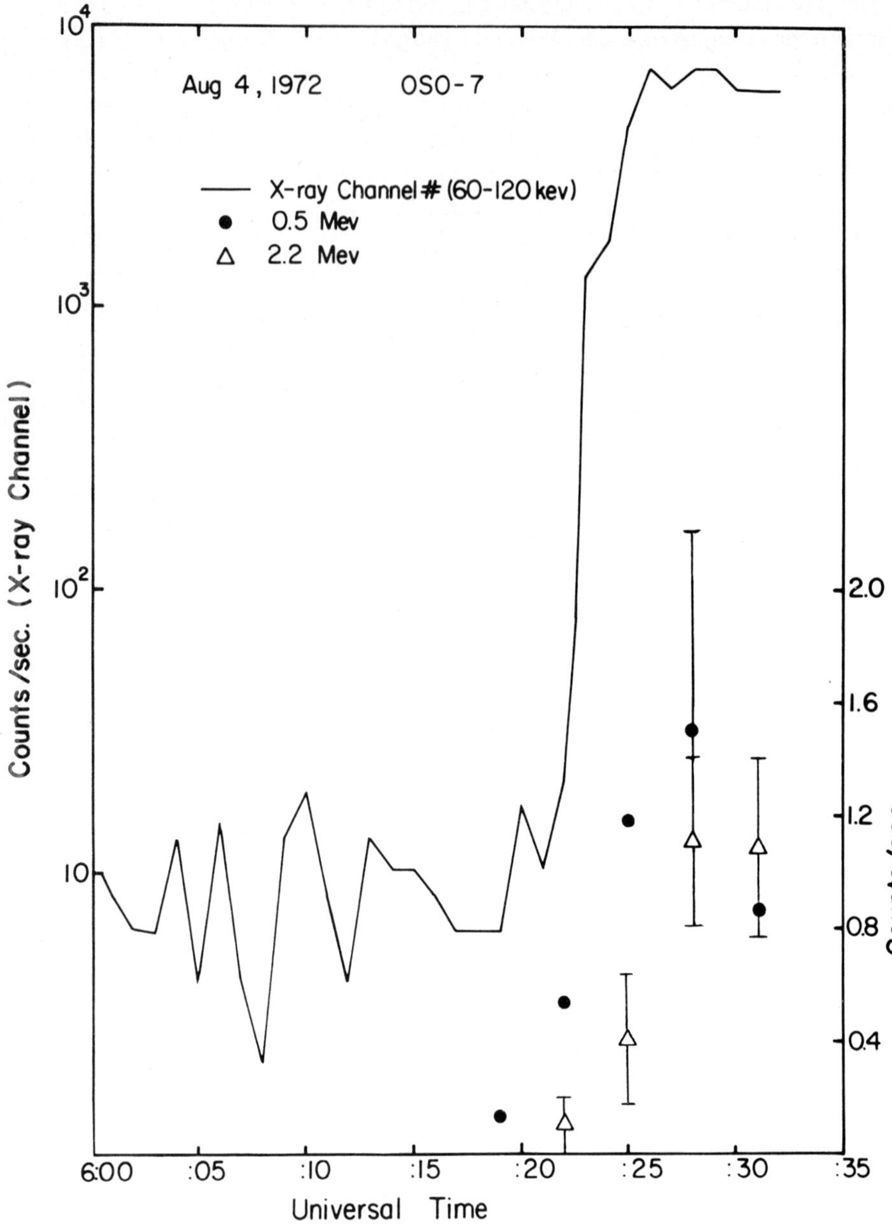

Fig. 4. The photopeak counting rates (right ordinate) shown for the 0.5 MeV and 2.2 MeV lines versus time during the August 4 event. The time profile (—) for the counting rate in one X-ray channel (60–120 keV) is shown (left ordinate) covering the period of satellite day from about 20 min before the optical onset and until the satellite went into eclipse.

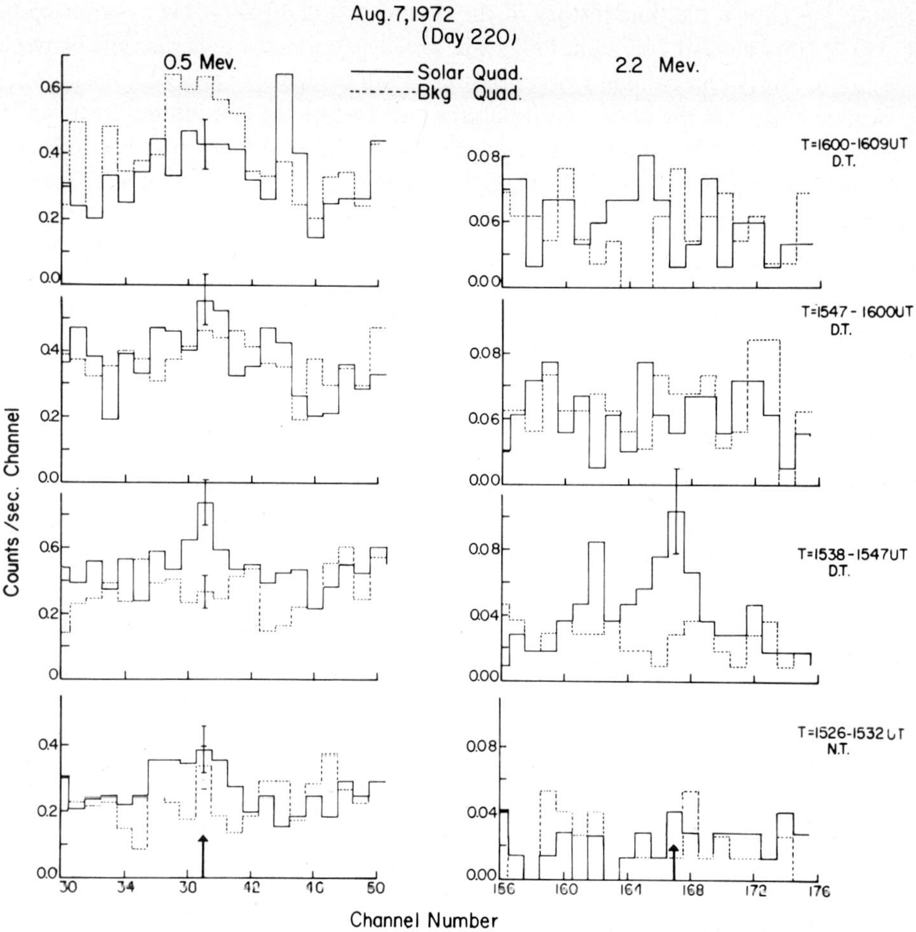

Fig. 5. Gamma ray pulse height spectra are shown for the energy regions 435 keV–610 keV (on the left) and for 2.1 MeV–2.4 MeV (on the right) on August 7, 1972. The lower spectrum was taken at nighttime just before the satellite Sunrise, but while the 3B optical flare was in progress. The second spectrum from the bottom was taken just after the Sunrise calibration and during the declining phase of the optical flare. The top two spectra were taken at later times in the same orbit during the satellite daytime. The accumulation time intervals are shown for each spectrum. ——— solar quadrant, ... background quadrant.

saturation occurs earlier as the mean energy of the X-rays is lower. The event seen in channel 3 (30–60 keV) begins to decay just before 0627, well in advance of the eclipse by the Earth. Also channel 4 (60–120 keV) does not saturate but reaches a plateau before being eclipsed. The question of atmospheric effects such as scattering and absorption at grazing incidence is discussed later. As will be shown, only scattering is a problem and the absorption by the atmosphere occurs within ~ 1 min from (0633–0634) UT as is also shown by the sharp cut off in the X-ray counting rates in Figure 3. At the lower part of Figure 3 is shown the total counting rate in all gamma ray channels as recorded with a rate-meter with a 30 s time resolution. At the top in

Figure 3 is shown the time history of the radio burst at 19000 MHz as reported in UAG 21 (Lincoln and Leighton, 1972). The excellent correspondence in time between the radio, X-ray and gamma ray continuum is self evident.

A time history of the photopeak counting rate (before the continuum) is shown in Figure 4 on an expanded time scale, for the 0.5 and 2.2 MeV lines, along with X-ray channel 4 (60–120 keV). After the OSO 7 returns to daylight at ~0705 UT, there is no residual detectable evidence for gamma ray lines at any energy.

3.2. August 7, 1972

The solar gamma rays observed during this event were, unfortunately, recorded after the rise and well after the maximum phase of the optical event since the satellite was in nighttime. Figure 5 shows a sequence of counting rate spectra which were taken at nighttime and during the satellite daytime period while the 3B Hα flare was still in progress. In this figure only portions of the energy spectrum around 0.51 and 2.2 MeV are shown. The spectra shown are the sum of several separate consecutive full scans, as indicated by the time interval. By comparison of the lowest spectrum (nighttime) with that immediately above it (daytime), it is evident that there are significant increases corresponding to photon energies of 0.5 and 2.2 MeV. In both cases the enhancement appears only in the solar quadrant spectra just after Sunrise. We wish to argue that these enhancements are also due to a direct solar flux, since it would be expected that a local satellite or atmospheric effect would produce an enhancement in both the solar and background quadrants for the detector's aspect with respect to Earth just after the night-to-day transition. A remark should be made about the upper spectrum in Figure 5, which shows an increase in the 0.5 MeV region in the background quadrant. At this time this quadrant was viewing the Earth with good aspect but at lower rigidity ~3.0 GV, and therefore, registered atmospheric radiation. Atmospheric fluxes of 0.51 MeV radiation seen from OSO 7 are being discussed separately (Dunphy *et al.*, 1973).

4. Justification as an Extraterrestrial Event

As is already evident, the 2B event on August 2, 1972, produced a copious ($\times 10^3$) enhancement in low energy (~30 MeV) protons at the Earth in the polar cap regions and outside the magnetosphere. It must be noted that since protons in this energy range are a very efficient producing agent for gamma ray lines, we must critically ask whether a local atmospheric or spacecraft background effect could explain the spectra just presented instead of direct solar emissions.

Since the gamma ray monitor has wide angular response (~100°) there may be some concern that the detector is recording atmospheric gamma rays produced by solar protons (the solar direction actually grazes the Earth's atmosphere during part of the time intervals in question). By a fortunate coincidence the satellite was in a region of high geomagnetic cut-off during both events. For the crucial time on August 4 the cut-off varied from 15.8 GV to 15 GV and on August 7 the cut-off

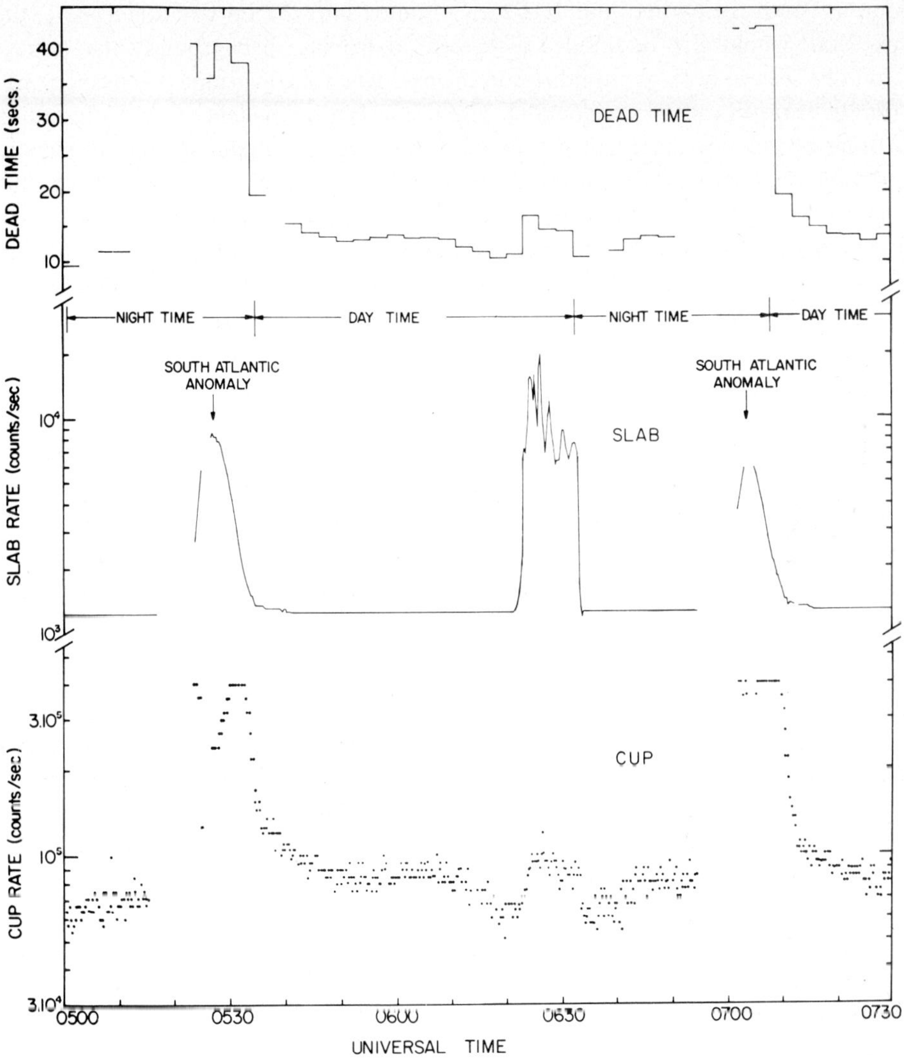

Fig. 6. In the lower part of the figure, the counting rate in the cup portion of the CsI(Na) anticoincidence shield is shown versus Universal Time on August 4, 1972. In the center portion of the figure is shown the corresponding rate in the slab portion of the anticoincidence shield and in the upper portion of the figure the measured deadtime for the gamma ray monitor is presented. The large increases near 0530 UT and 0700 UT are due to passage of the satellite near the South Atlantic anomaly. The interval for satellite daytime and nighttime are indicated near the center of the figure.

varied from 15 GV to 12 GV. The spacecraft was also well away from any trapped radiation during the event times.

The integrity of the events as true solar events can also be established by considering the solar-background quadrant spectra in Figure 2. We note that there is no similar set of spectra normally seen when the satellite monitor is in the similar aspect to Earth's atmosphere as for the flare time. Second, if there were coincidentally solar protons (from the earlier August 2 event) raining into the Earth's atmosphere, the spacecraft would also be flooded with them and then a background enhancement would be seen in both quadrants (as in the case when the instrument is in the South Atlantic trapped radiation anomaly). There is in fact no evidence for an excess flux of solar protons or precipitated trapped protons at the satellite during the time in question. This is shown most clearly by the shield counting rates during the time span surrounding the flare observations. Figure 6 shows a plot of the rates of the anticoincidence shield (i.e. its elements, the cup and slab) and the gamma ray spectrometer deadtime for some time before and after the event. No abnormal enhancement of the shield rates is evident during the time of interest. The small slab and cup enhancements during the event are in agreement with X-ray and gamma ray rates measured independently during the event. The variations shown at other times are due to the latitude effect and passage of the instrument through the radiation anomaly. The final point on this matter is the possibility of a 'hot spot' in the Earth's atmosphere (and no excess protons at the spacecraft) seen only by the solar quadrant. If this were so, then the background quadrant would have been seeing this 'hot spot' about fifteen minutes earlier (because of the monitor's broad angle). No such enhancement was seen, thus; such a phenomenon would have to be coincidentally short lived (~ 15 min) and localized.

5. Atmospheric Effects

There are two terrestrial atmospheric effects that must be considered in the correction of the gamma ray and X-ray spectra. Because of the grazing incidence geometry of solar radiation during part of the event and the broad angle response of the instrument, we must consider both the possibilities of absorption and scattering by the Earth's atmosphere. It should be noted though that absorption only would be important for correcting photopeak gamma ray intensities since scattering would serve to remove the line from the peak completely or make it asymmetric on the low energy side.

The basic geometric problem of absorption has been treated some time ago by Chapman (1931), who considered the absorption of monochromatic solar radiation incident on an exponential atmosphere with scale height H at grazing incidence. Chapman (1931) (Equation (5)) gives, for monochromatic solar radiation, the fractional intensities, s/s_0, remaining after absorption, for a detector at an altitude h above the Earth's surface basically in terms of this altitude h, the scale height H, the Sun's zenith angle χ and the monochromatic absorption coefficient $A\,(\mathrm{cm}^2\ \mathrm{gm}^{-1})$ in the atmosphere. The atmospheric density is assumed to be of the form $\varrho = \varrho_0 \exp(-h/H)$ where $\varrho_0\,(\mathrm{gm\ cm}^{-3})$ is the sea level density; therefore, we must use suitable values of these quantities to represent the atmosphere above 100 km where absorption could be significant before eclipse. Chapman's basic equation can be put in the following suitable form for our purpose;

$$\ln s/s_0 = -A\,[\varrho_0 H \exp(-h/H)\,\mathrm{Ch}(x, \chi)], \tag{1}$$

where

$$\mathrm{Ch}(x, \chi) = x \sin\chi \int_0^{\chi} \exp\{x(1 - \sin\chi/\sin\lambda)\} \operatorname{cosec}^2 \lambda \, d\lambda$$

is the so-called Chapman Function tabulated by Wilkes (1954), $x = a + h/H$ and the other quantities have been previously defined. The quantity in brackets [ ] in Equation (1) is clearly the total terrestrial atmospheric thickness in gm cm^{-2} between the detector and the Sun if A, the absorption coefficient for the gamma ray energy and time of interest, is expressed in units of cm^2 gm^{-1}.

TABLE I

UT	Atmospheric thickness (gm cm^{-2})
0630	6×10^{-9}
0631	1.5×10^{-6}
0632	1.8×10^{-2}
0633	1.6×10^{4}

The atmospheric thickness calculated from Equation (1) using reasonable values for constants are given below in Table I for a few crucial times.

6. Summary of Results

The flux values for various gamma ray lines are shown in Table II. These values, still based on the preliminary analysis, are obtained by determining the deadtime corrected photopeak counting rates above any continuum and dividing by the sensitivity for that energy.

The interpretation of the above observations implies production of discrete gamma ray lines by a variety of high energy processes taking place in the solar atmosphere. The probable sources of the gamma ray lines are positrons and neutrons produced in the nuclear reactions and high energy protons. Annihilation of positrons with electrons produces the 0.51 MeV annihilation line while the 2.23 MeV line results from the capture of neutrons by the hydrogen atoms. Inelastic scattering of fast protons on the less abundant C, N, O, and Ne nuclei produce a large number of weaker deexcitation gamma ray lines: 4.43 MeV ($C^{12}*$) and 6.13 MeV ($O^{16}*$) are likely to be the most prominent. We believe that the line features we observe in the 2B Hα flare are the annihilation line (0.51 MeV), the neutron capture line (2.23 MeV) and the deexcitation lines of $C^{12}*$ (4.43 MeV) and $O^{16}*$ (6.13 MeV). Ramaty and Lingenfelter (1973) have recently confirmed this general interpretation of these observations. The enhanced continuum shows the possibility of a large number of unresolved line features. At this state of the analysis of the spectra, we are not sure whether the

TABLE II

Associated flare and the time of observations	Tentative designations and preliminary solar flux at 1 AU Photons $cm^{-2} s^{-1}$			
	0.5 MeV	2.2 MeV	4.4 MeV	6.1 MeV
2B (Hα) 4 Aug 1972 0623–0632 UT (Before Hα max)	$(7 \pm 1.5) \times 10^{-2}$	$(2.2 \pm 0.2) \times 10^{-1}$	$(3 \pm 1) \times 10^{-2}$	$(3 \pm 1) \times 10^{-2}$
3B (Hα) 7 Aug 1972 1538–1547 UT (After Hα max)	$(3.4 \pm 1.1) \times 10^{-2}$	$(4.8 \pm 1) \times 10^{-2}$	$<2 \times 10^{-2}$	$<2 \times 10^{-2}$

continuum (particularly the high energy end) is due to electron bremsstrahlung or summation of unresolved line features.

It is appropriate to point out here that the gamma ray emission in solar flares could be of more complex character than has been assumed in the past (Lingenfelter and Ramaty, 1967). As discussed by Shima and Alsmiller (1970), the inelastic scattering of protons (50 MeV) on C produces a line at ~ 2 MeV in addition to the 4.43 MeV line, while even 15 MeV protons on O^{16} yield prominent lines at ~ 4.4 MeV and ~ 5.2 MeV in addition to the 6.13 MeV line. Thus the possibility that more than one source is contributing to the observed line features cannot be ruled out.

Acknowledgement

This work was supported by NASA under Contract Nas 5-11054. We are thankful for the contributions of several persons in the fabrication and data analysis phase of the experiment particularly; Dr P. R. Higbie, A. A. Sarkady, I. U. Gleske, S. Foss, Sue Crouteau, Margaret Simmons, Kishore and Dipika Patel, Ron Adams, P. Dunphy, C. Tsai, and Pat Ferguson. Dr Claus Reppin joined the group in November 1972 and his contribution to the later phase of the data analysis is much appreciated. Mary M. Chupp's help on the manuscript was invaluable.

References

Chapman, S.: 1931, *Proc. Phys. Soc.* **43**, 483.

Chupp, E. L., Forrest, D. J., and Suri, A. N.: 1973a, in R. Ramaty (ed.), Proceedings of the Symposium on High Energy Phenomena on the Sun, held at Goddard Space Flight Center, Greenbelt, Maryland, 28–30 September 1972, NASA Report X-693-73-193.

Chupp, E. L., Forrest, D. J., Higbie, P. R., Suri, A. N., Tsai, C., and Dunphy, P. P.: 1973b; *Nature* **241**, 333.

Dunphy, P. P.: 1974, Doctoral Thesis – University of New Hampshire – Department of Physics.

Forrest, D. J., Higbie, P. R., Orwig, L. E., and Chupp, E. L.: 1972, *Nucl. Instr. Methods* **101**, 567.

Higbie, P. R., Chupp, E. L., Forrest, D. J., and Gleske, I. U.: 1972, *IEEE Trans. Nucl. Sci.* **NS–19**, No. 1, 606.

Lincoln, J., and Leighton, Virginia and Hope I.: 1972, *World Data Center A, Report UAG-21*; U.S. Department of Commerce, NOAA, Environmental Data Service, Ashville, North Carolina, November.

Lingenfelter, R. E. and Ramaty, R.: 1967, in B. S. P. Shen (ed.), *High Energy Nuclear Reaction in Astrophysics*, Benjamin, New York.

Ramaty, R. and Lingenfelter, R. E.: 1973, in R. Ramaty (ed.), Proceedings of the Symposium on High Energy Phenomena on the Sun, held at Goddard Space Flight Center, Greenbelt, Maryland, 28–30 September 1972, NASA Report, X-693-73-193.

Shima, Y. and Alsmiller, R. G.: 1970, *Nucl. Sci. Enging.* **41**, 47.

Suri, A. N., Forrest, D. J., Chupp, E. L., and Reppin, C. R.: 1973, *Bull. A.P.S.* **18**, No. 4, paper JH1.

Wilkes, M. V.: 1954, *Proc. Phys. Soc.* **67**, 304.

TIME PROFILES AND PHOTON SPECTRA OF SOLAR HARD X-RAYS

H. F. VAN BEEK, L. D. DE FEITER, and C. DE JAGER

Space Research Laboratory of the Astronomical Institute at Utrecht, The Netherlands

Abstract. With a spectrometer aboard the ESRO TD-1A satellite, time profiles and photon spectra of solar X-rays have been observed in the energy range above 24 keV. Most of the X-ray bursts studied thus far appear to consist of numerous short-lived spikes with rise and decay times of a few seconds. Such spikes, called 'Elementary Flare Bursts' (EFB's) cluster together in large numbers and constitute the conventional high energy flare. The form of the photon spectra could be determined very accurately and was in all cases of the power-law type. For the August 4, 1972 flare, the photon spectra appear to be combinations of two power-law spectra; for the lower energies the spectral coefficient is smaller than for the higher energies. The transition energy is about 60 keV.

1. Instrumentation

The spectrometer was designed and manufactured at the Space Research Laboratory of the Astronomical Institute at Utrecht, Holland, and is incorporated in the ESRO TD-1A satellite, which was launched on March 12, 1972. It aims at measuring the spectral distribution and the intensity of hard X-rays emitted by the hot plasma which appears in a solar flare region. The energy range of the instrument in orbit is from 24 keV to 900 keV, split into twelve logarithmically spaced, differential energy channels. In order to be able to observe fast variations in the flux of the solar photons, the time resolution was chosen as high as possible: 1.2 s for the lower part and 4.8 s for the remaining part of the energy range. The TD-1A spacecraft is particularly suitable for carrying this instrument because of its stabilization, which implies that one axis is continuously pointed towards the Sun, and because of the orientation of its orbit, which is near polar and Sun-synchronous. The height of the orbit is 545 km.

The experimental techniques are based on the use of a scintillation counter, combined with two solid state detectors and a passive shielding against background radiation. Detection of the solar photons is performed by the scintillation counter, which has a sensitive area of 5 cm^2. The solid state detectors, mounted in front of the CsI (Na) scintillation crystal, are incorporated to detect those charged particles that enter via the entrance window and which might penetrate into the scintillation crystal. Events in this crystal due to such particles can thus be rejected electronically. The solid state detectors also fulfil an important task in the system of inflight calibration. This system operate with the help of two on board radioactive sources, ^{241}Am and ^{90}Sr, as references. It yields precise information about the most important instrumental characteristics during the flight. The lead-tin-copper shielding around the scintillation counter, ending in a collimator in front of the counter, has been incorporated to absorb background X-rays of various sources. It was designed to achieve a maximum shielding against Bremsstrahlung photons generated by radiation belt electrons.

D. E. Page (ed.), Correlated Interplanetary and Magnetospheric Observations, 533–543. *All Rights Reserved.*

Preflight calibrations were effectuated with the help of mono-energetic photons emitted by radioactive sources. The response characteristics of the scintillation counter, rather complicated owing to numerous photon scattering and absorption effects occurring inside and outside the scintillation crystal, could be described by the superposition of several elementary curves. Knowing the numerical values of the parameters describing these curves as a function of the photon energy, the detector responses to mono-energetic photons or to photons with a specific spectral distribution can be simulated with the help of a computer program. Examples of these responses are presented in Figure 1 and Figure 2. The very good agreement between the measured and simulated detector responses enables one to determine the spectrum of solar photons with high precision.

A detailed description of the instrument and the calibration procedures has been given elsewhere (Van Beek, 1973).

2. Observations

2.1. The March 21, 1972 Flare

As an example of a rather small flare, where short time variations of the photon intensity can clearly be seen, the March 21, 1972 flare is presented. The flare was of im-

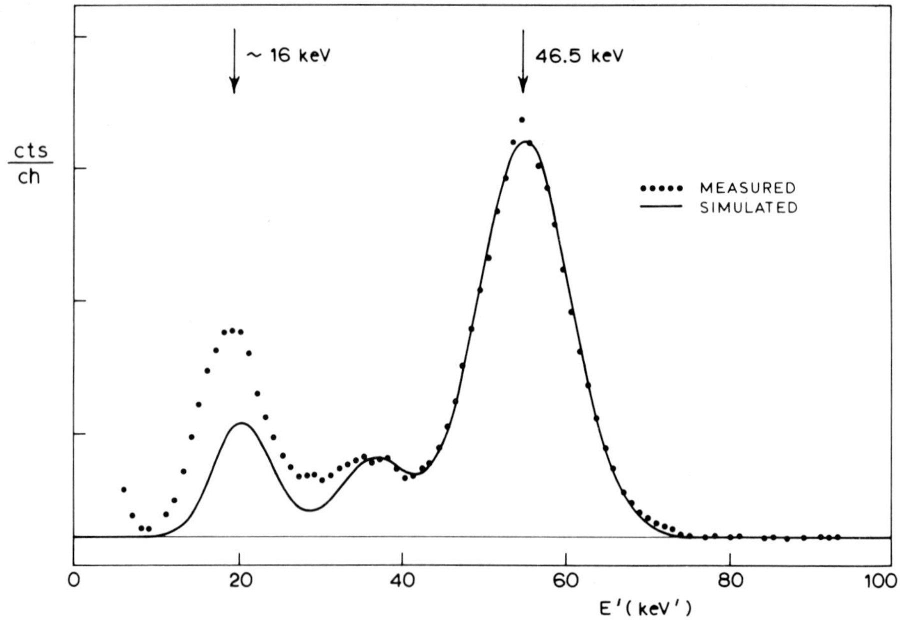

Fig. 1. Measured and simulated detector response to radiation emitted by a ^{210}Pb source. The difference between the pulse height distributions at about 16 keV is due to X-rays emitted by the source, which do not appear in the simulated response. Although the output of the scintillation counter is not proportional to the energy absorbed in the crystal, a linear scale along the *X*-axis is introduced. At 662 keV the 'normalized' energy, *E*′, corresponds to an absorption of an energy of 662 keV in the scintillation crystal.

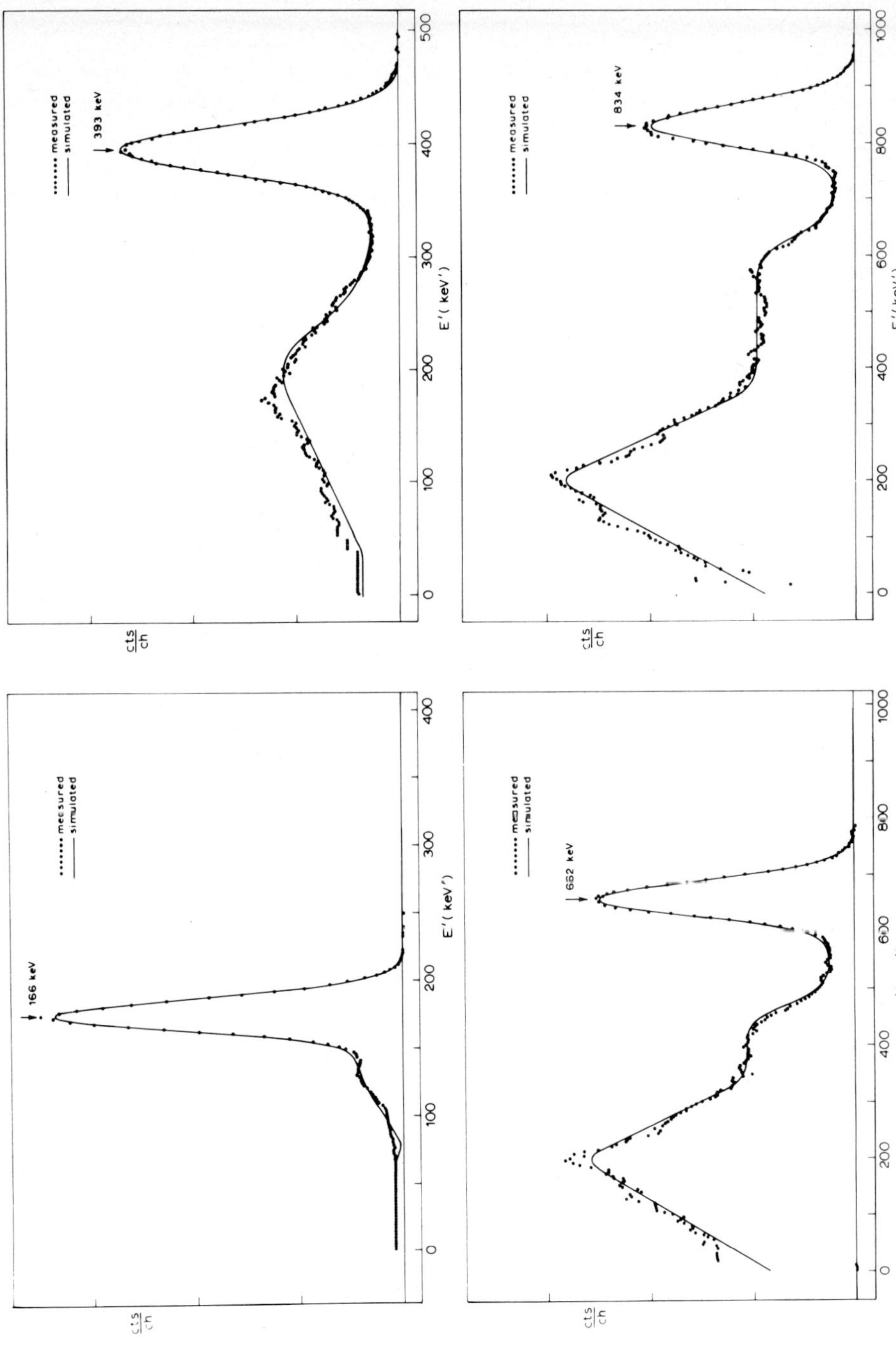

Fig. 2. Measured and simulated detector responses to mono-energetic radiation emitted by ^{139}Ce, ^{113}Sn, ^{137}Cs and ^{54}Mn sources.

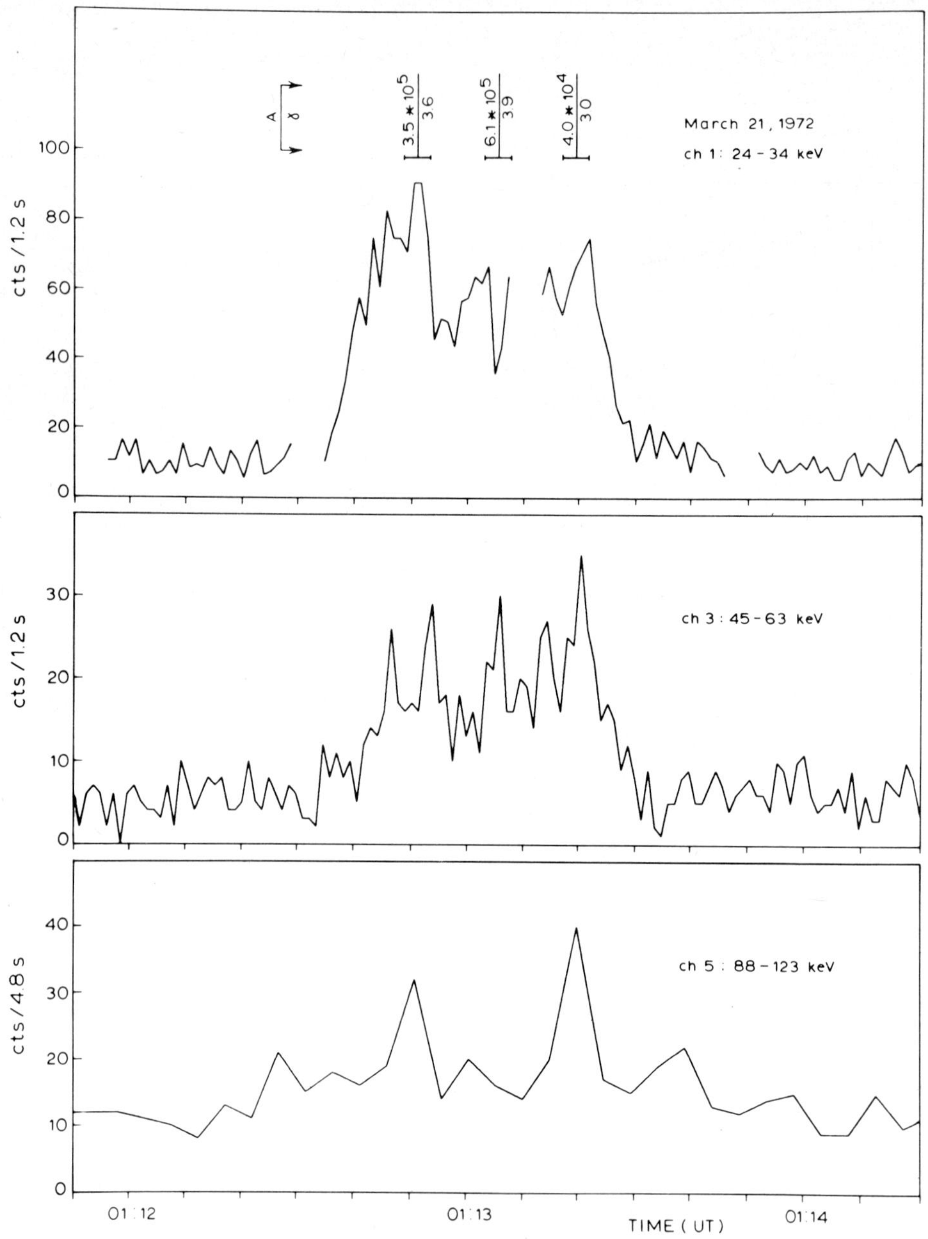

Fig. 3. Time profiles for the flare appearing on March 21, 1972. The parameters A and γ define the power-law photon spectra which yield the best fit to the observations. The periodical absence of channel 1 data every 38.5 s during 4.8 s is due to electronic subcommutation.

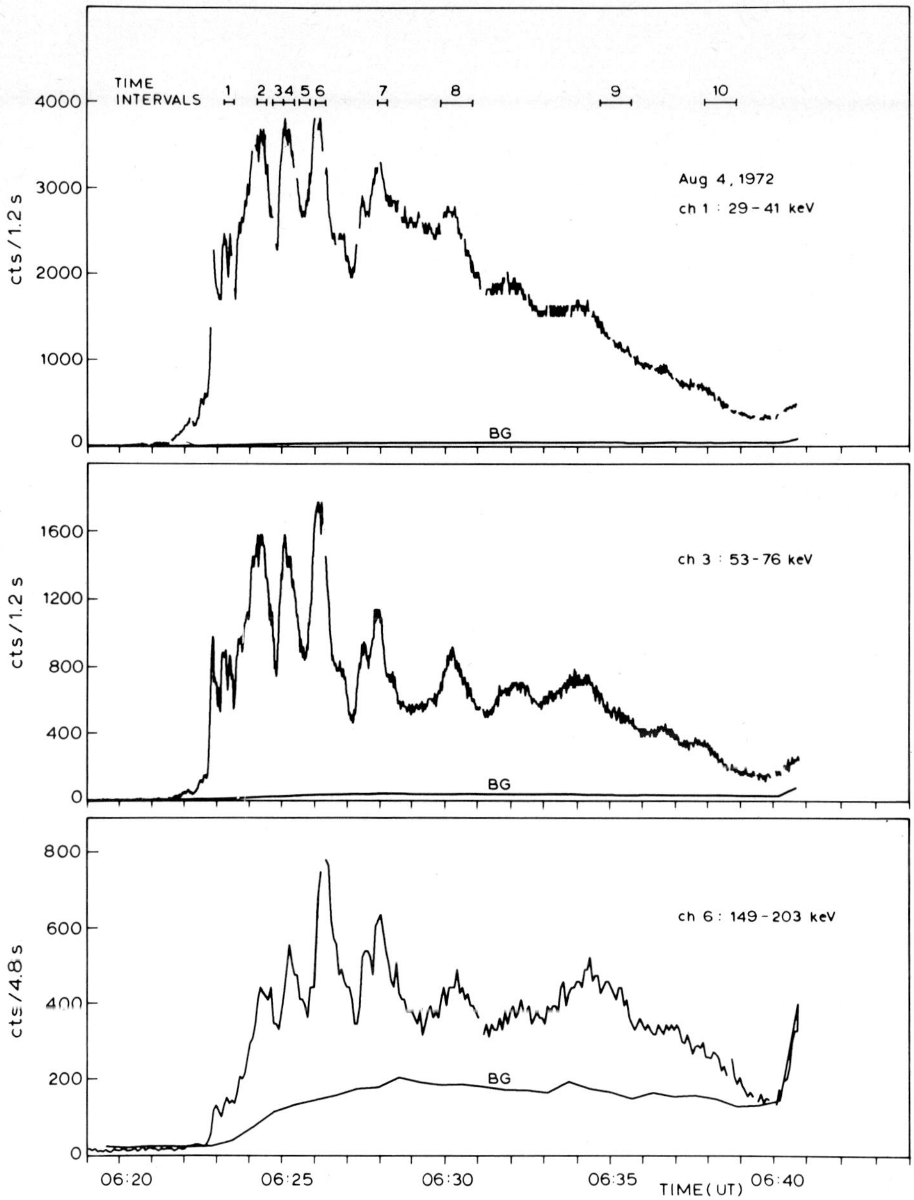

Fig. 4. Time profiles for the August 4, 1972 flare. The background count rates which have to be subtracted from the observations are indicated by BG. The numbers refer to the selected time intervals of the flare for which the solar photon spectra have been approximated by power-law spectra.

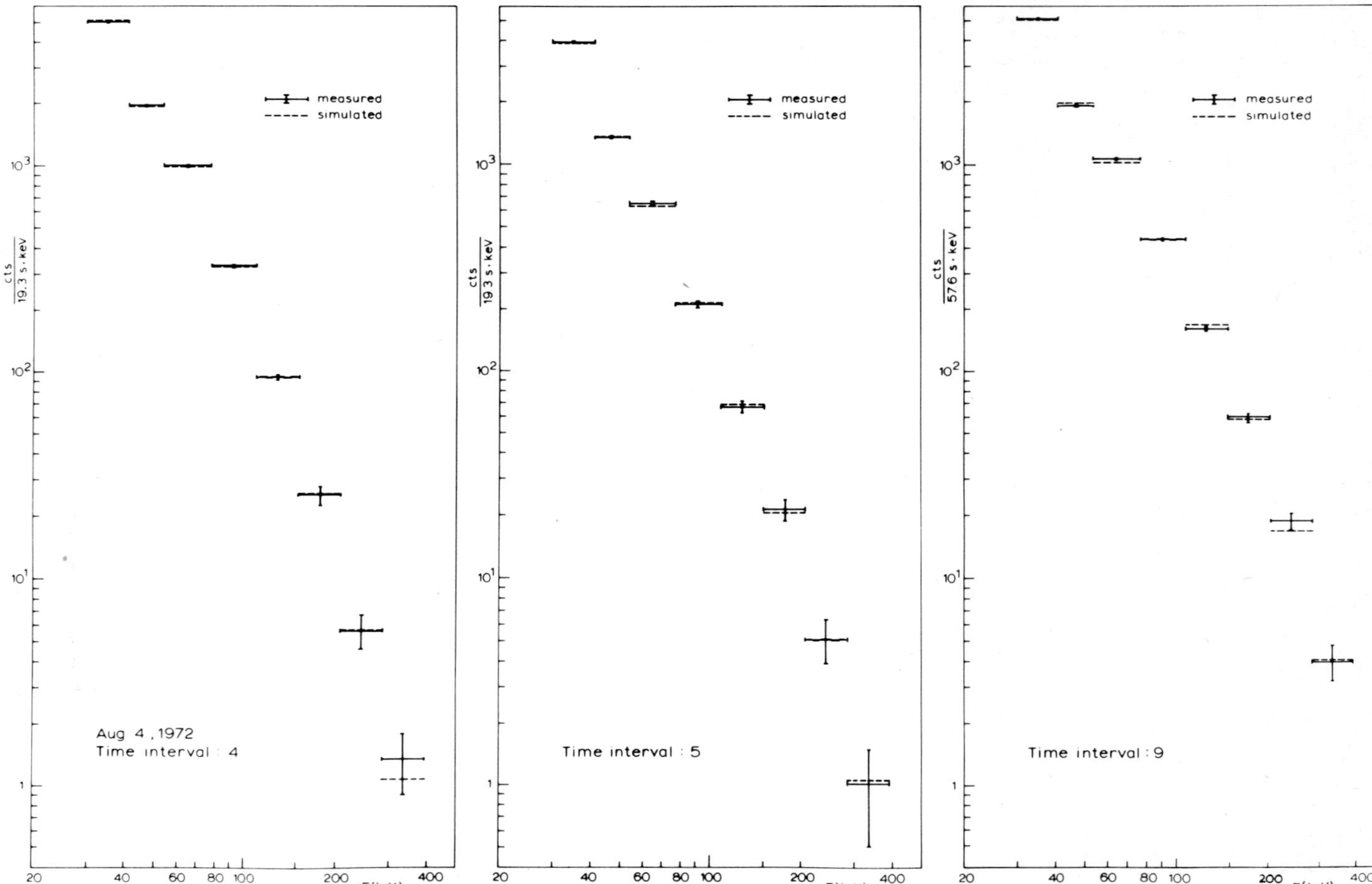

Fig. 5. Pulse height distributions measured during time intervals 4, 5 and 9 of the August 4, 1972 flare, showing the very good agreement of these distributions with those resulting from assumed combinations of two power-law spectra defined by the parameters A_1, γ_1, A_2 and γ_2 and connected to each other at the knee energy E_k. Vertical bars represent the statistical uncertainty in the measured values.

portance 1B, while the X-ray classification was M4 according to the classification scheme used by IUWDS, which refers to the soft X-ray intensity. The position on the solar disk was N08 E42. Time profiles of channel 1, 3 and 5 are presented in Figure 3. With the help of the computer program which simulates the detector response, the type of photon energy distribution which provides the best fit to the observed distributions could be determined. The latter could be approximated to a high degree of precision with power-law energy distributions ($dN/dE = AE^{-\gamma}$). For a few selected time intervals the parameters A and γ were determined using data integrated over time intervals marked in Figure 3 just below the parameter values. The accuracy for A is $\pm 8\%$ and for $\gamma \pm 0.2$.

Thanks to the good time resolution of the instrument it can be shown from the observations that the flares consist of a number of short-lived spikes with equivalent rise and decay times of about 2 s. A specific study of the short time variation will be given by van Beek *et al.* (1974). Earlier, Frost (1969) and Parks and Winckler (1969) reported quasi-periodic structures of 35 s and 16 s, respectively.

2.2. The August 4, 1972 Flare

The strong flare appearing on August 4, 1972 in the region N14 E09, was classified as a 3B, X5 event. The observed time profiles for channels 1, 3 and 6 are shown in

TABLE I

Parameters A_1, γ_1, A_2 and γ_2 and the knee energy E_k of the power-law spectra which yield the best fit to the photon spectra measured during time intervals 1 through 10 of the August 4, 1972 flare (Figure 4)

Time Interval	A_1 (photons/cm^2 s keV)	γ_1	A_2 (photons/cm^2 s keV)	γ_2	E_k (keV)
1	3.1×10^6	3.1	1.6×10^{10}	5.1	73
2	1.2×10^6	2.7	5.9×10^8	4.2	64
3	2.6×10^6	3.0	1.6×10^8	4.0	61
4	2.4×10^6	2.9	2.4×10^8	4.0	63
5	5.4×10^6	3.2	4.0×10^7	3.7	56
6	1.2×10^6	2.7	3.3×10^8	4.0	73
7	6.1×10^6	3.2	2.0×10^7	3.5	54
8	1.4×10^7	3.5	1.5×10^8	4.0	114
9	8.1×10^5	2.9	2.9×10^6	3.2	71
10	2.4×10^5	2.8	1.4×10^7	3.7	94

TABLE II

Solar flare August 4, 1972. Standard deviation (σ_{obs}) of the observed count rates (in percent) in comparison with the difference (in percent) between these count rates and those prescribed by the two component power-law approximations

Energy channels (keV)	Time interval					
	4		5		9	
	$\sigma_{obs.}$	p.l. appr.	$\sigma_{obs.}$	p.l. appr.	$\sigma_{obs.}$	p.l. appr.
29– 40	0.5	+0.4	0.6	−1.0	0.5	−1.4
40– 53	0.7	−0.5	1.0	+1.2	0.7	+3.0
53– 75	0.8	−2.5	1.2	−3.2	0.8	−4.3
75–107	1.8	−1.0	2.8	+0.8	1.5	−0.6
107–149	4.4	0.0	6.6	+3.3	2.8	+4.0
149–203	11	+1.7	14	−3.4	6.0	−0.5
203–279	19	+0.2	23	−2.3	10	−9.9
279–382	33	−23.3	49	+3.8	20	+3.3

Figure 4. As for other flares, the observations of the August 4 flare have been approximated by power-law photon spectra. Surprisingly, it appeared that the observations fit very accurately to a combination of two power-law energy distributions with a γ-discontinuity ('knee') in between. For ten selected short time intervals the parameters A_1 and γ_1, referring to the low energy part of these distributions, and A_2 and γ_2, referring to the upper part, are given in Table I together with the energy E_k of the knee. Since the A- and γ-values could be determined very accurately (errors are $\pm 4\%$ and ± 0.1 respectively), the E_k-values are defined with the fairly high precision of about 5 keV. The E_k-values do not vary considerably with time. However, there is a weak indication for an increase of E_k towards the end of the event, and for a higher E_k-value in the intensity 'peaks' rather than in the intensity 'valleys', as far as the first part of the flare is concerned.

Because the occurrence of the knee in the photon spectra might be physically very significant, the evidence of this knee and the accuracy of the power law fit will be illustrated in more detail.

Therefore, the observed pulse height distributions for the time intervals 4, 5 and 9 indicated in Figure 4 are presented in Figure 5 together with those resulting from assumed combinations of two component power-law distributions.

The discrepancies between the observed and simulated count rates are very small and are covered by the statistical uncertainties of the observations. In Table II they are compared with the standard deviations of the observed count rates.

It has been investigated whether thermal spectra, $dN/dE \sim e^{-E/E_0}$, or spectra of the form $dN/dE \sim e^{-\sqrt{E/E_0}}$ fit to the observations. Of course spectra of these forms can be fitted to the observations at two arbitrary energy values, but severe deviations, which are a multiple of the statistical uncertainties of the observations, occur at the remaining parts of the energy range. This can easily be checked by comparing the energy dependence of a power-law distribution, which yields such a very good fit, with those prescribed by the alternative distributions.

The knee in the photon spectrum must therefore be accepted as a reality.

3. Conclusions

3.1. Time profiles

The observations suggest a rejection of the concept of solar flares which considers the X-ray burst to result from one injection of high energy particles into the coronal plasma, followed by a gradual decrease of the energy of the particles in several ways until the initial preflare situation is restored. On the contrary, the March 21 flare appears to consist of numerous short-lived spikes with rise and decay times of about 2 s. Each spike has its own well-defined spectrum, which seems not to be correlated with that of the preceding or following spike. This is an indication that the spikes are individual physical phenomena; they do not simply result from variations of the intensity of the emitted radiation. The spikes have been called 'Elementary Flare Bursts' (EFB's). It should be noted that the March 21 burst gives an example of the way almost all weak bursts behave.

Remarkable is the constancy of the decay times over a wide photon energy range. According to Hoyng and Stevens (1972), the decay time is coupled to the ambient electron density by the expression:

$$t_{1/2}=\frac{10^8\, E_{\text{keV}}^{3/2}}{N_e}\left(\frac{3}{2^{2\gamma+1}}-1\right). \tag{1}$$

This is under the assumption that the loss of energy of the excited electrons is determined by electron-electron collisons. The absence of an $E^{3/2}$ energy-dependence can be clarified by assuming that the ambient electron density, N_e, is relatively high. The expected decay times according to the expression (1) become then short as compared with the observed decay times, and, hence, the observed profiles of the EFB's must be practically identical to the time profiles of the injection of accelerated electrons or the time profile of the electron acceleration process.

Also the photon fluxes of the August 4 flare exhibit statistically significant rapid variations within a few seconds as have been observed during the March 21 flare. This might be an indication that such a big flare event is also constituted of correspondingly short-lived EFB's. Here, the EFB's seem to be clustered together in such a way that quasi-periodic bursts are formed on a 40 s time base (time intervals 2, 4 and 6 in Figure 4).

3.2. Photon spectra

The knee in the photon spectra observed during the August 4 flare must also be present in the energy distributions of the solar electrons which cause the X-rays via the Bremsstrahlung mechanism. This is because the Bremsstrahlung mechanism does not have any cross-section discontinuity, which could give rise to such a knee.

If one would assume a production spectrum for the energetic electrons which is a simple power law in energy, the observed spectrum can result from either a removal of low-energy electrons or a removal of high-energy electrons. In the first case one could think of energy transfer to the ambient plasma electrons through electron-electron collisions; a knee is then very unlikely as compared with a rapid increase of the spectral steepness, going to lower energies. In the second case synchrotron losses could be called upon if such losses would not only occur at considerably higher energies. Another assumption would be that electrons with energies above E_k might more easily escape from the high density flare region and might enter a region with lower density where the Bremsstrahlung production is correspondingly reduced. This hypothesis is very unlikely because the gyro radii of electrons in the flare region are very small as compared with the size of the flare region. Hence, it must be concluded that the knee-effect is related to the properties of the acceleration mechanism; it might be an indication for a two stage acceleration process in flares as was suggested by De Jager (1969). Primary electron acceleration should occur up to energies not far above the knee energy, while secondary acceleration would yield electrons with higher energies and a different (for the August 4 flare steeper) energy distribution.

It is obvious, however, that the present problem, on the interpretation of the 'knee' in the differential photon spectrum, cannot be considered detached from the result described in Section 3.1, i.e. that the observed time profiles reflect the instantaneous injection profiles of energetic particles; the decay time of these particles is so short that they will not keep their energy during an observable time interval.

If there are two acceleration processes active in the flare plasma, secondary acceleration of a particle must occur immediately after the primary acceleration in a time interval short as compared with the decay time of the elementary flare bursts.

The observed shape of the photon spectra offers considerable problems to a theoretical interpretation. For example, there is the question why the shape of the spectrum around E_k has always been observed to be 'convex' and never 'concave'. Neither an intensity 'step' has ever occurred, which means that some kind of equilibrium mechanism equalizing the number of particles just below and above E_k, must be operative. The occurrence of convex shaped spectra is in fact a strong argument for the existence of a two-stage acceleration mechanism. It excludes the possibility of more or less isolated flare regions, each of them having its own electron energy distribution. Addition of the separate power law spectra of the generated X-rays would in all cases yield a concave shaped spectrum.

It is our impression that the present new observation will greatly reduce the number of applicable theories on particle acceleration processes in solar flares.

References

De Jager, C.: 1969, in C. de Jager and Z. Švestka (eds.), *Solar Flares and Space Research*, North-Holland Publ. Co., Amsterdam, p. 1.

Frost, K. J.: 1969, *Astrophys. J.* **158**, L159.

Hoyng, P. and Stevens, G. A.: 1973, 'Interpretation of a Hard Solar X-Ray Burst', in *Proceedings of the First European Astronomical Meeting* (to be published).
Parks, G. K. and Winckler, J. R.: 1969, *Astrophys. J.* **155**, L117.
Van Beek, H. F.: 1973, 'Development and Performance of a Hard Solar X-Ray Spectrometer', Thesis.
Van Beek, H. F., de Feiter, L. D., and de Jager, C.: 1974, 'Observations of Elementary Flare Bursts, and Their Interpretation', in *Space Res.* **14** (to be published).

PIONEER SOLAR PLASMA AND MAGNETIC FIELD MEASUREMENTS IN INTERPLANETARY SPACE DURING AUGUST 2–17, 1972

J. D. MIHALOV, D. S. COLBURN, H. R. COLLARD, B. F. SMITH, C. P. SONETT, and J. H. WOLFE

NASA Ames Research Center, Moffett Field, Calif., U.S.A.

Abstract. Solar wind plasma and magnetic field measurements from Pioneers 9 and 10 during August 2–17, 1972, reveal complex and large-amplitude variations on a one-hour time scale and numerous discontinuities. During this time period an approximate radial alignment of the two spacecraft as seen from the Sun occurred with heliocentric distances of 0.8 AU for Pioneer 9 and 2.2 AU for Pioneer 10, both at 45° east of the Earth's solar longitude. The peak hourly average solar wind proton bulk velocity measured at Pioneer 9 was 990 km s^{-1} during hour 0 UT of August 5. The peak hourly average proton number density was 62 cm^{-3} during hour 11 UT of August 3. The peak solar wind speeds are generally much reduced at Pioneer 10 compared with those observed at Pioneer 9. The peak 30-min average magnetic field magnitude was 85 γ during 1245–1315 UT of August 3. The Pioneer 9 data indicate passage of four fast forward interplanetary shocks, and one slow forward interplanetary shock.

1. Introduction

An unusual period of solar flare activity in early August, 1972, occurred at McMath Plage Region 11976. During August 2 to 10, two 3B and three 2B flares were reported (*Solar-Geophysical Data*, 1973). Intense and varied geophysical activity (cf. McKinnon, 1972) and high cosmic ray fluxes (cf. Pomerantz and Duggal, 1973) were observed at the Earth. Various spacecraft observations which also show unusually large fluxes of energetic solar particles and high-speed solar wind streams have also been given at the December 1972 and April 1973 American Geophysical Union meetings. This report presents interplanetary plasma data from the Ames Research Center solar wind plasma experiments on the Pioneer 9 and 10 interplanetary spacecraft, as well as interplanetary magnetic field data from the Ames Research Center magnetometer on Pioneer 9. The solar flares which probably were the source of the principal features in the plasma and magnetometer records are listed in Table I. Note that the two spacecraft were near 45° east of the Earth's solar longitude during this time period.

2. Experiment Details

The Pioneer-9 Ames solar wind plasma experiment utilizes 120° electrostatic deflection for energy per unit charge analysis, with three current collectors connected to electrometer amplifiers. Proton measurements are made at 30 logarithmically spaced intervals ranging from 210 to 15000 eV, for a proton energy/plate voltage difference ratio of 6. Similarly electron measurements are made at 14 logarithmically spaced intervals ranging from 13 to 1000 eV. The mean radius of the nearly hemispherical analyzer plates is 6 cm and their separation is $\frac{1}{2}$ cm. The solar wind azimuthal flow direction measurements are performed by sampling $2\frac{13}{16}$ deg wide sectors of the

D. E. Page (ed.), Correlated Interplanetary and Magnetospheric Observations, 545–553.

TABLE I

Large solar flares during early August 1972 (*Solar-Geophysical Data*, 1973)

Day, August 1972	Time, UT (onset)	Class	Approx. lat.	Approx. long.
2	0505	2B	13N	35E
2	1958	2B	14N	28E
4	0617	3B	14N	8E
7	1449	3B	14N	37W

spacecraft rotation. The spacecraft spin axis is perpendicular to the ecliptic plane. The polar flow direction measurements give the flow vector component approximately normal to the ecliptic plane, using the relative currents in the three collectors. The solar wind proton parameters presented here are obtained by a least-squares fit of the flight data to a convecting Maxwellian velocity distribution model with an isotropic temperature, using an instrument model derived from laboratory calibration data. This procedure is analogous to that used for the early Pioneer-6 and -7 Ames plasma probe data (Mihalov and Wolfe, 1971), although linear rather than non-linear least squares fits are done currently. The uncertainties in the plasma parameters for Pioneer 9 have not yet been completely evaluated but for similar instrument modes and solar plasma conditions they will be much smaller than those given by Mihalov and Wolfe (1971) for the Pioneer-6 and -7 data because the later instrument has significantly improved energy and angular resolution. The Pioneer-9 plasma experiment is almost identical to that on Pioneer 8 which has been described by Intriligator *et al.* (1969), and both experiments are similar in many respects to the Pioneer-6 and -7 plasma probes described by Wolfe and McKibbin (1968).

The Pioneer-9 Ames magnetometer is a triaxial fluxgate instrument with on board signal processing to preclude contamination of the measurement by vector aliasing (Sonett, 1966). Analog signals from the three sensors are prefiltered and then sampled at a high rate (27 s^{-1}) relative to the 1 Hz spin frequency. Digital spin demodulation is performed on the spin plane components by mathematical transformation to a non-rotating coordinate system using the Sun angle to obtain a high data rate representation of the field as would be measured by a non-spinning spacecraft. Digital filtering is then employed on each component to observe the Nyquist criterion prior to telemetry sampling. The filter constants are automatically adjusted to match any of 5 commandable telemetry bit rates. During the time of the magnetometer data presented here, the field sampling interval was 19 s, except for ~8 h on August 6-7 when it was 2.3 s. The digitization window is 0.4 γ (one gamma = 10^{-5} G) in each component.

The Pioneer-10 Ames plasma experiment is similar in many respects to the Pioneer-6 and -7 experiments described by Wolfe and McKibbin (1968). Two independent sets of quadrispherical 90° electrostatic deflection plates are used in the Pioneer-10 experiment for energy analysis, but data from the higher resolution set which uses 26 Bendix continuous channel multipliers as detectors are not given in this report. The

Pioneer-10 plasma analyzer views along the spacecraft spin axis which is directed toward Earth. The lower resolution assembly deflects incoming particles into five current collectors connected to electrometer amplifiers. The deflection plates have a mean radius of 12 cm and a 1 cm separation. The five current collectors provide a total angular acceptance of $\pm 70°$ from the normal to the instrument aperture. The central three collectors each have a 15° wide angular acceptance. The energy analysis for protons may be done at four different sets of energy intervals, the maximum number being 64 logarithmically spaced intervals ranging from 100 to 18000 eV for a proton energy/plate voltage difference ratio of 6. Similarly electrons are measured at 15 logarithmically spaced intervals that range from 2.1 to 500 eV. The solar wind flow directions are obtained from the combination of the relative currents on the five collectors together with identification of the particular 0.70° wide spacecraft roll sectors in which the solar wind is detected.

The Pioneer-9 spacecraft was launched November 8, 1968 into a heliocentric orbit with a 298 day period. This spacecraft is spin stabilized at 60 rpm. The Pioneer-10 spacecraft was launched March 3, 1972 on a trajectory which, with subsequent corrections, permits encounter with the planet Jupiter on December 4, 1973. This spacecraft is spin stabilized at about 4.7 rpm. The Pioneer-10 mission is discussed in NASA (1971), in which some description of the spacecraft and experiments is also given.

During the times of these observations the Pioneer-9 and -10 spacecraft were at heliocentric distances of 0.8 and 2.2 AU respectively, approximately 45° east of the Earth's solar longitude. Pioneer 9 was located from 110000 km to $\gtrsim$130000 km north of the ecliptic plane during this time, while Pioneer 10 was $11–12 \times 10^6$ km south of the ecliptic plane.

3. Description of Observations

The hourly average solar wind proton bulk velocity, number density and temperature observed at Pioneers 9 and 10 are given in Figure 1. These results are presented with an arbitrary delay of the Pioneer-10 data by 3 days with respect to the Pioneer-9 data to achieve on the figure an approximate alignment of the most prominent features. Assuming radial solar wind flow and considering the plasma speed observations alignment was at hour 08 UT of August 7 for Pioneer 9 which corresponds to hour 19 UT of August 11 for Pioneer 10.

At both Pioneers 9 and 10, features often observed in association with interplanetary disturbances are seen, in particular, fast velocity increases associated with increased proton temperatures (Wolfe, 1972). At the shocks in the data discussed here, the abrupt velocity increases are followed by peaks in the number density. An exception to this behavior is the event seen at Pioneer 9 at the end of August 4 which is not followed by a large density increase. A recent summary of theoretical calculations on post-shock conditions may be found in Dryer *et al.* (1972).

Another feature evident at both Pioneers is the gradual recovery of number densities over several days from low values that follow the high density peaks seen on

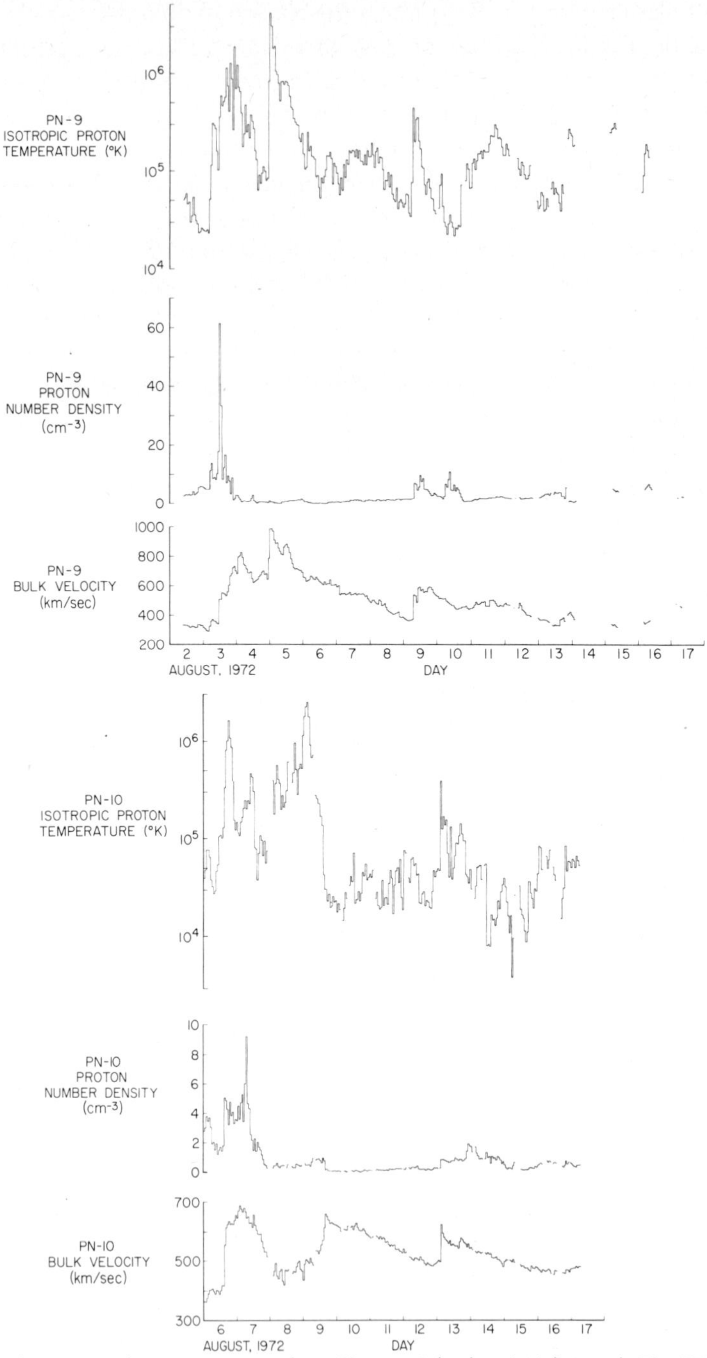

Fig. 1. Hourly average plasma parameters from Pioneers 9 (top) and 10 (bottom). The UT times refer to the Earth, so that a ~6 min correction must be applied to obtain times at Pioneer 9, and a ~13–15 min correction for Pioneer 10. The Pioneer 10 plot is arbitrarily delayed three days with respect to the Pioneer 9 plot, as indicated by the horizontal scales.

TABLE II

Interplanetary shocks observed by Pioneer 9 during early August 1972

Day, August 1972	Time, UT	Solar wind proton bulk velocity change (km s^{-1})	Interplanetary magnetic field magnitude change (γ)	Type
3	0440	55	5.6	Fast forward
3	1117	~ 210	55	Fast forward
4	2323	240	14	Fast forward
9	0707 ± 10 min	~ 145	~ 15	Fast forward
9	0958	85	~ -3	Slow forward

August 3 at Pioneer 9 and on August 7 at Pioneer 10 associated with the second shock of Table II. One also notes that during the first few days of these events, a 'background' proton temperature level which might be $\sim 10^5$ K from the Pioneer-9 records is $\sim 3 \times 10^4$ K at Pioneer 10. If an adiabatic, quiescent solar wind expansion with an $r^{-4/3}$ temperature gradient were assumed, a 10^5 K temperature at Pioneer 9 at 0.8 AU would imply a 3×10^4 K value at Pioneer 10 at 2.2 AU. Here r refers to heliocentric distance. We speculate that convective energy has been transformed into thermal energy during part of the disturbance, as the wind passed between 0.8 and 2.2 AU, because the Pioneer-9 high velocity peaks of 830 km s^{-1} at 03 h on August 4 and 990 km s^{-1} at 0 h and 01 h on August 5 appear to be replaced at Pioneer 10 by a single reduced amplitude peak of 690 km s^{-1} at 02 h on August 7.

The plasma bulk velocities of Figure 1 have not been corrected for the spacecraft velocities. The Pioneer-9 spacecraft velocity at this time is 36 km s^{-1} approximately normal to the solar direction in the direction of planetary motion and in the ecliptic plane. The Pioneer-10 spacecraft velocity at this time is 15 km s^{-1} in the antisolar direction and 17 km s^{-1} in the direction of planetary motion, parallel to the ecliptic plane.

Thirty minute averages of the Pioneer-9 magnetic field measurements during August 2 to 10 are given in Figure 2. The field latitude and longitude are given in the conventional ecliptic coordinate system with ϕ measured counterclockwise from the solar direction, looking from the north and θ measured from the ecliptic plane to the vector, with positive values corresponding to northward pointing field vectors. There is a peak in the average magnetic field magnitude at 85 γ during 1245 to 1315 UT on August 3. After the early part of August 4 the magnetic field magnitude is generally much smaller than 24 γ. After the early part of August 6 until August 9 the field magnitude is at essentially the pre-disturbance value, less than 10 γ.

4. Pioneer-9 Interplanetary Shock Observations

Five interplanetary shocks are observed in the Pioneer-9 plasma and magnetic field data. The times of these events, observed changes in magnetic field magnitude and

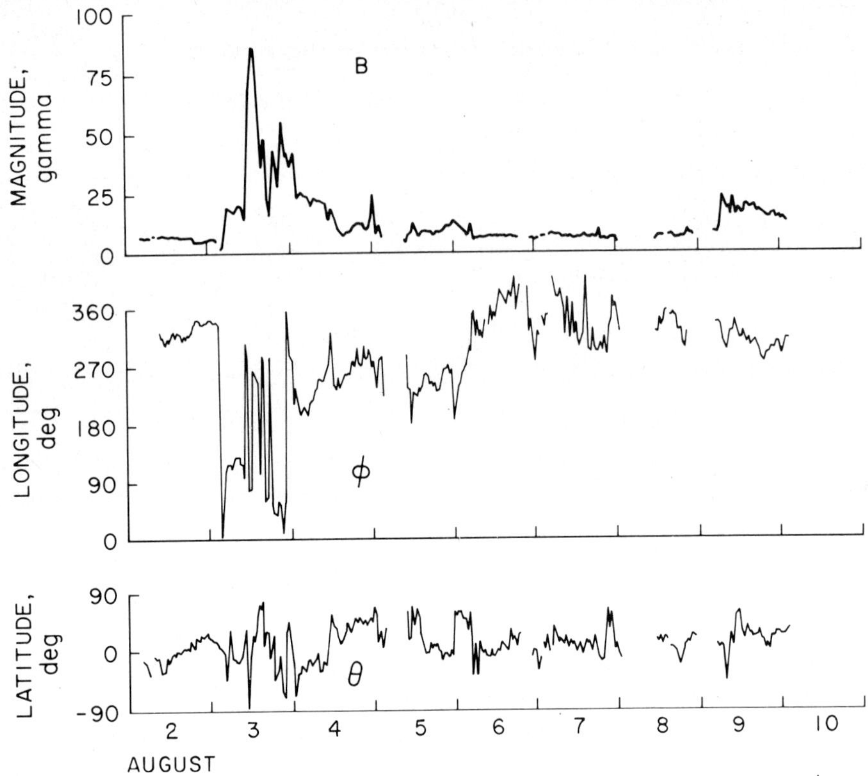

Fig. 2. Thirty-minute average magnetic field measurements from Pioneer 9. The UT times refer to the spacecraft.

TABLE III

Best-fit calculation (*Lepping and Argentiero*, 1971) for Pioneer-9 event of August 3, 1972, 1117 UT

Parameter	Average value	σ	Best estimate value
B_{1x} (gamma)	3.6	1.48	3.9
B_{1y}	−7.3	2.02	−6.1
B_{1z}	−19.4	1.66	−19.7
B_{2x}	−3.6	1.25	−3.6
B_{2y}	−15.0	5.76	−16.5
B_{2z}	−74.0	6.77	−73.3
W_x(km s^{-1})	−196	11	−194
W_y	52	15	61
W_z	−21	20	−26
n_1 (cm^{-3})	19.2	2.7	16.2
n_2	48.9	13	58.4
Δp(10^{-10} dyne cm^{-2})	112		−55

calculated proton bulk velocity, and the shock type (Colburn and Sonett, 1966) are given in Table II. The times are obtained from the magnetometer records and the identification to the minute is arbitrary in some of the cases with slower times on the order of a minute for the change in magnetic field magnitude. The shock types are deduced from examination of detailed data that is not presented here.

For about 45 min before the final event of Table II, extremely non-Maxwellian proton energy spectra are observed. These spectra appear to be composed of pre-shock plasma combined with a subsidiary stream of post-shock plasma of approximately $\frac{1}{5}$ the proton number density of the pre-shock plasma. The 'post-shock' plasma stream observed prior to the shock has a velocity that is lower than that of the first post-shock plasma spectrum but which is 100–150 km s^{-1} higher than that of the pre-shock plasma. In addition, the subsidiary plasma stream observed before the shock is at an azimuthal flow direction about 7 deg more from the west, the same as the post-shock plasma. He^{++} is observed in association with the subsidiary proton stream, while observation of the helium associated with the main pre-shock plasma is evidently obscured by the high energy side of the subsidiary stream proton distribution. These are the most intense non-Maxwellian plasma fluxes observed by Pioneer 9 during the entire 16 day time period discussed here.

The best-fit calculation of Lepping and Argentiero (1971) has been performed for the August 3, 1117 event, excluding an error cone angle calculation. The results are given in Table III. The notation of Table III follows that of Lepping and Argentiero (1971), with subscripts 1 and 2 denoting pre- and post-shock quantities. The coordinate system used is the conventional right handed, orthogonal solar ecliptic system with $+x$ in the solar direction and $+z$ northward. $\mathbf{W}=\mathbf{V}_2-\mathbf{V}_1$ where $\mathbf{V}$ refers to the plasma (proton) bulk velocity. B_{ij} and n refer to the jth component of magnetic field and the proton number density, respectively. $\Delta p=p_2-p_1$ where p refers to the thermal pressure, in this case due to protons and electrons alone. The solar wind helium is ignored in this calculation. σ here is an estimated uncertainty for each measured parameter, including statistical variance and systematic sources. The measured pre- and post-shock electron temperatures are 7×10^5 K and 9×10^5 K respectively.

The best-fit shock normal is $\hat{n}=(-0.938, 0.341, 0.064)$, with a 17° change from a shock normal based on average values. This shock normal is within 4° of the direction to the flare location from Pioneer 9 at the time of the event. An arbitrary preliminary 1.0 γ northward instrumental offset magnetic field has been assumed for these calculations. Based on the best-fit shock normal the measured Alfvén, whistler and sonic shock Mach numbers (Tidman and Krall, 1971) are 11, 17 and 2.6 respectively.

Various estimates of shock speed are possible. The expression given by Lepping and Argentiero (1971) yields

$$V_s=\frac{(n_2\mathbf{V}_2-n_1\mathbf{V}_1)}{n_2-n_1}\cdot\hat{n}\sim645 \text{ km s}^{-1}.$$

Evolutionarity (Colburn and Sonett, 1966) yields 754 km $s^{-1} < V_s \leqslant 788$ km s^{-1}. These two estimates for shock speed are subject to experimental uncertainties that may be as large as $\sim 10\%$ but a final evaluation has not yet been done. The transit time between Pioneers 9 and 10 yields a mean speed of 787 km s^{-1} along the direction between the locations of the two spacecraft at the times when this event was observed; this direction is $\sim 27°$ west and $\sim 7°$ south of the above shock normal. An indication of inaccuracy in the above shock normal is the negative calculated value for the thermal pressure difference Δp given in Table III. Perhaps a more correct shock normal exists which would yield better agreement with this measured parameter which is not used in the best-fit calculation reported here. Agreement between the measured and calculated thermal pressure differences might be supposed if a large excess of momentum flux ($\sim$200 erg cm^{-3}) due to waves or turbulence were present upstream from the shock. This large an amount of extra momentum flux would seem unlikely, as no more than $\frac{1}{30}$ this amount was inferred for any of 18 interplanetary shocks studied by Chao and Goldstein (1972), although only two of the 18 shocks they studied were of comparable Mach number to the event discussed here. Also, one would expect excess momentum flux due to waves or turbulence downstream rather than upstream from the shock.

For measured average proton number densities twice the values of Table III the best estimate Δp is 75×10^{-10} dyne cm^{-2}. The good agreement between the other measured average and calculated best estimate values of Table III is unchanged for this case. The Alfvén, whistler and sonic shock Mach numbers would be 16, 20 and 2.3 respectively for proton number densities increased by two times, for the 787 km s^{-1} shock speed between the two spacecraft projected onto the shock normal.

Note added in proof. Subsequent to preparation of this paper an apparent error in the Pioneer-9 proton number densities has been discovered which has not yet been conclusively determined. Some estimates indicate that the Pioneer-9 proton number densities should be multiplied by a factor of almost two.

References

Chao, J. K. and Goldstein, B.: 1972, *J. Geophys. Res.* **77**, 5455.
Colburn, D. S. and Sonett, C. P.: 1966, *Space Sci. Rev.* **5**, 439.
Dryer, M., Smith, Z. K., Endrud, G. H., and Wolfe, J. H.: 1972, *Cosmic Electrodyn.* **3**, 184.
Intriligator, D. S., Wolfe, J. H., McKibbin, D. D., and Collard, H. R.: 1969, *Planetary Space Sci.* **17**, 321.
Lepping, R. P. and Argentiero, P. D.: 1971, *J. Geophys. Res.* **76**, 4349.
McKinnon, J. A.: 1972, *NOAA Tech. Memo. ERL SEL-22*, U.S. Department of Commerce, N.O.A.A. Environmental Research Laboratories, Boulder, Colo.
Mihalov, J. D. and Wolfe, J. H.: 1971, *Cosmic Electrodyn.* **2**, 326.
NASA: 1971, *The Pioneer Mission to Jupiter, SP-268*, National Aeronautics and Space Administration, Washington, D.C.
Pomerantz, M. A. and Duggal, S. P.: 1973, *Nature* **241**, 331.
Solar-Geophysical Data (comprehensive reports): 1973, *No. 342-part II*, U.S. Department of Commerce, Boulder, Colo., p. 4.
Sonett, C. P.: 1966, in R. L. Smith-Rose (ed.), *Space Res.* **6**, Spartan Books, Washington, D.C., p. 280.

Tidman, D. A. and Krall, N. A.: 1971, *Shock Waves in Collisionless Plasmas*, Wiley-Interscience, New York.

Wolfe, J. H.: 1972, in C. P. Sonett, P. J. Coleman, Jr., and J. M. Wilcox (eds.), *Solar Wind, SP-308*, National Aeronautics and Space Administration, Washington, D.C., p. 170.

Wolfe, J. H. and McKibbin, D. D.: 1968, *Planetary Space Sci.* **16**, 953.

OBSERVATION OF INTERPLANETARY SHOCKS ON AUGUST 4, 1972

M. B. CATTANEO, P. CERULLI-IRELLI, L. DIODATO, A. EGIDI, and G. MORENO
Laboratorio per il Plasma nello Spazio, CNR, Roma, Italy

and

P. C. HEDGECOCK
Dept. of Physics, Imperial College of Science and Technology, London, U.K.

Abstract. In this report we describe plasma and magnetic field observations performed by the European satellite HEOS 2 during a series of large interplanetary shocks, which overtook the spacecraft on August 4, 1972.

1. Experimental Apparatus

The HEOS-2 mission has been described by De Waard *et al.* (1971). Details of the plasma experiment (S. 202), have been given by Coletti (1971) and De Waard (1971). Here we will only recall the following. The S. 202 experiment performs measurements of positive ions and electrons, as a function of flow direction and energy, in the range 10 eV q^{-1} to 40 keV q^{-1}. The plasma sensor consists of an electrostatic analyser with 2 Faraday cups and 2 channeltrons as particle detectors (Figure 1): the multiple detector system allows the simultaneous measurement of proton and electron fluxes.

The voltage applied to the deflector is kept constant during two complete spin revolutions and then varied in 6 steps: a complete energy spectrum is taken every 192 s. During the first spin revolution, 32 angular measurements are performed within 11.25° sectors, giving the following information for the four sensors:

(i) the integral flux, i.e. the sum of counts accumulated in the whole revolution;

(ii) the maximum flux, defined as the counts accumulated in the 11.25° sector, where the maximum flux is observed;

(iii) the direction corresponding to the sector, where the maximum is detected, with an uncertainty of ±5.6°.

During the subsequent revolution, the previous data are elaborated, supplying the width of the distribution, defined as the number of sectors in which the flux exceeds half the maximum.

After each complete energy spectrum, the position of all sectors is shifted of $\frac{1}{8}$ of their angular width, improving the precision of the directional measurements.

The following discussion is based only on the Faraday cup data. Channeltron data are not presented here due to a failure of their operation occurred before the considered events. As a consequence of the channeltron failure, the energy channels were rearranged by telecommand, in such a way that the maximum observable energy was reduced to ~4 keV.

2. Observations

Three shock waves associated with sudden commencements at ground, were detected

D. E. Page (ed.), Correlated Interplanetary and Magnetospheric Observations, 555–562.

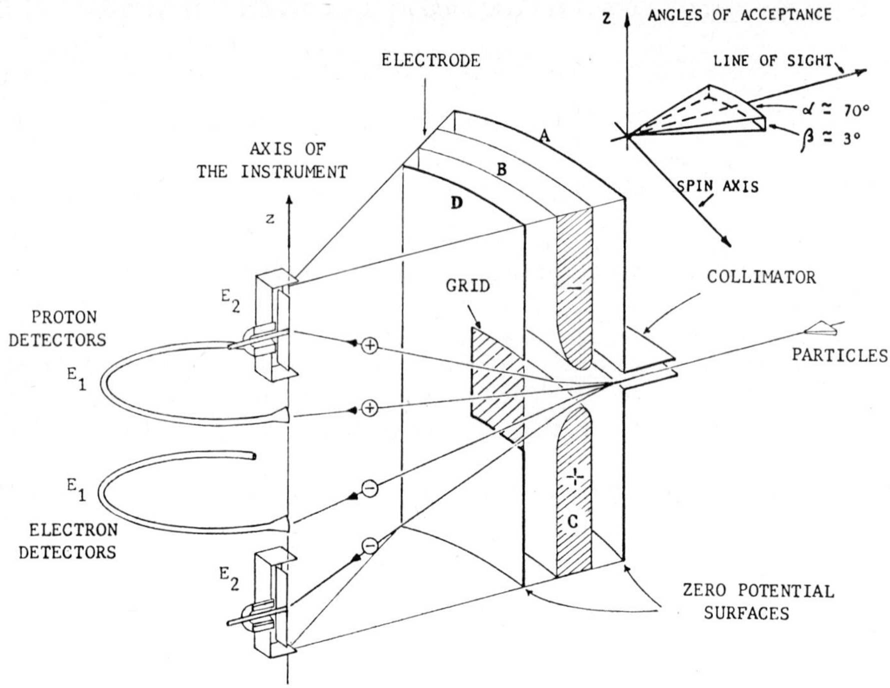

Fig. 1. Cross section of the electrostatic deflector used in the S. 202 plasma experiment.

by HEOS 2 on August 4, 1972 respectively at ~UT 0120, 0220 and 2100.

Figure 2 shows plasma observations in the period from August 3, UT 1200 to August 4, UT 0400, during which the first two shocks overtook the satellite. Only fluxes in some of the most interesting energy channels of protons and electrons are given in the figure.

For each channel, two lines are shown in the same graph, giving respectively the total electrometer counts over one revolution of the spacecraft (one count is of the order of 5×10^4 particles $cm^{-2}\ s^{-1}$) and the counts for the angular sector in which the maximum was detected. The θ plot, for the 580–870 eV proton channel, gives the direction from which the plasma is coming, measured in the equatorial plane of the satellite from the solar direction (the satellite spin axis was almost perpendicular to the ecliptic). For comparison, the magnetic field intensity measured by the S. 201 experiment is also shown. This experiment has been described by De Waard (1971).

During the considered period, a number of Earth's bow shock crossings are identified, using both plasma and magnetic data. As usual, magnetosheath traversals are characterized by: a higher magnetic field intensity, increase of the low energy electron flux, a proton thermalization (the 'maximum' flux becomes much smaller than the 'integral') and a change of the direction of the incoming protons. The last change is relatively small in the ecliptic plane, due to the position of the satellite ($X_{SE} = 6.8\ R_E$, $Y_{SE} = 5.9\ R_E$, $Z_{SE} = 36.1\ R_E$, at UT 1100).

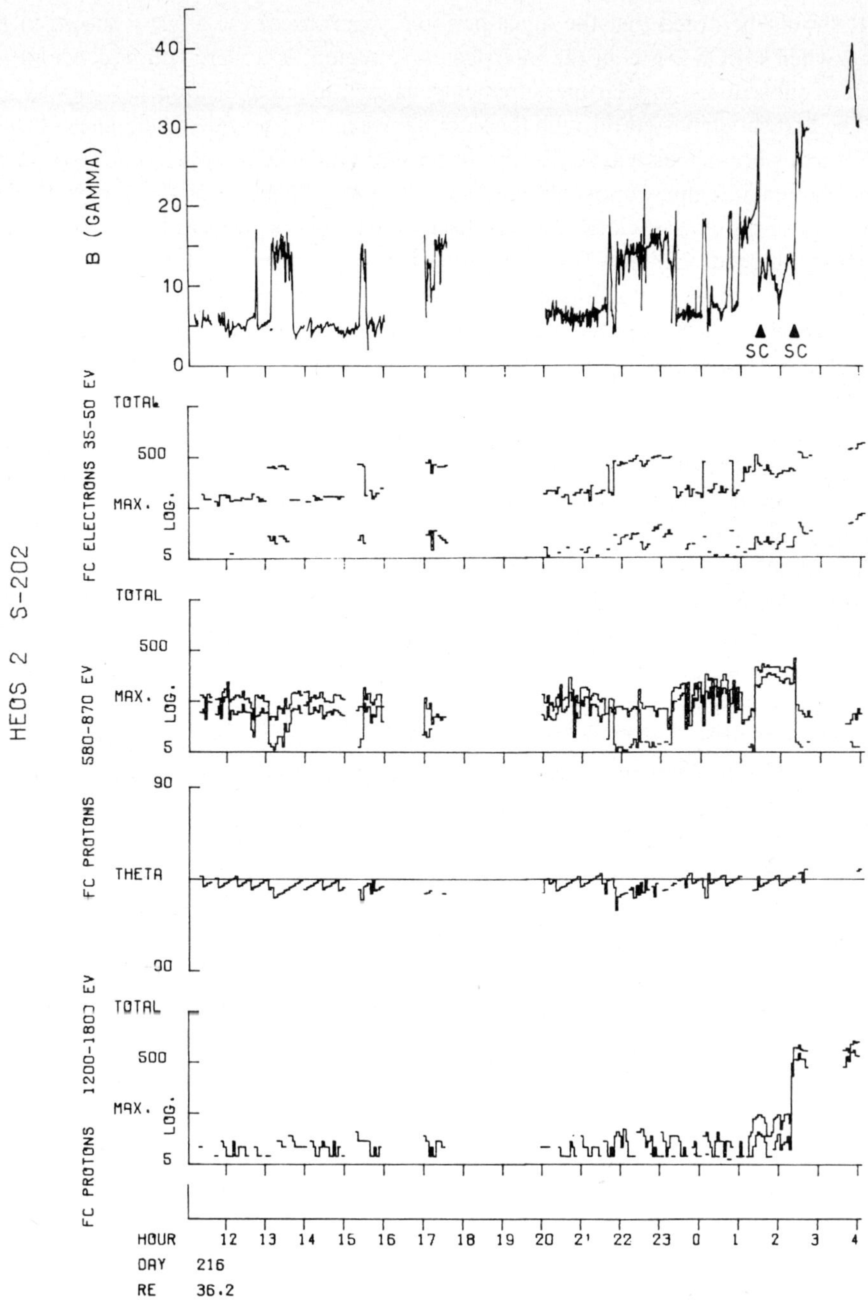

Fig. 2. Time history of plasma and magnetic field during the multiple shock crossings occurred on August 3 and 4, 1972. From the top are given: the magnetic field intensity, the electron flux in the 35–50 eV channel, the proton flux in the 580–870 eV and 1200–1800 eV channels. In each graph, the two lines refer to the 'integral flux' (upper bars) and to the 'maximum' (lower bars). The θ plot refers to the proton direction in the ecliptic plane in the energy range 580–870 eV.

It should be noted that the quasi-periodic variation of the angle θ shown in the plot, when HEOS 2 was in the interplanetary region, is a mere consequence of the shift of subsequent angular measurements, as mentioned in Section 1.

The first shock passed through the satellite, when it was inside the magnetosheath: the Earth's bow shock was suddenly compressed and HEOS 2 was found to be in the interplanetary region almost simultaneously to the detection of the shock. Plasma observations are better clarified, considering the magnetic field: at UT 0120, the field intensity, indeed, increases from values typical of the magnetosheath ($\sim 17\ \gamma$) up to almost 30 γ and then suddenly decreases to $\sim 10\ \gamma$. This situation prevents the computation of the plasma parameters on the two sides of the discontinuity.

Fuller information is obtainable for the second shock observed at UT 0220, when HEOS 2 was still in the interplanetary region. The jumps of the plasma parameters through the shock front are in this case the following:

Bulk velocity (V): from ~ 400 km s^{-1} to ~ 600 km s^{-1}.

Proton number density (N): from ~ 2 cm^{-3} to ~ 6 cm^{-3}.

Magnetic field (B): from $\sim 11\ \gamma$ to $\sim 29\ \gamma$.

The shock speed was of the order of 730 km s^{-1}.

After the shock crossing, B underwent a further increase reaching values larger than 40 γ. From $\sim$UT 0400 to $\sim$UT 1200 data are missing. Figure 3 shows the time history of protons and magnetic field in the following period. The analysis of the plasma data shown in this figure has been prevented by a very unusual phenomenon, never observed by S. 202 or by other plasma experiments, in the knowledge of the authors. The instrumental background, always constant within the measurement quantization, underwent large fluctuations returning to its normal values only at the end of August 4. The fluctuations were observed with opposite signs on both the positive ion and electron Faraday cups, the sum of their background levels being always constant. This suggests that the observed phenomenon is related to real flux of particles rather than to a failure of the electronics. In fact, Bostrom *et al.* (1972), using IMP-5 and IMP-6 data, found in this period fluxes of high energy protons ($E > 30$ MeV) 10^4 times larger than usual. Figure 4 shows that an extremely good correlation exists among the S. 202 noise level (proton Faraday cup) and the observed high energy proton fluxes.

In Figure 3 the abnormal background has been removed, by linearly interpolating in the time interval separating subsequent noise measurements (the noise level is sampled by the instrument only every ~ 17 min). This procedure is subject to large errors, as the variations of the noise level often exceed the counts due to the low energy protons. Therefore, no attempt has been made to derive the plasma parameters from these data.

The character of the plasma data from UT 1200 to UT 2100 seems to suggest that HEOS 2 was in that period inside the magnetosheath: protons are indeed observed in a wide range of energies with comparable intensities. On the other hand, the observed plasma thermalization could be related also to the interplanetary shock detected a few hours earlier. To distinguish between these two possibilities, we have

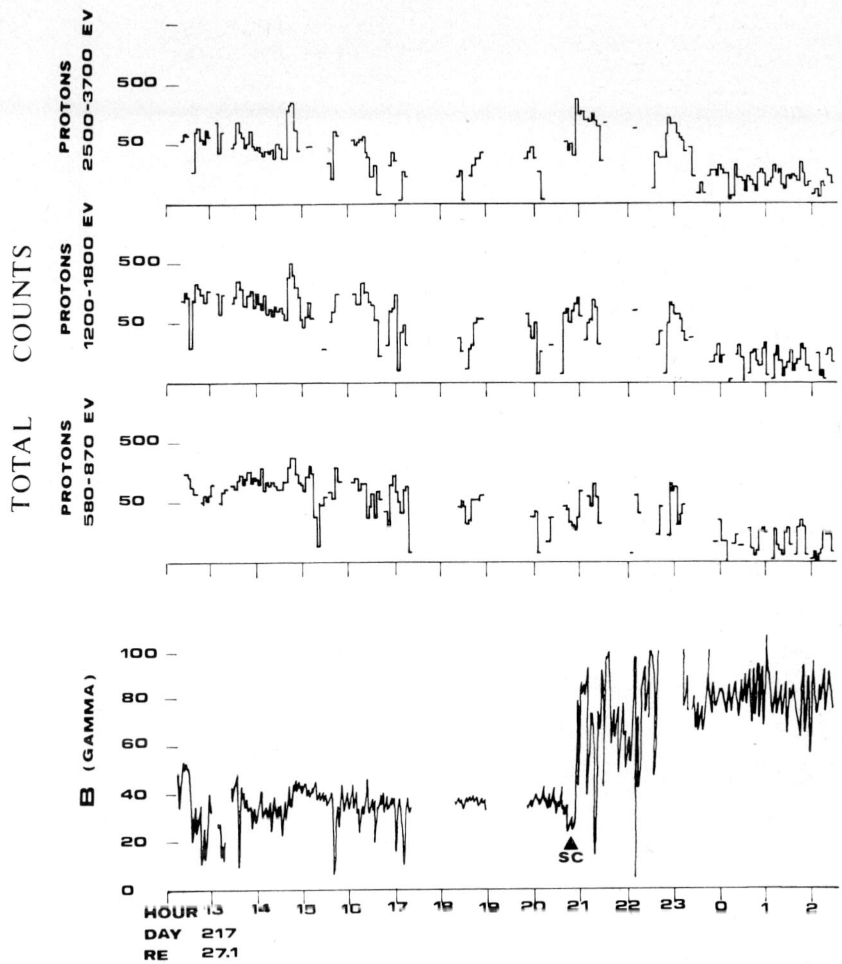

Fig. 3. Time history of protons and magnetic field on August 4–5, 1972. Only the 'integral' flux is shown for protons in the specified energy channels.

computed the position of the bow shock, on the basis of the plasma parameters measured at UT 0400, i.e. just before the beginning of the above mentioned data gap. The radial distance of the subsolar point is given by the equation (Spreiter *et al.*, 1968):

$$D = \frac{C}{\sqrt[6]{NV^2}} \left[1 + 1.1 \frac{(\gamma - 1) M^2 + 2}{(\gamma + 1) M^2}\right]$$

V is the plasma bulk velocity,
N is the proton number density,
γ is the ratio of specific heats, and
M is the magnetosonic Mach number.

The constant C was computed assuming the average position of the shock observed

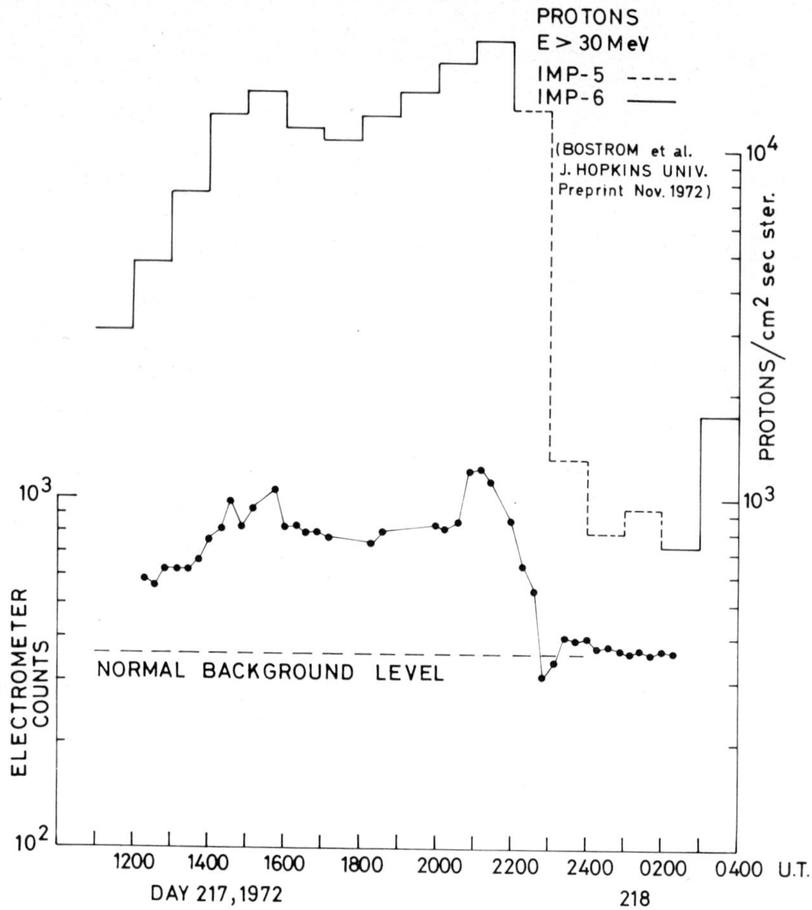

Fig. 4. Time history of the instrumental background, considered along with the high energy proton fluxes measured by IMP 5 and 6.

by HEOS 1 in 1969 and the average dynamical pressure and Mach number ($NV^2 = 6.3 \times 10^{15}$ cm^{-1} s^{-2}, $M = 4.7$) measured by the same spacecraft (Egidi *et al.*, 1970; Formisano *et al.*, 1973). Assuming $\gamma = \frac{5}{3}$ and the values of the parameters supplied by HEOS 2 ($NV^2 = 2.6 \times 10^{16}$ cm^{-1} s^{-2} and $M = 4.9$), the subsolar point of the bow shock should be located at 12.1 R_E from the Earth. To obtain the shock position at HEOS-2 orbit, the hypothesis was made that the shape of the surface (assumed to be axially symmetric about the Sun-Earth line) was the same observed by HEOS 1. The results are given in Figure 5 where the expected position of the bow shock is shown along with HEOS-2 orbit in the X_{SE}, $(Y_{SE}^2 + Z_{SE}^2)^{1/2}$ plane.

The first portion of orbit corresponds to the period considered in Figure 2 (different symbols have been used to indicate interplanetary and magnetosheath observations). After the data gap, the satellite should be just inside the magnetosheath according to the above computations (portion of orbit marked as 'possible magnetosheath', corresponding to the data shown in Figure 3).

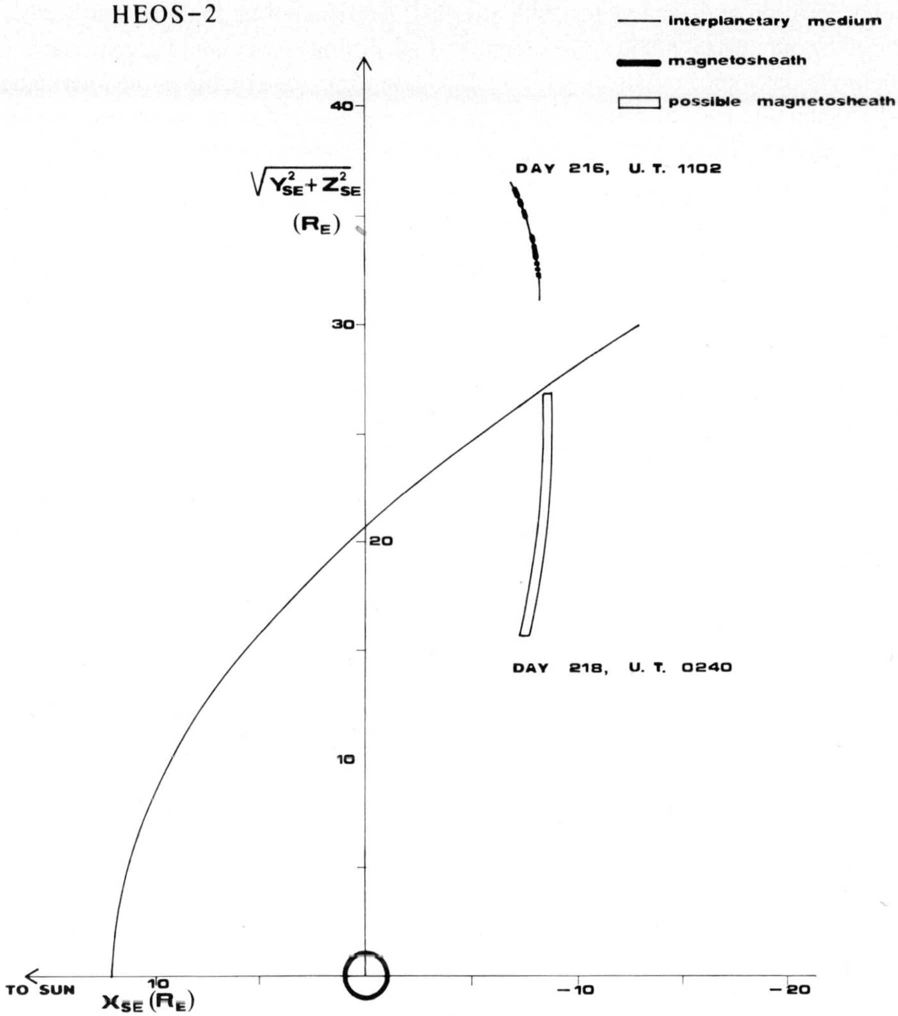

Fig. 5. HEOS-2 orbit from August 3, UT 1102 to August 5, 0240. The orbit gap corresponds to lack of data. The continuous curve represents the expected bow shock position on August 4, UT 0400.

At ~UT 2100 the largest shock overtook the satellite, as seen from the magnetic field intensity (Figure 3) which exhibits a sudden increase up to 80 γ. Simultaneous with the magnetic field variation, the proton energy increases, the maximum flux being observed in the highest energy channel (2500–8700 eV). Then the plasma signals decrease in all channels reaching almost the noise level a few hours later. The observations seem to indicate that the plasma bulk velocity is still increasing, reaching values outside the range of the instrument (protons become unobservable when their velocity exceeds 900 km s^{-1}). Such an interpretation is confirmed by Pioneer-9 data (World Data Center, Report UAG-21, November 1972): the bulk velocity, subsequently to the detection of the considered shock, appears to increase up to 1130 km s^{-1}, the largest value ever measured in the solar wind.

The location of the magnetosheath after UT 2100 can not be unambiguously determined by our measurements. The extremely high proton energy could suggest a strong compression of the bow shock which could leave again the satellite in the interplanetary region. On the other hand, the magnetic field intensity seems to be more typical of the magnetosheath: if the satellite was in the interplanetary space, the field intensity would be the largest observed until now.

Acknowledgements

We are indebted to Prof. A. Bonetti, Dr S. Cantarano, Dr A. Coletti, Dr R. Marconero, Dr G. V. Pallottino and to the other people of the Laboratorio per il Plasma nello Spazio of Rome who have participated in the development of the plasma experiment S. 202.

This research has been supported by the Consiglio Nazionale delle Ricerche of Italy.

The HEOS-2 magnetic field experiment was supported by the British Science Research Council.

References

Bostrom, C. O., Kohl, J. W., McEntire, R. W., and Williams, D. J.: 1972, 'The Solar Proton Flux – August 2–12, 1972', Johns Hopkins University, Silver Spring, Maryland, Preprint, Nov.

Coletti, A.: 1971, 'Descrizione dell'esperimento ESRO-S. 202 per la misura del plasma interplanetario col satellite HEOS 2', Laboratorio per il Plasma nello Spazio, Roma, Nota Interna LPS-71-18.

De Waard, J.: 1971, 'HEOS-A2 Satellite: Spacecraft Handling Plan', ESRO/ESTEC, Internal Note, Noordwijk, The Netherlands.

De Waard, J., Kooter, C. J., Summerhill, S., and Taylor, B. G.: 1971, *ELDO/ESRO Sci. Techn. Rev.* **3**, 185.

Egidi, A., Formisano, V., Palmiotto, F., Saraceno, P., and Moreno, G.: 1970, *J. Geophys. Res.* **75**, 6999.

Formisano, V., Moreno, G., and Amata, E.: 1973, *J. Geophys. Res.* (submitted).

Spreiter, J. R., Alksne, A. Y., and Summers, A. L.: 1968, in R. L. Carovillano (ed.), *Physics of the Magnetosphere*, D. Reidel Publ. Co., Dordrecht-Holland, p. 301.

RELATION OF CORONAL MAGNETIC STRUCTURE TO THE INTERPLANETARY PROTON EVENTS OF AUGUST 2–9, 1972

E. C. ROELOF, J. A. LEZNIAK, and W. R. WEBBER
University of New Hampshire, Durham, N.H., U.S.A.

and

F. B. McDONALD, B. J. TEEGARDEN, and J. H. TRAINOR
NASA/GSFC, Greenbelt, Md., U.S.A.

Abstract. In order to interpret the time history of the ⩾13.5 MeV protons observed on Pioneers 9 and 10 and IMP 5 from the flares of August 2, 4 and 7, we have estimated the coronal connection longitudes of the interplanetary field lines using the solar wind velocities measured on the spacecraft. Pioneer 9 was 46° east of the Earth-Sun line at 0.78 AU and Pioneer 10 was nearly on the same radial at 2.2 AU. We find reasonable agreement between the observed (preliminary) magnetic field polarity at Pioneer 9 and the equatorial chromospheric polarity (from Hα filtergrams) at the inferred connection longitudes. The protons appear to be released preferentially in a large magnetic neutral-line complex that surrounds the flare region (McMath 11976) from Carrington longitudes 0° to 60°. Although impulsive proton events were observed from all three flares by Pioneer 9, only the second two were seen at Earth, and Pioneer 10 never showed comparable impulsive response (the fluxes rose steadily August 3–6 and then decayed irregularly). These apparent descrepancies are resolved by comparison with magnetic neutral line structure. The long neutral line that passes through McMath 11976 and extends over more than 150° appears to separate regions of markedly different interplanetary proton fluxes.

1. Time Histories of ⩾13.5 MeV Proton Fluxes

During the series of solar energetic particle events August 2–13, 1972, the locations of the Pioneer-9 and -10 spacecraft were most fortuitous for the study of interplanetary propagation. Pioneer 9 was at 0.78 AU, about 47° east of the Sun-Earth line, while Pioneer 10 had reached 2.2 AU, but was nearly on the same radial as Pioneer 9, being about 45° East of Earth. Moreover, Pioneer 10 and Earth would be connected by an ideal Archimedean interplanetary field line if the solar wind velocity V were ~670 km s^{-1}. However, for lower velocities, the spiral through Pioneer 10 would pass to the west of Earth, and vice versa.

Protons ⩾13.5 MeV were measured on Pioneer 9 by the University of New Hampshire (UNH) detector (Webber and Lezniak, 1973) and also on Pioneer 10 by the GSFC/UNH (Goddard Space Flight Center) detector (Teegarden *et al.*, 1973). Comparable data were available from the GSFC experiment on IMP 5 (Explorer 41, McDonald, 1970) and kindly made available to us prior to publication. Although the IMP-5 threshold (⩾11.5 MeV) was not exactly the same as the Pioneer thresholds, its fluxes could be transformed to ⩾13.5 MeV using the spectral information derived from other energy channels of the IMP-5 detector. Three hour averages from these three data sets are shown in Figure 1 for the period August 1–15, 1972.

The three time histories are markedly dissimilar from August 2 through August 5 and from the middle of August 7 through August 9. Pioneer 9 observes impulsive injections from three of the seven flares of importance ⩾1B reported in McMath plage

D. E. Page (ed.), Correlated Interplanetary and Magnetospheric Observations, 563–571.

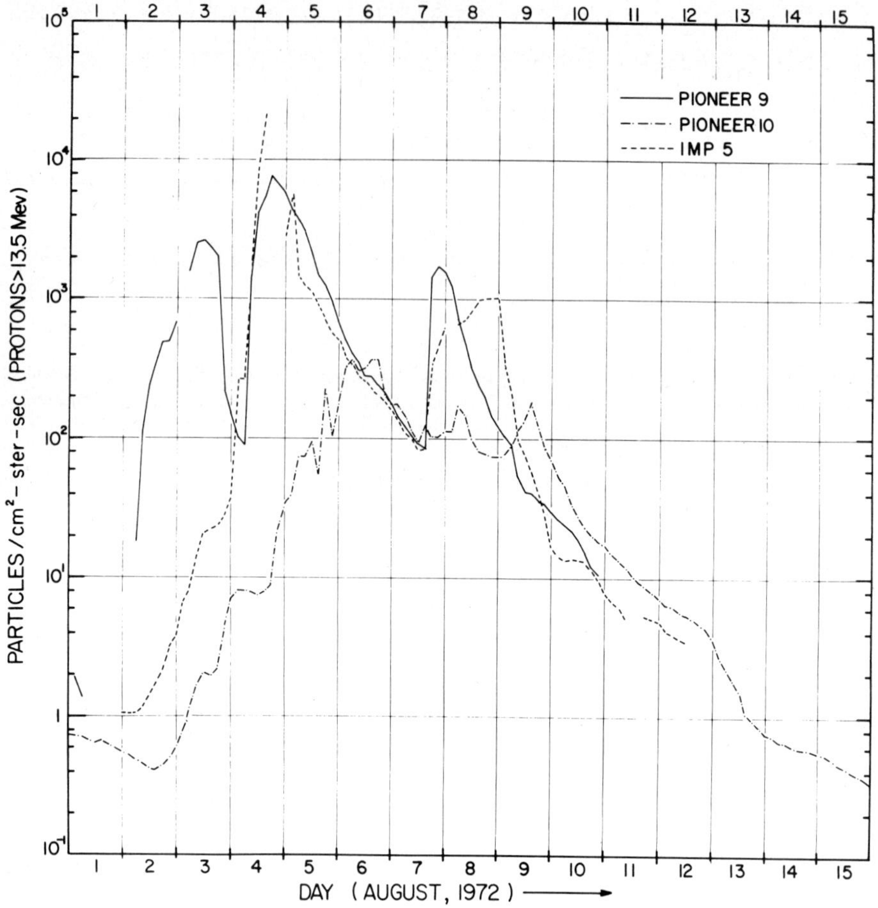

Fig. 1. Three-hour averages of proton fluxes ⩾13.5 MeV from Pioneers 9 and 10 and IMP 5 versus time. Fluxes from IMP 5 ⩾11.5 MeV have been corrected to ⩾13.5 MeV using spectrum estimates derived from other energy channels of the same detector.

region 11976 August 2–15 (1B, August 2, 0316 UT, N13 E35; 3B, August 4, 0530 UT, N14 E09; 3B, August 7, 1455 UT, N14 W37), although the maximum in the flux history of the August 2 flare is delayed until ~0900 on August 3. While fluxes on IMP-5 increase gradually on August 2, there is no impulsive response to the first flare, while the Pioneer 10 fluxes actually *decline* for most of the day before gradually increasing. For the second and third flares, both Pioneer 9 and IMP 5 show impulsive responses, but the Pioneer-10 flux still rises gradually until the middle of August 6 when it joins Pioneer 9 and IMP 5 in a common decay which ceases with the flare in August 7. However, the increases of August 7–8 have little in common other than their starting times. Pioneer 9 shows a 'classical' impulsive increase and decay, while IMP 5 has its maximum delayed one day from that of Pioneer 9 and followed by a much more rapid decay, and Pioneer 10 shows only two small irregular increases on August 8 and 9.

Because of these dissimilarities, the period of very small interplanetary gradients (radial *and* azimuthal) during August 6–7 and the comparable fluxes of the three detectors on August 10 are even more remarkable. The small gradients cannot be simply a coincidence because, as Figure 1 shows, the decay *rates* at the different spacecraft are also remarkably similar.

We shall show that the explanation of these striking features, both dissimilarities and similarities, lies in the large-scale structure of the interplanetary magnetic field and its connection into distinct regions of the coronal magnetic field. We shall not deal with structure on a time scale much less than a day, since a detailed analysis of the interesting smaller-scale structure in these events has been presented to this conference by our colleagues Van Hollebeke and McDonald (1973).

2. High Coronal Magnetic Field Line Connection Longitudes

Our explanation of the large-scale structure of the events is based on the assumption that ~ 15 MeV protons have negligible transport transverse to interplanetary magnetic field lines, so that the ordering of the data which we find then validates the assumption, after the fact. Thus if we know where the field line passing over the spacecraft connects into the high corona, we know the injection longitude of the energetic protons. We estimate the approximate connection longitudes of interplanetary field lines to the high corona by using the observed solar wind velocity to construct 'instantaneous' ideal Archimedean spirals, as first suggested by Snyder and Neugebauer (1966). Although the 'instantaneous' spirals themselves bear little resemblance to the actual field structure during disturbed times, Nolte and Roelof (1974a, 1974b) have presented theoretical and observational evidence that the *high coronal connection longitudes* of the spirals are approximately ($\sim 10°$) the same as those of the actual large-scale interplanetary field lines due to the cancellation of two competing effects: (1) corotation of the plasma with the coronal magnetic field inside the magnetohydrodynamic critical points (tending to shift the connection longitude *eastward* from that given by an ideal spiral from the center of the Sun); and (2) acceleration of the plasma in interplanetary space (yielding a *westward* shift from the spiral longitude). Krieger *et al.* (1973) have demonstrated that this 'extrapolated quasi-radial hypervelocity' (EQRH) approximation for the connection longitudes is sufficiently accurate to identify the equatorial source of a recurrent solar wind stream as a magnetically open coronal 'hole'. Moreover, Nolte and Roelof (1974b) have offered evidence that the EQRH approximation holds even during disturbed interplanetary conditions (with the exception of velocity peaks in the disturbance) and may be used to reconstruct the actual large-scale field configuration if data from several appropriately located satellites are available.

If ϕ_s is the solar (Carrington) longitude of the spacecraft at radius a, then the EQRH connection longitude ϕ_c for an observed solar wind velocity V is

$$\phi_c = \phi_s + \Omega a / V,$$

where Ω is the sidereal rotation rate of the Sun.

In Figure 2, the proton fluxes from the three spacecraft are plotted versus their high coronal connection longitudes (Carrington solar coordinates) calculated from the equation above. We have used solar wind velocity data presented to this conference and graciously provided to us prior to publication by Mihalov *et al.* (1974) for Pioneers 9 and 10 and by Grünwaldt *et al.* (1973) for HEOS 2 (near Earth). Since both the injection rates and interplanetary conditions are obviously changing rapidly

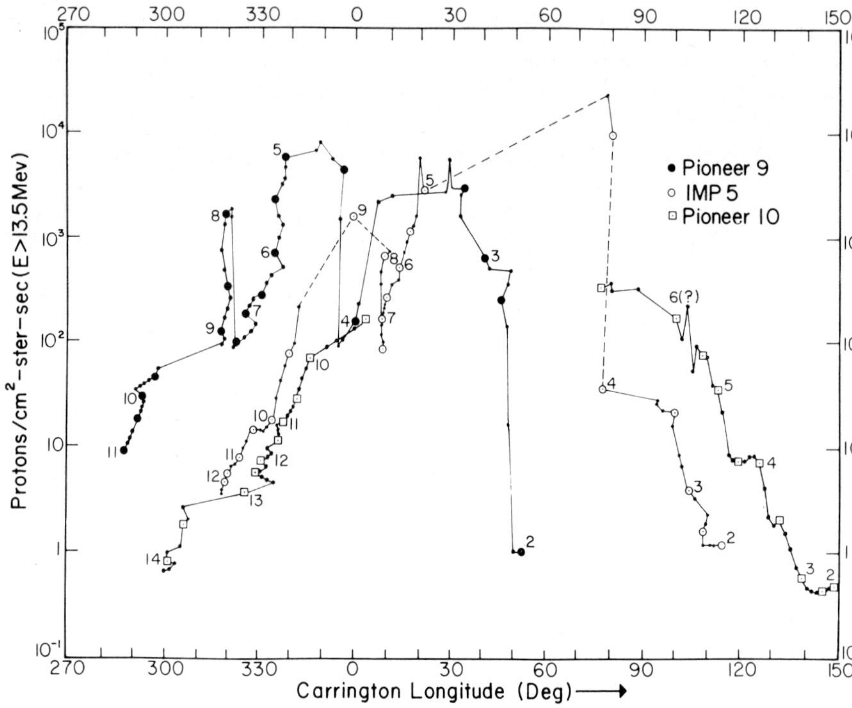

Fig. 2. Three-hour averages of proton fluxes ⩾13.5 MeV from Pioneers 9 and 10 and IMP 5 versus estimated Carrington longitude at which large-scale interplanetary field lines connect to the equatorial high corona. See Figure 1 and text. Solar wind used is from Pioneers 8 and 9 (Mihalov *et al.*, 1974) and HEOS 2 (Grünwaldt *et al.*, 1974).

with time, the interplanetary fluxes (measured at very different positions in space) are not necessarily simply related to the fluxes escaping from the corona. For instance, even if there were no injection gradient in coronal longitude, there might well be a longitude dependence in interplanetary propagation conditions. However, during the times when there is only a small interplanetary flux gradient while all three spacecraft measure similar decay rates (August 6–7 and 10–12), the simplest assumption is that both coronal *and* interplanetary gradients are small, and therefore that the coronal injection rate and interplanetary loss rate are nearly independent of longitude over those longitudes spanned by the coronal connections from the spacecraft.

3. Effect of Solar Wind Structure on ⩾13.5 MeV Proton Flux Histories

Connection longitudes are plotted from 3-h averages of solar wind velocities at each spacecraft. Identifying symbols are plotted every 12 h, and small dots every 3 h unless higher time resolution was desirable. The rising phases of large solar wind disturturbances are easily identified by an eastward longitude shift greater than the average solar rotation rate of ~3° every 6 h. For instance, there was a large solar wind velocity increase at Pioneer 9 after 1200 on August 3, and the broad front (>45° wide) of the disturbance reached Earth at the beginning of the next day (SSC, 0120 UT, August 4). Due to the velocity rise, the Pioneer-9 connection longitude shifted ~20° eastward in 3 h and that of Earth shifted ~15° (see Figure 2). It is likely that this shift distorted the Pioneer-9 proton flux profile, producing the second flux maximum at ~0900 UT, August 3. After this longitude shift the connection of Earth on August 4 lies only 25° west of the connection longitude that Pioneer 9 had on August 2, a fact that bears on the August 4 proton increase having an impulsive profile at both Earth and Pioneer 9, while that of August 2 was impulsive only at Pioneer 9.

The largest shift in connection longitudes was the result of the extremely high velocity increase late on August 4 observed at both Pioneer 9 and Earth (SSC, 2054 UT, August 4). Although the Pioneer-9 connection longitude shifted only 10° (because the solar wind velocity was already ~700 km s^{-1} prior to the rise to 1000 km s^{-1}), the Earth's connection shifted ~40° eastward. Then from August 5 through August 8, the connection longitudes 'dwell' during the long decay of the solar wind velocities, Pioneer 9 between 315° and 335° and Earth between 0° and 20°. Recall that the period of small interplanetary flux gradients was August 6–7, during these longitude 'dwells'.

The solar wind disturbances of August 3 and 4 arrive at Pioneer 10 ~1200 UT, August 6, but in a highly attenuated form (Mihalov *et al.*, 1973) since the peak velocity of 1000 km s^{-1} at Pioneer 9 has dropped below 700 km s^{-1} at Pioneer 10. This strong deceleration between 0.78 AU and 2.2 AU clearly invalidates the EQRH approximation during these strongly disturbed conditions so no connection longitudes are calculated for Pioneer 10 from 1200 UT, August 6 until the recovery of the rarefaction is completed at 1200 UT, August 9. The connection longitudes earlier on August 6 are also somewhat in doubt, since the disturbance is then approaching Pioneer 10. During this interval, the connection has moved eastward from 80° to 5°, and then from August 10 to August 12 Pioneer 10 has almost the same connection longitude as Earth (320°–345°) and lies only ~30° from that of Pioneer 9 (290°).

The final solar wind increase occurred on August 9 at Pioneer 9 and Earth (SSC 0036 UT), shifting connection longitudes eastward ~15° at both spacecraft. The velocity increase 4 days later at Pioneer 10 (~0300 UT, August 13), produces a 20° eastward connection shift and is well-associated with those increases on August 9 at the inner spacecraft since the transit time over 1.4 AU agrees with the peak velocities at both Pioneers of ~600 km s^{-1}. This disturbance terminates the distorted impulsive proton flux peak at Earth, and produces a factor of 2–3 drop in fluxes at Pioneer 9 (on August 9) and Pioneer 10 (on August 13) when the connection longitudes for *both* the latter spacecraft shift from ~320° to 300°. Moreover, August 10–12 is the second period noted in which all three spacecraft exhibit comparable proton fluxes and similar decay rates. This decay rate is not likely to be affected by gradients in coronal injection rates, because the connection longitudes of both Pioneer 10 and Earth 'dwell' between 315° and 345° on these days, just as the Pioneer 9 connection 'dwelled' at these longitudes August 5–9 (so that the 'classical' profile of the August 7 event observed at Pioneer 9 was probably undistorted by the usual coronal gradients).

4. Control of Proton Transport by Coronal Magnetic Structures

We may summarize our discussion of Figure 2 in three items as follows:

(1) Proton fluxes in the interplanetary medium are lower by a factor >100 from connection longitudes west of 90° compared to those east of 60°. Compare fluxes and connection longitudes of Pioneer 9 and IMP 5 on August 3 and of Pioneer 10 and IMP 5 on August 5 in Figure 2.

(2) Proton fluxes in the interplanetary medium are comparable for connection longitudes east of 60° and west of 330°.

(3) A small but readily discernable reduction (by a factor of ~2–4) of interplanetary proton fluxes is found between connection longitudes ~300° as opposed to ~330°. Note the drop in Pioneer 9 fluxes on August 9 and Pioneer 10 fluxes on August 13 when the shift is made for either spacecraft from 330° to 300°.

We know that there are magnetic structures in the low equatorial corona that correspond well with the longitude structures found in the proton fluxes. Since it has recently been argued (Roelof and Krimigis, 1973) that interplanetary structure in ~1 MeV proton fluxes corresponds to chromospheric magnetic neutral lines as defined by features in Hα filtergrams (McIntosh, 1972), we present in the upper panel of Figure 3 an Hα synoptic chart from Carrington solar rotation 1590 prepared by S. Wayland under the supervision of P. S. McIntosh of the National Oceanic and Atmospheric Administration. Dark filaments are cross-hatched and other magnetic neutral lines (filament channels, fibril patterns, etc.) are light lines, while magnetic neutral lines inferred from fine structure, continuity, etc. are dashed lines. Polarities (checked against Mt. Wilson magnetograms) are indicated by pluses (field out) and minuses (field in).

Since preliminary Pioneer 9 magnetic field polarities were graciously provided us

---

Fig. 3. *Upper panel:* Hα filtergram of low coronal magnetic neutral line structure prepared by S. Wayland under the supervision of P. S. McIntosh of the National Oceanic and Atmospheric Administration. Dark filaments are cross-hatched, and other magnetic neutral lines (filament channels, fibril patterns, etc.) are light lines, while neutral lines inferred from fine structure, continuity, etc., are dashed lines. Polarities (checked against Mt. Wilson magnetograms) are indicated by pluses (field out) and minuses (field in).

Lower panel: Proton fluxes ⩾13.5 MeV from Pioneers 8 and 9 and IMP 5 at the beginning of each day (August, 1972); (Inset) Pioneer 9 interplanetary magnetic field polarity from three-hour periods described as positive (+), negative (−) and mixed (/), plotted at their estimated high coronal connection longitudes (see text and Figure 2).

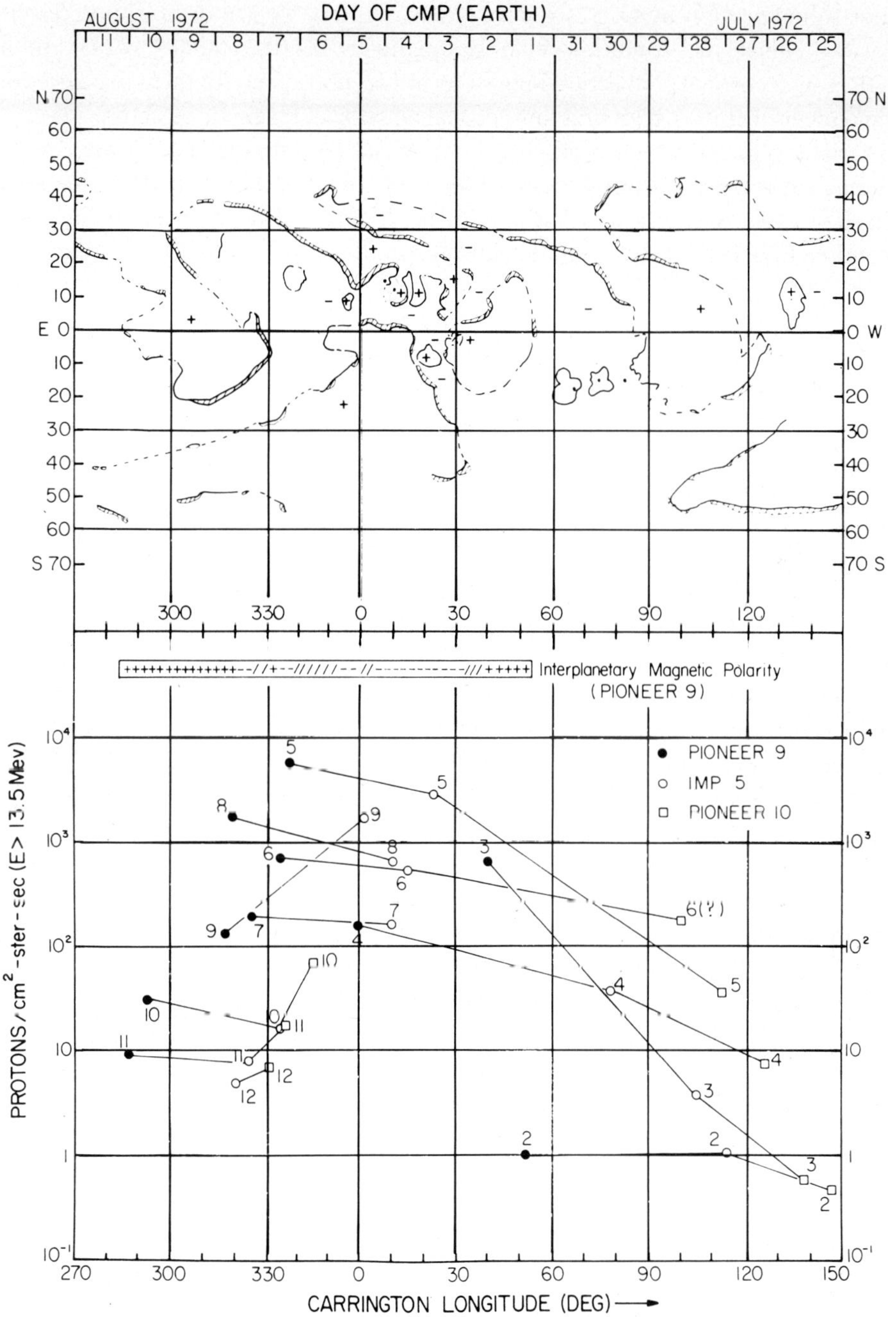
DAY OF CMP (EARTH)
AUGUST 1972
JULY 1972
11 10 9 8 7 6 5 4 3 2 1 31 30 29 28 27 26 25
N 70
E 0
S 70
70 N
0 W
70 S
300 330 0 30 60 90 120
+++++++++++++--//+--//////--//-----------///+++++
Interplanetary Magnetic Polarity
(PIONEER 9)
PIONEER 9
IMP 5
PIONEER 10
PROTONS/cm² -ster-sec (E > 13.5 Mev)
CARRINGTON LONGITUDE (DEG)
270 300 330 0 30 60 90 120 150

prior to publication by D. S. Colburn (NASA/Ames), we can demonstrate directly that two interrelated conditions hold for longitudes 55° (through 0°) to 285°: (i) the EQRH high coronal connection longitudes are accurate; and (ii) there is a close correspondence between high and low coronal equatorial magnetic structure.

Assigning positive (+), negative (−) or mixed (/) polarities to 3-h periods (excluding those periods during the rise phases of solar wind disturbances), the interplanetary polarities were compared with predominant Hα polarities at the corresponding EQRH longitudes and the sub-spacecraft latitude (~NO2). The interplanetary polarities are plotted in the lower panel of Figure 3 and the contingency table for the correlation is given in Table I wherein mixed polarities were assigned a random correlation. The correlation coefficient is high (0.67) and significant ($\chi^2 = 22$), as would be anticipated from a direct comparison of the equatorial polarities in Figure 3.

TABLE I

Contingency table for inferred Hα and EQRH-mapped interplanetary magnetic field polarities (Pioneer 9).

Solar \ Interplanetary	+	−	
+	20.5	2.5	23
−	6	21	27
	26.5	23.5	50

In order to better visualize the interplanetary gradients, the proton fluxes at the beginning of each day are replotted from Figure 2 onto the lower panel of Figure 3, and the points from the three spacecraft on each day are interconnected. Items (1) and (2), above, are easily verifiable from this simplified plot.

When we look to the low coronal magnetic structure, the significant longitudes of 90°, 60°, 330° and 300° all correspond to well-marked equatorial neutral lines. Moreover (with the exception of 60°) all crossings are of the *same* neutral line, namely the long sinuous filament that runs through McMath plage region 11976 (20°–30°, N10), the site of all the large August flares.

It therefore appears that ⩾13.5 MeV protons were only inefficiently transported to the high coronal structure west of the neutral line which crossed the equator at 90° (item 1) but were injected into the interplanetary medium across longitudes from 330° to 60° with much more comparable efficiency (item 2). This latter domain corresponds to the large negative polarity region most intimately associated with the flare site, McMath 11976. The longitude of reduced fluxes ~300° (item 3) is separated from the main negative cell by a positive region.

The actual magnetic structure of the high corona cannot be inferred from low coronal polarity data alone, although it is reasonable to assume that current sheets extend upward above the Hα neutral lines to preserve the corresponding polarity changes observed in the interplanetary magnetic field. If so, such structures are an intrinsic part of the mechanism that controls the high coronal distribution and interplanetary propagation of energetic solar flare particles.

Acknowledgements

The work at the University of New Hampshire was supported by Pioneer 9 data analysis grant NGR 30-002-087 and one of us (E.C.R.) acknowledges support from Department of Commerce Contract NOAA 03-3-022-41.

References

Grünwaldt, H., Montgomery, M. D., and Rosenbauer, H.: 1973, personal communication.

Krieger, A. S., Timothy, A. F., and Roelof, E. C.: 1973, *Solar Phys.* **29**, 505.

McDonald, F. B.: 1970, in V. Manno and D. E. Page (eds.), *Intercorrelated Satellite Observations Related to Solar Events*, D. Reidel Publ. Co., Dordrecht-Holland, p. 34.

McIntosh, P. S.: 1972, *Rev. Geophys. Space Phys.* **10**, 837.

Mihalov, J. D., Colburn, D. S., Collard, H. R., Smith, B. F., Sonett, C. P., and Wolfe, J. H.: 1974, this volume, p. 545.

Nolte, J. T. and Roelof, E. C.: 1974a, *Solar Phys.* (in press).

Nolte, J. T. and Roelof, E. C.: 1974b, *Solar Phys.* (in press).

Roelof, E. C. and Krimigis, S. M.: 1973, *J. Geophys. Res.* **78**, 5375.

Snyder, C. W. and Neugebauer, M.: 1966, in R. J. Mackin and M. Neugebauer (eds.), *Solar Wind*, Pergamon, New York, p. 25.

Teegarden, B. J., McDonald, F. B., Trainor, J. H., Roelof, E. C., and Webber, W. R.: 1973, *Astrophys. J. Letters* **185**, L155.

Van Hollebeke, M. A. and McDonald, F. B.: 1973, 'Propagation of Solar Particles Through the Major Disturbances in the Events of August 1972', 7th ESLAB Symposium (Saulgau) (not published).

Webber, W. R. and Lezniak, J. A.: 1973, *J. Geophys. Res.* **78**, 1979.

MeV ELECTRONS, PROTONS AND ALPHA PARTICLES OBSERVED AUGUST 2–12, 1972

D. E. PAGE, V. DOMINGO, and K.-P. WENZEL

Space Science Department (ESLAB), European Space Research and Technology Centre, Noordwijk, The Netherlands

1. Introduction

Our experiment on the ESRO satellite HEOS 2 monitored electrons, protons and alpha particles during the August 1972 solar particle events. Measurements for the particle types and energies as shown in Table I are reported:

TABLE I

HEOS 2 monitored electrons, protons and alpha particles during August 1972 solar particle events

Particles	Energy windows			
Electrons	0.45– 0.76	0.76– 1.18	1.18– 1.77	1.77– 3.5 MeV
Protons and alpha particles	9.3 –12.2	12.2 –17.0	17.0 –27.7	27.7 –35.7 MeV nucleon^{-1}

The detector system used to make the measurements has been explained in detail by Köhn *et al.* (1972). It consists of a 500 μ surface barrier silicon detector placed in front of a 6 mm lithium drifted detector, both being housed in a plastic scintillator anticoincidence cup. The half opening angle of the telescope is 33° and the geometric factor which is slightly energy dependent is about 1.1 cm^2 sterad. The telescope looks perpendicular to the spin axis of HEOS 2, measuring electrons in eight sectors and protons and alpha particles in four sectors.

The orbit of HEOS 2 in early August 1972 is shown in Figure 1 where the coordinates are Geocentric Solar Ecliptic. The period of the orbit is 5.2 days and the orbital plane turns 1° per day relative to the Earth-Sun line. Projections of successive apogee positions on the *X*, *Y* plane are shown for the period since the HEOS-2 launch on 31 January 1972.

2. The Particle Intensities Recorded

The dramatic intensities recorded in the first few days of August 1972 are generally referred to as the 'August' event (World Data Centre A, 1973). However, Figures 2 and 3 show that in fact it might be more appropriate to refer to the July 19 event, because since that date the particle intensities encountered had been far above background and remained so until mid-September 1972.

D. E. Page (ed.), Correlated Interplanetary and Magnetospheric Observations, 573–585. All Rights Reserved.

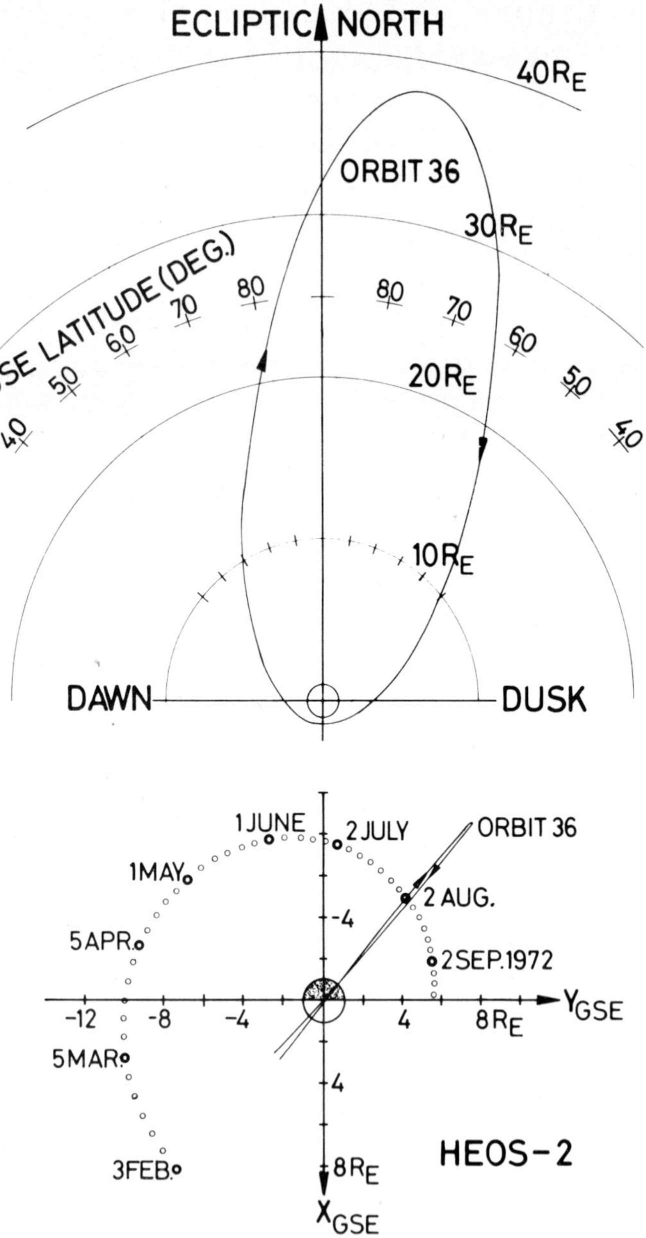

Fig. 1. HEOS-2 orbit 36 projected on the *Y*-*Z* plane (top) and on the *X*-*Y* plane (bottom). The positive *X* direction is toward the Sun. The open circles in the bottom part show the movement of apogee from February till September 1972.

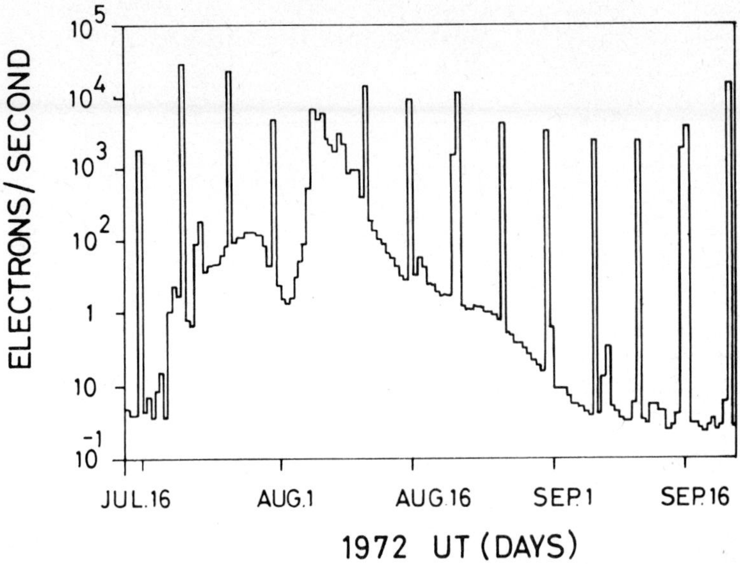

Fig. 2. Count rate of 0.45–3.5 MeV electrons (12-h averages) from July 14 till September 21, 1972. The regular spikes at 5.2 day intervals are produced by the crossing of the Earth's radiation belts.

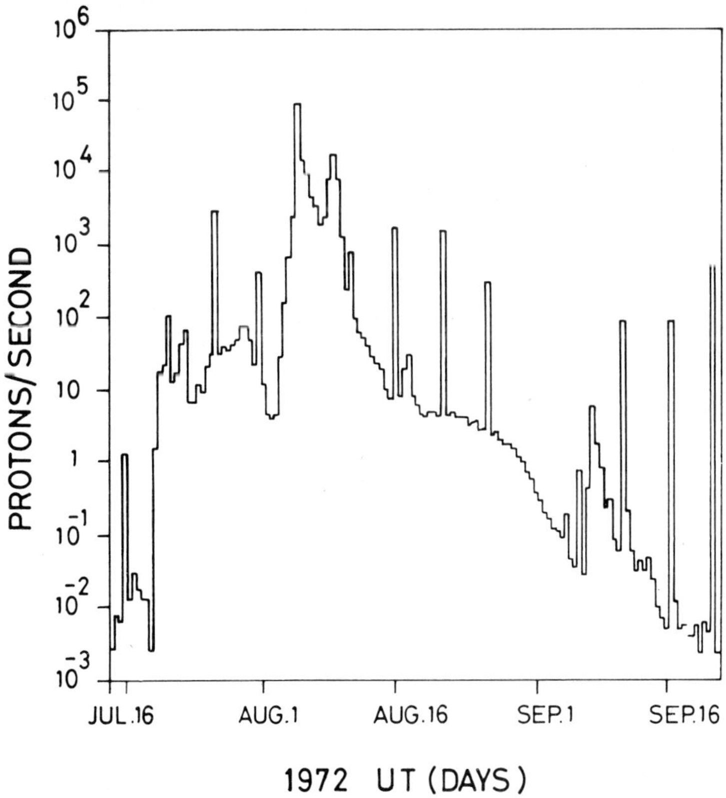

Fig. 3. Count rate of 9.3–35.7 MeV protons (12-h averages) from July 14 till September 21, 1972.

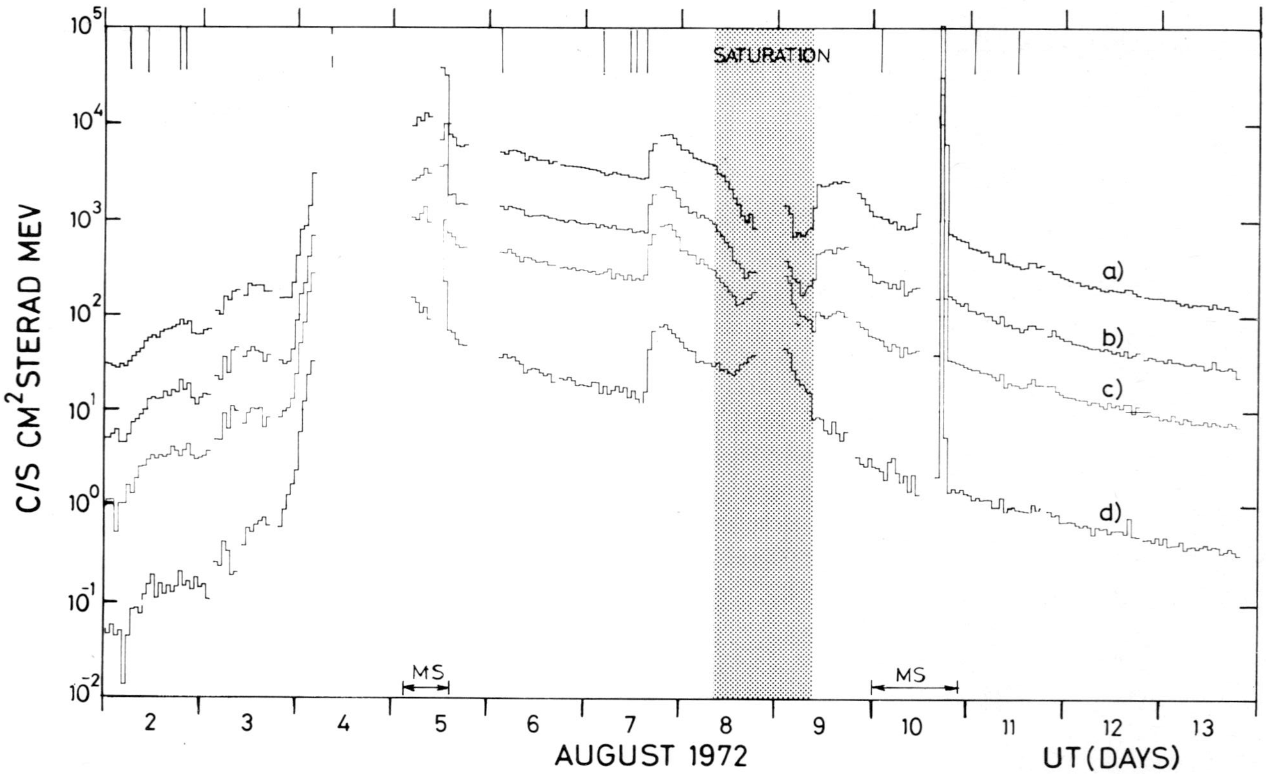

Fig. 4. Electron fluxes averaged over one-hour periods. The energy intervals are (a) 0.45–0.76 MeV, (b) 0.76–1.18 MeV, (c) 1.18–1.77 MeV, (d) 1.77–3.5 MeV. (For the electron measurements there is some uncertainty in geometric factors and energy intervals.)

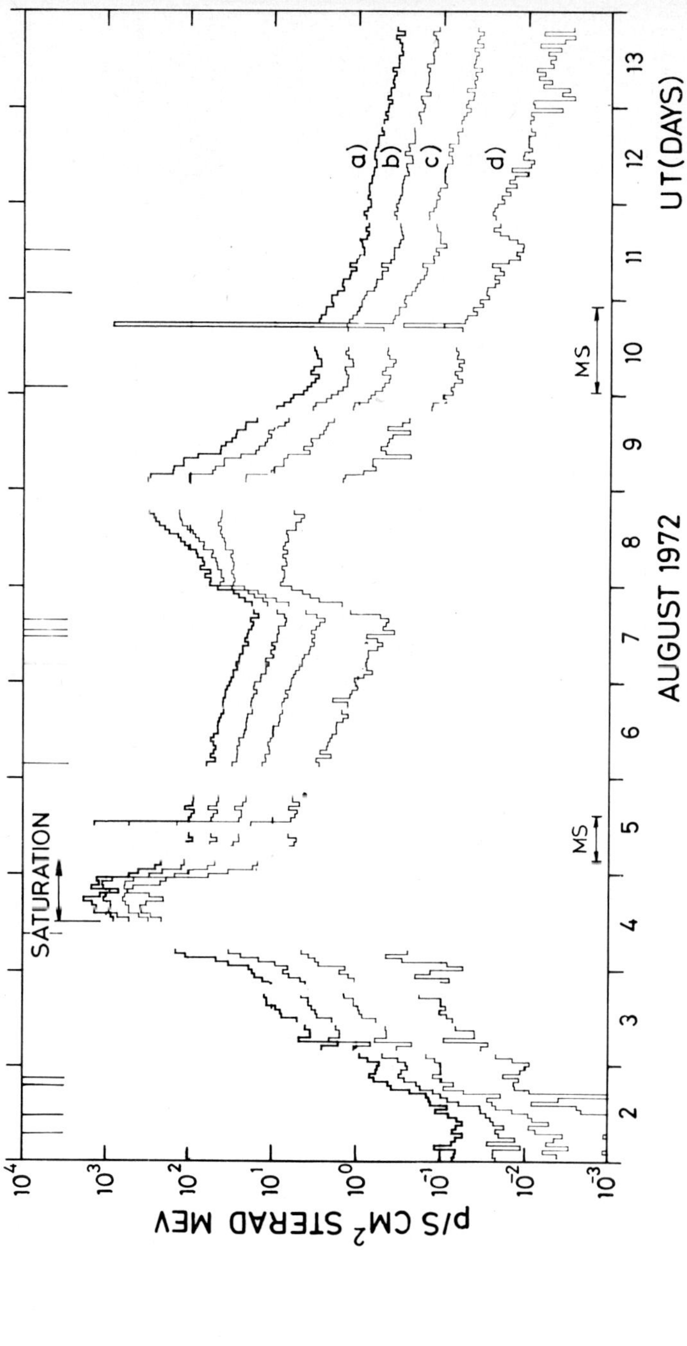

Fig. 5. Proton fluxes averaged over one-hour periods. The energy intervals are (a) 9.3–12.2 MeV, (b) 12.2–17.0 MeV, (c) 17.0–27.7 MeV, (d) 27.7–35.7 MeV.

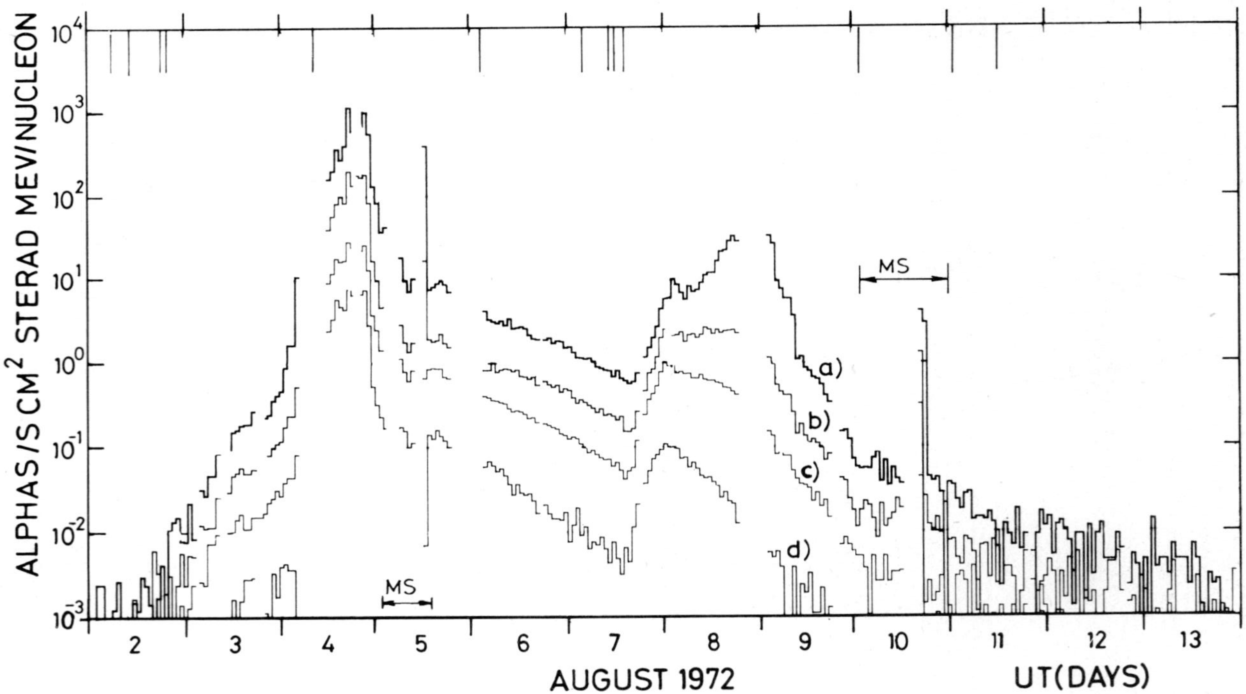

Fig. 6. Alpha particle fluxes averaged over one-hour periods. The energy intervals are (a) 9.3–12.2 MeV nucleon^{-1}, (b) 12.2–27.0 MeV nucleon^{-1}, (c) 17.0–27.7 MeV nucleon^{-1}, (d) 27.7–35.7 MeV nucleon^{-1}.

Figures 4, 5, and 6 show respectively the electron, proton and alpha particle intensities measured between August 2 and August 13 in those energy windows listed in the relevant figure captions. HEOS 2 passes through the radiation belts every 5.2 days and the particle intensity spikes corresponding to magnetospheric crossings can be seen in all of Figures 2 to 6.

During the peak intensity period from 1200 UT till 2340 UT on August 4 the electron measurements have been disregarded and the proton measurements are almost certainly underestimates because of saturation of the electronics. The alpha particle fluxes are probably nearly correct throughout. For the same reasons the electron values between August 8, 1000 UT and August 9, 1000 UT have been decreed to exist in a 'grey' area.

The problem arises most severely for electrons and less significantly for protons for the following reason. During the alpha measurement period the gain of the experiment amplifiers is set at a minimum and the alpha pulses pass through preset discriminator windows which have fixed upper and lower levels. During the proton measurement period an intermediate gain setting allows proton pulses to pass through the same preset windows, and in this case alpha particles are rejected because the pulses they produce are too big, while electrons are not registered because 'electron' pulses do not reach the lower threshold of the window. However, during the electron measurement period when the amplifiers are operating at maximum gain all pulses pass through to the window where proton and alpha pulses are too large to be accepted. Thus in the electron measurement period all produced pulses combine to clog the electronics. It might be possible to make corrections for proton and alpha contamination of the electron measurement if the fluxes of these particles were known. However, the measurements actually recorded are valid for the coincidence mode only (i.e. for protons with energy >9.3 MeV) and the very high flux of lower energy protons producing pulses in the front detector is unknown.

3. Some Features of the Intensity Plots August 2–13

The relatively slow and step wise rise of the particle intensity during August 2 and 3 was typical of the effects to be expected at the Earth arising from a series of east limb solar flares. The peak flux was reached for all particle types late on August 4. The 9.3–12.2 MeV nucleon^{-1} alpha peak was about 10^3 counts s^{-1} cm^{-2} sterad^{-1} (MeV nucleon^{-1})$^{-1}$ compared with the August 2 value of around 10^{-3}. The 9.3–12.2 MeV proton flux increased by at least 2×10^4 from its August 2 value of around 8×10^{-2} counts s^{-1} cm^{-2} sterad^{-1} MeV^{-1} and the 0.45–0.76 MeV electron flux increased during the same period by at least a factor of 10^3. However, the electron and proton upper limits are probably very considerable underestimates because of saturation effects described earlier.

The sharp flux spike seen by many observers in the early hours of August 5 was not recorded by our experiment because of a gap in spacecraft data recovery.

From about mid August 5 (UT times) till mid August 7 there was a linear decay of all particle types. The stability and linearity (on a log scale) of this is particularly interesting in view of the very disturbed solar and interplanetary conditions existing immediately before and after. From Figures 4, 5 and 6 it would appear that alphas disappeared faster than protons which in turn decayed faster than the electron fluxes. This linear decay period will be discussed in more detail below.

On August 7 there occurred what appeared to be a 'classical' prompt onset particle event arising probably from the 3B flare at N 14 W 38. This onset will be described below in the context of the interplanetary conditions as deduced from the evidence of the linear decay period of August 5–7.

The flux build-up of the lower energy protons and alphas continued till the sudden commencement around midnight on August 8, but by this time the higher energy channels had begun to decay. With the sudden commencement all energies of all particle types showed a flux decrease.

4. The Flux Decay Period, August 5–7

Here the time constants of the decay and the anisotropies observed for the different particle species are studied to provide information about the interplanetary conditions. The answers obtained at an early stage from our particle data agree rather well with more recently published directly measured interplanetary data. This would indicate that the connecting theory was probably correct.

4.1. The Anisotropy

Figure 7 shows the integrated proton (9.3–35.7 MeV) and electron (0.45–3.5 MeV) anisotropies calculated for the period 1200 UT on August 6 till 1200 UT on August 7. It can be seen that about 10% more protons come from the direction of the Sun than from other directions, while about 7% more electrons appear to come from the anti-Sun direction. This result appears valid throughout the period and the degree of anisotropy is not a function of the flux value. This is mentioned because of the fear that proton saturation of the electronics might affect the result in such a way that minimum electron flux would be recorded just where the protons showed a maximum.

If the measurement is valid it is of considerable interest because it would indicate that while protons were still being convected out from the Sun past the Earth, an electron particle density gradient (e.g. McCracken *et al.*, 1971) had been set up beyond the Earth giving at the Earth a resultant net movement of electrons from the anti-Sun direction. The interplanetary magnetic field at this time was essentially radial (HEOS-2 magnetometer data provided by P. C. Hedgecock) so that particle anisotropies precisely in the Sun or anti-Sun directions were to be expected. Because of this radial field direction it is not possible to identify the different 'stages' of the event (defined by McCracken *et al.*, 1971) by the anisotropy direction, since vector addition of convective and diffusive (pressure gradient) terms always yields a resultant along the Sun-Earth line.

If it can be assumed that the proton population has reached a state of equilibrium

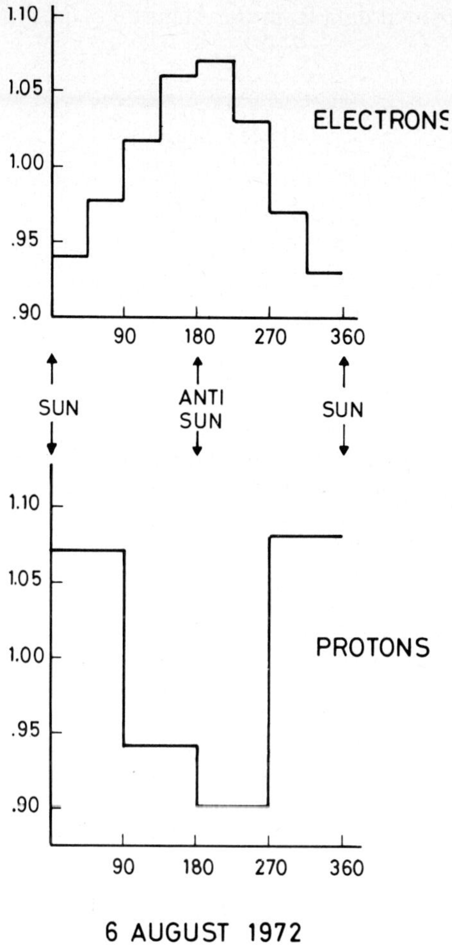

Fig. 7. Electron (0.45–3.5 MeV) and proton (9.3–35.7 MeV) anisotropies summed over the period 1200 UT August 6 till 1200 UT August 7. Electron measurements are available in 8 directions, proton measurements in 4 directions.

where convection is dominant the following analysis may be applied following the method of McCracken *et al.* (1971) and Forman (1970):

$$\delta = (2 + \alpha\gamma)\frac{V}{v}$$

where

V is the solar wind velocity; v is the proton velocity; $\alpha = (T + 2\,M_0C^2)/T + M_0C^2$; T is kinetic energy of the proton; M_0 is the proton rest mass; γ is the exponent in the proton differential spectrum; and δ is the anisotropy in percent.

Then using the experimental data from the August 6–7 decay period

$$\delta \simeq 10$$
$$\gamma \simeq 2.5$$
$$v \simeq 0.4 \times 10^{10} \text{ cm s}^{-1} \text{ (10 MeV proton).}$$

This gives

$$V \simeq 625 \text{ km s}^{-1}.$$

This solar wind speed derived from particle data agrees rather well with the direct measurements (Mihalov *et al.*, 1974).

4.2. The Decay Time Constant

Using still the flux decay period of August 6–7 as described above and the notation of McCracken *et al.* (1971) the following is valid if variations due to heliographic longitude changes are neglected.

$$\ln \frac{U_1}{U_2} = -K(t_1 - t_2)$$

where

U_1, U_2 are the cosmic ray particle densities at times t_1 and t_2 respectively and

$$K = \frac{2\, V_c(2 + \alpha\gamma)}{3r} = \frac{1}{\tau}$$

V_c is the convective velocity given to the cosmic ray particle by the solar wind; and r is the distance from the Sun.

Taking now t_1 as 1200 UT on August 6 and t_2 as 1200 UT on August 7 and using the experimentally measured particle intensity values at these times we obtain Table II.

TABLE II

Events on August 6–7

	γ	V_c
Protons	2.5	340 km s^{-1}
Alphas	4.0	350 km s^{-1}
Electrons	3.0	270 km s^{-1}

As pointed out by McCracken *et al.* (1971) for other events the velocity imparted by the solar wind is around one half of its own velocity. The proton and alpha V_c values are essentially the same. Although all the experimental data used are somewhat approximate the V_c for electrons having a lower value may be significant and would apparently be consistent with the concept of some electrons moving back toward the Sun at this time.

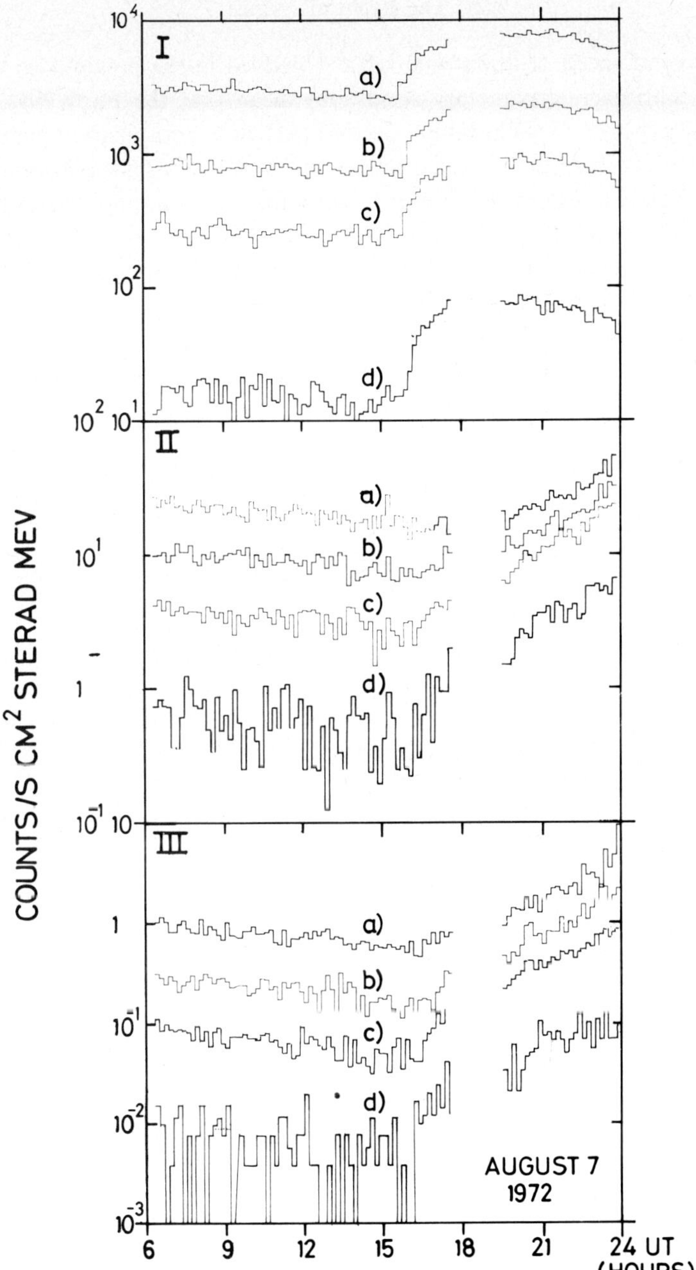

Fig. 8. Onset of the event of August 7 (10-min averages). I, electrons; II, protons; III, alpha particles. The (a), (b), (c) and (d) labels are as in Figures 4, 5 and 6.

5. The Event of August 7

The solar wind speed of about 600 km s^{-1} derived in the previous section would under equilibrium interplanetary conditions mean that the Earth was connected magnetically to a 38° W solar longitude. The particle event starting around 1530 UT on August 7 was therefore expected to be a 'prompt' event since it appeared to arise from a solar flare at N 14 W 38. Figure 8 shows the event onset for electrons, protons and alpha particles.

The MeV electrons arrived at 1540 UT, the 10 MeV $nucleon^{-1}$ alphas at about 1710 UT and the 10 MeV protons later than 1740 UT. It is difficult to reach conclusions about different transit times because of the uncertainty in selecting the event zero point. In an attempt to do this the response of the Geiger counter when it

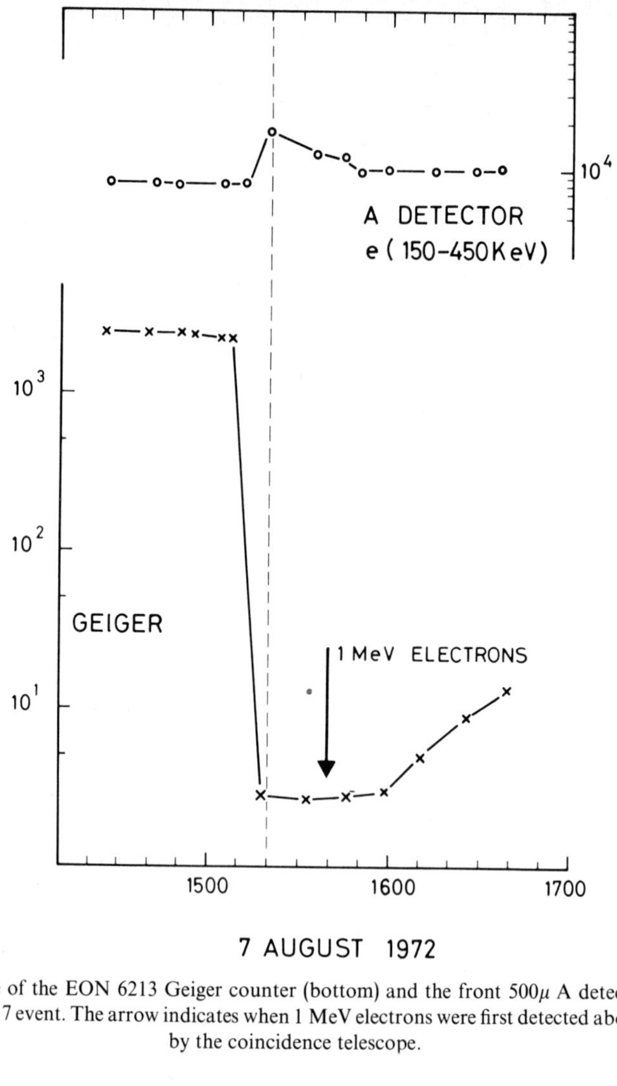

Fig. 9. Response of the EON 6213 Geiger counter (bottom) and the front 500μ A detector (top) at the start of the August 7 event. The arrow indicates when 1 MeV electrons were first detected above background by the coincidence telescope.

looked at the Sun was examined (Figure 9). The dip in Geiger count rate arises because of saturation and in this case the saturation was probably due to solar X-rays in the wavelength range 2–12 Å (Van Allen, 1967). The arrival of 1 MeV electrons is indicated by the arrow in Figure 9.

An interesting feature on which the authors have been unable to find any comment in the literature is the response when it looked toward the Sun of the front solid state 'A' detector shown in the upper part of Figure 9. This detector is a 500 μ surface barrier silicon detector with its light insensitive surface uppermost in the telescope. This surface is covered by 120 $\mu g\ cm^{-2}$ of aluminium and is further shielded by 10 μ of Kapton foil. On the basis of range vs energy considerations the detector should respond mainly to electrons between about 150 and 450 keV. Thus at first sight we appeared to have an exciting example of very prompt field aligned low energy electrons (e.g. Lin, 1970; Frank and Gurnett, 1972) and this explanation cannot be ruled out. However, the time coincidence with the Geiger response and the X-ray data in the literature (Van Beek *et al.*, 1974) lead us to believe that sufficient of the X-rays generated in this very intense event were converted in the front detector to come above our detection threshold. This emphasizes the value of considering more than one detector response in studying any event (Shea and Smart, 1974).

Acknowledgements

The assistance is gratefully acknowledged of Drs D. Köhn and B. G. Taylor and of engineers T. Iversen, G. Korfmann and R. Roger.

References

Frank, L. A. and Gurnett, D. A.: 1972, *Solar Phys.* **27**, 446.
Forman, M. A.: 1970, *J. Geophys. Res.* **75**, 3147.
Köhn, D., Page, D. E., Taylor, B. G., and Wenzel, K.-P.: 1972, *ESRO/ELDO Sci. Techn. Rev.* **4**, 19.
Lin, R. P.: 1970, *J. Geophys. Res.* **75**, 2583.
Mihalov, J. D., Colburn, D. S., Collard, H. R., Smith, B. F., Sonett, C. P., and Wolfe, J. H.: 1974, this volume, p. 545.
McCracken, K. G., Rao, U. R., Bukata, R. P., and Keath, E. P.: 1971, *Solar Phys.* **18**, 100.
Shea, M. A. and Smart, D. F.: 1974, this volume, p. 381.
Van Allen, J. A.: 1967, *J. Geophys. Res.* **72**, 5903.
Van Beek, H. F., De Feiter, L. D., and De Jager, C.: 1974, this volume, p. 533.
World Data Centre A, Collected data reports on August 1972 solar-terrestrial events (ed. by Helen E. Coffey), Report UAG-28, Boulder, 1973.

SOLAR PARTICLE OBSERVATIONS DURING THE AUGUST 1972 EVENT

L. J. LANZEROTTI and C. G. MACLENNAN
Bell Laboratories, Murray Hill, N.J. 07974, U.S.A.

Abstract. Observations of solar electron, proton, and alpha particles made with the Explorer-41 satellite at 1 AU during the period of intense solar activity in August 1972 are presented. The intensities of low energy particles are the largest measured at the Earth during the 20th solar cycle; the peak fluxes of 500 keV protons are more than a factor of 10 times the intensity of those measured during the event of April 10, 1969. Some comments are also made on the intensities of interplanetary particles measured at 1 AU during a two-week period prior to the August interval as well as on the intensities of the ground-level events during the August period.

1. Introduction

This paper represents a report of the observations of low energy solar cosmic ray protons, alpha particles, and electrons made at 1 AU during the period of large solar activity in McMath region 11976 in August 1972. The solar particle observations were made with a semiconductor detector telescope instrument (Lanzerotti *et al.*, 1969) flown by Bell Laboratories on the Explorer-41 satellite.

Prior to the beginning of the 'August events' on August 2, the interplanetary space in the vicinity of the Earth had been populated for approximately two weeks with significant fluxes of low energy solar particles. As shown in Figure 1, the interplanetary particle fluxes sharply increased in intensity beginning on July 19. Although they fluctuated in intensity to some extent, the fluxes of observed particles, particularly of electrons and protons, remained appreciably above background intensity through August 1 (cf. Figure 1). It is difficult to determine whether or not the sources of interplanetary particle fluxes observed prior to August 2 resulted only from active regions on the visible solar disk. It is quite likely that some portion of the fluxes, particularly those measured beginning on July 24–25, may be a resultant of flares (on the invisible disk) from the future McMath region 11976. Such a wide spread in solar azimuth of solar-flare particles would not be unusual (e.g., McCracken *et al.*, 1971).

2. Ground Level Events

Aspects of the changes in galactic and solar particle fluxes as recorded on the earth by ground-level detectors during the passage of McMath region 11976 have been reported by Pomerantz and Duggal (1973), Mathews and Lanzerotti (1973), and Swinson (1973). Measurements from August 1 through August 13 made with super neutron monitors at Calgary (114.1° W, 51.1° N; altitude 1128 m; vertical cutoff rigidity $P_T = 1.09$ GV) and Sulphur Mountain (115.6° W, 51.2° N; altitude 2283 m; $P_T = 1.14$ GV) were reported by Mathews and Lanzerotti (1973) and are shown in Figure 2. Plotted at the bottom of Figure 2 are values of the solar wind speed recorded

D. E. Page (ed.), Correlated Interplanetary and Magnetospheric Observations, 587–596. All Rights Reserved.

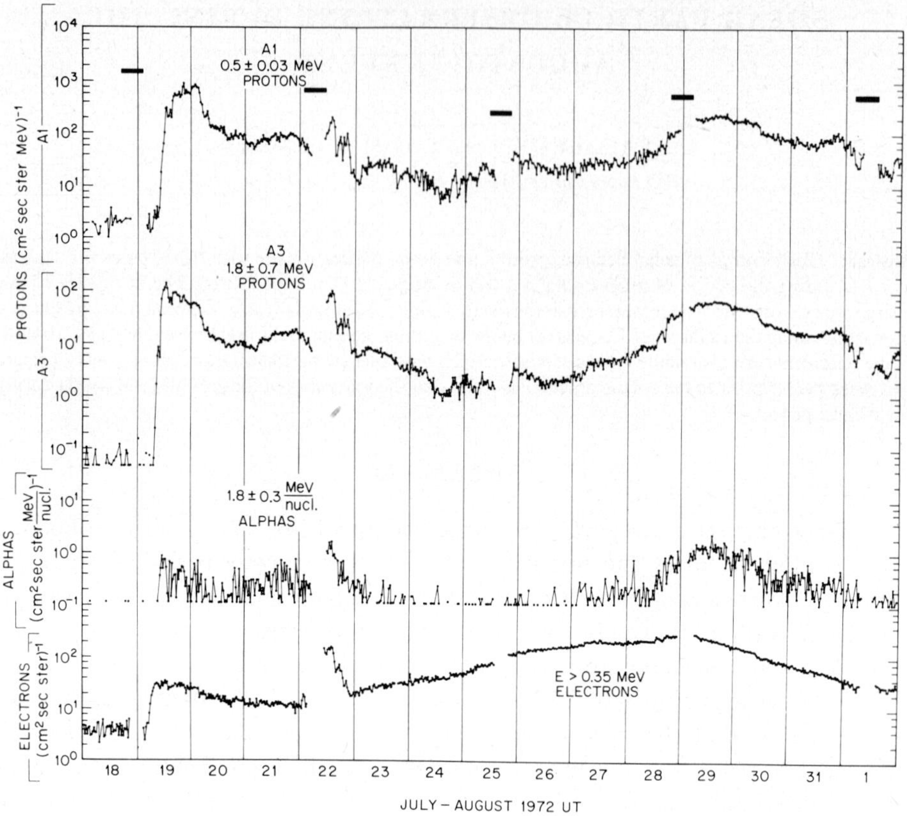

Fig. 1. Interplanetary proton, alpha particle, and electron fluxes for a two-week period prior to the August 1972 solar active period.

at various times during the August period by an instrument on Pioneer 9 and reported in *Solar-Geophysical Data* (1972). During the measurement period Pioneer 9 was ~0.78 AU from the Sun and ~52° east of the Earth-Sun line.

The first flare-produced sudden commencement, recorded on the earth at ~0220 UT, was accompanied by a Forbush decrease measured at both neutron monitor stations shown in Figure 2. The ground level event (GLE) on August 4 occurred several hours prior to the observation of the second sudden commencement at the Earth at ~2054 UT. It is interesting to note that the 2054 UT sudden commencement at the Earth was observed at about the same time as an interplanetary shock at Pioneer 9. The solar wind velocity measured at Pioneer 9 at this time was ~1100 km s^{-1}.

Following the second sudden commencement on August 4, one of the largest Forbush decreases in recorded cosmic ray history was measured at the Earth (Pomerantz and Duggal, 1973). The beginning of the recovery of this Forbush decrease was accompanied by a neutron monitor enhancement during the period ~0300–0500 UT on August 5. The GLE on August 7 occurred during the main recovery phase of the Forbush decrease.

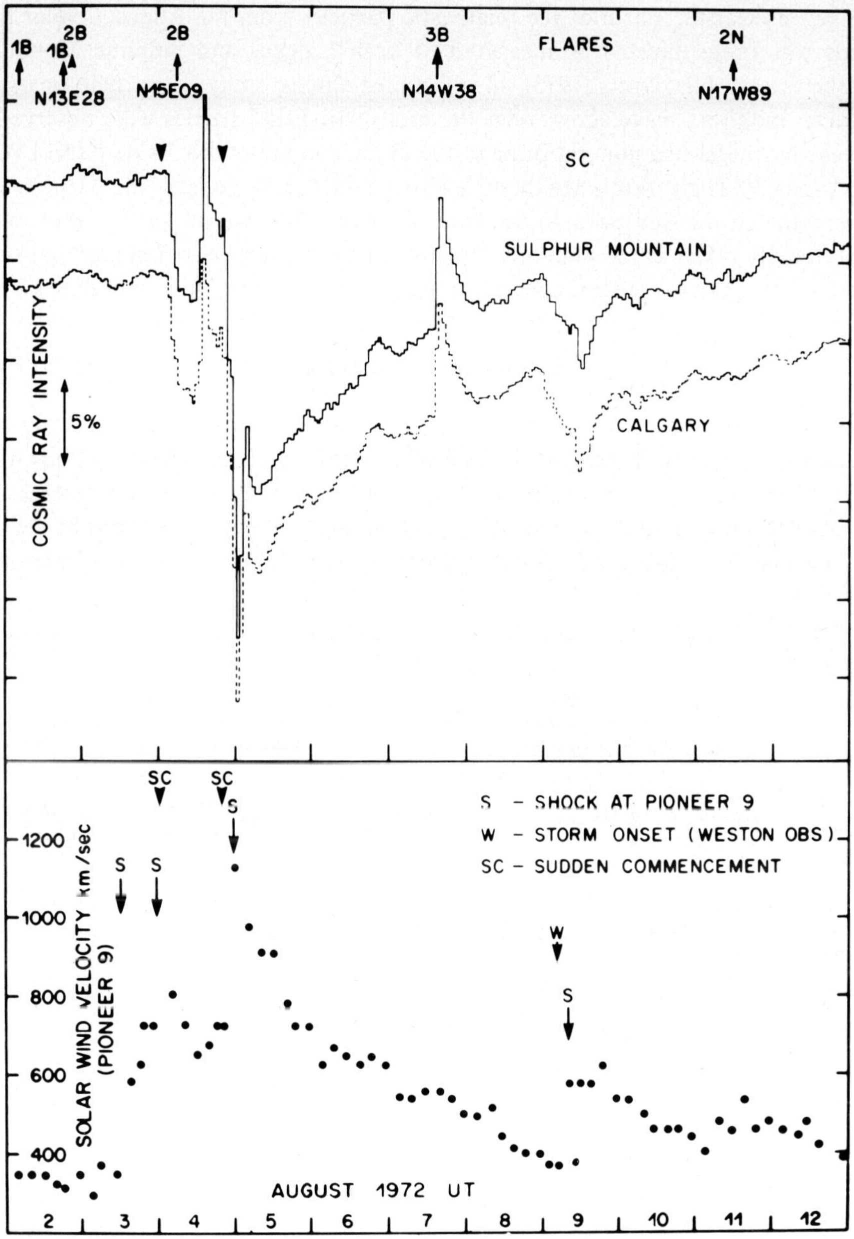

Fig. 2. Super neutron monitor measurements from Calgary and Sulphur Mountain during the August events. Plotted at the bottom are solar wind velocity measurements from Pioneer 9. (Figure from Mathews and Lanzerotti, 1973.)

The onset at the Earth of the relativistic particles from the August 7 solar flare event was quite unusual as measured at both Calgary and Sulphur Mountain. Mathews and Lanzerotti (1973), in a detailed timing study of particle onsets in neutron monitors, have shown that the first relativistic particles were observed at the earth at least five minutes prior to the Hα maximum of the 3B flare (N14 W38) on August 7. Their results are the first direct evidence that relativistic particles can be present on the Sun prior to the Hα maximum (the time of which is very often traditionally taken as the time of release of the initial flare particles) and that these particles are perhaps present coincident with the time of the white-light flare.

3. Satellite Measurements

The half-hourly averages of relativistic electrons measured on Explorer 41 during 2–11 August are plotted in Figure 3. The black bars above the data indicate periods when the satellite was at an altitude of $<10\ R_E$. The electron fluxes began to increase at ~0900 UT on August 2 and reached a peak intensity about one hour after the first sudden commencement on August 4. A decrease in the fluxes was followed by another

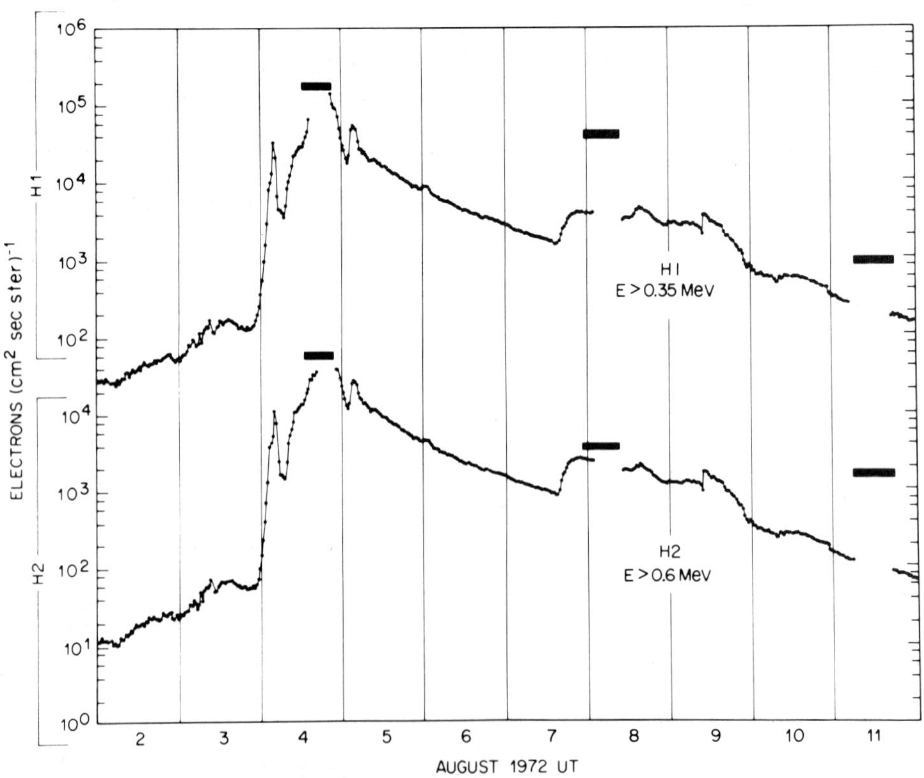

Fig. 3. Solar electron fluxes measured on Explorer 41 during the August events. The solid bars indicate passage of the spacecraft through the magnetosphere.

increase which apparently peaked during the spacecraft perigee pass through the magnetosphere. An interplanetary electron enhancement was measured from ~0300–0500 UT on August 5. The electron fluxes from the August 7 flare exhibited the typical (for a western-hemisphere flare) rapid particle onset to a peak intensity. On August 11, four days after the last large flare, the electron fluxes were still about a factor of twenty above background intensity.

Lanzerotti *et al.* (1973) have used the time to maximum flux of the electrons measured in the August 7 event to estimate the scattering mean free path for the electrons within 1 AU. They obtained a value of $\lambda_r \sim 0.013$ to ~ 0.016 AU for electrons of energy ~1 MeV to ~300 keV, respectively. These are among the lower values of λ_r reported for electrons from a number of impulsive solar particle events studied by Lanzerotti *et al.* (1973). The λ_r-values from the electron observations appear contrary to the expectations of rapid particle transit (large λ_r) from the flare site when the results of the rapid arrival at the Earth of the particles constituting the GLE are considered (see previous section). This consideration is currently being investigated in more detail to understand interplanetary particle transport during the August period.

Low energy solar proton fluxes (half-hour averages) measured in three differential energy channels are plotted in Figure 4. The low energy proton flux profiles on 2–4 August are similar to the electron flux profiles until the time period following the first sudden commencement. After ~0700 UT the proton fluxes remain depressed at least until the satellite enters the magnetosphere. Upon exiting the magnetosphere, at ~2200 UT, the interplanetary proton fluxes are at a greatly enhanced level.

A substantial enhancement is measured in the proton fluxes from ~0300 to ~0500 UT on August 5. The low energy particles from the August 7 flare are not observed until about the time that the satellite exits the magnetosphere on August 8. The time-intensity behavior of the proton fluxes in the interval around ~0000 UT August 9 suggests that significant trapping of low energy protons occurs behind the interplanetary shock that originated from the August 7 flare. In fact, the lowest energy protons (0.5 ± 0.03 MeV) are only observed two to three hours before the shock arrival.

Alpha particle fluxes (half-hour averages) in three differential energy channels are plotted in Figure 5 for the interval August 2–11. The alpha particle time-intensity profiles closely resemble the proton flux profiles of Figure 4. The fluxes in the highest energy channel appear to increase in intensity prior to the spacecraft entrance into the magnetosphere on August 4. The large flux enhancement on August 5 is observed in the alpha fluxes as it was in the proton and electron fluxes (cf. Figures 3 and 4). The alpha particle flux profiles around 0000 UT August 9 resemble the proton flux profiles at that time (cf. Figure 4).

Half-hour averages of the alpha particle-to-proton flux ratios for the August 2–11 period are plotted in Figure 6. Plotted at the bottom are proton fluxes (3.4 ± 0.9 MeV) from Figure 3. The two alpha-to-proton ratios in the upper portion of the figure are

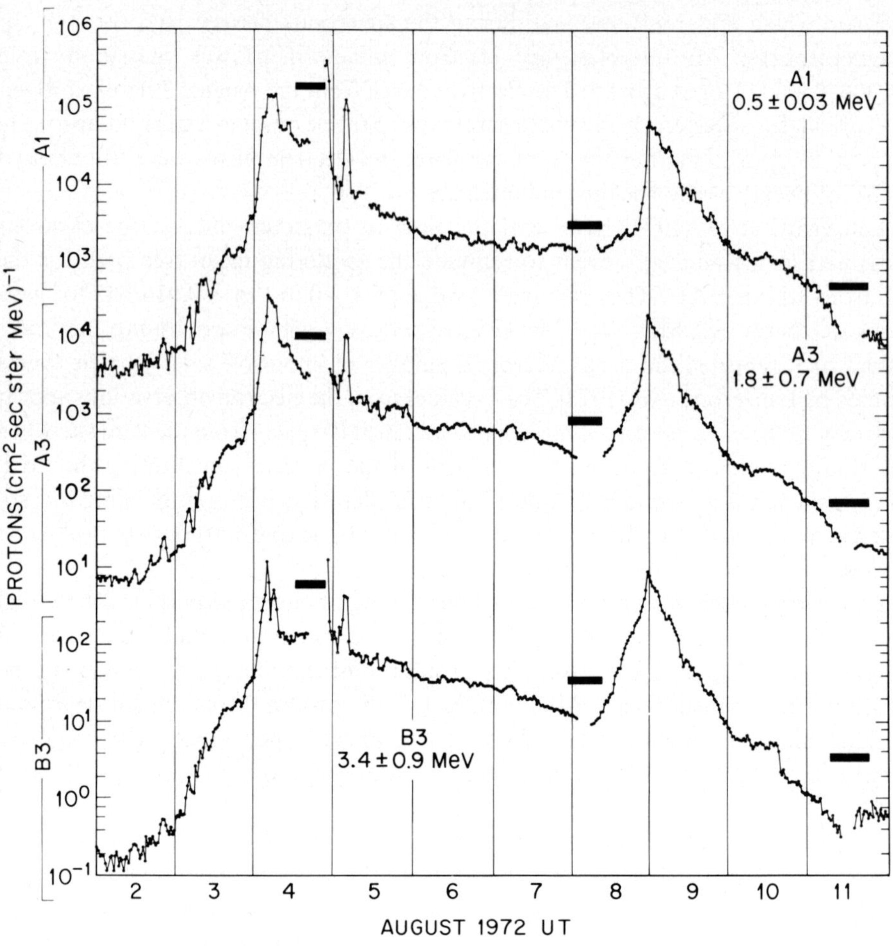

Fig. 4. Low-energy solar proton fluxes measured on Explorer 41 during the August events. The solid bars indicate passage of the spacecraft through the magnetosphere.

plotted for particles of equal velocity v (equal energy per nucleon); the third set of ratios are for particles of equal energy per charge. A large increase in the latter ratio is observed at the time of the flux increase on April 4. Increases in the alpha-to-proton ratios in both representations are observed during the flux enhancement on August 5; i.e., relatively more alpha particles than protons are observed during this period. Large changes in the alpha-to-proton ratios for equal energy per nucleon particles are observed at ~0000 UT August 9. At this time no changes are observed in the particle fluxes compared as to equal energy per charge, a typical behavior of the ratios across an interplanetary shock disturbance (e.g., Lanzerotti and Robbins, 1969; Lanzerotti, 1973, 1974; Scholer *et al.*, 1972) which has not as yet been satisfactorily explained (e.g., Lanzerotti, 1974).

Details, in fine-time resolution, of the unusual flux enhancement on August 5 are

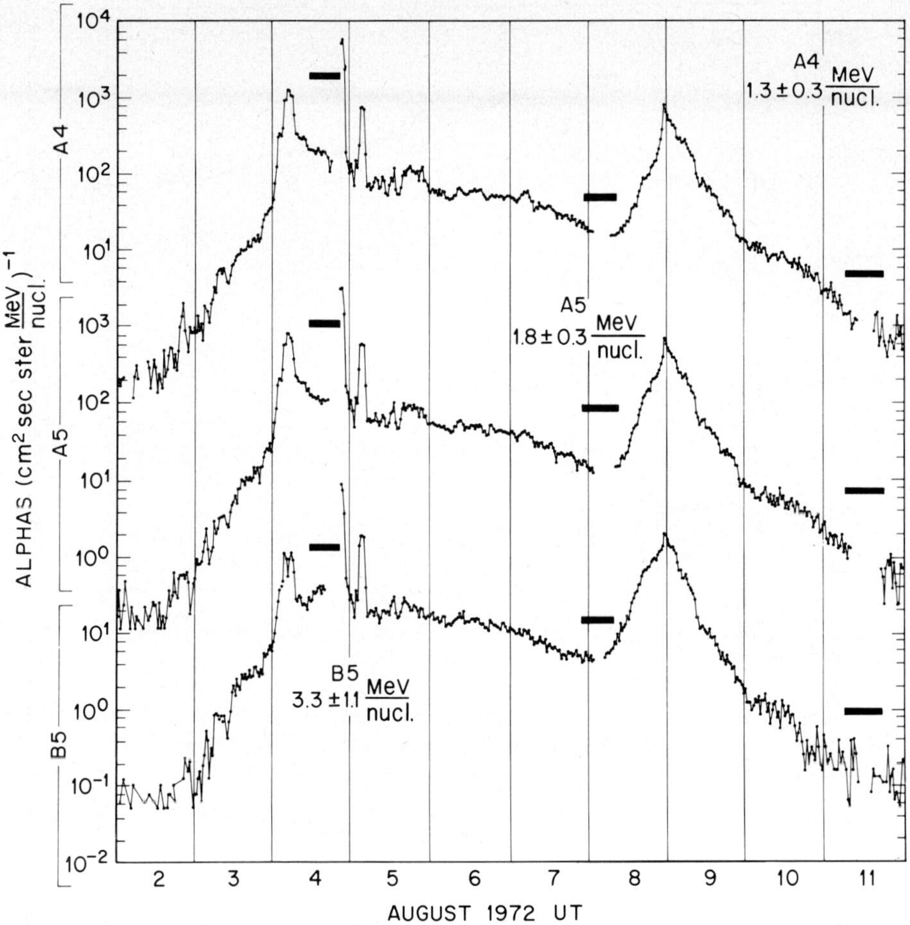

Fig. 5. Low-energy solar alpha particles measured on Explorer 41 during the August events. The solid bars indicate passage of the spacecraft through the magnetosphere.

shown in Figure 7. All individual counting rate values that were measured (consisting of 9.28 s spin-averaged rates) have been plotted for the interval 0200–0700 UT. The lowest-energy proton channel was sampled approximately once per minute; the electron channel was sampled approximately three times per minute, and the remainder two channels (protons and alphas) were sampled approximately once per five minutes. An unfortunate telemetry data gap prevented a precise determination of the event onset; nevertheless, it is possible to determine that the electron fluxes increased approximately five minutes before the increase in the nuclei fluxes. There is some evidence that the alpha particles increased before the protons, suggesting a weak energy dependence to the event onset for the nuclei. The fluxes during the two hours duration of the event were quite constant; the flux decreases at ~0500 UT were

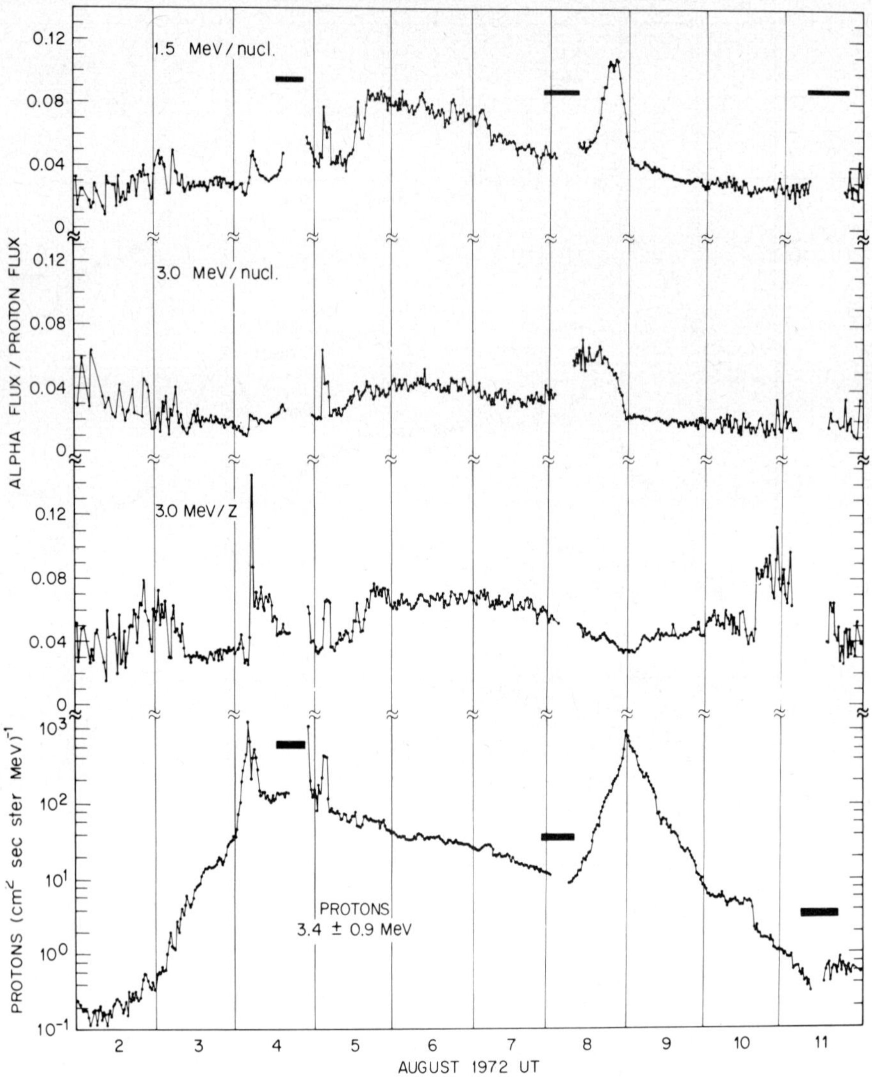

Fig. 6. Half-hourly averages of the alpha-to-proton ratios determined from measured particle fluxes during the August events. Plotted at the bottom is the time-intensity profile of 3.4 MeV solar protons.

approximately energy independent. The electron flux level after the event was enhanced approximately a factor of two over the level measured prior to the event.

4. Discussion

The low energy particle fluxes from the series of August 1972 flare events, as observed at 1 AU at the Earth, were determined by both the locations of the flares on the solar disk as well as by the interplanetary conditions prevailing during the entire

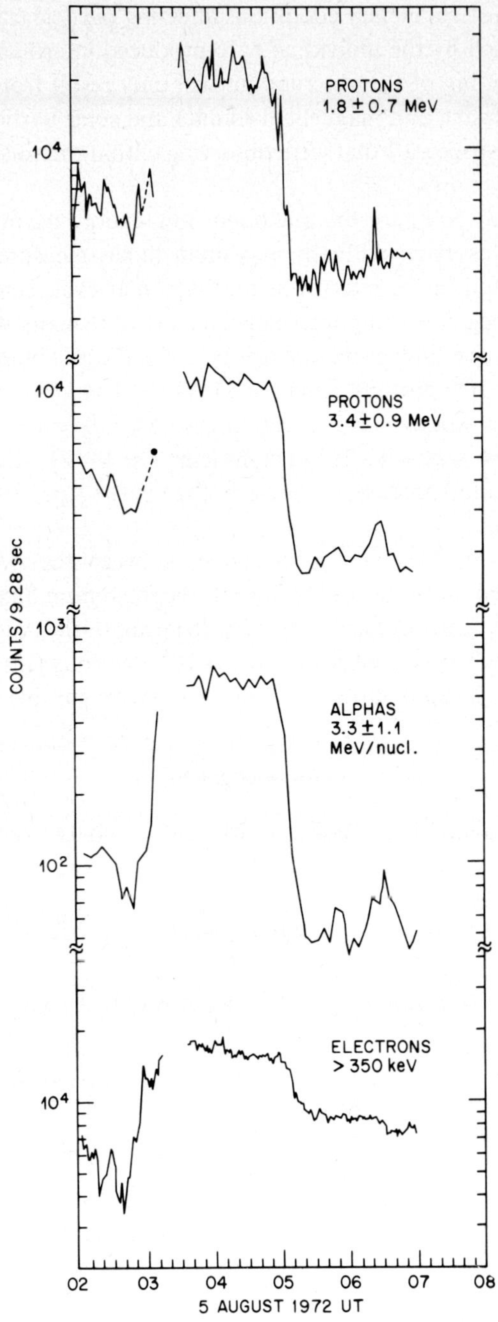

Fig. 7. Detailed time-intensity plot of particle fluxes during the unusual particle intensity enhancement on August 5.

active period. The interplanetary conditions affecting particle transport appeared to be largely determined by the individual flare-produced interplanetary shock waves. Ultimate understanding of the observations will only result from a future synthesis of interplanetary plasma and magnetic field data and solar particle data obtained at the locations of all spacecraft that were observing within the solar system during the time of the event.

It is interesting to compare the maximum low energy particle flux magnitudes measured during this event with the maximum fluxes measured during the next-largest event observed in the interval of the 20th solar cycle since May 1967 (when Bell Laboratories began making interplanetary measurements with instruments on Explorer 34). In the east limb event of April 10, 1969, the maximum flux of ~ 600 keV protons was $\sim 3 \times 10^4$ protons $(\mathrm{cm}^2\ \mathrm{s\ sr\ MeV})^{-1}$. The peak flux of ~ 500 keV protons during the August 1972 event, measured following the magnetosphere passage on August 4, was $\sim 4 \times 10^5$ protons $(\mathrm{cm}^2\ \mathrm{s\ sr\ MeV})^{-1}$; this is a flux greater than a factor of ten more intense, at the peak, than in the April 1969 event (cf. Figure 2, Lanzerotti, 1974).

The comparison of electron flux intensities between the two events is not so dramatic. At the peak of the April 1969 event, the maximum flux of >300 keV electrons was $\sim 8 \times 10^4$ electrons $(\mathrm{cm}^2\ \mathrm{s\ sr})^{-1}$; at the peak of the August 1972 event, the maximum flux of >350 keV electrons was $\sim 10^5$ electrons $(\mathrm{cm}^2\ \mathrm{s\ sr})^{-1}$. Thus, the peak electron fluxes in both events are more nearly comparable than were the peak proton fluxes.

Acknowledgments

We thank J. B. Reagan, J. B. Blake, T. Mathews, and D. Venkatesan for useful discussions about various aspects of the August 1972 event.

References

Lanzerotti, L. J.: 1974, this volume, p. 345.
Lanzerotti, L. J.: 1973, in D. Venkatesan (ed.), *Proc. Solar-Terrestrial Relations Conf.*, Univ. of Calgary, p. 283.
Lanzerotti, L. J. and Robbins, M. F.: 1969, *Solar Phys.* **10**, 212.
Lanzerotti, L. J., Lie, H. P., and Miller, G. L.: 1969, *IEEE Trans. Nucl. Sci.* **NS-16**, 343.
Lanzerotti, L. J., Venkatesan, D., and Wibberenz, G.: 1973, *J. Geophys. Res.* **78**, 7986.
Mathews, T. and Lanzerotti, L. J.: 1973, *Nature* **241**, 335.
McCracken, K. G., Rao, U. R., Bukata, R. P., and Keath, E. P.: 1971, *Solar Phys.* **18**, 100.
Pomerantz, M. A. and Duggal, S. P.: 1973, *Nature* **241**, 331.
Scholer, M., Hovestadt, D., and Hausler, B.: 1972, *Solar Phys.* **24**, 475.
Solar Geophysical Data, 1972, No. 337, Part 1 (U.S. Dept. of Commerce, Boulder, Colorado, September.)
Swinson, D. B.: 1973, *J. Geophys. Res.* **78**, 1707.

INTERPLANETARY PARTICLE FLUXES OBSERVED BY OV5–6 SATELLITE

G. KENNETH YATES and LUDWIG KATZ
Air Force Cambridge Research Laboratories, Bedford, Mass. 01730, U.S.A.

and

BACH SELLERS and FREDERICK A. HANSER
Panametrics, Inc., Waltham, Mass. 02154, U.S.A.

Abstract. The OV5-6 (1969–046B) satellite launched 23 May 1969 and with 2.9 R_E perigee, 19.3 R_E apogee and 30.6° inclination, has observed numerous solar events since its launch. The proton-alpha telescope aboard this satellite is capable of measuring protons in 8 energy bins between 1 and 100 MeV and alphas in 3 energy bins between 20–100 MeV. Time histories of fluxes of various channels and energy spectra are presented for the August 2–15, 1972 series of events. This multiple event included a more than 10^4 factor increase in the proton and alpha particle fluxes.

Satellite OV5-6 (International designation 1969-046B) measures solar fluxes of protons and alpha particles. In August 1972, the perigee was 2.9 R_E and the apogee 19.3 R_E. The orbital period was nearly 52 h (3117.6 min). Apogee was about 40° east of the Sun. For the period reported here, the first apogee occurred at 1659 UT on August 2.

The proton-alpha particle detector on OV5-6 consists of two totally depleted silicon surface barrier detectors in a telescope configuration. The detectors each have a 2 cm^2 area and are separated by 2.54 cm. The outer one is 200 μ thick and the inner one 750 μ. The outer detector is shielded from light by 0.6 mil of aluminum foil. In the coincidence mode of operation, the telescope has a geometric factor of 0.52 cm^2 sr, with a detection cone of 30° half angle. The average angle of detection is 17°. A coincidence is set by an energy loss window on the first detector and a threshold on the second detector. The resulting coincidences detect protons and alpha particles (principally) in the following ranges: protons 5.3 to 8 MeV, 8 to 17 MeV, 17 to 40 MeV, and 40 to 100 MeV; alpha-particles 20 to 32 MeV, 32 to 68 MeV, and 68 to 100 MeV. The telescope cycles sequentially through these seven ranges, each range being counted for ten seconds and then read out. The complete cycle is completed in approximately two minutes. The telescope looks in the equatorial plane of the satellite. The satellite spin axis is stable and is directed toward 0^h and 40^m RA and 32° declination in celestial coordinates. In August 1972 the Sun appeared thirty-five degrees below (−35°) the satellite's equator. Since the spin period at this time was 4.5 s, the telescope accumulated counts in a given particle energy range for approximately two satellite revolutions.

The telescope also has a calibration mode which allows some lower energy proton data to be obtained during high flux periods from a single detector. Each coincidence count period is followed by a two-second count period from one of the two detectors. The differential energy loss in the 200 μ detector and the integral energy loss in the

D. E. Page (ed.), Correlated Interplanetary and Magnetospheric Observations, 597–602. All Rights Reserved.

750 μ detector are measured. The detectors alternate in readout. Thus, the singles rate cycle lasts about four minutes. The instrument aperture revolves about 130° during the two-second count time. For this mode the 200-μ detector has a geometric factor of 3.2 cm^2 sr and provides information on protons in the energy ranges 1.25 to 1.41 MeV, 141 to 1.79 MeV, 1.79 to 2.71 MeV, 2.7 to 4.8 MeV, and 4.8 to 5.6 MeV. During the peak fluxes on August 4, the normal mode of operation of the telescope gave unreliable data for a period of several hours. However, the inner 750 μ detector did give reliable counts during this period, since it has a smaller geometric factor than the front detector and is better shielded.

Figure 1 shows 30-min averages of the proton fluxes as a function of time. Most of the gaps are due to the omission of data while the satellite was in the magnetosphere at perigee. However, it was at the peak between 2100 and 2200 h on August 4 that the high particle flux saturated the coincidence (telescope) mode. This gap was bridged by singles mode results from the 750-μ detector. The singles rates before and after the gap can be compared with the lower but overlapping energy interval measured in the coincidence mode. A smaller peak occurs 0300 to 0550 h on August 5. These peaks are probably associated with the passage of interplanetary shock waves. The

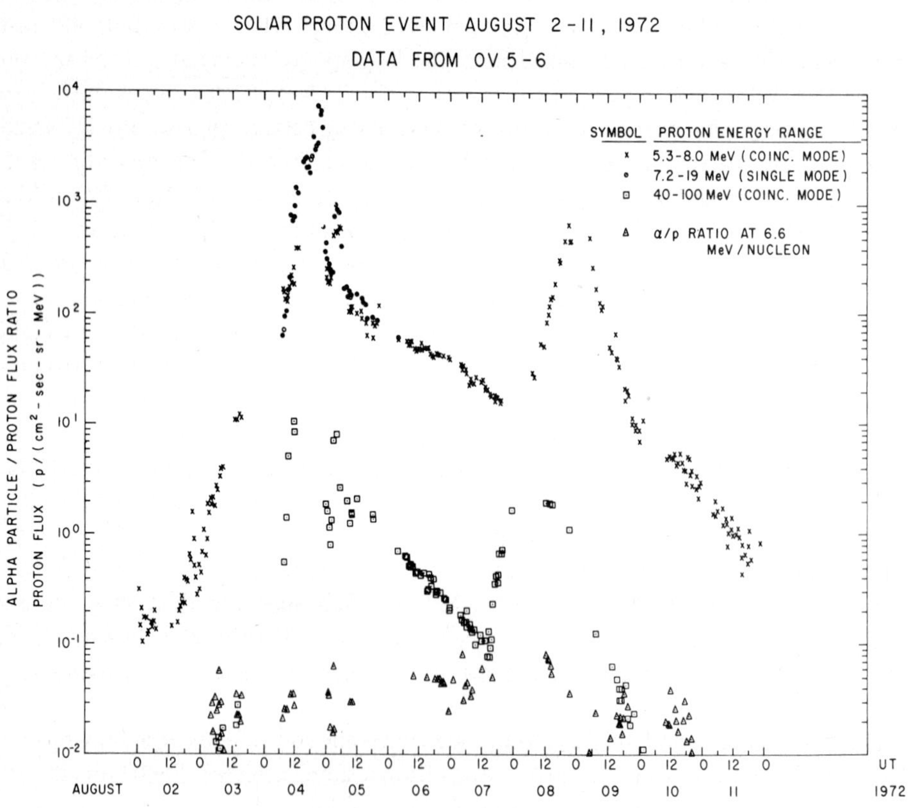

Fig. 1. Time history of proton fluxes and alpha particle/proton flux ratios.

decay time in the August 5 to 7 period is about 22 h. The flux peaks associated with the August 7 event can be seen. The subsequent decay time is about 15 h for the lower energy protons. The alpha particle/proton flux ratios at 6.6 MeV nucleon^{-1} are also shown.

The remaining figures present various proton energy spectra observed throughout the period. The singles (calibration) mode proton data were averaged over the satellite spin to make them comparable to the coincidence mode data. The lower energy data have been corrected for the contribution by high energy protons and for background from a weak alpha (Am^{241}) source normally used to check the gain of the detector electronics. Where they are larger than the symbol size, one sigma statistical uncertainties are shown by vertical bars.

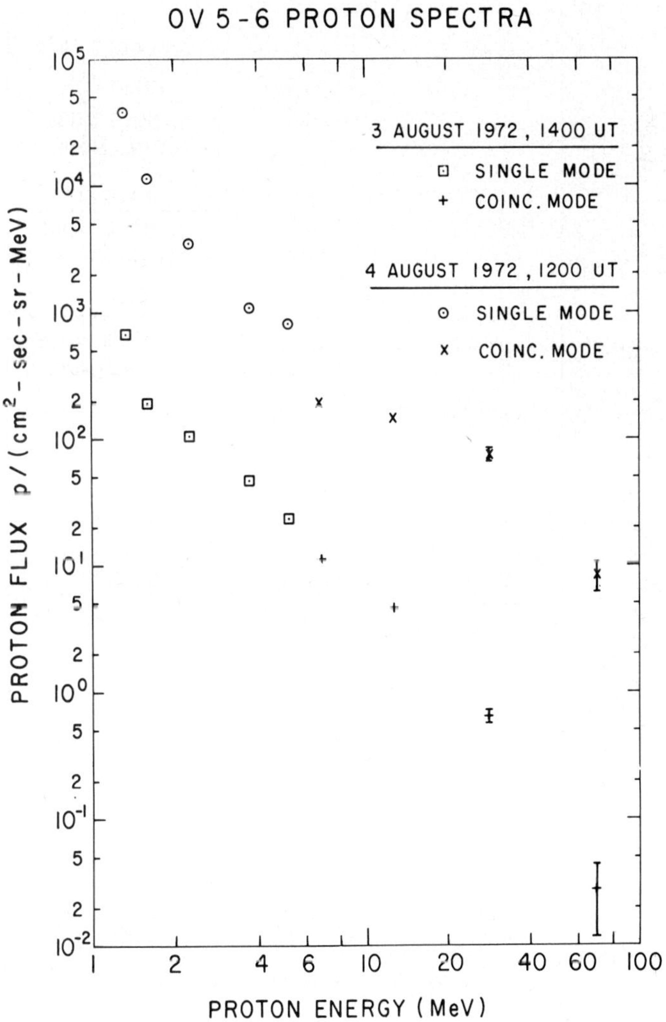

Fig. 2. Proton energy spectra during the early flux increase.

Figure 2 shows two energy spectra preceding the August 4 peak. The complicated nature of this series of events is illustrated by the fact that the increase of proton flux from the August 3 to 4 was accompanied by hardening of the energy spectrum. This indicates solar particle injection subsequent to the start of the flux increase. During the peak fluxes, saturation prevented the measurement of energy spectra. In Figure 3, the energy spectra before, during and after the secondary peak of early August 5 are shown. Lower energy protons predominated the increase and as is to be expected the net result is a softer spectrum. Figure 4 shows the continued decay of the particle flux and softening of the spectra through the next two days.

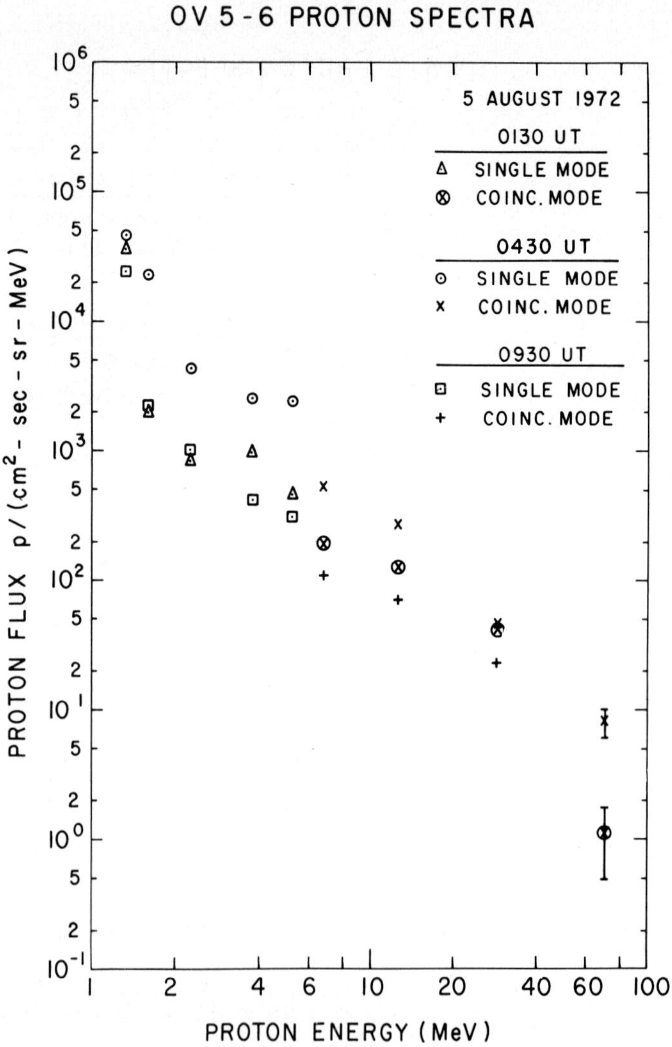

Fig. 3. Proton energy spectra before, during and after the secondary flux peak of August 5.

The spectra during the August 8 peak and subsequent flux decay are illustrated in Figure 5. The earlier spectrum was measured when the 5.3–8.0 MeV proton flux was at its maximum. The flattening of this spectrum below 6 MeV is attributed to the

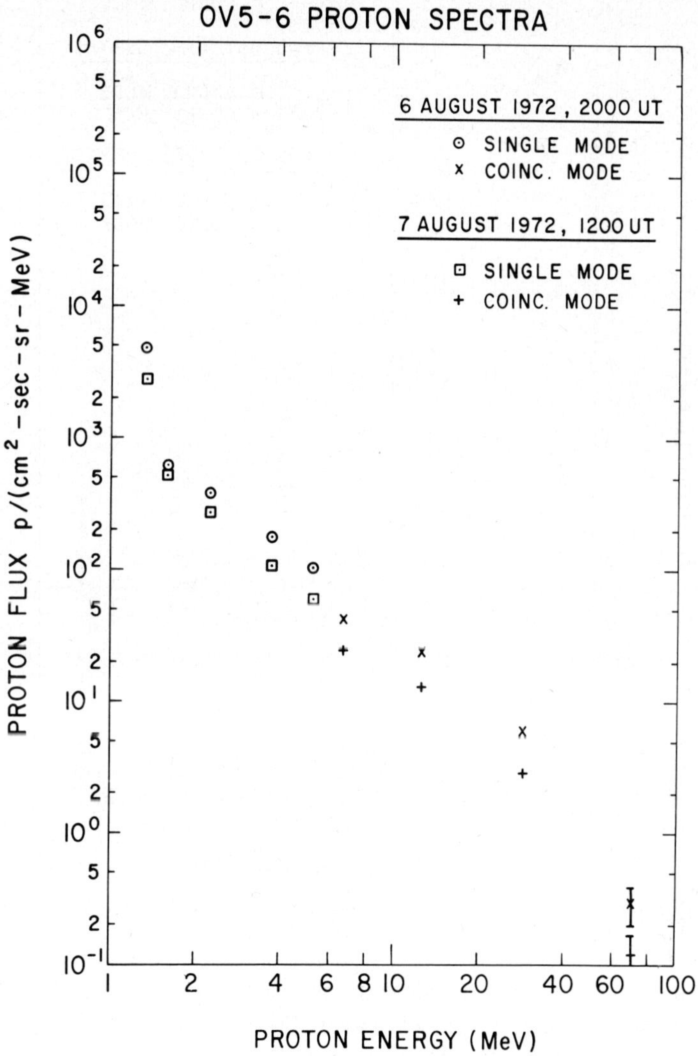

Fig. 4. Proton energy spectra during the flux decay of August 6–7.

fact that the lower energy fluxes were still increasing with time. Although not shown here, this was indeed observed. After the peak of these lower energy fluxes, the final decay was accompanied by a slight softening of the spectra.

Of the two major flux increases, the first and larger was more complicated. The

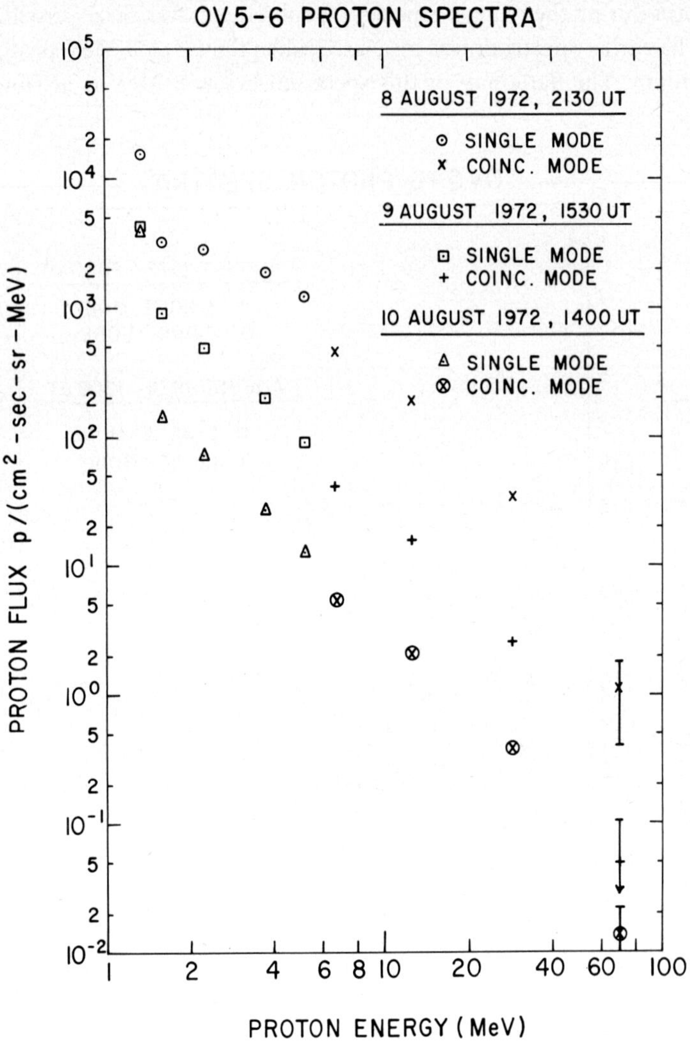

Fig. 5. Proton energy spectra during and following the August 8 proton flux maximum.

evidence supports the near coincidence of more than one solar injection followed by perturbation by interplanetary shock waves. The second large increase follows the usual pattern of earlier arrival and decay of the higher energy particles.

TEMPORAL VARIATIONS OF SOLAR PARTICLE SPECTRA (AUGUST 4–7, 1972)

G. KREMSER, G. PFOTZER, E. KIRSCH, and H. SPECHT
Max-Planck-Institut für Aeronomie, Institut für Stratosphärenphysik, 3411 Lindau, Germany

and

W. RIEDLER* and K. ZIRM
Institut für Nachrichtentechnik, Technische Hochschule Graz, Austria

Abstract. Solar particle fluxes were measured with balloon-borne instrumentation at Kiruna (Sweden) during the solar particle events of August 1972. Data from these balloon flights and from the satellite Explorer 41 were used to investigate the temporal variations of the energy spectrum during the event following the flare at 0620 UT on August 4, 1972. It was found that the slope of the energy spectrum increased linearly in the latter part of the event. Extrapolation of these linear variations to the time of the flare yielded power-law differential energy spectra with an exponent $1.8 \leqslant \gamma \leqslant 2.3$ or exponential rigidity spectra with a characteristic rigidity 85 MV $\leqslant P_0 \leqslant$ 110 MV. These values are in agreement with the corresponding parameters of time-integrated fluxes from Explorer 41. Implications of these results on the source energy spectrum of the solar particles and their propagation in interplanetary space are discussed.

1. Introduction

During the outstanding solar particle events of August 1972 eight balloons were launched from Kiruna with detectors for auroral X-rays and charged particles. A first report on these measurements was given by Kremser *et al.* (1973). The present paper describes the results on temporal variations of the energy spectrum during the event following the flare at 0620 UT on August 4, 1972. This event was the most spectacular one of the whole sequence of events (Pomerantz and Duggal, 1973).

The balloon payloads contained a scintillation counter (NaI (T1), 1 in. diameter, 1 in. height) for X-ray measurements and three Geiger counters in a threefold coincidence telescope arrangement for measurements of charged particles. It is not possible to separate protons, electrons and α-particles. However, at balloon altitudes (atmospheric depth of about 10 g cm^{-2}) solar protons are the main component of the particle flux (Pfotzer, 1965).

The method to derive the particle spectra is based on range measurements during the ascent of the balloon (Heristchi *et al.*, 1969). From the range measurements first the zenith angle distribution of the particles in the atmosphere, then the particle fluxes themselves, can be derived. The kinetic energy of the protons corresponding to the different pressure levels is obtained from a range-energy relationship (Pfotzer, 1965). In this way the energy spectra of solar particle fluxes can be obtained provided that no drastic changes in particle flux and spectrum occur during the ascent.

* Also with Geophysical Observatory, Kiruna, Sweden.

D. E. Page (ed.), Correlated Interplanetary and Magnetospheric Observations, 603–607.

2. Observations

2.1. Representation of Energy Spectra

The energy spectra to be discussed are presented in Figure 1. The balloon measurements cover the energy range from 65 MeV up to about 600 MeV. They have been supplemented by low-energy proton measurements (> 10, > 30, > 60 MeV) made on-board Explorer 41, which were published in *ESSA Solar Geophysical Data* (1973).

The horizontal bars in the data points indicate the energy interval over which the intensities were averaged. The vertical bars represent an estimate of the statistical uncertainty. The fluxes are given in relative units, as a normalisation factor had to be introduced to adjust the satellite data to the balloon measurements. This could not be done in a completely sufficient manner for all spectra. Therefore only the slope of

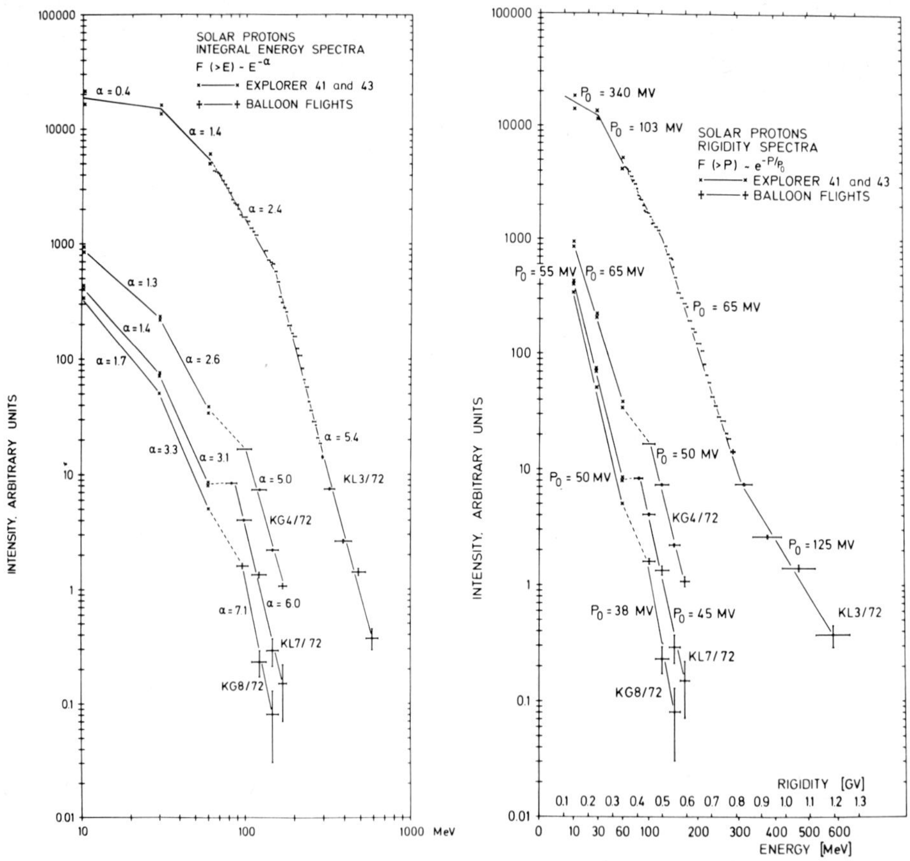

Fig. 1. Solar particle spectra obtained from range measurements during the ascent of balloons and from satellite data (see text). The balloon measurements were made during the following periods:
KL 3/72: Aug. 4, 1972, 1810–1920 UT;
KG 4/72: Aug. 5, 1972, 1910–1940 UT;
KL 7/72: Aug. 6, 1972, 1330–1400 UT;
KG 8/72: Aug. 6, 1972, 1855–1920 UT.

the spectra will be discussed in the following. The spectra are represented in two different ways. For the left part a power-law fit was used, i.e.

$$F(>E) \sim E^{-\alpha}.$$

For the right part we tried an exponential fit of the rigidity spectrum, i.e.

$$F(>P) \sim e^{-P/P_0}.$$

As can be seen, it is not possible to approach the measured energy spectrum over the whole energy range with a constant value of α or P_0, respectively. Furthermore the spectra steepen with time. A comparison of both representations shows that the data at lower energies (<60 MeV) are fitted better by a power-law spectrum. At energies between about 80 and 170 MeV both fits are equally good. At higher energies (>150 MeV) the power-law fit seems to be better.

2.2. Temporal Variations of Energy Spectrum

In Figure 2 we have plotted as function of time the exponent γ of the power-law representation of the differential energy spectrum and the characteristic rigidity P_0 of the exponential representation of the rigidity spectrum. Again data from the balloons and Explorer 41 were used.

Both parameters vary rather irregularly until the morning hours of August 5. Then the exponent γ increases almost linearly with time, but more rapidly for higher particle energies than for lower ones. The rigidity P_0 decreases also linearly, but the decrease does not depend on energy.

Extrapolation of the linear parts of the curves shows that for γ these lines intersect approximately at the onset of the flare at 0620 UT on August 4. At this time the values of γ range between 1.8 and 2.3. No such intersection is observed for P_0. The lines are rather close together and yield values of P_0 between 75 MV and 95 MV at the onset of the flare.

3. Discussion

It is tempting to regard the values of γ and P_0 found by extrapolation as parameters of the source energy spectrum. One possibility to check this assumption is to compare them with the parameters of the time-integrated spectra calculated by Boström *et al.* (1972). Doing this one obtains values of γ between 1.8 and 2.7 and values of P_0 between 85 MV and 110 MV, which are in acceptable agreement with the extrapolated values. This result also shows that it is not possible to decide whether the proton source spectrum was of a power-law or an exponential form. It seems, however, that all protons observed during the period August 4–7 were produced by the same flare.

This is an important result, as at 1830 UT on August 4, about 13 h after the flare, protons with energies up to at least 600 MeV were still present near the Earth. Neutron monitor recordings at high magnetic latitudes indicate additional proton fluxes with even greater energies from about 14 UT until 21 UT (Pomerantz and Duggal,

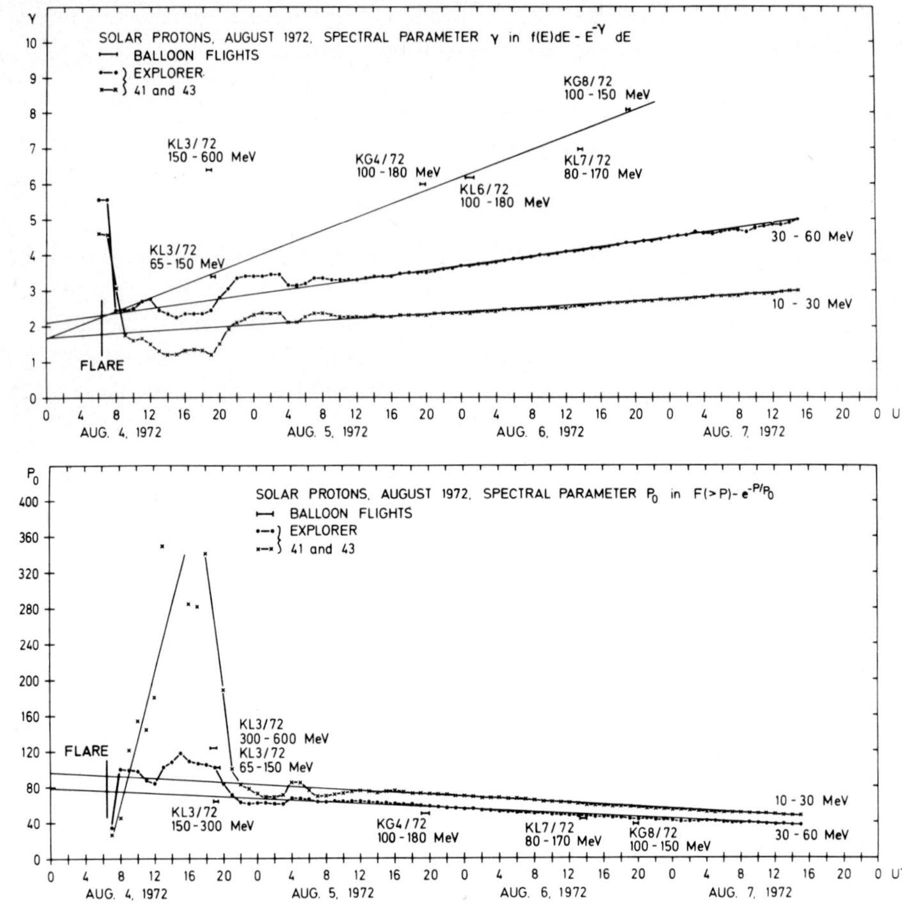

Fig. 2. Temporal variations of the exponent γ of the differential energy spectrum and the characteristic rigidity P_0 of the rigidity spectrum (see text).

1973). As no flare was observed at this time, we must assume that these relativistic protons had been produced already by the large flare at 0620 UT. This long delay time is rather unusual.

One explanation might be found in the interplanetary field measurements obtained onboard Pioneer 9 (46° east of the Earth-Sun line at that period), which show that on August 4 the interplanetary magnetic field had a nearly tangential direction with field strengths up to 20 γ. Such fields are able to trap even relativistic particles (1200 MeV protons have gyro radii of about 2×10^{-3} AU in a field of 20 γ). The trapping region might have been convected with the solar wind from the Sun and be responsible for the long delay of relativistic protons.

The low-energy protons (< 100 MeV) reached their flux maximum earlier. As they have gyro radii of the order of 10^{-4} AU, it may be assumed that they were scattered more often by field irregularities. Therefore they might have leaked out more easily from the trapping region.

References

Boström, C. O., Kohl, J. W., McEntire, R. W., and Williams, D. J.: 1972, 'The Solar Proton Flux – August 2–12, 1972', Preprint, The Johns Hopkins University, Applied Physics Laboratory.

ESSA Solar Geophysical Data, 1973, No. 342, Part 2.

Heristchi, Dj., Kangas, J., Kremser, G., Legrand, J. P., Masse, P., Palous, M., Pfotzer, G., Riedler, W., and Wilhelm, K.: 1969, *Ann. IQSY* **3**, 267, MIT Press.

Kremser, G., Pfotzer, G., Kirsch, E., Riedler, W., Specht, H., and Zirm, K.: 1973, *Kleinheubacher Ber.* **15**, 405.

Pfotzer, G.: 1965, in B. Maehlum (ed.), *High Latitude Particles and the Ionosphere*, Logos Press, London, p. 167.

Pomerantz, M. A. and Duggal, S. P.: 1973, *Nature* **241**, 331.

POLAR RIOMETER ABSORPTION DATA AUGUST 1–16, 1972

P. STAUNING

Ionosphere Laboratory, Meteorological Institute, c/o Technical University, DK 2800 Lyngby, Denmark

Abstract. The present paper contributes a summary of data on cosmic noise absorption in the polar ionosphere during the major polar cap absorption events of August 2–15, 1972. The data are obtained from a net of 10 riometers at 6 observatories in Greenland and at the Faroe Islands. The observatories are located within the northern polar cap and extends over a range of latitudes from $L=4.5$ to the geomagnetic pole. Observations related to the occurrence of solar flares are discussed and the riometer absorption data are compared to the fluxes of high energy solar protons measured by the Explorer-41 satellite.

1. Introduction

The locations of the riometer observatories that have supplied the data presented below are indicated in the map of the north polar region shown in Figure 1. Contours of constant L at 100 km's height have been plotted in the map to illustrate the observatory positions relative to the geomagnetic field.

Geographic and geomagnetic coordinates and 100 km L value for each observatory and receiving frequencies for the various riometers are shown in Table I. Further information on observatories and ionospheric measurements in Greenland may be found in Ionlab Report B55 (1972).

The 30- and 50-MHz riometers at the Geopole observatory are ARI, mark II versions while the remaining 20- and 30-MHz instruments are Ionlab, model II riometers. All riometers use 3-element linearly polarized Yagi antennas.

During interference-free conditions, data samples up to appr. 13 dB of absorption are believed to be fairly accurate. Above 13 dB the accuracy rapidly deteriorates. In

TABLE I

Location of riometer observatories

Observatory	Geographic		Geomagnetic		100 km	Riømeter
	Latitude	Longitude	Latitude	Longitude	L value	frequencies
Geopole	76.55	291.35	88.25	2.67	181.48	20, 30, 50
Godhavn	69.26	306.49	80.05	30.20	20.41	20, 30
Sdr. Strømfjord	67.02	309.28	77.60	34.78	14.42	31
Godthåb	64.19	308.27	74.99	29.93	10.89	30.2
Narssarssuaq	61.17	314.59	71.35	37.12	7.31	20, 30.2
Thorshavn	62.03	353.24	65.38	85.20	4.47	30

D. E. Page (ed.), Correlated Interplanetary and Magnetospheric Observations, 609–623. All Rights Reserved.

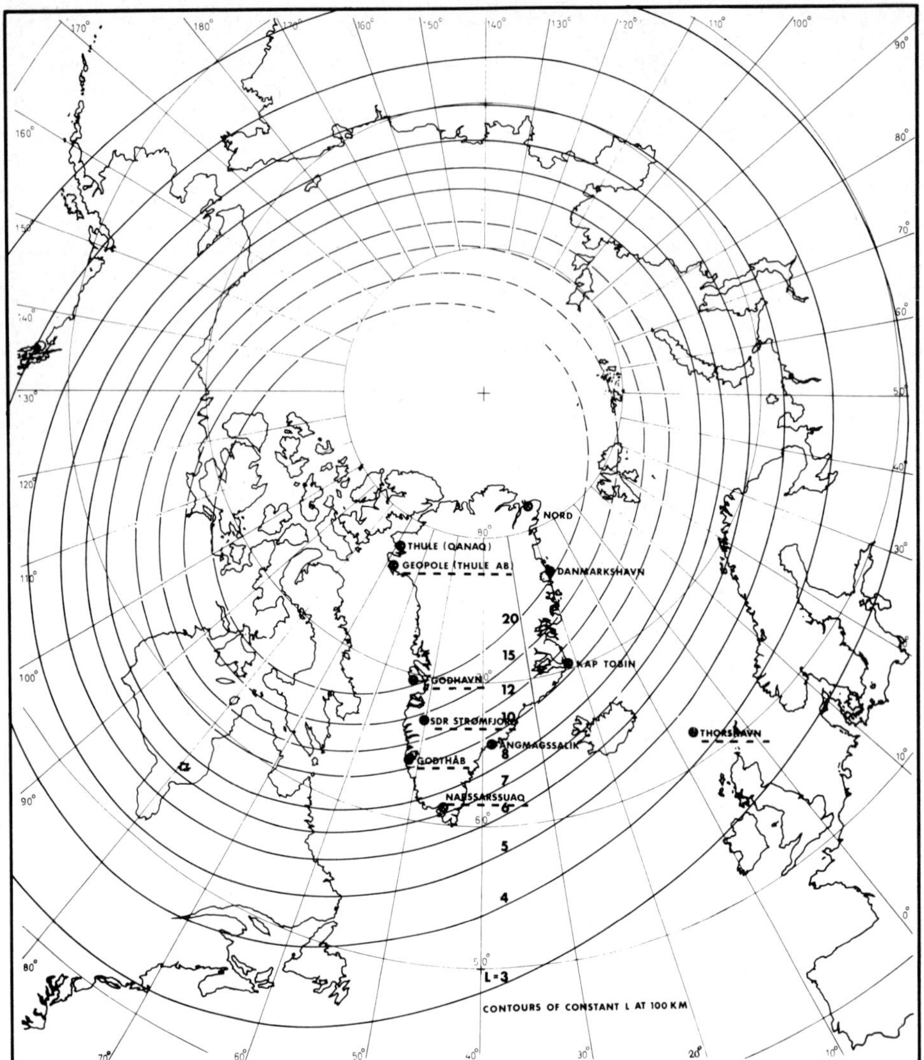

Fig. 1. Map of the North Polar region showing the location of riometer observations in Greenland and at Faroe Islands. The underlined observatories were in operation during the August 1–16, 1972 period.

addition to the obvious difficulties in the scaling of small recorder deflections, the riometers may fail to track the very weak cosmic noise signals received.

Such effects of excessive absorption are seen in all 20- and 30-MHz records during the day hours of August 4 and also in the 20-MHz records during part of the day hours of August 8. In addition to the above reservations it should be noted, that various kinds of local or propagated interference may influence the riometer readings.

A useful indication of data reliability may be obtained from comparisons of absorption values measured either at different frequencies from one location or measured at neighbouring observatories. Apart from the above intervals of excessive absorption the mutual consistency is quite good during the August 1–16 period for all except the 50-MHz riometer at Geopole. The absorption values derived from this riometer are somewhat questionable.

2. Data Presentation

The riometer data are presented in Figures 2–6 in the form of diagrams of superimposed traces representing 10-min average absorption values from all 10 riometers. Equivalent absorption values derived from hourly averages of the high energy proton flux are also plotted in the diagrams. All traces are marked with different symbols.

The riometer data have been 'normalized' to a reference frequency $f_0 = 30$ MHz by using the relation:

$$A(f_0) = A(f) \cdot (f/f_0)^n,$$

where $A(f)$ is the absorption (in dB) measured at the frequency f. The exponent n was calculated from absorption data for observatories, where both 20- and 30-MHz riometers were operating. From these calculations an average value of $n = 1.60$ was selected for normalization of all data.

The values of equivalent 30-MHz day-time absorption have been derived from the Explorer-41 data on solar proton fluxes above 10 MeV, 30 MeV and 60 MeV (*Solar-Geophysical Data*, vol. 342, II). The conversion of solar proton fluxes into ionospheric absorption values is based on the frequently used relation:

$$A_{eq} = m \cdot J(E > E_0)^{1/2},$$

where $J(E > E_0)$ is the 2-π omnidirectional flux of protons with energies greater than E_0. The conversion factor m is discussed below.

The bars drawn in the upper part of the diagrams show the occurrence of solar flares (*Solar-Geophysical Data*, vol. 343, II). Their importance and the sequence number for the active region is noted at each bar.

The occurrence of geomagnetic sudden commencements (sc) is indicated along the horizontal UT-time axis.

3. Summary of Observations

The main feature of the riometer absorption traces during the August 1–16 period is the occurrence of several intense polar cap absorption events related to the time varying fluxes of high energy protons from multiple solar flares.

All traces show the daily modulation produced by the varying solar zenith angle. The ratio of night-time absorption minima to day-time maxima varies with latitude from appr. 0.9 at the northernmost location, Geopole, to appr. 0.1 at the southernmost observatory, Narssarssuaq.

The absorption trace from Torshavn at an L value of 4.5 is strongly modulated by variations in the latitude of geomagnetic cut-off for the solar protons of relevant

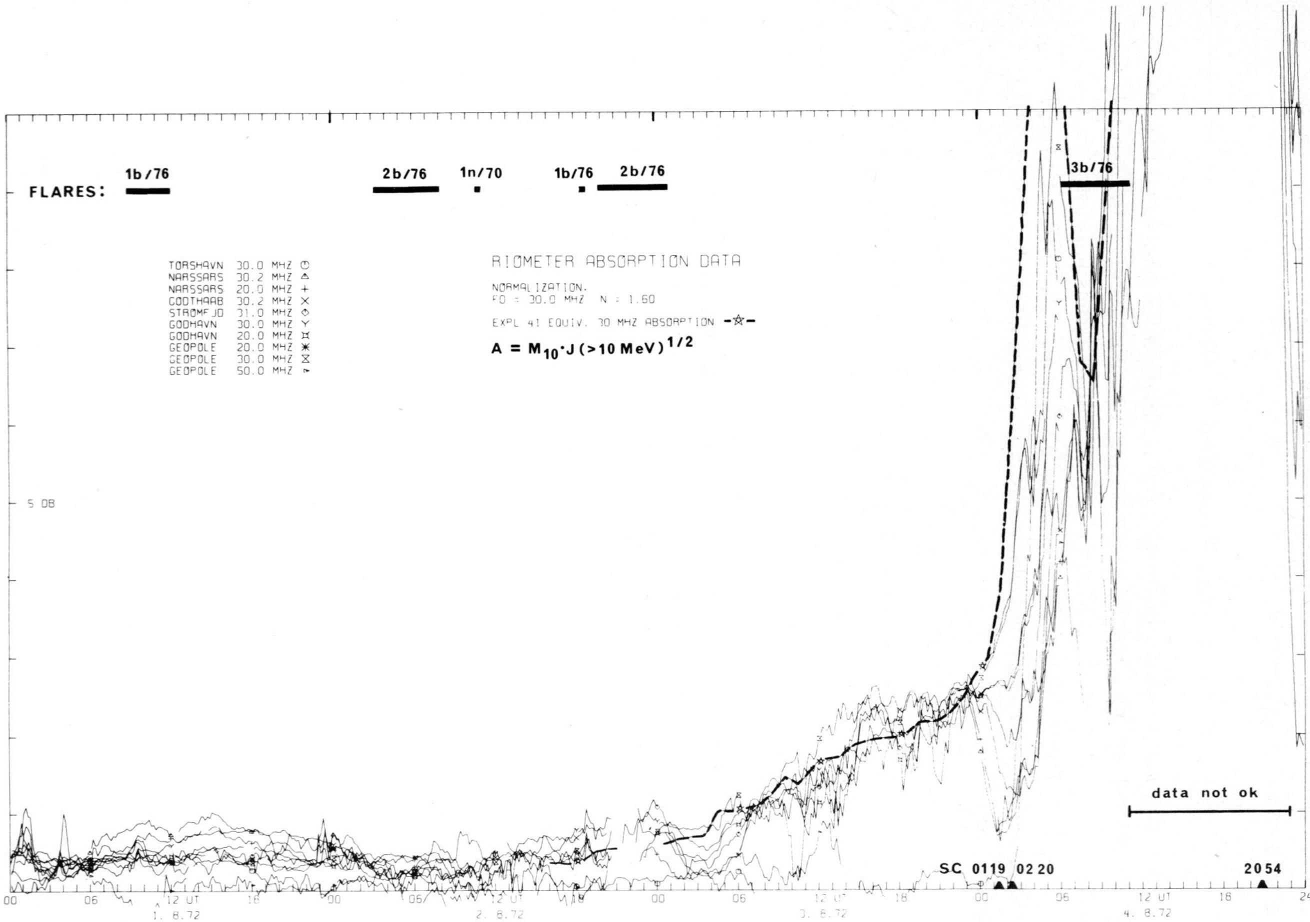

Fig. 2. Plot of riometer absorption data for August 1–4, 1972. All absorption values are normalized to $f_0 = 30$ MHz. A curve for the equivalent absorption intensities calculated from Explorer-41 data on solar protons is included. The relation used for the conversion is noted in the diagrams. The occurrences of solar flares and geomagnetic sudden commencements are also indicated in the diagrams.

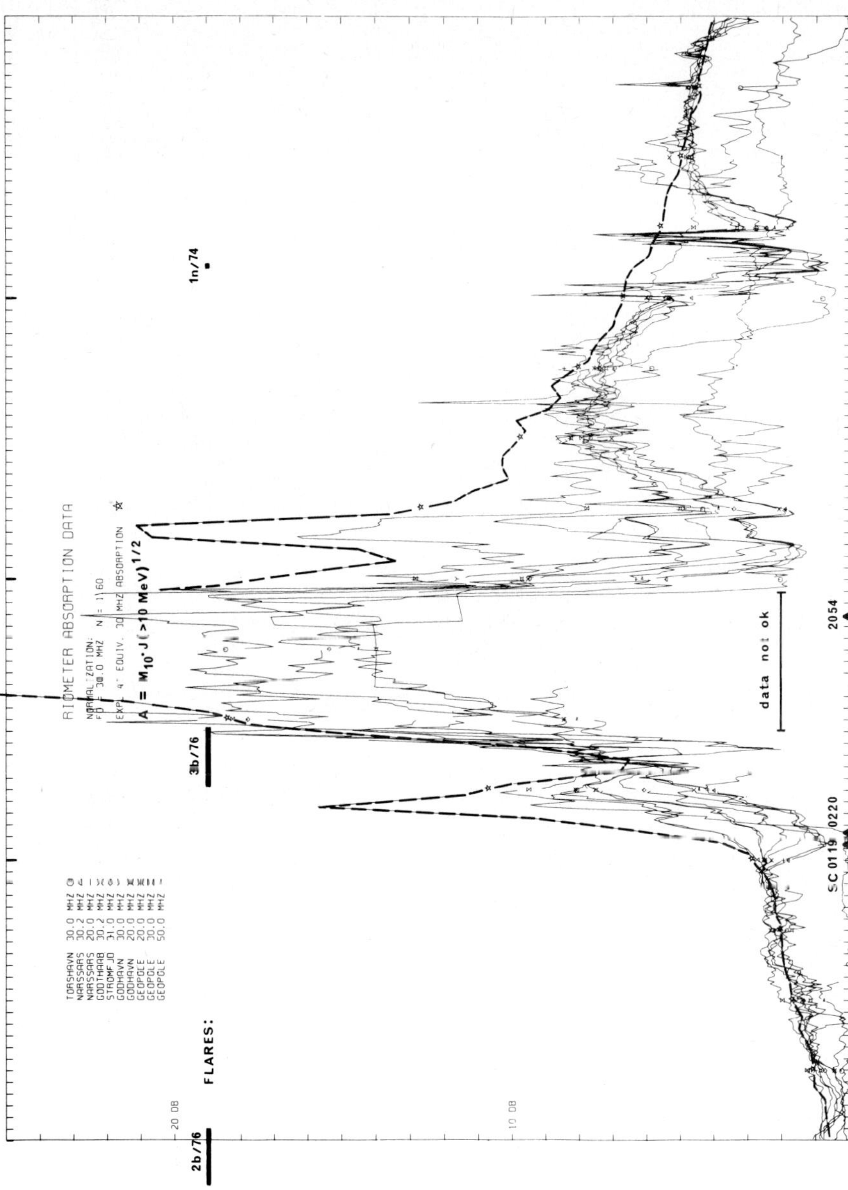

Fig. 3. Riometer absorption data and equivalent absorption intensities derived from solar proton fluxes for August 3–6, 1972.

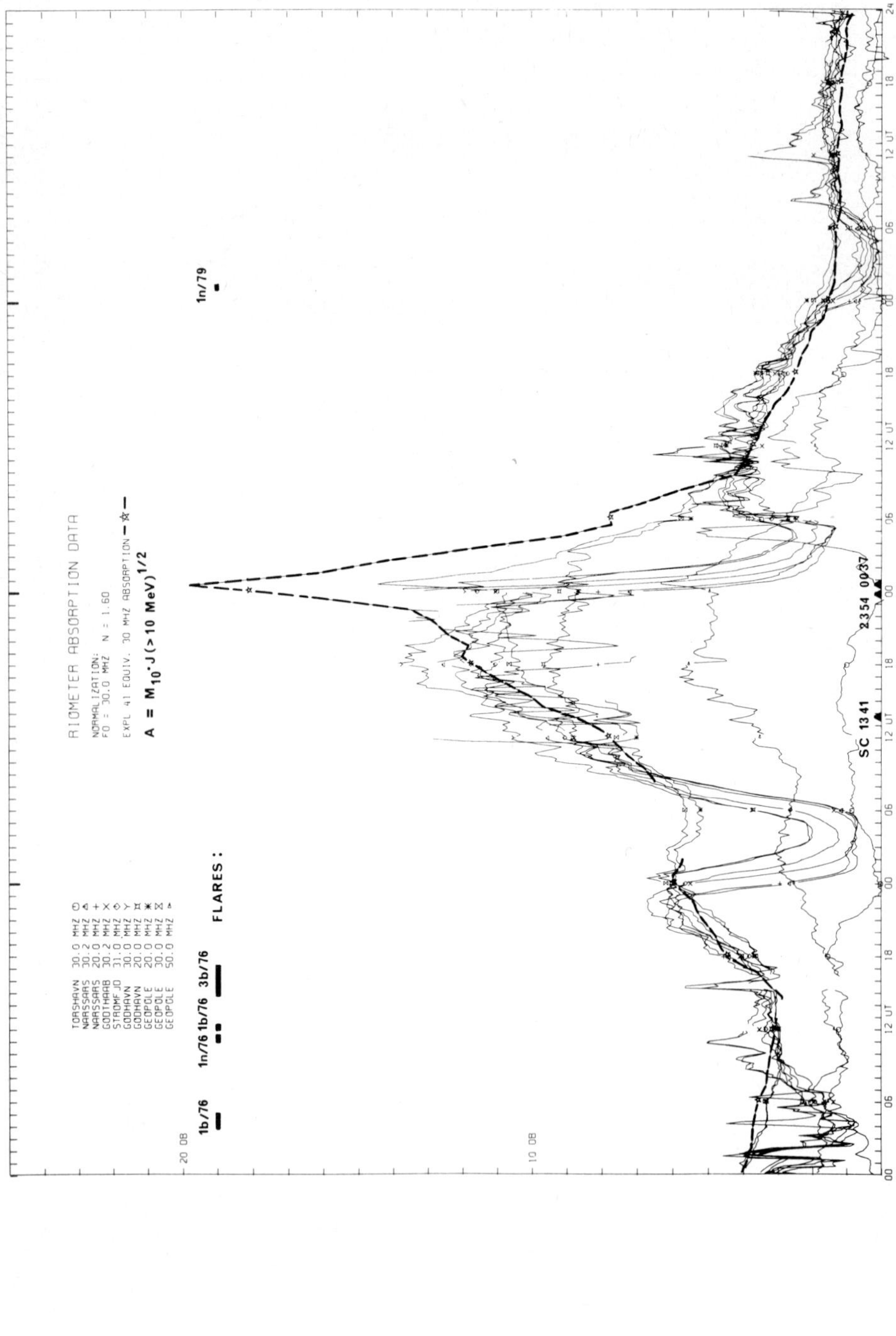

Fig. 4. Riometer absorption data and equivalent absorption intensities derived from solar proton fluxes for August 7–10, 1972.

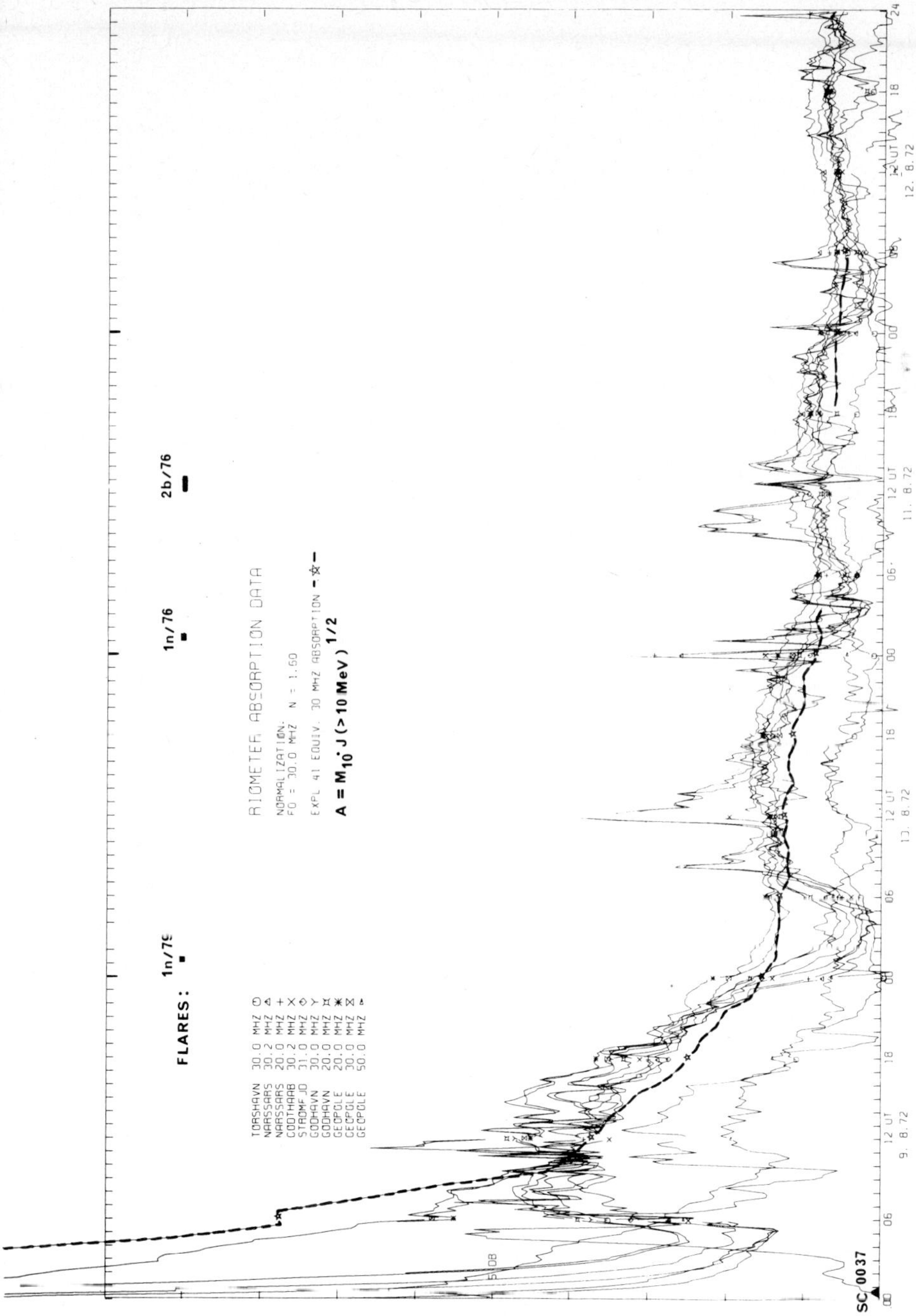

Fig. 5. Riometer absorption data and equivalent absorption intensities derived from solar proton fluxes for August 9–12, 1972.

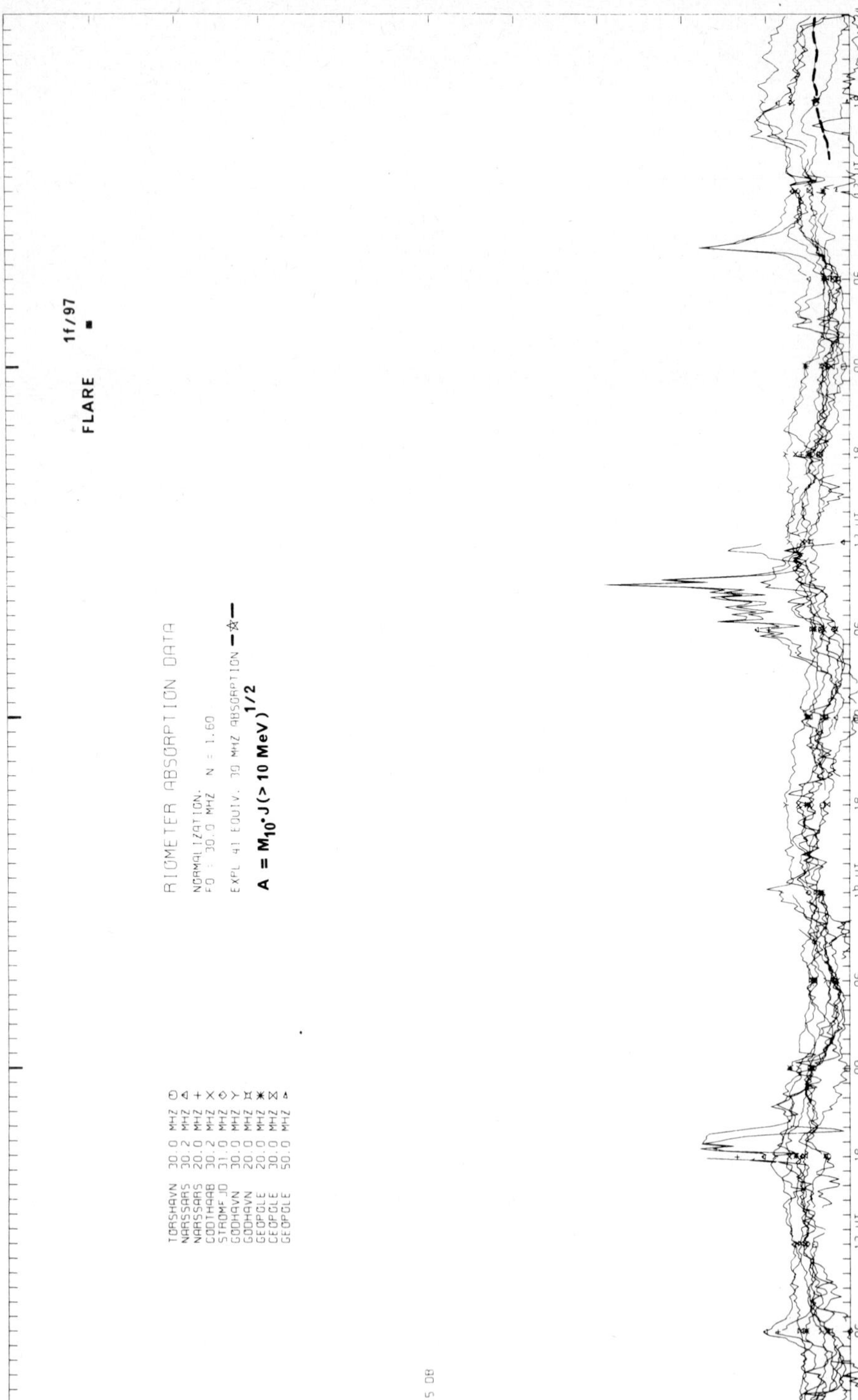

Fig. 6. Riometer absorption data and equivalent absorption intensities derived from solar proton fluxes for August 13–16, 1972.

energies (1–100 MeV). The other observatories are located poleward of the proton cut-off boundary.

Disturbances in the form of sudden cosmic noise absorption events (SCNA) related to solar X-ray emissions and interference events related to solar radio emissions are also observed in the data.

In addition to events related directly to solar emissions the absorption traces also show the occurrence of substorm events. The substorm-related absorption adds to the absorption associated with the solar proton flux producing the large and rapid fluctuations observed around local midnight (at appr. 03. UT) as well as the more slowly varying excursions seen in the absorption traces during local morning (from appr. 06 to 12 UT). The observations of flare-associated absorption events will be briefly discussed in the following. The discussion refers to the absorption diagrams in Figures 2–6.

(i) The flare of importance 1b in the activity region no. 76 occurring 0855–1218 UT on August 1 produces no significant effects in the present set of riometer recordings.

(ii) The flare 2b/76 at 0315–0805 UT on August 2 occurs during local night for the riometer observatories considered here and the immediate effects are not seen in the absorption traces.

The flare 1n/70 at 1040–1110 UT produces no significant event in the riometer data.

The flare 1b/76 at 1840–1855 UT produces a small SCNA event ending shortly after 1900 UT.

The flare 2b/76 lasting from 1955 on August 2 to 0110 UT on August 3 produces an immediate event of SCNA and interfering solar radio noise.

A PCA event related to the high energy solar particles emitted during the flares of August 2 is progressing in the evening of this day and develops further during August 3. The average polar noon Absorption is appr. 2.5 dB at 30 MHz on this day.

After an extended onset phase, the riometer absorption increases considerably just after the sudden commencements at 0119 and 0220 on August 4. At this time the Earth probably intersects the plasma stream confining protons of lower energies. The enhanced solar wind pressure also moves the cut-off boundary equatorward of Torshavn at $L = 4.5$ and this location is then fully exposed to the PCA-event for some time. The intensity of this PCA event peaks at appr. 05 UT i.e. during local night or early morning for the present riometer sites. The maximum observed absorption value at 30 MHz is 10.4 dB for Geopole at 0545 UT which corresponds to an equivalent day-time value of appr. 12 dB.

From this time the PCA event related to the flares of August 2 decreases rapidly but during its initial decay phase the next major solar event starts.

(iii) The flare 3b/76 at 0620–1130 on August 4 produces an event of SCNA and interfering solar radio noise starting at local Sunrise at the various observatories.

From appr. 09 UT the PCA develops rapidly. At appr. 11 UT the intense absorption causes the cosmic noise signals to drop below detection level for all 20- and 30-MHz riometers.

The 30 MHz absorption probably exceeds 20 dB during all day-hours.

The riometers do not recover until Sunset at appr. 23 UT. During the early hours of August 5 the absorption decreases rapidly but from appr. 0300 to 0530 a peculiar enhancement is observed. The absorption intensity increases rather abruptly by a factor 2 and stays at the high level during appr. 2 h. Then the absorption intensity suddenly drops to the pre-enhancement level.

From appr. 06 UT on August 5, through August 6 and part of August 7 the PCA event related to the flare of August 4 decays steadily. Polar average 30 MHz noon absorption values are 7,5 dB, 4,8 dB and 3,2 dB respectively for these days.

(iv) The flare 1n/74 at 0240–0300 UT on August 6 is possibly responsible for the short-lasting absorption enhancement observed at the southernmost observatories from appr. 03 to 06 UT and at the northernmost observatory (Geopole) from 0430 to 0500 UT. This event, however, might also be related to some unusual substorm activity.

(v) The flares 1b/76 at 0350–0510 UT, 1n/76 at 1055–1145 UT and 1b/76 at 1155–1235 UT on August 7 produce no immediate event of significant intensity in the riometer data.

The flare 3b/76 at 1445–1720 UT on August 7 produces an immediate and rather intense event of SCNA and interfering solar radio noise lasting from appr. 1515 to 1630 UT.

From appr. 16 UT the PCA related to the flares of this day gradually develops. The onset phase is rather extended and the absorption increases gradually until the sudden commencements at 2354 UT on August 8 and 0037 UT on August 9.

A remarkable feature observed during the onset phase is the very clean appearance of the absorption traces during the night of August 7/8. The variation of absorption with solar zenith angle is very distinct and the measured absorption values are nicely placed within the dynamical range of the instruments. In terms of geomagnetic activity, this night is unusually quiet. From 00 to 12 UT on August 8 K_p varies between 0^+ and 1^+, and there is no indication of substorm activity in the riometer data. This situation is favourable for a study of D-region processes provided data on the precipitating solar proton fluxes are made available.

At the sudden commencement at 0037 on August 9 the absorption intensity peaks at a value of appr. 15 dB. This time is close to local sunset and the equivalent daytime absorption intensity could be a few dB higher.

From the time of peak absorption the PCA decays steadily. The absorption decrease is rather abrupt in the morning hours of August 9. From noon on August 9 and through August 10, 11 and 12 the absorption intensity decreases gradually. Noon 30 MHz absorption values are 3,7 dB, 1.7 dB, 0.9 dB and 0.6 dB respectively on these days.

(vi) None of the flares 1n/79 on August 10 and 1n/76 on August 11 produce significant events in the present set of riometer data.

(vii) The flare 2b/76 at 1215–1320 UT on August 11 produces an immediate event of SCNA and solar radio noise. There is no indication of a succeeding PCA event in the absorption data.

(viii) The flare 1f/97 on August 16 produces no significant event in the riometer data.

Absorption events related to substorm activity are observed during local night (around 03 UT) and morning (06–12 UT) on most days. The substorm-related absorption is normally most intense at the southernmost observatories as they are located close to the auroral zone.

Rather intense events of substorm activity are seen in the night and morning hours (00–12 UT) of August 6, 7, 9, 10 and 11. The substorm absorption intensity is rather weak during the first days of August and also surprisingly small during the night of August 7/8 and the morning of August 8 as earlier mentioned.

4. Conversion of Proton Fluxes

The fluxes of high energy solar flare protons measured by integral proton detectors on Explorer 41 have been converted into equivalent polar cap absorption values in order to relate the intensities of solar protons to the plotted riometer absorption traces. The conversion is based on the model presented by Potemra (1972). From a daytime ionization model for the D-region he derives a relation between polar cap riometer absorption A and the integral proton flux $J(E > E_0)$ of the form:

$$A = m \cdot J(E > E_0)^{1/2}.$$

A power law proton spectrum of the form $J(>E) \sim E^{-\gamma}$ was assumed to derive the conversion factor m as a function of the energy exponent γ and the threshold energy E_0.

The threshold energy $E_0 = 10$ MeV for the most sensitive of the three solar proton detectors on Explorer 41 is used in the present case. At this value of E_0 the variations of m with γ are rather small according to the model.

Equivalent 30-MHz absorption values have been calculated using a fixed value of $m = 0.1$ dB cm $s^{1/2}$. This value is a fair approximation to the theoretical value of m, when the spectral exponent is close to $\gamma = 1.0$.

When the functional relation between m and γ is used, an uncertainty arises in the proper definition of the spectral exponent, when the observed proton fluxes above 10, 30 and 60 MeV fail to fit a simple power-law spectrum.

In one approach the exponent $\gamma_{10/30}$ is given the value that makes the fluxes measured by the >10-MeV and the >30-MeV detectors fit a power-law. The corresponding values of m are denoted $M_{10/30}$.

In another approach an exponential expression of the form

$$\ln[J(>E)] = a \cdot (\ln E)^2 + b \cdot \ln E + c$$

is first derived from the observed proton intensities. An equivalent exponent γ_{10} is then derived from fitting a power-law spectrum to the exponential spectrum at $E_0 = 10$ MeV. The values of m derived from this approach are called M_{10}.

For all three cases the equivalent absorption intensities calculated from the proton intensities have been plotted in the riometer absorption diagrams to check the correspondence between computed and observed polar absorption values. The best fit was found when $m=M_{10}$ was used. This case is illustrated in Figures 2–6.

To derive a numerical measure of the accuracy of the fit, hourly averages of the mean polar absorption intensity were calculated from the absorption values measured by 5 polar 30-MHz and 31-MHz riometers during local day-time i.e. from 12 to 22 UT. The period of excessive absorption from 11 UT to 23 UT on August 4 and intervals of flare activity were excluded from the data set. A few samples from the southernmost observatories were omitted due to substorm activity. From the data for the August 3–10 period, 67 samples of hourly averages of mean polar cap absorption were obtained (of 80 possible day-time values).

For each of the three sets of equivalent absorption values the rms deviation from the measured polar cap absorption was calculated. The results are shown in Table II. The table also shows the corresponding figures found in the analysis made by Potemra. His samples of individual satellite passes are to some extent statistically comparable with the samples of hourly averages derived here.

One might note that the average polar absorption intensity was appr. 3 dB during the events considered by Potemra, while the average absorption is 5.0 dB for the samples selected during the August 3–10 period. With this difference in mind, the best fit obtained here between Explorer-41 proton data and polar cap absorption is comparable to the optimum fit obtained by Potemra using data from the near-Earth polar orbiting satellite 1963 38c.

The variations in the actual conversion factor with energy exponent γ is illustrated in Figures 7 and 8 where the observed ratio M of the average polar riometer absorption values to the square root of the >10-MeV integral proton flux is plotted versus γ_{10} and $\gamma_{10/30}$ respectively. The relation between m and γ devised by Potemra for $E_0=10$ MeV is also plotted in the diagrams.

It is seen that using the exponent γ_{10} gives a more even distribution of data points around the curve corresponding to the theoretical model than found when using $\gamma_{10/30}$.

TABLE II

Comparison of results obtained from linear fits of 30 MHz day-time riometer absorption to the square root of integral solar proton fluxes. Explorer 41 detected the proton intensities outside the magnetosphere, while 1963 38c measured the proton fluxes at low altitudes above the polar caps.

	Number of samples	Threshold E_0 MeV	m dB cm $s^{1/2}$	rms deviation, dB
This study		10	0.100	0.80
Aug. 3–10, 72	67	10	$M_{10/30}$	1.27
(Explorer 41)		10	M_{10}	0.62
Potemra (1972)		2.2	0.040	1.11
all events	93	8.4	0.084	0.38
(1963 38c)		25	0.18	1.42

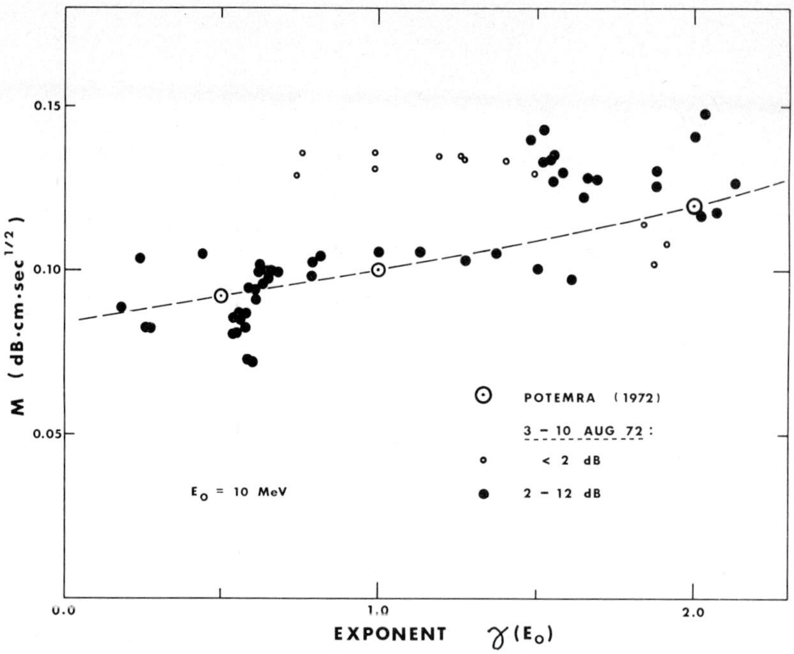

Fig. 7. Plot of the ratio of hourly riometer absorption values to the square root of the integral solar proton flux above 10 MeV versus spectral exponent for the protons. The spectral exponent is derived from the slope of the energy spectrum at $E_0 = 10$ MeV. The plot comprises samples from August 3–10, 1972. The dotted line indicates the theoretical relation (Potemra, 1972) between the conversion factor and the spectral exponent.

It is, however, evident from both diagrams, that a fixed value of $M = 0.10$ is a fair approximation in most cases.

5. Conclusions

The riometer data considered here are believed to provide a comprehensive illustration of polar cap absorption intensities related to the solar flare events of August 1–16 except during the interval of extremely intense solar proton fluxes from 11 to 23 UT on August 4.

The absorption summary plots presented above show a considerable amount of details related to immediate effects of solar flares and to substorm activity. For investigations of such events a more detailed time resolution is required but the plots provide a reliable identification of their occurrence.

The observed day-time polar cap absorption intensities are closely related to the fluxes of high energy solar protons measured by the >10-, >30- and >60-MeV integral detectors on the Explorer-41 satellite. The best fit between absorption values and proton data is obtained, when the fluxes measured by the >10-MeV integral detector and a power-law spectral exponent referring to $E = 10$ MeV are used in the theoretical model devised by Potemra (1972). For this case the rms deviation cal-

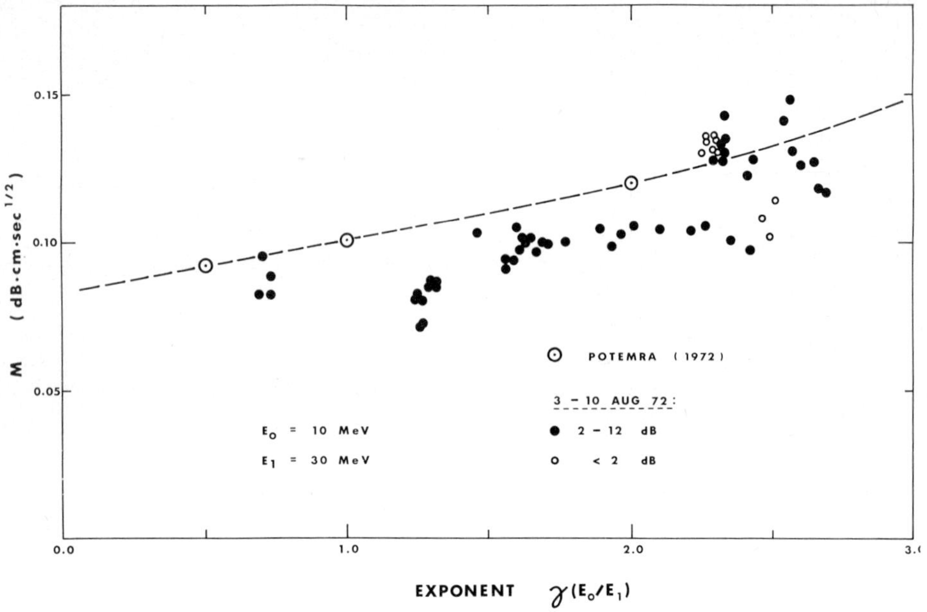

Fig. 8. Plot of the ratio of hourly riometer absorption values to the square root of the integral solar proton flux above 10 MeV versus spectral exponent for the protons. The spectral exponent is calculated from the ratio between >10-MeV and >30-MeV proton intensities. The plot comprises samples from August 3–10, 1972. The dotted line indicates the theoretical relation (Potemra, 1972) between the conversion factor and the spectral exponent.

culated from samples of hourly averages of the proton flux and the mean polar cap absorption intensity is 0.62 dB during the August 3–10 period. The samples represent an average absorption intensity of 5.0 dB.

This is comparable to the fit obtained in studies of the relation of polar cap absorption intensities to solar proton fluxes measured by the near-Earth polar orbiting satellite 1963 38 (Potemra, 1972) and also to the fit obtained using measurements of solar protons from the geostationary ATSI satellite (Potemra and Lanzerotti, 1971) during polar cap absorption events.

Acknowledgements

The riometers in Godthåb and Sdr. Strømfjord are installed at communication stations operated by the Greenland Technical Organization (GTO). The Geopole observatory at Thule AB is operated by Air Force Cambridge Research Laboratories (AFCRL). The remaining riometers are installed at observatories operated by the Danish Meteorological Institute. The computer processing of riometer data was made at the Northern Europe University Computing Center (NEUCC). The support and assistance received from these organizations is gratefully acknowledged.

I wish in particular to thank the many individuals responsible for the daily operation of riometer installations in Greenland and at the Faroe Islands. The cooperation with and the assistance received from B. Christensen, B. Ramsberg and other members of the staff at the Ionosphere Laboratory is greatly appreciated.

References

Neble Jensen, V. and Taagholt, J.: 1972, Danish Ionospheric Measurements in Greenland, Ionlab Rept B55.

Potemra, T. A.: 1972, *Radio Sci.* **7**, 571.

Potemra, T. A. and Lanzerotti, L. J.: 1971, *J. Geophys. Res.* **76**, 5244.

Solar-Geophysical Data, 1973, No. 342, part II.

Solar-Geophysical Data, 1973, No. 343, part II.

PART 7

FUTURE STUDY TOPICS

SOLAR ACTIVITY AND THE WEATHER

LEIF SVALGAARD

Institute for Plasma Research, Stanford University, Stanford, Calif., U.S.A.

Abstract. Some new evidence that the weather is influenced by solar activity is reviewed. It appears that the solar magnetic sector structure is related to the circulation of the Earth's atmosphere during local winter. About $3\frac{1}{2}$ days after the passage of a sector boundary the maximum effect is seen; apparently the height of all pressure surfaces increases in high latitudes leading to anticyclogenesis, whereas at midlatitudes the height of the pressure surfaces decreases leading to low pressure systems or to deepening of existing systems. This later effect is clearly seen as an increase in the area of the base of air with absolute vorticity exceeding a given treshold. Since the increase of geomagnetic activity generally is small at a sector boundary it is speculated that geomagnetic activity as such is not the cause of the response to the sector structure but that both weather and geomagnetic activity are influenced by the same (unknown) mechanism.

1. Introduction

The idea that solar activity influences weather and climate on the Earth is nearly as old as the concept of solar activity itself. The Italian Jesuit priest Riccioli suggested in 1651 that the temperature of the Earth falls with increasing spottedness of the Sun. Unfortunately in 1801, William Herschel came to the opposite conclusion from an examination of wheat prices in England. This inconclusiveness has been the mark of the whole field of relations between solar activity and weather ever since. Part of the reason for this is the complexity and variability of both solar and terrestrial parameters over a wide range of time scales. Different regions of the Earth such as oceans and continents may also show different responses to solar events. One author investigates surface pressure patterns in response to solar flares; another examines rainfall and sunspot numbers. Their results can hardly be compared and may even be contradictory to results obtained from other geographical regions. After some initial optimism in the beginning of the century serious research in the solar activity-weather field has since been discouraged, particularly among meteorologists.

Recently, however, a number of rather convincing findings have appeared linking the Earth's weather to solar and interplanetary parameters. For example the IUGG Symposium convened by E. R. Mustel in Moscow during August 1971 summarized many new researches in this field. On a whole it seems that direct solar activity influence on weather is gaining growing acceptance in the Soviet Union, whereas the field still is somewhat suspect to Western researchers. The fact that no plausible mechanism has yet been advanced to explain how solar activity possibly can influence the weather is undoubtedly the main reason for this lack of respectability. A famous case in point, of course, was the slow acceptance of the relationship between solar activity, auroras and magnetic storms. From energy considerations Lord Kelvin concluded in 1892 that it was impossible that terrestrial magnetic storms are due to magnetic action of the Sun or to any kind of dynamical action taking place anywhere near the Sun.

D. E. Page (ed.), Correlated Interplanetary and Magnetospheric Observations, 627–639.

The existence of the solar wind overcomes Lord Kelvin's objections because the magnetic field of the Sun is brought out to the Earth by the outflowing plasma. But the energy content of the solar wind is many orders of magnitude less than that of the total radiation which is the driving force of the atmospheric circulation; so again from energy considerations, it is hard to see how solar activity can have any measurable influence on the lower atmosphere. Without a plausible theory it is then difficult to organize and interprete the empirical evidence. On the other hand some really firm evidence might stimulate the search for a physical mechanism. In this paper I will discuss a recent investigation (Wilcox *et al.*, 1973) of the apparent response of the atmospheric circulation to the solar magnetic sector structure.

2. Solar Magnetic Sector Structure

The large-scale solar photospheric magnetic field is extended outward from the Sun by the radially flowing solar wind (Wilcox, 1968). Seen from the Earth, the interplanetary magnetic field can be divided into sectors such that within each sector the field polarity is either toward the sun or away from the Sun. Adjacent sectors having opposite field polarities are separated by a very narrow boundary. Most commonly there are four such sectors, each having a width of about 90° in solar longitude, and the whole sector structure rotates past the Earth with a period of approximately 27 days. While the Earth may be within the same sector for several days, it usually takes only a few hours down to a few minutes for a sector boundary to sweep past the Earth. Hence the time of passage of a sector boundary may be used as a precise timing signal to examine the atmospheric response to the sector structure using the superposed epochs technique. The zero days (or key-days) were defined as the times at which the sector boundaries tabulated by Wilcox and Colburn (1972) swept past the Earth. These are boundaries for which the polarity of the interplanetary magnetic field observed by spacecraft near the Earth – but still outside the Earth's magnetosphere – was in one direction (away from or toward the Sun) for at least four days before the boundary, and in the opposite direction for at least four days after the boundary. Similarly defined sector boundaries for the year 1970 (Wilcox and Colburn, 1973) were added to the list of well-defined boundaries. A total of more than 100 sector boundaries were available for analysis. Due to the strict requirement that each boundary must be clearly defined by *observations* only about one-fourth of all possibly existing boundaries went into the analysis.

3. The Hemispheric Vorticity Area Index

The atmospheric parameter in the investigation was the vorticity area index of Roberts and Olson (1973). A basic quantity in hydrodynamics is *circulation*, involving movement around a closed path within a fluid. Circulation is defined as the line integral of the velocity of the fluid around this closed path. *Vorticity* may be defined as circulation per unit area. For a circular disk rotating rigidly with angular velocity

ω the circulation around the outer boundary is $C = 2\pi R \cdot \omega R$, where R is the radius of the disk. Thus $C = \pi R^2 (2\omega)$ or area times 2ω, so that the vorticity of a disk rotating as a solid is twice the angular velocity of rotation. Since the Earth is rotating, it must have a vorticity; the Earth's atmosphere corotates with the Earth, and so its mean vorticity is simply 2Ω, where Ω is the angular velocity of the Earth. However, local variations in atmospheric vorticity occur, and these are of prime significance for both climate and weather. Strictly speaking we have components of vorticity for the three axes of a Cartesian system. Since the vorticity about the rotation axis is by far the most important, we only consider that from now on.

A distinction must be made between relative vorticity, which is the vorticity of the airflow relative to the Earth's surface, and absolute vorticity. The latter is simply the sum of the relative vorticity ζ and the vorticity f of the Earth itself: $\zeta_A = \zeta + f$.

Now it can be shown (e.g., Pedlosky, 1971) that for a rotating air column in the atmosphere the absolute vorticity is conserved for that volume of air. The distribution of absolute vorticity over the surface of the Earth is highly uneven. There are numerous centers or regions with rather high vorticity separated by areas of low absolute vorticity. These high vorticity centers move across the surface with the weather systems but conserve their value of high absolute vorticity. If the pressure is falling in one of these centers its area increases to conserve volume, and the area over which the high vorticity is observed increases accordingly. The high vorticity centers are seen in connection with low pressure troughs associated with increased storminess and precipitation. A deepening of these troughs leads to an increase of the area with high absolute vorticity; and it is possible to define an index indicating the overall severeness of the low pressure systems. This vorticity area index is the total area over which the absolute vorticity exceeds a given treshold.

The vorticity area index can be computed from maps of the height of constant pressure surfaces using the geostrophic wind approximation. These maps prepared twice a day, at 0^h UT and at 12^h UT for the part of the northern hemisphere north of 20° N – were used to compute the vorticity area index for the years 1964–1970. The original vorticity area index of Roberts and Olson was defined as the area (in km^2) over which the absolute vorticity exceeded 20×10^{-5} s^{-1} plus the area (also in km^2) over which the vorticity exceeded 24×10^{-5} s^{-1}. For virtually all low pressure troughs there was some area over which the vorticity exceeded the lower value, and for the major troughs there was nearly always some area over which the higher value was exceeded. The pressure level used to calculate the vorticities was the 300 mbar level at about 9000 m height. The area was summed over the entire hemisphere north of 20° N, and two values of the index was obtained per day. Due to systematic (small) differences between the 0^h UT map and the 12^h UT map each vorticity index value was smoothed according to $V_i = (V_{i-1} + 2V_i + V_{i+1})/4$ to remove the systematic variation.

4. Vorticity Area Index Response to Sector Structure

Figure 1 shows the average response of the hemispheric vorticity area index to the sector structure sweeping past the earth during the winter months, November to March. A total of 54 sector boundaries was used to calculate the average response. On the average the vorticity is decreasing until 1 day after the boundary, and then increasing until a maximum is reached about $3\frac{1}{2}$ days after the boundary. A similar analysis covering the summer months showed little or no effect. This may not be too surprising because the atmospheric circulation patterns are quite different for summer and winter. Also it appears to be one of the most persistent empirical evidence emerging from most other investigations that any influence of solar activity or solar parameters on the weather tends to be strongest during the winter season.

The vorticity area index is influenced by many physical effects. The superposed epoch analysis presented here emphasizes one particular effect – the solar magnetic sector. Since the phase (zero day) of the analysis is fixed by the time at which a sector boundary passes the Earth, physical effects on the vorticity index related to the sector structure will tend to be reinforced in this analysis, and other physical effects on the vorticity index will tend to occur at random phases and therefore to be averaged out.

The physical significance and reproducibility of the result shown in Figure 1 have been investigated by dividing the data sample into two parts in three different ways, and performing the same superposed epoch analysis separately on each part. Figure 2 shows the result of analysing separately the boundaries at which the field polarity

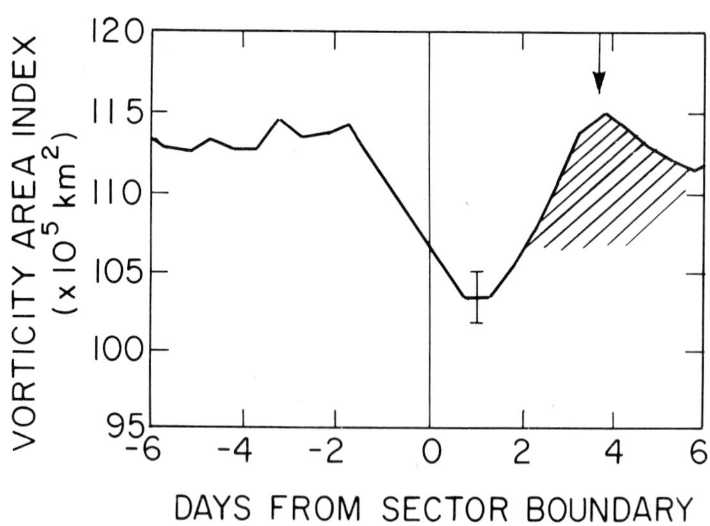

Fig. 1. Average response of the vorticity area index to the solar magnetic sector structure. Sector boundaries were carried past the Earth by the solar wind on day 0. The analysis includes 54 boundaries during the winter months November to March in the years 1964 to 1970. The standard error of the mean (error bar) was calculated after subtracting a 27-day mean centered on each sector boundary, to remove long-term trends. The increase in the vorticity area index $3\frac{1}{2}$ days after the boundary is shaded.

changed from toward the Sun to away from the Sun and the boundaries where the field polarity changed from away to toward. Also shown are the results of separate analyses of the boundaries occurring in the first half of the winter (November 1 to January 15) and those occurring in the last half (January 16 to March 31). Finally, the boundaries occurring during the years 1964 to 1966 and those occurring during 1967 to 1970 have been analyzed separately. We note, that the analyses performed on these various subsets of the data all gave essentially the same results, so that the result shown in Figure 1 is highly reproducible and is quite independent of the way the data was subdivided. In all cases do we see a decline of the vorticity area index until 1 day after the boundary followed by a rise in the index to a maximum about $3\frac{1}{2}$ days after the sector boundary passed the Earth.

The results shown in Figures 1 and 2 were based on the vorticity area index calcu-

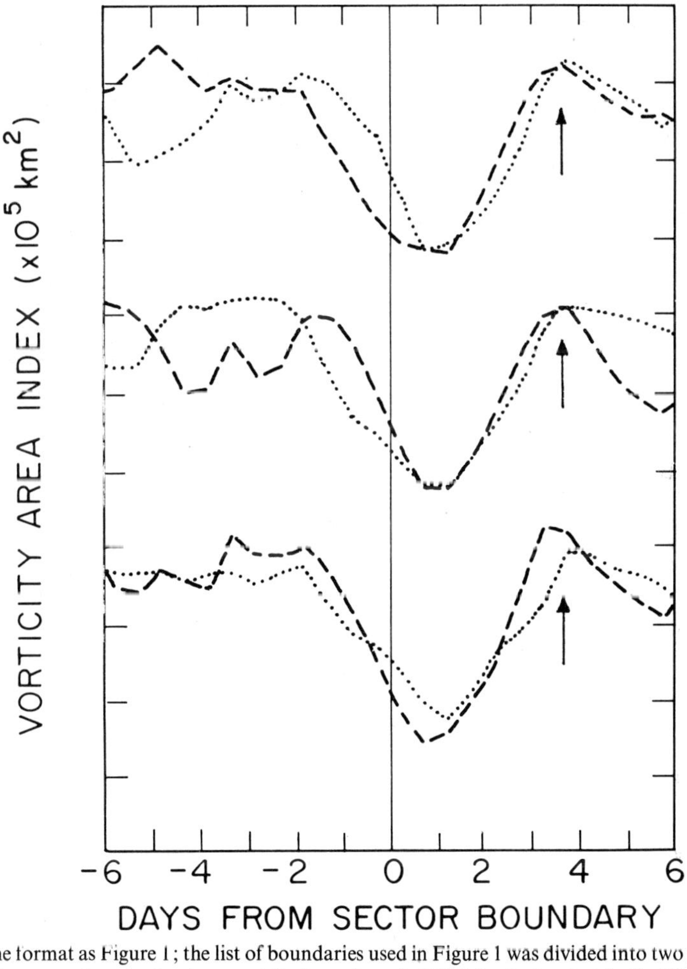

Fig. 2. Same format as Figure 1; the list of boundaries used in Figure 1 was divided into two parts according to (top) the magnetic polarity change at the boundary, (middle) the first or last half of the winter, and bottom) the yearly intervals 1964–1966 and 1967–1970. The curves have been arbitrarily displaced in the vertical direction, but the scale of the ordinate is the same as in Figure 1.

lated for the 300 mbar level, and using the original definition of the index as the sum of two areas corresponding to two different vorticity tresholds. Using only one treshold gives essentially the same results and it was felt an unnescessary complication to proceed with the analysis with the two tresholds. The final vorticity area index may now be characterized by two parameters: (i) the pressure level for which it is computed, and (ii) the treshold or discriminator value used.

To check if the results found at the 300 mbar level could be reproduced at other levels in the atmosphere, the analysis was repeated for each of the so-called standard pressure levels ranging from 850 mbar up to 10 mbar. The vorticity discriminator was first set to the fixed value $20 \times 10^{-5}\ \mathrm{s}^{-1}$ for all levels. Figure 3 demonstrates that the

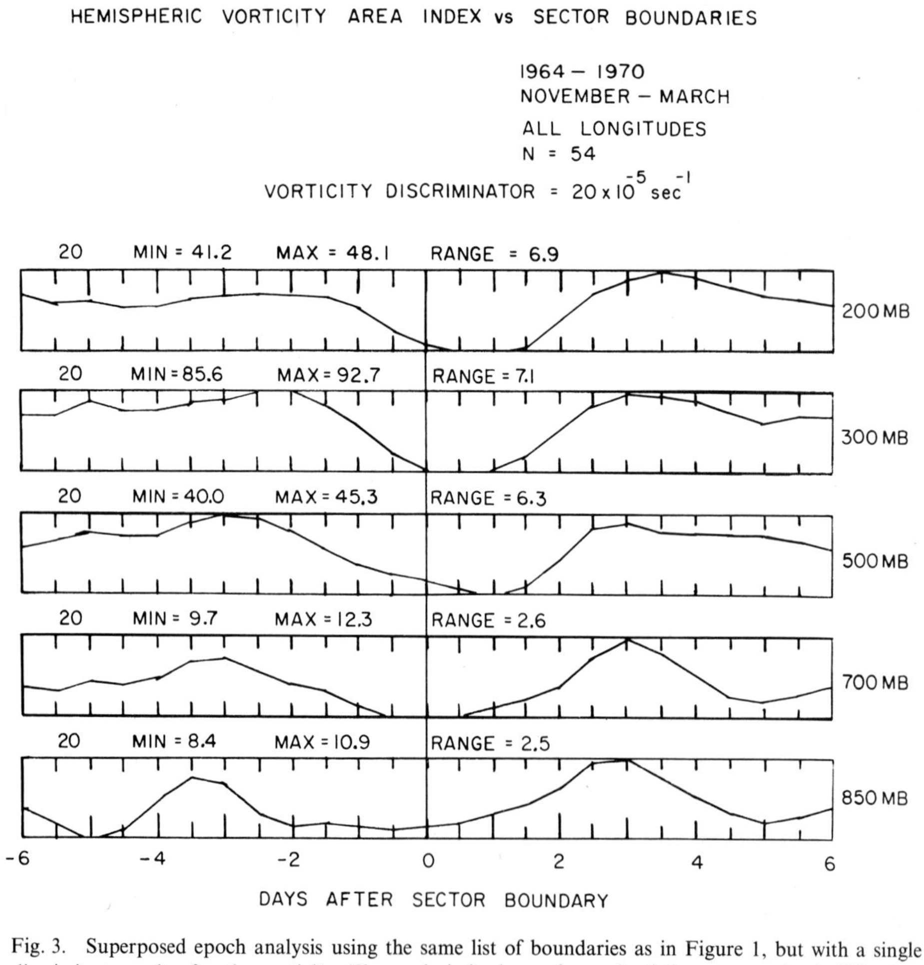

Fig. 3. Superposed epoch analysis using the same list of boundaries as in Figure 1, but with a single discriminator value for the vorticity. The analysis is shown for each of the pressure levels 850 mbar, 700 mbar, 500 mbar, 300 mbar, and 200 mbar. The vertical scale is for each strip given by the maximum and minimum value of the vorticity area index (in units of $10^5\ \mathrm{km}^2$).

effect is present at all levels in the troposphere. It could not be seen at stratospheric levels above 100 mbar. This may simply be a consequence of the fact that the atmospheric circulation is quite different in the stratosphere and in the troposphere. It is quite noteworthy that the effect can be observed as low as at the 850-mbar level. The persistent feature in Figure 3 is the maximum 3–4 days after the boundary. This maximum starts to develop at the same time at all levels, whereas there may be some indication that the decline of the indexvalue starts first at the lowest level and becomes effective later and later as we go to higher levels. It is tempting to ascribe the slower decay as due to decreased friction with altitude.

From Figure 3 it is evident that there is quite a variation of the absolute vorticity from one level to the next; the vorticity area index changes from a minimum of $\approx 10 \times 10^5$ at 850 mbar to $\approx 90 \times 10^5$ km^2 at 300 mbar. Since the atmosphere is basically a *thin* structure, air masses generally have much larger horizontal dimensions compared to their vertical size. Therefore the analysis was finally repeated varying the vorticity discriminator at each level such as to make the areas nearly equal to 40×10^5 km^2 at all levels. The result is shown in Figure 4. We note that again we get essentially the same result as in the previous analysis.

The reproducibility of our results when the parameters of the analysis are varied, for all levels of the troposphere and for different subsets of the data strongly suggests that the effect is real. The fact that the times of the sector boundaries were published long before the analysis precludes wishful selection-effects. Finally the whole analysis is completely objective and may be repeated by anybody using only published data.

5. Pressure and Sector Structure

In a follow-up study of the atmospheric response to the solar sector structure Svalgaard *et al.* (1973) analyzed atmospheric pressure rather than vorticity. Instead of analyzing the pressure at a given height it is costumary in meteorology to examine the height of a given pressure level; many equations in dynamic meteorology are simpler with pressure as an independent variable instead of height. Using the same 54 sector boundaries as zero days in a superposed epoch analysis the quantity investigated was the average height of a given constant pressure level over a latitude belt 10° wide. As for the vorticity area index significant systematic variations of the average heights were found in response to the passage of a sector boundary. In general the heights increased a few days after the boundary at high latitudes, whereas they decreased at mid-latitudes.

A single quantity which incorporates both of these changes of height is the height difference between the levels of constant pressure in the 40°–50° N latitude belt and in the 60°–70° latitude belt. In the free atmosphere there is generally a low pressure area around the north pole in the winter time and a region with high pressure in the subtropics. The height difference defined defined above is a measure of the slope of the surface of constant pressure slanting somewhat downwards toward the pole. This slope is then positive and of the order of 400 m on the average. The response

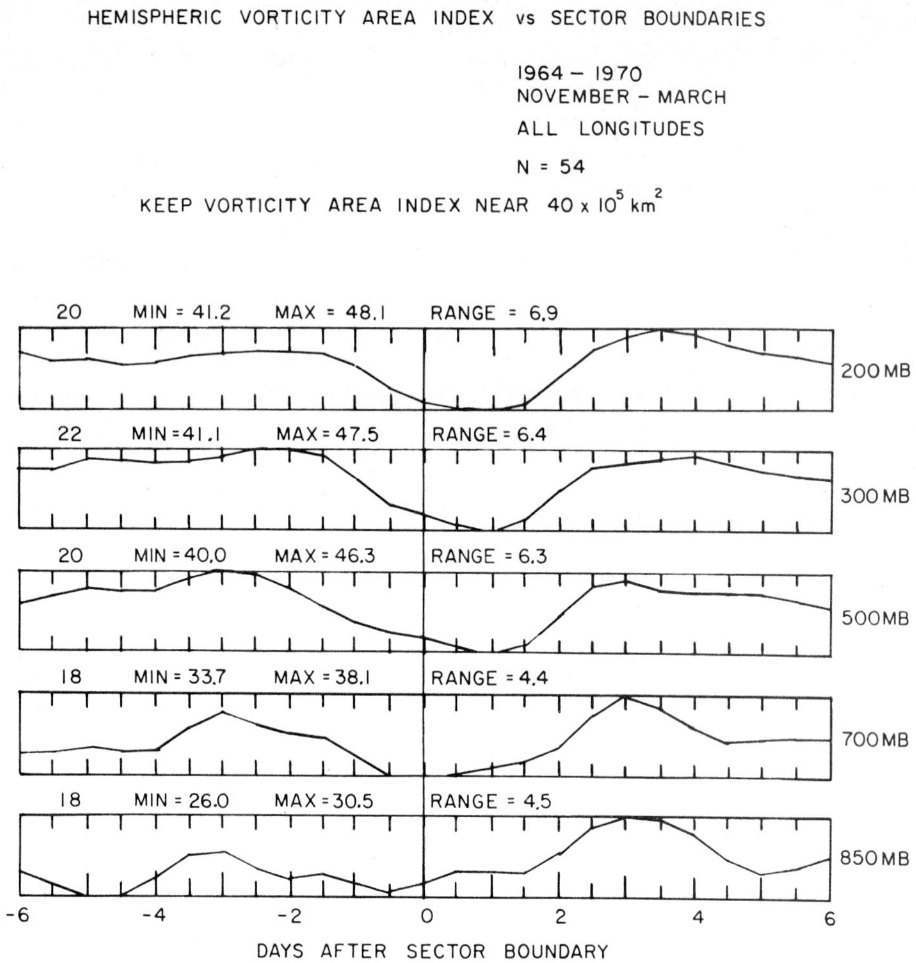

Fig. 4. Same format as Figure 3; the vorticity discriminator has been varied to bring the area index close to 40×10^5 km^2 and is given at the left end of each strip.

of the height difference to passage of a sector boundary is shown in Figure 5 for all standard levels up to 30 mbar. A persistent feature of the response is the decrease in height difference starting at the time of passage of the boundary and being maximal 3–4 days after the boundary. This decrease corresponds to a 'flattening' of the height profile along a meridian. In interpreting Figure 5 two circumstances must be taken into account. First, there is the pronounced seasonal variation of the slope as mid-winter is approached, and secondly it must be remembered that there is often another sector boundary about 7 days before a sector boundary, such that 3–4 days *before* the zero-day of our analysis we may see the effects of the preceding boundary. There are 32 boundaries in the first half of the winter but only 22 in the last half, so that residual trends become apparent: increasing height difference with time at strato-spheric levels, but decreasing in the troposphere.

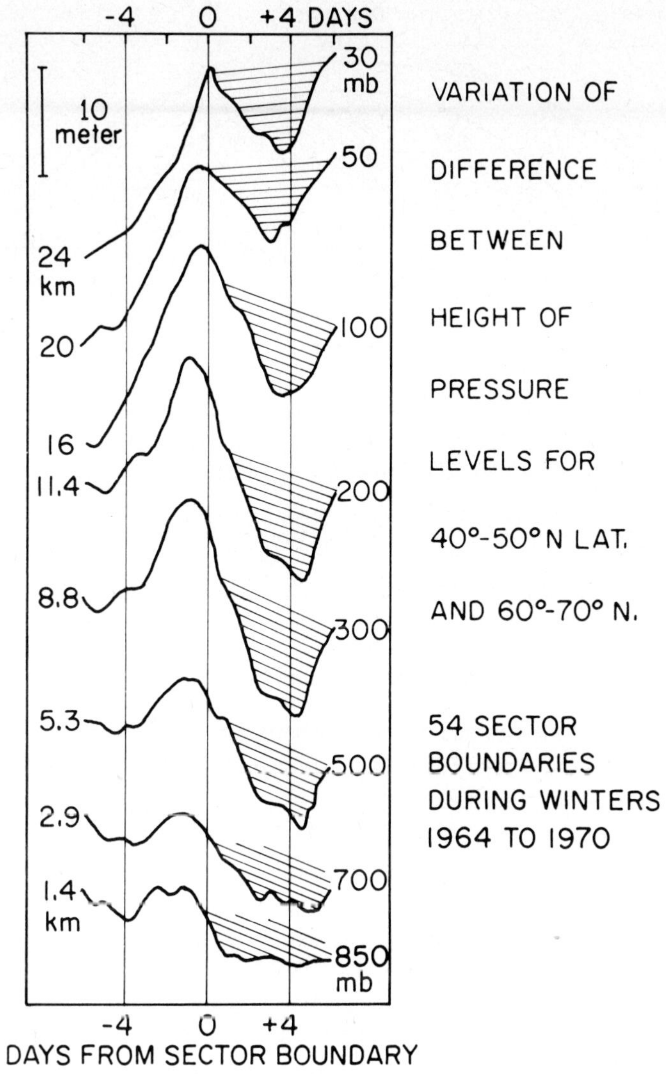

Fig. 5. Superposed epoch analysis of the difference between the height of constant pressure levels for the two latitude belts 40°–50°N and 60°–70°N. The same sector boundaries as in Figure 1 are used. The approximate height of each level is also given.

A crude, schematic picture of the physical changes in the pressure distribution after the passage of a sector boundary is given as Figure 6. Following a sector boundary the height of all pressure surfaces increases at high latitudes leading to anticyclogenesis (high pressure system), whereas at midlatitudes the height of the pressure levels decreases leading to low pressure systems or deepening existing ones with an overall result of increased storminess and precipitation. This picture is quite consistent with the vorticity analysis which indicated an increase in absolute vorticity about $3\frac{1}{2}$ days

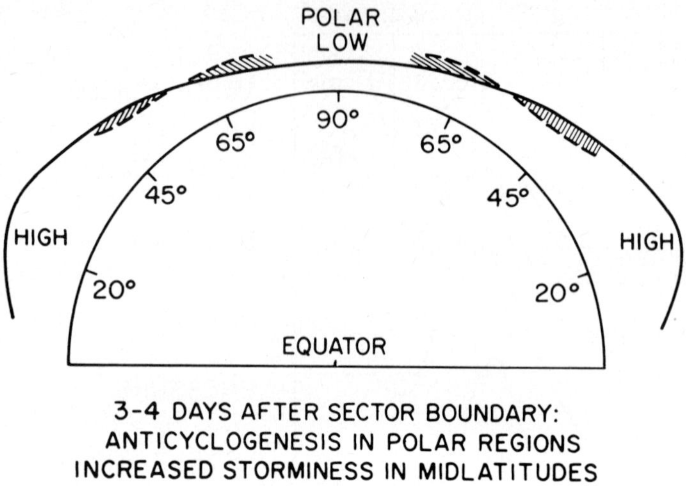

Fig. 6. Schematic of the change in meridional profile for a pressure level in response to passage of a sector boundary.

after the boundary with the same consequences at midlatitudes where most of the vorticity centers are.

6. Geomagnetic Activity and Sector Structure

It is well known that the passage of a sector boundary is associated with a transient increase in geomagnetic activity as measured for example by the K_p index. However, this increase is usually quite small; K_p increases on the average from a minimum of 1_+ just before the boundary to about 2_+ or 3_- a day after the boundary. Numerous authors have searched for atmospheric responses to strong geomagnetic activity and the claims are legio. The effects reviewed here are not particularly associated with strong or not even moderate geomagnetic disturbances. On the other hand, it may well be that many of the effects claimed to follow geomagnetic activity are related to the sector structure. Solar proton flares occur most often near solar sector boundaries. Geomagnetic sudden commencements show a marked tendency to occur near sector boundaries. High solar wind velocities, densities and interplanetary magnetic field strengths also are often observed near sector boundaries. It is not inconceivable that at least some of the many past investigations relating geomagnetic activity and the weather actually correlated sector structure and weather instead.

Rather, the relationship between the sector structure and the weather reviewed in the present paper suggests that it is not geomagnetic activity as such that may influence the lower atmosphere but that geomagnetic activity and some weather changes are results of the same basic mechanism which still waits to be uncovered.

Acknowledgements

This work was supported in part by the Office of Naval Research under Contract

N00014-67-A-0112-0068, by the National Aeronautics and Space Administration under Grant NGR 05-020-559, and by the Atmospheric Sciences Section of the National Science Foundation under Grant GA-31138.

References

Pedlosky, J.: 1971, in W. H. Reid (ed.), *Mathematical Problems in the Geophysical Sciences*, vol. 1, American Mathematical Society, Providence.
Roberts, W. O. and Olson, R. H.: 1973, *J. Atmospheric Sci.* **30**, 135.
Svalgaard, L., Wilcox, J. M., Roberts, W. O., and Olson, R. H.: 1973 (in preparation).
Wilcox, J. M.: 1968, *Space Sci. Rev.* **8**, 258.
Wilcox, J. M. and Colburn, D. S.: 1972, *J. Geophys. Res.* **77**, 751
Wilcox, J. M. and Colburn, D. S.: 1973 (in preparation).
Wilcox, J. M., Scherrer, P. H., Svalgaard, L., Roberts, W. O., and Olson, R. H.: 1973, *Science* **180**, 185.

Appendix

List of observed and Well-defined Sector Boundaries

The date, sign change (+away, −toward), and time (in 3-h intervals) is given for all observed sector boundaries with at least four days of opposite field polarity on each side of the boundary. The notation 8-1 means that the boundary occurred between the last 3-h interval of that day and the first 3-h interval of the next day.

Observed and well-defined sector boundaries

Year	Day of Year	Sign	Date	Time
1962	253	+, −	Sept 10	8–1
	269	−, +	Sept 26	3–4
	281	+, −	Oct 8	4–5
	293	, +	Oct 20	8–1
1963	336	−, +	Dec 2	8 1
	346	+, −	Dec 12	4–3 (gap)
	354	−, +	Dec 20	1–2
1964	007	+, −	Jan 7	7–8
	016	−, +	Jan 16	2–2 (gap)
	023	+, −	Jan 23	3–4
	035	+, −	Feb 4	2–3
	284	−, +	Oct 10	6–7 (1 day gap)
	291	+, −	Oct 17	7–8
	297	−, +	Oct 23	6–8 (1 day gap)
	306	+, −	Nov 1	5–6
	312	−, +	Nov 7	2–1 (gap)
	320	+, −	Nov 15	5–6
	325	−, +	Nov 20	3–2 (gap)
	332	+, −	Nov 27	7–8
	341	−, +	Dec 6	4–5
	345	+, −	Dec 10	8–1

Year	Day of Year	Sign	Date	Time
	349	−, +	Dec 14	8–1
	361	+, −	Dec 26	1–2
1965	002	−, +	Jan 2	1–2
	008	+, −	Jan 8	1–2
	012	−, +	Jan 12	2–3
	032	+, −	Febr 1	8–1
	125	+, −	May 5	4–5
	153	+, −	June 2	8–1
	161	−, +	June 10	2–3
	230	−, +	Aug 18	7–6 (gap)
	235	+, −	Aug 23	5–7 (gap)
	259	−, +	Sept 16	2–3
1966	001	+, −	Jan 1	6–7 (1 day gap)
	032	+, −	Febr 1	4–5
	043	−, +	Febr 12	2–3
	062	+, −	March 3	3–4
	067	−, +	March 8	2–3
	089	+, −	March 30	2–3
	099	−, +	April 9	1–2
	127	−, +	May 7	8–1
	249	−, +	Sept 6	5–6
	257	+, −	Sept 14	6–7
	276	−, +	Oct 3	6–7
	285	+, −	Oct 12	2–3
	303	−, +	Oct 30	5–6
	312	+, −	Nov 8	4–5
	331	−, +	Nov 27	7–8
	338	+, −	Dec 4	3–4
1967	001	+, −	Jan 1	7–8
	013	+, −	Jan 13	3–4
	018	−, +	Jan 18	2–3 (1 day gap)
	081	−, +	March 22	7–8
	216	−, +	Aug 4	5–6
	242	−, +	Aug 30	2–3 (1 day gap)
	249	+, −	Sept 6	6–7
	270	−, +	Sept 27	3–4
	276	+, −	Oct 3	1–2
	297	−, +	Oct 24	2–3
	324	−, +	Nov 20	4–5
	338	+, −	Dec 4	5–6
1968	001	+, −	Jan 1	6–5 (gap)
	028	+, −	Jan 28	8–1
	042	−, +	Febr 11	3–4
	057	+, −	Febr 26	6–7
	070	−, +	March 10	4–5
	083	+, −	March 23	5–6
	096	−, +	April 5	7–8
	112	+, −	April 21	3–4
	123	−, +	May 2	1–2
	138	+, −	May 17	5–6

Year	Day of Year	Sign	Date	Time
	185	−, +	July 3	3–4
	191	+, −	July 9	8–1
	199	−, +	July 17	4–5
	207	+, −	July 25	4–5
	213	−, +	July 31	7–8
	226	−, +	Aug 13	7–8
	234	+, −	Aug 21	2–3
	263	+, −	Sept 19	2–3
	290	+, −	Oct 16	5–6
	318	+, −	Nov 13	2–3
	334	−, +	Nov 29	6–8 (gap)
	345	+, −	Dec 10	2–3
	359	−, +	Dec 24	6–7
1969	006	+, −	Jan 6	5–6
	023	−, +	Jan 23	8–1
	033	+, −	Febr 2	5–6
	050	−, +	Febr 19	2–3
	090	+, −	March 31	6–7
	110	−, +	April 20	7–1 (gap)
	119	+, −	April 29	3–4
	127	−, +	May 7	6–3 (gap)
	132	+, −	May 12	8–2 (gap)
	138	−, +	May 18	6–7
	147	+, −	May 27	1–2
	165	−, +	June 14	3–4
	192	−, +	July 11	2–3
	202	+, −	July 21	5–6
	219	−, +	Aug 7	6–7
	248	−, +	Sept 5	3–4
	303	−, +	Oct 30	8–1
	330	−, +	Nov 26	5–6
	343	+, −	Dec 9	1–2
	356	−, +	Dec 22	7–8
1970	040	−, +	Febr 9	7–8
	067	−, +	March 8	8–1
	120	−, +	April 30	3–4
	131	+, −	May 11	6–7
	158	+, −	June 7	6–7
	243	+, −	Aug 31	8–5 (gap)
	309	−, +	Nov 5	3–4
	328	+, −	Nov 24	3–4

THE JOVIAN RADIATION BELT

D. B. BEARD and J. L. LUTHEY

Dept. of Physics and Astronomy, The University of Kansas, Lawrence, Kan., U.S.A.

Jupiter is a copious source of radiowaves having a wavelength of 3–75 cm. The radiation is 20% or so linearly polarized perpendicular to the planet rotation axis, does not vary much in intensity over the entire range of observed frequencies, and changes slowly in intensity over a time interval of the order of a few years. Chang and Davis (1962) have shown that the source is a belt of very energetic electrons trapped in the dipole field of the planet radiating by synchrotron emission. Spatially resolved observations at different frequencies have suggested to us a method by which the electron number density and energy can be determined everywhere in the equatorial plane.

Figure 1 taken from Berge (1966) illustrates the general radiation pattern observed with widely spaced antenna arrays. One can also represent the data as radiation

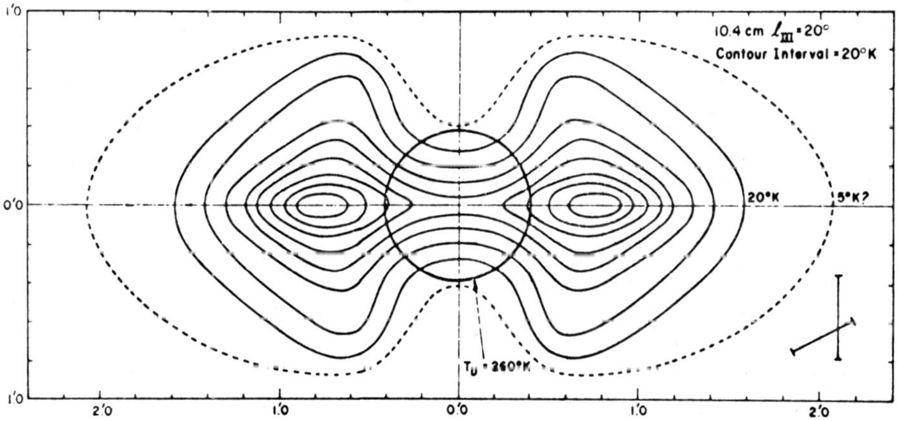

Fig. 1. Temperature contour map of Jupiter. The 10.4-cm brightness distribution determined for $l_{III} = 20°$. The scale, in minutes of arc refers to a distance of 4.04 AU (Berge, 1966).

received from strip scans in the equatorial plane, that is, radiation received from narrow strips parallel to the rotation axis and spaced out along the planet equator. This is shown in Figure 2 for observations by Berge at 10.4 cm wavelength and by Branson at 21.3 cm (Branson 1968). The contribution from a planet surface temperature of 260° has been subtracted out from Berge's data. We divided up the region around the planet into cylindrical shells $\frac{1}{4}$ Jupiter radii thick as shown in Figure 3 and

D. E. Page (ed.), Correlated Interplanetary and Magnetospheric Observations, 641–647. *All Rights Reserved.*

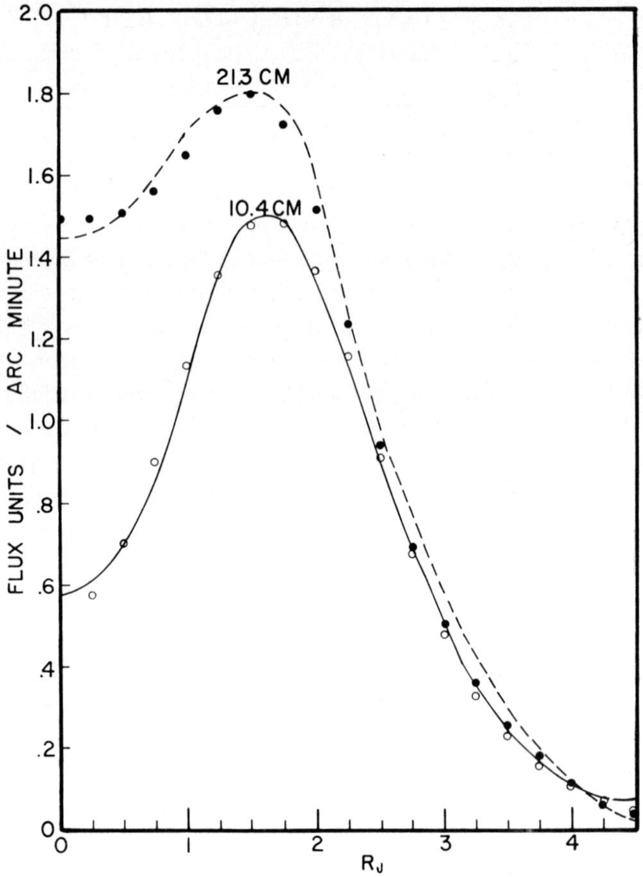

Fig. 2. Equatorial strip scans of Jupiter's decimetric radio emission at two wavelengths. The 10.4-cm radiation has had a 260° planet surface radiation subtracted out (Berge 10.4 cm, Branson 21.3 cm). The points are the calculated strip scans from the source emissivities in Figure 4.

assumed that the source intensity in each shell was constant throughout the shell except for effects of pitch angle distribution.

In this way we were able to relate the total intensity S seen in each strip to the source intensity D at each frequency in each concentric cylindrical shell.

$$S_1 = V_{11} D_1$$

$$S_2 = V_{22} D_2 + V_{21} D_1 \text{ etc.}$$

We then took account of the finite antenna resolution for the observations and arrived at a matrix connecting the observations with the source intensities

$$S_i = \sum_k F_{ik} D_k .$$

We adjusted the D's to give a best fit to the observations S and obtained source intensities at each frequency as a function of equatorial distance from the planet

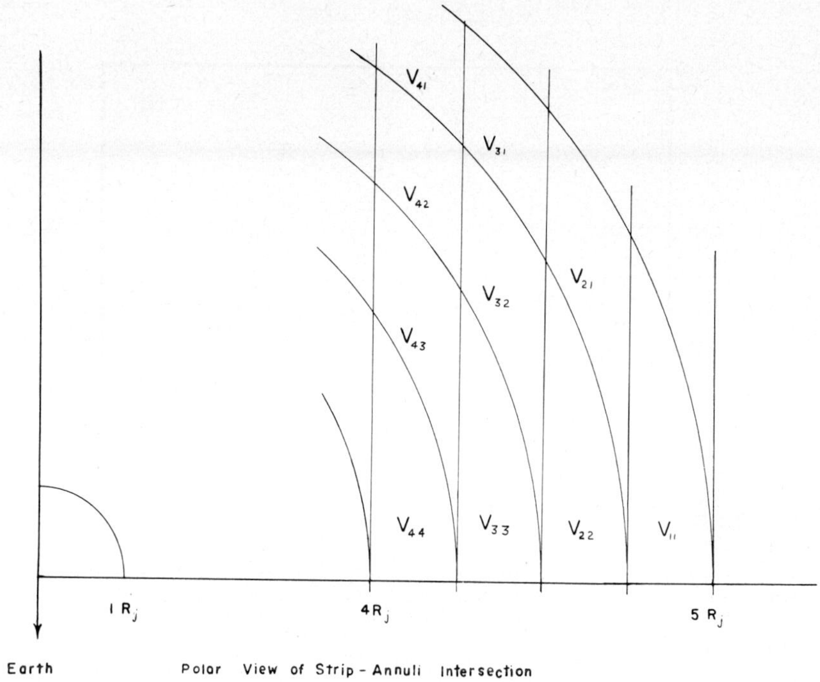

Fig. 3. Equatorial plane projection illustrating the geometry used in obtaining source emissivities from the strip scans.

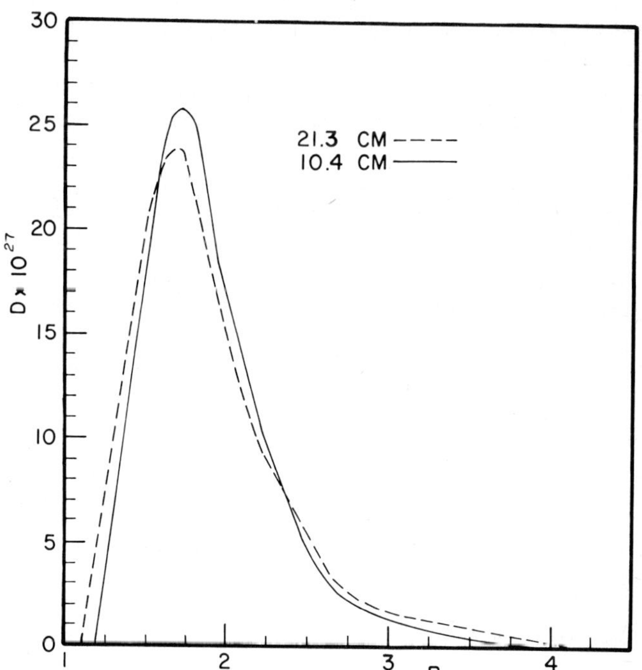

Fig. 4. Source emissivities (radiation per volume of source) as a function of equatorial distance for two wavelengths.

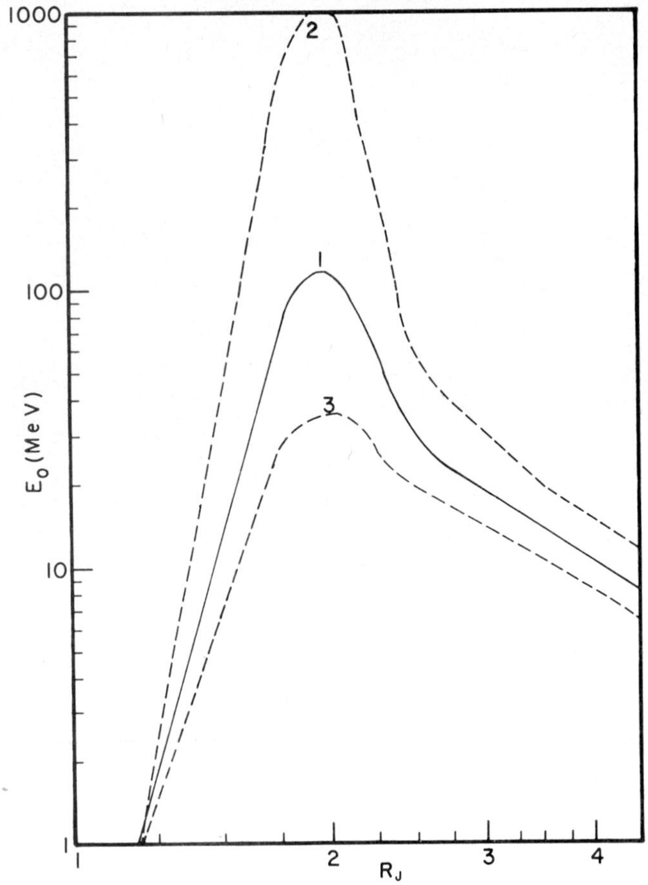

Fig. 5. Electron temperature as a function of equatorial distance. The dashed curves are (2) raising the intensity of the 10.4 cm radiation relative to the 21.3 cm by 10% (3) lowering it by 10%.

shown in Figure 4. The S's from these assumed D's are shown as open and closed circles in Figure 2.

The theoretical radiation from energetic electrons in a magnetic field must be integrated over the assumed isotropic pitch angle distribution and over its energy distribution. We assumed both a Maxwell Boltzmann energy distribution e^{-E/E_0} with E_0 a yet to be determined electron temperature and a power law energy distribution $E^{-\gamma}$ with γ an undetermined parameter. In this way we obtained the theoretical source intensity per unit volumes as a function of frequency, field, energy parameter, and electron density

$$D_{\text{theo}} = Nf(\omega, H, E_0)$$

or

$$D_{\text{theo}} = Nf(\omega, H, \gamma).$$

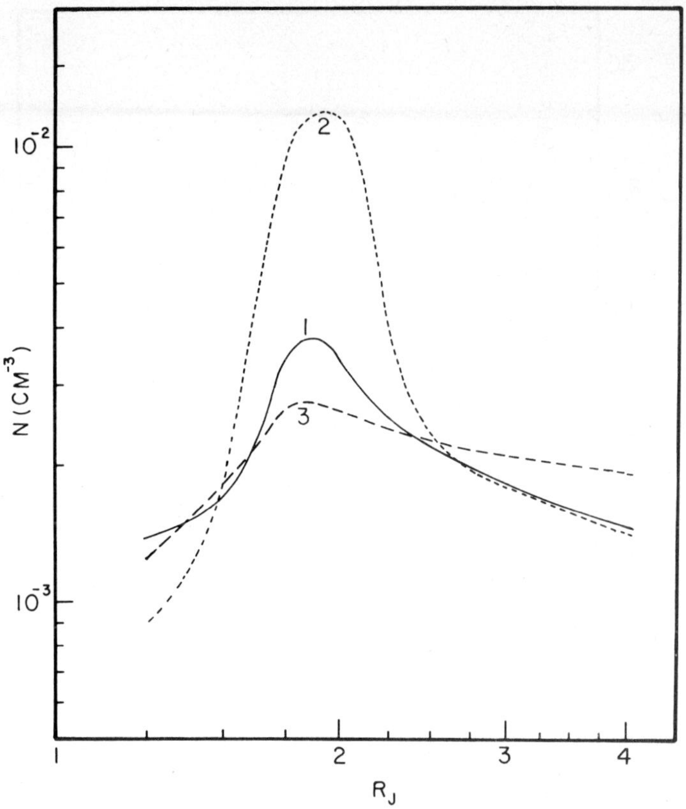

Fig. 6. Electron number density as a function of equatorial distance. The dashed curves result from changing the relative intensity for the two wavelengths by 10% as stated in the caption of Figure 5.

The magnetic field H is for a dipole field; observations of long wave-length cyclotron radiation indicate a surface equatorial field of 7 G which we assumed for the magnetic field strength, i.e. $H = (7/r^3)$ G.

By comparing the theoretical emission to the two sets of data at two different frequencies we are able to solve for the two unknowns, N and E_0. Our results are shown in Figures 5–7. The dashed curves are obtained by changing the intensity of one observation relative to the other by 10%. This illustrates the central importance of having available accurate relative intensities at different frequencies. The work to date suffers in that the two measurements were taken at different epochs nearly four years apart and the intensity may very well have changed by as much as 10% in that interval. We are now analyzing data taken at the same epoch and expect to have a more reliable result soon.

The time in which an energetic electron loses half its energy to radiation is given by

$$\tau_{1/2} \sim 1/E_0 H^2 .$$

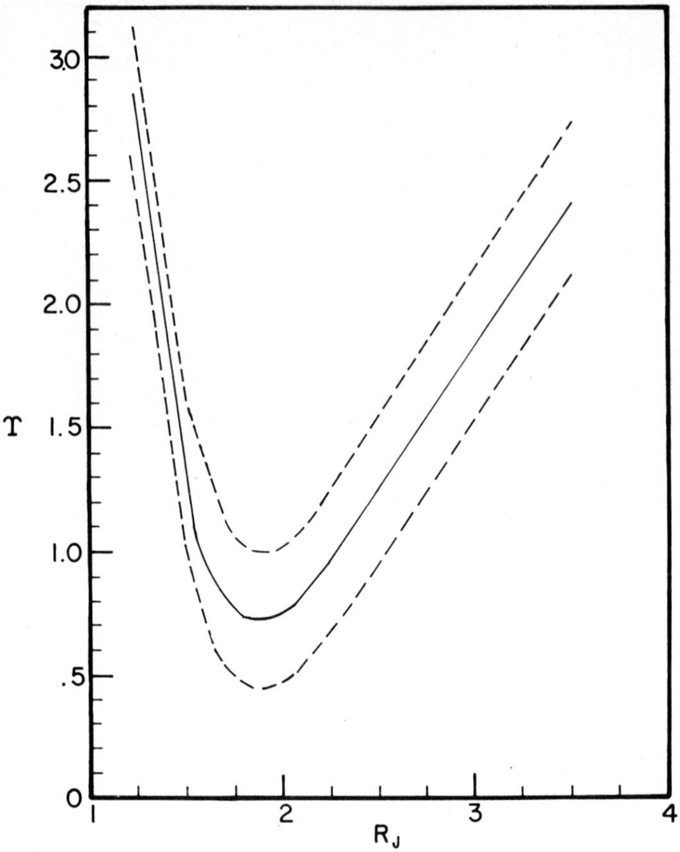

Fig. 7. The exponent γ in the energy spectrum $E^{-\gamma}$ as a function of equatorial distance. The dashed lines result if the relative intensities are changed by 10%.

The half-life for our assumed H and the E_0 we found is shown in Figure 8. Note that the half-life is very long beyond the radiation peak and doesn't seem to affect anything. The half-life is very short inside of the peak and appears to be the reason the electron energy decreases so fast near the planet. Note further that the half-life is only a few months at and within the peak. This means that the injection time is of the order of months indicating a surprisingly fast source.

It looks like electrons having an energy of tens of kilovolts in the magnetosheath diffuse in conserving the first adiabatic invariant and have their energies increased to tens of MeV at 2 R_J. They then radiate copiously and lose their energy as they diffuse farther in.

If electrons can do this, protons may do it also. In the proton case the particles do not lose their energy by radiation. This presents the possibility that there may be a

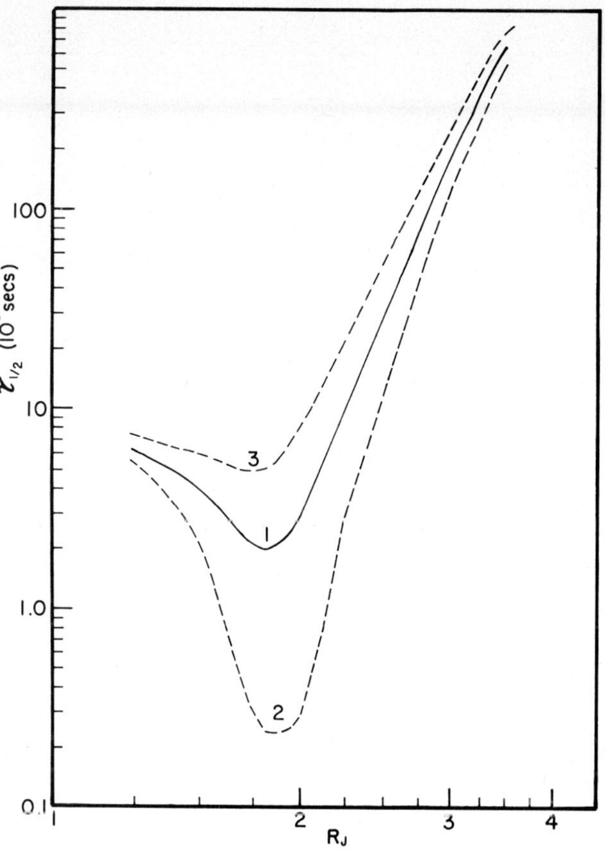

Fig. 8. The half-life of the energetic electrons as a function of equatorial distance. The dashed curves result from changing the relative intensities by 10% as stated in the caption to Figure 5.

very dense population of high energy protons of the order of 100 MeV close to the planet.

References

Berge, G. L.: 1966, *Astrophys. J.* **146**, 767.
Branson, N. F. B. A.: 1968, *Monthly Notices Roy. Astron. Soc.* **139**, 155.
Chang, D. B. and Davis Jr., L.: 1962, *Astrophys. J.* **136**, 567.

THE DESIGN AND DEVELOPMENT OF A SPACE LABORATORY TO CONDUCT MAGNETOSPHERIC AND PLASMA RESEARCH

ALAN ROSEN

TRW Systems Group, Redondo Beach, Calif., U.S.A.

1. Introduction

During the past two years a study was performed by the Space Science Group at TRW Systems to identify the outstanding magnetospheric and plasma physics problems of the Eighties, to define the experimental programs required to solve these problems, and finally to design a plasma physics space laboratory to be housed in the space shuttle to carry out these programs. This study was performed for Marshall Space Flight Center of the NASA under the direction of Mr William T. Roberts.

The unsolved magnetospheric problems and the experiment concepts were obtained from replies to a worldwide solicitation of prominent space and plasma physicists. A questionnaire, together with a brief description of possible shuttle mission capabilities, was circulated to well over 200 active investigators all over the world. The replies consisted of 206 experiment concepts, comments and ideas. These replies were studied by three scientific working groups and a science advisory board under the auspices of the National Aeronautics and Space Administration. Eight areas of investigation were identified during the course of the study. A magnetospheric and plasma physics laboratory, suitable for inclusion in the space shuttle payload bay was designed so that the experiment program identified during the course of the study could be carried out. The laboratory design and mission schedule are based on the data gathered and requirements defined by the scientists and magnetospheric physics investigators responding to the questionnaire.

2. Scientific Problems and Experimental Requirements

The experimental requirements for each area of investigation were deduced from the correspondence file generated from the worldwide solicitation. For each reply a format sheet was developed which gave a one-page synopsis of the experiment objectives and instrumentation requirements. These sheets were than assembled according to eight areas of research and published as a synopsis volume for further use in the study and as reference for others interested in plasma physics experiments in space. The eight areas break down into two broad disciplines of research:

1. Geophysical Studies of the Dynamics and Structure of the Magnetosphere
   1. Wave Characteristics
   2. Wave Particle Interactions
   3. Magnetospheric Modifications

D. E. Page (ed.), Correlated Interplanetary and Magnetospheric Observations, 649–662. *All Rights Reserved.*

4. Beam-Plasma Interactions
5. Energetic Particles and Tracers

II. Studies in the Field of Plasma Physics
1. Plasma Physics in Space
2. Wake and Sheath Studies
3. Propulsion and Devices

The eight areas of research were useful in finding the relationships and interrelationships between experiments and delineating various groups performing common experiments in terms of common interest. The demarcation between areas is not sharp and there is a broad overlap in experiments and interest in certain areas.

The following sections outline the scientific problems in each area, the experimental requirements that must be met to carry out research in that area, and the broad constraints imposed on a space system by these requirements.

2.1. Wave-particle interactions – particle acceleration and scattering

The significant unsolved problems identified in the areas of wave characteristics, wave-particle interactions, and beam-plasma interactions relate to dynamic processes within the magnetosphere. Some of the major unanswered questions are: Do electromagnetic waves cause sufficient pitch angle scattering to account for the observed precipitation and stable trapping limits? How are triggered emissions, chorus, and micropulsations generated? What determines ionospheric reflection and transmission for VLF? How effective are electrostatic waves in energizing precipitated particles? Are these phenomena linear, quasi-linear or non-linear?

The area of wave characteristics includes experiments emphasizing the study of wave propagation and damping characteristics, plasma instabilities, and wave-wave interactions. Most of the experiments in this area are aimed at studies of linear and non-linear dispersion relations and damping and growth of waves. The suggested experimental techniques involve the use of large transmitting antennas operating at frequencies below the ion plasma frequency, receivers placed remote from the sources, and diagnostic equipment to measure the dispersion of the sheath in the intervening region. The receivers and transmitters must be designed to measure resonances, long delay echos, non-linear effects and three-wave interactions, and the generation of low frequency electromagnetic waves from within the plasma at ULF.

The experiments in the Beam-Plasma Interactions area emphasize the interactions of beams with the ambient plasma. This category includes artificial auroras, beam instability and turbulence, return currents, neutralization, collisionless dissipation and acceleration mechanisms. The suggested experimental techniques involve the use of accelerators that generate versatile beams with known characteristics; and diagnostic equipment to measure electromagnetic and electrostatic waves generated by plasma instabilities and the particle motions affected by the wave fields. Waves, or particles, generated by the space laboratory must be measured on remote platforms placed on long booms or from subsatellites remote from the laboratory. The generation of natural magnetospheric wave modes such as chorus, high frequency electrostatic emissions, VLF and ELF hiss, and triggered emissions, require long antennas and remote platforms for viewing. Many of the observations in this area would provide important data if performed in conjunction with active experiments involving the injection of cold plasma in regions beyond the plasma pause, the emission of powerful electromagnetic waves, and correlated ground observations of precipitated particles and waves.

Experiments in the area of wave-particle interactions emphasize the interactions of certain plasma particles with locally generated waves. Such experiments utilize wave modes extending over a frequency range of fractions of Hz to many MHz and the interactions of interest generally take place near the lines of force or on field lines passing through the source. In the low density space plasma, the effects of organized groups of particles are manifest by their collective oscillations, or plasma waves. The wave-particle interactions play an important role in controlling the dynamics of the medium. Measurements to be performed generally include cyclotron resonance instabilities, pitch angle diffusion, acceleration, turbulent resistivity and the generation of electromagnetic waves by phased electron or proton gun arrays. The equipment required to perform experiments in this area includes transmitting antennas, phased electron and ion accelerators used as sources, and diagnostic equipment placed in a region remote from the sources, either on extended booms, or a subsatellite located at a known position from the source.

2.2. Magnetospheric Configuration, Particle Trajectories, and Convection

The experiments in the area of Energetic Particle and Tracer experiments are designed to shed light on the configuration of the geomagnetic field and the processes that provide stable or quasi-stable trapping of energetic particles. Some of the significant, unsolved problems in this area revolve around the questions of: How do solar wind particles enter the magnetosphere? In what regions are they energized? What is the distribution of electric fields in space? What is the dependence of convection patterns on ionospheric conditions? Do particles with different Z/A values experience the same acceleration and loss mechanisms?

The suggested experimental technique involves ion releases and subsequent detection and tracing of particle orbits and life histories, a study of field line topology, and parallel and perpendicular electric fields. Included in this category are shaped charges fired from the space laboratory, large releases to carry out magnetospheric modification experiments, and particle accelerators carried on the laboratory to provide variable ion beams with known characteristics. Sensors, remote from the laboratory, either on a long boom or on a subsatellite are required in order to detect and study the effects of the intervening magnetosphere. Studies of large scale field configurations, particle trajectories, and convection, must also be performed in regions remote from the space laboratory. The requirement for long booms and subsatellites is reflected in each of the suggested experiments in this area.

2.3. Magnetospheric modifications

Some of the more dramatic experiments were suggested in the area of magnetospheric modification. These are experiments designed to produce large scale perturbations in the magnetosphere or ionosphere and to identify the underlying mechanisms. Experiments in this area include studies of radiation belt precipitation by changing wave growth rates, generation of high power VLF waves to trigger precipitation events, and ionospheric heating involving parametric instabilities and RF heating. Also included in this area are studies of the response of the ionosphere to controlled fluxes of suprathermal particles, modification of ionospheric conductivity, artificial mid-latitude SAR red arcs, and models of solar flare radiation mechanisms. Some examples of suggested experimental techniques involve the injection of fresh particles into trapped orbits with a particle accelerator to provide beams with known characteristics, the generation of high powered VLF waves by means of a VLF antenna, and diagnostic equipment preferably placed remote from the source to measure the state of the magnetosphere over a large volume of space. In conjunction with these experiments it is necessary to perform correlated ground observations of precipitated particles and wave reflection and transmission properties.

2.4. Plasma physics in space

Experiments in this category are essentially those laboratory type experiments that can be performed in space to take advantage of large volumes, high vacuum conditions, weightlessness, and the ambient plasma present in the magnetosphere. Besides plasma physics, this general category includes wake and sheath studies and studies of propulsion techniques and other plasma devices.

In the area of plasma physics, the proposed experiments involve studies of levitron type confinement devices utilizing deployed magnets, studies of the motion and configuration of spinning conducting fluids, studies of neutral gas plasma beam interactions, and the generalized ohm's law. The equipment required to perform experiments in this area include, besides the conventional plasma diagnostic equipment, those special purpose components that would be required to generate large magnetic fields and MHD phenomena.

Wake and sheath experiments are designed to study the wake or sheath around orbital bodies. The experiments suggested in this area include studies of the validity of current theories describing the size and shape of the perturbed regions, the potential distributions within the regions and the Cherenkov cones in the wakes. Studies of the stability of the wake and sheath with different surface materials and different shapes and different biasing and the effects of wake and sheath on antenna impedance and particle probes will also be undertaken. Included in this area are experiments involving generalized terella simulations with large magnets. The suggested experimental techniques involve the deployment of known targets and the use of suitable diagnostic instruments mounted on a remote maneuverable platform about the deployed target. A remote maneuverable boom may be used to study the characteristics of the wake and sheath in the region close to the target. At distances large compared to the target diameter, remote and maneuverable subsatellites carrying the diagnostic equipment would be advantageous.

The area of propulsion and devices includes experiments that have applied goals such as ion thruster studies, or the application of $v \times B$ forces. Also included in this area are studies to develop new diagnostic techniques and resolve outstanding problems connected with the use of standard diagnostics on unmanned rockets and satellites. The suggested experimental techniques involve use of Langmuir probes, Faraday cups, dc electric field and magnetic field probes to measure small plasma drifts, the doppler shifts, and electric and magnetic fields. Experiments with an ultimate propulsion application include studies of MPD arcs in large volumes and plasma beam/ambient plasma interactions.

3. The Basic Requirements of a Plasma Physics and Environmental Perturbation Laboratory (PPEPL)

In order to perform the array of experiments identified in the previous sections, a plasma physics laboratory must include electron and ion accelerators, long maneuvering booms, large gimbaled platforms, subsatellites, arrays of transmitters and receivers and the standard diagnostic complement of a plasma physics laboratory. With such a laboratory the same payload subsystems can be used repeatedly to perform different experiments. An example of the common or simultaneous usage of the major subsystems is illustrated in Table I. This table was deduced from the synopsis volume of experiment concepts that was described in Section 2. A black dot indicates that at least 75% of the experimenters in a given area had a firm requirement for the subsystem (booms, subsatellites, accelerators, transmitters, or plasma diagnostic array). An open circle indicates that at least 75% felt that the subsystem was desirable, but not essential to the performance of the experiments. It was determined that the simultaneous use of four of the five subsystems was required by over 75% of the experiment concepts, and the simultaneous use of all five subsystems was desired in approximately 80% of the experiment concepts. Thus it is desirable to include all five major subsystems in a plasma physics laboratory. In the following sections the physical description of each of the major subsystems of the plasma physics laboratory will be presented.

The plasma physics laboratory could be incorporated in the space shuttle using the dedicated laboratory approach or the modular birdcage approach. In the modular approach, all the major subsystems of the plasma physics laboratory would be integrated into one or more removable modules. These modules could then be integrated into the space laboratory, which, in turn, is carried by the space shuttle.

3.1. Long-boom assemblies and subsatellites

The requirement that active perturbing and sensing instruments be remotely located from the large shuttle system and from each other dictates that a baseline plasma physics facility include long-boom assemblies and subsatellites. The synopsis volume

TABLE I

Space physics PPEPL experiment sub-areas; Simultaneous usage of the major subsystems

	Major instrument and mission rqmts.					
	Maneuverable Long Booms	Subsatellites	e^- and Ion accelerators	Transmitting and receiving antenna	Plasma diagnostic array	Other
Geophysical studies of the dynamics and structure of the magnetosphere						
Wave characteristics (35 Experiment concepts)	●	●	○	●	●	
Wave-particle interactions (18 Experiment concepts)	●	●	●	●	●	
Magnetospheric modification (21 Experiment concepts)	●	●	●	●	●	Correlated ground observations
Beam-plasma interactions (23 Experiment concepts)	●	●	●	○	●	
Energetic particles and tracers (20 Experiment concepts)	●	●	●	○	●	Ion injection and release capability
Studies in the field of plasma physics						
Plasma physics in space (15 Experiment concepts)	●	○	○	●	●	Large magnets
Wake and sheath (29 Experiment concepts)	●	●			●	Deployment of large targets
Propulsion and devices (22 Experiment concepts)	●		●		●	

● Indicates that 75% of the experiment concepts in that area have a firm requirement for the subsystem.

○ Indicates that 75% of the experiment concepts would benefit from the subsystem, however, the subsystem is *not* a firm requirement.

of experiment concepts was used to perform an analysis of the extent to which remote deployment methods were required. In all, 166 experiment concepts were examined to determine the degree to which the scientific objectives of each experiment could be fulfilled by the addition of a varying number of long maneuverable booms or a subsatellite. Figure 1 shows the results of this examination. Each of the five plots represents a different fraction of the experiment objective that is fulfilled. If only 20% of the experiment objectives need be fulfilled, then 140 experiments could be carried out with the one boom configuration. It is shown in Figure 1 that approximately half of the experiments could have 80% of their scientific objectives fulfilled with a two boom configuration and two-thirds of the experiments could have 60% of their scientific objectives fulfilled with such a configuration.

The addition of subsatellites to the PPEPL greatly increases the depth to which most of the experiments can be carried. For example, whereas half of the 166 experi-

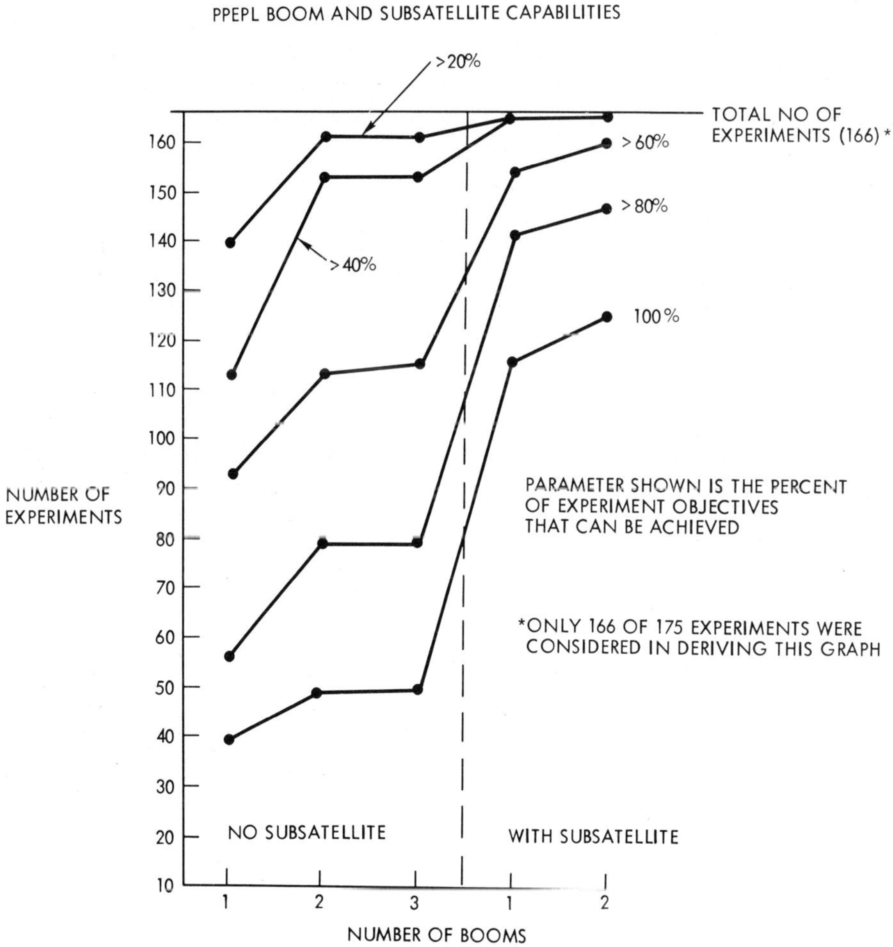

Fig. 1.

ments could achieve 80% of the experiment objectives with a two boom configuration, with the addition of a subsatellite more than 80% of the experiments achieved 80% of their experiment objectives.

Subsatellites increase the distance over which measurements can be performed and thereby broaden the spatial domain over which experiments can be carried out. With subsatellites it is possible to study characteristics of long wave lengths, plasma waves, provide correlations with plasma parameters at a point remote from the shuttle, and perform studies at magnetic conjugate points.

3.2. Accelerators, Transmitters and Plasma Diagnostic Array

A number of high powered electron and ion beam guns and plasma accelerators have been considered in the course of the study. These guns should be mounted on a pallet, external to the pressurized, manned plasma physics laboratory, exposed to the space environment. The requirements placed on the accelerators by the experiments include stability and pointing, beam width, beam current and energy. In order to cover the broad range of energies and species, several types of guns may be required, but all may operate from a common power supply. For example, one electron accelerator being suggested contains a congruent flow, cylindrical geometry, Pierce-type gun. The gun perveance is 10^{-6} A $V^{-3/2}$ yielding a 1-A output at 10 kV; the gun operational range is 10 kV to 50 kV. The likely element for the proton gun is a Von Ardenne source (duoplasmatron) with an initial output of 1 A at 50 kV. The abundance of H^+ is 85% of the ion flow for hydrogen operation. Operation on all other noble gases and on N_2 is possible. The plasma accelerator is a magneto plasma dynamic arc. This plasma device is inherently adaptable to pulsed operation and possesses a wide range of possible output power. Such a device would allow high energy plasma deposition onto a field line in milliseconds. A low energy (5–20 eV) electron gun may also be mounted at the end of a boom, for boom-to-boom transmission studies.

The accelerators should be coupled with the plasma detection system located on the gimbaled platform at the end of one of the long booms. Normally, one of the guns will be oriented in a given direction while the boom mounted equipment is used to measure the characteristics of the beam.

The plasma diagnostic array should consist of a platform which provides common power, data and control interfaces to accommodate a variety of diagnostic instruments requiring pointing. These instruments will be mounted on the platform and used in the analysis of ambient as well as perturbed plasma parameters. Typical instruments to be found on the platform include spectrometers, photometers, and particle sensor clusters. The platform will also be used as a 'standard' environment to test various new instrument concepts and compare the performance of various diagnostic instruments.

A high powered transmitter should be mounted on the pallet. It will be used to modulate long sounding antennas (up to 330 m per element) at frequencies up to ~20 MHz and amplitudes up to 20 kV. The transmitters operate in conjunction with the wave detection instrumentation located on the gimbaled platform at the end of the boom or with sensors on a subsatellite.

4. Versions of the PPEPL

A layout of a typical laboratory is shown in Figure 2. The laboratory consists of a pressurized, manned module, and a pallet or platform. Labels indicate the major equipment and instrumentation. A summary of the basic laboratory feature selected for this layout is presented in Table II.

Near the far end of the pallet are mounted two articulated 50-m deployable booms. The two 50-m booms are shown deployed in Figure 3, along with the 1000 ft VLF/RF transmitting antenna. One of the booms is a passive, or diagnostic, boom while the other boom is used as a perturbing source. The typical complement of equipment for a two boom configuration is given at the bottom of Table II.

The plasma diagnostic instrument array is mounted on a gimbaled platform at the end of the diagnostic boom. This array includes antennas, particle detectors, magnetometers, and other equipment to be used for many experiments, particularly those in the areas of waveparticle and beam-plasma interactions, wave characteristics, devices, and wake and sheat measurements. The layout of this platform is shown in Figure 4.

Two subsidiary 5-m booms are deployed from the above described platform for isolation of electric and magnetic field sensors from the extensive instrumentation on the platform. A 30-m dipole receiving antenna is also deployed from the platform.

The second 50-m boom shown in Figure 3 acts as a perturbation source. As such,

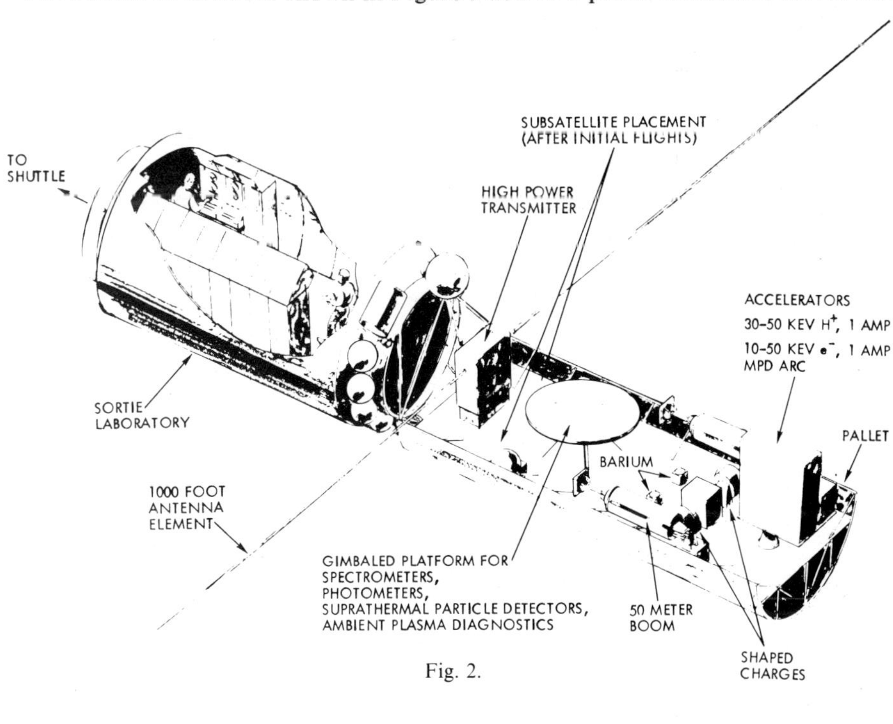

Fig. 2.

TABLE II

Summary of basic laboratory features

- Two 50-m booms; two maneuverable subsatellites
- No cryogenic systems
- Photometer array on gimbaled platform; no spectrometers
- Standardized accelerators
  - 30–50 keV protons, up to one Ampere (on pallet)
  - 10–50 keV electrons, up to one Ampere (on pallet)
  - 5–20 eV electron gun (on boom)
  - High power MPD plasma accelerator (on pallet)
- High powered transmitter only for $f \gtrsim 10^5$ Hz; 1000 ft dipole elements
- Low power transmitter for VLF and below; boom-to-boom transmissions
- Shaped charges, barium canisters will be carried if safety considerations permit
- Complete diagnostic packages

Typical complement for a two-boom configuration

Boom No. 1
- Plasma diagnostic array

Boom No. 2
- Targets for wake-sheath studies; variable sizes, biases, surface properties
- Low energy (5–20 eV) electron gun for measurements of $\mathbf{E}\|$ to $\mathbf{B}$, study of low energy beam-plasma streaming instabilities
- Acoustic (electrostatic) plasma wave generator (parallel grids) for boom-to-boom transmission experiments

balloons or acoustic wave generators, for example, would be deployed from this boom and the effects measured on the second, or diagnostic, boom.

Mounted at the far end of the pallet are high-power electron-ion beam guns complete with power supply. An example of the configuration of the power supply for the ion accelerator is shown in Figure 5. The guns themselves are of several types, but they operate from a common power supply.

On the opposite end of the pallet nearest the sortie lab is mounted a variable transmitter and power supply with associated dipole antenna. Figures 6 and 7 illustrate the configuration of a suitable transmitter and motor drive system. The dipole antenna may be extended to about 330 m per element once the sortie lab and pallet are deployed.

A gimbaled platform some eight feet in diameter is located about halfway between the antenna and the electron-ion beam guns. This gimbaled platform contains optical and particle detectors requiring pointing. These sensors are used for a number of experiments, especially those in the areas of beam plasma interactions, magnetospheric modifications, and energetic particles and tracer experiments.

The pallet provides sufficient area to accommodate other experiment items. For example, canisters containing lithium, barium, or other chemicals may be mounted on the pallet and ejected to carry out ionospheric wind studies, field line tracing, and electric field investigations. In a similar manner, canisters containing inflatable 'wake bodies' may also be ejected, as may maneuverable subsatellites.

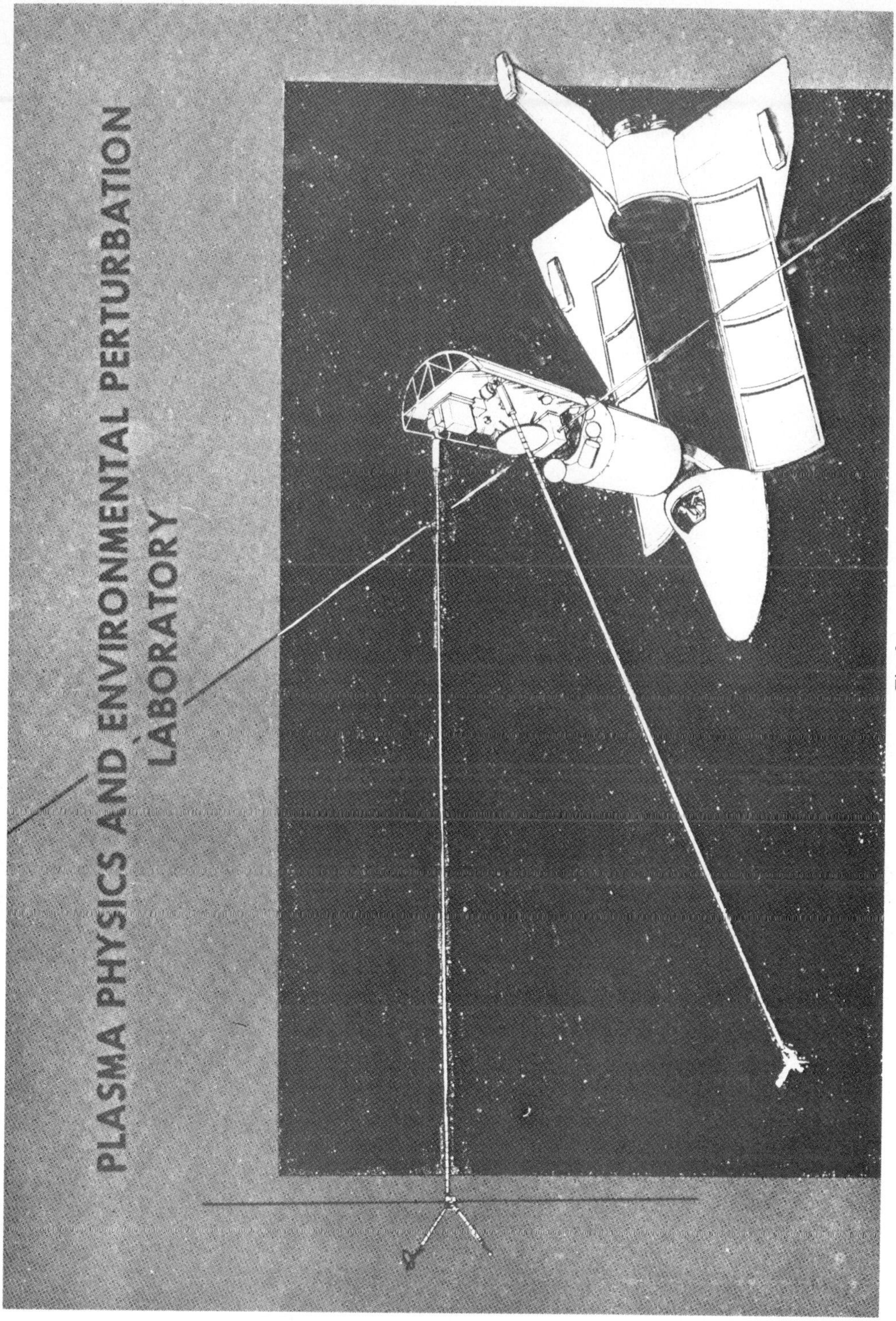
PLASMA PHYSICS AND ENVIRONMENTAL PERTURBATION
LABORATORY

Fig. 3.

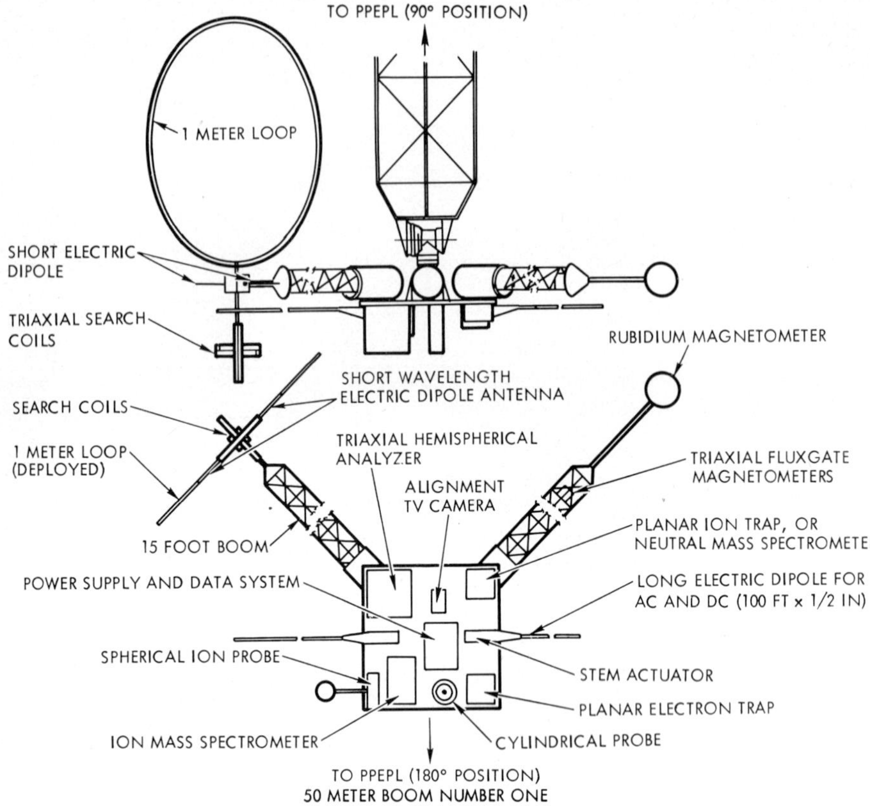

Fig. 4.

Inside the pressurized portion of the laboratory are located the control and display consoles for the sensors, the transmitters and receivers for the RF and VLF experiments, the electron and ion beam control and display consoles, a computer and spectrum analyzer for near real time data evaluation, additional power supplies, general work areas, and recorders.

In light of current NASA austerity guidelines, the entire PPEPL was designed so that only relatively minor manufacturing changes would be required in adapting commercial units to the PPEPL. Almost all display units chosen are commercially available and the design of the control units reflects standard earth laboratory equipment. At present the quality assurance and reliability requirements for experiment equipment on the shuttle are not defined. The goal, however, is to minimize these requirements so that only minor modifications to commercial equipment is necessary.

5. Problem Areas

There are a number of problem areas that have been identified in the process of designing a plasma physics laboratory. In general these problems do not represent in-

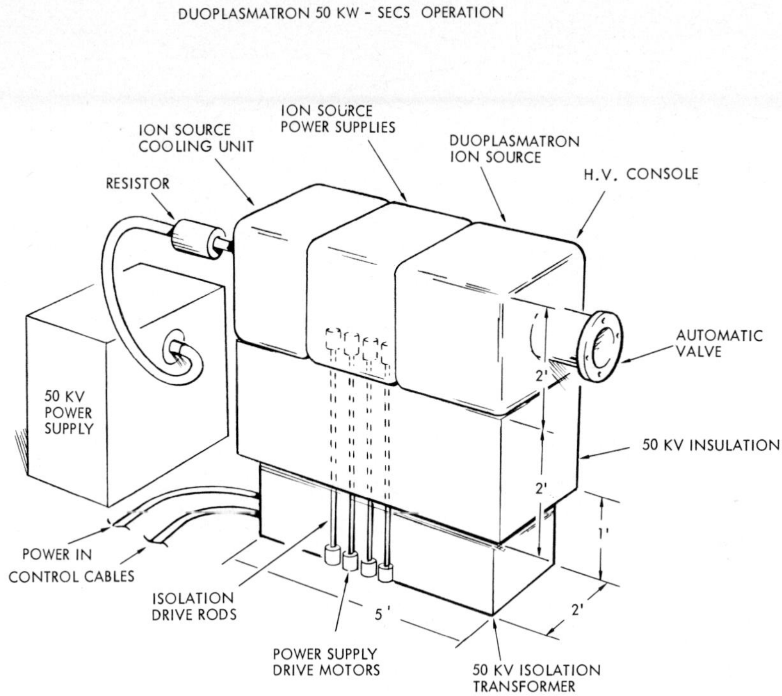
DUOPLASMATRON 50 KW - SECS OPERATION
ION SOURCE POWER SUPPLIES
ION SOURCE COOLING UNIT
DUOPLASMATRON ION SOURCE
H.V. CONSOLE
RESISTOR
AUTOMATIC VALVE
2'
50 KV POWER SUPPLY
50 KV INSULATION
2'
1'
POWER IN
CONTROL CABLES
ISOLATION DRIVE RODS
5'
2'
POWER SUPPLY DRIVE MOTORS
50 KV ISOLATION TRANSFORMER

Fig. 5.

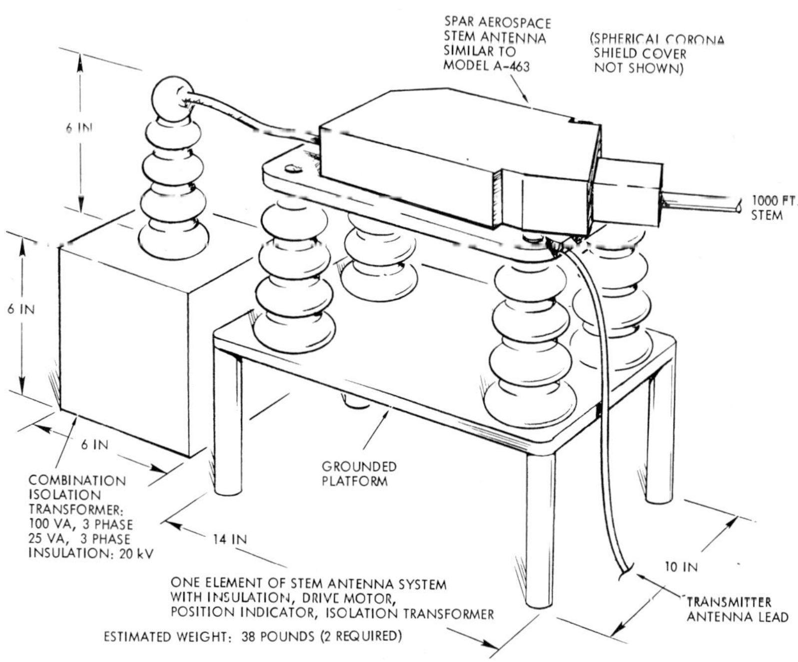
SPAR AEROSPACE STEM ANTENNA SIMILAR TO MODEL A-463
(SPHERICAL CORONA SHIELD COVER NOT SHOWN)
6 IN
1000 FT. STEM
6 IN
6 IN
GROUNDED PLATFORM
COMBINATION ISOLATION TRANSFORMER: 100 VA, 3 PHASE 25 VA, 3 PHASE INSULATION: 20 kV
14 IN
10 IN
ONE ELEMENT OF STEM ANTENNA SYSTEM WITH INSULATION, DRIVE MOTOR, POSITION INDICATOR, ISOLATION TRANSFORMER
TRANSMITTER ANTENNA LEAD
ESTIMATED WEIGHT: 38 POUNDS (2 REQUIRED)

Fig. 6.

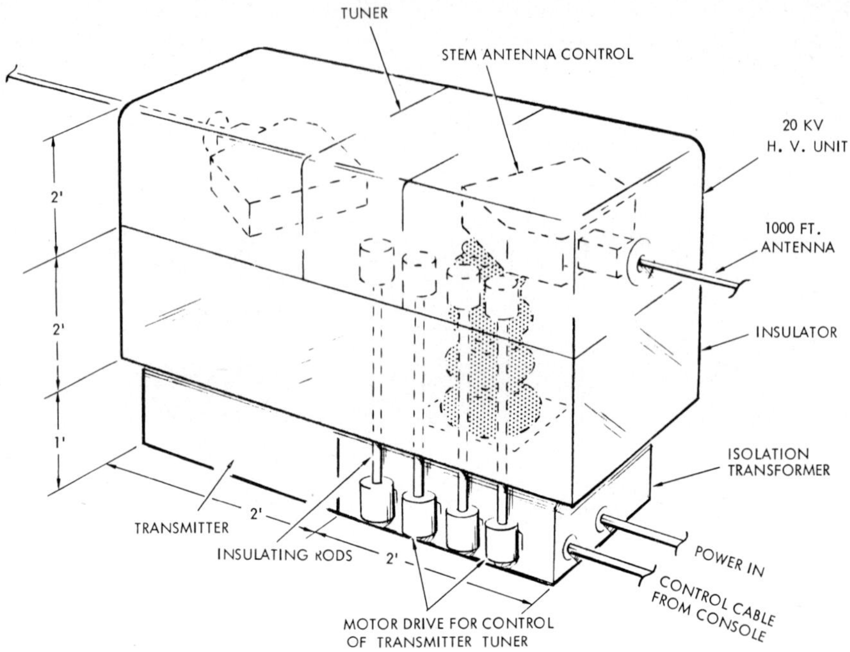

Fig. 7.

surmountable obstacles to the concept of a plasma physics laboratory; however they require further study. Some of the problems that are being studied include electromagnetic interference, outgassing and contamination, cooling of high voltage supplies, and reflected light problems, particularly from deployed booms and antennas. The design of a suitable accelerator system represents an area with special problems, such as space charge forces, stable neutralization, purity of proton beam, cathode contamination by outgassing, and electrostatic and magnetic 'contamination' for the low energy gun. There are a large number of 'system' problems associated with the design and development of a plasma physics laboratory; however, all such problems can be solved in a satisfactory manner if a decision is made to proceed with a plasma physics laboratory.

Acknowledgements

This work was performed for Marshall Space Flight Center under Contract NAS8-28047.

ASTROPHYSICS AND SPACE SCIENCE LIBRARY

1. C. de Jager (ed.), *The Solar Spectrum. Proceedings of the Symposium held at the University of Utrecht, 26–31 August, 1963*. 1965, XIV+417 pp.
2. J. Ortner and H. Maseland (eds.), *Introduction to Solar Terrestrial Relations. Prodeedings of the Summer School in Space Physics held in Alpbach, Austria, July 15–August 10, 1963 and Organized by the European Preparatory Commission for Space Research*. 1965, IX+506 pp.
3. C. C. Chang and S. S. Huang (eds.), *Proceedings of the Plasma Space Science Symposium, Held at the Catholic University of America, Washington, D.C., June 11–14, 1963*. 1965, IX+377 pp.
4. Zdeněk Kopal, *An Introduction to the Study of the Moon*. 1966, XII+464 pp.
5. Billy M. McCormac (ed.), *Radiation Trapped in the Earth's Magnetic Field. Proceedings of the Advanced Study Institute, held at the Chr. Michelsen Institute, Bergen, Norway, August 16–September 3, 1965*. 1966, XII+901 pp.
6. A. B. Underhill, *The Early Type Stars*. 1966, XIII+282 pp.
7. Jean Kovalevsky, *Introduction to Celestial Mechanics*, 1967, VIII+427 pp.
8. Zdeněk Kopal and Constantine L. Goudas (eds.), *Measure of the Moon. Proceedings of the Second International Conference on Selenodesy and Lunar Topography held in the University of Manchester, England, May 30–June 4, 1966*. 1967, XVIII+479 pp.
9. J. G. Emming (ed.), *Electromagnetic Radiation in Space. Proceedings of the Third ESRO Summer School in Space Physics, held in Alpbach, Austria, from 19 July to 13 August, 1965*. 1968, VIII+307 pp.
10. R. L. Carovillano, John F. McClay, and Henry R. Radoski (eds.), *Physics of the Magnetosphere. Based upon the Proceedings of the Conference held at Boston College, June 19–28, 1967*. 1968, X+686 pp.
11. Syun-Ichi Akasofu, *Polar and Magnetospheric Substorms*. 1968, XVIII+280 pp.
12. Peter M. Millman (ed.), *Meteorite Research. Proceedings of a Symposium on Meteorite Research held in Vienna, Austria, 7–13 August, 1968*. 1969, XV+941 pp.
13. Margherita Hack (ed.), *Mass Loss from Stars. Proceedings of the Second Trieste Colloquium on Astrophysics, 12–17 September, 1968*. 1969, XII+345 pp.
14. N. D'Angelo (ed.), *Low-Frequency Waves and Irregularities in the Ionosphere. Proceedings of the 2nd ESRIN-ESLAB Symposium, held in Frascati, Italy, 23–27 September, 1968*. 1969, VII+218 pp.
15. G. A. Partel (ed.), *Space Engineering. Proceedings of the Second International Conference on Space Engineering, held at the Fondazione Giorgio Cini, Isola di San Giorgio, Venice, Italy, May 7–10, 1969*. 1970, XI+728 pp.
16. S. Fred Singer (ed.), *Manned Laboratories in Space. Second International Orbital Laboratory Symposium*. 1969, XIII+133 pp.
17. B. M. McCormac (ed.), *Particles and Fields in the Magnetosphere. Symposium Organized by the Summer Advanced Study Institute, held at the University of California, Santa Barbara, Calif. August 4–15, 1969*. 1970, XI+450 pp.
18. Jean-Claude Pecker, *Experimental Astronomy*. 1970, X+105 pp.
19. V. Manno and D. E. Page (eds.), *Intercorrelated Satellite Observations related to Solar Events. Proceedings of the Third ESLAB/ESRIN Symposium held in Noordwijk, The Netherlands, September 16–19, 1969*. 1970, XVI+627 pp.
20. L. Mansinha, D. E. Smylie, and A. E. Beck, *Earthquake Displacement Fields and the Rotation of the Earth. A NATO Advanced Study Institute Conference Organized by the Department of Geophysics, University of Western Ontario, London, Canada, June 22–28, 1969*. 1970, XI+308 pp.
21. Jean-Claude Pecker, *Space Observatories*. 1970, XI+120 pp.
22. L. N. Mavridis (ed.), *Structure and Evolution of the Galaxy. Proceedings of the NATO Advanced Study Institute, held in Athens, September 8–19, 1969*. 1971, VII+312 pp.
23. A. Muller (ed.), *The Magellanic Clouds. A European Southern Observatory Presentation: Principal*

Prospects, Current Observational and Theoretical Approaches, and Prospects for Future Research. Based on the Symposium on the Magellanic Clouds, held in Santiago de Chile, March 1969, on the Occasion of the Dedication of the European Southern Observatory. 1971, XII + 189 pp.

24. B. M. McCormac (ed.), *The Radiating Atmosphere. Proceedings of a Symposium Organized by the Summer Advanced Study Institute, held at Queen's University, Kingston, Ontario, August 3–14, 1970.* 1971, XI + 455 pp.
25. G. Fiocco (ed.), *Mesospheric Models and Related Experiments. Proceedings of the 4th ESRIN-ESLAB Symposium, held at Frascati, Italy, July 6–10, 1970.* 1971, VIII + 298 pp.
26. I. Atanasijević, *Selected Exercises in Galactic Astronomy.* 1971, XII + 144 pp.
27. C. J. Macris (ed.), *Physics of the Solar Corona. Proceedings of the NATO Advanced Study Institute on Physics of the Solar Corona, held at Cavouri-Vouliagmeni, Athens, Greece, 6–17 September 1970.* 1971, XII + 345 pp.
28. F. Delobeau, *The Environment of the Earth.* 1971, IX + 113 pp.
29. E. R. Dyer (general ed.), *Solar-Terrestrial Physics 1970. Proceedings of the International Symposium on Solar-Terrestrial Physics, held in Leningrad, U.S.S.R., 12–19 May 1970.* 1972, VIII + 938 pp.
30. V. Manno and J. Ring (eds.), *Infrared Detection Techniques for Space Research, Proceedings of the Fifth ESLAB-ESRIN Symposium, held in Noordwijk, The Netherlands, June 8–11, 1971.* 1972, XII + 344 pp.
31. M. Lecar (ed.), *Gravitational N-Body Problem, Proceedings of IAU Colloquium No. 10, held in Cambridge, England, August 12–15, 1970.* 1972, XI + 441 pp.
32. B. M. McCormac (ed.), *Earth's Magnetospheric Processes. Proceedings of a Symposium Organized by the Summer Advanced Study Institute and Ninth ESRO Summer School, held in Cortina, Italy, August 30–September 10, 1971.* 1972, VIII + 417 pp.
33. Antonin Rükl, *Maps of Lunar Hemispheres.* 1972, V + 24 pp.
34. V. Kourganoff, *Introduction to the Physics of Stellar Interiors.* 1973, XI + 115 pp.
35. B. M. McCormac (ed.), *Physics and Chemistry of Upper Atmospheres. Proceedings of Symposium Organized by the Summer Advanced Study Institute, held at the University of Orléans, France, July, 31–August 11, 1972.* 1973, VIII + 389 pp.
36. J. D. Fernie (ed.), *Variable Stars in Globular Clusters and in Related Systems. Proceedings of the IAU Colloquim No. 21, held at the University of Toronto, Toronto, Canada, August 29–31, 1972.* 1973, IX + 234 pp.
37. R. J. L. Grard (ed.), *Photon and Particle Interaction with Surfaces in Space. Proceedings of the 6th ESLAB Symposium, held at Noordwijk, the Netherlands, 26–29 September, 1972.* 1973, XV + 577 pp.
28. Werner Israel (ed.), *Relativity, Astrophysics and Cosmology, Proceedings of the Summer School, held 14–26 August, 1972, at the BANFF Centre, BANFF, Alberta, Canada.* 1973, IX + 323 pp.
39. B. D. Tapley and V. Szebehely (eds.), *Recent Advances in Dynamical Astronomy. Proceedings of the NATO Advanced Study Institute in Dynamical Astronomy, held in Cortina d'Ampezzo, Italy, August 9–21, 1972.* 1973, XIII + 468 pp.
40. A. G. W. Cameron (ed.), *Cosmochemistry. Proceedings of the Symposium on Cosmochemistry, held at the Smithsonian Astrophysical Observatory, Cambridge, Mass., August 14–16, 1972.* 1973, IX + 173 pp.
42. D. E. Page (ed.), *Correlated Interplanetary and Magnetospheric Observations. Proceedings of the Seventh ESLAB Symposium, held at Saulgau, W. Germany, 22–25 May, 1973.* 1974, XIV + 662 pp.
45. C. B. Cosmovici (ed.), *Supernovae and Supernova Remnants. Proceedings of the International Conference on Supernovae, held in Lecce, Italy, May 7–11, 1973.* 1974, XVII + 387 pp.

In preparation:

41. M. Golay, *Introduction to Astronomical Photometry.* 1974, in press.
43. R. Giacconi and H. Gursky (eds.), *X-Ray Astronomy.*
44. B. M. McCormac (ed.), *Magnetospheric Physics. Proceedings of the Advanced Summer Institute, held at Sheffield, U.K., August 1973.*